CSR, Sustainability, Ethics & Governance

Series Editors

Samuel O. Idowu, London Metropolitan University, London, UK

René Schmidpeter, Cologne Business School, Cologne, Germany

In recent years the discussion concerning the relation between business and society has made immense strides. This has in turn led to a broad academic and practical discussion on innovative management concepts, such as Corporate Social Responsibility, Corporate Governance and Sustainability Management. This series offers a comprehensive overview of the latest theoretical and empirical research and provides sound concepts for sustainable business strategies. In order to do so, it combines the insights of leading researchers and thinkers in the fields of management theory and the social sciences—and from all over the world, thus contributing to the interdisciplinary and intercultural discussion on the role of business in society. The underlying intention of this series is to help solve the world's most challenging problems by developing new management concepts that create value for business and society alike. In order to support those managers, researchers and students who are pursuing sustainable business approaches for our common future, the series offers them access to cutting-edge management approaches.

CSR, Sustainability, Ethics & Governance is accepted by the Norwegian Register for Scientific Journals, Series and Publishers, maintained and operated by the Norwegian Social Science Data Services (NSD).

More information about this series at http://www.springer.com/series/11565

Eduardo G. Pereira · Rochelle Spencer ·
Jonathon W. Moses
Editors

Sovereign Wealth Funds, Local Content Policies and CSR

Developments in the Extractives Sector

Springer

Editors
Eduardo G. Pereira
Siberian Federal University
Krasnoyarsk, Russia

The University of West Indies
St. Augustine Campus
St. Augustine, Trinidad and Tobago

University of São Paulo
São Paulo, Brazil

Jonathon W. Moses
Department of Sociology and Political Science
Norwegian University of Science and Technology
Trondheim, Norway

Rochelle Spencer
College of Arts, Business, Law and Social Sciences
Murdoch University
Murdoch, WA, Australia

ISSN 2196-7075 ISSN 2196-7083 (electronic)
CSR, Sustainability, Ethics & Governance
ISBN 978-3-030-56091-1 ISBN 978-3-030-56092-8 (eBook)
https://doi.org/10.1007/978-3-030-56092-8

This Springer imprint is published by the registered company Springer Nature Switzerland AG
The registered company address is: Gewerbestrasse 11, 6330 Cham, Switzerland

Foreword

The Stuff of the Stars

Who Owns Natural Resources?

Many of the creation stories of the world's great (and smaller) religions tell us that we did not create the natural resources that we often claim to own. Instead, we were also created as part of the natural world and given custodianship rather than ownership of natural resources.

Astrophysicists also tell us that within twenty minutes of the birth of our Universe, 90% of all of the lithium was created in an event called the Big Bang Nucleosynthesis (BBN). All of the elements heavier than lithium were subsequently created in stars that began to form 180 million years later. Iron and related elements (chromium, manganese, copper and nickel), the so-called iron peak, are the ejecta of supernovas where oxygen and silicon fused to create these elements (and still do). Gold comes from the collision of neutron stars. Tin and lead are from red giants (like our sun will one day become) and on it goes. Carbon the foundation of life as we know it is, with a tinge of cosmic irony, from dying stars. We are truly the stuff of stars (as Carl Sagan famously said).

According to astrophysicists, all of the metals on our Earth arrived here on interstellar and solar winds and coalesced into this cooling cauldron of molten metal that we now call home. That all happened between 13.8 billion and 4.5 billion years ago. Magmatic and tectonic forces then distributed those atoms in our planet to be moved around, subducted, erupted and ultimately placed on our cooling crust. The hydrocarbons that form the basis of fossil fuels began being laid down half a billion years ago.

It is in these distant stories of creation and cosmological history that we, human beings, find ourselves often arguing about who owns and who should have right of access to harness what atoms. It is the hubris of the construct of our human societies that lays claim to "ownership" of the stuff that is the stuff of stars—materials which we were given custodianship of, to harness and to maintain for the benefit of all.

While this book does not deal in astrophysics or theology or take a view measured in billions of years, it does directly challenge the ideas of "ownership" of natural resources. It casts an eye to more recent histories and suggests that in our current configuration of societies and nation-states, we need to re-examine who "owns" the natural resources and who has the right to exploit and prosper from them. The theological origins of planet Earth and haphazard cosmological game of stellar and planetary formation both suggest no basis for humans to claim ownership of natural resources. The equally haphazard and often historically dubious incidences that led to the drawing of maps that formed national boundaries as we know it today also cast doubts on the "rights" of current residents in specific countries and communities to claim ownership of materials found beneath their land.

Drs Spencer, Pereira and Moses have worked together virtually on this volume. They have never met in person (in this COVID-19 constrained world) but have nonetheless cooperated and collaborated closely to bring together 44 authors to examine a dozen jurisdictions. This is a significant achievement in its own right. Taken together this compendium explores three mechanisms to manage the ownership of atoms: sovereign wealth funds (SWFs), local content policies (LCP) and corporate social responsibility (CSR). The strength of the volume is found in both the depth and breadth of analysis coupled with the diversity of professional perspectives and how the chapters invite us to undertake a comparative policy approach to the myriad of "solutions" described. The value of the book is found in the ways that it provokes our own thinking (and hopefully doing) to seek inclusive economic growth that brings enduring benefit to local communities. The very communities who are the closest to the atoms and are often are the last to benefit and often benefit the least.

It is an important contribution to an urgent issue and we commend it to you, the reader.

David Doepel
STM Chair, Africa Research Group, Murdoch University
Perth, Australia

Prof. Kevin Chika Urama
FAAS, Senior Director, African Development
Institute, African Development Bank Group
Abidjan, Côte d'Ivoire

Contents

List of Figures

Non-Renewable Resource Revenue Savings and Distribution in Canada: Alberta

Norway's Sovereign Wealth Fund

Local Content Policy in Indonesia Oil and Gas Industry

Norwegian Local Content Policy

Corporate Social Responsibility in the Mining Sector in Canada

Conclusion

List of Tables

Corporate and Environmental Responsibility Among Russian Oil Companies

Corporate Social Responsibility in the Mining Sector in Canada

Conclusion

Introduction

Jonathon W. Moses, Rochelle Spencer, and Eduardo G. Pereira

Abstract The analogy that follows includes 35 chapters, 16 cases and 12 very different countries. This chapter provides a brief overview of these cases and introduces the concepts and literatures associated with three main tools for managing a modern resource economy: sovereign wealth funds, local content policies and corporate social responsibility.

Keywords Sovereign wealth funds · Local content policy · Corporate social responsibility · Resource curse · Resource rents

"The revenue of the state is the state". These words, from Burke's (1790) *Reflections on the Revolution in France*, sum up the anthology you hold in your hands. The states in this volume are beholden to one type of extractive industry or another: their main source of revenue comes from the exploitation of rare, valuable and non-renewable resources. Many of these states find themselves in the developing world, with relatively weak political institutions, and they hope that their natural resource wealth can help them to develop further.

J. W. Moses (✉)
Department of Sociology and Political Science, Norwegian University of Science and Technology, Trondheim, Norway
e-mail: jonathon.moses@ntnu.no

R. Spencer
Centre for Responsible Citizenship and Sustainability, Murdoch University—Perth, Perth, WA, Australia
e-mail: rochelle.spencer@murdoch.edu.au

E. G. Pereira
Siberian Federal University, Krasnoyarsk, Russia
e-mail: egp.portugal@gmail.com

The University of West Indies, St. Augustine Campus, St. Augustine, Trinidad and Tobago

University of São Paulo, São Paulo, Brazil

E. G. Pereira et al. (eds.), *Sovereign Wealth Funds, Local Content Policies and CSR*, CSR, Sustainability, Ethics & Governance,
https://doi.org/10.1007/978-3-030-56092-8_1

By depending on a single source of revenue, states take on three types of vulnerabilities. First, they become very sensitive to the needs of the provider (e.g. International Oil, Gas and Mining Companies, or IOGMCs) to secure a steady supply of resource wealth, often at the expense of local community priorities and indigenous land rights. This segues to the second vulnerability: states risk creating or exacerbating social tensions and corporate-community conflict over the social and environmental impacts of extractives activity. Third, states become exposed to markets that tend to be very volatile. This suite of vulnerabilities highlights that sustainability in extractive resources has long been a contentious issue owing to the depletive character of the extractives industry and the associated social and environmental impacts that result from extractive activities.

The dangers of these sorts of vulnerabilities can be illustrated with a simple illustration. Imagine two kings and two kingdoms, of relatively equal size: one king ($King_1$) controls a kingdom that sits on an enormous mountain of gold. The other king ($King_2$) rules a kingdom that is barren of natural wealth. Your first thought, at this point in the story, is probably: Wow, lucky $King_1$! After all, ruling a kingdom filled with gold is like winning the lottery: what can be better than swimming in money? While this is a common first impression—it is far too often wrong. As we shall soon discover, natural resource wealth is seldom a manna from heaven. The kingdom's very wealth and inflated prices make it difficult to export other goods and services and (as a consequence) all of the kingdom's energy is directed at securing more gold. *The revenue of the state is the state.*

Because the golden king has access to enormous wealth, he is (more or less) free to behave as he likes: he can afford a large army and he can use that wealth to extend the borders of his kingdom, and/or to rule over his people with an iron fist. When $King_1$ threatens to invade the other kingdom, $King_2$ must find a means of defence. Without recourse to an independent source of financing (i.e. gold), $King_2$ is forced to tax the people of his kingdom to fund his defence. This puts $King_2$ at the mercy of his people: in exchange for their taxes, the people can demand something in return, such as constraints on sovereign authority (think Magna Carta). Because $King_2$ receives his revenue from the people, the people are able to hold him accountable. Because $King_1$ pulls his revenue from the ground, his biggest constraint is the size of his gold holdings. *The revenue of the state is the state.*

This fanciful example is used to depict, very briefly, the main challenges—both economic and political—facing states that have found wealth in non-renewable natural resources. The rest of this book shines further light on these challenges.

Natural resource wealth holds the promise of raising incomes and living standards—in both developed and developing contexts alike. Extractives industries contribute to GDP, increase tax revenues, export earnings and employment, and the wealth that these industries generate can contribute to sustainable development by enriching local communities, including those in rural and remote regions (where these industries are often located). This is the hope offered by the extractives sector. Still, there is now a rising knowledge of environmental sustainability as a premise for economic and social sustainability (Folke et al. 2016). The "planetary boundaries" framework (Rockström et al. 2009; Steffen et al. 2015) demonstrates that we are already transgressing four of the nine identified planetary boundaries, hence moving out of the safe operating space for humanity. At least three of these boundaries are of high significance for the extractives industries; climate change, biological diversity and land-system change. This book does not get into these overarching issues as such, but some of these challenges are discussed in, e.g. the Chapter "CSR in the Norwegian Context".[1]

Notwithstanding and perhaps despite these celebrated spoils of the extractives sector, many resource-dependent countries prove unable to fully benefit from their natural resource wealth. Too often, poor communities living adjacent to extractives operations are denied the benefits of economic progress owing to historical legacies of colonialism and dispossession; poverty persists; jobs may be hard to come by; the industry can become "enclavic", providing limited backward and forward linkages with other industries or enterprises; and the environment can be degraded, impacting the livelihoods and cultures of subsistence farmers and indigenous communities whose connection to land is trivialised in order to prioritise extractives industries. The wealth pouring into the country can also breed corruption and raise inequalities. These unwanted outcomes risk perpetuating adverse social, cultural and environmental legacies and may create fertile grounds for violence and conflict. In fact, on average, resource-dependent economies score lower on the human development index than less endowed countries. Out of the 3.5 billion people living in resource-dependent countries, approximately 887 million live below $1.25/day (UNDP 2014).

It does not have to be this way. Some countries have managed their extractives industries successfully: they have diversified their economies, ensured backward and forward linkages, tackled corruption and conflict and invested in human and physical capital. The experiences of these states show that it is possible for natural bounty to be managed in a way that facilitates good governance and maintains international competitiveness. By surveying these experiences, we can better understand the threats that resource wealth brings to political, cultural and economic sustainability, and the tools that states have to minimise or avoid those threats.

This is exactly what we propose to do in the book that follows: we consider how twelve very different states employ three of the most important tools for managing a modern resource economy: *sovereign wealth funds*, *local content policies* and *corporate social responsibility*. To better understand how these tools are wielded, this introductory chapter provides a brief and general introduction to the challenges hinted at above: the unique qualities of natural resource wealth; the sort of socio-economic imbalances this wealth can generate, and the sundry ways that states manage those challenges.

[1]For further reading, we suggest Hickel (2019) and Nilsen (2019).

1 Kingdoms of Gold

Each of the case studies in this anthology is struggling with the wealth generated from the process of extracting raw materials from the earth (and subsequently selling them to consumers). The extractives industries are those that remove metals, mineral and aggregates from the Earth, for example, through the process of oil and gas extraction, mining, dredging and quarrying.

These extractives industries have the potential to drive growth, development and poverty reduction (Spencer 2018; Blowfield and Dolan 2014). Indeed, these industries already play a dominant role in the economic, social and political lives of 3.5 billion people living in 81 countries (World Bank 2019c). If managed well, the revenues and experiences from these industries could reduce poverty, increase prosperity and even minimise inequality, while protecting the surrounding communities and environment (World Bank 2019a). In practice, however, most of these states (from the global North and global South) do a poor job in managing their resource wealth, resulting in both resource dependency and weak governance.

To understand the difference between good and poor management practices, we need to begin by recognising that there is something special about the extractives industry. This industry is not like the textile, agricultural or service industries. Whether it rests on gold or gas (or anything in between), the extractives industries rely upon non-renewable natural resources, often in high demand. This matters, on at least three counts.

First, natural resources are *owned by the people* in common. Public ownership in natural resources is secured by centuries of common law/experience, but also by international agreements, such as the 1962 UN Resolution on Permanent Sovereignty over Natural Resources and Article 1 of the International Convention on Civil and Political Rights. Whether it is because of historical convention, national (constitutional) and/or international law, it is almost everywhere recognised that subsurface petroleum and minerals are owned by the people, and the responsibility for managing these resources lies with their representatives, i.e. national governments.[2] As we shall see, in some exceptional situations in, e.g. in certain parts of Canada and several parts of the USA, the private ownership of mineral rights is still allowed. But it is a mistake to think that such North American practices are normal or commonplace in the world at large.

Second, because these resources are non-renewable and limited by nature, and because they are in high demand, they tend to be associated with a significant rent—the so-called *resource rent*.[3] A resource rent is an extraordinary value derived from

[2] As is clearly evident in the chapters on Alaska ("Alaska's Petroleum Industry, Institutions and Sovereign Wealth Fund", "Alaska's Tug of War on Land Rights" and "Alaska's Corporate Social Responsibility: The Economics of the Corruption Case of VECO"), federal systems complicate this simple picture, as the interests of state residents may not correspond with the interest of the national citizenry.

[3] This understanding of rent was initially introduced by Ricardo (1817), but subsequently developed by George (1879).

exploiting a natural resource (fixed in quantity)—after accounting for all costs and the provision of normal returns (including profit). We hasten to underscore that the rent is not created by risk-taking, skill or expertise, and has nothing to do with profit: it is an *unearned surplus* that results from restricted access to a valuable commodity. The rent is what is left over after all the input costs are covered, and a normal return on investment is paid off.

The size of the resource rent is determined by many different factors, none of which are associated with commercial prowess. For one, the rent is determined by the productivity of the process of extracting/producing the resource, relative to the least efficient producer in the market. In some contexts (think Kuwait), it is relatively easy to get oil out of the ground—and the rent comes easily. In other countries (think Norway), it is very expensive to get the oil out of the ground, and the corresponding rent will be relatively smaller. In this instance, the difference in rent is determined by geological luck: Kuwait is simply luckier than Norway, with regard to the ease of accessing its oil reserves.

Daniel Yergin, in his award-winning book, *The Prize*, provides a very useful example:

> [I]n the late 1940s, oil was selling for around $2.50 a barrel. Some grizzled stripper-well operator in Texas might only make a 10 cent profit on his oil. But in the Middle East it only cost 25 cents a barrel to produce oil. Deducting 50 cents for other costs, such as transportation, and allowing a 'profit' of 10 cents on the $2.50 barrel, that would still leave a very large sum—$1.65 on every barrel of Middle Eastern oil. That sum would constitute rents. Multiply it by whatever the rising production numbers, and the money added up very rapidly. (Yergin 2009: 414)

The size of the rent is also determined by the market price of the resource, which is—in turn—determined by supply. Much of the value in natural resources comes from the fact that they are rare and limited. That they are rare in nature is easy to understand, but they are also limited because states limit commercial access to these resources.

In the early years of the US petroleum industry, it was relatively costless for new actors to enter the petroleum market, and the price of oil fluctuated wildly. New discovery areas were immediately flooded with wildcat producers, and the price of oil (and the poorly managed reserves) fell precipitously. This price volatility became unbearable, even for liberal, free-market states in Southern USA, so state authorities in Texas and Oklahoma, for example, began to limit access to the resource with an eye at stabilising its price.[4] Further access restrictions were introduced in response to increased concerns about inefficient use of the resource and potentially detrimental environmental effects. Whatever its initial motivation, we know that access to non-renewable natural resources is limited by both nature and government.

In particular, and as we shall see below, states use a plethora of licensing arrangements to restrict access in all the extractives industries. Restricting access in this way allows political authorities to increase the size of the rent, by limiting access

---

[4]See, e.g., Yergin (2009: Chaps. 12 and 13). These examples came to influence the subsequent creation of OPEC. See Yergin (2009: 495), but also Rabe (1982).

to potential competition. When governments limit access to natural resources (e.g. through a system of contracts, licences or concessions), they increase the potential rent—and hence, the resource becomes even more valuable. As the resource rent is a function of the underlying resource (not the result of some clever manipulation on the part of the contractor), and because the size of the rent is a function of restricted access, the rent belongs to the owner of the resource and the creator of the rent (i.e. nature and government—both of which belong to the people).

Third, the global price on natural commodities is notoriously *volatile*. As we have just seen, this is one of the reasons that governments initially became involved in restricting access to these resources. For example, in just two years, around 2014, the price of a barrel of crude Brent oil halved, from about $117 in 2013 to about $56 in 2015. You can see that significant drop in Fig. 1, which illustrates the long-term price volatility in the oil market, using overlapping price measures. Figure 2 shows similar long-term price developments in three aggregate commodity indices: energy (including crude oil, natural gas and coal), metals and minerals, and precious metals. As the prices of these important natural resources rise and fall, so too do the fortunes of countries that rely upon them.

When a country leans heavily on a single natural resource, the price of which is volatile, it is exposed to a number of challenges. First of all, state revenues will vary significantly, along with the price of the resource. This is a problem, in that the

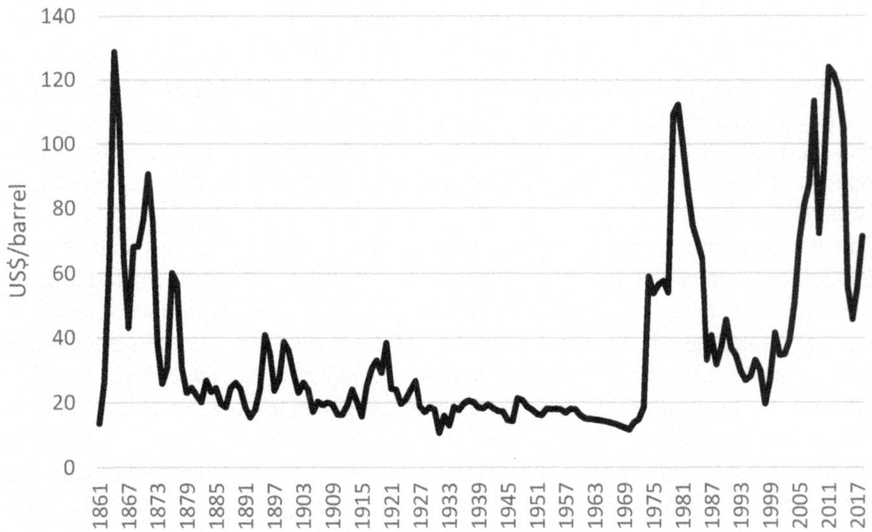

Fig. 1 Crude oil prices, 1861–2018.
Source BP (2019).
Note In 2018 US dollars (deflated using the Consumer Price Index for the USA). Different measures were use to cover the different periods: 1861–1944 uses the US average; 1945–1983 uses the price of Arabian light posted at Ras Tanura; and in 1984–2018, we used the Brent dated data

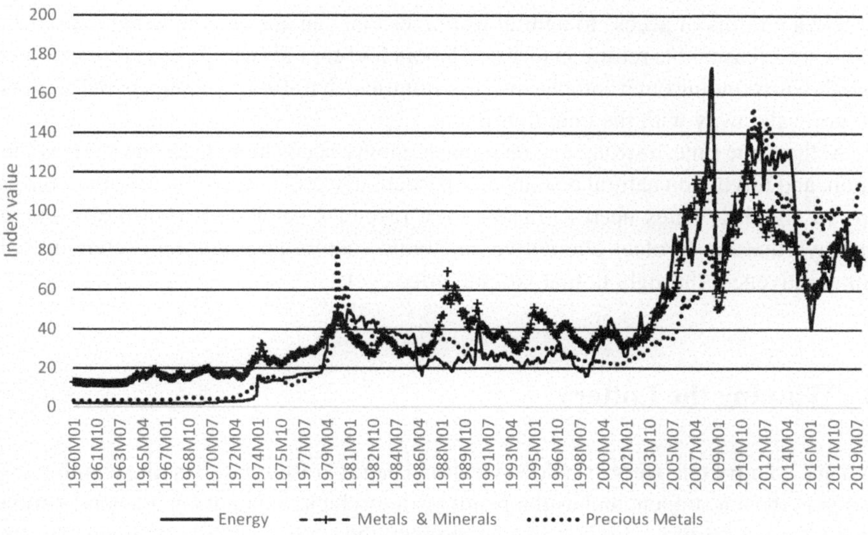

Fig. 2 Commodity price indices, monthly, 1960 to present.
Source World Bank (2019b).
Note "Pink Sheet". Monthly indices based on nominal US dollars, 2010 = 100. See source for further clarification on the composition of the indices, e.g. the "energy" index is made up of coal (4.7%), crude oil (84.6%) and natural gas (10.8%)

state's spending needs are more constant (consider education or defence). In other words, the state needs to find a way to smooth out the revenue flow, such that it can continue to pay for necessary state activities, even when the resource price is low. Second, when the market price of a natural resource is high, the industry will draw to it the country's productive factors (capital, labour), leaving fewer qualified people and investment capital to satisfy the needs of other sectors. As a result, these countries become more and more dependent on a single source of economic activity. As long as the price of the single resource remains high, this might not appear to be a problem. But when that price falls, as it inevitably does, it brings along with it the jobs and revenues that are connected to the related industry—and with them the fortunes of the state. Absent of economic alternatives (having been drained of labour and capital), it becomes very difficult for any country to adjust to the new (low-price) market conditions.

In short, the challenge of resource management is twofold.[5] The first is a need to secure the resource rent for its rightful owners. As the resources belong to the people, and governments contribute to the rent by limiting access to the natural resource, it is essential that private actors are not allowed to abscond with the rent (by this, we mean corrupt government, but also IOGMCs). When all is said and done, states need

[5]In this volume, we set aside the other, arguably more common, aspects of resource management (e.g. how to maximise efficiency, protect the environment, etc.), in order to focus on the political and economic effects.

to provide terms of access to natural resources that can attract competent expertise; allow for a reasonable return on invested labour and capital (while protecting workers and the surrounding environment and communities), but still ensure that private actors do not walk away with the underlying rent.

At the same time, a strong and resilient economy needs more than one leg to stand upon, and relying on natural resources is particularly problematic, as the price trends are so volatile. States need to manage the revenues generated from the resource in ways that can protect alternative economic sectors and maintain international competitiveness throughout that volatile price cycle.

2 Winning the Lottery

When a country uncovers its hidden wealth in natural resources, it is like winning the lottery. All of a sudden, and as the result of dumb luck, a country finds itself sitting on a valuable resource. Like the lottery winner, the state is initially bombarded with consultants and advisers, anxious to help them spend their new-found wealth and/or lend them additional money. Lottery winners tend to assume that their life starts anew: they need not worry about future revenues streams, and they finally have the money they need to fix the roof or educate the kids. In the absence of a plan, a lottery winner might quit her job, buy her cousin a new home, and begin to spend extravagantly…until all the money is gone. The same can easily happen with a state that has discovered natural resource wealth: the promise of future wealth makes it easy to secure loans, to throw money at problems (both new and old), and to distance oneself from previous (less lucrative) forms of employment.

When a commercially viable natural resource is first discovered, it is understandable that most of the early attention is focused on the technical challenge of locating the resource and getting it to market: countries spend a great deal of money acquiring the expertise necessary to make this happen. After all, the extractives sectors tend to be very capital and knowledge-intensive: the resources they chase do not float on the surface waiting to be picked up. But this challenge is relatively easy to overcome: the expertise is out there to secure, it is just a matter of attracting its attention and finding the money to pay for it. While petroleum engineers are often loath to admit it, the *real* challenge for the country happens only after this technical miracle has been performed: what to do with all the money that results?

Two threats are particularly menacing: one mostly economic, the other mostly political. The first has to do with maintaining international competitiveness and a balanced economy. The second concerns the nature of so-called rentier states—states that depend upon the rents from their natural resources. Both threats can play havoc with good governance. After all, it is not without reason that Juan Pablo Pérez Alfonzo—Venezuela's former oil minister and the main driver behind the creation of OPEC—referred to oil as the "Devil's excrement" in a 1975 speech (Starr 2007).

The problem of excess money has many different names and can take many different forms. Whether we refer to it as a resource (or oil) curse, or a paradox of plenty, the problem is one that has long haunted countries.[6] The initial work on the resource curse showed how those countries that relied on natural resource exports tended to have lower per capita growth rates (see e.g. Sachs and Warner 1995). Subsequent work extended the argument to more political effects (such as increased corruption, inequality and internal conflict). The argument is on full display in a 2003 World Bank study of the extractive industries:

> Data on real per capital gross domestic product (GDP) reveal that developing countries with few natural resources grew two to three times faster than resource-rich countries over the period 1960–2000. Of 45 countries that did not manage to sustain economic growth during this time, all but six were heavily dependent on extractive industries, a majority of them also experienced violent conflict and civil strife in the 1990s. (World Bank 2003: 12)

Over time, the nature of the relationship has changed, so the economic curse seems less pronounced now than it was in the past (Ross 2012). Today, it would seem that economic growth is neither faster nor slower in resource-rich states (relative to the rest of the world). While this is an improvement on the past, it is hardly what we should expect.[7] Why does not this increased wealth transfer to greater economic activity, more jobs, and higher standards of living?

The main problem is one of economic capacity: the domestic economy simply cannot absorb the new capital in an efficient or effective manner, and this inability to adjust bloats the surrounding economy, making it harder for it to compete on the international stage. This particular part of the problem is what we refer to as Dutch Disease: where the new wealth from natural resources floods the local economy, creating inflation and a real appreciation of the national currency. As a result, the price of local goods and services rises across the board (as the injection of new money drives up food costs, housing costs, wages…). This, in turn, makes the goods and services produced in the country more expensive, and hence less competitive, internationally.

In addition, the wealth being generated in the natural resource sector will attract skilled workers and dormant capital away from traditional centres of economic activity. This sort of internal factor flow makes it difficult to sustain alternatives to the natural resource sector: it pushes up wages and returns across the country—crowding out investment and jobs elsewhere. As prices increase across the economy,

---

[6]The «Paradox of Plenty» comes from a 1997 book title by Terry Lynn Karl. The "resource curse" was first coined by Auty (1993) to describe how natural resource wealth was not generating the effect on economic growth that economists expect. Subsequent econometric work by Sachs and Warner (1995) and a 1999 article in *World Politics* by Michal Ross brought his work to even broader audience. The notion of a "rentier state" was first introduce by Mahdavy (1970) with reference to the political challenges associated with Middle Eastern states that rely heavily on resource rents; whereas the term, "Dutch Disease" was first introduced in a 1977 *Economist* article referring to a decline in the Dutch manufacturing sector after that country's natural gas discoveries in the 1950s.

[7]Expectations are a large part of the problem. The defining characteristic of developing countries is a lack of investment capital (or excess labour, relative to the domestic capital supply). As resource wealth introduces an abundance of capital to countries that were previously starved for investments (but pregnant with labour), economists expect significant economic growth from the injection of new capital.

the traditional sectors can no longer maintain their price competitiveness with others in the global economy.[8]

From these economic challenges, we see how easy (and dangerous) it is for a country to become overly dependent upon the extractives sector. This economic dependence brings with it political challenges as well—and these challenges can be even more troubling for the general population. The problem is that the state is less likely to concern itself with the needs of the people when its main paymaster resides elsewhere. To put it bluntly: after the discovery of natural resource wealth, the keys to the state's treasury lie with the extractives industry—not with the people, themselves.

Having direct access to a strong revenue stream from natural resource extraction/production, state officials can avoid the unenviable task of taxing their citizens. Knowing which side of their bread is buttered, these officials are increasingly sympathetic to the needs of the industry that fills their coffers, and increasingly callous to the needs of their citizens. Resource wealth allows governments to ignore the needs of their constituents: they can become more corrupt, and they are more susceptible to political violence, even civil conflicts. With their treasuries full, governments can pursue stick and carrot policies. On the one hand, officials can buy-off any political opposition, by offering generous social policies or subsidising important fuel/food costs. If that does not work, government officials can use their wealth to build up a domestic security force to protect governors from the governed, and to further repress the people. In the end, following Burke, the revenue of the state becomes the state.

3 Managing the Resource

Because natural resources are commonly owned by the people, their governments are responsible for managing them in an efficient and just manner. The first task of the government, in this regard, is to find, extract and get these resources to market. In so doing, states need to present terms and conditions that are generous enough to attract competent skills and contractors, but fair enough to ensure that the contractors do not walk off with the resource rent. The "generosity" tends to be higher for the so-called first comers as they take higher risks and expect higher rewards. As we will see in the Chapters "The Experiences of Managing the Heritage and Stabilisation Fund in Trinidad and Tobago and the Sovereign Wealth Fund Guyana", "The Development and Implementation of Local Content in the Extractive Industries in Trinidad and

[8]Consider a country that has previously relied on agricultural exports, but then uncovers a mineral Eldorado. All of a sudden, capital is flooding into the country, inflating prices across the board (as we saw above). Wages and returns on investment grow higher in the mining sector, relative to the agricultural sector, so that workers and capital are drawn from agriculture and into mining. The agricultural export industry finds it increasingly difficult to attract new workers and investors, and the industry must raise their prices (wages) accordingly. As a result, the price of the country's agricultural exports rises, relative to the global competition, making it more difficult to compete/sell. Over time, the agricultural industry falls further into decline.

Tobago and Guyana" and "The Approach to Corporate Social Responsibility in the Extractive Industries in Trinidad and Tobago and Guyana", this was the case in Guyana over a decade ago as not many oil and gas companies were keen to venture into those waters as no discovery was found nor any infrastructure in place. But it is no longer the case as Guyana became one of the largest finds in recent years with its first production started in 2020. Nevertheless, it is difficult to find a balance for such "first comers" as an oil and gas contract should last for many years and host countries should avoid expropriation or forced renegotiation.

The international community is increasingly aware of the scope of the challenge, and several international organisations help countries manage their natural resource wealth in a way that can maximise the national benefit from the sector, contribute to economic growth and reduce poverty. While many of the programs offer out-of-the-box state improvement courses [e.g. increase effective governance and transparency, promote inclusive growth, protect both people and the environment (World Bank 2019a)], new global standards are being developed—*inter alia*—to promote the open and accountable management of oil, gas and mineral resources, the use of improved stakeholder engagement, the safety of extractives operations. For example, the Extractive Industries Transparency Initiative (EITI) was established to publish information on the extractives industry value chain. This initiative allows us to follow the money: how these revenues make their way through the government, and how they benefit (or not) the public. Currently, the work of the EITI is supported by a broad coalition of government, companies and civil society, and 52 countries have agreed to implement the EITI Standard (EITI n.d.).

For our purposes, there are basically two different sorts of tools used to manage natural resources: the licensing regime and the revenues regime. How these two regimes are combined determines the size of what is often called the "government take", or the public's share of the resource rent.

The first step a government must take is to decide how to get the resource to market, i.e. to secure a commercial partner. Generally speaking, two paths avail themselves. On the one hand, the state can keep the process in-house, using state-owned enterprises (SOEs), for example, a national oil, gas or mining company (NOGMC) to extract and bring the resource to market. Alternatively, the government can hire in private (often international) expertise, e.g. an International Oil, Gas or Mining Company (IOGMC) to do the work. In either case, the government issues a permit or licence, often allocated using some sort of bidding/procurement process, to the chosen firm so that it can access and extract the natural resource.

Governments provide access to their natural resources by way of a system of contracts, or licences, and the form of these contracts tends to vary over time and by national contexts (see Moses and Letnes 2017: 86–91 for a discussion). Although the scope of these licensing agreements can vary significantly, they all determine how rents and costs will be distributed between the host government (and/or its NOGMC), and any interested private companies (e.g. IOGMCs). In the early years of the industry, following American experience, countries tended to rely on concession or royalty-based contract forms; but in the late 1960s, following Indonesia's example (see the Chapter "Overview of Extractive Resources Management in Indonesia"), more and more countries came to employ production sharing agreements (PSAs) or contracts (PSCs). More recently, varying forms of service contracts (SCs) have grown in frequency. Each of these contract types offers varying levels of control, risk and reward, and there is much overlap among them. The most relevant point is that these licensing agreements are used to control access to a valuable non-renewable resource: they are, in effect, licences to print money (i.e. access the resource rent).

The second tool is used to ensure that the underlying rent remains with the people. In order to ensure that the licensed contractor does not abscond with the rent, the state needs to appropriate that rent, and it must do so in a way that does not undermine the willingness (and profit) of private interests to do the job, and/or undermine the country's competitiveness in international markets. The form of appropriation can either be political (e.g. ensuring local content), or economic (e.g. through fees, royalties and taxes) in form. The latter is often referred to in terms of "government take".[9]

The size of the government take is only partly determined by the nature of the contract. Just as important is the system of revenue collection used by the state. After all, states might demand a share of equity in any venture, royalties,[10] taxes/fees/bonuses, etc. (See Moses and Letnes 2017: 91–95 for a discussion). Some of these taxes and fees are used to incentivize the industry to act in particular ways (e.g. to reduce harmful emissions); others are used to secure the resource rent. In practice, states use a plethora of tools to secure their share of the resource rent, and the particular mix of tools can vary across countries and time.[11]

Finally, as the nature and size of the rent can change over time, it is important for political authorities to adopt a system that is flexible enough to adjust to the changing conditions and allows for updates in ways that can facilitate the obsolescing bargaining mechanism (Vernon 1971). By spreading out the allocation of licenses/contracts (over time), and by embracing shorter licensing/contractual

---

[9]There is no agreed-upon means for measuring or comparing the level of "government take" across countries. Indeed, most measures ignore the political take and focus on the economic value, often measured as if it was some sort of "taking" from the international oil company! See, e.g., Johnston (2007) for a description of the different means used to estimate government take and Johnston (2008: 34) for a useful graphic depiction of variations in government take, across countries. For an alternative metric, see API (2012).

[10]A royalty is levy where the owner is entitled to a given percentage of the wealth being produced. In other words, a royalty is a payment to an owner for the use of property (here natural resources)—it is designed to compensate an owner for use of her/his asset.

[11]For example, in Norway, the authorities today rely on six sources of income from the petroleum sector: dividends from its majority ownership share in the national oil company (Equinor); state direct ownership in sundry production licences (so-called SDFI); area fees; a NOx tax; a CO_2 tax; and taxes on petroleum companies. The latter consists of an ordinary company income tax of about 27% (which all companies pay) plus an additional 50% tax on petroleum firms, to secure the rent. See Moses and Letnes (2017: 100–103).

periods (but sufficiently long for investors to secure their rightful returns and establish efficient production routines), the political authorities can ensure that the resource rent remains with the people.[12]

4 Three Popular Tools of Resource Management

With this general background in place, we can now turn to the focus of this anthology: how states employ three of the most popular tools in resource management. Each of these tools is aimed at different aspects of the aforementioned challenges, and each has their own strengths and weaknesses. This section introduces these three popular tools and the role they can play in the larger toolbox.

4.1 Sovereign Wealth Funds

Sovereign wealth funds are the new darlings of resource management. At the most general level, a sovereign wealth fund (SWF) is simply a state-owned investment fund. Traditionally, SWFs were built from central bank reserves, currency operations, privatisation drives, etc., but we are concerned with states that build SWFs from the revenues generated by their extractives industries.[13]

Although there are many variants (as we shall see!), there are basically two main motivations for resource-rich states to start and employ a sovereign wealth fund: as a tool to protect the economy from the price volatility that tends to characterise these sectors; and as a means to save for the future or current needs. Table 1 provides an overview of different types of SWFs, but where the savings/spending types of funds are further sub-divided into three objectives (the bottom three rows).

4.1.1 Savings/Spending Funds

The bottom three rows of Table 1 reveal several of the different ways that SWFs can be used as a savings/spending fund. A core concern of any government official responsible for managing a non-renewable resource is clear: the resource will

[12]Investors are often concerned by the lack of political and/or fiscal stability among the large number of oil and gas provinces around the world. This is why it is fairly common for host governments to sign bilateral investment treaties (BIT) and/or add stabilisation provisions in their relevant contracts order to offer further assurances to investors that future governments will not change the key terms of the said deal. However, it is outside the scope of this book to analyse these matters.

[13]There are good introductions to the international variance in SWFs found in the Chapters "Overview of Extractive Resources Management in Indonesia" and "Sovereign Wealth Funds and Impact Investing in Australia". See also the forthcoming edited volume by Okpanachi and Tremblay (2020).

Table 1 Sovereign wealth funds types and objectives

Classification of sovereign wealth funds				Legal basis
Objectives		Type	Investor	
Stabilisation	Protect and stabilise the budget from market volatility in revenue/exports and/or earn greater returns than foreign exchange reserves. Smooth revenues and expenditures, from fluctuations in commodity prices	Stabilisation funds	Central bank or monetary authority	• Constitutive law • Fiscal law • Constitution • Company law • Other laws and regulations
Capital preservation	Future liability funds preserving the real value of capital to meet future, contingent liabilities	Pension reserve funds	Intergenerational sovereign investor	
Capital maximisation	Increase savings for future generations. Liability profiles are multi-generational, usually have long investment horizons with higher target returns. They favour relatively diversified and low-risk portfolios and illiquid assets such as infrastructure, public equity and private equity and real estate	Future generation funds	Liability sovereign investor	
	Manage a nation's foreign reserves and earn greater returns than foreign exchange reserves; and assist monetary authorities dissipate unwanted liquidity	Reserve investment funds	Liquidity sovereign investor	

(continued)

Table 1 (continued)

Classification of sovereign wealth funds				Legal basis
Objectives		Type	Investor	
Socio-economic development	Fund social and economic development, and/or sustainable long-term capital growth. Characterised by long-term investments, especially in infrastructure and private equity. They invest in illiquid and strategic assets which can stimulate economic development	Strategic development SWF	Strategic development SWF	

Source SWFInstitute.org https://www.swfinstitute.org; Bortolotti et al. (2015)

eventually be depleted. Hence, any oil (or gold, or diamonds) that we take out of the ground today will not be available for future generations. This introduces three problems, each of which can be addressed with a SWF.

First, we like to think of resources in the ground as money in the bank. The problem is that untapped natural resources do not earn any interest. Once a state's natural resources are extracted and sold, then the wealth of the people is (in effect) spent. In the process, the country is simply draining its resource pool (and wealth) and leaving nothing in return. A SWF can take the revenues generated from the sale of natural resources and invest them in the market, generating a return on that investment. In the process, a state can turn its fixed pool of capital into a dynamic stream of revenue.

Second, when a natural resource remains in the ground, its value is shared by current and future generations. Once it is brought to market, however, its value is captured by the current generation, at the expense of future generations. This raises concerns about inter-generational equity. If we think of natural resources as the people's "family jewels", then we need to think of ways to reimburse future generations when selling the family jewels. A SWF can do this, by squirrelling away the resource money, to be used by future generations—for example, as a means to transition to a new economic footing, once the resource is depleted.

Third, the savings inherent to a SWF can either be used as an engine of investment (if the money is invested at home in necessary projects), or as a bulwark against Dutch Disease (when invested abroad). The rationale for using a SWF to affect domestic investments is somewhat convoluted, as the money is probably better spent through traditional government channels, where the spending process will be exposed to more political scrutiny.[14] When the money is invested abroad, however, the government can protect the local economy from the inflationary and currency appreciation affects that we associate with Dutch Disease. This brings us to the second type of SWF: as a buffer or stabilisation fund.

4.1.2 Buffer Funds

Sovereign wealth funds can also be used to stabilise or buffer the national economy from the price volatility that characterises these markets. This can be done in two ways. First, a country that relies on natural resource wealth is susceptible to the swings in production (in some years, there will be major finds, in other years, there will be nothing), as well as significant variation in the market price for their resource. As we have already noted, this is challenging for states, whose expenditure needs are much more stable: the state still needs to protect, educate and inspire its people, even if there were no new discoveries that year, or the global price of their resources is falling.

To compensate for the cyclical nature of resource revenues, a country can invest the resource revenues and then allow the return on these investments to slowly (and predictably) return to the state (or the people directly). This is consistent with one form of a savings fund, as described in the previous section. But here, our focus is on the means by which the money is repatriated to the domestic economy. By investing its money in this way, the government can plan and prepare for how to use the revenues in a way that does not overwhelm the domestic economy. This buffers the state economy from the ups and downs of the extractives sector.

This type of buffer fund is also useful to immunise against Dutch Disease. When this is the objective, the state can invest its revenues offshore (beyond the home economy), in effect protecting the local economy from the sort of inflationary/appreciatory effects that we associate with Dutch Disease. When all the resource revenues are deposited in an offshore account, the returns on these investments can be transferred back to the state (or people) in a steady and predicable rate.

The second way to buffer the national economy is to invest the money in a way that can counter-balance, or at least water-down, the price volatility inherent to the extractives sector. After all, as the SWF grows, the returns from the country's investments can be ploughed back into the fund, allowing for new sources of revenue (not related to the original natural resource wealth). If the money was invested in IT, for example, then the returns on IT investments are unlikely to corelate with the price of oil (for example). Hence, returns on investments may increase even as the price on the natural resource is falling. A SWF allows the state to develop a more diversified

[14] When governments are suspected of being corrupt or inefficient, it is often tempting (if not always accurate) to think of a SWF as an alternative budget, one that can be created with better means of control and oversight. But SWFs that are poorly integrated with the budget can easily lead to a loss of overall fiscal control and problems of coordinating expenditures. In practice, corrupt officials have not found it difficult to bypass controls and drain the SWFs for their own private gain.

portfolio and with it a more diversified source of revenue—which can be used in a counter-cyclical fashion.

For this to happen, it is important to isolate the money in the SWF, so it is not just the result of a government budget surplus. Most states take in their resource funds, decide how to spend them, and then send whatever remains to a SWF. This approach tends to facilitate wasteful spending, as it is tempting to use more of this money than may be prudent. Worse, this approach can be destabilising as well. As the price of the resource will vary substantially from year to year, letting it flood into the government's budget introduces a great deal of uncertainty and volatility in an important source of government financing. To counter this, the government can set up a system such that the resource money goes directly to an offshore SWF, and the government budget is only granted an annual share of the expected return from that fund (see, e.g., the Chapter "Norway's Sovereign Wealth Fund", on Norway).

While more and more states are employing SWFs, it is doubtful whether they are appropriate in many contexts. Developing countries suffer from an acute lack of investment capital, and resource wealth can be used to fill the void. The challenges for these countries are to bring that money into the country in a way that does not overwhelm the economic capacity of the existing economy—so the capital can be utilised efficiently. It would be a shame if this needed capital is hidden away in a foreign account (susceptible to future theft/corruption), if it could be spent at home to develop domestic economic capacity (without increasing inflation and an appreciation of the currency).

In any case, it is quite challenging for any country to save funds for future generations whenever there are high demands to deal with critical issues on present generations experiencing poverty and a lack of basic needs. However, such decisions might be much easier in a developed economy like Norway or Canada in comparison with Nigeria or Guyana. Nevertheless, most governments tend to face challenges when implementing SWFs for various reasons:

1. transparency on the amount of money they collect into the funds,
2. maintain a constant flow of income inside the fund so it can continue to increase,
3. avoid large and/or constant withdrawals from the fund, or
4. wisely manage such funds.

4.2 Local Content Policies

A second tool used by state officials to manage their revenue wealth is the use of local content policies (LCPs). LCPs are used by states to secure a larger part of the resource rent: they are a means to help spread the wealth of the resources to a wider swatch of the local economy, as they spread out the allocation of jobs and government revenues. States introduce LCPs in an effort to create jobs, promote local enterprises, support local communities, and to secure new types of skills and technologies.

LCPs tend to be aimed at different goals, depending upon the state's particular needs, capabilities and contexts. Some states will wish to maximise local employment creation, others will want to expand domestic ownership and entrepreneurial activity or encourage inward direct investment, and still others might want to focus on building up a domestic supply industry, providing local skills, transfer of technology and technical competence.

None of this is easy. Even defining local content can be a significant challenge: "local" can be operationalised at various levels of aggregation (regional, national, community) and content might refer to workers, firms (their location, ownership, tax registration status), share of economic activity, or even the type of knowledge.

In addition, the capacity of states to use LCPs has been severely restricted by changes in international trade and investment rules. For example, membership in the WTO brings constraints in the form of several agreements that limit the state's capacity to employ LCPs, including the agreements concerning Trade-Related Investment Measures (TRIMS), the General Agreement on Trade in Services (GATS), the Agreement on Government Procurement (GPA) and the Agreements on Subsidies and Countervailing Measures (ASCM).[15] International funding agreements and bilateral investment treaties often include restraints on the use of local procurement tools and require lenders to abide by the Model Law on Procurement of Goods, Construction and Services (see Ssennoga 2006: 222); and international contracts and trade agreements are making it increasingly difficult to use national regulations as a means to prioritise local producers by introducing ISDS (investor state dispute settlement) clauses that allow investors/corporations to sue if a government's actions result in reduced profits.

Still, states have a need and desire to encourage local production, and their ability to control access to natural resources provides them with the leverage to do so. This leverage can be applied by prioritising their NOGMCs, by enacting national legislation, and even by employing specified contract terms.

In many states, NOGMCs are the strongest drivers of local content policies. It is for this reason that states tend to prioritise them when allocating their licences. After all, NOGMCs are more likely to hire local employees, train local workers and choose local contactors as their suppliers. Other states enact laws that formally require contractors to employ local goods, services and/or workers. For example, consider Article 27.1 of Angola's Law No. 10 on Petroleum Activities (2004):

> Licensees, the National Concessionarie (i.e., Sonangol) and its associates, and any other entities which cooperate with them in carrying out petroleum operations shall:
>
> (a) acquire materials, equipment, machinery and consumer goods of national production, of the same or approximately the same quality and which are available for sale and delivery in due time, at prices which are not more than 10 percent higher than the imported items including transportation and insurance costs and custom charge due;
>
> (b) contract local services providers, to the extent to which the services they provide are similar to those available on the international market and their prices, when subject to

[15]See Moses and Letnes (2017: 139–143) for a longer discussion about the nature of international constraints on LCPs.

> the same tax charges, are no more than 10 percent higher than the prices charged by foreign contactors for similar services. (Sonangol 2004)

As with SWFs, LCPs are not a panacea. They often result in less efficient and/or more expensive options than what can be provided by IOGMCs. Policymakers need to employ careful cost–benefit analyses to ensure that the political benefits from local content enhancement outweigh the potential loss in the form of securing a less efficient provider. LCPs can also fail, especially when poorly designed. After all, there is no point in forcing Licensees/Contractors to use local suppliers if there are no local suppliers able to provide the needed goods and services; this will only exacerbate the Dutch Disease. It is essential that policymakers establish targets that are within the reach and/or capabilities of local providers and the needs of IOGMCs, or have a plan for expanding those local capabilities.

Thus, there are a number of challenges for LCPs and practices. The first issue that any country should consider before implementing LCPs is to conduct a fair assessment of the current reality of its capabilities and then set reasonable targets to be achieved with a doable progression. When governments fail to conduct such assessments or to provide strategies to reach such goals, negative results ensue either with exorbitant increases in supply costs/prices and/or lack of compliance, resulting in potential fines, which might scare away investors. In this case, it is relevant to note that LCPs tend to focus on a compliance versus penalties approach rather than positive incentives to encourage higher performance. Another key element for any LCPs is to boost the capabilities of the competent authorities so they can monitor and provide oversight instead of relying on the industry to do so without a proper compliance system in place. In addition, it is paramount to invest in education and transfer of technology, so the local industries are able to secure the added value rather than exporting the low value commodity and/or relying on foreign technology. Furthermore, diversifying the economy and creating SWF might help to avoid the so-called resource curse. Finally, local content policies should have a beginning and an end as at some point the "locals" should be able to compete on the same level nationally and internationally.

4.3 Corporate Social Responsibility

Corporate social responsibility (CSR) is different from the other two tools in that this tool is not typically part of an explicit government policy.[16] Indeed, CSR is better understood as a response to ineffective government policy. For that reason, it may be more controversial than the other two instruments, as it has the potential to undermine government competencies and transfer even more power to powerful multinational actors (at the expense of the state). This instrument can also allow weak

[16] Although in some countries it is, e.g. India.

governments to continue to be so when the corporate sector addresses development and infrastructure needs within the country via their CSR activities.

There is remarkably little consensus about what constitutes CSR.[17] Indeed, economists have long argued about whether companies should be engaged in CSR at all, as it threatens to undermine shareholder value. Friedman (1970), for example, argued that the main social responsibility of business is to increase its profits. However, in the current climate of the United Nations' Agenda 2030 and its associated sustainable development goals, there is a greater emphasis and expectation placed on the private sector globally to enact responsible business for progressing a more sustainable and equitable pathway forward (Spencer 2018).

From an analytical perspective, it is difficult to measure the effect or effectiveness of CSR, or even to put a finger on what it is, exactly. The former is complicated by the fact that corporations are seldom willing to share the sort of data required to test and control for the effect of their corporate engagement; the latter is evident in the many faces of CSR. As Frynas (2009: 5) notes, CSR can take many different faces:

- business ethics and morality;
- corporate accountability;
- corporate citizenship;
- corporate giving and philanthropy;
- corporate greening and green marketing;
- environmental responsibility;
- human rights;
- responsible buying and supply chain management; and
- socially responsible investment.

In the most thorough study of CSR in an extractive sector (oil), Frynas (2009: 105) finds that the most socially responsible actors tend to be national oil companies, which invest heavily in social programs as part of their governments' broader social policy objectives. But NOGMCs are often derided in the broader literature for being less "efficient" than private companies, who often enjoy the luxury of ignoring the social and environmental consequences of their actions. In this light, CSR can be a double-edged sword: when employed by government-owned enterprises, it is seen to be inefficient; when done by multinational corporations (IOGMCs), CSR is seen as progressive.

CSR seems to be driven by two different developments. The first is a perceived need to right past wrong doings. In effect, CSR is the industry's response to blood on its hands. IOGMCs work in politically sensitive situations and under trying political conditions. In order to secure contracts under corrupt regimes, they are often forced to become corrupt themselves. In short, these companies face severe ethical challenges, and CSR is a strategy for dealing with those ethical challenges: it offers a means to insure against growing reputational costs, and to secure greater influence over the regulations that govern their industry (by encouraging voluntary, rather than mandatory or regulatory responses). "For the companies claiming to have a broader social

[17]For an influential introduction to the subject matter, see Carroll (1979, 1991).

responsibility, e.g. a 'force for good' (BP), and 'building a better world' (Shell), the paradox of plenty directly affects their legitimacy as corporate citizens" (Skjærseth et al. 2004: 18).

The second driver behind CSR has been the demise of the state: that is, the state's decreased capacity to supply the benefits and protections of traditional citizenship (see, e.g., Matten and Crane 2005). The reasons for this demise are many and include globalisation and the rise of market fundamentalism (read market liberalisation and deregulation). Whatever the reason, states now enjoy less scope to provide and defend social, political and civil rights (compared to, say the 1970s). This has created a vacuum, into which firms (and CSR) have flowed, often unwillingly.

The result has created a paradox, where "government failure to deliver effective governance in the form of environmental protection and economic and social development often results in both government and local communities seeking to shift the burden of delivering such benefits to oil and gas companies" (Wagner and Armstrong 2010: 144). In this new context, firms are embracing CSR as a cheap way to secure a social licence to operate or a reputational insurance (Minor and Morgan 2011; Spense 2011), avoid the spotlight effect (Spar 1999) of poor publicity, and/or increase the likelihood of securing future licences (and their rent!).

In short, there is a growing fear that CSR can undermine the effort for greater revenue transparency and effective governance. In this regard, it is part of a larger movement to move away from government regulation and capacity. It is for this reason that the former US Secretary of Labour, Reich (2008) has argued that CSR can do significant damage as it removes the incentives for government to enact laws and to undertake actions that maximise social welfare. More recently, at the end of 2019, Reich declared "corporate social responsibility is the second-biggest con of 2019 (Donald Trump remains in first place)" (Reich 2019).

In the end, there are clearly a number of challenges involving CSR from its definition to its need to manage expectations. However, it is clear that extractive resource companies tend to deal with resources, which most likely belong to the public and often are located in poor and remote areas. These investors should find a way to extract such valuable resources with broader support (i.e. not only from the host government but also with the support of local and wider public communities as well as from their shareholders and financiers). This is why their CSR policies and practices might assist them to secure a so-called social licence to operate; otherwise, they might find serious challenges to conduct their business. Nevertheless, a serious concern that extractive resource companies could face is the lack of governmental presence within their business operations. A number of countries do not have a fair distribution of resources nor wealth among its regions and population. This is why an area might be rich in resources with poor economic development at the same time as such rent might be moved to the capital and/or abroad. In these cases, extractive resource companies might be expected to intervene as if they were the state and build schools, roads, hospitals, etc. even though this is outside their expertise. Some companies might prefer to pay the relevant funds to the government so they can build it. But in other cases, the money might not be allocated as it should be. Resource companies might add information on CSR investments but it is less clear what the real benefits and impact that such "investments" might add to communities and society as a whole. In any case, the main challenge with such a complex topic is not even who put them in place but rather who will staff and maintain them.

5 Overview

In the book that follows, we compare a wide variety of cases and how they employ and react to these three tools: SWFs, LCPs and CSR. These cases vary significantly, in that they are written by authors with very different professional backgrounds (lawyers, engineers, political scientists, anthropologists, economists, activists…), and they come from very different national contexts and industries (sometimes minerals and metals, sometimes, petroleum, sometimes both and some external). Even the political contexts vary in terms of both level of aggregation (some of the cases are states within larger federal systems), and level of development.

Given this wealth of examples, we have decided to divide them into three different groups, and let the chapters speak for themselves. The first grouping compares how the sundry cases employ (or do not employ) a sovereign wealth fund. The second grouping looks at how the same group of cases employs local content policies. The third and final grouping looks at the use of corporate social responsibility in each of these twelve states.

Within each grouping, we list the cases in order of their relative wealth. This, we hope, will facilitate comparisons. This is no easy task, as the cases sometimes include more than one nation state, or are based on sub-national actors (such as provinces or states). But if we consider how each nation state is ranked in the recent World Human Development Index, as is done in Table 2, we can then list the cases in terms of their national ranking. In those cases, e.g. Alaska, where the case is a member state to a larger federal union, it is presented in the order of its federal ranking (i.e. the USA).

In this way, the reader will be free to choose how to organise his/her reading. Some readers may only be interested in SWFs and will elect to read that section to grasp the breadth of experiences we report. Other readers may be interested in a particular country, say Indonesia, and wish to choose the thematic chapters from each section for that country (reading the Indonesian chapter on SWF ("Overview of Extractive Resources Management in Indonesia"); the Indonesian Chapter on LCP ("Local Content Policy in Indonesia Oil and Gas Industry") and the Indonesian Chapter on CSR ("Practice of Corporate Social Responsibility (CSR) in Extractives Sector in Indonesia").

The main body of the text is organised to display the variety of ways that different states employ these three tools, and our concluding chapter will return to the lessons generated by this variance. However you decide to read the book that follows, we are sure you will find a whole new layer of wealth.

Table 2 Rank chapters by HDI (2019)

Country	Rank	Human Development Index (HDI) (value)	Life expectancy at birth (years) SDG3	Expected years of schooling (years) SDG 4.3	Mean years of schooling (years) SDG 4.6	Gross national income (GNI) per capita (PPP $) SDG 8.5
Nigeria	158	0.534	54.3	9.7	6.5	5086
Kenya	147	0.579	66.3	11.1	6.6	3052
Guyana	123	0.670	69.8	11.5	8.5	7615
Indonesia	111	0.707	71.5	12.9	8.0	11,256
Brazil	79	0.761	75.7	15.4	7.8	14,068
Iran	65	0.797	76.5	14.7	10.0	18,166
Trinidad and Tobago	63	0.799	73.4	13.0	11.0	28,497
Russian Federation	49	0.824	72.4	15.5	12.0	23,036
US	15	0.920	78.9	16.3	13.4	56,140
Canada	13	0.922	82.3	16.1	13.3	43,602
Australia	6	0.938	83.3	22.1	12.7	44,097
Norway	1	0.954	82.3	18.1	12.6	68,059

Source UNDP (2019)

References

API [American Petroleum Institute]. (2012). Composite index: Global rating and ranking fiscal systems for oil and natural gas upstream investment. Press release. American Petroleum Institute. Accessed January 5, 2020. https://www.api.org/~/media/files/news/2012/12-march/composite-index-onepager.pdf.

Auty, R. M. (1993). *Sustaining development in mineral economies: The resource curse thesis*. London: Routledge.

Blowfield, M., & Dolan, C. (2014). Business as a development agent: Evidence of possibility and improbability. *Third World Quarterly, 35*(1), 22–42.

Bortolotti, B., Fotak, V., & Megginson, W. L. (2015). The rise of sovereign wealth funds: Definition, organization, and governance. In S. Caselli, G. Corbetta, & V. Vecchi (Eds.), *Public private partnerships for infrastructure and business development*. New York: Palgrave Macmillan.

BP. (2019). *BP Statistical Review of World Energy 2019*. Online at https://www.bp.com/en/global/corporate/energy-economics/statistical-review-of-world-energy.html. Accessed January 5, 2020.

Burke, E. (1790). *Reflections on the revolution in France*. Online at https://constitution.org/eb/rev_fran.htm. Accessed December 28, 2019.

Carroll, A. B. (1991). The pyramid of corporate social responsibility: Toward the moral management of organizational stakeholders. *Business Horizons* (July–August), 39–48.

Carroll, A. B. (1979). A three-dimensional conceptual model of corporate social performance. *Academy of Management Review, 4*(4), 497–505.

EITI [Extractive Industry Transparency Initiative]. (n.d.). *Who we are.* Online at https://eiti.org/who-we-are. Accessed January 5, 2020.

Folke, C., Biggs, R., Norström, A. V., Reyers, B., & Rockström, J. (2016). Social-ecological resilience and biosphere-based sustainability science. *Ecology and Society, 21*(3), 41. https://doi.org/10.5751/ES-08748-210341.

Friedman, M. (1970, September 13). A Friedman Doctrine—The social responsibility of business is to increase its profits. *New York Times Magazine.*

Frynas, J. G. (2009). *Beyond corporate social responsibility: Oil multinationals and social challenges.* New York: Cambridge University Press.

George, H. (1992 [1879]). *Progress and poverty.* New York: Robert Schalkenbach Foundation.

Hickel, J. (2019). The contradiction of the sustainable development goals: Growth versus ecology on a finite planet. *Sustainable Development*, 1–12.

Johnston, D. (2008). Changing fiscal landscape. *Journal of World Energy Law and Business, 1*(1), 31–54.

Johnston, D. (2007). How to evaluate the fiscal terms of oil contracts. In M. Humphreys, J. D. Sachs, & J. E. Stiglitz (Eds.), *Escaping the resources curse* (pp. 53–88). New York: Columbia University Press.

Karl, T. L. (1997). *The paradox of plenty: Oil booms and petro-states.* Los Angeles: University of California Press.

Mahdavy, H. (1970). The patterns and problems of economic development in Rentier States: The case of Iran. In M. A. Cook (Ed.), *Studies in economic history of the Middle East from the rise of Islam to the Present Day* (pp. 428–467). London: Oxford University Press.

Matten, D., & Crane, A. (2005). Corporate citizenship: Toward an extended theoretical conceptualization. *The Academy of Management Review, 30*(1), 166–179.

Minor, D., & Morgan, J. (2011). CSR as reputation insurance: Primum non nocere. *California Management Review, 53*(3), 40–59.

Moses, J. W., & Letnes, B. (2017). *Managing resource abundance and wealth. The Norwegian experience.* New York: Oxford University Press.

Nilsen, H. R. (2019). The hierarchy of resource use for a sustainable circular economy. *International Journal of Social Economics, 47*(1).

Okpanachi, E., & Tremblay, R. (Eds.). (2020). *The political economy of natural resource funds.* Springer: Forthcoming.

Rabe, S. G. (1982). *The road to OPEC. United States relations with Venezuela, 1919–1976.* Austin: University of Texas Press.

Reich, R. B. (2008). *The case against corporate social responsibility.* Goldman School Working Paper Series, No. GSPP08-003. August 1.

Reich, R. (2019). The biggest business con of 2019: Fleecing workers while bosses get rich. *The Guardian.* https://www.theguardian.com/commentisfree/2019/dec/29/boeing-amazon-business-ethics-robert-reich. Accessed December 29, 2019.

Ricardo, D. (1817). *The works and correspondence of David Ricardo, Vol. 1: Principles of political economy and taxation.* Online Library of Liberty.

Rockström, J., Steffen, W., Noone, K., Persson, Å, Chapin, F. S., III., Lambin, E., et al. (2009). Planetary boundaries: Exploring the safe operating space for humanity. *Ecology and Society, 14*(2), 32.

Ross, M. L. (2012). *The oil curse. How petroleum wealth shapes the development of nations.* Princeton: Princeton University Press.

Ross, M. L. (1999). The political economy of the resource curse. *World Politics, 51*(2), 297–322.

Sachs, J. D., & Warner, A. M. (1995). *Natural resource abundance and economic growth.* NBER Working Paper 5398.

Skjærseth, J. B., Tangen, K., Swanson, P., Christiansen, A. C., Moe, A., & Lunde, L. (2004). *Limits to corporate social responsibility: A comparative study of four major oil companies.* Lysaker, Norway: Fridtjof Nansen Institute.

Sonangol, E. P. (2004). Republic of Angola LAW No. 10/04 of 12 November 2004. Petroleum Activities Law. Translation provided by Miranda, Correia, Amenoeira and Associates. Accessed December 11, 2016. https://www.eisourcebook.org/cms/files/attachments/other/Angola%20Petrloleum%20Activities%20Law,%202004.pdf.

Spar, D. (1999). Foreign investment and human rights. *Challenge, 42*(1), 59–85.

Spense, D. R. (2011). Corporate social responsibility in the oil and gas industry: The importance of reputational risk. *Chicago-Kent Law Review, 86*(1), 59–85.

Spencer, R. (2018). Corporate social responsibility as sustainable development: A critical anthropology of development perspective. In M. Brueckner, R. Spencer, & M. Paull (Eds.), *Disciplining the undisciplined: Perspectives on responsible citizenship, corporate social responsibility and sustainability* (pp. 73–88). UK: Springer.

Ssennoga, F. (2006). Examining discriminatory procurement practices in developing countries. *Journal of Public Procurement, 6*(3), 281–349.

Starr, A. (2007). Caracas: Living large on oil. *The American Scholar*. March 1. Online at https://theamericanscholar.org/letter-from-caracas/#.Xg-rgfx7mUk. Accessed January 3, 2020.

Steffen, W., Richardson, K., Rockström, J., Cornell, S. E., Fetzer, I., Bennett, E. M., et al. (2015). Planetary boundaries: Guiding human development on a changing planet. *Science, 347,* 6223. https://doi.org/10.1126/science.1259855.

The Economist. (1977, November 26). The Dutch disease.

UNDP. (2014). *Development and extractive industries*. Issue Brief. January. Online at https://www.undp.org/content/dam/undp/library/Poverty%20Reduction/Extractive%20Industries/Issue-Brief-Extractive-Industries.pdf. Accessed March 10, 2020.

UNDP. (2019). *2019 Human Development Index Ranking*. Online at https://hdr.undp.org/en/content/2019-human-development-index-ranking. Accessed December 30, 2019.

Vernon, R. (1971). *Sovereignty at Bay: The multinational spread of US Enterprises*. New York: Basic Books.

Wagner, J., & Armstrong, K. (2010). Managing environmental and social risks in international oil and gas projects: Perspectives on compliance. *Journal of World Energy Law and Business, 3*(2), 140–165.

World Bank. (2003). Striking a better balance—The World Bank Group and extractive industries: The final report of the extractive industries review. *Open Knowledge Depository*. Last modified December. Accessed May 4, 2015. https://openknowledge.worldbank.org/handle/10986/17705.

World Bank. (2019a). *Extractive industries*. October 11 version. Online https://www.worldbank.org/en/topic/extractiveindustries/overview.

World Bank. (2019b). *Monthly indices based on nominal US dollars, 2010 = 100, 1960 to present*. World Bank Commodity Price Data (The Pink sheet). Updated 3 December 2019. Online at https://pubdocs.worldbank.org/en/561011486076393416/CMO-Historical-Data-Monthly.xlsx. Accessed January 1, 2020.

World Bank. (2019c). *Extractives industries: Overview*. Updated 11 October 2019. Online at https://www.worldbank.org/en/topic/extractiveindustries/overview.

Yergin, D. (2009). *The prize: The epic quest for oil, money and power*. New York: Free Press.

SWF

Public Wealth Management and Distribution in the Extractive Industry in Nigeria

Chilenye Nwapi

Abstract Few issues surrounding public wealth management in Nigeria enjoy broad geo-political agreement and support. The most vexed question is that of which level of government ought to have control over extractive resources. Behind the resource control debates is the question of the appropriate resource revenue sharing formula the country ought to adopt. Under current arrangements, states receive 13% of the revenues generated from natural resources located within their territory, leaving the federal government with control of 87% of the country's resource wealth. Agitations by oil-producing states for state resource ownership and control have been raging for decades but have not resulted in any change in resource ownership and control. Although the federal legislature has the constitutional authority to change the current sharing formula to give states more share than they currently have, it has not yet exercised that authority. In order to ensure that extractive companies pay adequate revenue to the state, the government imposes an assortment of fiscal instruments, including: royalties, petroleum profits tax, capital gains tax and value-added tax. In order to ensure that revenues generated from the resources benefit both present and future generations, the country operates a sovereign wealth fund. The operation of this fund has been riddled in managerial controversy, even its constitutionality has boggled the minds of Nigerian legal pundits. This chapter analyses these issues, proffering views on how they may be better understood and approached.

Keywords Excess Crude Account · Resource wealth · Sovereign wealth fund · Derivation

---

C. Nwapi (✉)
Canadian Institute of Resources Law, Calgary, Canada
e-mail: enyenwapi@yahoo.com

E. G. Pereira et al. (eds.), *Sovereign Wealth Funds, Local Content Policies and CSR*, CSR, Sustainability, Ethics & Governance,
https://doi.org/10.1007/978-3-030-56092-8_2

1 Introduction

Extractive resource revenue management is perhaps the sorest spot in Nigeria's body politic since the 1960 independence. Conflict after conflict over the negative environmental and health impacts of extractive resource projects, over the underdevelopment and marginalisation of the subnational region (the Niger Delta region) where the resources are located, over the distribution of the wealth from the resources between this region and the rest of the country, and over the lack of accountability of public officers responsible for managing revenues from the resources have been the defining characteristics of the political economy of resource extractivism in Nigeria (Nwapi 2010). The explanation for this should not, however, be laid open to one-sided generalisations promulgated in the resource curse literature regarding the link between natural resources and conflict and underdevelopment, for there is strong evidence that its roots lie more deeply in structural distortions that for several decades have cut off the extractive economy (particularly the oil economy) from other sectors of the national economy (see Obi 2018). This is not to say, however, that the resource curse theory has no explanatory value for understanding the Nigerian resource tragedy. It is rather to highlight what a number of Nigerian scholars have suggested that the Nigerian situation is much more mixed and complex than the resource curse theory wants us to believe (see, e.g., Obi 2018). There is, for instance, what Finkel (2017: 178) calls the "chauvinistic" attitude and "bunker mentality" of many public agency personnel as a factor militating against the effective functioning of regulatory institutions around the world. These two pathologies manifest quite strongly in Nigeria in the form of ethnicism and distributive politics, whereby public officers (elected or appointed) view their main mission as that of grabbing their own personal and ethnic share of the common wealth instead of helping to create and sustain it (see Ikpeze et al. 2004).

This chapter discusses Nigeria's experience with the management of its extractive resource wealth. It addresses questions surrounding resource ownership, the fiscal regime for generating resource revenue, the distribution of resource revenue among the different levels of government and the establishment and management of natural resource funds. The chapter highlights the complexity of resource wealth management in Nigeria and provides thoughts on contemporary debates around the subject.

Following this introduction, Sect. 2 presents an overview of the development of the extractive industry in Nigeria. Section 3 analyses the constitutional framework for extractive resource wealth management in Nigeria. Section 4 discusses the fiscal regime for extractive resource revenue generation in the country, while Sect. 5 analyses the country's efforts to establish a sovereign wealth fund (SWF). Section 6 concludes the discussion.

2 The Development of the Extractive Industry in Nigeria

Nigeria is rich in oil, gas and solid mineral (i.e., mining) resources. While oil and gas resources are located in commercial quantities in the Niger Delta region, solid minerals are spread all over the country. Some of the key solid minerals are coal, tin, gold, lead/zinc, iron ore, limestone and gemstone. Although the history of solid mineral development in Nigeria dates decades further back than oil and gas development (Maduaka 2014),[1] the latter has dominated the country's extractive industry since the mid-1970s. Increasing oil discovery led to increased production, which in turn combined with rising oil prices to render Nigeria almost completely dependent on oil, as federal budgets were funded almost exclusively from oil revenue (Nwapi 2010; Frynas 2001). This was at the expense of the solid mineral sector which, until then was, together with agriculture, the mainstay of Nigeria's economy (Aigbediion and Iyayi 2007). Today, oil and gas resources account for about 80% of government revenue and more than 90% of foreign exchange earnings (Agbaeze et al. 2015). Nigeria holds the enviable honour of Africa's largest oil producer after Libya and the tenth largest in the world, with approximately 37 billion barrels of proven oil reserves (Central Intelligence Agency 2017). It is Africa's largest natural gas producer and the ninth largest in the world, with about 187 trillion cubic feet of proven natural gas reserves (Central Intelligence Agency 2017).

In contrast, Nigeria's solid mineral industry is vastly underdeveloped. Until 1995, the mineral sector had no standalone ministry, but was married with power to form the Ministry of Mines and Power. Even then, the mines arm of the marriage was treated like the "junior partner" (Eyre 2007: 28). This was in spite of the fact that in other African countries, such as Botswana, Ghana, South Africa and Zambia, solid mineral development was generating massive revenues and was the mainstay of their economies, similar to oil in Nigeria (Eyre 2007). This is not to say, however, that the government was not interested in developing the sector. Rather, the revenues coming from oil and gas were so humongous that the government lost focus on other economic sectors. With dwindling oil revenue, however, due mainly to decreasing demand for oil and oil price volatility, the government has been compelled to see the need to diversify the economy, and the solid mineral sector has been identified as a priority sector because of its potential to bring significant economic benefits to the country.

Like most countries, Nigeria's extractive sector development is private sector-driven although this is historically more so with solid minerals than with oil. While all extractive resources belong to the state, the government's policy thrust is to create an enabling environment for the sector to thrive. The government grants oil concessions or licences to third parties to exploit oil within defined acreages while reserving for itself the right to participate directly in oil and gas development. The two main participation arrangements used by the government are joint-venture agreements

[1]Oil was officially discovered in commercial quantities in 1956 while commercial mining of tin ore began as far back as 1904, and coal was discovered in commercial quantities in 1909. By 1919, the Nigerian Geological Survey was already established.

and production-sharing contracts. The national oil company—the Nigerian National Petroleum Corporation (NNPC)—serves as the government's representative in all participation arrangements. While the NNPC is supposed to focus on the commercial interests of the government, it has often engaged in policy-making and regulation although much more so previously than now (Ministry of Petroleum Resources 2017). Authority to grant oil licences is vested in the Minister of Petroleum Resources, while technical issues of policy, regulatory control, fiscal control and licencing and the day-to-day monitoring of oil and gas operations are the responsibilities of the Department of Petroleum Resources (DPR)—a technical unit of the Ministry. The Petroleum Act, 1969, does not specify how licences are to be awarded, leaving the Minister with the discretion to decide. Discretionary allocation was the official licencing policy of the government until around 2000 when the government adopted competitive bidding with the intention of promoting transparency. However, since the Petroleum Act is silent on the method of awarding licences, discretionary allocation remains a legally valid method of licencing. Where bids are to be conducted, they are conducted by the DPR based on its established guidelines.

In the case of solid minerals, however, the role of the government is limited to that of regulator and administrator because there are no government participation arrangements in the industry. The government awards mineral titles based on a "first come, first served" principle. Under section 9 of the Mines and Minerals Act, 2007, however, the Minister of Solid Minerals and Steel Development may make regulations requiring that specified exploration licences and mining leases be awarded based on competitive bidding. All other mineral titles, such as a reconnaissance permit, a small-scale mining lease and a quarry lease, are awarded solely on the first-to-apply principle; authority to make an award is vested in the Minister. Where competitive bidding is to be conducted, the Mining Cadastre Office is charged with its administration and conducts the bid based on guidelines established under the Mines and Minerals Regulations 2011 (Regulation 24).

Nigeria has faced significant challenges in developing its extractive sector. The challenges range from lack of revenue accountability, environmental degradation, community–industry conflicts and weak institutions. The oil and gas sector is still regulated primarily by the 1969 Petroleum Act, the oldest existing petroleum statute of any major oil-producing country in the world. A Petroleum Industry Bill (PIB) was introduced in 2008 to carve a new regulatory path for the sector but was stalled by politics and industry lobbying. Several versions of the bill were produced without success in the legislature. In 2016, the government adopted a new policy strategy that not only decoupled the gas policy from the oil policy, but that also split the PIB into five smaller bills: the Petroleum Industry Governance Bill (PIGB), the Petroleum Industry Fiscal Framework Bill, the Petroleum Industry Downstream Administration Bill, the Petroleum Industry Revenue Management Framework Bill and the Petroleum Industry Host Communities Bill. The PIGB was passed by the federal legislature in mid-2018. However, the President declined to sign the bill due to concerns over certain sections of the bill that seemed to cut his decision-making power over petroleum development. Although under section 58(4) of the Nigerian Constitution, the two chambers of the federal legislature can vote to pass the bill into law if the President fails to sign it after a period of 30 days have passed since it was submitted to him, neither chamber has yet called for a vote.

Major developments in the solid mineral sector include the passage of a new mining statute in 2007 (the Nigerian Minerals and Mining Act), which established a Mining Cadastre Office (an independent agency responsible for the management and administration of mineral titles), the adoption of a new mining policy in 2008 (the Nigerian Minerals and Metal Policy 2008) and the adoption of a policy roadmap for the sector in 2015 (Ministry of Solid Minerals and Steel Development 2016). In 2016, the government committed 30 billion naira (approximately $82 million) to boost the development of the sector, with priority for minerals considered of strategic importance to the economy (coal, bitumen, iron ore, barites, gold and lead/zinc) (Fayomi 2018). In the third quarter of 2018, mining's contribution to overall Gross Domestic Product (GDP) was 10.55% (National Bureau of Statistics 2018a; b). Although this was lower than its contribution in four previous quarters, it represented a significant increase from 2016 levels since the 1980s (Ministry of Mines and Steel Development 2016). It represents a visible sign that the steps undertaken so far to boost the sector are bearing positive results.

3 The Constitutional Framework for the Management of Extractive Resource Wealth

Nigeria is a federation made up of three tiers of government, namely: the federal, state and local governments, with a federal capital territory (FCT). The FCT is freestanding in the sense that, although under the ultimate control of the federal government, akin to a state, it has a distinct political system composed of the executive, the legislature and the judiciary. In fact, given the distinct status of the FCT, it can be said that the country operates a four-tier rather than a three-tier system.[2] However, it continues to be regarded as a three-tier mainly because the FCT is almost akin to a state and that is how it will be regarded in this chapter.

There is a widespread belief among Nigerians that Nigeria's federal structure is brutally flawed by an over-concentration of political and economic power in the centre (see Kalu 2008). This power imbalance is treated as the original sin in Nigeria's political historiography and has given rise to a chant of "restructuring" as the most important step towards the country's survival as a nation (see Abutudu 2010). That is to say, unless the federation is restructured to reduce the concentration of power in the federal government, all efforts to address the country's myriad of socio-economic and political problems must prove vain. While the exact nature of restructuring needed remains unsettled in the national debate, the most controversial issue is control over natural resources. Under the current constitutional arrangement [section 44(3)], ownership of natural resources is vested in the federal government; there is no state

[2]The 1999 Constitution recognises the distinct standing of the federal capital territory in the distribution of resource revenue.

or individual ownership. The federal government owns the resources, subject to the requirement that it shall manage them for the benefit of the entire country.

Management power over the resources flows from ownership, and this is fully established by the constitutional allocation of legislative power. There are two principal heads of legislative power under the 1999 Constitution: the Exclusive Legislative List (ELL) and the Concurrent Legislative List (CLL). The ELL contains items exclusively allocated to the federal government. The CLL deals with matters on which both federal and state governments are granted power to legislate. In the case of a conflict between federal law and state law enacted pursuant to the CLL, the doctrine of paramountcy applies to give supremacy to federal law. Items that cannot be directed into either the ELL or the CLL recede automatically into what, by constitutional convention, has become known as the Residual List (Ingelson and Nwapi 2014).

By virtue of section 251(1)(n) of the 1999 Constitution, power to regulate "mines and minerals (including oil fields, oil mining, geological surveys and natural gas)" and matters incidental to these items belong to the ELL and therefore are under exclusive federal jurisdiction. The federal government's power is all-inclusive. It covers not only the establishment of the policy, legal, fiscal and institutional regimes for the exploration and exploitation of the resources but also the establishment of the regime for managing their externalities. However, there are state laws and even local council bylaws that impose certain fees or levies on companies (including extractive companies) operating within their territories. But those fees or levies are treated under federal fiscal laws as nothing other than part of extractive companies' cost of doing business, like any other expense an extractive company makes, deductible for the purposes of calculating an extractive company's taxable profits. However, since most of the activities related to extractive resource development take place in local areas under the jurisdiction of states, states have taken an interest in at least protecting their environment by enacting laws and establishing institutions to deal with environmental matters within their territories. These state laws and institutions, however, only complement the powers of the federal government to manage the negative fallout of extractive resource development and do not supplant or compete with them, since the exclusive power of the federal government to regulate extractive resource development is settled law. Thus, even state environmental laws are very limited in their scope of application.

While the federal government has all-inclusive power to regulate extractive resource development, the regime for the distribution of revenue generated from the resources among the three tiers of government is dealt with substantially by the constitution. Section 162(1) of the constitution provides for the establishment of a "Federation Account" into which shall be deposited all revenue accruing to the government of the federation, except revenues from personal income taxes of military and police personnel, personnel of the ministry of foreign affairs and residents of the FCT. Revenue deposited in the Federation Account is to be allocated to the three tiers of government based on a distribution formula established by an act of the federal legislature. In establishing the distribution formula, however, the federal legislature must be guided by the following revenue allocation principles: "population, equality of States, internal revenue generation, land mass, terrain as well as population density".[3] Revenue accruing to local government councils in the Federation Account is to be allocated to the relevant state for the benefit of its local governments, and each state must deposit the revenue in a "State Joint Local Government Account". Thus, local governments do not receive their allocations directly from the Federation Account, and the exact amount they receive is determined by an act of the federal legislature.[4]

Regarding revenues from natural resources, the Constitution stipulates that "the principle of derivation shall be constantly reflected in any approved formula as being not less than thirteen per cent of the revenue accruing to the Federation Account directly from any natural resources".[5] Thus, the distribution formula for natural resource revenue is based on the principle of derivation. The derivation principle stipulates that a state's share of natural resource revenue depends on how much of the resources in question is derived from the territory of that state (Akinola and Adesopo 2011; Suberu 2010). Its underlying philosophy is that resource extraction generates externalities that disproportionately affect residents of the states where the resources are located; therefore, providing those states a percentage of the revenues generated from those resources would serve to compensate the residents for the negative impacts. Under current arrangements, states receive 13% of the revenues generated from resources located within their territory. This means that the federal government controls 87% of Nigeria's resource wealth. However, the 13% established by the Constitution is the minimum that states can be allocated. The federal legislature has authority to increase it, but it has not yet exercised it.

The issue of how much of the revenue from extractive resources that the resource-bearing states should receive is one of the most controversial issues in Nigeria. Historically, resource wealth distribution has been governed, at different points in time, mainly by two principles: the derivation principle (already explained) and the needs principle. The needs principle allows distribution to be made based on the responsibilities of each tier of government and the financial expenditures and obligations of the various governments (Adango 2015). Needs is accordingly an equalising principle that is based on the rationale that all citizens should have equal claim to the national wealth regardless of the region of the country in which they reside (Adango 2015). Whether derivation or needs, the actual distribution formula has historically gone through a "back-and-forth process" (Appiagyei-Atua 2005). It went from 100% in 1953 (pre-independence period) to 50% in 1960, 45% in 1970, 40% in 1975, 0% in 1979 in favour of a Special Account for mineral producing areas, 2% in 1982, 1.5 in 1984, 3 in 1992 and 13% since 1999 to date (Akinola and Adesopo 2011). The thinking of oil-producing states is that the changing revenue distribution formula has its roots in the fact that oil is found mainly in areas inhabited by the minorities who are disadvantaged in the political power play among the major ethnic groups and who (at least until relatively recently) were unable to pose a real threat to

[3] 1999 Constitution, section 162(2).

[4] 1999 Constitution, section 162(5)–(8).

[5] 1999 Constitution, section 162(5)–(8).

national unity (Oguine 1999). There is much force in this thought, for during the early post-independence period when national revenue came mainly from commodities (such as cocoa, groundnuts and palm oil) produced in areas inhabited by the major ethnic groups, not only was derivation the generally accepted principle for allocation of national revenue, the percentage of revenue distributed to the commodity-bearing states was as high as 45% (Omorogbe 2002; Oguine 1999). The debate is a continuing one and is at the heart of the intensified quest for restructuring.

4 The Fiscal Regime in Nigeria's Extractive Industry

The fiscal regime for petroleum consists of several fiscal instruments that can be basically divided into: taxes, resource rents and royalties, state participation and incentives. The regime for solid minerals is distinct from that of petroleum although some of the fiscal instruments applicable to petroleum are equally applicable to solid minerals. The major difference is that there is no state participation in the solid mineral sector. Also, some taxes that are applicable to petroleum do not apply to solid minerals, the most important example being petroleum profit tax (PPT). Mainly for want of space and the fact that petroleum fiscal regime has generated a lot of debate in Nigeria, the following discussion focuses on petroleum. It would suffice, however, to note that solid mineral companies in Nigeria are subject to the following taxes: companies income tax (CIT), personal income tax (for individuals, cooperatives and partnerships engaged in mining business), royalties, withholding tax, capital gains tax (CGT), value-added tax (VAT) and annual surface rent. In the discussion of the petroleum fiscal regime, it is beyond the scope of this chapter to evaluate the strengths and weaknesses of the various fiscal instruments. After all, it is not the theoretical impact of each of the instruments, viewed in isolation, that matters, but the net impact of all of them, as that is what determines a company's decision to invest in a country as well as the government's revenue (Wilde 2016).

4.1 Taxes

Petroleum companies operating in Nigeria are subject to several taxes, including petroleum profit tax (PPT), company income tax (CIT), capital gains tax (CGT) and value-added tax (VAT). PPT is the principal tax and is established under the Petroleum Profit Tax Act, 2004 (PPTA). Section 2 of the Act defines petroleum operations as:

> the winning or obtaining and transportation of petroleum or chargeable oil in Nigeria by or on behalf of a company for its own account by any drilling, mining, extracting or other like operations or process, not including refining at a refinery, in the course of a business carried by the company engaged in such operations, and all operations incidental there to and sale of or any disposal of chargeable oil by or on behalf of the company.

PPT payments are made either in cash or in-kind, depending on the operating contract between the company and the government. The applicable tax rate is 85% (subject to certain incentives) or 65.75% during the first five years of their operations to enable them to recover their capital expenditure.

However, associated gas producers are subject to the CIT under the Company Income Tax Act, 2004, rather than to the PPT and are assessed at the rate of 30% of taxable profits. Thus, a company must segregate its associated gas profits from its other profits in order to determine tax payable under the CIT. On the other hand, CGT is payable, under the Capital Gains Tax Act, 2004, on capital gains that accrue to a company when the company disposes its chargeable assets (regardless of the location of the assets) and is payable at the rate of 10%. Under the Value-Added Tax Act, 2004, VAT is charged at a flat rate of 5% on supplies of taxable goods and services.

4.2 Resource Rents and Royalties

Every corporation involved in the production of oil and gas is statutorily obliged to pay royalty. Royalty payment is governed principally by the Petroleum Act, 1969 and the Petroleum (Drilling and Production) Regulations, 1996. Chargeable oil for the purposes of determining royalty is calculated by ascertaining the quantity of crude oil produced on a field by field basis and reducing that quantity by deducting the quantities used for production operations, the quantities used for re-injection and the quantities lost through evaporation. Royalty is thus paid on net crude oil production. The royalty rate depends on whether the oil is produced onshore or offshore. For onshore areas, it is 20% while the rate for offshore depended, until a 2019 amendment of the Deep Offshore and Inland Basin Production Sharing Contract Act, on the water depth. Depths up to 100 m water depth was 18.5%; up to 200 m water depth 16.5%, from 201 to 500 m water depth 12.5%; from 501 to 800 m water depth 8%; from 801 to 1000 m water depth 4%; and beyond 1000 m water depth 0% (Lawal 2013). Under the 2019 amendment, however, introduced a mix of production and crude price-based royalty system. It specifies a baseline royalty of 10% for crude oil and condensates produced in the deep offshore (greater than 200 m water depth) and 7.5% for the Frontier and Inland Basin. In addition, a royalty based on the price of crude oil, condensate and natural gas at the relevant time will apply when the crude price exceeds 20/barrelinagraduatedmanner: 0 0 and 20/barrel; 2.5 20 up to 60/barrel; 4 60 and up to 100/barrel; 8 150/barrel; and 10% when it is above $150/barrel.

4.3 State Participation

State participation is regarded as part of the fiscal regime because one of its main purposes is to enable the country to generate maximum revenue from petroleum resources. The Nigerian government participates directly in oil and gas development through its national oil company, the Nigerian National Petroleum Corporation (NNPC). The two major participation arrangements are joint-ventures (JV) and production-sharing contracts (PSC). Under a JV, the NNPC and the oil company contribute to fund the oil operations according to the proportion of their equity holdings in the JV. Companies in JV with the NNPC are assessed to tax at the rate of 65.75% of chargeable profits during the first five years of operation of the JV and thereafter at the rate of 85% (KPMG 2014). Due to NNPC's inability to fund its own share of JV operations, the JV option is no longer popular. The preferred arrangement is the PSC, under which the NNPC is the sole concessionaire, and engages a company to conduct the petroleum exploration at its own financing risks. If it is successful, the company will recover its costs when commercial production begins but will lose everything if the exploration is not successful. The company is assessed to tax at the rate of 50% of chargeable profits (KPMG 2014).

4.4 Incentives

As a response to declining levels of exploration and production in the 1980s, which reduced petroleum revenue, the Nigerian government entered into a Memorandum of Understanding (MOU) with their joint-venture partners under which it provided several incentives to the partners in exchange for certain work commitments (Ayoade 2010). It appears, however, that the MOU schemes were cancelled around 2000, the incentives system has taken on a life of its own under the PPTA and is composed of deductions, investment tax allowance/tax credit and tax holidays.

Section 10 of the PPTA provides for allowable deductions before taxes are charged. Such allowable deductions include: rent incurred by a petroleum company in respect of land or buildings occupied under an oil-prospecting licence or an oil-mining lease for disturbance of surface tights or the like; royalties, non-productive rents; expenses incurred for repair of premises, plant, machinery, or fixtures used for carrying on petroleum operations; etc. (Idubo 2015). In 1996, the Nigerian Supreme Court expanded the scope of allowable deductions by interpreting the meaning of "petroleum operations" under the PPTA.[6] Shell had claimed for deductions for "foreign exchange losses, Central Bank commissions and educational scholarship expenses". This claim was rejected by the Federal Board of Inland Revenue. The Court held that this claim should be allowed because the expenses "were 'incidental to petroleum operations' and were 'wholly, exclusively, and necessarily' incurred for this purpose". It is not necessary that the expense be directly related to petroleum operations; it suffices that the petroleum company has a "statutory or contractual obligation to incur [the] expense" (Ayoade 2014: 189). In June 2018, however, the Federal High Court held that payments for gas flaring activities were not deductible because they were penalties and not expenses (Anderson Tax 2018).

[6] *Shell v Federal Board of Inland Revenue*, (1996) NWLR (Pt 466) 285.

Capital Allowance relates to expenses on equipment, pipelines and storage facilities, buildings and drilling costs. The applicable rate of Capital Allowance is 20% of the cost of the qualifying asset for the first four years and 19% for the fifth year (Wahab and Diji 2017). Capital Allowance deduction is restricted: for an accounting period, a company's tax cannot be less than 15% of the tax which it would have paid had no Capital Allowance been granted it (Wahab and Diji 2017). A petroleum investment tax allowance relates to new investments in assets and is available where a petroleum producing company executes a PSC with the NNPC. Section 22 of the PPTA allows companies that incur capital expenditure for petroleum operations a tax allowance of 5% for onshore operations and 10% for operations in territorial waters and the continental shelf.

Tax holidays are given to companies with "pioneer status" for the first year of their production operations. The company must have incurred capital expenditure of not less than 5 million Naira (Ghebremusse 2014).

4.5 Indirect Taxes

Indirect taxes are taxes levied on corporations that are not tied to the corporations' income or profits, but typically to fund specific government programmes. An example of indirect tax aimed at funding a specific government programme is provided under the Tertiary Education Trust Fund Act, 2011. The Act mandates all companies registered in Nigeria to pay annually 2% of their assessable profits into the Tertiary Education Trust Fund [section 1(2)], which shall be used "for the rehabilitation, restoration and consolidation of tertiary education in Nigeria" [section 3(1)]. The tax is assessed and collected by the Federal Inland Revenue Services, and the fund is managed by a Board of Trustees. Also, companies are required to contribute 3% of their total annual budget to the Niger Delta Development Commission to help fund development in the Niger Delta region of the country.[7]

4.6 The New Petroleum Fiscal Policy

In 2016, the Ministry of Petroleum Resources released a draft of National Petroleum Fiscal Policy whose main objective was to separate the fiscal requirements for oil from those for gas (Ministry of Petroleum Resources 2016). The new policy is based on six key principles: (1) right pricing (pricing should be fixed by market forces), (2) sustainability, (3) non-consolidation/non-recovery of gas costs from oil income (gas project costs should not cross-subsidise oil projects), (4) distinct fiscal treatments for oil and gas, (5) upstream incentives for gas-for-development investments and (6) midstream incentives (incentives to attract midstream investments).

[7]Niger Delta Development Commission (Establishment) Act, 2000, section 14(2)(b).

The new policy proposes a National Hydrocarbons Tax in place of the current PPT. Although it retains the current Companies Income Tax, the policy places more emphasis on royalties than on tax and proposes to replace the current royalty system, which is based on water depth, with a new system that is based on price and volume of production. The implication is that even if a company operates in an area with more than 1000 m water depth, it will still pay royalty, whereas under the current system no royalty is payable. The afore-mentioned 2019 amendement to the Deep Offshore and Inland Basin Production Sharing Contract Act was an implementation of this policy.

The policy also proposes the elimination of several deductions and reliefs currently in place, such as petroleum investment allowance, investment tax allowance, investment tax credit, gas flare deductibility and caps recoverable costs incurred both locally and abroad. Overall, the prime objective of the policy is to increase government revenue from oil and gas resources. The policy is expected to be implemented through the Petroleum Industry Fiscal Framework Bill.

5 Managing the Resource Wealth for Present and Future Generations

Until 2011, resource wealth management in Nigeria was based on political arrangements between the federal government and the federating units. Historically, the first proposal was made during the military era in the mid-1980s for the establishment of a stabilisation fund. The idea was for the government to save additional money when oil price exceeded US$16 per barrel and to draw from the savings when oil price fell below that price (World Bank 2003). The proposal was implemented in 1989. However, the government lacked the fiscal restraint to manage the fund, as it made withdrawals from the fund even in times of oil price boom in violation of the spirit of the fund (World Bank 2003). The fund could thus not be sustained, but the government renewed its efforts in 2004 by establishing another stabilisation fund, called the Excess Crude Account (ECA), to help the country meet its budget deficits and contribute to local infrastructure development. This political arrangement which, like its predecessor, had no constitutional or legal parentage allowed the federal government, the custodian of the Federation Account, to draw upfront into the account national oil revenue excess of the budgetary benchmark price, so that only revenue within the benchmark would be shared between the three tiers of government based on the existing formula (Ugwuibe 2012).

From the very beginning, the ECA was riddled in controversy due mainly to its lack of legal backing. It came under severe attack from a number of governors who felt that the federal government was illegally withholding revenues constitutionally meant to be shared by the three tiers of government (Ezeani 2012). A fusillade of lawsuits challenged its legality (Ekokoi 2015). Observers raised concerns about its management: there was no systematic way to track the operations of the account—how money came in and how money went out—due to absence of "rules of practice governing deposits, withdrawals and investments" (Natural Resource Governance Institute 2017). In the 2017 Resource Governance Index, the ECA was ranked as the most poorly managed sovereign wealth fund (SWF) in the world (Natural Resource Governance Institute 2017).[8] But its greatest obstacle was lack of political will to save in times of bumper harvest (Nigerian Extractive Industries Transparency Initiative 2017). Increased public pressure led to the enactment of the Nigeria Sovereign Investment Authority (NSIA) Act, 2011, apparently to fix the ECA. At the time of its enactment, Nigeria was one of only three OPEC countries without a legally backed SWF (Eguzozie 2011).[9]

The NSIA Act not only purports to address the questions surrounding the legality of the ECA. It also purports to provide rules for the establishment of a structured, central storehouse of resource revenues and rules for the collection, allocation, management and disbursement of the funds, based on long-term national development goals and strategies, with potential positive implications for transparency and accountability. This is mainly what distinguishes it from the ECA, apart from its statutory backing and wider scope. Also, in the inaugural edition of the African Sovereign Wealth Fund Index released in June 2018, which ranked African sovereign wealth funds based on four indicators: governance and disclosure, size, domestic investment mandate and source of funding, Nigeria was ranked first (Nanfuri 2018).

Below is a discussion of the main features of resource revenue management under the NSIA Act. As will be seen, although the Act is a significant step forward on the part of Nigeria to design a mechanism to manage its resource wealth, its capacity to achieve inclusivity in resource wealth management has been called into question (Olawuyi and Onifade 2017). In addition, the Act has ended up in the same constitutional hot water from which the ECA could not recover.

5.1 Main Features of the NSIA Act

The NSIA Act establishes a savings account. Its policy backdrop consists of building a savings base for the benefit of future generations of Nigerians, boosting the development of infrastructure in critical sectors, and in insulating Nigeria's economy from oil-price volatility-induced external shocks (Ugwuibe 2012). What is saved, however, is not a percentage of oil revenue, but the difference between budgeted earning and actual earning, which means that when actual earning is equal to or below budgeted earning, nothing would be saved. However, the Act provides other sources of funding

[8]The index assessed how 81 resource-rich countries governed their extractive resources (oil, gas and minerals).

[9]The others were Ecuador and Iraq. Most observers believe that Nigeria joined the league of countries with SWFs only with the enactment of the NSIA Act. This belief, however, ignores the fact that the ECA was in itself a type of SWF, and despite its legal problems and management deficiencies it continued to exist de facto as such.

for the SWF outside resource revenue. Section 30 of the Act provides for "Residual Funds" from the Federation Account above the "Budgetary Smoothing Amount" to be transferred to NSIA. Since the Federation Account does not hold only resource revenue, resource revenue is not the only source of funding for the SWF. However, as resource revenue is the major source of funding for the Federation Account, it is the major source of funding for the SWF. Discussed below are the main features of the SWF established under the NSIA Act.

5.2 The Funds

The Act establishes three SWFs: the Future Generations Fund (FGF), the Nigeria Infrastructure Fund (NIF) and the Stabilisation Fund (SF). The FGF is a diversified investment portfolio for the benefit of the future generations of Nigerians. The NIF is a portfolio dedicated to the development of "critical infrastructure" to facilitate the attraction of foreign investment and to promote "economic diversification and growth". The SF is an investment portfolio intended to provide supplemental stabilisation funding for the country in times of need, that is, when other funds set aside for fiscal stabilisation have been exhausted or are insufficient for that purpose (NSIA Act, s 4(1)). Thus, the SF provides the country a buffer against external shocks.

Ownership of the funds is vested jointly in the federal government, the state governments, the FCT and the local and area councils [NSIA Act, s 32(1)]. They hold the funds on behalf of the Nigerian people. None can transfer or assign or otherwise dispose of or encumber any of its interests in the funds [NSIA Act, s 32(2) and (3)]. As one commentator has wryly argued, this latter provision is intended to prevent state governors from encumbering their interests in the funds by using them as collateral to secure foreign debts (Ezeani 2012). Foreign debt acquisition has become customary among Nigerian state governors (Debt Management Office 2018; National Bureau of Statistics 2018a, b).

5.3 Administration of the Funds

The administration of the three funds is vested in NSIA, which is further vested with the authority to receive, manage and invest money on behalf of the funds and to reinvest portions of the profits and proceeds generated through investments of the funds.[10] The Act also established a Governing Council (headed by the President) whose function is only advisory.[11] Allocation of money to the funds is to be determined by the Board of Directors of NSIA established under section 15 of the Act. This need not be in equal proportion. What is required is that none of the funds shall

[10] NSIA Act, s 4(2).

[11] NSIA Act, s 7.

be allocated less than 20% of any available funding.[12] Currently, the allocation is as follows: FGF: 40%, NIF: 40% and SF: 20% (Hove and Ncube 2014).

5.4 *The Investment Framework*

The NSIA Act provides an investment framework for each of the funds. The framework mandates NSIA to develop an investment plan for the FGF and the NIF as well as a reinvestment plan for the proceeds, interests and dividends derived from each fund. There is no explicit requirement for such a plan to be developed for the SF, but this would seem to be implicit in the provisions for SF since an investment plan would at all times need to be developed. For the FGF, "the investment plan may be subject to strict short-term, tailored confidentiality restrictions".[13] Investments in the NIF are to focus on domestic infrastructure development in such areas as power, gas, transport, agriculture, healthcare, housing and water resources, using both direct investments and public–private partnerships.[14] The Act allows NSIA to make financial investments with money in the NIF pending when it becomes practical to make infrastructure investments [section 41(2)]. NSIA may enter into infrastructure co-investment arrangements with private companies. Social infrastructure investments in under-served sectors or regions that are likely to yield less economic returns are limited to 10% of the funds in the NIF [section 41(5)]. Realised proceeds, interests and dividends from each fund are to be reinvested in new or existing assets for that fund (sections 40, 44). The investment allocation for the SF is split between hedge assets and growth assets [section 47(1)].

In all cases, NSIA's investments shall align with the country's investment policy and priorities. NSIA may keep its investment plans strictly confidential for a short time to preserve its ability to effectively make strategic investments (sections 39(2), 41(3)).

5.5 *Withdrawal of Funds from the SWF*

The NSIA Act does not establish strict withdrawal conditionalities. Section 34 of the Act authorises the Board, with the approval of the Governing Council, to declare that funds be distributed from the "uninvested and uncommitted available funds" realised from the profits of NSIA, provided that NSIA has set aside sufficient funds to meet its anticipated operational costs before payment of any declared distributions, however, which shall not be more than 60% of NSIA's profits at the time of distribution. Such

[12]NSIA Act, s 31.

[13]NSIA Act, s 39(2).

[14]NSIA Act, s 41(1).

declared funds are to be paid into the Federation Account for distribution to the owners of the SWF according to their respective contributions.[15]

The Act contains withdrawal provisions specific to the SF. Section 47 gives the Minister of Finance authority to direct NSIA to utilise the fund's capital and assets to supplement available resources to stabilise the economy. The Minister only needs to demonstrate "urgency" and satisfy NSIA that the Budgetary Smoothing Amount has been depleted and that actual revenue from hydrocarbons is lower than projected revenue. The Minister may direct NSIA to release from the SF an amount equal to the difference between actual and projected hydrocarbons revenue.[16]

5.6 The Constitutional Fragility of the NSIA Act

The passage of the NSIA Act and its eventual take off was more the product of a political arrangement than a legal solution to the problems faced by the ECA. Right from the beginning, state governors believed that the Act was unconstitutional for violating the constitutional stipulations on revenue distribution in light of existing Supreme Court decisions. Although following political discussions with the federal government, the governors had decided to support the Act's enactment, when faced with increased financial burdens arising from the passage of a new minimum wage law that increased workers' salaries across the country, they withdrew their support for the law and sought to truncate the eventual establishment of NSIA and the transfer of any funds to it (Ugwuibe 2012). The governors threatened to challenge the Act in court if the federal government went ahead with any transfer of funds. They had earlier consented to the establishment of the fund but later claimed that the Act did not accurately reflect the substance of what they consented to (Ezeani 2012). In late 2011, however, the governors reached a settlement with the federal government to proceed with the establishment of NSIA and the transfer of $1 billion seed money drawn from the ECA (Ugwuibe 2012; Ezeani 2012). The problem with the political settlement–that is, the manner in which it was done–is that the constitutional question still remains.

The constitutional question is whether money deposited (or to be deposited) in the Federation Account can be transferred to another account instead of distributed among the three tiers of government. Most analysts have cited section 162(3) of the Constitution, which provides that all money in the Federation Account shall be distributed among the three tiers of government, to argue that it cannot be done (Ezeani 2012). This view, however, ignores the fact that the same section 162(3) authorises the federal legislature to prescribe the "terms" and the "manner" in which money in the Federation Account shall be distributed. While this does not allow the federal legislature to unilaterally authorise the transfer of any money in the Federation Account to another account, such a transfer would seem to be constitutional if

[15] NSIA Act, s 35.

[16] NSIA Act, s 48.

freely agreed to by all the three tiers of government. However, it is essential that the agreement properly involves the three tiers of government without leaving out any tier. This is where the NSIA Act fails.

Thus, the problem with the earlier political arrangement that established the ECA was that it did not properly involve the three tiers of government. If money in the Federation Account belongs jointly to the tiers, there is nothing in the constitution or in the principle of fiscal federalism that prohibits the tiers from entering into an agreement to save a portion of that money in a joint account, such as an SWF. It does not violate the constitutional requirement that the money shall be distributed among the tiers because each tier is simply contributing a freely agreed portion of its share of the distribution to the SWF. Such an agreement would be lawful and binding if it followed the constitutionally laid down procedure for making fiscal decisions especially on the part of state governments. In this respect, the state governors must obtain the approval of their state legislatures before entering into such an agreement with the federal government. This is mandated by section 120 of the Constitution, which provides as follows:

1. All revenues or other moneys raised or received by a State (not being revenues or other moneys payable under this Constitution or any Law of a House of Assembly into any other public fund of the State established for a specific purpose) shall be paid into and form one Consolidated Revenue Fund of the State.
2. No moneys shall be withdrawn from the Consolidated Revenue Fund of the State except to meet expenditure that is charged upon the Fund by this Constitution or where the issue of those moneys has been authorised by an Appropriation Law, Supplementary Appropriation Law or Law passed in pursuance of section 121 of this Constitution.
3. No moneys shall be withdrawn from any public fund of the State, other than the Consolidated Revenue Fund of the State, unless the issue of those moneys has been authorised by a Law of the House of Assembly of the State.
4. No moneys shall be withdrawn from the Consolidated Revenue Fund of the State or any other public fund of the State except in the manner prescribed by the House of Assembly.

It is clear from the above that revenues accruing to a state shall generally be paid into a Consolidated Revenue Fund of that state [subsection (1)] and that a law of the state legislature is required to authorise the state executive branch to allow money to be withdrawn from the Fund [subsections (2)–(4)].[17] Since the consent given by state governors to the passage of the NSIA Act and the establishment of the SWF did not follow this constitutionally laid down process, the Act is unlikely to pass constitutional muster.

The popular view is that a constitutional amendment is required to legalise the Act (see Olawuyi and Onifade 2017; Ekokoi 2015); this is not necessary. While it

[17] Whether money from the Federation Account is first paid into the Consolidated Revenue Fund of the state before a portion of it is transferred by the state to the agreed SWF, or that portion of the money is transferred directly from the Federation Account to the SWF, makes no difference provided that the transfer is backed by a law of the state legislature.

is true that "the Constitution does not provide for the SWF" (Olawuyi and Onifade 2017: 331), it need not provide for it for the fund to be legally established; moreover, there is nothing in the Constitution that prohibits its establishment. It is submitted that the NSIA Act can be regularised without any constitutional amendment. What is required is for state governors to obtain statutory authorisations from their legislatures to allow the transfer of the already agreed upon portion (under the NSIA Act) of their constitutionally allocated revenue to the SWF. In addition, since local government councils have a constitutional interest in the Federation Account,[18] each state must consult its local government councils to obtain their consent to the establishment of the SWF. Failure to do so would render the SWF open to constitutional challenge by local governments.

Other analysts have argued that the NSIA Act is fully consistent with section 80(1) of the Constitution which provides as follows:

> All revenues or other moneys raised or received by the Federation (not being revenues or other moneys payable under this Constitution or any Act of the National Assembly into any other public fund of the Federation established for a specific purpose) shall be paid into and form one Consolidated Revenue Fund of the Federation.

It is argued that this provision "recognises the existence of revenues payable under an Act of the National Assembly into different public funds established for a specific purpose" and which are not payable into the Consolidated Revenue Fund of the Federation (CRFF) and that "[t]he SWF is one of such funds" (Odude 2008). Opponents of this view have argued, however, that the CRFF is an account that belongs exclusively to the federal government, whereas the Federation Account belongs to the Federal Republic of Nigeria (Amadi and Obutte 2018). That is to say, all money due to the federal government, including money due to it from the Federation Account, is paid into the CRFF (Amadi and Obutte 2018). There appears to be no judicial authority in support of this interpretation. The Constitution itself does not define the terms "Consolidated Revenue Fund of the Federation" and "Federation Account". However, it defines "Federation" to mean "the Federal Republic of Nigeria",[19] thereby suggesting that the above interpretation of section 80(1) is wrong. However, section 120 of the Constitution also establishes the Consolidated Revenue Fund of the State, into which shall be paid "[a]ll revenues or other moneys raised or received by a State" (with exceptions similar to those established for CRFF under section 80(1)). This companion provision strongly suggests that the CRFF belongs exclusively to the federal government regardless of the definition of Federation provided under the Constitution.[20] It follows that the constitutional question remains. And unless and until it is resolved, the sustainability of the SWF established under the NSIA Act will, as one scholar has suggested, "rely on the strength of the

[18]Section 162 of the 1999 Constitution.

[19]Section 318(1) pf the 1999 Constitution.

[20]It must be pointed out that outside the definition of "Federation" provided in the Constitution, the multiple contexts in which the Constitutional draftsperson uses the term that makes it unclear whether it is referring to the federal government or to the Federal Republic of Nigeria.

political deals cut between the state governors and the federal government, which from previous experiences appear to be shaky agreements" (Ugwuibe 2012: 57).

5.7 *Transparency and Accountability Mechanisms Under the SWF*

The NSIA Act [section 4(2)(d)] explicitly adopts a key prescription of the Santiago Principles by specifically enjoining NSIA to "implement best practices with respect to management independence and accountability, corporate governance, transparency and reporting on performance". To implement this principle, section 12 of the Act contains reporting and disclosure provisions, requiring NSIA to provide written yearly reports to the Governing Council "on the assets, liabilities, redemptions, realisations, sales, general performance by asset class and significant trends affecting [its] investment objectives". Also, under section 36, NSIA is required to keep proper books of accounts regarding its transactions and businesses in accordance with internationally recognised financial reporting standards and generally accepted accounting principles in Nigeria. And for every financial year, NSIA shall deliver its Annual Report to the President, the Minister of Finance, the Central Bank Governor, the National Economic Council, the National Assembly and the House of Assembly of each state [section 37(1)]. The Annual Report shall be made accessible to the public. There are also provisions mandating NSIA to publish its investment plans, policies and procedures (see, e.g., section 39(2) of the Act). These provisions are commendable and in 2016 Nigeria's SWF was adjudged the most transparent in Africa, scoring 9/10 in the Linaburg-Maduell Transparency Index administered by the Sovereign Wealth Institute (Diallo et al. 2016).

Yet, the Act contains a number of other provisions that undermine NSIA's capacity to effectively implement the Santiago transparency prescription. For instance, the withdrawal rules give too wide discretion to political actors. For instance, section 47(2) of the Act authorises the Minister of Finance to direct NSIA to "utilise capital and assets" in the SF to supplement available resources to stabilise the economy. The criteria set out for giving such direction is vague. The Act only requires the Minister to demonstrate "urgency", provided also the "Budgetary Smoothing Account" has been depleted. It does not provide any guidance on what conditions could be regarded as urgent (Olawuyi and Onifade 2017). There is no procedural mechanism in the Act to ensure that the Minister's demonstration of urgency can be evaluated by the public. In addition, the Act does not stipulate preconditions for withdrawal of money from the FGF. Also, section 34 gives the Board authority to declare that uninvested and uncommitted available money in the funds be distributed without establishing preconditions for making such a declaration other than that the declaration shall be approved by a resolution of the Governing Council. In fact, the granting of this authority to the Board, when combined with the authority granted to the Minister of Finance to direct that money in the SF be utilised, makes if difficult to know who has final authority over the utilisation of money in the SWF. These lacunae have negative implications for transparency and accountability in the management of the SWF.

5.8 The Question of Inclusivity

The NSIA Act opens up an important question about how a fiscal mechanism should be designed and implemented to address the issue of over-concentration of power at the centre, believed to be one of the most critical problems in Nigeria's federalism. In more practical terms, this question has to do with whether all stakeholders in the funds are adequately involved in decision-making regarding the management of the funds.

There is a lack of adequate representation of stakeholders in the Governing Council of NSIA. The Council consists of the President and each of the state governors, the federal Attorney General, the Minister of Finance and the Minister in charge of national planning, the Governor of the Central Bank, the Chief Economic Advisor to the President, Chairman of the Revenue Mobilisation, Allocation and Fiscal Commission, and four representatives of private sector. There is no representation for the FCT, Local Government and Area Councils, which by virtue section 32(1) of the Act hold ownership interests in the funds. In addition, the FCT, Local Government and Area Councils are excluded from the entities to whom NSIA shall submit its Annual Report. Scholars have also rightly pointed out the lack of stakeholder consultation during withdrawal of money from the fund (Olawuyi and Onifade 2017). The Minister of Finance determines when funds are to be withdrawn from the SF and instructs NSIA accordingly.[21] There is no requirement for the Minister to consult with the states let alone the FCT and local government and area councils before determining that funds should be withdrawn.

Another important inclusivity issue is whether the SWF effectively addresses the needs of both present and future generations of Nigerians or whether one generation has been sacrificed for the needs of the other. Applying the theory of functional distributive justice, it has been argued that a broader range of stakeholder representation in the management of the SWF is needed to cater for the needs of present and future generations (Olawuyi and Onifade 2017). One way to approach this is to appoint people of diverse age and social groups onto the Board, the Governing Council and the management team of NSIA. Under current arrangements, the Governing Council shall have two representatives of the youth and regard shall be had to gender equity in appointing persons to the Council.

Perhaps a better approach to the intra- and inter-generational equity question is to first look holistically at the funds established under the NSIA Act to ascertain whether if effectively implemented, the interests of present and future generations would be catered for, and then to ask whether there are adequate provisions in the

[21] NSIA Act, s 47(2).

Act to ensure their effective implementation. It would appear that the three funds established by the Act fully contemplate both intra- and inter-generational equity. The SF is especially suited for meeting the needs of the present generation while the FGF is especially designed to meet the needs of future generations. The NIF, however, would serve both present and future generations in that the infrastructure that would be developed, such as roads, rails, dams and healthcare systems, would outlive the current generation. Are there adequate provisions in the Act to ensure the effective implementation of the three funds? While there are obvious deficiencies in the Act, as pointed out in the preceding sections, it would appear that much depends on political will. So far, NSIA has been adjudged very positively on transparency, governance and disclosure, size, domestic investment mandate and source of funding by two independent ranking institutions. This is a positive sign for the realisation of the objectives of the Act.

5.9 The Coexistence of the ECA and the SWF

Contrary to expectations, the enactment of the NSIA Act did not put an end to the existence of the ECA. Both the ECA and the SWF established under the NSIA Act are being operated concurrently. The reason for their coexistence is hard to explain. From a legal standpoint, the NSIA Act says nothing about the continued existence of the ECA. Since the ECA was not established by statute, a judicial pronouncement would suffice to terminate its existence. Otherwise, what was required to bring it to an end was the same political arrangement that brought it into existence in the first place. Analysts believe it was due to "political struggles" between officials at the highest levels of government that the ECA was not "subsumed" in the SWF (Investopedia 2018). Initial seed money deposited in the SWF was $1 billion drawn from the ECA (Odude 2008). The natural expectation then was that the ECA would be fully absorbed into the SWF but eight years later that is not yet the case. The prudence of managing both accounts concurrently has been rightly disputed (see Odude 2008). However, it does not seem that any more moneys have been directed to the ECA since the NSIA Act was enacted. Instead, withdrawals have been progressively made from the Account and distributed among the three tiers of government. If this trend continues, then the ECA would be automatically shut down once all moneys in it have been drawn.

6 Conclusion

Public wealth management in Nigeria is riddled with controversy. Few issues surrounding it enjoy geo-political agreement. The most controversial is the question of who should have control over the resources. The Constitution vests ownership in the federal government, the propriety of which remains a vexed question in Nigerian public debates. Behind the resource control debates is the question of the appropriate resource revenue sharing formula for the country. Derivation is the current operative principle. Under current arrangements, states receive 13% of the revenues generated from resources located within their territory. The federal government therefore controls 87% of the resource wealth. Agitations by oil-producing states for state resource ownership and control have been going on for decades but have not resulted in any constitutional change in resource ownership. The current 13% derivation was fixed under the 1999 Constitution. Although the federal legislature has constitutional authority to increase it, it has not yet exercised that authority. The debate around resource revenue distribution is a continuing one. It is centred exclusively on oil rather than on both oil and mining, making it appear as though the %derivation applies to oil alone, whereas it applies to all natural resources. However, this is arguably because the amount of shareable revenue from mining is low. As mining's contribution to GDP continues to rise, the solid mineral sector would likely assert its place in the debate.

The fiscal regime for generation of revenue from extractive resources is characterised by an assortment of instruments intended to ensure that extractive companies pay adequate revenue to the State. They include royalties, PPT, CIT, CGT and VAT. Revenues generated from these sources are distributed among the three tiers of government in line with the derivation principle. Various government administrations have recognised the need for a resource-dependent country to "save for the rainy day" to help it withstand external shocks induced by oil and commodity price volatility. However, these funds were badly mismanaged due to lack of fiscal discipline on the part of government.

The constitutionality of the NSIA Act remains a live issue in Nigeria. While this chapter supports the view that the Act is potentially unconstitutional, it is not in agreement with the view that a constitutional amendment is required to save the Act. An agreement between the three tiers of government to establish an SWF is fully compatible with the constitution and the principle of fiscal federalism. What is needed to save the Act is for state governors to obtain the necessary statutory approvals from their state legislatures and to consult with their local government councils to obtain their consent to the transfer of funds from the Federation Account to the SWF. It is hoped that state governors will one day take this step to lay to rest the constitutional question.

Key lessons learned from the management of extractive resources in Nigeria include: (1) effective management of natural resources within a federation requires cooperation among the federating units; (2) resource revenue management frameworks must have sound legal backing; (3) transparency and accountability are key to the effective resource revenue management; and (4) political will to save is essential for the success of SWFs.

References

Abutudu, M. (2010). Federalism, political restructuring, and the lingering national question. In S. Adejumobi (Ed.), *Governance and politics in post-military Nigeria* (pp. 23–60). New York: Palgrave Macmillan.

Adango, Z. (2015). The principle of derivation and the search for distributive justice in the Niger Delta region of Nigeria: The journey so far. *Journal of Law, Policy and Globalization, 41,* 115–133.

Agbaeze, E. K., et al. (2015). Resolving Nigeria's dependency on oil—the derivation model. *Journal of African Studies and Development, 7*(1), 1–14.

Aigbedion, I., & Iyayi, S. E. (2007). Diversifying Nigeria's petroleum industry. *International Journal of Physical Sciences, 2*(10), 263–270.

Akinola, S. R., & Adesopo, A. (2011). Derivation principle dilemma and national (dis)unity in Nigeria: A polycentric planning perspective on the Niger Delta. *Journal of Sustainable Development, 4*(5), 251–263.

Amadi, L., & Obutte, P. C. (2018). The treasury single account and the search for effective revenue management in Nigeria's oil and gas sector. *Afe Babalola University: Journal of Sustainable Development Law and Policy, 9*(2), 101–124.

Anderson Tax. (2018). *Federal High Court rules that payments for gas flaring activities are not tax-deductible.* https://andersentax.ng/federal-high-court-rules-that-payments-for-gas-flaring-activities-are-not-tax-deductible/. Accessed February 20, 2019.

Appiagyei-Atua, K. (2005). Self-determination v state sovereignty: A critique of the African Commission's decision in the Ogoni case. In I. Law & I. Peoples (Eds.), *Joshua C & Niamh W* (pp. 313–314). Leiden: Martinus Nijhoff.

Ayoade, M. A. (2008–2010). Nigerian petroleum profits tax and financial incentives: *50-year requiem? University of Ghana Law Journal* 24:171–204.

Central Intelligence Agency. (2017). *The World Factbook 2017*. https://www.cia.gov/library/publications/the-world-factbook/rankorder/2244rank.html. Accessed February 20, 2019.

Debt Management Office. (2018). *States, FCT and federal governments' external debt stock as at 30th June, 2018*. https://www.dmo.gov.ng/debt-profile/sub-national-debts/2543-states-and-federal-government-s-external-debt-stock-as-at-30th-june-2018/file. Accessed January 20, 2019.

Diallo, B., et al. (2016). Sovereign wealth funds and long-term investments in Sub-Saharan Africa. *World Bank Policy Research Working Paper 7903*, https://documents.worldbank.org/curated/en/736301480964634718/pdf/WPS7903.pdf.

Eguzozie, B. (2011). How sovereign wealth fund can drive Nigeria's growth, infrastructure development. *Business Day Online*. https://www.businessdayonline.com/NG/index.php/analysis/features/21470-how-sovereign-wealth-fund-can-drive-nigerias-growth-infrastructure-development. Accessed February 10, 2019.

Ekokoi, S. E. (2015). Legal and constitutional evaluation of the Nigerian sovereign wealth fund. *Journal of Sustainable Development Law and Policy, 5*(1), 101–128.

Eyre, J. M. (2007). *An economic analysis of natural resources sustainability for the mining sector component*. Nigeria, World Bank. https://openknowledge.worldbank.org/bitstream/handle/10986/7941/561780ESW0whit10Box349489B01PUBLIC1.pdf?sequence=1&isAllowed=y. Accessed February 22, 2019.

Ezeani, J. C. (2012). An overview of the Nigerian sovereign investment fund. *The sovereign wealth fund initiative, The Fletcher School, Tufts University*. https://sites.tufts.edu/sovereignet/files/2017/08/Archives-An-Overview-of-the-Nigerian-Sovereign-Investment-Authority.pdf. Accessed December 20, 2018.

Fayomi, O (2018). *Repositioning the Nigerian mining sector through improved regulatory and incentives framework*. SPA Ajibade & Co. https://www.spaajibade.com/resources/wp-content/uploads/2018/06/NIGERIA-MINING-SECTOR-INCENTIVES-IN-PERSPECTIVE-Niyi.pdf. Accessed February 20, 2019.

Finkel, A. M. (2017). Beyond best-in-class: Three secrets to regulatory excellence. In C. Coglianese (Ed.), *Achieving Regulatory Excellence* (pp. 166–187). Washington, DC: The Brookings Institution.

Frynas, J. G. (2001). Corporate and state responses to anti-oil protests in the Niger Delta. *African Affairs, 100*(398), 27–54.

Ghebremusse, S. Z. (2014). *Assessing the petroleum fiscal regimes of Nigeria, Ghana, and Cameroon.* LLM Thesis, University of Toronto. https://tspace.library.utoronto.ca/bitstream/1807/67935/1/Ghebremusse_Sara_201411_LLM_thesis.pdf.

Hove, S., & Ncube, M (2014). *Sovereign wealth funds as a driver of African development.* quantum global. https://quantumglobalgroup.com/wp-content/uploads/2017/10/Sovereign-Wealth-Funds-as-a-driver-of-African-development.pdf. Accessed February 24, 2019.

Idubor, R. (2015). Appraising taxation and the Nigerian oil industry. *Journal of Law, Policy and Globalization, 37,* 188–197.

Ikpeze, N.I., Soludo, C.C., & Elekwa, N.N. (2004). Nigeria: The political economy of the policy process, policy choice and implementation. In C. Soludo, O. Ogbu & H. Chang (Eds.), The politics of trade and industrial policy in Africa: Forced consensus? (pp. 342–343). Trenton, NJ and Asmara, Eritria: Africa World Press.

Ingelson, A., & Nwapi, C. (2014). Environmental impact assessment process for mining, oil and gas projects in Nigeria: A critical analysis. *Law, Environment and Development Journal, 10*(1), 35–56.

Investopedia, (2018). *Excess Crude Account (ECA).* https://www.investopedia.com/terms/e/excess-crude-account.asp. Accessed February 20, 2019

Kalu, N. K. (2008). *State power, autarchy, and political conquest in Nigerian federalism.* Lanham: Lexington Books.

KPMG. (2014). *Nigeria's oil and gas industry brief.* https://www.blog.kpmgafrica.com/wp-content/uploads/2016/10/Nigerias-oil-and-gas-Industry-brief.pdf. Accessed March 1, 2019.

Lawal, T. K. (2013). Taxation of petroleum profit under the Nigeria's Petroleum Profit Tax Act. *International Journal of Advanced Legal Studies and Governance, 4*(2), 1–17.

Maduaka, A. G. (2014). *Contributions of solid mineral sectors to Nigeria's economic development.* Masters Thesis, Institute of Graduate Studies and Research, Eastern Mediterranean University, Cyprus. https://i-rep.emu.edu.tr:8080/xmlui/bitstream/handle/11129/1684/MaduakaAda.pdf?sequence=1. Accessed February 20, 2019

Ministry of Mines and Steel Development. (2008). National minerals and metals policy. https://www.minesandsteel.gov.ng/wpcontent/uploads/2016/04/National_Minerals_and_Metals_Policy.pdf

Ministry of Mines and Steel Development. (2016). Roadmap for the growth and development of the Nigerian mining sector. https://www.minesandsteel.gov.ng/wp-content/uploads/2016/09/Nigeria_Mining_Growth_Roadmap_Final.pdf

Ministry of Mines and Steel Development. (2016). *Roadmap for the growth and development of the Nigerian mining industry.* http://www.minesandsteel.gov.ng/wp-content/uploads/2016/09/Nigeria_Mining_Growth_Roadmap_Final.pdf. Accessed January 20, 2019

Ministry of Petroleum Resources. (2017). *National petroleum policy: Nigerian government policy and actions.* www.petroleumindustrybill.com/wp-content/uploads/2017/07/National-Petroleum-Policy-Approved-by-FEC-in-July-2017.pdf. Accessed February 1, 2019

Ministry of Petroleum Resources. (2016). *National Petroleum Fiscal Policy—Nigerian Government's Policy and Actions.* https://www.lawyard.ng/wp-content/uploads/2017/03/Draft-National-Petroleum-Fiscal-Policy.pdf. Accessed January 20, 2019

Natural Resource Governance Institute. (2017). *Resource governance index 2017.* https://resourcegovernance.org/sites/default/files/documents/2017-resource-governance-index.pdf. Accessed January 20, 2019

Nanfuri, P. (2018). Ghana ranks third in sovereign wealth fund index.*Joy Business.* https://www.myjoyonline.com/business/2018/june-5th/ghana-ranks-third-in-sovereign-wealth-fund-index.php. Accessed January 19, 2019

National Bureau of Statistics. (2018). *Nigerian gross domestic product report (Q3 2018).* file:///C:/Users/Owner/Downloads/GDP_Report_Q3_2018.pdf. Accessed January 26, 2019

National Bureau of Statistics. (2018). *Nigerian domestic and foreign debt—Q4 and full year 2017*. https://www.proshareng.com/admin/upload/reports/11049-NIGERIANDOMESTICANDFOREIGNDEBT2017-proshare.pdf. Accessed January 20, 2019

Niger Delta Development Commission (Establishment) Act. (2000).

Nigerian Extractive Industries Transparency Initiative. (2017). *The case for a robust oil savings fund for Nigeria.* NEITI Occasional Paper Series, Issue 2. https://www.proshareng.com/admin/upload/reports/FinalNEITIOPSonOilSavingsFund.pdf. Accessed January 17, 2019

Nwapi, C. (2010). A legislative proposal for public participation in oil and gas development in Nigeria. *Journal of African Law, 54*(2), 184–211.

Obi, C. (2018). Nigeria: The role of civil society in the politics of oil governance and revenue management. In I. Overland (Ed.), *Public brainpower: Civil society and natural resource management* (pp. 201–216). Basingstoke: Palgrave Macmillan.

Odude, F. (2008). The sovereign wealth fund should subsume the Excess Crude Account, *Financial Nigeria.* https://www.financialnigeria.com/the-sovereign-wealth-fund-should-subsume-the-excess-crude-account-blog-350.html. Accessed December 20, 2018

Oguine, I. (1999). Nigeria's oil revenue and the oil producing areas. *Journal of Energy Natural Resources Law, 17*(2), 111–120.

Olawuyi, D. S., & Onifade, T. T. (2017). Promoting functional distributive justice in the Nigerian sovereign wealth fund system: Lessons from Alaska and Norway. *Nigerian Yearbook of International Law, 1,* 317–354.

Omorogbe, Y., et al. (2002). The legal framework for public participation in decision-making on mining and energy development in Nigeria: Giving voices to the voiceless. In D. N. Zillman (Ed.), *Human rights in natural resource development: Public participation in the sustainable development of mining and energy resources* (pp. 549–587). New York: Oxford University Press.

Shell v Federal Board of Inland Revenue. (1996). NWLR (Pt 466) 285.

Suberu, R. (2010). The Nigerian federal system: Performance, problems and prospects. *Journal of Contemporary African Studies, 28*(4), 459–477.

Ugwuibe, C. (2012). *Strengthening the Nigerian sovereign investment authority: A policy analysis of the Nigerian Excess Crude Account and the Nigerian Sovereign Investment Authority Act.* MA thesis, University of California Los Angeles. https://cloudfront.escholarship.org/dist/prd/content/qt5zn1h1d6/qt5zn1h1d6.pdf. Accessed January 11, 2019

Wahab, L., & Diji, C. J. (2017). Comparative analysis of Nigeria petroleum fiscal systems using royalty and tax optimization models to drive investments. *Oil and Gas Research, 3*(3), 1–14.

Wilde, D (2016). *Key issues in natural resource taxation and revenue management in the commonwealth.* Economic Paper 96, The Commonwealth Secretariat Economic Paper Series. https://www.oecd-ilibrary.org/docserver/9781848599543-en.pdf?expires=1544711656&id=id&accname=id36621&checksum=BD588FB061D411C9CC4CF3DB7C739B0F. Accessed February 27, 2019.

World Bank. (2003). *Nigeria: Policy options for growth and stability.* Report No. 26215-NGA, Washington, DC.

Public Wealth Management and Distribution in Kenya's Extractives Sector

Kate Wanza Mavuti, Helen Hoka Osiolo, and Caroline Wanjiru Kariuki

1 Background

Countries that are endowed with extractive resources such as oil and gas can have tremendous impact to the economy if the emanating wealth is well managed, allocated, and distributed. Good governance of extractive resources creates employment thus improving well-being among national citizens. At the same time, for governments, this may generate new revenue streams and fund basic government services and stimulate further economic growth.

Despite this importance, extractive resources from many African countries remain poorly managed with their ensuing wealth unequally distributed. According to EU-UN partnership (2012), a raft of challenges contributes to new conflicts as well as act as impediments to the peaceful resolution of existing ones including: poor engagement of communities and stakeholders; inadequate benefit sharing mechanisms and strategies, mismanagement of extractive resources, inadequate institutional and legal frameworks, inadequate natural resources management and distribution, and poor transparency and accountability among the relevant state and related agencies.

K. W. Mavuti (✉)
Strathmore Extractives Industry Centre (SEIC), Strathmore University School of Law, Nairobi, Kenya
e-mail: kmavuti@strathmore.edu

H. H. Osiolo
Center for Research in Applied Economics, Strathmore University Institute of Mathematical Sciences, Strathmore University, Nairobi, Kenya
e-mail: hosiolo@strathmore.edu

C. W. Kariuki
Strathmore Institute of Mathematical Sciences, Strathmore University, Nairobi, Kenya
e-mail: cwkariuki@strathmore.edu

E. G. Pereira et al. (eds.), *Sovereign Wealth Funds, Local Content Policies and CSR*, CSR, Sustainability, Ethics & Governance,
https://doi.org/10.1007/978-3-030-56092-8_3

The debate on who owns and who manages extractives resources is evident in many developing countries and even more so for those countries with new discoveries of extractives resources. Though there are diversified extractives resources across countries, the fiscal regimes that determine how the revenues from extractives projects are shared between government, companies, and communities also vary depending on the fiscal tools. The International Monetary Fund reports that distinct fiscal regimes for extractives industries are critical in addressing key tax challenges. For example, extractives taxes that offer an attractive tax base/substantial rents based on efficiency and equity measures, pervasive uncertainty in commodity prices as well as in relation to geology, input cost, political risk, and asymmetric information, where the private sector is considered better informed than government when it comes to undertaking exploration and development. The government, on the other hand, is better informed with regard to its own future fiscal intentions, high sunk costs including the long production periods, extensive involvement of multinationals in some countries, and of state-owned enterprise in others, raising questions about how to address tax and the sharing of benefits from extractives (IMF 2012). Royalties, taxes, production sharing systems, and bonuses are some of the fiscal tools available to establish a fiscal regime to govern extractives resources (Natural Resource Governance Institute 2015).

This chapter seeks to identify the main extractives resources in Kenya; the fiscal regime within its extractives industry and who owns the resources; and how extractives resources are managed and distributed. The chapter also sheds light on Kenya's proposed Sovereign Wealth Fund, which aims to manage and invest income from current oil, gas, mineral, and other natural resources for future generations. While two bills have been proposed to parliament, neither has been approved, and Kenya remains without a Sovereign Wealth Fund.

2 Ownership of Extractive Resources in Kenya

The 2010 constitution of Kenya indicates under Article 62(1)(f) that public land in Kenya includes minerals and mineral oils. Further, Article 62(3) provides that public land, including the definition above, shall vest in and be held by the national government in trust for the people of Kenya. This provision states that this land shall be administered by the National Land Commission. What this essentially means is that minerals and mineral oils are owned and controlled by the state for the benefit of the citizenry. This constitutional provision of course reaches into the private ownership of property, fair compensation, and/or reallocation to land owners when minerals or mineral oils are found on their land.

Further, Article 69 (1) directs the state to guarantee sustainable exploitation, utilization, management, and conservation of the environment and natural resources, and to ensure that the accruing benefits are shared equitably, ensuring the resources are utilized for the benefit of the people of Kenya. In Kenya's nascent oil and gas industry, the matter of equitable distribution has seen intensive debate; whereby, the industry related legislation, that is, the Petroleum Act 2019, the Mining Act 2016, and associated regulations, seeks to provide guidance on the matter.

Of further importance is Article 201 of the constitution, which provides that the burdens and benefits of the use of resources shall be shared equitably between present and future generations. Again, Kenya's supreme law unequivocally prescribes the equitable sharing and distribution of national wealth, including natural resources wealth, not only among the national and county governments for the development of the current population, but also for the benefit of future generations. This principle, together with transparency and accountability, forms part of the principles of public finance in Kenya.

Kenya's extractives industry has experienced substantive legal, policy, and institutional reform in line with the Constitution 2010. The Mining Act 2016, specifically, Article 6(1) (a–c), states that every mineral in its natural state in, under, or upon land in Kenya or in or under a lake, river, stream, or water courses in Kenya, or in the exclusive economic zone and an area covered by the territorial sea or continental shelf, is the property of the Republic and is vested in the national government in trust for the people of Kenya. This law applies to all minerals, which are detailed in the First Schedule of the Mining Act, 2016 except for petroleum and hydrocarbon gases. Further, in April 2016, the Ministry of Mining launched the Mining and Minerals Policy 2016, which underpins the Mining Act, 2016. The policy ensures that key issues related to sustainable exploitation of natural resources such as community engagement, environmental issues, and the benefits from mining are addressed. In line with the provisions of the Constitution of Kenya 2010, the Mining Act, 2016 bestows the ownership of minerals with the national government in trust for the people of Kenya.

Likewise, the Petroleum Act 2019, which came into effect on March 28, 2019, vests all petroleum existing in its natural condition within Kenya and its continental shelf, with the national government in trust for the people of Kenya. This Act provides a framework for the contracting, exploration, development, and production of petroleum and repeals the Petroleum (Exploration and Production) Act 1984. The new Act also gives effect to relevant Articles of the Kenyan Constitution, which apply to upstream, regulation of midstream, and downstream petroleum operations. The Act provides further that the national government may participate in any phase of upstream operation and create a conducive environment for investment in petroleum operations and infrastructure development, including formulation of guidelines in collaboration with relevant national government agencies on development of petroleum investments and to disseminate them among potential investors.

Echoing the state duty enshrined in the Constitution, the Act also mandates the national government to ensure that petroleum operations and infrastructure development are carried out for the benefit of the people of Kenya. In its effort to promote petroleum operations and investments, the national government shall facilitate access to land for exploration activities in accordance with the Constitution and any other written law. In Kenya, negotiations over land rights can at times be highly charged due to the historical legacy of contention over land ownership in Kenya. In addition, and in many cases, areas where extractives resources are discovered are mainly used for agriculture and livestock production, residential or communal holdings, recreation, transportation, and at times commercial activities, further exacerbating tensions.

The Kenyan legal regime over the past few years can be said to be alive to the special and specific circumstances of land use and land ownership in Kenya while seeking to promote sustainable exploitation of the natural resources found therein. However, certain issues such as regulatory authorities and their scope, acquisition, compensation, and relocation still require more legislative attention and review. Unlike in other jurisdictions, where ownership of hydrocarbons or other natural resources is directly tied to private land ownership, the Kenyan legal regime is clear that the national government owns the resource, both onshore and offshore.

Additionally, the Energy Act, 2019 also came into effect on March 28, 2019 and repealed the Energy Act, 2006. This Act consolidates the laws relating to energy, provides for national and county government functions in relation to energy, promotes renewable energy, promotes the exploration, recovery, and commercial utilization of geothermal energy, provides regulation of midstream and downstream petroleum and coal activities, among others (The Energy Act 2019). In relation to renewable resources, Article 73 of the Energy Act, 2019 confers all unexploited renewable energy resources under or in any land in the National Government of Kenya.

3 Main Extractives Resources in Kenya

Kenya is abundant in resources that largely remain untapped. According to the Economic Survey (2019), the data on the contribution of mining and quarrying to GDP is captured alongside that of forestry and logging, fishing and aquaculture, and water supply to form the environment and natural resource sector. Though environment and natural resources as a sub-sector contributed to about 3.7% of GDP in 2014, this growth dropped to 3.2% of GDP in 2018. For mining and quarrying, the growth has been steady at 0.8% in the same period. This was attributed to larger significant growth rates in other industries than with or within the environment and natural resources.

For mining and quarrying, key resources documented include: soda ash, fluorspar, salt, crushed refined soda, carbon dioxide, diatomite, gold, coal, gemstones (cut), gemstones (rough), and titanium ore minerals (illeminte, rutile, and zireon). Other minerals that are not captured in the Economic Survey include limestone, manganese, gypsum, and niobium.

Kenya's petroleum sector on the other hand is described as a nascent sector. According to the Ministry of Mining and Petroleum, the key petroleum projects include oil and gas exploration, development of a crude oil pipeline, geoscientific data acquisition and management to further inform the sector, enhancement of liquified petroleum gas uptake, and the eradication of fuel adulteration, dumping and illegal LPG refilling. Unlike the presence of mineral occurrence across the eight former geographical regions in Kenya excluding Nairobi, petroleum resources are available in very few regions. In particular, Kenya has four (4) petroleum exploration basins and these are: Lamu Basin, Anza Basin, Mandera Basin, and Tertiary Rift Basin. Kenya's oil reserves in the Northern County of Turkana, which were discovered in 2012, have been estimated to be at 560 million barrels of oil (mnbl), which have been described as 2C, signifying the best estimate of contingent reserves.

Oil and gas exploration in the country began in 1956 with the discovery of commercially viable crude oil discovered in March 2012 with the discovery well—Ngamia 1 Well, in Lokichar Basin in Turkana County. To date, the National Oil Corporation reports that over 86 wells have been drilled with a majority within the Tertiary Rift. In June 2018, Kenya launched the Early Oil Pilot Scheme (EOPS) reportedly geared towards gathering technical data for the design and preparation of the Field Development Plan (FDP) and to test the international market for Kenyan Crude. In August 2019, Kenya reportedly flagged off the exportation of a cargo of approximately 200,000 barrels from the port of Mombasa, lifting the oil from the EOPS into the international market. The transportation of crude oil has, however, seen its fair share of challenges mainly due to poor infrastructure in certain areas and stoppages due to protests mainly by community members highlighting issues of insecurity, employment, and access to information, particularly about the beneficial sharing of accruing benefits.

In Kenya, with the petroleum sector being a nascent yet fast developing sector, the longer established mining industry affords us with comparative as well as complimentary information. In this light, the Mining and Minerals Policy of 2016 identify key factors that hinder the sector's performance, which may also apply to the petroleum or the oil and gas sector. To begin with is the legal regime; previously, the mining sector operated under outdated legal frameworks (the Mining Act Cap 36 of 1940). The Act did not consider several minerals that are mined and quarried on commercial basis. It also failed to provide clear guidelines on procedures and time line for licensing and supervision that resulted in cases of investors making speculations on when to sell or hoard minerals. A similar set of circumstances existed in the petroleum sector before the enactment of the Petroleum Act 2019.

Further, the linkages between mineral marketing and value addition are lacking. This is due to several factors such as inadequate expertise, under development of the mineral processing industry, a lack of appropriate technology, and high energy costs that contribute to the low level of value addition to Kenya's minerals. A lack of sector strategies for the marketing, promotion, and value addition to minerals domestically, regionally, and internationally of Kenya as a preferred mining destination limits the full benefit from its mineral wealth. Further, efforts to reduce environmental degradation and promote sustainable exploitation are hampered by wanting alignment of regulatory polices between both the mining and petroleum sectors and the environmental sector. Gender and labor issues have been identified in both sectors as hampering sustainable development. For example, child labor particularly in artisanal and small-scale mining, lack of access and control over land resources for women, patriarchal decision-making processes in both the workplace and beyond and either lack of or inadequate access to relevant information to facilitate sensitization of the wider community. Finally, despite the presence of a well structured, clear and enabling fiscal regime, there have still been calls to ensure transparency, accountability, and predictability throughout the entire fiscal regime and benefit and revenue collection, allocation, sharing, and management processes.

4 Fiscal Regime Within the Extractives Industry

The level of development of the extractives resource industry is largely defined by the fiscal regime. While most governments would want to enhance development of their extractives sector through fiscal regimes that are internationally competitive, they must do so by balancing the interest of the host community; often being difficult to achieve. Fees, royalties, and taxes are the common fiscal instruments applied to both petroleum and mining resources in Kenya.

The fiscal system used in Kenya's petroleum sector can be said to be a royalty/tax system and production sharing hybrid contract. There are two main fiscal systems employed in extractives: The concessionary or royalty tax system and the contractual-based systems where production sharing contracts (PSCs) and service agreements are classified. The choice on which fiscal system a country uses depends on the laws, policies and regulations, and objectives of that country in the exploitation of its natural resources. Thus, there is no one-size-fits-all system or preferred fiscal system in the extractives sector.

The royalty tax system is characterized by factors such as the oil and gas companies are contracted to explore for and produce oil and gas and where there are commercial funds; upon production, title of the resource passes to the company who in turn pays royalties on the resource produced. The companies also pay taxes on the profits from the sale of the resource. On the other hand, with the contractual systems, title over the resource remains with the state. Depending on the type of contract, the resource company receives its compensation via a share of the actual production. As mentioned above, the Kenyan scenario leans heavily toward the PSC system; however, the companies are still mandated by law to pay taxes on their profits/income. While the mining sector is required to make the payment of royalties to the state authorities, the petroleum sector in Kenya does not pay royalties.

4.1 Production Sharing Contracts (PSCs)

According to the Petroleum Act, there are areas delineated for petroleum exploitation known as "blocks." Section 16 of the Act provides that the licensing regime for these blocks for upstream operations shall be through an executed petroleum agreement or a non-exclusive exploration permit for the purpose of obtaining geological, geophysical, and geochemical information. The petroleum agreements or PSCs are entered into between the state and the company and are negotiated after the conclusion of bidding rounds. The Petroleum Act provides a model for the petroleum agreement known as the Model Production Sharing Agreement (Model PSC), which is prescribed in the Schedule of the Act. Just as the Petroleum Act 2019 repealed the previous Act, the accompanying Model PSC repeals all other versions contained in previous legislation. The Model PSC provides parties with a country-specific template for further negotiation. It is within this PSC that Kenya and its contractors agree on fiscal terms and conditions for the lifespan of the agreement based on both domestic laws and regulations as well as international norms and best practices. Section 53 of the Act bolsters this and provides that contractors (or oil and gas companies) shall pay the state all taxes, relevant fees, and levies in such manner as is stipulated in the petroleum agreement. Of note here is that despite the devolution of government in Kenya with the introduction of county governments introduced by the 2010 Constitution, the fiscal repository for oil and gas exploration activities remains with the national government. However, the Act makes reference to revenue sharing of oil and gas proceeds in Section 58, which provides that 'the national government's share of the profits derived from upstream petroleum operations shall be apportioned between the national government, the county government and the local community'. We shall discuss this further in the next section on revenue allocation.

4.2 *Profit Petroleum Sharing*

The hybrid system in Kenya's petroleum PSC system is prescribed by clause 37 of the Model PSC, 2019, which provides that profit petroleum shall be shared between the state and the contractor on a quarterly basis, according to the value of the R-Factor in respect of a specific contract area. This steers the country away from the previous system where production share was determined by a daily rate of production. The R-factor is defined as a ratio of cumulative receipts from the sale of petroleum to cumulative expenditures. This ratio is initially zero (during exploration there is no sale of petroleum while there may be considerable expenses) and gradually grows with time. An R-factor less than 1.0 would mean that total expenditure exceeds total receipts or revenue. An increase in the R-factor means an increase in the state's profit share or revenue.[1]

The Kenyan Petroleum Act thus provides the following as the application of the R-factor for the share of profit petroleum Table 1.

4.3 *Fees*

An applicant or a holder of mineral right, a mineral dealer's license or a diamond dealer's license are obliged to make payment of fees or charges as prescribed by the

[1]Contracts for Petroleum Development, Part 2, World Bank https://siteresources.worldbank.org/INTOGMC/Resources/cambodia_oil_gas_newsletter_8.pdf.

Table 1 Profit petroleum sharing formula for every calendar quarter

R-factor	Government share (%)	Contractor share (%)
Less than 1.0	50	50
Equal to or greater than 1.0 and less than 2.5	65	35
Equal to or greater than 2.5	[75]	[25]

Source Clause 37 of the Model PSC, 2019

Kenyan Gazette notices toward application fees, report filing fees, fees for access to geological data, and fees for access to public registers. Similarly, in petroleum, contractors or a holder of a petroleum agreement must pay annual fees prescribed by the Petroleum Act to include surface fees, training fees, and "such other fees that may be prescribed."

4.4 Royalties

As mentioned above, Kenya's petroleum sector does not make use of the royalty system. However, in mining, mineral right holders are obliged to make payments of royalty to the state as per the prescribed rates, manner, and within a specified period. Even so, mineral samples removed for purposes of testing, shall not be subject to royalty unless they exceed the maximum value stipulated in regulations. Royalties payable according to the Mining Act shall be distributed as follows: seventy percent to the national government; twenty percent to the county government; and ten percent to the community where the mining operation occurs.

This is in contrast to the petroleum sector where rather than royalties being shared among national and county governments and the community; instead, it is the national government's share of profits derived from upstream operations, or what is otherwise known as the government's share of profit petroleum.

4.5 Ring Fencing

This is a financing model that aims to block expenses incurred to be used against revenues derived from the same license area. Ideally, ring fencing limits the ability to utilize losses from one license area against another. With regard to Kenya's petroleum sector, ring fencing is applicable in the upstream sector; losses from one block cannot be used to reduce the taxable income of another.[2] Expenditure incurred by a contractor in a license area can only be offset against income derived from the same license area.

[2]Taxation in the Upstream oil and gas sector, Oraro and Company Advocates, https://www.oraro.co.ke/2018/09/14/taxation-in-the-upstream-oil-and-gas-sector/.

4.6 Corporate Tax

Corporate companies are subjected to a direct tax on profits as stated in the Kenya Income Tax Act, which provides information on the determination of taxable income and rates of taxation. The Act stipulates a normal rate of 30% for resident companies and 37.5% for non-resident companies. However, the Income Tax Act does not presently provide specific rules for determining the value of sales of hydrocarbons. In its ninth Schedul, however, does provide for specific deductions of expenses incurred in the exploration, development, and production phases in determining the taxable income:

- *Exploration costs*: including capital expenditure incurred in undertaking exploration operations shall be fully deductible for tax purposes in the year in which incurred.
- *Development expenditure*: excluding plant and machinery and social infrastructure is depreciated for tax purposes at a rate of 20% per annum (straight line) commencing the year after the asset is brought into use and the year in which production commences.
- *Operating costs*: including geological and geophysical (G&G) and intangible drilling costs are fully deductible in the year incurred.

The Income Tax Act requires oil and gas companies to withhold tax at a rate of 10% on dividends paid. In the repealed Petroleum Act, this provision appeared to conflict with the provisions therein, which stated that all income taxes including that on dividends are carved out of the government's share of production hence a "deemed" income tax. This position has, however, been remedied by the Petroleum Act 2019, which under Clause 39 of the Model PSC provides that the government's share shall be exclusive of all taxes payable by the contractor therein mandating the contractor to remit the income tax.

4.7 Capital Gains Tax

In Kenya, capital gains tax, which had been suspended since 1985, was reintroduced effective January 1, 2015. Gains arising on the disposal of all qualifying assets by oil and gas companies (except share and license interest disposals, as discussed in 3.3 above) are subject to capital gains tax at the rate of 5%.

4.8 Transfer Pricing

This covers taxes levied against transfers of interest in either oil, gas, and mining agreements/licenses. Currently, Kenya has a broad transfer pricing regime with no specifications for the oil and gas sector.

4.9 Value Added Tax (VAT)

The VAT Act No. 35 of 2013 stipulates that registration for VAT is compulsory, where the annual turnover is expected to be KES 5 million and more in taxable supplies. VAT is considered a consumption tax levied on designated goods and services with a standard applicable rate of 16%. The supply or importation of the following goods shall be considered exempt supplies upon recommendation from the cabinet secretary. The taxable supplies, exclude motor vehicles imported or purchased for direct and exclusive use in geothermal, *oil or mining prospecting or exploration*, by a company granted prospecting or exploration license in accordance with Geothermal Resources Act (Cap. 314A), production sharing contracts in accordance with the provisions of Petroleum (Exploration and Production) Act (Cap. 308) or mining license in accordance with the Mining Act (Cap. 306). These goods are also subject to levies such as 1.5% Railway Development Levy (RDL) paid on all goods imported into the country and may vary as stipulated in the Finance Act of 2019.

4.10 Withholding Tax on Natural Resource Income

The resident withholding tax rates for royalty or natural resource income are stipulated at five percent of the gross amount payable. Further, the following payments made by a licensee to a non-resident shall be subject to withholding tax: (i) 20% on royalties or natural resources; (ii) 10% on dividends; (iii) 15% on interest payments; and (iv) 12.5% on management or professional fees. Withholding tax on service fees paid to non-resident subcontractors for services provided to the contractor are at a rate of 5.625%. However, this does not apply where the non-resident has a permanent establishment in Kenya; in which case, the permanent establishment is liable to corporation tax at the non-resident rate of 37.5%.

4.11 Tax Losses

Losses estimated under the tax rules may be carried forward against income from the same source for a maximum of ten years, including the year in which the losses arise. It is only for natural resources where losses can be carried back indefinitely.

4.12 Rehabilitation Expenditure

Amounts spent on rehabilitation incurred by a licensee is subject for deduction for tax consideration. Conversely, if such amounts are removed and reimbursed to the licensee, it is subject to tax. The approved decommissioning plan is also subject to income tax.

5 Revenue Allocation and Distribution in Kenya's Extractives Sector

The Mining Act, 2016 states that the holder of a mineral right shall pay royalty to the state according to the prescribed rates set by the government. Mineral royalties provide monetary compensation to the people of Kenya, as owners of the minerals until they are won, for the loss of Kenya's non-renewable asset (The Mining (Royalty) Regulations 2017). The Mining Act, 2016 provides for sharing of revenue from royalties between the national government, county government, and local communities. Article 183(5) of the Act declares that the royalties payable shall be apportioned as follows: (i) 70% to the national government; (ii) 20% to the county government; and (iii) 10% to the community where the mining operations occur.

In relation to resources under the petroleum sector, the Petroleum Act, 2019 provides for sharing of revenue from upstream petroleum operations to ensure that the county governments and local communities benefit directly from exploitation of petroleum resources located in their counties and sub-counties. Upstream petroleum operations involve the process of exploration, development, and production of crude oil and natural gas. In particular, Section 58 of the Act stipulates that the national government's share of the profits will be shared out as follows: (i) 75% to national government; (ii) 20% to the county government; and (iii) 5% to the local community, payable to a trust fund managed by a board of trustees established by the county government in consultation with the local community. The Act also mandates parliament to review these percentages within ten years.

The Energy Act, 2019 also provides for revenue sharing between the national government, county government, and local communities in relation to geothermal energy. Specifically, Section 85(3) of the Act stipulates that royalties received by the national government from geothermal energy shall be paid into the treasury of the national government and distributed between the national government, county government and the local community as follows: (i) 20% to the county government; (ii) 5% to the local community, payable through a trust fund managed by a board of trustees established by the local community and (iii) the remaining 75% shall be treated as national revenue to be dealt with in accordance with Article 203 of the Constitution of Kenya.

6 Managing Kenya's Extractive Resources Though a Sovereign Wealth Fund

A Sovereign Wealth Fund (SWF) is a state-owned investment fund or entity that is commonly established from revenues obtained from natural resources. This revenue is professionally managed and invested in equity, debt, property, or other areas of potential growth so as to benefit current and future generations. Kenya's Constitution 2010 provides for a SWF as a Constitutional Public Fund to be established by an Act of Parliament under Article 206(1)(a) (Constitution of Kenya 2010). In July 2013, the Presidential Taskforce on Parastatal Reforms was appointed by the President of Kenya with the aim of addressing sectoral challenges while achieving government policy priorities including developing a framework for the establishment of a Sovereign Wealth Fund. The creation of a SWF for Kenya is a policy choice that is consistent with Kenya's Vision 2030 objectives of achieving greater intergenerational equity and prudent management and investment of future natural resources revenue (GoK 2013). There have been two proposed Sovereign Wealth Fund Bills in Kenya. These are the National Sovereign Wealth Fund (NSWF) Bill 2014 and the Kenya Sovereign Wealth Fund (KSWF) Bill 2019.

7 The National Sovereign Wealth Fund (NSWF) Bill 2014

The proposed National Sovereign Wealth Fund (NSWF) Bill 2014 aims to secure income from current oil, gas, mineral, and other natural resources for future generations. This will be accomplished through investing in a diversified portfolio of medium and long-term investments. The fund will aid in building a savings base to be used for national development, stabilize the economy from excess volatility in revenues or exports, and enhance intergenerational equity for future generations. According to the NSWF Bill 2014, the initial start-up capital of the fund shall be 10 billion Kenyan shillings. Additional sources of financing for the fund shall include capital from the privatization of state corporations, dividends from state corporations, oil, gas, and mineral revenues allocated to the national government and revenue from other natural resources.

The NSWF Bill 2014 proposes the establishment of a professionally managed mixed commodity and non-commodity fund with three main funds within it. Thirty percent of the fund deposits will be directed to the Stabilization Fund, forty percent to the Infrastructure and Development Fund, and thirty per cent to the Future Generations Fund. These proportions are subject to periodic reviews by the NSWF Council, who provide advice and general direction about the funds. The main objective of the Stabilization Fund is to insulate the economy from the impact of volatility in revenues, including mineral and petroleum revenues. The Infrastructure and Development Fund aims to provide funding of infrastructure in Kenya for economic and social development, while the Future Generations Fund aims to provide funding for future generations of Kenya with revenue accrued from minerals, petroleum reserves, and exploitation of other exhaustible natural resources. Moreover, deposits in the fund will not be used for the government's day-to-day operations but will be invested to generate returns for the state.

In relation to the proposed management of the fund, the National Sovereign Fund Board of Trustees will be established and entrusted with the responsibility of managing the fund. The Board shall consist of a chairperson appointed by the president, with the approval of Parliament; the Principal Secretary in the Ministry responsible for the National Treasury; the Principal Secretary in the Ministry for planning; three persons appointed by the President; the Chief Executive Officer of the fund; and the Corporation Secretary (The National Sovereign Wealth Fund Bill 2014). The National Sovereign Wealth Fund Council will provide advice and general direction to the Board. The Council shall comprise the President, the Cabinet Secretary to the National Treasury, the Cabinet Secretary responsible for Economic Planning, the Cabinet Secretary responsible for Mining, the Cabinet Secretary responsible for Energy and Petroleum, the Attorney General, the chairperson of the Board, and the Chief Executive Officer of the Board (The National Sovereign Wealth Fund Bill 2014).

7.1 The Kenya Sovereign Wealth Fund (KSWF) Bill 2019

Further to the National Sovereign Wealth Fund (NSWF) Bill 2014, the Kenya Sovereign Wealth Fund (KSWF) Bill 2019 was proposed and was open for comments from stakeholders and the public in February 2019 via press release by the National Treasury. This Bill is currently scheduled for submission to the National Assembly in Parliament for approval. According to Bauer, Olan'g, and Mihalyi (2019) from the Natural Resource Governance Institute, the proposed KSWF Bill incorporates elements that promote effective, accountable, and transparent management of natural resource revenues. Key elements of the Bill include clear objectives, clear deposit rules, significant public disclosure requirements, and competitive and transparent selection of external managers.

Similar to the NSWF Bill 2014 proposal, the KSWF Bill 2019 also recommends the establishment of three distinct components of the fund: the stabilization component, the infrastructure development component, and the *urithi* (Swahili word meaning inheritance) component:

- The purpose of the *stabilization component* is to insulate expenditure of the national government from fluctuations in resource revenues and manage shocks that may affect macro-economic stability;
- The *infrastructure development component* shall provide funding for public sector infrastructure development priorities to foster a stronger and more inclusive growth and development; and
- The purpose of the *urithi component* is to build a savings base for future generations by providing an endowment to support development when the revenue from minerals and petroleum is depleted, distributed wealth across generations, and generated an alternative stream of income to support expenditure on capital projects for future generations.

These three components have found clearer and better defined objectives, which is an improvement from previous drafts of the Bill. This seeks to avoid the problem of ambiguity and potential mismanagement of funds in the specific components.

Further, Section 8 of the Bill provides that at least fifteen percent of the funds in the holding account will be channeled to the stabilization component; at least sixty percent to the infrastructure development component; and at least ten percent to the urithi component. An additional source of funds for each of the components shall be the investment incomes earned on the respective components.

In addition, the KSWF Bill 2019 states that transfers to the stabilization component shall cease when it grows to twenty percent of GDP, and the share shall be utilized to service national debt or may be distributed to either of the two other components. In situations, where there is a windfall in resource revenues, the windfall may be used for debt servicing to reduce national debt; transferred to the stabilization component; transferred to the infrastructure development component to provide basic services such as education, health care; or transferred directly to Kenyan citizens through tax cuts (Kenya Sovereign Wealth Fund 2019).

The management of the KSWF shall be assigned to a Board, which shall consist of a chairperson appointed by the President on nomination by the Cabinet Secretary, the Principal Secretary to the National Treasury or an alternate appointed by the Principal Secretary in writing; the Principal Secretary responsible for matters relating to petroleum or an alternate appointed by the Principal Secretary in writing, Governor of the Central Bank or an alternate appointed by the Governor in writing, three other persons appointed by the Cabinet Secretary from outside of government, and the Chief Executive Officer who shall be an ex-officio member with no rights to vote. The Bill provides clear and comprehensive Board mandates and procedures; this too has been hailed a considerable improvement on previous drafts of the Bill. With the management of the KSWF outlined under the Bill, it is proposed that the administration of the fund be governed by the Public Finance Management Act as well as the KWSF Bill 2019.

However, the KSWF Bill 2019 has been critiqued for having inadequate independence and oversight. This is because all board members are either directly or indirectly nominated by the Office of the President, including the auditor-general who audits the fund. Bauer et al. (2019) have recommend ensuring at least three board members are nominated by non-executive bodies, e.g., parliament or professional associations. Further, as drafted, the KSWF Bill 2019 proposes a Board that includes the Principle Secretary of Petroleum but not the Principle Secretary of Mining, yet the Bill speaks of funds obtained from mineral and petroleum resources.

Other areas for improvement are identified as: a need to reduce asset management risks because the Bill does little to provide processes that cushion against or mitigate high-risk investments; a need to improve the aspect of transparency and accountability of the management and administration of the fund, more specificity for instance the need for guidelines as to what constitutes "significant depletion" of mineral and petroleum resources; and a general need to highlight principles of sustainable development. The investment management issue is particularly a concern when dealing with the urithi component; of note, the Bill does not provide for "withdrawal" from the component but rather "transfers" which shall be invested in accordance with the provisions of the Bill. The Bill does not go into detail about these transfers, which may need more detailed procedures when regulations are rolled out. Many of these areas of improvement might be considered in upcoming drafts as many stakeholders have headed the call by the Treasury to provide comments on the Bill. At the time of publication of this edited book, the authors remain vigilant for the revised versions.

7.2 Kenya Sovereign Wealth Fund Draft Policy 2019

There is also the Kenya Sovereign Wealth Fund (KSWF) Draft Policy 2019. The National Treasury has called for comments from stakeholders. In its current draft, the policy outlines its objectives to guide the utilization of natural resource income for inclusive growth and development and to secure income from current resources for future generations. The Draft Policy states that the scope covers revenues from the petroleum and mining sources payable to the national government. The scope of the KSWF Bill and Draft Policy has been a recurrent area of concern and issue with stakeholders requiring a clear and unambiguous provision as to whether they cover *all* mineral and petroleum revenues or only *surplus revenues.* As currently drafted, the KSWF Bill 2019 and the KSWF Draft Policy 2019 provide that they cover mineral and petroleum resources payable to the national government presupposing that this means *all* mineral and petroleum revenues. One of the obvious issues that arises is the matter of either complementarity or conflicting provisions with the Mining Act 2016 and the Petroleum Act 2019 with regard to the management of mining and petroleum revenues.

8 Conclusion

Kenya's public wealth management systems for the extractives sector have seen and continues to see much positive growth and experience legislative review to reflect such growth. The Constitution of Kenya 2010, the Petroleum Act 2019, the Mining Act 2016 as well as other relevant pieces of legislation and policies continue to propel the country further into sustainable and effective management of mineral and petroleum resources for the benefit of the current population as well as future generations. A 2019 Bill proposes a Sovereign Wealth Fund for Kenya, but has not yet been approved at the time of this writing. However, there is still much room for improvement in the legislation and regulations as the landscape of the Kenyan extractives sector changes.

Kenya has made an admirable effort to propose ways in which Kenya's resources can be managed thus ensuring economic stability and guaranteeing future generations benefit from Kenya's current natural resources. There remain high expectations across stakeholder groups for the principles of responsible leadership, good governance, and sustainable development outlined under the Constitution of Kenya 2010 to emanate from the management of extractives wealth.

References

Bauer, A., Olan'g, S., & Mihalyi, D. (2019, March 6). *Kenya proposes transparent, but risky, new sovereign wealth fund*. Retrieved from Natural Resource Governance Institute website: https://resourcegovernance.org/blog/kenya-proposes-transparent-risky-new-sovereign-wealth-fund.

Deloitte. (2014). *Kenya's petroleum fiscal regime*. https://www2.deloitte.com/content/dam/Deloitte/ke/Documents/tax/tax_Kenya_petroleum_fiscal_regime_2014.pdf.

Deloitte. (2016). *Oil and gas taxation in Kenya*. https://www2.deloitte.com/content/dam/Deloitte/global/Documents/Energy-and-Resources/gx-er-oil-and-gas-taxguide-kenya.pdf.

EU-UN Partnership. (2012). Toolkit and guidance for preventing and managing land and natural resources conflict. *EU and The United Nation Interagency Framework Team for Preventive Action*.

GoK. (2013).*Report of the presidential taskforce on Parastatal Reforms*. Retrieved from https://www.cofek.co.ke/Report%20of%20The%20Presidential%20Task%20force%20on%20Parastatal%20Reforms.pdf.

Government of the Republic of Kenya, GoK. (2010). *Constitution of Kenya*. Government Printers, Nairobi: Kenya.

Government of the Republic of Kenya, GoK. (2014).*The national sovereign wealth fund bill, 2014*.

Government of the Republic of Kenya, GoK. (2016a). *Draft National Extractive Policy*. Ministry of Mining, Nairobi: Kenya.

Government of the Republic of Kenya. (2016b). *Mining and minerals policy. The popular version*. Retrieved from https://www.idlo.int/sites/default/files/pdfs/highlights/Kenya%20Mining%20Policy%20Popular%20Version-LowRes.pdf.

Government of the Republic of Kenya. (2016c). *The mining (royalty) regulations, 2017*.

Government of the Republic of Kenya, GoK. (2018). *The Tax Laws (Amendment Bill, 2018*. Government Printers, Nairobi: Kenya.

Government of the Republic of Kenya. (2019a). *The energy act, 2019*. , Pub. L. No. Kenya Gazette Supplement No. 29 (Acts No. 1).

Government of the Republic of Kenya, GoK. (2019b). *Petroleum Act No. 2 of 2019*. Government Printers, Nairobi: Kenya.

Government of the Republic of Kenya, GoK. (2019c). *Mining Act No. 12 of 2016*. Government Printers, Nairobi: Kenya.

Government of the Republic of Kenya, GoK. (2019d). *Economic survey*. Government Printers, Nairobi: Kenya.

Government of the Republic of Kenya, GoK. (2019e). *The Finance Act No. 23 of 2019*. Government Printers, Nairobi: Kenya.

Government of the Republic of Kenya, GoK. (2019f).*Draft Kenya sovereign wealth fund bill, 2019.*

Government of the Republic of Kenya, GoK. (2019g). *Draft Kenya sovereign wealth fund policy, 2019.*

Harraz, H. Z. *Petroleum laws: Types of international petroleum contracts* (Slideshare) https://www.slideshare.net/hzharraz/topic-4-types-of-petroleum-contracts-agreement

International Monetary Fund, IMF. (2012). *Fiscal Regimes for extractive industries: Design and implementation*. IMF, Washington D.C: USA

Kenya National Bureau of Statistics. (2019). *Economic survey 2019*. Retrieved from Kenya Ministry of Energy. (2018). The State Department of Petroleum. Retrieved from Ministry of Energy website: https://energy.go.ke/?p=328

Ministry of Petroelum and Mining website. https://www.petroleumandmining.go.ke/contact-us/

Muchiri, G. *Raxation in the upstream oil and gas sector*. https://www.oraro.co.ke/2018/09/14/taxation-in-the-upstream-oil-and-gas-sector/.

National Bureau of Statistics website. https://www.knbs.or.ke/?wpdmpro=economic-survey-2019

Natural Resource Governance Instituted, NRGI (2015). *Fiscal Regime design: What revenues government will be entitled to collect*. RGI Reader.

Petroleum Economist. (2019). *Kenya enters the oil business*. https://www.petroleum-economist.com/articles/politics-economics/africa/2019/kenya-enters-the-oil-business

Sigei, G. C. *The taxation of the extractives industry (mining, oil and gas), 4th annual tax summit*. https://summit.kra.go.ke/wp-content/uploads/2018/10/Taxation-of-the-Extractive-Industries-Mining-oil-and-gas.pdf.

Tullow Oil Kenya website: https://www.tullowoil.com/operations/east-africa/kenya

United Nations Conference on Trade and Development, UNCTAD. (2012). *Extractive industries: Optimizing value retention in host countries*. UNCTAD, Geneva: Switzerland.

Wanyagathi Maina, A. (2019). The Kenyan tax regime for the oil and gas sector: An international tax perspective to policy and practical challenges. *Revista de Derecho Fiscal n.* 14, 203–219, enero-junio de.

World Bank. (2007). *Contracts for Petroleum Development, Part 2*. https://siteresources.worldbank.org/INTOGMC/Resources/cambodia_oil_gas_newsletter_8.pdf

Overview of Extractive Resources Management in Indonesia

Rivana Mezaya, Yudo Anggoro, Wisnu Jaluakbar, and Wulan Asti Rahayu

Abstract Since the first drill that struck oil in 1884, the extractive industry has been the key contributor to the Indonesian economy. Being blessed with natural resources, the Indonesian Constitution strictly mandates these resources to be used for the welfare of the people with the Government acting as the representative of the people, placed as the guardian of these natural resources. The regulatory regimes over the two main extractive sectors, oil and gas as well as minerals and coal, have experienced changes over the years. These changes correspond to the political changes that the country was experiencing, from being a newly independent nation to a reformed democracy with a focus of decentralising power.

1 Introduction

Since the first drill that struck oil in 1884, the extractive industry has been the key contributor to the Indonesian economy. Being blessed with natural resources, the Indonesian Constitution strictly mandates these resources to be used for the welfare of the people with the Government acting as the representative of the people, placed as the guardian of these natural resources. The regulatory regimes over the two main extractive sectors, oil and gas as well as minerals and coal, have experienced changes over the years. These changes correspond to the political changes that the country

R. Mezaya (✉) · Y. Anggoro · W. A. Rahayu
School of Business and Management, Bandung Institute of Technology, Bandung, Indonesia
e-mail: rivana.mezaya@sbm-itb.ac.id

Y. Anggoro
e-mail: yudo.anggoro@sbm-itb.ac.id

W. A. Rahayu
e-mail: wulan_asti@sbm-itb.ac.id

W. Jaluakbar
SKK Migas, Taskforce of Upstream Oil and Gas Business Activities, Jakarta Selatan, Indonesia
e-mail: wjaluakbar@skkmigas.go.id

E. G. Pereira et al. (eds.), *Sovereign Wealth Funds, Local Content Policies and CSR*, CSR, Sustainability, Ethics & Governance,
https://doi.org/10.1007/978-3-030-56092-8_4

was experiencing, from being a newly independent nation to a reformed democracy with a focus of decentralising power.

Nevertheless, the Constitutional mandate did not change, granting ownership of natural resources to the State, and the Government as the Mining Authority. However, third-party involvement in the sector has always been deemed to be important especially with considerations of requirements of capital and technology. Arguably, regimes of Production Sharing Contracts for Oil and Gas, and Contracts of Works for Mineral and Coal, are the answer to this. After decades of growth and the success of the extractives industries, peaking in the 1980s and 1990s, Indonesia's production started to decline particularly in response to low numbers of exploration permits and further investment. This, coupled with rapidly increasing demands, led to Indonesia becoming an importer of oil. This trend came at a time where political and regime changes were made. Despite the trend, Indonesia still has significant reserves of resources. However, while it is yet to maximise the management of its natural resource profits, it has improved its management of profit sharing between the Central and Regional Governments. It is yet to explore and adopt other fund management approaches such as establishing funds to further boost the economic impact of its natural resources. In this chapter, we examine how extractive resources are regulated in Indonesia, but first we offer an overview of sovereign wealth funds internationally.

2 Overview of Sovereign Wealth Funds

Sovereign Wealth Funds (SWF) are a government-saving fund separate from the State's budget and managed to make some particular purpose investment to gain maximum returns (Devlin and Brummit 2007). Butt et al. (2008) (cited in Alhashel 2015) define SWFs not only from the purpose point of view, but also from the source of the Fund itself. They state that SWFs are the state-owned investment vehicle that invests globally in various types of assets ranging from financial to real to alternative assets. These investment vehicles are usually funded by commodity export revenues or the transfer of assets directly from official foreign exchange reserves. In some cases, government budget surpluses and pension surpluses are also transferred into SWFs. Lowery (2007) (cited in Griffith-Jones and Ocampo 2008) provides a sharp definition of SWF and highlights the way SWFs must be managed separately from official foreign exchange reserves. SWFs must be invested for long term sustainable profit and target high risk, high return investment so that it can grow the wealth of the State (Kimmitt 2008; IMF 2007, cited in Griffith-Jones and Ocampo 2008).

From these definition statements, we can conclude that there are several critical success factors for the sustainability of SWFs. The first essential factor is the source of funds must be able to guarantee the continuity of foreign exchange surplus. Second, the economic purpose is a fundamental reason why a nation develops a SWF. The third factor is the way SWFs are managed, which includes who has the primary role in making an investment decision and the relationship of a SWF to the State budget.

The last factor is how the SWFs are invested and whether the investment results build better wealth of state or social development.

Over the past few years, the existence and rapid development of Sovereign Wealth Funds has attracted increasing attention. Presently, the total number of SWFs operated globally is around 80 and those are valued at \$8 Trillion (2018) (SWFI 2018, cited in Bahoo et al. 2019), which increased 2.5 times compared to 5 years before (around \$3 Trillion in 2013) (Truman 2009: IFSL 2009, cited in Gould 2010). The total funds raised in SWF around the world exceed hedge funds and private equity firms. At 2017, SWF doubled the overall hedge fund amount and tripled the size of private equity funds (Megginson and Gao 2019). SWFs have existed since 1953 when the first one was established in Kuwait under the name Kuwait Investment Authority to manage the country's oil revenue surplus. Three years later, the Pacific Island of Kiribati developed its own SWF called the Revenue Equalisation Reserve Fund (RERF). Following these two initial countries, in the 1970s, many oil-exporting countries, particularly in the Middle East, set up their own SWFs.

According to Aizenman and Glick (2009), the increasing price of commodities, mainly oil, is one reason behind the recent high accumulation of foreign assets in SWFs. Thus, it is not surprising that the number of SWFs tends to increase sharply in periods when the oil price is high. The second reason behind the high number of foreign assets in SWFs is the hoarding of foreign exchange excess by non-natural resources countries for precautionary motives where they transfer some parts of them to particular investment vehicles to gain high profit. Based on Griffith-Jones and Ocampo (2008), commodities, mainly oil and gas, recently accounted for approximately 70% of the total value of SWFs. However, Bahoo et al. (2019) observed that commodities only contribute to around 55%, while 33% are accounted for by non-commodities, and 12% held by other types of assets. One country with an active SWF in Southeast Asia is the non-commodity nation of Singapore. Hunt's (2016) (cited in Megginson and Gao 2019) analysis of Singapore's SWF is that it is one of the most effective funds internationally, able to increase its state domestic social economy under the Temasek Holding and Singapore Investment Corporation. These two funds contribute to 2% of Singapore's GDP and 12–15% of the government's annual operating budget. Hence, through its SWF, the Singapore Government can better finance its social expenditure without increasing the tax burden on its citizens, thereby improving social welfare.

On the other hand, oil-producing countries that manage their revenue under SWFs are dominated by developing countries. Examples of developing nations that have successfully initiated SWFs from the extractives industry include Algeria, Azerbaijan, Botswana, Brunei, Chile, Gabon, Iran, Kazakhstan, Kiribati, Kuwait, Libya, Nigeria, Oman, Qatar, Sudan, Saudi Arabia, UEA. Timor Leste and Trinidad Tobago. There are two fundamental economic purposes for developing SWFs from natural resources revenue. The first one is the Stability Fund, which aims to reduce the impact of natural resources revenue volatility (linked to the pro-cyclical pattern of export price or volume). This means the government will accumulate resources when the rates of export are high and pay when they are low (Griffith-Jones and Ocampo 2008). The second economic purpose is the Saving Fund, which is intended to store wealth for consumption or investments of future generations. Timor Leste and Kiribati are two countries in the Pacific area, which are focusing on the savings purpose rather than the stability purpose. However, in the implementation process, these two purposes cannot be distinguished.

Some oil-producing countries that utilise SWFs to manage government revenue from exhaustible natural resources to improve country or state development have aided intergenerational equity and macroeconomic stability, while others have struggled to bring about improvements in wellbeing (Gould 2010). The United Arab Emirates and Saudi Arabia show the highest SWFs with the source of funding stemming from the extractives industry. Those two nations successfully developed their country by using the revenue from the extractives industry due to the huge number of oil-products. On the contrary, there are some countries in the Pacific, which aim for similar things. Table 1 describes those country efforts in managing SWFs coming from natural resources.

From Table 1, two main causes lead to the decreased performance of SWFs value. First, integrating SWFs and the State budget without a proper and clear rule on withdrawal procedure, as was experienced by Kiribati and Papua New Guinea (before the year 2000). The second is the wrong investment strategy, which was experienced by Nauru. Across those Pacific countries, Timor Leste is an excellent example in managing SWFs. This country provides regular reports so that the SWFs are well-monitored. Even though the investment strategy needs to be improved by doing a diversification, current strategy has led to an increase in economic growth.

In terms of fund management, the key factors that differentiate SWFs from other investors are their independent arrangements, ownership, management and control systems. SWFs are directly or indirectly owned and controlled by the government or government representatives (Bahoo et al. 2019). Some countries tend to manage SWFs directly via the central banks and/or finance minister (Truman 2009). Norway's SWF stems from the government pension fund-global and is managed by the government indirectly through a national bank. Still, it is controlled by the Norwegian parliament.

SWFs investment strategy is aimed at minimising exposure from the oil and mining industry. The conventional investment strategy will rely on a financial instrument or bilateral relationship on financing the infrastructure development in the neighbouring country. Another alternative of SWFs investment strategy, considered relatively safe, is to allocate the funding on the developed country instrument, which cover financial instruments such as bonds and property investment. Recently, PWC (2019) argue that SWFs investment must be diversified into new technology development by financing the new technology start-up.

Several factors must be well-evaluated to have an excellent SWF performance, which can give sustainability improvement for state wellbeing and country economic growth. One of them has a steady source of funds that can provide continuous supply for SWFs. The precise economic purpose also being a critical point that must be decided before managing SWFs, especially when the fund is coming from the extractives industry in which we have to deal with price volatility. The third important factor is in defining the role of SWFs management. Robust procedure on the use of SWFs to

Table 1 SWFs management in Pacific Countries

Country	Source of Funds	SWFs name	SWFs management and investment	impact
Kiribati	Phosphate mining	Revenue Equalisation Reserve Fund	Since economic is dominated by public sector, SWFs integrated with state budget (fiscal surpluses a re added to the fund and any budget shortfalls are financed via drawdowns, the real per capital value is kept constant	increased drawdowns and equity market falls, combined significant domestic inflation due to high international food and fuel price in 2008 have reduced the rea1 per capita value of RERF
Nauru	Phosphate mining	Nauru Phosphate Royalties Trust (NPRT)	Poor NPRT management by investing in international property rather than fixed-income assets or equity composed of four funds: LT Investment Fund, The Lane Owners' Royalty Trust Fund, The Housing Fund, and the Rehabilitation Fund	Investment loses at the same time when phosphate price was falling down in 90's caused the fiscal deficits, as government revenue fel1 and expenditure continued to increase
Timor Leste	Economy is dominated by an offshore oil and gas sector; non-oil economy: subsistence agriculture and government consumption and investment	The Petroleum Fund established in August 2005 to manage royalty and tax revenue from oil and gas	BPA and the government publish quarterly report to have a transparency on fund performance and currently on the process of having monthly pubic report. Being a member of International Forum of SWF A1l investments are putting on developed country government bonds, 10% of the fund may be invested in equities-not enough to maintain the domestic purchasing power, need investment diversification	Transfer from the Petroleum Fund Finance around 90% of the National Government's budget. Increase annual economic growth to an average 9% over 2007–2009 There may be strong case for investing more in Timor Leste's physical and human capital. With such a low capital base, the social return form such expenditure could be higher than the financial returns the Petroleum Fund can achieve (Collier et al. 2009)

(continued)

Table 1 (continued)

Country	Source of Funds	SWFs name	SWFs management and investment	impact
Papua New Guinea	Dominated by resource extraction, primarily gold, oil, and copper	PNG Mineral Resource Stabilisation Fund (MRSF) established in 1974	Tax, Royalty, and Dividend Payments from all mining and oil enterprises were placed in the MRSF rather than in consolidated revenue. Aim of the fund was to reduce the impact of mining revenue volatility on the budget (Parsons and Vincent 1991)	The Fund's Capital then used to finance the budget. The MRSF was held by in the Bank of PNG (central bank), not invested, and was managed by selected departmental secretaries. The Fund was effectively close in 1999, with final balance used to reduce the country's ballooning debt

Source https://treasury.gov.au/publication/economic-roundup-issue-1-2010/economic-roundup-issue-1-2010/managing-manna-from-below-sovereign-wealth-funds-and-extractive-industries-in-the-pacific

cover state budget can avoid a significant decrease in the value. However, in terms of investment strategy, some countries such as Singapore and Norway have proven that being flexible and diversification will empower the SWFs impact on state development. In the end, the ability for SWF manager to dance in the rain instead of waiting for the storm to pass will decide the success of Sovereign Wealth Funds management.

3 Ownership of Extractive Resources in Indonesia

Extractive resources in Indonesia is owned by the State and considered public goods. This principle is enshrined in Article 33 of the 1945 Constitution of the Republic of Indonesia ("**Constitution**"), which states that "*branches of production which are vital to the State and affect the livelihood of the majority of the people shall be controlled by the State*" and "*land, waters and natural resources therein shall be controlled by the State and shall be used for the maximum welfare of the people*". Under these clauses, the rights of ownership to the natural resources are given to the State with the purpose to be used for the welfare of the Indonesian people. This Article embodies the principles of "economic democracy" in which the economy is organised as a collective endeavour, and prioritises welfare of the people. Through this provision, the right to mineral resources is separated from the rights to the land.

The implementing laws of Article 33 of the Constitution relating to extraction of natural resources were issued in 1960 identifying two categories of the mining sector: (1) minerals, and (2) oil and natural gas. Minerals were regulated through

Law No. 37 of 1960 while oil and gas were regulated through Law No. 44 of 1960. These laws were consistent with the Constitution, granting the rights to the natural resources to the State, not to the holder or owner of the land, while providing an opportunity for the private sector to extract natural resources through the granting of economic rights from mining activities (although for Oil and Gas, this early law gives authority to mine or mining rights to State Enterprises, which could then contract with third party (private corporations) as contractors).

These laws experienced changes throughout the years, and the latest changes happened after Indonesia went through a reformation era starting in 1998, which saw more efforts to democratise the economy through liberalisation and lessen the role of the Government in business operations. These newer laws are Law No. 22 of 2001 concerning Oil and Gas ("**Oil and Gas Law**") and Law No. 4 of 2009 concerning Mineral Resources and Coal Mining ("**Mining Law**"). Nevertheless, the newer laws still maintain the Constitutional mandate regarding ownership.

The term "State" as owner of natural resources based on the Constitution is further defined in implementing regulations. For Oil and Gas, the Oil and Gas Law regulates that "*The control by the state as... shall be conducted by the Government as the holder of* ***the Mining Authority***" with Government being the Central Government of the Republic of Indonesia. While the upstream business activities are opened for private sector participation through a Cooperation Contract (ref. Article 6 paragraph (1) of the Oil and Gas Law), the Cooperation Contract shall at least contain the following requirements:

a. the ownership of natural resources shall remain in the hands the Government up to the point of transfer;
b. the management control of operations shall be by the Implementing Body;
c. all capital and risk shall be borne by the Business Entity or Permanent Establishment.

From these provisions, it can be concluded that for upstream activities:

- Mineral rights are with the State;
- Mining rights are with the Government through the Ministry of Energy and Mineral Resources; and
- Economic rights are given to Contractors/Oil Companies.

In this context, Mineral Rights are the rights that deal with the ownership of the minerals in the ground (in situ); Mining Rights are the rights to bring the minerals to the surface; and Economic Rights deal with the ownership of the minerals once they have been mined (Utomo 1999). Under the Cooperation Contract, mineral rights, mining rights, and also the economic rights are vested in the state. The investor only realises economic rights based on its working interest share of production at the point of export once commercial production commences (Machmud 2002).

Meanwhile in the Minerals and Coal sector, the 2009 Mining Law no longer adopts contractual regime as in the Oil and Gas Law. The Mining Law regulates that minerals and coals are exploited through a licencing regime. The mineral rights are controlled by the Government while the economic rights are with those who have obtained the licence (*Izin Usaha Pertambangan* or *IUP*). This 2009 Mining Law is a departure from the previous regime where minerals and coal were exploited through Contracts of Works where private corporations entered into contracts directly with the Government to be given a concession to a specific area of work. Even though it is a concession-like contract, the Contract of Works is not a common concessionary contract since it is considered *rights in personaam* not *rights in rem,* which can be treated as a property and mortgaged (see Kusumaatmadja 2013).

The concessionary-like regime in the mineral and coal sector required royalty from the private corporation deriving from the awarding of concession rights over a certain area and the natural resources within. The title to the natural resources from the Government passed at the point of extraction (different with a common concession contract where title passes when the concession is given even before extraction). This is a different approach from the Oil and Gas Law that adopts a production sharing contract approach where title passes to the investor to its share of production at the point of export. The investor assumes all the pre-production risk and recovers both cost and profit share from production, in predetermined proportions, once commercial production from the contract area commences (Machmud 2002).

4 Main Extractive Resources in Indonesia

Indonesia is rich in natural resources, particularly in hydrocarbons and mining. The extractive industries contribute 7% to the Gross Domestic Products ("**GDP**"), totalling 21% of export, amounting to 10% of government revenue and providing 1.2% of total employment in the country. Table 2 lists Indonesia's extractive resources.

4.1 Oil and Gas

Indonesia was a significant oil-exporting country that joined OPEC in 1962. Its oil peaked in 1995 where the production was 1,624,000 barrels of oil per day. In the following decade, it experienced a decline in oil and gas production attributed largely to the lack of exploration activities and investment in the sector. The 1998 Asian economic crisis as well as the fundamental reform in government that followed contributed to this lack of investment since it provided a lack of certainty for investors. Indonesia made fundamental changes to its Oil and Gas regime in 2001 through the Oil and Gas Law, and to its mining regime in 2009 through the Mining Law. It also changed its geothermal regime twice in a decade, in 2006 and 2015 through the Geothermal Law.

This decline in production happened while Indonesia experienced a rapid increase in its domestic consumption. By 2003, Indonesia had become a net-importer of oil. Nevertheless, Indonesia still has a significant amount of oil reserves. Indonesia sits

Table 2 Indonesia's extractive resources

Commodity	Reserves	Unit	Significance
Oil	7390	Million stock tank barrels	Indonesia ranks as the world's 27th largest in terms of crude oil reserves
Gas	150	Trillion standard cubic feet	Indonesia ranks at the world's 14th largest in terms of natural gas reserves
Gold	4248	Metric tonnes	Indonesia ranks as the world's 5th largest in terms of gold reserves
Lead	1.6	Metric tonnes	Indonesia ranks as the world's 2nd in terms of lead reserves
Zinc	7	Metric tonnes	
Nickel	577	Million metric tonnes	
Bauxite	180	Million metric tonnes	
Silver	4104	Metric tonnes	
Copper	4161	Million metric tonnes	
Coal	31.4	Billion metric tonnes	Indonesia ranks as the world's 10th largest in terms of coal reserves
Tin	0.7	Metric tonnes	

Source EITI Indonesia, https://eiti.org/indonesia#extractive-industries-contribution

at number 26 of the top crude oil-exporting countries in 2017, with export value in the amount of USD 5.2 billion or 0.6% of the total export value in the world for crude oil. While for gas, Indonesia sits at number 8 with an export value of USD 8.9 billion or 3.8% of the total world's export.

4.2 Minerals and Coal

Indonesia has a significant amount of minerals and coal reserves. It is an important player in the global mining industry with significant production of coal, copper, gold, tin and nickel. It is the second-biggest coal exporter after Australia, exporting USD 17.9 billion worth of coal in 2017 or 16.1% of the world's production. It holds 3% of the world's coal reserves and is ranked 10 for the world's largest reserve.

Based on the data reconciliation in the coal sector at the end of 2018, the Ministry of Energy and Mineral Resources stated that Indonesia has 116 billion tonnes of coal resources and 37 billion tonnes of coal reserves, which can last for 76 years based on the latest production target of 486 million tonnes in the year 2018. In 2018, the Mining Industry contributes 4.98% to the Indonesian GDP and made up 16% of Indonesian exports.

5 Fiscal Regime in the Extractive Resources Industries in Indonesia

The fiscal regime in Indonesia varies depending on the commodity, such as

a. For the Oil and Gas sector, the fiscal regime is Production Sharing Contract (PSC) adopting an in-kind basis approach;
b. For the Mining sector, the fiscal regime is Royalty and Tax (R/T) which adopts cash mechanism.

5.1 Oil and Gas

Fiscal regimes relating to the extractive resources are mostly designed to balance the interests amongst labour and capital as well as land/resource owners. The Government as land/resource owner also plays a role as the policymaker who has an interest in capturing as much economic rent as possible through various levies, taxes, royalties and bonuses. At the same time, the Government needs to capture this economic rent efficiently while considering gains for the private sector, which plays a role as a contractor to extract these resources. The private sector is exposed to the high risks in the extractive resources industries; thus, the profit margin should be large enough to accommodate the high risks.

The Government has tried to capture the economic rent from the upstream activities through the following:

a. signature bonuses or production bonuses during the production phase;
b. royalties/first tranche petroleum;
c. production sharing;
d. taxes;
e. state participation (indirect, through state-owned enterprises).

Below is the illustration for the allocation of revenues and profit in the Cost Recovery scheme (Fig. 1).

Based on the considerations above, the Government created a fiscal system to address an important question of whether an exploration or development is feasible.

The Oil and Gas Law differentiates upstream business activities (exploration and exploitation) and downstream business activities (processing, transport, storage and commerce) and stipulates that upstream activities are controlled by the Government through "Joint Cooperation Contracts" between the business entity and the executing agency. Downstream activities are controlled by business licences issued by the regulatory agency for downstream businesses.

After previously adopting a concessionary system, Indonesia first introduced the Production Sharing Contract (PSC) to the world as a fiscal regime in the upstream oil and gas sector in 1961. The Indonesian PSC model was adopted by many countries, including Chile, Guatemala, Israel, Ivory Coast, Egypt, India, Peru, Libya, Malaysia,

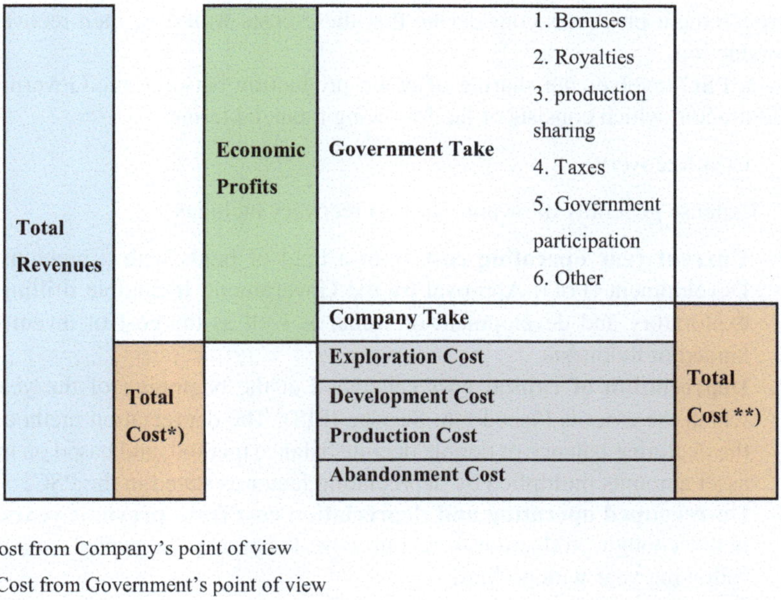

Fig. 1 Cost recovery scheme

Syria, Trinidad, Oman and Sudan (Machmud 2002).[1] The PSC was designed with consideration that oil and gas is a high risk as well as capital and technology-intensive business, thus needing private sector participation. To address this, the sector needs a system that is flexible to changes in the economy but also able to provide long-term certainty. The issue of ownership is also one of the key considerations in selecting the PSC regime compared to the concessionary regime. Under a concessionary system, the Contractor has the title to mineral resources (e.g. crude oil) produced against which it pays royalties and taxes. Under the PSC system, the Government retains title to the mineral resources as mandated by the Constitution.

Under the PSC system, the Government and Contractor agree to split the production by a certain percentage measured in terms of revenue. Operating costs are recovered from production through Cost Recovery as defined by the PSC, and the Contractor has the right to take and separately dispose its shares of oil and gas (with title to the hydrocarbons passing at the point of export or delivery). The PSC sets the share or take of the Contractor in a fixed percentage of gross production (before tax). It is calculated after recovery of the Contractor's costs under the cost recovery scheme. Production sharing should not be confused with profit sharing. Profit sharing is often not advantageous to owners of the resources, as they have no control at all over cost (Kusumaatmadja 2013). While in the production sharing system, the Government still controls cost through approval mechanisms of plans of developments and other

[1] Tengku Nathan Machmud, The Indonesian Production Sharing Contract: An Investor's Perspective, p. 39.

procurement processes considering that these costs would be then recovered after production.

A PSC involves the sharing of crude production between the Government and Contractor, which consists of the following financial terms:

a. Cost Recovery

Expense generally allowable for cost recovery includes:

1. **Current year operating cost** from a field of fields with a previous Plan of Development (PoD) Approval by the Government, intangible drilling cost on exploratory and development of wells, as well as the cost of inventory when landed in Indonesia.
2. **Depreciation of capital cost** calculated at the beginning of the year during which the asset is Placed into Service (PIS). The depreciation method is either the declining balance or double decline balance method, and based on individual asset amounts multiplied by depreciating factor as stated in the PSC.
3. **Un-recouped operating and depreciation cost from previous years.** If there is not enough production to recoup cost, these may be carried forward to the following year with no limit.

b. Investment credit

An investment credit (IC) is available on direct development and production capital cost as negotiated and approved by Government. The credit ranging from 17%–55% of the capital cost of development, transport and production facilities which were historically available. The IC must be taken in oil and gas in the first years of production but can generally be carried forward to later years.

The Ministry of Energy and Mineral Resources (MEMR) issued Ministerial Regulation No. 08/2005 on the Marginal Oil Field incentive programme. The regulation provides contractor with an additional cost recovery of 20% when working with marginal oil field. This treatment is likely to be applied similarly to an investment credit (i.e. cost recoverable but taxable).

c. Equity share

Any production that remains after IC and cost recovery is split between the Government and the Contractor. The Indonesian PSC has evolved through five "generations" with the main change being the production sharing split. The second and third generations of PSC, which were issued after 1976, removed the earlier cost recovery cap of 40% of revenue and confirmed an after tax oil equity split of 85/15 for Government and the Contractor, respectively.

Below is the comparison between generations of PSCs in Indonesia (Table 3).

To most recent, PSC generation is based on the Oil and Gas Law of 2001, which founded a fundamental change in the Oil and Gas management. Pertamina, the National Oil Company, which previously acted on behalf of the Government in negotiating with Contractors no longer played that role. The Government established a separate public entity, BP Migas (now SKK Migas or the Taskforce for Upstream Oil

Table 3 Generations of PSCs in Indonesia

	New Contracts %	PSC Gen IV (incentive package 1988–1989)	PSC Gen III (1984)	PSC Gen II (1978)	PSC Gen I (1966)
FTP		20%		–	–
Cost recovery				CR 100% (no limit)	40% CR limit
Equity to be split (before tax)					
Government			(oil) 71.1538% (gas) 42.3077–32.6923%	(oil) 65.9091% (gas) 31.8181%	65%
Contractor			(oil) 28.8462% (gas) 57.6923–67.3077%	(oil) 34.0909% (gas) 68.1818%7	35%
DMO allocation				25% after 60 months	25%
DMO Fee		10% of export price		US$ 0.2/bbl	Full price for first 5 years of production, US$ 0.2/bbl thereafter
IC		17%	17%	20%	–
Tax			48%	56%	Paid by contractor
Share of production after tax					
Government		Frontier: <50 MBOPD = 80% 50–50 MBOPD = 85% >150 MBOPD = 90%	85% (oil) 70–65% (gas)	85% (oil) 70% (gas)	
Contractor		Frontier: <50 MBOPD = 20% 5–150 MBOPD = 15% >150 MBOPD = 10%	15% (oil) 30–35% (gas)	15% (oil) 30% (gas)	

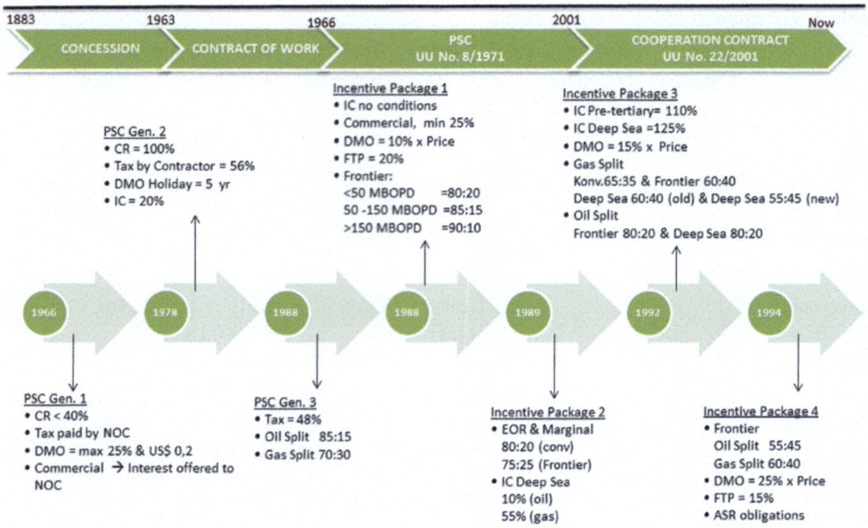

Fig. 2 Changes in Fiscal Regime in Indonesia. *Source* SKK Migas

and Gas), which is tasked with the management of natural resources and enters into contracts with Contractors on behalf of the Government. Pertamina with a separate law became a limited liability wholly owned State-owned Enterprise, which acts as a Contractor in PSCs. Since 2012, SKK Migas became a quasi-part of the Ministry of Energy and Mineral Resources based on the Constitutional Court decision to return the management and supervisory function of the Oil and Gas industry to the Government through the Ministry of Energy and Mineral Resources, specifically the Directorate General of Oil and Gas.

There were also several changes in the fiscal regime, as illustrated below (Figs. 2 and 3).

Under the PSC system, the production is split in a way as illustrated by the chart below:

The arrangement above enables the Government, without risking its own resources in the beginning, to attract Contractors with capital and technological capabilities required to extract the natural resources, but still maintain ownership of the resources at the end.

In 2017, the Government introduced the Gross Split scheme in oil and gas upstream contracts. The new scheme was introduced in the Minister of Energy and Mineral Resources Regulation No. 8 of 2017 ("MEMR 8/2017"), which came into force in January 2017, and is further regulated by its subsequent revisions. The new scheme departed from the cost recovery mechanism in the previous PSC to a direct split between Government and Contractor's take of the production (before tax) regardless of Contractor's operational costs. The Gross Split scheme was first applied to the Offshore North West Java in January 2017 in the new PSC where the state-owned Pertamina took over the contract after the previous PSC expired.

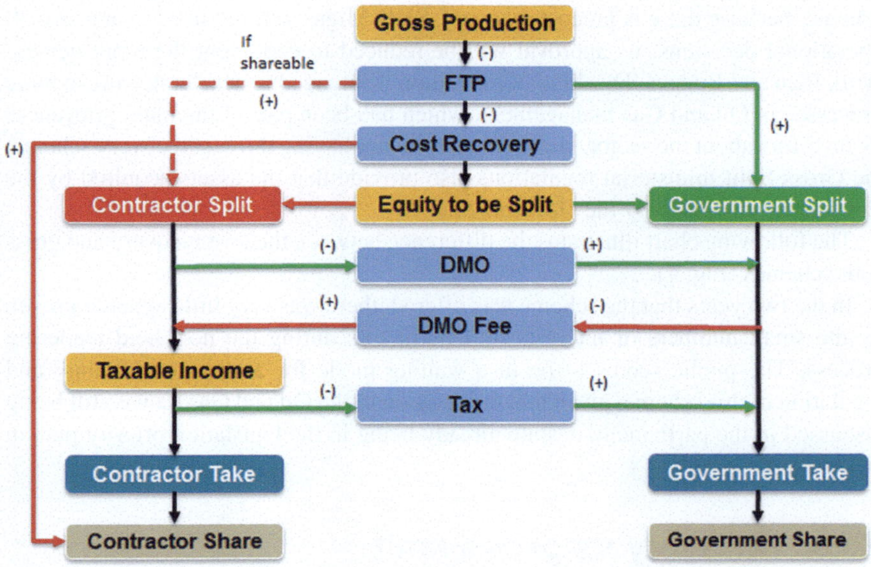

Fig. 3 Production sharing contract in Indonesia. *Source* SKK Migas

Table 4 Gross split scheme

	Crude oil (%)	Natural gas (%)
Government take	57	52
Contractor take	43	48

The Gross Split has two main components that are (a) the base split component; and (b) the adjustment component that allocates the shares as follows (Table 4).

The Gross Split scheme also takes into account two additional components, which are: (a) variable component and (b) progressive component. Variable components consider factors of field development stage, field location, depth of reservoir, availability of supporting infrastructure, reservoir type, CO_2 content, H2S content, gravity, domestic content during field development and production methods (water injection or enhanced oil recovery) to adjust the split between Government and Contractor. While the progressive components are to be adjusted monthly by SKK Migas considering additional shares available for the Contractors based on a reversed sliding scale based on cumulative production over a certain period. The higher the cumulative production, the lower the additional share.

The Government sees that this scheme does not depart from what is required of extractive resources management as regulated in the Constitution and in the Oil and Gas Law (Nurtjahyo 2017). The Gross Split scheme maintains ownership title of the mineral resources in the Government until point of delivery while it maintains certain control of operational management through SKK Migas. The Contractor is still responsible to provide capital and bear all the development risks. Under the scheme, because there is no cost recovery, SKK Migas will not need to approve all operational decisions; its approval will be reduced to endorsing the more general Work Plan and Budget. This kind of relaxation is intended to accelerate bureaucratic processes in Oil and Gas management, which had been part of the main grievances by investors about the sector. However, despite not having the cost recovery scheme, the Gross Split ministerial regulations also provide that the assets acquired by the Contractor are owned by the Government.

The following chart illustrates the difference between the cost recovery and gross split schemes (Fig. 4).

In the two years that the scheme was offered, there was very little appetite proven by the small numbers of interests by Contractors during the new field tendering process. The public seems to be in a waiting mode for a stronger certainty and regulation of this scheme, given that the revision to the Oil and Gas Law is still being discussed in the parliament, despite already being in the legislation priority plan in 2014–2019.

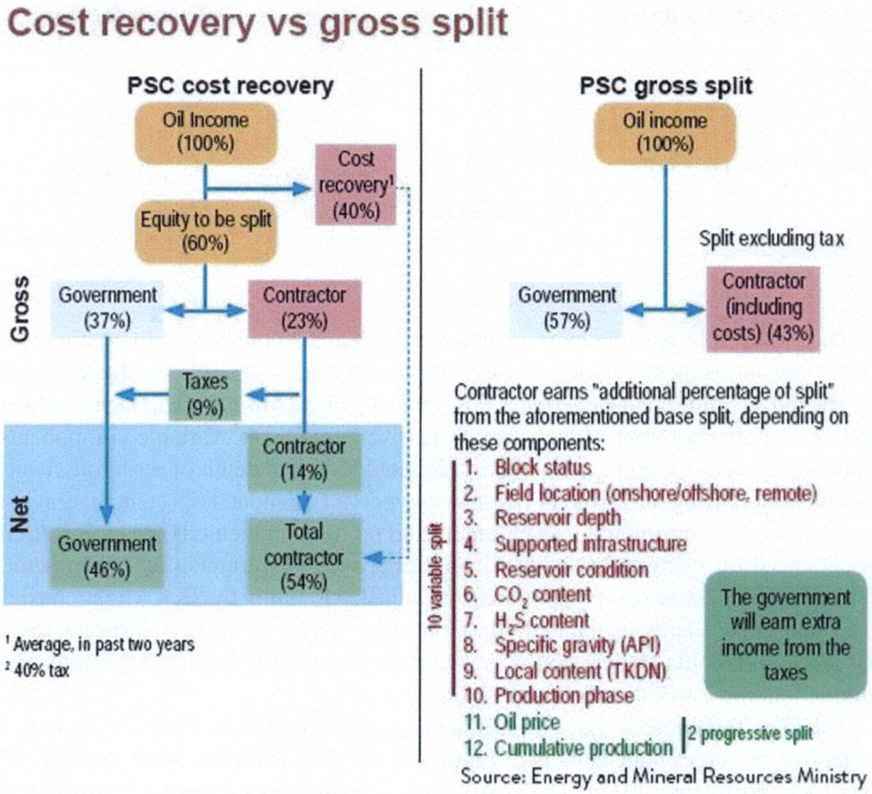

Fig. 4 PSC versus gross split. *Source* Ministry of energy and mineral resources

5.2 Mining

The Mineral and Coal sector is regulated by the 2009 Mining Law and its implementing regulations. It adopts a licencing requirement for extraction by private corporations in the form of Mining Licence (*Izin Usaha Pertambangan* or IUP) or Special Mining Licence (*Izin Pertambangan Khusus* or IUPK). Existing Contracts of Works was to be converted into IUP or IUPK accordingly.

All IUP or IUPK holders are required to pay production royalties at varying rates, depending on the mining scale, the production level and the mining commodities price (PWC 2019). The following is production loyalty rates applicable to IUP and IUPK holders (Table 5).

In addition to the above rates, holders of IUPK are subject to an additional royalty of 10%. Other than these royalties, holders of IUP and IUPK are also subject to certain taxation regime applicable to the mining sector as regulated by Government Regulation No. 37 of 2018 concerning the Treatment of Taxation and/or Non-Tax State Revenues in the Mineral Mining Business.

Another fiscal obligation imposed to the mining sector is the export duties that are tied to the progress of a certain licence holder's progress in building a refining facility/capacity. This is due to a new regulation under the 2009 Mining Law that bans exports of ore. However, the Government relaxed this provision by allowing some exports of ore subject to export licence and duties while continuing to impose obligations to the exporters to build refining facilities/capacities domestically.

Table 5 Production royalty rate

Commodity	Production royalty rate (%)
Coal • Open pit • Underground	3–7 2–6
Nickel	4–5
Zinc	3
Tin	3
Copper	4
Iron	3
Gold	3.75
Silver	3.25
Iron Sand	3.75
Bauxite	3.75

6 Management of Funds from Extractive Resources in Indonesia

Under the Indonesian treasury system, all State revenues are managed by the Ministry of Finance. The Ministry of Energy and Mineral Resources as the Ministry responsible for the Oil and Gas and Mining sectors will be allocated some budget to execute its regulatory and supervisory capacity through the State Budget managed by the Ministry of Finance.

At the time this chapter is written, there is no sovereign or dedicated fund for extractive resources in Indonesia. State revenue from extractive resources becomes part of the ordinary State revenue and is used to fund development and operational programmes as planned in the annual State Budget.

However, Indonesia is in the process of reforming its Oil and Gas Law, a process it has been undergoing since 2011. In the draft of the revised Oil and Gas Law,[2] there is a chapter on Oil and Gas Fund. Further, regulations on this matter are still unclear, and are mandated to be governed through subsequent Government Regulation. The Academic Background section of this draft mentioned the need for the Oil and Gas Fund and look at the practices of Norway, Timor Leste, Thailand, Kazakhstan, and Ghana for guidance.

In 2015, the debate about the Oil and Gas Fund took centre stage in the context of establishing a Renewable Fund. In the 2015 draft of the revised Oil and Gas Law, it is expressly stated that one of the purposes of the Oil and Gas Fund is to support renewable energy development in Indonesia.[3] However, this expressly stated purpose is not found in the 2018 draft, which states that the Oil and Gas Fund is to be used for activities relating to the replacement of oil and gas reserves through exploration, development of oil and gas infrastructure, as well as oil and gas research and development.[4]

The Oil and Gas Fund will consist of certain percentages of[5]:

1. First Tranche Petroleum (FTP);
2. Bonuses payable to the central government based on contracts and/or the Oil and Gas Law; and
3. Any levy payable to the state based on laws and regulations.

---

[2]House of Representative Oil and Gas Law Draft of September 2018, as referred to in Arvirianty 2019.

[3]House of Representative Oil and Gas Law Draft of 2015.

[4]Article 63 (2), Chapter IV, Oil and Gas Law Draft, September 2018.

[5]Article 63 (3), Chapter IV, Oil and Gas Law Draft, September 2018.

This was already a departure from the previous drafts where the contribution to the Oil and Gas Fund was to be accrued from a certain portion of total commercial production specially set aside, outside of the portion of the Government or Contractor.[6]

The revised draft Oil and Gas Law mandates the making of Government Regulations to further regulate the Oil and Gas Law. This means that even after the enactment of the revised Law, there will still be some time before the Oil and Gas Fund can be established, pending the enactment of said Government Regulations. Since the Law will unlikely regulate a timeline for the Government Regulations, it would be unpredictable. As a reference, the first Government Regulation enacted under the old Oil and Gas Law was only enacted three years after, with the latest one only issued 10 years after.[7]

In the meantime, some local governments have tried to establish their own Oil and Gas Fund in the form of putting aside some of their portion of oil and gas revenue sharing fund (*dana bagi hasil/DBH*) into an endowment fund to be used to fund development programmes in their specific localities such as in Bojonegoro and Musi Banyuasin. However, at the time this chapter is written, there has been no successful establishment of such an endowment fund. There were concerns relating to committing the amount of funds for a long period of time while their investment plan is not clear yet.

7 Distribution of Funds from Extractive Resources to Regions and Other Entities in Indonesia

After the reform of 1998, the Indonesian Government focused on decentralisation of power to the regions, which also led to a fairer, more proportionate, and efficient distribution of economic benefits, including the distribution of benefits from extractive resources. The decentralised government structure in Indonesia is divided into three main categories: (1) Central Government, (2) Provincial Government and (3) District (and/or City) Government. By law, the Central Government only holds power over six matters: foreign policy, defence, security, judiciary, monetary and fiscal and finally religious affairs (Law No. 32 of 2004 on Decentralisation). Other matters are reserved for Provincial and District Governments, depending on the scope.

The reformed benefit-sharing mechanism to honour the commitment to decentralisation is reflected in Law No. 33 of 2004 concerning Fiscal Balance between Central and Regional Governments and its implementing regulations through Government

[6]House of Representative Oil and Gas Law Draft of 2015, "*jumlah tertentu dari hasil total produksi komersial yang disisihkan secara khusus di luar bagian Pemerintah Pusat dan kontraktor*", translated as "*a certain amount of the total commercial value of the production will be set aside outside of Government and Contractor's take*".

[7]UU 22/2001 on Oil and Gas, PP 35/2004 on Upstream Oil and Gas Activities, PP 79/2010 on Cost Recovery.

Regulation No. 55 of 2005 concerning Fiscal Balance Funding. The Fiscal Balance Regulations adopt a distribution mechanism known as Revenue Sharing Fund (*Dana Bagi Hasil*/DBH). DBH comes from the revenue post at the National Budget, which is allocated to the regions based on a certain percentage to fund regional needs in order to implement decentralised powers (Article 1 point 20 of Law No. 33 of 2004 and Article 1 point 9 of Government Regulation No. 55 of 2005). The Natural Resources DBH include DBH from the forestry, fisheries, mining, oil, gas and geothermal sectors.

In principle, tax, royalties and income directly linked to extractive resources in Indonesia are collected by the Central Government. This income will be blended in the National Treasury without specific earmarked uses. The Central Government will then distribute some of this income to the Regional Governments (provinces or districts), both those where the extractive resource originated from (the producing regions) and those who do not produce the extractive resource (the non-producing regions) to ensure equity for all Indonesians. Income received by the Central Government that is used to be the basis of distribution to regional governments is the Net Operating Income.

DBH is allocated based on the following two principles:

1. "By Origin", whereas the producing region receives a bigger percentage of revenue sharing while the other regions in one province receive an aggregate percentage.
2. "By Actual", whereas the revenue shared to the regions, both the producing region and otherwise, is based on realisation of tax and non-tax revenue in the same year.

The source of DBH that will be distributed to the regions comes from taxes and extractive resources income. DBH from taxes include those part of the land and building tax, levies on the acquisition of rights over land and buildings (BPHTB), and income tax.

Below is DBH mechanism that comes from tax (Table 6).

DBH that comes from natural resources is distributed based on the following criteria:

1. Forestry, in the form of:
   a. Forest Utilisation Business Permit Fee (*Iuran Izin Usaha Pemanfaaan Hutan*/IIUPH)
   b. Forest Royalties (*Provisi Sumber Daya Hutan*/PSDH)
   c. Reforestation Fund
2. Mining, in the form of:
   a. Land Rent
   b. Royalty from Exploration and Exploitation
3. Fisheries, in the form of:
   a. Commercial Fisheries Charges

Table 6 Revenue sharing fund mechanism

No.	Tax	Central Gov. (%)	Reg. Gov.	Proportion			
				Province (%)	Producing district (%)	Other districts (%)	Upah Pungut (%)
1	Land and building tax	10	90	16.2	64.8		9
2	Levies on the acquisition of rights over land and buildings (BPHTB)	20	80	16	64		
3	a. Domestic individual income tax (Art. 25 dan Art. 29)	80	20	8	8.4	3.6	
	b. Income tax Art. 21	80	20	8	8.4	3.6	

   b. Fisheries Levy

4. Oil extraction
5. Gas extraction
6. Geothermal production, in the form of:

   a. Government take
   b. Land rent and royalty.

Below is the distribution of natural resources DBH (Table 7).

Aside from the above distribution of funds, the Provinces of Aceh and West Papua are given an additional 55% for Oil DBH and 40% for Gas DBH due to their special autonomy status, which should be earmarked to fund education and health programmes (Ministry of Finance 2017).

Even though the distribution of fund to the regions is already regulated by percentage, there is still criticism by the regional governments regarding certainty of income from extractive resources. This is because the actual distribution is conducted only after the annual financial audit, which overlaps with the State budgeting cycle for the next year. This means that regional governments often have to use inaccurate assumptions of projected income to budget for the next year's spending.[8] Another criticism regarding the portion for the producing regions is that some regions argue that the central government takes too much of the income from extractive resources

[8]https://eiti.ekon.go.id/eiti-untuk-perbaikan-pengalokasian-dana-bagi-hasil-dbh/.

Table 7 Natural resources revenue sharing fund

No.	Types of revenue sharing	Central Gov. (%)	Reg. Gov. (%)	Proportion		
				Prov. (%)	Producing district (%)	Other districts (%)
1	*Forestry*					
	a. IHPH	20	80	16	64	–
	b. PSDH	20	80	16	32	32
	c. Reforestation	60	40	–	40	–
2	*Mining*					
	a. Land rent from producing district	20	80	16	64	–
	b. Land rent from producing province	20	80	80	–	–
	(c) Royalty from producing district	20	80	16	32	32
	(d) Royalty from producing province	20	80	26	–	54
3	*Fishery*	20	80	Distributed in aggregate to all regions in Indonesia		
4	*Oil*					
	a. From producing district	84.5	15.5			
			15	3	6	6
			0.5[a]	0.1	0.2	0.2
	b. From producing province	84.5	15.5			
			15	5	–	10
			0.5 [a]	0.17	–	0.33
5	*Gas*					
	a. From producing district	69.5	30.5			
			30	6	12	12
			0.5[a]	0.1	0.2	0.2
	b. From producing province	69.5	30.5			

(continued)

Table 7 (continued)

No.	Types of revenue sharing	Central Gov. (%)	Reg. Gov. (%)	Proportion		
				Prov. (%)	Producing district (%)	Other districts (%)
			30	6	12	12
			0.5[a]	0.1	0.2	0.2
6	*Geothermal*	20	80	16	32	32

[a]Earmarked for primary education

while already taking control of income tax for both corporations and individuals.[9] These criticisms have been subject of review by the Constitutional Court, the only court authorised to review laws passed by the parliament, upon challenges by various civil society organisations as well as regional governments. However, the Constitutional Court has consistently rejected those challenges and thus the laws on balancing of State revenue amongst the Central and Regional governments still stand.

8 Conclusion

Since its independence in 1945, Indonesia has experienced changes in government, which inevitably modifies its ways of regulating its natural resources and the relevant industries. All the while, the fundamental principle of natural resources as public goods to be used for the welfare of the people did not change. This principle was strengthened in recent years through various Constitutional Court decisions, which in essence provided guidance for the Government as regulator of this sector. The challenges lie with the Government on how to maximise the impacts of the natural resources industry to the Indonesian economy. Strategic initiatives should be taken with considerations that although Indonesia has significant natural resource reserves, it is depleting and competing for investments. The Government has attempted to adopt new approaches in order to further investment in the extractive industry, such as the new gross split scheme in oil and gas, or obligation to process ore domestically in the mineral and coal sector to add value to the industry as well as converting contracts of works to mining licences for regulatory purposes. However, these new policies are not without their challenges since investment in the natural resources sector requires a degree of certainty and is made for a relatively long period of time. The gross split scheme has yet to be accepted by the industry, shown by low interest for fields requiring the adoption of the gross split scheme, and no new fields is yet to be contracted with this scheme. The obligation to process ore in the mineral and coal sector and the conversion from contract to licencing regime were met with various challenges to the Constitutional Court and the Supreme Court, and resulted

[9]https://eiti.ekon.go.id/eiti-untuk-perbaikan-pengalokasian-dana-bagi-hasil-dbh/.

with no less than five revisions to the Government Regulation regulating this matter. From this experience, changes need to be applied gradually and in close consultation with all stakeholders, particularly the industry, to avoid any significant hindrance in the implementation stage, and any potential breach of existing obligations to the investors.

While continuing to reform its natural resources sector, Indonesia also experiences challenges from the change to decentralisation where local governments have more authority to issue licences, including environmental and several mining licences. These relatively new authorities resulted in extra layers of bureaucracy for the sector. The Extractive Industries Transparency Initiative (EITI) Indonesia noted that there were at least 2522 non-clear and clear (CNC) mining licences, partly due to the implementation of local authorities in licencing, which sometimes created overlapping licences across regions. Even though the intention was to lessen the gap between local citizens with its ownership of mineral resources, governance on implementation by local authorities needs to be improved to create certainty for investment.

In dealing with the revenues from its natural resources, Indonesia should continue to explore the establishment of a sovereign fund to further optimise the economic impacts of its natural resources. Especially considering that there might be changes in the near future that require heavy investment from the Government, such as the move to renewable energy and the scarcity of financing for the extractives sector. It should also continue to explore better and fairer ways to redistribute profits from natural resources exploitation to the regions, continuing its already positive effort to do so since the reformation twenty years ago. These new strategies should be considered and taken sooner rather than later to ensure that natural resources are truly leveraged for the maximum welfare of the people, as its Constitution consistently mandates.

References

Aizenman, J., & Glick, R. (2009). Sovereign wealth funds: Stylized facts about their determinants and governance. *International Finance, 12*(3), 351–386.

Alhashel, B. (2015). Sovereign wealth funds: A literature review. *Journal of Economics and Business, 78*, 1–13.

Arvirianty, Anastasia. RUU Migas Digodok, Ini Bocorannya, 24 January 2019, available at https://www.cnbcindonesia.com/news/20190124101358-4-52136/ruu-migas-digodok-ini-bocoran-rancangannya, accessed on 29 November 2019.

Bahoo, S., Alon, I., & Paltrinieri, A. (2019). Sovereign wealth funds: Past, present and future. *International Review of Financial Analysis*, 101418.

Butt, S., Shivdasani, A., Stendevad, C., & Wyman, A. (2008). Sovereign wealth funds: A growing global force in corporate finance. *Journal of Applied Corporate Finance, 20*(1), 73–83.

Collier, P., Van Der Ploeg, R., Spence, M., Venables, A. J. (2009) Managing resource revenues in developing economies. *IMF Staff Papers, 57*(1), 84–118.

Constitution of the Republic of Indonesia and Explanatory Memorandum attached to the 1945 Constitution, available at https://www.dpr.go.id/jdih/uu1945, accessed on 29 November 2019.

Constitutional Court of the Republic of Indonesia. Decision No. 36/PUU-X/2012, 2012, available at https://www.bphn.go.id/data/documents/36-2012.pdf, accessed on 29 November 2019.

Devlin, W., & Brummitt, B. (2007). A few sovereigns more: The rise of sovereign wealth funds. *Economic Round-up*, 119.

Extractive Industries Transparency Initiative (EITI) Indonesia. Country Overview, available at https://eiti.org/indonesia#extractive-industries-contribution, accessed on 29 November 2019.

Erlangga, Doni. EITI untuk Perbaikan Pengalokasian Dana Bagi Hasil, 2017, avalaible at https://eiti.ekon.go.id/eiti-untuk-perbaikan-pengalokasian-dana-bagi-hasil-dbh/, accessed on 29 November 2019.

Gould, M. (2010). Managing manna from below: Sovereign wealth funds and extractive industries in the Pacific. *Economic Round-Up, 1*, 63.

Griffith-Jones, S., & Ocampo, J. A. (2008). Sovereign wealth funds: A developing country perspective.

House of Representative, Republic of Indonesia. Draft Law on Revision to Law No. 22 of 2001 on Oil and Gas, 2015, available at https://www.dpr.go.id/doksileg/proses2/RJ2-20150626-022308-2742.pdf, accessed on 29 November 2019.

IMF. (2007). Global Financial Stability Report, Annex 1.2, September 2007, Washington DC, International Monetary Fund.

International Financial Services London (IFSL). (2009). Sovereign wealth funds 2009, March. https://dio.org/www.ifsl.org.uk/upload/CBS_Sovereign%20Wealth%20Funds%202009.pdf.

Kimmitt, R. M. (2008). Public footprints in private markets, *Foreign Affairs, 87*, 1. January/February 2008.

Kusumaatmadja, M. (2013). Indonesia's national policy on offshore natural resources: Some legal issues. *Indonesian Journal of International Law, 10*(4).

Lowery, C. (2007). Remarks on sovereign wealth funds and the international financial system. US Department of the Treasury.

Machmud, T. N. (2002). The Indonesian production sharing contract: An investor's perspective. *Journal of Energy and Natural Resources Law, 20*(3).

Megginson, W. L., & Gao, X. (2019). The state of research on sovereign wealth funds. *Global Finance Journal*, 100466.

Ministry of Energy and Mineral Resources, Republic of Indonesia. Rekonsiliasi Data, Sumber Daa Batubara Indonesia Kini 166 Miliar Ton, Cadangan 37 Miliar Ton, 2018, available at https://www.esdm.go.id/id/media-center/arsip-berita/rekonsiliasi-data-sumber-daya-batubara-indonesia-kini-166-miliar-ton-cadangan-37-miliar-ton, accessed on 29 November 2019.

Ministry of Finance, Republic of Indonesia. Handbook on Allocation of Natural Resources Revenue Sharing Fund, 2017, available at https://www.djpk.depkeu.go.id/wp-content/uploads/2017/06/Buku-Pegangan-Perhitungan-Alokasi-DBH-SDA.pdf, accessed on 29 November 2019.

Nurtjahyo, Prahoro. Expert Staff to the Minister ESDM on Investment and Infrastructure Development on Answering Doubts on Gross Split, 2017, available at https://www.esdm.go.id/assets/media/content/content-menjawab-keraguan-terhadap-gross-split-tanggapan-atas-opini-dr-mad jedi-hasan-potensi-permasalahan-dalam-gross-split.pdf, accessed on 29 November 2019.

Parsons, D., & Vincent, D. (1991). High stakes: Mineral and petroleum development in papua new guinea. Institute of National Affairs Discussion Paper No. 49. Port Moresby.

PWC. Mining in Indonesia: Investment and Taxation Guide, 2019, available at https://www.pwc.com/id/en/energy-utilities-mining/assets/mining/mining-guide-2019.pdf, accessed on 29 November 2019.

Republic of Indonesia, Law No. 37 of 1960 on Mining.

Republic of Indonesia, Law No. 44 of 1960 on Oil and Gas.

Republic of Indonesia, Law No. 22 of 2001 on Oil and Gas.

Republic of Indonesia, Law No. 4 of 2009 on Mineral and Coal Mining.

Republic of Indonesia, Law No. 32 of 2004 on Decentralization.

Republic of Indonesia, Law No. 33 of 2004 on Fiscal Balance Between the Central and Regional Governments.

Republic of Indonesia, Government Regulation No. 55 of 2005 on Revenue Sharing.

Republic of Indonesia, Government Regulation No. 35 of 2004 on Upstream Oil and Gas Activities.

Republic of Indonesia, Government Regulation No. 79 of 2010 on Cost Recovery.

Truman, E. M. (2009). A blueprint for sovereign wealth fund best practices. *Revue d'economie Financiere, 9*(1), 429–451.

Utomo, S. P. (1999). Butir-Butir dalam Penyusunan Rencana Undang-undang tentang Usaha Pertambangan Minyak dan Gas Bumi.

The Social Fund: A Brazilian Sovereign Wealth Fund

Hirdan Katarina de Medeiros Costa and Isabela Morbach Machado e Silva

Abstract In conjunction with expectations of an increase in revenue from the exploration of oil and gas after the discovery of Brazil's pre-salt layer, important issues emerged on the best methods of administering public revenue in the benefit of current public interest and futures generations. In this context, the Executive Branch sent to the National Congress a proposal for a new regulatory framework for the formulation and implementation of public policies in the energy sector. As part of this policy a Social Fund of petroleum (i.e. a Sovereign Wealth Fund) was created in order to create a source of resources for social and regional development, through programs and projects aimed at combating poverty and stimulating development. In light of this, this chapter will describe the main aspects of Brazil's Social Fund.

Keywords Public finance law · Non-renewable natural resources · Social fund and sovereign wealth fund

1 Introduction

The possibility of exploiting abundant oil in the Brazilian pre-salt layer has raised the possibility of several economic effects for Brazil (Rodrigues and Sauer 2015). The volume of revenue generated by the oil activity can bring development as well as economic losses due to market currency imbalance and economic dependence on these revenues. This phenomenon is labelled as "resource curse" or "paradox of plenty" (Auty and Gelb 1986; Conway and Gelb 1988; Gelb et al. 1988; Auty 1988, 1991, 2003, 2005; Bacon and Tordo 2006; Sachs and Warner 1995, 1997, 1999, 2001). In order to minimise such problems, Sovereign Wealth Funds (SWF)

H. K. de Medeiros Costa (✉) · I. M. M. e Silva
University of São Paulo, IEE, São Paulo, Brazil
e-mail: hirdan@usp.br

I. M. M. e Silva
e-mail: sabelamorbach@usp.br

E. G. Pereira et al. (eds.), *Sovereign Wealth Funds, Local Content Policies and CSR*, CSR, Sustainability, Ethics & Governance,
https://doi.org/10.1007/978-3-030-56092-8_5

have been used as a tool for achieving distributive justice, as well as energy justice (considering resource policy taxation) (Heffron 2018; Heffron and McCauley 2017).

Classified as a SWF, Brazil created a Social Fund (hereinafter referred to as SF). The central challenge of the SF was to design a legal and financial mechanism to convert this temporary and volatile source of income, that is, the economic exploitation of oil and gas, into a source of income that is regular and stable for current and future budgetary activities (Costa 2010, 2018, de Medeiros Costa 2012; Silva and de Medeiros Costa 2019).

It is part of this challenge to protect the SF against discretionary changes promoted by the Executive and Legislative Branches of the government—for example, by avoiding the use of the SF for addressing possible public deficit problems (including budget deficits)—but at the same time, to use these same resources for the promotion of social development (Jordan et al. 2013).

With the possibility of changing Brazil's status to be among the largest reserves of oil and gas in the world, the Executive Branch was obliged to discuss the question of extractives resources and to define new arrangements for their allocation. These include: defining objectives and targets, outlining public policies and economic planning, taking into account the harms of oil exploration, observing other producing countries and cases of occurrence of the resource curse phenomenon. In this context, the Executive Power sent to the National Congress a proposal for a Regulatory Framework, formed by four bills, which gave rise to Law 12.276/2010, Law 12.304/2010 and Law 12.351/2010.

The proposals brought three main innovations for the formulation and implementation of public policies in the Brazilian energy sector, since (i) a new regulation was established for the exploitation of petroleum in the area of pre-salt—the system of production sharing; (ii) a new public company was created, responsible for the management of oil and gas production and sales distribution contracts in the area of pre-salt, called the Brazilian Petroleum and Natural Gas Administration Company—Pre-Sal Petróleo SA (PPSA); and (iii) the creation of a Sovereign Wealth Fund called the Social Fund, with an accounting and financial remit, directly linked to the Presidency (de Medeiros Costa et al. 2017a).

The Social Fund was created by Law 12.351/2010, which provided for general regulation about exploration and production of oil, natural gas and other fluid hydrocarbons under the production sharing regime in areas of the Pre-Salt and in strategic areas, defining its purpose, objectives, structure and sources of resources, among other measures. Therefore, this chapter describes the SF, analyses its structure and draws out lessons to be learned.

2 The History of the Brazilian Social Fund (SF)

In Brazil, the Oil Law and Law No. 7990/89 currently provide for the ways the 'government takes' should be divided between federal entities and agencies of the federal government (Costa and Santos 2013). Historically, the legal provisions did not contain specific requirements as to how the revenues should be applied by these entities and agencies (de Medeiros Costa 2012). Law 12.734/2012 changed those provisions (Costa 2018). Indeed, there were only a few restrictions to the management of government oil and gas revenues in Brazil, namely:

- Law 7.990/89, providing for the distribution of the share of royalties provided in the concession contract, representing 5% of production, in Article 8, provides a prohibition to invest such funds to pay debts, except government debts, and for the payment of permanent staff (Costa and Santos 2013), and
- Resolution No. 43/2001 of the Senate, providing for domestic and foreign credit transactions of the government entities, in Article 5, VI, that states the prohibition to anticipate 'government take' revenue concerning the period after the term of office of the Head of the Executive Branch; note that these prohibitions do not apply to capitalization of pension funds or to repay debt with the government.

The absence of a specific regulation of expenditure of government oil revenues could increase the occurrence of the oil curse in Brazil, because in the case of states and municipalities, expenditure can be concealed by macroeconomic policies subject to federal decisions. Costa and Santos (2013) found indications of a resource curse in the municipal executive data for the localities studied where they observed a failure to adhere to Law 7.990/89, with public expenses being directed toward paying personnel. Also, they found that the oil industry in the municipalities did not observe social and economic human rights, nor were the municipal executives transparent, thereby hindering the monitoring of, and popular participation in, the administrative process.

In view of the large oil and gas reserves discovered in the pre-salt layer in deep waters off the Brazilian coast, on August 31st, 2009, the Government sent to the Congress Bill No. 5.940/09, creating a Pre-Salt Social Fund (SF) for the purpose of regulating government funds relating to the exploration and production of oil and gas. Then, in 2010, this Congress Bill became Law 12.351 that created the SF (de Medeiros Costa et al. 2010). Historically, according to this Congress Bill No. 5.940/09, the accounting and financial nature of the Social Fund is directly linked to the Presidency, in order to provide a regular source of funds for the performance of projects and programs in the areas of poverty alleviation and the development of education, culture, science and technology and environmental sustainability (Brasil 2010a; de Medeiros Costa et al. 2010, 2017a).

It was determined that the following funds are allocated to the Social Fund: (i) the portion of the signature bonus amount to be allocated by production sharing contracts; (ii) the portion of the royalties to which the Government is entitled, upon deduction of those to its specific agencies, as established in the production sharing contracts; (iii) the income resulting from the sale of oil, natural gas and other hydrocarbon fluids of the Government, as defined by law; (iv) the results of financial investments on their cash and cash equivalents; and (v) other funds that may be allocated by law (de Medeiros Costa et al. 2010, 2017a; Costa 2018).

3 The Characteristics of the Social Fund

As it was conceived, the Social Fund can be considered a sovereign wealth fund, with allocation characteristics and intergenerational purposes (i.e., savings formation and promotion of development) (Costa 2018; Silva and de Medeiros Costa 2019). There is a clear linkage of resources, provided in the Brazilian Constitution itself, which authorizes the creation of these kind of funds.

The Social Fund can still be classified as a sovereign wealth fund, according to the classification proposed by the International Monetary Fund, which distinguishes five modalities, depending on their dominant characteristics: stabilization funds, savings funds, investment funds, development funds and pension funds (Backer 2016). According to Scaff[1] (2013), the SF in Brazil is closer to being a savings fund with its focus on future generations.

4 The Social Fund and Its Management

Law 12.351/2010 states that a Deliberative Social Fund Board (DSFB) will be created with the duty to decide on the priority and the allocation of resources recovered from the fund, which will involve representatives of civil society and the Federal Government, with the composition, duties and operation established in the Executive Branch (Brasil 2010b).

However, according to Law 12.351/2010, the Social Fund will be actually managed by the Financial Management Committee of the Social Fund (FMCSF), whose composition and operation will also be established by the Executive Branch. The FMCSF will have the duty to define: (a) the amount to be annually recovered from the SF with its financial sustainability being ensured; (b) the expected minimum return; (c) the type and level of risk that may be undertaken in the making of investments; (d) the minimum and maximum percentages of funds to be invested in the country; (e) the minimum and maximum percentages of funds to be invested abroad; (f) the minimum and maximum percentages of funds to be invested by the economic activity or sector; and (g) the minimum capitalization to be achieved (Brasil 2010b).[2]

Regarding the control system, Law 12.351/2010 defines that financial statements and results of the Social Fund's investments will be developed and assessed every six months, the Ministry of Finance will submit on a quarterly basis to the Congress

---

[1]Scaff (2013: 518) says: "And the FS is a typical Fund made up of incomes derived from the exploitation of petroleum resources or, in the nomenclature of the IMF, a savings fund for future generations".

[2]In order to promote the investment in assets in Brazil and abroad, the Federal Government, with funds from the Social Fund, could also participate, as the sole member, of a specific investment fund for such purpose. Payment of the shares of the fund will be authorized by an act of the Executive Branch, after opinion of the FMCSF is heard. The investment fund will set out operational matters of administrative and financial management and the prudential rules of investment supervision.

a report performance, and a future decree of the Executive Branch will define the other supervisory rules of this fund (Brasil 2010b).

Indeed, by analysing the proposal, the level of discretion granted to the Executive Branch is of great concern. First, we can see that the Social Fund is virtual, as it will be created within the normal budgetary process of the Government, being directly linked to the Presidency. There is no independent institution responsible for the fund management: the composition and operation of both the DSFB and FMCSF will be established by the Executive Branch, which, in principle, will keep the control of these two agencies in the hands of the Government, i.e., such agencies will not be given political independence or institutional autonomy.[3]

5 The Social Fund and Its Analysis

There are no limits to the types of investments that may be made by the Fund. The lack of limitations to investments in oil and gas projects, for example, can result in great risk due to the lack of encouragement to economic diversification. Similarly, the limitations of location, so important to prevent the appreciation of the real exchange rate, and further inflationary pressures resulting from the expansion of local aggregate demand are addressed only indirectly. Moreover, the FMCSF has the duty of defining the minimum and maximum percentages of funds to be invested in Brazil or abroad.

According to Gelb and Auty, the direct involvement of the Government in the allocation of revenues from oil and gas, in the way proposed in Bill No. 5.940/09, can lead to poor allocation of these funds. Gelb and Auty claim that specific regulation should be created to bind the government revenues from oil and gas, protecting the management of these revenues from the discretion of the Executive Branch, in order to force long-term investments such as in infrastructure. This would be in addition to the creation of funds for promotion and development, thereby allowing for the competitiveness of other industries.

It is a logical reasoning that government oil and gas revenues are temporary revenues and belong not only to the present generation but also to future generations. Therefore, their inclusion in government accounts as if they were taxes, without a specific regulation as to their condition, necessarily leads to mismanagement of these

[3]Bucheb (2004), *The autonomy of the regulatory agencies and the job security of their leaders*, jus navigandi. The essential requirements for independence or political and institutional autonomy are: (1) job security of the leaders: impossibility of dismissal, except for a serious misconduct determined by due process of law; (2) fixed term of office; (3) appointment of officers backed by political parties; (4) impossibility of administrative appeal to the Ministry to which it is bound: no hierarchical instance to review its acts, except judicial review; (5) Management autonomy: not being bound to any government level; and (6) Establishment of own sources of funds to the agency, if possible generated from the very exercise of the regulatory activity.

funds, generating significant losses to the present generation and even greater losses to future generations.[4]

Under the Brazilian Government's proposal, it is not known whether the Social Fund will be primarily a stabilization or a savings fund. The Fund's goals, according to the Bill, among others, are to: (i) provide regular source of funds for social development, in the form of projects and programs in the areas of poverty alleviation and development of education, culture, science and technology and environmental sustainability; and (ii) mitigate the income and price fluctuations in the national economy, resulting from changes in the income generated by production activities and exploration of oil and other non-renewable resources, which are characteristics of stabilization funds; and (iii) provide long-term public savings through revenue earned by the Government, which are a characteristic of a savings fund.

The vagueness of the legislation is also reflected in the choice of applying discretionary rules, not linked ones. The law lists in Article 1 the areas that will benefit from the creation of the Social Fund. The specific criteria for investment of funds, however, will be defined by the FMCSF, and the decision on the priority and the allocation of the funds recovered from the Social Fund is the prerogative of the DSFB.

All such decisions made by the Government may result in future losses to society. Although there is a control system of the Fund's decisions, as proposed in the Bill, the questioning of government decisions after the event is superfluous, since the scope of the review, judicial or not, of discretionary administrative decisions is limited and its legal possibility is even questioned.

6 Brazilian Law 12.351/2010

When the Government sent the Bill 5.940/2009 related to creation of a Pre-Salt Social Fund to the National Congress in 2009, it argued that the Fund should be created because of three crucial points (MME 2009). All of them were dealing with the nature of the hydrocarbon industry revenues. Those points are found in Item 2 of the Memorandum 119/2009 (MME 2009). The first and second points are the finiteness of the natural resource itself, which leads to temporal limitation of the income derived from oil exploration and the volatility of international oil and natural gas prices, characterized by factors that go beyond the simple supply-demand equation (MME 2009). The last point relates to Dutch Disease, because the purpose is to mitigate the volatility of prices and of incomes into the national economy (MME 2009). The third item of the mentioned Memorandum relates to the benefits of the exploitation of these resources for future generations: "Governments must act to prevent only the current generation enjoying the benefits of the exploitation of finite resources. For this it is necessary that the oil and gas wealth will be transformed into active

[4]The referred loss will be suffered over a period of time and by means of methods of transmitting the resource curse.

Table 1 Brazilian rate of illiterates—aged 15 or older (in 2010)

Urban area	Rural area	Total
7.3%	23.2%	9.6%

Source IBGE (2010a, b)

enjoyment which can be extended in time, even after the oil and/or gas has run out" (MME 2009).

When the Law 12351/2010, which actually created the SF was approved by Congress in December 2010, there was a prediction in Article 47, §2° which stated that "50% of the total revenue earned by the Fund should be invested in programs for public education development, with a minimum of 80% for basic and child education" (Brasil 2010b). However, former President Luis Inacio Lula da Silva (Lula) vetoed the Article on the revenues from oil and natural gas exploration focused on royalties. Thus, discussion continued in the Brazilian National Congress (Brasil 2010b).

Despite this veto, Congress continued to argue that the revenues should be directed to education. They said the best way to benefit from oil wealth over time, even after the oil and gas have run out, is to invest those revenues in education (de Medeiros Costa et al. 2010; ADPM 2019). They considered the Brazilian weakness in education has been causing also a deficit in other areas, such as the shortage in skilled workers (de Medeiros Costa et al. 2010; ADPM 2019). In fact, the data from IBGE (2010a, b) shows high levels of illiteracy for 15 years old people and older (Table 1).

Also, according to the IPEA (2012), the labour productivity in Brazil is historically low, showing very little growth over the years. Measured in terms of gross domestic product (GDP) for employed persons, labour productivity in Brazil is three times lower than in South Korea, four times lower than in Germany and five times lower than in the United States. In Brazil, college enrolments are low (IPEA 2012).

Although the debate in Congress had gone on, former president Lula wrote in the consideration of his veto: "the Social Fund is a long term savings in order to ensure the intergenerational benefits arising from the exploration of pre-salt. In this context, it is not appropriate to establish in advance which areas to be prioritized among those already considered, including education. For this reason, the Law created the Advisory Board of the Social Fund, which will be the interface with the demands of society, and that it will allow adjusting, over time, the definition of its funds rescued" (Brasil 2010a). Thus, in the final text of the Law 12.351/2010, all of the areas were equally sharing the investments. Article 47 remained with the following areas: education; culture; sports; public health; science and technology; the environment; and mitigation and adaptation to climate change (2010a).

However, in relation to the Social Fund, the Provisory Measure (PM) 592/2012, edited by the former Brazilian President Dilma Rousseff, changed Lula's decision, and the third paragraph of Article 47 of Law 12.351/2010 stipulates that 50% of the proceeds of return on the Social Fund capital must be dedicated to education, according to later regulation (Brasil 2012a).

The remaining 50% should be allocated to programs and projects in the areas of culture, sports, public health, science and technology, the environment and mitigating and adapting to climate change. Also, this PM included Article 50-B in Law 9.478/97 stating that revenues from royalties and special participation relating to contracts signed after 3th December of 2012 must be directed to education (Brasil 2012a).

7 Brazilian Law 12.734/2012 and a New Path

The new rules of the distribution of royalties focus on future contracts under the production sharing system. After all, the Law 12.734/2012 was approved at the end of 2012. From the new royalties' rate on oil production, 22% of the offshore royalties and 15% of onshore royalties will be allocated to the Pre-Salt Social Fund (Brasil 2012b) (Table 2).

Considering the relationship between the distribution and the destination of those royalties, there is no doubt that the amount they will represent may influence choices and decisions to the Brazilian Government (Brasil 2012b). The SF was created as a mechanism to help avoid Dutch Disease in Brazil, which could be caused by big oil discoveries in the pre-salt area, as well as for serving as an intergenerational fund. It is clear that Law n. 12.734 did not change these purposes.

Although the SF will receive 22% of the royalties share of the offshore production, the exact amount of the financial resources that will go to education and the other social areas remains vague. All scenarios of the SF are based on hypothesized percentages, however, the social demand in Brazil is based on facts and needs. Moreover, some Congressmen questioned why this amount should be directed only to education while the need for healthcare remains a critical issue (Jornal do Brasil 2012). Also, it was not clear, for example, what percentage of revenue from pre-salt will be used in the early years for the capitalization of the SF (Oliveira 2010). From 2012 to 2018, the SF's revenues have increased mainly because of pre-salt production (Portal do Brasil 2016; Petrobras 2016) (see Graph 1).

The Law 12.734 did not change the structure of the SF, which was defined by Law 12.351 and Financial Ministers will monitor the Fund. Thus, the influence of Ministers within the social areas in this Fund could be questioned. On the other hand, there

Table 2 Distribution of royalties among the different beneficiaries (% of total royalties)

Beneficiares	Onshore (%)	Offshore
Producer States	20	22
Producer Municipalites	10	5
Municipalities impacted by oil and natural gas loading and unloading facilities operations	5	2
Special Fund of Non-producer States	25	24.5
Special Fund of Non-producer Municipalites	25	24.5
Social Fund of the Pre-Salt	15	22

Source Law 12734/2012 (Brasil 2012b)

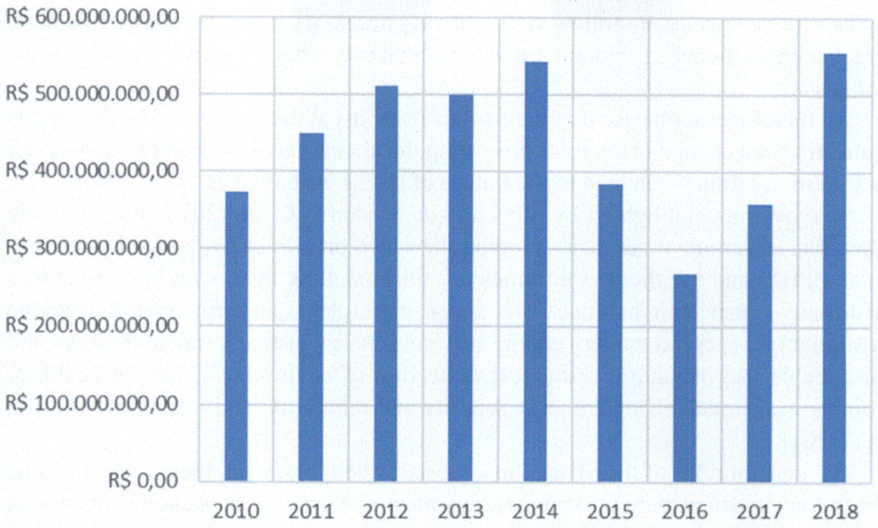

Graph 1 PSSF's revenues [US$ (1 US$ = 3,364 R$ Brazilian currency (average to 2018). Available at: https://economia.acspservicos.com.br/indicadores_iegv/iegv_dolar.html)]. *Source* Inforoyalties (2019)

were concerns about the Presidential behaviour related to the SF, because historically the Federal Government has been using its financial reserves as an instrument to guarantee the primary surplus (de Medeiros Costa et al. 2010; Costa 2018).

8 Conclusion and Lessons to Be Learned

As noted, the Brazilian Social Fund is indeed a sovereign wealth fund, its purposes are mixed, combining the functions of a stabilization fund and an intergenerational savings fund. It is the exclusive property of the Federal Government, as a legal person under domestic public law. Brazilian states and municipalities do not participate in its domain or management.

These circumstances pertaining to the ownership of the SF appear to move in the opposite direction of successful SWFs. The Federal Government should not be considered highly dependent in terms of oil and mineral revenues. On the other hand, at other government levels, such as the Brazilian states and municipalities dependent on oil resources that did not create their own funds during promising economic periods, they are in a worrisome situation, to say the least (Costa and Santos 2013).

The mere existence of legislation to carefully regulate the funds is perhaps not as important as the enforcement of these rules. It is necessary that the Fund is well established as per the management and investment policies recommended by those in charge. Nevertheless, more important than the system informing who is responsible for such policies and what penalties to impose in cases of noncompliance, is to make sure those rules are followed. It is always necessary to ascertain which de facto practices are efficient, accountable and beneficial over what the legislative framework suggests.

We therefore emphasise that the low levels of fiscal discipline in Brazil—as it is indicative of a change in the rules governing the allocation of the Fund's values even before its creation—can lead to the failure of the SF's objectives.

Moreover, as highlighted by Silva and de Medeiros Costa (2019: 166), that the Brazilian government can have adopted the seven principles explained by Heffron et al. (2018) and use them as a framework for modelling the Social Fund to ensure distributive justice in practice. The seven principles speak to: natural resource sovereignty; access to modern energy services; energy justice; prudent, rational and sustainable use of natural resources; protection of environment, human health & combatting climate change; energy security and reliability; and resilience (Heffron et al. 2018).

The general rules of the SF are largely established. However, they do not become discipline. In Brazil there is a strange phenomenon of laws with not enough or enough enforcement. Such unstable levels of enforcement have not contributed to the social and economic development of the country. Instead, an institutional environment of convenience has been created to the detriment of legal and institutional security. This is a concrete risk and the SF, about which authorities, academia and organized society should address and it begs the following question: how to make the Brazilian Social Fund a concrete instrument and not only a formal device?

Sovereign wealth funds can take on several roles or goals. Within their applications, SWFs can be a fundamental tool for better national savings management, especially for those nations exposed to international economic volatility. SWFs also can be an important tool against the "resource curse" or the "paradox of plenty". In the Brazilian context, in which SWFs are federal property and they play dual roles (i.e. a mix of stabilization and integrational fund), some lessons can be garnered:

1. The fact that Brazil—as a federal union—is not considered as highly dependent on oil and mineral revenues, weakens the enforcement of its SWF. On the other hand, subnational states are highly dependent on those revenues. They could establish their own SWFs in order to save for the future;
2. The legal and political enforcement matters in order to evolve and strengthen SWFs;
3. Fiscal Discipline is perhaps more important than the establishment of detailed legislation.

Acknowledgements We gratefully acknowledge support of the RCGI—Research Centre for Gas Innovation, hosted by the University of São Paulo (USP) and sponsored by FAPESP—São Paulo Research Foundation (2014/50279-4) and Shell Brasil, and the strategic importance of the support given by ANP (Brazil's National Oil, Natural Gas and Biofuels Agency) through the R&D levy regulation.

References

ADPM. Congressmen approved royalties to education. Available at https://www.adpmnet.com.br/index.php?option=com_content&view=article&id=242:senado-aprova-proposta-que-destina-75-dos-royalties-do-petroleo-a-educacao-e-25-para-a-saude&catid=12&Itemid=107. Accessed June 16, 2019.

Auty, R. M. (1988). Oil-exporters' disappointing diversification into resource-based industry: the external causes. *Energy Policy 13*, 230–242.

Auty, R. M. (1991). Resource-based industry in boom, downswing and liberalization: Mexico. *Energy Policy, 11,* 13–23.

Auty, R. M. (2003). Third time lucky for Algeria? Integrating an industrializing oil rich country into the global economy. *Resour. Policy, 29,* 37–47.

Auty, R. M. (2005). Transition reform in the mineral-rich Caspian region countries. *Resource Policy, 27,* 25–32.

Auty, R., & Gelb, A. (1986). Oil windfalls in a small parliamentary democracy: Their Impact on Trinidad and Tobago. *World Development, 14*(9), 1161–1175.

Bacon, R., & Tordo, S. (2006). Experiences with oil funds: Institutional and financial aspects. Energy Sector Management Assistance Program (ESMAP). Washington D.C., USA: The International Bank for Reconstruction and Development/The World Bank.

Backer, L. C. (2016). Sovereign wealth funds as regulatory chameleons: The Norwegian sovereign wealth funds and public global governance through private global investment. Available at https://litigation-essentials.lexisnexis.com/webcd/app?Action=DocumentDisplay&crawlid=1&doctype=cite&docid=41+Geo.+J.+International'l+L.+425&srctype=smi&srcid=3B15&key=97e14f1968976e98161bc39dcb00c887. Accessed June 16, 2019.

Brasil, Presidência da República Federativa do. (2010a). Casa Civil. Subchefia para Assuntos Jurídicos. Mensagem n. 707, DE 22 de dezembro de 2010. Available at https://www.planalto.gov.br/ccivil_03/_Ato2007-2010/2010/Msg/VEP-707-10.htm. Accessed June 16, 2019.

Brasil, Presidência da República Federativa do. (2010b). Lei 12.351, de 22 de dezembro de 2010. Available at https://www.planalto.gov.br/ccivil_03/_Ato2007-2010/2010/Lei/L12351.htm. Accessed June 16, 2019.

Brasil, Presidência da República Federativa do. (2012a). Provisory Measure 592, de 03 de outubro de 2012. Available at https://www.planalto.gov.br/ccivil_03/_Ato2011-2014/2012/Mpv/592.htm. Accessed June 16, 2019.

Brasil, Presidência da República Federativa do. (2012b). Lei 12.734, de 30 de novembro de 2012. Available at https://www.planalto.gov.br/ccivil_03/_Ato2011-2014/2012/Lei/L12734.htm. Accessed June 16, 2019.

Bucheb, J. A. (2004). The autonomy of the regulatory agencies and the job security of their leaders. Revista Jus Navigandi, ISSN 1518-4862, Teresina, ano 9, n. 210, 1 fev. 2004. Disponível em: https://jus.com.br/artigos/4783. Accessed June 16, 2019.

Conway, P., & Gelb, A. (1988). Oil windfalls in a controlled economy. *Journal of Development Economics, 28,* 63–81. ((North-Holland)).

Costa, H. K. M. (2012). *O princípio da Justiça Intra e Intergeracional como elemento na destinação das rendas de hidrocarbonetos: temática energética critica na análise institucional brasileira.* São Paulo: Universidade de São Paulo.

Costa, H. K. M. (2018). *Royalties de Petróleo, Justiça e Sustentabilidade* (1st ed.). Rio de Janeiro: Synergia.

Costa, H. K. M., & Santos, E. M. (2013). Institutional analysis and the—resource curse—in developing countries. *Energy Policy, 63,* 788–795.

Costa, H. K. M., Simões, A., & Santos, E. M. (2017a). Integral sustainability as driving force for paradigmatic change in human lifestyle. *Sustentabilidade Debate, 8,* 100–110.

Costa, H. K. M., Simoes, A. F., Dos Santos, E. M., & Silva, I. M. M. (2017b). Modificações Legislativas E Impactos Nos Royalties E Na Participação Especial Destinados Aos Órgãos Da Administração Direta. *Revista de Políticas Públicas da UFMA, 21,* 959–981.

Costa, H. K. M., et al. (2010). Uma análise do Projeto de Lei n° 5.940/2009: Fundo Soberano do pré-sal. In *Rio Oil & Gas: Expo and Conference, 2010, Rio de Janeiro. Rio Oil & Gas* (pp. 1–12). Rio de Janeiro: IBP.

Gelb, A. H., et al. (1988). *Oil windfalls: Blessing or curse?* Oxford: Oxford University Press.

Heffron, R. J. (2018). The application of distributive justice to energy taxation utilising sovereign wealth funds. *Energy Policy, 122,* 649–654.

Heffron, R. J., & McCauley, D. (2017). The concept of energy justice across the disciplines. *Energy Policy, 105,* 658–667.

Heffron, R. J., Ronne, A., Bradbrook, A., Tomain, J. P., & Talus, K. (2018). A treatise for energy law. *The Journal of World Energy Law & Business, 11*(1), 34–48.

IBGE (Instituto Brasileiro de Estatística). (2010a). Social indicators. Available at https://www.ibge.gov.br/home/presidencia/noticias/noticia_visualiza.php?id_noticia=2019&id_pagina=1. Accessed June 16, 2019.

IBGE (Instituto Brasileiro de Estatística). (2010b). Pesquisa Industrial Mensal, time series. Available at https://seriesestatisticas.ibge.gov.br/lista_tema.aspx?op=0&no=8. Accessed June 16, 2019.

Inforoyalties. (2019). Indicadors. Available at https://inforoyalties.ucam-campos.br/informativo.php. Accessed June 16, 2019.

IPEA (Instituto de Pesquisas Estatíticas Aplicadas). (2012). Qualification of labor is analyzed. Available at https://www.ipea.gov.br/portal/index.php?option=com_content&view=article&id=16476. Accessed June 16, 2019.

Jordan, L., de Medeiros Costa, H. K., Santos, E. M., Simoes, A. F. (2013). The Brazilian Sovereign Fund and the Brazilian Social Fund: A measure of the efficiency of Brazilian Sovereign Wealth Funds in fighting the resource curse and the Dutch disease. In *4th Meeting of Latin-American Energy Economics, 2013, Montevideo, Uruguay. Proceedings of the 4th ELAEE*, April 8–9, 2013—Montevideo—Uruguay (pp. 1–20). Montevideo: ELAEE4.

Jornal Do Brasil. (2012). Dilma fará 'análise exaustiva' da lei dos royalties. Available at: https://www.jb.com.br/pais/noticias/2012/11/08/dilma-fara-analise-exaustiva-da-lei-dos-royalties/. Accessed June 16, 2019.

MME (Ministério de Minas e Energia). (2009). Projeto de Lei 5940/2009. Available at https://www.mme.gov.br/documents/10584/1657891/Projeto_de_Lei_5940.pdf/383d7cd5-ed89-400c-b513-b7bf843003e1. Accessed June 16, 2019.

Oliveira, R. (2010). Valor Econômico (SP): Superávit primário com e sem petróleo. Available at https://www.ipea.gov.br/portal/index.php?option=com_content&view=article&id=6608&Itemid=75. Accessed June 16, 2019.

Petrobras. (2016). Nossos novos recordes no pré-sal. Available at https://www.petrobras.com.br/fatos-e-dados/nossos-novos-recordes-no-pre-sal.htm. Accessed June 16, 2019.

Portal Brasil. 2016. Pré-sal representou um terço da produção nacional de petróleo em janeiro. Available at: https://www.brasil.gov.br/economia-e-emprego/2016/04/pre-sal-representou-um-terco-da-producao-nacional-de-petroleo-em-janeiro. Accessed June 16, 2019.

Rodrigues, L. A., & Sauer, I. L. (2015, December). Exploratory assessment of the economic gains of a pre-salt oil field in Brazil. *Energy Policy, 87*, 486–495. https://doi.org/10.1016/j.enpol.2015.09.036

Sachs, J. D., & Warner, A. M. (1995). Nature resource abundance and economic growth. NBER Working paper series, Cambridge, MA.

Sachs, J. D., & Warner, A. M. (1997). Nature resource abundance and economic growth, Harvard University, Working Paper, revised version.

Sachs, J. D., & Warner, A. M. (1999). The big push, natural resource booms and growth. *Journal of Development Economics, 59,* 43–79.

Sachs, J. D., & Warner, A. M. (2001). The curse of natural resources. Natural Resources and Economic Development. *European Economic Review, 45,* 827–838.

Scaff, F. F. S. (2013). Royalties decorrentes da exploração de recursos naturais não renováveis. Tese de Livre-Docência em Direito Financeiro. Universidade de São Paulo, Mimeo.

Silva, I. M. M., & Costa, H. K. M. (2019). Brazilian social fund: The lessons learned from the norway fund experience. *Energy Policy, 129,* 161–167.

Public Wealth Management and Distribution in Iran

Zoha Abdolalizadeh

1 Introduction

Iranian officials and planners have always been concerned about the optimum use of the country's oil and gas revenues, as well as maintaining stability in the way those revenues are applied to the national budget. Various methods were implemented in pre-revolutionary development programs, including the establishment of a Planning and Budget Organization in 1948 to ensure the efficient use of these resources (Arab-mazar and Noormohammadi 2016). Using oil revenues to build production and infrastructure capacity was an important incentive for the establishment of this organization; but from the very beginning, the government was forced to spend part of the natural resource proceeds (Hadi-zonooz 2010). As oil revenues increased, corresponding investment in public utilities was also required, and the public infrastructure budget became greater. A shortfall in tax revenue resulted in an even greater need to rely on the proceeds from oil sales.

For many years, another part of the income was invested in for-profit production to prevent companies from the need to offset their losses with more oil revenues. If they were profitable, the country's production capacity and tax revenue would increase.

The devastating impact of oil revenue fluctuations on Iran's economic stability is another important issue. With the rise in global market oil prices over recent decades and an increase in national income, Iranian politicians have turned oil dollars into Iranian Riyals. This has resulted in an increase in the economic development of a specific sector at the expense of a decline in others and injected a temporary boom to the country's economy, with accompanying inflation. The situation presented various complications, including the "Dutch disease" and "rentier state", where resources

Z. Abdolalizadeh (✉)
Senior Legal Specialist, Dana Energy Company, Tehran, Iran
e-mail: zoha.abd@gmail.com

E. G. Pereira et al. (eds.), *Sovereign Wealth Funds, Local Content Policies and CSR*, CSR, Sustainability, Ethics & Governance,
https://doi.org/10.1007/978-3-030-56092-8_6

are rented externally (Sadatrasoul et al. 2015). However, unstable oil prices fell after a few years, and the swollen economy (fuelled by deflated demand, and huge government spending on new hires and half-finished projects) became the centrepiece for offsetting the deficit. There was no other choice than to borrow from the central bank, which brought a new wave of inflation.

This strategy hit industrial and agricultural productivity, which had already lost production potentials because of cheap imports during the boom. Inflation rose in both sectors, although as a result of different issues. With foreign exchange earnings rising, the monetary base swelled with a net increase in the central bank's foreign assets; and because of the government's budget deficit during the recession, the monetary base also rose through the net increase in government debt to the central bank.

The idea of a foreign exchange reserve account was adopted and articulated in Article 60 of the Islamic Republic of Iran's Third Development Plan (2000–2005), in order to stabilize the country's economy (Abbasi 2008). This account was also maintained in the subsequent Fourth Development Plan (2005–2010). The overall policies of the Fifth Development Plan (2011–2016), under Paragraph 22, emphasized a shift towards oil and gas revenues with the creation of a National Development Fund. To implement these policies, the Statute of the National Development Fund was adopted in Article 84 of the Fifth Development Plan, and the foreign exchange account maintained by new duties in Article 85 of the Fifth Plan.

Here, we first describe how the financial relationship between the government and Iran's oil formed a basis for a foreign exchange reserve account in the Third Development Plan (2000–2005) and its performance in this, as well as the Fourth Development Plan (2005–2010). The Statute of the National Development Fund, together with the new face of the Reserve Account in the Fifth Development Plan (2011–2016), is outlined, alongside the strengths and weaknesses of the strengths and weaknesses of all the Plans.

2 Establishment of Foreign Exchange Reserve Account in Iran

There are now more than 60 National Oil Funds around the world that operate under different mechanisms. Creating foreign exchange reserves or oil funds is one strategy undertaken by countries to manage oil wealth. These funds are either used to solve the problems created by oil revenue fluctuations (stabilization funds), or part of the revenue is saved for future generations (savings funds). Both goals can also be pursued in tandem.

Petroleum funds are created for the same purpose, which is to help governments deal with the challenges and problems arising from oil revenues. There are generally two purposes: first, to create a stabilization fund to balance the budget. This anticipates and confronts the problem of revenue volatility. Second, attending and being proactive to the future. In this case, a savings fund would be created to set aside a portion of oil revenues for future generations. The philosophy behind this is based on the presumption that oil is a national asset. For this reason, oil revenues can only be used for investment whether inside or outside the country, while allocating them to current and consumption costs is not permitted.

In 1996, the Iranian government faced greater projected foreign exchange earnings, and a reform bill was submitted to parliament to spend more than 6000 billion Rials on the approved budget. However, in 1997, the government had to submit amendments to parliament to borrow this amount from the central bank because of a lack of oil revenue (PRC 2001). This experience forced policy-makers at the time think about how to design a system of stability for the Iranian economy. Prior to ratification of the Third Development Plan, experts at the Ministry of Petroleum proposed the establishment of a foreign exchange fund which would store surplus income from the government's proposed production ceiling, to be used for investment.

The proposal did not take into account how to prevent unbudgeted fluctuations, and the impact of the oil revenue trap. This trap occurs when the means of production and investment are diverted to an industry that brings economic prosperity. However, the result negatively impacts other parts of the economy, leaving it undiversified and vulnerable if the oil sector undergoes a downturn. Even though it was flawed, the proposal was therefore accepted by the Planning and Budget Organization as Article 60 of the Third Plan. Under it, the government was obliged in 2001 to introduce the Foreign Exchange Reserve Account of Crude Oil Revenue and the Rial Reserve Account at the Central Bank of the Islamic Republic of Iran. Its mechanism specified that the surplus of foreign exchange earnings higher than the Plan's anticipated figures would be reserved in this account, and whenever oil revenues experienced a decline, withdrawals would be made, enabling the currency to be converted into Rials. The Plan prohibited the government from using the account to compensate its tax or any other kind of income reduction. However, after the Rial revenues for each year's budget were forecasted, short-term lending from the remaining currency was permitted for production and investment activities (Haji-Mirzaee 2006).

Instead of predicting oil prices, those creating the Third Plan considered managing prices to adjust annual budgets. For example, no matter what the price of oil would be in 2001, about $13 billion oil revenue was projected for the annual budget, which would be offset by the failure to meet the reserve requirement. With the sharp rise in oil prices on the world market less than six months after the Third Development Plan was announced in 2000, MPs amended Article 60. The Rial Reserve Account was removed, and only one account was designated the Foreign Exchange Account from crude oil export. The amendment also changed the figures for oil export earnings forecasted in 2000 and 2001. It additionally allowed the government to spend up to 50% of the foreign exchange reserve account on investment and to provide part of the credit needed for production and entrepreneurship projects. As well, non-governmental sectors were provided with a portion of the credits for industrial, mining, agricultural, transportation and engineering services.[1]

In 2000, the Executive Regulation of Article 60 of the Plan was adopted. A Board of Trustees was established, and this consisted of the Chairman of the Management and Planning Organization, the Minister of Economic Affairs and Finance, the Chairman of the Central Bank, and four delegates elected by the President. At least two ministers were put in place to enforce the law and make decisions on matters such as determining how the rate of return was to be calculated, and when concessional facilities were to be repaid. They would also prioritize plans for when these facilities would be used.

According to this regulation, proposed projects would benefit from foreign exchange facilities after operating banks had conducted technical and economic reviews, and once approval was granted by the relevant ministries. The acquisition period of the facility for investments and experimental operations was up to three years, and the maximum loan repayment term was five. The method of depositing foreign exchange funds into the reserve account was also specified. In order to ensure public funding, the central bank was obliged to deposit the equivalent of 100% of the foreign exchange earnings from the crude oil export to the relevant public revenue account during the first four months of the year. And from the fifth month to the ceiling of the year's currency creation obligation, it had to deposit the equivalent of one-twelfth of the said ceiling into the public revenue account. The remainder was to be deposited into the foreign exchange reserve account, and after fulfilling the annual budget, all amounts received would be deposited into the currency account. Article 5 of the regulation provided that foreign exchange facilities users shall repay the principal and the interest accrued in accordance with national foreign exchange regulations. The granting of foreign exchange facilities to projects approved under the Law of Attraction and Protection of Foreign Investments of Iran (joint ventures) was authorized under the same Article.

The interesting point in this regulation is the Central Bank's requirement to pay interest on the maintenance of the account, referred to in Article 7 (Mardoukhi 2005). The Foreign Exchange Reserve Account Board set a maximum of 7.5% interest per annum on customer lending facilities. The operating bank was required to deposit 3% of the interest received in the same currency from the customer into the foreign exchange reserve account. The facility rate for industrial restructuring plans, which included at least 50% of available machinery and existing equipment, was minus 0.25%, and the facility rate for investment projects in deprived provinces was minus 0.5%. The deduction from interest accrued to accelerate the utilization and repayment over the planned period was 10% for six months, and 15% for one year.

[1]"Overview of Foreign Exchange Reserve Accounts Out of Crude Oil Exports, related to the Modification Plan of Section B: Amendments to the Third Development Plan Act".

3 Foreign Exchange Reserve Account Input Resources and Expenses

Foreign Exchange Reserve account resources included

1. The difference between the real income and the share of oil revenues allocated to the annual state budget in the Third and Fourth Plans;
2. Receiving the principal and benefit of the facility paid from the account; and
3. Revenue from interest received from the Central Bank.

Expenses included

1. Withdrawal from the account, equivalent to the deficit of crude oil export earnings forecast for the year;
2. Withdrawal from the account, equivalent to up to 50% of the balance to pay off loans to non-government investors and entrepreneurs; and
3. Other options according to related laws.

4 Foreign Exchange Reserve Account in the Fourth Development Plan (2005–2010)

The Fourth Development Plan was designed with a focus on accelerating economic growth, empowerment, and expanding the private sector. The account was given a special place in the Plan, and Article 1 assigned to it. This Plan did not differ greatly from its predecessor, except that the government was now allowed to provide at least 10% of available resources to Agribank (Keshavarzi Bank), in the non-governmental sector. Agribank was to invest and distribute this amount to justified agricultural schemes, and to provide working capital to the private sector for export expansion projects. (Hadi-zonooz and Pileh-foroush 2013)

Similar to Article 60 of the Third Plan, the Fourth Plan allowed the government to invest up to 50% of the account balance for investment. It could provide part of the credit required for industrial, mining, agricultural, transportation projects and services (including tourism, etc.), to non-governmental companies who had been approved on technical and economic grounds by the relevant specialized ministries (Khalesi and Farhadikia 2009).

This permission was protected as a duty of government in Article 28 (2) of the Law "On Amendments to the Law of the Fourth Plan of Economic, Social and Cultural Development of the Islamic Republic of Iran and the Implementation of the General Policies of Article Forty-Fourth of the Constitution", and the Government was required to adopt policies to allocate 40% of the account balance from the previous year to the non-governmental sector. In addition, if there was a demand, accompanied by technical and economic justification plans in the private sector, the government would be obliged to allocate funds to applicants from this balance. The general rule was that the share of the private sector in any given year shall not be less than 40% of the withdrawal from the account in that year. Paragraph (3) of the same Article also enabled the Board of Trustees of the Account and the Central Bank of Iran to increase the share of foreign exchange facilities to the private sector, part of the Foreign Exchange Reserve Account, or the Central Bank of the Islamic Republic of Iran to be used as a deposit with operating banks and to open foreign exchange credit lines. This share could additionally be used for external and other payment facilities (Mahdavian 2007).

Implementing Regulations of Article 1 of the Fourth Development Plan was approved by the Cabinet in 2005. It was similar to the Article 19 by-law of the Third Development Plan Act, with minor modifications. Article 7 of the by-law allowed the Board of Trustees to pay the remaining 30% of the account balance to state-owned banks as a shareholder in the private financing syndicate, in accordance with the Law on Non-Usury Banking approved in 1984, to be used to increase the capacity of the facility. In Note (2) of Article 8, users of foreign exchange facilities were required to repay principal and interest in the same currency as the facility granted. Article 11 reduced the number of Board Trustees from seven to five, and the President's representatives were removed.

Other structural changes associated with the account included the cabinet's decision to amend the Implementing Regulations in 2008. With the amendment of the Implementing Regulations of Article 1 of the Fourth Development Plan, the functions of the Board of Trustees were delegated to the "Economic Commission" of government, and in all the provisions of this code, the term Board of Trustees was replaced by "Economic Commission".

The performance of the account has been directly related to the government's fiscal policy since its establishment. Prior to the Third Plan, oil revenues were largely allocated to the general government budget. In spite of repeated amendments to Article 60 and the establishment of the Foreign Exchange Account during the Third Plan period, government withdrawals from the account led to a significant share of the oil export revenue being applied to the general government budget. The government increasingly withdrew funds from the account during the Fourth Plan period.

The public budget's dependency on oil revenues was about 68% in 2000 when the account was established. This amount gradually declined and reached its lowest level in 2004 (43.3%). However, it has increased since 2005,[2] reaching its highest level (69.6%) in 2006 (Abbasi 2008).

It is noteworthy to calculate the real share of oil revenues in Iranian government expenditures. Since 2005, the accounts have been used for government expenditures outside the national budget. Also, the amounts paid to the National Iranian Oil Company to produce crude oil have not been seen as a total government cost and as a result are not included in the percentage of public budget dependency on oil revenues. The general budget's dependency on oil revenues was actually higher.

[2]This happened after 2004 governmental election in which the right won the election.

5 Account Performance in the Third and Fourth Development Plans (2000–2010)

The provision designed to stabilize the Plans was revised more than all other articles. These revisions and withdrawals were made when oil revenues were less than expected in the Third Plan, but also when the government faced a surplus in oil export revenues during the Plan's executive years. This experience continued with the implementation of Article 1 of the Fourth Development Plan. From 2000 to 2009, about $164 billion was deposited into the account, and over $160 billion withdrawn in total. The notable point is that the government expenditure was far greater than funds granted to facilities; more than 90% of total withdrawals were made by the government during that time, with the remainder spent on facilities to the private sector (Hadi-zonooz and Pileh-foroush 2013). This reflects the fact that the mere existence of a foreign exchange reserve account has not led to governmental financial discipline.

During the execution years of the Third Plan (2000–2005), about $30 billion was deposited into the foreign exchange reserve account, and about $20 billion withdrawn, $4 billion of which paid to private sector facilities. The numbers for the Fourth Plan (2005–2010) were $134 billion deposited, and $140 billion withdrawn, including $12 billion for private sector facilities. It should be noted that although about 10% of the account balance was allocated for the private sector, about 50% of the instalments were reportedly delayed.

6 Foreign Exchange Reserve Account and National Development Fund Under the Fifth Development Plan (2011–2016)

The account in the Third and Fourth Plans had two tasks:

1. To stabilize the budget in the event of a decrease in oil revenues and prevent the damage of oil shocks to the balance of the country's economy; and
2. To save part of the oil revenue for future generations by converting these funds into a productive investment.

Assigning this account with these two roles simultaneously led the parliament to allow the government to withdraw funds from it to accelerate implementation of development plans, as well as to address some budgetary problems (Kianpour 2010). It seems there are specific reasons why the account's performance fell short, and why it could not fulfil its tasks. It failed to support the country's economic development, which included deficiencies in the Planning and Budget Law, and in decision-making mechanisms (Hadi-zonooz 2009).

The following are some of the account's bugs:

1. The possibility of withdrawal from the account to support the government's general budget;
2. A lack of independent institutions for policy-making, implementation and monitoring of resources and costs;
3. Interference of tasks (stabilizing and developmental roles);
4. Sequential reforms and amendments of the relevant laws, and non-adherence of the executive and legislative powers to statutory approvals in different periods;
5. The lack of transparency of banking responsibilities, and interference with the responsibilities of the account's trustees regarding large projects; and
6. The impossibility of an independent audit.

Building on the unsuccessful Foreign Exchange Reserve Account experience, the National Development Fund structure in the Fifth Development Plan was designed to

1. Establish the existence of the account to continue only for the purpose of balancing the country's general budget and to maintain its stability; and
2. Set up a National Development Fund in order to convert part of the oil proceeds into productive economic resources and capital.

7 Foreign Exchange Reserve Account in the Fifth Development Plan (2011–2016)

In the bill of the Fifth Development Plan, it was proposed to convert the Foreign Exchange Reserve Account into a National Development Fund. But eventually with parliamentary scrutiny, the continued use of the Foreign Exchange Reserve Account with a narrower task description was adopted in Article 85 of the Fifth Development Plan.

With Article 60 of the Third Plan and Article 1 of the Fourth Plan, only surplus crude oil revenues were transferred to the account, but in Article 85 of the Fifth Plan, government revenues from the export of petroleum products and gas exports could also be deposited into it, in addition to revenues from crude oil and gas condensate exports (either in cash or barter).

Another important feature of the Fifth Development Plan, contained in Article 85, was the government's responsibility to keep track of the collection of instalments, as well as the interest accrued from facilities granted from the account and deposited into it. In addition, given the billion-dollar account's obligations and the lack of sufficient resources to fulfil these obligations, this Plan stipulated that the account would bear its remaining obligations to the non-governmental, private and cooperative sectors, and any new commitment was forbidden.

8 National Development Fund Under the Fifth Development Plan and Its Statute

Paragraph 22 of the General Policies of the Fifth Development Plan emphasized that the National Development Fund was established with the aim of changing the pervasive approach of using oil and gas and their revenues for public financing and budgeting, to productive economic resources instead. To implement this policy, Article 76 of the government's proposed bill was allocated to the National Development Fund. This article, set out in the form of a by-law and passed by the cabinet, outlined the Fund's charter and other key decisions for managing the fund.

Fund's statute begins with this statement that indicates: "The National Development Fund of Iran, hereinafter referred to as the "Fund", is established to turn a portion of the revenues originated from selling oil, gas, gas condensate and oil products to durable wealth and productive economic investments as well as preserving the share of future generations from the oil and gas resources and products."[3] It should be noted that the statute does not indicate principal matters such as the main purposes of the fund (stabilization or saving), its obligations and related enforcements. The uncertainties have severely affected the fund's procedures and function. For instance, despite the fact that statute determines the fund's spending, the restrictions and regulations have been neglected and deviated in many cases. To clarify, here, the specific spending terms are reviewed. The statute confines the fund's expenditure in six items as following:

1. to grant financial facilities to the private and cooperative sector as well as the economic entities owned by public non-governmental institutions, to be invested in production and development with economic, financial, and technical feasibility,
2. to grant financial facilities to Iranian private and cooperative companies awarded international tenders for the export of engineering and technical services, through the fund's resources or syndicated facilities,
3. to grant purchase credit to the buyers who purchase Iranian goods and services in the export-target markets of Iran,
4. to invest in foreign monetary and financial markets,
5. to grant financial facilities to foreign investors on competitive and economic viability grounds in order to attract and protect investors in Iran in accordance with Principle 80 of the Iran's constitution,
6. to pay for expenses of the fund.[4]

The statute sets some restrictions including the ban on using fund to cover costs, acquisition of capital assets and repaying debts of the government in any form and shape, which has been violated from the very beginning of the operation of fund. Besides, there is a ban imposed on the facilities, which should be paid in foreign

---

[3]http://en.ndf.ir/About-NDF/Articles-of-Association.

[4]http://en.ndf.ir/About-NDF/Articles-of-Association#111628-i---the-funds--spending.

currencies, indicating that the recipients are not allowed to convert the foreign currency into Rial in the domestic market.[5]

The subsequent articles explain the details of shares of the resources in fund. As mentioned, the principal resource of fund is composed of at least equivalent to 30% of the revenues from exporting oil (crude oil, gas condensates, gas and oil products) during the period of the Fifth Development Plan (2011–2016), while its amount is determined in the annual budget. Besides, at least 20% of the value of swap of the mentioned items should be allocated to the fund. These shares should increase on an annual basis by three percent. In addition, there are another resources including fifty percent of the cash balance of reserve account at the end of each year since 2011, resources available from international money markets under the permission of the Board of Trustees in accordance with relevant rules and regulations, net profit of the fund during the financial year, interest of the fund's resources with the Central Bank of Iran, not less than the average of interest rate of the Central Bank of Iran's deposits in foreign markets, with calculation and payment every three months, and twenty percent of resources subject matter of Part D of Paragraph 4 of the 2011 Annual Budget Law.[6]

Key features of the National Development Fund were

1. Separate elements, including the Board of Trustees, the Executive Board, and the Supervisory Board for policy-making, implementation, and oversight;
2. Decision-making to be made by a nine-member board of trustees with a decisive majority (in addition to the head of the Iran's Chamber of Commerce, Industries and Mines, and the head of the Iran's Cooperative Chamber as a supervisor without voting right);
3. The Board of Trustees to be required to include its decisions in the official gazette of the country as well as in one of the major newspapers;
4. The Board of Directors to have five members selected by the Board of Trustees and appointed by the President. Dismissal of the members of the Board and acceptance of their resignation to be conditional on confirmation of a two-thirds majority ballot by the trustees;
5. The Supervisory Board to be consisted of: the president of the Supreme Audit Court of Iran, the president of Inspectorate General of the country, and the president of the General Audit Office. This board is required to submit their supervisory report every six months to Parliament and to the Board of Trustees (Hadi-zonooz and Pileh-foroush 2013);
6. In the first year of the program, Fund Resources would include 20% of the revenues derived from exports of crude oil and gas condensates and petroleum products, whether in cash or barter. The share of the deposits would also increase by 3% points each year;

[5]http://en.ndf.ir/About-NDF/Articles-of-Association#111628-i---the-funds--spending.

[6]According to this paragraph, petroleum ministry shall deposit any revenue from export of crude oil directly into the Treasury accounts of the country as a whole, after deducting the buy-back reimbursements. The Central Bank is obliged to deduct twenty percent (20%) of this account on a monthly basis as the share of National Development Fund and deposit it into the relevant account.

7. The most important tasks of the Fund were the provision of facilities and loans to the private sector, cooperatives and enterprises owned by public non-governmental organizations, and to invest in international financial and monetary instruments;
8. It was forbidden to use the fund's resources for spending credits and the acquisition of capital assets as well as the repayment of government debt in any form; and
9. The loans were to be in foreign currency, and investors did not have the right to exchange them into Rials on the domestic market.

9 Performance of the National Development Fund of Iran

From the approved foreign exchange facilities, only 28% has been definitely paid. Concerning the composition of accepted projects, it can be said that in total more than 42% of projects accepted belong to public non-governmental organizations. However, according to the paragraph "d" of the statute of the Fund, the total amount of funds allocated from banks' resources to non-governmental public institutions (such as municipalities, social security and foundations) and subsidiaries and affiliates shall not exceed 20% of Fund resources (NDFI 2018).

By the end of the last year, the Fund has also signed agency contracts in Rials with a total of 314 billion Riyals with 25 state-owned and privately owned banks, out of which about 50% are allocated to plans and projects submitted by operating banks (NDFI 2018) Although Rial payments have been made mainly with legal authorizations, the purpose of the Fund of the fund is to provide facilities only in foreign currency, and according to the Fund's Statute, investors using these facilities are not allowed to convert the currency into Rials in the domestic market.

10 Conclusion

Iran's Fifth Development Plan (2011–2016) set out to change the way oil funds are organized in Iran, but as the National Development Fund's performance illustrates, in practice, shifting from an account to a fund resulted in no significant progress (Pile-Foroush et al. 2018).[7]

Under the current situation, the Iranian government's financial decision-making system lacks integration and coherence. It cannot prioritize and pursue the goals of Development Plans by defining policies that are cost-effective and implementable within a certain revenue ceiling. If oil revenues cannot be properly managed because institutional conditions are needed, which includes an integrated budgetary system, a new mechanism (such as the oil fund) will sooner or later suffer the same bugs as

[7] Pile-Foroush et al. (2018).

the current, established financial system, and it will not solve institutional problems (Shirkhani et al. 2010).[8]

One reason for this is the lack of attention to Iran's medium-term expenditure framework. It seems the prerequisite for success would be to set a ceiling table for all government spending, with a realistic forecast based on the average of the past 12 years, together with current oil prices, and an anticipation of those prices three years ahead. An additional requirement would be the fundamental institutionalization of the budgetary system and an optimal performance of the Iranian oil fund. By setting a ceiling and placing restraints on spending, the next necessary correction would be to show the non-oil budget deficit in the budget document. This simple, yet key, recommendation does not apply for most oil-producing countries.

If politicians have short-term horizons, they take their own interests into account. In the face of demands from various groups in society, they will opt to spend more on oil revenues in the short run, both in legal and even illegal ways. In other words, if the interests of a regulatory system are always that the immediate use of petroleum resources is required, the oil fund mechanism will not succeed. This feature is one of the most important factors affecting the failure of Iran's Foreign Exchange Reserve Account or the National Development Fund.

All in all, if costs are not taken into account realistically, and compulsory percentages (e.g. 30% annually) are taken into account as a share of the oil fund—regardless of the country's economic conditions—it will base its economic realities on fictitious and unrealistic percentages. And the government will use all means to justify its use of the fund's resources, as happened to the National Development Fund of Iran.

The requirements for achieving the goals of the National Development Fund are as follows:

- Due to the lack of quantitative tables in the Fifth Development Plan (2011–2016), the amount of annual oil consumption in the general budget of the country is determined and the surplus of oil resources deposited into the reserve account after deducting the share of the National Development Fund, and the use of the public budget. As a result, there is always the possibility of economic instability due to the excessive injection of oil into the country's economy.
- Following the positive oil shock, it has been the government's national policy to maximize oil revenues. To alleviate the negative effects of this policy, it is necessary for the National Development Fund Trustees to calculate the capacity of private sector to invest domestically, to the extent of the absorb capacity of the national economy in order to inject the resources of the Development Fund into the country's economy, and invest its surplus in international financial markets.
- Adding out-of-budget mechanisms to a non-transparent and inefficient financial system can remove oil funds from the public eye and provide a platform for a small group of decision-makers to prioritize. Experience in other countries shows that transparency and accountability are one of the essential foundations for the success of these funds (Karimifard 2011). Although, it should be taken into account that

[8] Shirkhani et al. (2010).

the issue of transparency of oil revenues, which can be made more precise by the creation of financial rules such as the oil fund, may be reversed and reduced in the absence of the necessary institutional conditions (Pile-Foroush et al. 2018).

- The success of the National Development Fund requires understanding the complexities and difficulties of getting rid of the structure of a rentier political economy. Governmental financial discipline is one of the basic conditions to achieve this.

References

Abbasi, E. (2008). Review of foreign exchange account performance during third development plan. *Biannual Program and Budget, 106,* 101–127.

Arab-mazar, A., & Noormohammadi, K. (2016). A critical analysis of the economic goals of development plans in Iran (before and after the Islamic revolution). *Social Studies and Research in Iran, 5*(17), 19–42.

Hadi-zonooz, B. (2009). *Iran's economy, comments and critics.* Tose'e Publication.

Hadi-zonooz, B. (2010). *Investigation of economic planning system in Iran (before the Islamic revolution).* Research Center of Parliament.

Hadi-zonooz, B., & Pileh-foroush, M. (2013). Foreign exchange reserve account and national development fund in Iran; Differences, functions and legal principles. *Development Strategy, 40,* 75–107.

Haji-Mirzaee, M. (2006). The necessity of establishing an oil fund in Iran. *Economic Research and Policy, 15*(44), 1–46.

Karimifard, H. (2011). Assessment of the National Revenue Fund from the sale of non-renewable natural resources, with emphasis on the Islamic Republic of Iran Development Fund. *Parliament and Strategy, 18*(68), 147–184.

Khalesi, A., & Farhadikia, A. (2009). *The pathology of first to fourth development plans after the Islamic revolution.* Vice President of Strategic Planning and Supervision's Publication.

Kianpour, S. (2010). Analysis of Iran's foreign exchange accounts compared to the experiences of other countries' oil funds. *Journal of Business Reviews, 43,* 78–95.

Mahdavian, M. H. (2007). The review on function and structure of foreign exchange deposit account. Strategic Research Center, Second Bulletin.

Mardoukhi, B. (2005). Oil wealth management and foreign exchange reserve accounts in the third plan. *Economy and Society, 1*(5), 1–8.

NDFI. (2018). National Development Fund performance report. Published by National Development Fund of Iran.

Pile-Foroush, M., Pasbani, A., & Chashmi, A. (2018). Impact of financial rules on the performance of oil funds: A comparison of Iranian and Norwegian Funds. *Parliament and Strategy Magazine, 25*(93), 237–280.

PRC. (2001). A look into foreign exchange deposit account. Parliament Research Centre, Report No. 15.

Sadatrasoul, M., Gholamian, M., & Shahanaghi, K. (2015). Investigating the effect of selected sustainable development indicators on credit allocation: The case of National Development Fund of Iran. *Iranian Journal of Management Studies (IJMS), 8*(3), 453–477.

Shirkhani, M., Barari, A., & Pouzesh-Shirazi, H. (2010). The curse theory or how to manage oil revenues; Case study: A comparison of Norway and Iran. *Journal of the Faculty of Law and Political Science, 40*(2), 115–134.

The Experiences of Managing the Heritage and Stabilisation Fund in Trinidad and Tobago and the Sovereign Wealth Fund Guyana

Alicia Elias-Roberts and Indira Rampaul-Cheddie

Abstract This chapter provides an analysis of the legal and regulatory framework of both Trinidad and Tobago and Guyana's sovereign wealth funds. It discusses the legal regime of ownership of the mineral estate in Trinidad and Tobago. It discusses the procedure to obtain the Exploration and Production (Public Petroleum Rights) Licence and the Exploration and Production (Private Petroleum Rights) Licence, as well as the system of private petroleum leases in Trinidad and Tobago. The establishment and management of the Heritage and Stabilisation fund in Trinidad and Tobago is examined. In the second part of the paper the ownership of the petroleum resources in Guyana is discussed and some of the main features of Guyana's Natural Resource Fund Act 2019 is evaluated. Also, a critical analysis of the legal framework in Guyana is undertaken and lessons from Trinidad and Tobago which can be useful to Guyana are outlined.

Keywords Sovereign wealth fund · Stabilisation fund · Natural resource fund · Oil and gas law · Trinidad and Tobago · Guyana

[1] See Government of the Republic of Trinidad and Tobago Ministry of Energy and Energy Industries (2015); See also Trinidad and Tobago Extractive Industries Transparency Initiative Report (2016).

[2] The country is 1841 sq. miles and the population of Trinidad and Tobago is about 1.4 million according to the Government of Trinidad and Tobago, Central Statistical Office's Population, Social and Vital Statistics Division, https://cso.gov.tt/news/tt-population-reaches-1-4-million/, accessed 31 August 2019.

[3] *Heritage and Stabilisation Fund Act*, no. 6 of 2007, Law of Trinidad and Tobago.

[4] Ghany (2016).

A. Elias-Roberts (✉)
Faculty of Law, University of the West Indies, St. Augustine, Trinidad and Tobago
e-mail: aeliasroberts@gmail.com

I. Rampaul-Cheddie
Ministry of Energy and Energy Industries, Port of Spain, Trinidad and Tobago

E. G. Pereira et al. (eds.), *Sovereign Wealth Funds, Local Content Policies and CSR*, CSR, Sustainability, Ethics & Governance,
https://doi.org/10.1007/978-3-030-56092-8_7

1 Introduction

Trinidad and Tobago has over 100 years of experience in oil and gas exploration.[1] This small twin island State often experiences changes in political leadership, which results in varied priorities within the government and different application and interpretation of laws and policies.[2] The current Heritage and Stabilisation Fund of Trinidad and Tobago (hereinafter "HSF") is a sovereign wealth fund, which was established by the *Heritage and Stabilisation Fund Act*,[3] No. 6 of 2007 with effect from March 15, 2007. It was previously known as the Interim Revenue Stabilisation Fund (IRSF), which was set up in 2000. All of the proceeds in the IRSF were, in fact, transferred to the Fund in 2007. According to one academic,[4] this brief history demonstrates that the concern for the protection of the economy of Trinidad and Tobago has been shared on a bipartisan basis, where one administration had launched the IRSF in 2000 and after general elections and a change of government another administration had established the current Fund in 2007 as the replacement for the IRSF and transferred the proceeds from one account to the other.

In contrast to Trinidad and Tobago, the 2015 oil discoveries in the offshore areas in Guyana position the country as an emerging frontier oil and gas producer.[5] Guyana is set to be one of the few countries to have a Sovereign Wealth Fund (SWF) in place before "First Oil."[6] Since commencing exploratory drilling in Guyana, oil major ExxonMobil has discovered an estimated 10 billion barrels of oil equivalent of recoverable oil in the Stabroek deep water exploration block about 120 miles offshore.[7]

This chapter will provide an analysis of the legal and regulatory framework of both Trinidad and Tobago and Guyana's sovereign wealth funds. It will begin by explaining ownership of the mineral estate in Trinidad and Tobago. It will explain the hybrid system of public and private petroleum rights, which currently exists in Trinidad and Tobago and a brief history of the system. This chapter will also discuss the procedure to obtain the Exploration and Production (*Public* Petroleum Rights) Licence and the Exploration and Production (*Private* Petroleum Rights) Licence, as well as the system of private petroleum leases. The establishment of the Heritage and Stabilisation fund in Trinidad and Tobago will be examined and the management and distribution of the fund will be analysed. Next, ownership of the petroleum resources in Guyana will be discussed and some of the main features of Guyana's Natural Resource Fund Act 2019[8] will be evaluated. Before concluding with a critical analysis of the management of the sovereign wealth fund, the fiscal regime in Trinidad and Tobago will be evaluated and lessons from Trinidad will be outlined which can be useful to Guyana.

---

[5] Krauss (2017).

[6] Government of the Cooperative Republic of Guyana (2019).

[7] Myers (2018).

[8] *Natural Resource Fund Act* (no. 12 of 2019), Laws of Guyana.

2 Ownership of the Mineral Estate and the Extractive Resources in Trinidad and Tobago

Ownership of the mineral estate with regard to oil and gas is provided for under the *Petroleum Act and Regulations*.[9] This is the main statute governing petroleum operations in Trinidad and Tobago. Under the *Petroleum Act*, public petroleum rights are vested in the State and are exercisable by the President, while private petroleum rights are exercisable by the owner thereof, subject to the Act and Regulations, and any Rules and Orders made thereunder. Prior to the commencement of the *Petroleum Act*, exploration and production operations were regulated by oil mining leases. Historically, persons were not required to obtain licences to conduct certain petroleum operations. Other operations were conducted by virtue of a licence, grant or oil mining lease. The *Petroleum Act* introduced a system of licencing for petroleum operations. Note that the first types of licences under the Act, which we shall refer to as the older model licences, were issued until 2005 and thereafter a newer model of licences were issued.

One unique feature of the ownership of the mineral estate or petroleum resources in Trinidad and Tobago is that the country has a hybrid system of public and private petroleum rights. The *Petroleum Act*, section 2 provides a definition for private petroleum rights and public petroleum rights. It defines "private petroleum rights" as rights to petroleum that are not public petroleum rights and "public petroleum rights" as rights to petroleum in its natural condition in strata existing in State Lands and submarine areas.[10] Sub-section (3) of section 2 provides a cut-off date and states that all mineral rights are possessed by the State in grants of land after 30th Jan 1902. It states as follows: "(3) In this Act a reference to State Lands shall be read and construed as including a reference to the mineral rights in all lands by whomsoever possessed, the subject of a grant by the State after 30th January 1902."[11] The historical explanation for this hybrid system and the cutoff date is explained in the following excerpt from the speech of Mr. Bhoolai, Member of Parliament, in the Hansard Debate on the Petroleum Bill in 1969 where he states as follows:

> After 1902 surface land in this country was sold without mineral rights. Lands that were sold up to the year 1901… I am speaking with authority because I own them; my father bought them and I own them in the oil area too; and I think the Minister is fully aware of it himself, because he comes from that oil area and he owned plenty land there once. These lands that were sold after 1902 they have absolutely no oil rights and no mineral rights. And according to this bill the Government are treating these lands that are already sold and are owned by the public, with Houses and good roads and lots and villages established; with good cocoa cultivation, as Crown Lands of this country according to the rights and privileges given to the petroleum company whosoever might come.[12]

---

[9] *Petroleum Act* (no. 46 of 1969), Cap. 62:01, Laws of Trinidad and Tobago; see also the *Petroleum Regulations* (Legal notice 5 of 1970), made pursuant to section 29 of the Petroleum Act, Laws of Trinidad and Tobago.

[10] Ibid., *Petroleum Act*, section 2.

[11] Ibid.

[12] "Hansard Debate on the Petroleum Bill, No. 34 of 1969", *Debates of the Senate,* Volume 9, (1969) Hansard Parliamentary Report, Trinidad and Tobago.

Section 6 of the *Petroleum Act* proscribes any person from engaging in petroleum operations on land or in a submarine area, unless he first obtains a licence as provided for in the Act or the Regulations.[13] The interpretation section of this Act defines "petroleum operations" as the operations relating to the various phases of the petroleum industry, including, inter alia, "…exploring for and producing… petroleum." Section 3 of the *Petroleum Regulations* provides for the issue of two licences to govern exploration and production activities in Trinidad and Tobago: (i) the Exploration and Production (*Public* Petroleum Rights) Licence and (ii) the Exploration and Production (*Private* Petroleum Rights) Licence. The Private Petroleum Rights Licence is granted by the Government over lands in which the petroleum and mining rights are owned by private individuals. Trinidad is one of less than ten countries in the world in which there exists such private ownership. To date, only two of this type of Exploration and Production (*Private* Petroleum Rights) Licence have been issued in Trinidad. One such licence has been granted to Trinidad Exploration and Development Limited (TED).[14]

Petroleum activities over areas that are subject to public petroleum rights are carried out either under the authority of an Exploration and Production (Public Petroleum Rights) Licences ("Public E&P Licences") granted by the Minister, or by the Minister entering into a Production Sharing Contract ("PSC") with the contractor.[15] Another unique feature of the Petroleum laws in Trinidad and Tobago is that the Act provides that production sharing contracts shall prevail over the Act where there is a conflict. Whether this provision is constitutional has not been challenged. Section 6(4) provides as follows:

> Where a production sharing contract is entered into under subsection (3), so much only of this Act and the Regulations as are not excluded by the contract shall apply to any person carrying on petroleum operations under such contract, and where any provision of this Act or the Regulations is modified by the contract for the purposes of such contract, this Act and the Regulations shall be read and construed accordingly, and **where there is any conflict or variance with reference to any matter between the provisions of the contract and this Act or the Regulations, the provisions of the contract shall prevail. (bold for emphasis).**

Additionally, it should be noted that the regulations expand on the operation of public and private petroleum rights and the licenses. The Minister is charged with the general administration of the Act, and in the exercise of his powers and the performance of his duties, he is required to confirm with any general or special directions given to him by the Cabinet. Any decision made or action taken by the Minister in

[13]Section 6 of the *Petroleum Act* stipulates as follows:

(1) No person shall engage in petroleum operations on land or in a submarine area, unless he first obtains a licence as provided for in this Act or the Regulations.
(2) A person who contravenes this section is liable on summary conviction to a fine of thirty thousand dollars and in the case of a continuing offence, to a further fine of one thousand, five hundred dollars for every day during which the offence continues.

[14]Lynch (1997).

[15]See, *Petroleum Act* Chap. 62:01, section 6(3).

the exercise of his powers and the performance of duties in accordance with the Act and the regulations shall be deemed to be made or taken by the Government and is binding thereon. The Minister may also delegate by written instrument to any public officer or agency of the government any of his powers or functions under the Act, as specified in the instrument. Such delegation does not prevent the exercise of the said functions by the Minister. Every delegation is revocable at will.

The procedure to acquire a licence requires the applicant to determine whether the petroleum rights attached to a particular piece of land are private or public. In order to conduct petroleum operations involving private petroleum rights, the company would need to obtain two documents: a private oil mining lease and a licence. The owner of the Private Petroleum Right must transfer their right.[16] Minerals rights in an oil and gas lease differ from surface rights. In the event that one concludes that one is dealing with private petroleum rights, the next step would be to determine who the owner of the right is. The Petroleum Regulations provides that "the term for which an Exploration and Production (Private Petroleum Rights) Licence may be granted shall be twenty years, subject to renewals for successive periods of twenty years."[17]

Unlike with private petroleum rights, a person interested in exploring areas that are subject to public petroleum rights need only deal with Ministry of Energy and Energy Industries as these rights are vested in the State to obtain the Exploration and Production (Public Petroleum Rights) Licences or a Production Sharing Contracts (PSC). The Public E&P Licences is a type of concession agreement and is the direct descendent, in the international arena, of the original Drake Lease in Pennsylvania.[18] It is the first type of contract used internationally and still persists today. However, this system lost popularity as nationalism grew or countries sought to take charge of their natural resources.[19]

An Exploration and Production (Public Petroleum Rights) Licence is a permission granted by a government to explore for and produce petroleum on lands in which the petroleum and mining rights are owned by the government or in a submarine area. Section 38 of the Act allows a person carrying out exploration and production operations a grace period of twelve months in which to become licenced under the Act. However, time period was not strictly enforced and existing oil mining leases continued in effect even after this twelve-month period. This was an unacceptable situation, as the two regimes could not reasonably subsist side by side

[16]For a detailed discussion on the procedures to obtain a private petroleum rights licence see Rampaul (2003).

[17]Regulations section 13 (2). Note that more details about the regulations of the Exploration and Production (Private Petroleum Rights) Licence are included in a *Cabinet Minute* headed "Guidelines for Use in assessing Applications for Exploration and Production (Private Petroleum Rights) Licences". The Guidelines state that the Ministry of Energy and Energy Industries shall negotiate each licence on a case-by-case basis in accordance with the Petroleum Act and the guidelines laid out in the said minute. The Guidelines specify that the "minimum acreage to be covered by such a licence be no less than 500 acres." The reason for this is to avoid inefficient and unsafe operations on small holdings.

[18]Yergin (1991).

[19]Ibid., 165–541.

without conflict. This situation caused particular difficulties especially with regard to issues such as transparency, unitisation of blocks, and requests by holders of oil mining leases to expand their licenced areas.

In 1990, section 38 of the *Petroleum Act* was amended in an attempt to more effectively administer the conduct of exploration and production activities by phasing out the oil-mining leases. Section 38 provided that persons conducting petroleum operations under oil-mining leases would be deemed to be licensees in respect of those petroleum operations until licenced as such under the Act. It was the intention behind this amendment that new licences would be issued to replace oil-mining leases and these new licences would be more in conformity with the new licencing regime despite the fact that they would have been issued on terms and conditions appropriate to and as reasonably close as possible to the oil-mining leases that they were replacing. This process was not fully undertaken at the time, and so the oil-mining leases, now deemed licences by virtue of Section 38 of the Act, continued in full force and effect. These oil-mining leases have been slowly replaced over the years in an ad hoc fashion until recently when most of the oil-mining leases held by the Petroleum Company of Trinidad and Tobago were surrendered and replaced by new model licences.

The provisions of the oil-mining leases and older model licences were well drafted and very comprehensive but were found in the early 2000s to be archaic in some ways. For instance, they contained an obligation to properly abandon and decommission at the end but did not include for any mechanism such as places funds into an escrow account to provide finances for the abandonment of wells and decommissioning of facilities on cessation of operations or for the remediation of pollution. Thus, the oil-mining leases did not always meet the government's modern aims and objectives in an ever-evolving industry. Accordingly, the new model licences have been formulated to do so.

The oil-mining leases contributed to a heritage of pollution and poorly abandoned wells that Trinidad now must deal with. The older model licences and the new model licences place an obligation on the licensee to establish an escrow account in the name of the Minister to accumulate cash reserves for use as a contingency fund for the eventual abandonment of wells in the Licenced Area and decommissioning of facilities used for petroleum operations under licence.[20] The new model licence expands the escrow account to include remediation of pollution arising from petroleum operations carried out under the licence during the term of the licence. The Minister may at his sole discretion access funds from the escrow account if the Licensee fails to effect environmental clean-up, in light of an accident to the satisfaction of the Minister.

The Licensee is obligated to pay twenty-five cents in the currency of the United States of America per barrel of oil equivalent produced into the escrow account. In computing the relevant production, Natural Gas Production is added to Crude Oil Production after converting to barrels of crude oil on an Energy Equivalent Basis,

[20] See Government of the Republic of Trinidad and Tobago Ministry of Energy and Energy Industries (2019a).

which is defined as the equivalent of Natural Gas in barrels of Crude Oil with 5,800 standard cubic feet of Natural Gas being equivalent to one barrel of crude oil. The Licensee is obligated to maintain this account at a level considered adequate by the Minister to fulfil its purpose at any time.[21]

The oil mining leases and old model licences do not contain clauses regarding the use of local goods and services in their operations. The more recent licences require that the licensee comply with the provisions of the *Local Content and Local Participation Policy and Framework*.[22] The licensee is required to keep records of their use of local goods and services for the purpose of inspection and audit by the Minister and to submit reports on local content to the Minister on a quarterly basis. This change is in keeping with the government's policy to develop local participation in the energy sector thereby fostering the development of local expertise in the petroleum industry and ensuring a wider distribution of the revenue accruing to the country from the energy sector. In the new model licences, the government has given recognition and significance to the necessity for the development of local human resources. Accordingly, the new licences have clauses requiring contributions towards training and research and development and the provision of scholarships. The oil-mining leases had no such provisions and older model licences did not have such comprehensive clauses in these areas.

The oil-mining leases and old model licences did not contain many bonuses to be paid to the government. The new model licences contain a signature bonus, production bonuses, and a technical equipment bonus. The signature bonus is usually payable within ten days of execution of the licence, it represents a reasonable sum based on the project as a whole. The production bonuses are payable on first attainment of a sixty-consecutive day average at or in excess of certain production levels. This is a mechanism employed to ensure an equitable distribution of supernormal profits. The technical equipment bonus is payable either in cash, technical assistance, or technical equipment.

The older licences tended to treat with environmental issues very scantily. As mentioned earlier, this has led to a heritage of pollution. As part of the approach to dealing with the existing pollution, licensees are required to pay an environmental bonus within ten days of the effective date of the licence; this money is to be used by the government in the remediation of existing sites of pollution. Also, certain licensees are required to conduct a review of the well files and other related and available documents, as well as physical site inspection of the existing wells with a view to determining, to the extent reasonably possible, the abandonment and environmental status of the existing wells and the licensed area within eighteen months of the effective date of the licence. Further, the licensee is required each year to implement a programme of environmental remediation of identifiable sites of chronic oil pollution in the licensed area. This programme is to be approved by the Minister.

---

[21] Ibid.

[22] Government of the Republic of Trinidad and Tobago Ministry of Energy and Energy Industries (2004).

The new licences have more stringent relinquishment provisions than the oil-mining leases and older model licences. This is in keeping with the objective of encouraging exploration activity and avoiding acreage lying idle for long periods of time. The oil-mining leases provided for an initial thirty-year term with an option to renew for a further term not exceeding thirty years from the expiration of the initial subject to the same covenants, provisions, and agreements as were therein contained with the exception of the covenant for renewal. They traditionally contained no relinquishment provisions. Both the new and the older model licences provide that all acreage that does not form part of a field must be surrendered by the end of the sixth year of the licence. Additionally, in the new licences, the licensee is required to surrender fifty per cent of the licensed area by the end of the fourth year of the license.

3 Establishment of a Dedicated Fund for Extractive Industries

As illustrated above, Trinidad and Tobago has a long history of managing oil and gas exploration. Over the years, the laws were amended several times to make accommodation for new circumstances and the model contracts and licences used in the industry welcomed new iterations over the years, which included changes to update the legal, regulatory, and contractual relationship between the government and contractors. With regard to the Heritage and Stabilisation Fund of Trinidad and Tobago, as mentioned above, this Fund in its current form was established in 2007 but was previously known as the Interim Revenue Stabilisation Fund, which was set up in 2000. This fund was introduced after over 80 years of petroleum operations in Trinidad and Tobago.[23] The primary objectives of the fund are to save and invest surplus petroleum revenues derived from production business and to support and sustain public expenditures during periods of revenue downturn and to provide a heritage for future generations of the nation. According to Section 3(1) of the Act, the purposes of the fund are to:

(a) Cushion the impact on or sustain public expenditure capacity during periods of revenue downturn whether caused by a fall in prices of crude oil or natural gas;

(b) Generate an alternate stream of income so as to support public expenditure capacity as a result of revenue downturn caused by the depletion of non-renewal petroleum resources; and

(c) Provide a heritage for future generations of citizens of Trinidad and Tobago from savings and investment income derived from the excess petroleum revenues.[24]

The resources of the fund consist of moneys transferred from the Interim Revenue Stabilisation Fund, petroleum revenues deposited into the fund in accordance with

[23] See Government of the Republic of Trinidad and Tobago Ministry of Energy and Energy Industries (2019b).

[24] *Heritage and Stabilisation Fund Act, supra* n. 3, section 3(1).

section 13 of the Act, and assets acquired and earned from investments. Section 13 of the Act provides that where petroleum revenues collected in each quarter of any financial year exceed the estimated petroleum revenues for that quarter of the financial year by more than ten per cent, the currency of the United States of America equivalent of the excess revenue shall be withdrawn from the consolidated fund and deposited to the fund in accordance with section 14(1) of the Act. It further provides that where petroleum revenues collected in each quarter of any financial year exceed the estimated petroleum revenues for that quarter of a financial year but do not exceed such estimated revenues by at least ten per cent, the Minister may direct that the currency of the United States of America equivalent of all or part of the excess revenue shall be withdrawn from the consolidated fund and deposited to the fund in accordance with section 14(1) of the Act.

4 Management of the Heritage and Stabilisation Fund in Trinidad and Tobago

The International Working Group of Sovereign Wealth Funds, which was established in 2008 created a set of *Generally Accepted Principles and Practices* known as the *Santiago Principles.*[25] These principles are aimed at ensuring good governance, accountability, and prudence by Sovereign Wealth Funds. The Trinidad and Tobago Fund was built on the principles thereby created. This group subsequently became the International Forum for SWFs (IFSWF) and lends assistance to members in implementing the Santiago Principles.

The President, on the advice of the Minister of Finance appoints a Board of Governors for the fund. The Board comprises five members with expertise in matters of finance, investment, economics, business management, or law. The Board must always include an officer of the Central Bank; and the Ministry of Finance. Members of the Board are to be appointed for a term of three years and are eligible for reappointment. The Board has the responsibility to determine by resolution, the governance structure and the operational and investment guidelines of the fund based on prudential standards used by the Central Bank for investments of a similar nature. It is also responsible for the management of the fund and must review the performance of the fund from time to time.[26]

Section 10 of the *Heritage and Stabilisation Fund Act*, 2007, however, provides for the Board to delegate its management responsibility to the Central Bank of Trinidad and Tobago. As Manager of the fund, the Central Bank is responsible for the management of the assets and other resources of the fund in accordance with the Act and the prudent investor standard of an investment manager, engaged in the asset management profession.[27] It also oversees the investment of the assets and other resources

[25] International Working Group of Sovereign Wealth Funds (2008).

[26] *Heritage and Stabilisation Fund Act, supra* n. 3.

[27] Ibid.

of the fund in accordance with the *Heritage and Stabilisation Fund Act,* 2007 and the operational and investment guidelines developed by the Board. It manages the selection and retention on behalf of the fund appropriate third-party service providers, such as, attorneys-at-law, auditors, and advisors in order to carry out competently, the mandate specified in the instrument of delegation and the selection of an appropriate global custodian for the fund. Its duties include the maintenance of records and documentary support for all investments, receipts, disbursements, and other transactions relating to the management of the fund in accordance with prevailing accounting practice. It must ensure the submission of quarterly reports to the Board on the holdings, performance, and risk of the fund no later than one month after the end of each quarter and an annual report of the fund to the Board no later than two months after the end of the financial year which must contain audited financial statements and an investment report on the performance of the fund. The Board decides on the investment objectives and approves the manner in which the funds are to be invested by the Central Bank.

The management of the fund's portfolio is done by skilled asset managers engaged by the Central Bank, with the approval of the Board of Governors. Additionally, the Central Bank hired a global custodian to perform the role of safe keeping assets, trade settlement and reporting. In February 2008, the Central Bank commenced a rigorous external manager selection process which resulted in the engagement of eight managers to manage the four investment mandates.[28]

It should be noted that there is an Operational and Investment Policy guideline for the management of the fund, which was developed by the Central Bank with the assistance of the World Bank and approved by the Board of Governors on July 30, 2009. It provides the framework for the management of the fund by the Central Bank. The Board is required to review the operational and investment policy at least every three years.[29]

The parliament of Trinidad and Tobago has ultimate oversight of the fund through the review of annual reports and audited financial statements. The Act stipulates that the Board shall submit a quarterly investment report and an annual investment report to the Minister on the operation and performance by the fund. It also provides that the Minister may request such a report, and it must be submitted to the Minister within one month of a request made. Further, within four months of the end of the financial year, the Minister is obliged to cause the audited financial statements in respect of the fund to be laid in Parliament. These financial statements are prepared in accordance with generally accepted accounting practices and international accounting standards adopted by the Institute of Chartered Accountants of Trinidad and Tobago. It must be highlighted that because the fund is a public account, it is audited annually by the Auditor General.

[28] HSF Annual Investment Report for the Period ended September 30, 2017, https://www.finance.gov.tt/wp-content/uploads/2019/08/HSF-Annual-Report-2017.pdf, accessed August 29, 2019.
[29] Ibid.

5 Distribution of the Heritage and Stabilisation Fund Across Different Regions and State Entities

The Minister is responsible for approving deposits and withdrawals from the HSF in accordance with the legislation and reports annually to the parliament. The circumstances in which withdrawals may be made are clearly laid out in the Act. Section 15 states that where the petroleum revenues collected in any financial year fall below the estimated petroleum revenues for that financial year by at least ten per cent, withdrawals may be made from the fund. Withdrawal may be the lesser of either sixty per cent of the amount of the shortfall of petroleum revenues for that year or twenty-five per cent of the balance standing to the credit of the fund at the beginning of that year. It further states that the amount withdrawn from the fund must be deposited into the consolidated fund within forty-eight hours of such withdrawal. It is stipulated however that no withdrawal may be made from the fund in any financial year, where the balance standing to the credit of the fund would fall below one billion dollars in the currency of the United States of America, if such withdrawal were to be made.

To date, the Government has made 14 contributions to the fund amounting to US$2,554.6 million. For the first seven years of the fund's existence, with the exception of financial year 2009, the Government made deposits into the fund. Since the decline in energy prices in 2014, the conditions necessary for the Government to make deposits to the fund were not met and as a result, no further contributions were made.[30] The first drawdown occurred in May 2016 while the second occurred in March 2017 with the two withdrawals totalling US$627.6 million.[31]

On 13 May 2016 the Government of Trinidad and Tobago made its first drawdown of US$375.05 million from the fund. At that time, the balance in the fund was US$5.796 Billion, and, after the withdrawal, the balance was US$5.420 Billion. It was stated in the 2016 mid-year budget review, that any budget deficit would be financed through a combination of borrowing and a drawdown from the HSF. On 16 March 2017, Cabinet approved a drawdown from the fund in the amount of the US equivalent of TT$1,712,200,000.00 (US$251 million). This drawdown was used for the financing of the 2017 budget, in particular, the Development Programme, also known as the Public Sector Investment Programme (PSIP).[32]

Further, in the 2017 Budget Statement, it was stated that "in 2017, core revenue, defined essentially as revenue from taxation, royalties and customs duties is only projected to be of the order of $37 billion, $20 billion less than just two years ago [2015]. This leaves a fiscal gap in 2017 of over $16 billion, which must be financed by a combination of borrowings, and drawdowns from the Heritage and Stabilisation Fund, and one-off sources of income, such as the sale of assets, dividends from state

[30] HSF Annual Investment Report for the Period ended September 30, 2017, https://www.finance.gov.tt/wp-content/uploads/2019/08/HSF-Annual-Report-2017.pdf, accessed August 29, 2019.

[31] Ibid.

[32] See Government of the Republic of Trinidad and Tobago Ministry of Finance Press Release (2017).

enterprises, repayment of past lending and so on". In a press release[33] on 16 March, 2017 the government of the Republic of Trinidad and Tobago stated that "As the country continues to experience severe revenue shortfalls as a result of depressed petroleum prices, the HSF will be carefully used by the Government to ensure the country's financial stability".

6 Fiscal Regime in Trinidad and Tobago

The fiscal regime in Trinidad and Tobago is governed by various pieces of legislation as well as contracts and licences. The key legislation through which revenues are derived from the upstream petroleum sector is the *Petroleum Act and Regulations*.[34] This legislation governs the conduct of petroleum operations. Under *the Act*, companies are required among other things to pay a royalty that is stipulated in the licence as well as contribute to the Petroleum Impost, which is used to cover the administrative costs of the Ministry of Energy.

Royalty rates vary based on policies in existence at the time of the execution of the licence. For crude oil, the rate ranges from 10 to 12.5% of the Field Storage Values. According to the website of the Ministry of Energy and Energy Industries[35] "up until 1989, the Field Storage Value was based on the Royalty Lease Evaluation 1 Method (RLE1). This method provides for a price for crude oil that was determined by the values of the crude oil fractions (light oils, diesel and fuel oil) less a percentage for refining and handling charges. For licences signed from 1989, the Field Storage Values are determined using international market prices of reference crudes. In the case of natural gas, the royalty rate ranges from 0 to 15% of the value of natural gas.

Exploration and Production Licences were the main contractual arrangements used during the period 1900–early 1970s. However, given the rapid development of the sector, better administration of the contractual arrangements was necessary. Therefore, in 1974, the first two Production Sharing Contracts (PSCs), for acreage off the east coast of Trinidad, were signed. These earlier PSCs did not provide for cost recovery, they allowed government a share of production based on production levels and were also ring-fenced.

In 1995, with the adoption of the World Bank PSC Model by Government, the PSC was extensively expanded with enhanced contractual terms and conditions. These included provisions for cost recovery, relinquishment, abandonment, shares of Profit Petroleum to the government that were based on both price and production levels, minimum work obligations during the exploration period, procedure to encourage the development of natural gas markets and financial obligations such as signature bonus, research and development, training of nationals and technical equipment bonus. Like

---

[33]Ibid.

[34]*Petroleum Act supra* n. 9.

[35]See Government of the Republic of Trinidad and Tobago Ministry of Energy and Energy Industries (2019c).

the earlier PSCs, these continued to be ring-fenced and assured the Government of a steady revenue stream. In addition, under these PSCs the Contractor's tax liabilities were paid by the Government out of its share of profit petroleum. Simultaneously, similar type provisions were slowly being introduced in the E&P licences.

A review of the petroleum fiscal regime undertaken in 2005, and this led to the introduction of a new styled PSC, referred to as a "taxable PSC" that comprised three major features. Firstly, Government received a Share of Profit Petroleum in lieu of some taxes viz Supplemental Petroleum Tax, Royalty, Petroleum Impost and Petroleum Levy. Contractors were therefore exempt from payment of the aforementioned taxes but were required to pay all other taxes namely, Petroleum Profits Tax, Unemployment Levy, Green Fund Levy and Withholding Tax directly to the Ministry of Finance; this represented a departure from the earlier models in which the government paid these taxes on behalf of the Contractor. Secondly a windfall profits feature was introduced to capture higher shares of profit petroleum as petroleum prices increased. Thirdly, consolidation of the new PSCs, by type either deep water or land/shallow marine was permitted. This was to promote multi-block development and facilitate investment by consortia and in so doing minimise their exposure to risks.

Also included were provisions for re-openers, accessibility of natural gas supplies for both the domestic and export markets, improved funding procedures for abandonment, and assignments and transfers. A special incentive that provides for an uplift of 40% on the drilling of exploration wells in the deep water was also introduced. In 2010, the legislation regarding the "taxable PSC" was repealed and the 1995 model PSC reintroduced with the following changes: cost recovery levels fixed at 50, 55 and 80% for shallow, average and deep-water areas respectively; financial obligations are also fixed and the only two biddable items are the Government's profit share and minimum work program; and signature bonus is no longer compulsory."[36]

The *Petroleum Production Levy and Subsidy Act*,[37] was passed in 1974 with the objective of buffering large increases in petroleum product prices and provide a general level of market stability. This Act established a Petroleum Products Subsidy Fund to be managed by the Minister of Finance. Subject to the Act and to any Regulations and Orders made thereunder, the Minister of Finance, acting upon the advice of the Minister is authorised to cause advances to be made from the fund for the purpose of subsidising the prices at which petroleum products are sold by persons carrying on marketing business in accordance with price-fixing orders made by the Minister under section 31 of the *Petroleum Act.* In 1992, the Act was amended to place a ceiling on each company's gross levy payments of not more than 3% (later increased to 4%) of its value of its gross income derived from the sale of crude.[38] An inclusion was also made of those companies, previously exempt with production

[36]See ibid., *Tax Laws* https://www.energy.gov.tt/for-investors/fiscal-regime/tax-laws/, accessed 29 August 2019.

[37]*Petroleum Production Levy and Subsidy Act* Chap. 62:02, Laws of Trinidad and Tobago.

[38]See *Petroleum Profits Tax* (Part 1 of the Act), Chap. 75:04, Laws of Trinidad and Tobago; and *Supplemental Petroleum Tax* (Part 11), Introduced by Act 5 of 1981, Laws of Trinidad and Tobago.

level of less than 3500 barrels of oil per day. Any excess levy payments above the cap are to be made by the Government.

Taxation of Petroleum companies is governed by the provisions of the *Petroleum Taxes Act*,[39] Chap. 75:04, which applies to and is applicable to all companies engaged in petroleum operations. This is defined as hydrocarbon compounds; "petroleum operations" means the operations related to the various phases of the petroleum industry and includes natural gas processing, exploring for, producing, refining, transporting and marketing petroleum or petroleum products or both, and manufacturing and marketing of petrochemicals; but does not include mining operations involving the extraction of petroleum from bituminous shales, tar sands, asphalt or other like deposits.

Under the Act, two main taxes are paid by petroleum companies. These are *Petroleum Profits Tax* (Part 1 of the Act) and *Supplemental Petroleum Tax* (Part 11). The Petroleum Profits Tax (PPT) is applicable to all oil and gas producers as well as refinery operators and is applied to the net profits (chargeable income) from operations. The calculation for the net profit is derived by deducting from the gross income all operating expenses, capital allowances and other allowable deductions. According to the Ministry of Energy and Energy Industries, deductions for oil and gas producers include royalties, Supplemental Petroleum Tax, Petroleum Levy/Impost, decommissioning/abandonment costs and management fees paid to non-resident companies (limited to 2% of expenditure). Other special allowances are granted for signature and production bonuses, dry holes, work-overs, qualifying sidetracks, heavy oil and exploration costs (the latter available for the years 2014–2017). The current applicable tax rate charged on producers as well as refinery operators is 50% (reduced to 35% from income year 2011 for deep water operations only). Over the years, amendments have been made to the PPT as market conditions changed. According to the Ministry of Energy and Energy Industries, the last change was in 2014, when increased allowances were granted on capital expenditure.[40]

The Supplemental Petroleum Tax (SPT) was introduced by Act 5 of 1981 and has been amended on several occasions. The SPT is imposed on income generated from production of crude oil net of royalty and over-riding royalty. Prior to 2005, SPT was levied on the gross income from the disposals of crude oil (not natural gas income) less certain allowances based on expenditure incurred in specified exploration and development activities. Although the tax was imposed on crude oil sales, companies involved in both oil and gas activities benefitted from the allowances since they were broadly applied to exploration and development field activities. This significantly contributed to the development of the natural gas industry in Trinidad and Tobago.

Over the years, the SPT rates varied for marine and land operations and for licences or contracts that were agreed prior or post 1988. In 2006, SPT rates for deep-water operations were fixed as those for land operations post 1988. SPT rates were also based on a sliding scale for prices ranging from US$15.00 to $49.50 per barrel, thereafter the rate remained fixed. As time progressed and as economic and industry

---

[39] See Ibid., *Supplemental Petroleum Tax*.

[40] See *Tax Laws, supra* n. 35.

related factors warranted, several amendments were made to this tax. During the period 2011–2013 incentives in the form of discounts/tax credits were introduced to further stimulate the production of crude oil.

According to the Ministry of Energy and Energy Industries, companies also pay 5% of the chargeable income before loss relief plus any exempt income under the *Unemployment Levy Act*,[41] and the monies obtained are applied to assist in the Government's social programmes. The Green Fund Levy equates to 0.3% of the gross sales or receipts and is paid under the *Miscellaneous Taxes Act*.[42] The purpose of this levy is the restoration and preservation of the environment.

7 Ownership of the Petroleum Resources in Guyana

Guyana is the only English-speaking country in South America. It is a former British colony and has a common law legal system. Property in petroleum and the mineral estate in Guyana is vested in the State, which has the exclusive right of searching for petroleum.[43] The *Petroleum (Exploration and Production) Act* (1986)[44] and *Petroleum (Exploration and Production) Regulations* (1986)[45] govern and regulate exploration, exploitation, conservation, and management of petroleum existing in its natural condition in land in Guyana, including the territorial sea, continental shelf and exclusive economic zone in Guyana. The Act sets out the procedures for the application and grant of a petroleum prospecting licence and a petroleum production licence and for matters connected therewith.[46] The Act also sets out the conditions concerning the duration, renewal, cancellation, and other functions relevant to the petroleum licences as well as procedures for notification of a commercial discovery of oil.

A petroleum prospecting licence usually lasts for ten years. It consists of an initial period of four years with provisions for two further periods, which is at the option of the licensee. Relinquishment is mandatory and the licensee is required to relinquish blocks or a percentage of the licenced area at given intervals during the terms of the licence. These mandatory relinquishments exclude the discovery areas. Regarding the petroleum production licence, this licence is issued if a commercial

41 *Unemployment Levy Act,* Chap. 75:03, Law of Trinidad and Tobago.

42 *Miscellaneous Taxes Act,* Chapter 77:01, Laws of Trinidad and Tobago.

43 *Petroleum (Production) Act*, Cap. 65:05, Laws of Guyana, section 2.

44 *Petroleum (Exploration and Production) Act* (no. 3 of 1986), Cap. 65:10, Laws of Guyana available at https://parliament.gov.gy/documents/acts/8170-act_no._3_of_1986_petroleum_(exploration_and_production)_act_1986.pdf accessed 28 October 2019.

45 *The Petroleum Regulations* (Legal notice 5 of 1986), made pursuant to section 70 of the Petroleum (Exploration and Production) Act, (no. 3 of 1986), Cap. 65:10, Laws of Guyana.

46 Supra (n 4) Petroleum (Exploration and Production) Act, section 20: Application for a prospecting licence; section 21: Grant of prospecting licence; section 30: Notification of discovery of oil; section 31: Notification of discovery of oil in commercial quantities; section 34: Application for a petroleum production licence; and section 35: Grant of petroleum production licence.

petroleum discovery is made. This licence has an initial period of twenty years, with a single renewal period not exceeding ten years, which can be applied for, if necessary.[47] Production sharing agreements between the Government of Guyana and oil companies for licences granted to and held by oil companies are provided for under the Act. The terms and conditions of the petroleum prospecting licence are usually established through negotiations. The Act envisages that a production sharing agreement not inconsistent with the Act, will document any such settled terms and conditions to be included in licences granted under the Act. The Minister Responsible for Petroleum is authorised by the Act to conclude such agreements. The Minister responsible for finance may apply for an order to modify the tax laws in respect of a license as follows: "The Minister assigned responsibility for finance may, by order, which shall be subject to affirmative resolution of the National Assembly, direct that any or all of the written laws mentioned in subsection (2) shall not apply to, or in relation to, a licensee where the licensee has entered into a production sharing agreement with the Government of Guyana".[48]

8 Main Features of Guyana's Natural Resource Fund

While the HSF in Trinidad and Tobago is managed by a five-member Board of Governors as well as skilled asset managers engaged by the Central Bank, with the approval of the Board of Governors, Guyana's Natural Resource Fund Act 2019 mandates the establishment of several committees including a twenty-two member Public Accountability and Oversight Committee (POC), an Investment Committee and a Macro-Investment Committee.[49] Membership of the POC does not require expertise in financial investments and financial portfolio management. Members for the POC includes a nominee from the Private Sector Commission, a nominee from each of the 10 Regional Democratic Councils, a nominee from the consortium of civil society organisations and community-based organisations, which represent youth, etc.[50] The functions of this committee include monitoring and evaluating whether the fund has been managed in accordance with the principles of transparency, good governance and international best practices including the Santiago principles.[51]

The Minister of finance is responsible for the management of this fund and shall be assisted by a Senior Investment Advisor and Analyst.[52] There is also to be established a six-member Investment Committee to be appointed by the Minister. The nominees shall include a nominee of the Minister of Finance, a nominee of the Leader of the Opposition, a nominee of the Minister with responsibility for the administration of the

[47] Ibid., sections 39 and 40.

[48] Ibid., section 51(1).

[49] *Natural Resource Fund Act, supra* n. 8.

[50] Ibid., section 7(1).

[51] Ibid., section 6(2).

[52] Ibid., section 11.

petroleum sector, etc. It should be pointed out that the Minister of Natural Resources previously had responsibility for the administration of the petroleum sector, but the administration of the sector is now under the Office of the President and there is no ministry responsible for the administration of the petroleum sector. Section 13(1)(b) of the Act is therefore unclear where it states that the Committee will include "a nominee of the Minister with responsibility for the administration of the petroleum sector". Note that unlike the POC, the members of this Committee must have at least 10 years' experience and expertise in financial investments and financial portfolio management.[53]

Under section 20(1) of the Act a Macroeconomic Committee shall also be established, which shall be responsible for advising the Minister on the "Economically Sustainable Amount". This committee shall consist of five members appointed by the Minister. Members include a representative of the Ministry of Finance, who shall be Chairperson of the Committee, a nominee by the Leader of the Opposition, a representative of the Bank nominated by the Governor of the Bank, a nominee of the Private Sector Commission, and a leading international expert in macroeconomics identified and approved by Cabinet.

Section 21 of the Act indicates the various streams of revenues, which shall be paid into the fund. Deposits into the fund shall come from a range of petroleum revenue sources. Interestingly, deposits are not limited to petroleum revenues, but the Minister may deposit excess mining and forestry revenues into the fund. This fact might explain the name of the fund, which is wider than petroleum and refers to natural resources. With regard to withdrawals from the fund, section 22(1) stipulates that the maximum amount that may be withdrawn from the fund in a fiscal year shall be a portion of the fund and shall not exceed the total withdrawal from the fund approved by the National Assembly for that fiscal year in accordance with Section 28. The Act includes several technical and vague terms regarding the concept of "economically and fiscally sustainable". What this means is ambiguous and this might be deliberate to create room for flexibility. Section 24 provides that the economically and fiscally sustainable amount for a fiscal year shall be the lesser of the following amounts: (a) the economically sustainable amount for the fiscal year; and (b) the fiscally sustainable among for that fiscal year.

Section 25 of the Act goes on to provide that the economically sustainable amount is the maximum amount that can, in the opinion of the Minister after taking into account the recommendations of the Macroeconomic Committee established under section 20, be withdrawn from the fund for the next ensuing fiscal year without diminishing the competitiveness of the Guyana's economy. It must be pointed out that the use of the "economically and fiscally sustainable" concept is subject to varying interpretations and can lead to uncertainty.

[53] Ibid., section 13.

9 Conclusion and Critical Analysis of the Legal Framework and Management of the Sovereign Wealth Funds in Trinidad and Tobago and Guyana

According to an International Monetary Fund Report, which was submitted in 2012, before the first withdrawal was done from the HSF in Trinidad and Tobago, it was noted that in light of the likely changes to the HSF, consideration should be given to the proportion of the HSF that would be set aside for stabilisation purposes versus that for the future generation, determining new deposit and withdrawal rules and amending the reporting timeframe to allow for improved operational efficiency.[54] The International Monetary Fund Report also made the recommendation for the Government of Trinidad and Tobago to decide whether the HSF should continue in its present form or whether it should be split. According to the 2014 report of the HSF board of directors under the chairmanship of Dr Ralph Henry, there is an interesting sentence in the chairman's foreword which states: "After careful consideration, the board recommended that the fund should not be split formally into two funds at this time–i.e. a Heritage Fund and a Stabilisation Fund".[55] This comment is at odds with the Prime Minister's statement in 2015 that the Government intends to separate the Heritage from the Stabilisation in order to create two funds.[56] This is an important issue in the current HSF management, which needs to be clarified.

Another important development regarding management of the fund, which must be highlighted concerns the way the first withdrawal was made. It was widely reported in the local media in Trinidad and Tobago that the Government had "dipped" into the Heritage and Stabilisation Fund and no official announcement was made when the withdrawal was done. The manner in which the information about the withdrawal from the fund was made public generated controversy. It was reported that an investigative reporter was able to "blow the cover on what would have been an otherwise secret withdrawal" and after the first story, the media followed the report and it was publicised widely.[57] Another reportedly troubling aspect of the whole scenario is that the amount withdrawn had allegedly far exceeded what the Prime Minister had told the nation would have been withdrawn.[58]

From Trinidad and Tobago's short history and management of the HSF, there are a few important unanswered questions. First, what was the thinking of the HSF board in 2014 for not delinking Heritage from Stabilisation? Second, what is the Government's current position regarding this issue? Third, why was there no public announcement from the Ministry of Finance regarding the withdrawal of the funds? The need for more transparent discussion and scrutiny of the HSF after the first

---

[54] See International Monetary Fund (2012); and Williams (2008).

[55] *Trinidad and Tobago HSF Annual Report 2014,* at p. 4, available at https://www.finance.gov.tt/wp-content/uploads/2015/07/HSF-2014-Annual-Report.pdf, accessed 29 August 2019.

[56] Singh (2015).

[57] See Dr. Hamid Ghany, *supra* n. 4.

[58] *Ibid.*

withdrawal has been made and the issue regarding the separation of the fund need to be critically discussed. It is also recommended that clear rules for public notification regarding withdrawal from the funds must be established and enforced, which would help to improve the overall operational efficiency of the fund.

Turning to the Natural Resource Fund in Guyana, when compared to Trinidad and Tobago's HSF, there is more involvement of public interest groups in the monitoring of the fund. This appears to be an attempt to include the principles of transparency, good governance, and international best practices including the Santiago principles in the monitoring of the fund. Many of the Committees established and their functions outlined under the Natural Resource Fund Act in Guyana are similar to the bodies established under more recent sovereign wealth management legislation in developing countries, like Ghana's Petroleum and Revenue Management Act 2011.[59] One will have to wait and see whether having several committees and not requiring expertise in financial investments and financial portfolio management for membership in some committees will affect how they achieve the overall goals of the funds. Since this is a newly assented Act in Guyana, you will have to wait and assess how the POC committee functions to determine whether this is a good model to adopt. On another note, turning to the main function of the Macroeconomic Committee, it must be emphasised that the use of the "economically and fiscally sustainable" concept in the Act is subject to varying interpretations and can lead to uncertainty.

Concerns with the Natural Resource Fund in Guyana were highlighted by the Natural Resource Governance Institute (NRGI). NRGI released a report in which they urged the government of Guyana to address significant underlying problems with the proposed fund design. The main problems identified included rules governing how much money enters the fund and how much the government can withdraw for budgetary spending. "Guyana could end up in a situation where it is saving money and earning low interest while it borrows at a higher interest rate", a NRGI official said. "And in this way, the country ends up losing money".[60] The caution from NRGI that "Borrowing on the back of future natural resource revenues is a big risk for new oil producers like Guyana", should be taken seriously so that the country will avoid the "presource curse", which leads to accumulation of unsustainable debt followed by economic crises.[61]

References

Ghany, H. (2016, June 12). Heritage and Stabilization Fund usage. *Trinidad and Tobago Guardian*. https://www.guardian.co.tt/article-6.2.355013.4cf60100a3. Accessed August 31, 2019.

Government of the Cooperative Republic of Guyana. (2019, March 29). Department of Public Information 'Sovereign Wealth Fund assented to—Ministry of Finance. https://dpi.gov.gy/sovereign-wealth-fund-assented-to/. Accessed October 28, 2019.

[59] *Petroleum and Revenue Management Act* (2011), Act 218, Laws of Ghana.

[60] Natural Resource Governance Institute (2018).

[61] Ibid.

Government of the Republic of Trinidad and Tobago Ministry of Energy and Energy Industries. (2004). *Local content and local participation policy and framework'*. https://www.energy.gov.tt/wp-content/uploads/2013/12/Local-Content-and-Local-Participation-Framework.pdf. Accessed August 31, 2019.

Government of the Republic of Trinidad and Tobago Ministry of Energy and Energy Industries. (2015, June). White Paper on National Minerals Policy. https://www.energy.gov.tt/wp-content/uploads/2014/01/White-Paper-on-National-Minerals-Policy-June-2015.pdf. Accessed July 4, 2019.

Government of the Republic of Trinidad and Tobago Ministry of Energy and Energy Industries. (2019a). *Model production sharing contract*. Available at https://www.energy.gov.tt/wp-content/uploads/2013/12/The-Model-Production-Sharing-Contract-May-2019.pdf. Accessed August 28, 2019; and *Model JOA Trinidad Shallow Water Bid Round* (2018). Available at https://www.energy.gov.tt/wp-content/uploads/2013/12/Model-Joint-Operating-Agreement.pdf. Accessed August 31, 2019.

Government of the Republic of Trinidad and Tobago Ministry of Energy and Energy Industries. (2019b). *Historical facts on the petroleum industry of Trinidad and Tobago*. https://www.energy.gov.tt/historical-facts-petroleum/. Accessed August 28, 2019.

Government of the Republic of Trinidad and Tobago Ministry of Energy and Energy Industries. (2019c). *Tax Laws*. https://www.energy.gov.tt/for-investors/fiscal-regime/tax-laws/. Accessed August 29, 2019.

Government of the Republic of Trinidad and Tobago Ministry of Finance Press Release. (2017, March 17). Drawdown from the Heritage and Stabilisation Fund. https://news.gov.tt/content/drawdown-heritage-and-stabilisation-fund#.XWgKj0xFxjo. Accessed August 29, 2019.

International Monetary Fund. (2012, June). *Trinidad and Tobago: Selected issues*. IMF Country Report No. 12/128 (June 2012) at 12

International Working Group of Sovereign Wealth Funds. (2008, October). *Sovereign Wealth Funds: Generally accepted principles and practices "Santiago Principles"*. https://www.ifswf.org/santiago-principles-landing/santiago-principles. Accessed August 23, 2019.

Krauss, C. (2017, January 13). With a major oil discovery, Guyana is poised to become a top producer. *The New York Times*. https://www.nytimes.com/2017/01/13/business/energy-environment/major-oil-find-guyana-exxon-mobile-hess.html. Accessed June 14, 2019.

Lynch, J. (1997). Private rights to the subsoil in Trinidad and Tobago—A review of the grant of an exploration and production (Private Petroleum Rights) license to Trinidad Exploration and Development Company Ltd. *AIPN Advisor Newsletter*.

Myers, K. (2018, September 5). Westwood Global Energy Group, Westwood Insight 'Latest Guyana discovery opens the way to a new 10 billion barrel oil province and transformation for one of South America's smallest countries. https://www.westwoodenergy.com/news/westwood-insight/latest-guyana-discovery-opens-the-way/. Accessed June 14, 2019.

Natural Resource Governance Institute. (2018, November 13). Analysts suggests improvements to Guyana's proposed Sovereign Wealth Fund. https://resourcegovernance.org/news/analysts-suggest-improvements-guyana-proposed-sovereign-wealth-fund. Accessed October 28, 2019.

Rampaul, I. N. (2003). Privately owned petroleum rights in Trinidad and Tobago. *OGEL, 1*(2).

Singh, R. (2015, June 2015). Kamla slams Rowley recovery plans. *Trinidad and Tobago Guardian*. https://50tt.guardian.co.tt/news/2015-12-31/kamla-slams-rowley-recovery-plans. Accessed August 29, 2019.

Trinidad and Tobago Extractive Industries Transparency Initiative Report 2016 https://www.tteiti.org.tt/wp-content/uploads/TTEITI-Report-2016.pdf. Accessed July 4, 2019.

Williams, E. S. (2008, February 11–12). The challenge of abundance: Benefits from establishing a Sovereign Wealth Fund. Presentation by Governor, Central Bank of Trinidad and Tobago, at the *Sovereign Funds Roundtable*, Miami, Florida, USA.

Yergin, D. (1991). *The prize—The epic quest for oil, money, & power* (pp. 26–34). New York: Simon & Schuster.

Russian Sovereign Wealth Fund

Svetlana B. Globa

Abstract This contribution considers the experience of creating and operating sovereign wealth funds in Russia, analyses the basic principles of the functioning of funds, the sources of their formation, as well as the processes of managing funds. The experience of reforming the Stabilization Fund in Russia and its transformation into the Reserve Fund and the National Welfare Fund (NWF) is considered. Finally, the contribution explores the problems of increasing the significance and effectiveness of involving the resources of sovereign wealth funds in solving the most significant issues of the structural development of the national economy of Russia.

Keywords Russia · Sovereign wealth funds (SWF) · Russian national wealth fund (NWF) · Petroleum fund

1 Introduction

At present, the basic law determining the procedure for using the Russian subsoil is the Law of the Russian Federation "On Subsoil" dated 02.21.1992 (as amended on 02.08.2019) No. 2395–1. According to Article 1.2 of this law, subsoil within the borders of the territory of the Russian Federation, including underground space and minerals, energy and other resources contained in the subsoil, are State property. Issues of ownership, use and disposal of subsoil are jointly administered by the Russian Federation and its subjects.

Subsoil plots may not be subject to purchase, sale, donation, inheritance, contribution, pledge or alienated in any other form. Due to the fact that the subsoil is in State ownership, they can be provided only on the basis of the right to use. In accordance with the legislation of the Russian Federation, a subsurface user may be a legal entity, an entrepreneur, or a foreign citizen, if this is not prohibited by the legislation of the

S. B. Globa (✉)
School of Business Management and Economics, Siberian Federal University (SFU), Krasnoyarsk, Russia
e-mail: SGloba@sfu-kras.ru

E. G. Pereira et al. (eds.), *Sovereign Wealth Funds, Local Content Policies and CSR*, CSR, Sustainability, Ethics & Governance,
https://doi.org/10.1007/978-3-030-56092-8_8

Russian Federation. Exploration and mining of minerals is carried out on the basis of a license issued on a competitive basis, according to which users of the subsoil have the exclusive right to extract and develop minerals in accordance with the legislation of the Russian Federation. Minerals and other resources extracted from the subsoil under the terms of a license may be in federal State ownership, property of the subjects of the Russian Federation, municipal, private and in other forms of ownership.

In the Russian Federation, the use of subsurface resources is made on a paid basis, there are three types of payments by the subsoil user to the State budget—one-time payments for the use of minerals, regular payments, a fee for participation in the auction, and the law also allows the establishment of additional types of fees for the use of natural resources of the Russian Federation. Thus, at present, the Russian Federation has a two-tier system for regulating the use of the country's subsoil. The first level is federal, the second level is the level of subjects of the Russian Federation. Such a system makes it possible to achieve holistic economic regulation of the use of mineral resources taking into account the interests of the regions of the Russian Federation, while also making it possible to replenish both the federal budget and the budget of the subject of the Russian Federation on whose territory minerals are located.

Russia's natural resource potential is over 20% of world reserves. This provides Russia with a special place among industrialised countries. Russia ranks first in the world in gas reserves; second place in oil production; third place in coal reserves, first place in explored iron ore reserves, second in tin, and third in lead. In addition, metal ores are mined in the country: iron, nickel, copper, aluminium, polymetals, chromium, tungsten, gold, silver. Non-metallic ores are also diverse: phosphatites, apatites, talc, asbestos, mica, potassium and common salts, diamonds, amber, precious and semiprecious stones. Building materials are also widespread: sand, clay, limestone, marble, granite, cement raw materials and more.

As one of the world's largest oil-producing and gas-producing countries, Russia receives significant income from the use of its resources. Some of the oil and gas revenues are withdrawn in the budget system of the country in the form of taxes and fees. Currently, in Russia, tax legislation in the oil and gas sector is mainly fiscal. More than 40% of tax revenues are made up of special taxes paid by oil and gas companies—a tax on mining operations, on hydrocarbons (oil, combustible natural gas, gas condensate) and export customs duty on crude oil, natural gas, oil products.

According to Article 96.6 of the Budget Code of the Russian Federation, oil and gas revenues of the federal budget include federal budget revenues from tax on the extraction of minerals in the form of hydrocarbons (oil, natural combustible gas from all types of hydrocarbon deposits, gas condensate from all types of hydrocarbon deposits); export customs duties on crude oil; export customs duties on natural gas and export customs duties on goods derived from oil.

The Russian Tax on mining operations is a special form of financial relations between users of the subsoil and the State in the form of paying a tax depending on the type of resource extracted and geographic location.

The tax on mining operations was introduced on 1 January 2002 by including the 26th chapter in the Tax Code of the Russian Federation. The object of taxation is minerals. Payers of tax on mining operations are recognised organisations and individual entrepreneurs who are users of the subsoil. Tax on mining operations is distributed between budgets of various levels. Moreover, at present, over 98% of all taxes on the use of natural resources and 100% of all income from foreign economic activity (including export duties on oil, oil products, etc.) go to the federal budget, and not to the regional ones.

Reserve funds are created in those States whose budgets are highly dependent on market factors, as a rule, world commodity prices. In addition, some countries accumulate assets in such funds for the period when the subsoil is depleted. The Reserve Fund performs two functions. First, the Fund can be used to cover the State budget deficit at the time of unfavourable conditions. Second, during a period of high raw material prices, the Fund allows accumulating excess export earnings and preventing the development of "Dutch disease" of the economy.

The thesis of excess export earnings may seem paradoxical. The growth of export earnings leads, as a rule, to the rapid strengthening of the national currency. In itself, such a strengthening is not a threat to the economy, however, constant fluctuations in exchange rates following fluctuations in price conditions create macroeconomic instability and do not allow companies to choose a specific strategy—to adapt to low or high exchange rates. In addition, in conditions of high administrative barriers and strong monopoly (which is typical for most resource exporting countries), an increase in export earnings leads to an increase in inflation.

In addition to purely economic tasks, the Reserve Fund fulfils the political task of preventing the rapid growth of government spending. Government spending, as a rule, cannot be quickly reduced following a fall in revenue. As a result, during periods of unfavourable conditions, this can lead to large State budget deficits, failure to fulfil the promised social obligations and default on public debts. Such consequences are much more destructive for the economy than fluctuations in the volume of the State budget per se.

One of the key elements of macroeconomic policy in Russia since 2004 is the management of oil and gas revenues using various budget rule designs aimed at saving part of the income in sovereign funds. The savings mechanism, budget rules and the system of sovereign funds have changed many times over the past 15 years.

2 Stabilization Fund

The Stabilization Fund in Russia was established in 2004. State revenues from oil and gas production and export (in terms of export duties and mineral extraction tax) were directed to it when the world oil price exceeded a specially defined "cut-off price". That is, the State budget receives funds as if the oil price was equal to the "cut-off price", and this goes to the Stabilization Fund. Initially, the cut-off price was set at $20USD per barrel, then was raised to $27USD [1].

The Stabilization Fund's assets were directly controlled by the government represented by the Ministry of Finance, not the Federal Assembly through the federal budget and performed the following functions:

- was a built-in stabiliser;
- provided additional sustainability to public finances;
- could serve as a source of financing investment programs that ensure a reduction in budget expenditures for future periods.

When creating the Stabilization Fund in Russia, there was a situation where the yield of the Fund was significantly lower than the cost of servicing the country's external debt. At first glance, this paradoxical situation is explained by the following. First, it was not possible to agree on early repayment of external debt with all countries (for example, Germany could not technically do this, since bonds with many owners were issued on the external debt of the Russian Federation). Second, if the Stabilization Fund is spent on debt repayments, then there will remain a risk of a budget deficit in a period of low commodity conditions. Losses when placing funds in low-yield instruments are, in fact, an "insurance premium" when "insuring" the State budget against a deficit.

Since 1 February 2008, the Stabilization Fund has been divided into two parts: the Reserve Fund (3069 billion rubles) and the National Welfare Fund (782.8 billion rubles) [2].

3 Reserve Fund

Since 2008, oil and gas revenues are accounted for separately from other federal budget revenues. Oil and gas revenues of the federal budget are formed from:

- a tax on the extraction of minerals in the form of hydrocarbon raw materials (oil, natural combustible gas from all types of hydrocarbon deposits, gas condensate from all types of hydrocarbon deposits);
- export customs duties on crude oil;
- export customs duties on natural gas;
- export customs duties on goods derived from oil.

A certain part of these oil and gas revenues in the form of an oil and gas transfer is annually allocated to finance federal budget expenditures. The value of the oil and gas transfer is approved by the federal law on the federal budget for the next financial year and planning period.

The volume of oil and gas transfer is set as a percentage of the forecasted gross domestic product (GDP) for the corresponding year [3]:

- in 2008—6.1%;
- in 2009—5.5%;
- in 2010—4.5%;

- from 2011 onwards—3.7%.

After the formation of the oil and gas transfer in full, oil and gas revenues went to the Reserve Fund. The normative value of the Reserve Fund is approved by the federal law on the federal budget for the next fiscal year and the planning period in the absolute amount, determined on the basis of 10% of the GDP projected for the corresponding year. After filling the Reserve Fund to the specified size, oil and gas revenues are sent to the National Welfare Fund [1].

As of 1 October 2008, the Reserve Fund amounted to 3555.19 billion rubles [3].

The Reserve Fund was formed from:

- oil and gas revenues of the federal budget in an amount exceeding the amount of oil and gas transfer approved for the corresponding financial year, provided that the accumulated volume of the Reserve Fund does not exceed its normative value; and
- income from the management of the Reserve Fund.

The Reserve Fund of the Russian Federation, in which a significant part of oil and gas revenues was accumulated, ended with the beginning of 2018.

It represented that part of the proceeds from the export of oil, oil products, gas and condensate, which exceeds the approved oil and gas transfer for the new fiscal year (now it is defined as 3.7% of the GDP forecast by the authorities of the Russian Federation) [3]. If the normative value of the Reserve Fund is exceeded, the remaining income was credited to another national fund—the NWF. The normative (maximum) value was initially determined as 10% of the GDP of the Russian Federation projected for the next fiscal year. In addition to oil and gas revenues, another source of funds was income from the management of fund reserves.

The amount of funds deduction increased as the price of oil increased: from 2004 to 2006, revenues were transferred to the unified Stabilization Fund at a price of a barrel above $20USD, from 2006—over $27USD per barrel. According to the plan of the authorities, the existence of the Reserve Fund contributed to the fulfilment of government expenditure obligations and stabilisation of the economy (covering the budget deficit), reducing its dependence on fluctuations in oil and gas revenues. NWF was planned primarily to balance the Pension Fund of the Russian Federation.

According to the amendments to the Budget Code adopted on 30 September 2010 (Law No. 245-ФЗ) from 1 January 2010 to1 January 2015, the maximum amount of the Reserve Fund was not established, and oil and gas revenues were allocated to the federal budget for expenses. Replenishment of the Fund after a 3-year break occurred only in 2012. Also, from 1 January 2010, the Reserve Fund of the Russian Federation ceased to be replenished with income from the management of its funds—until February 2016, and these revenues were directed to the Federal Treasury.

The management of the Reserve Fund was vested in the Ministry of Finance, but part of these powers was allowed to delegate to the Central Bank. According to the government decree dated 29 December 2007 No. 955, funds should be managed in order to ensure:

- their safety; and
- long-term extraction of stable income from the allocation of reserves.

In the short term, it was allowed to obtain negative financial results in the process of managing funds. The order established by the government obliged to allocate funds:

- in a foreign currency;
- in assets nominated in foreign currency with a long-term credit rating not lower than "Aa3" according to Moody's classification;
- in assets nominated in foreign currency with a rating not lower than "AA–" according to Fitch-Ratings and Standard & Poor's.

The Reserve Fund was part of the gold and foreign exchange reserves of Russia, which was administered by the Ministry of Finance, and not the Central Bank. The order of the Ministry approved the currency structure of stocks [3]:

- Euro—45%;
- US dollar—45%;
- pound sterling—10%.

Foreign currency was credited to the accounts of the Bank of Russia, which paid interest equivalent to the return on the assets mentioned. Due to the high conservatism of investments (possibly caused by the financial crisis of 2008, which happened only a few months after the beginning of the independent existence of the Reserve Fund), the return on investments was approximately at the level of currency inflation. A balanced portfolio of stocks and bonds would allow the Bank to receive many times more income.

On 1 February 2018, the Reserve Fund of the Russian Federation ceased to exist, merging with the NWF. Thus, the Reserve Fund lasted exactly 10 years with an accuracy of a day.

On 1 December 2017, the proceeds from finding the funds in the Central Bank of the Russian Federation over the past year in the amount of 652 million rubles were transferred to the treasury, the remainder of foreign currency from accounts in the Central Bank—$7.62 billion, €6.71 billion and £1.10 billion—exchanged by 1.042 trillion rubles and used to cover the budget deficit [3].

In fact, the Reserve Fund today exists under the name of the National Welfare Fund of Russia, the volume of which on 1 January 2018, according to official data from the Ministry of Finance, amounted to $65.15 billion USD, or 3.75 trillion rubles.

4 National Welfare Fund of Russia (NWF)

The National Welfare Fund of Russia (NWF) was formed on 1 February 2008, after the separation of the Stabilization Fund. On 1 October 2008, the amount of the NWF was 1,228.88 billion rubles [4]. According to Article 96.10 of the Budget Code of the Russian Federation, "The National Welfare Fund is a part of the federal budget funds that are subject to separate accounting and management in order to ensure co-financing of voluntary pension savings of citizens of the Russian Federation, as well as to ensure the balance (deficit) of the budget of the Pension Fund of the Russian Federation". At the same time, "the objectives of the management of the Reserve Fund and the National Welfare Fund are to ensure the safety of these funds and a stable level of income from their placement in the long term" (Article 96.11 of the Budget Code of the Russian Federation).

The National Welfare Fund is formed by:

- oil and gas revenues of the federal budget in an amount exceeding the amount of oil and gas transfer approved for the corresponding financial year, if the accumulated amount of the funds of the Reserve Fund reaches (exceeds) its standard value; and
- income from managing the funds of the National Wealth Fund.
  - On 14 October 2008, a package of laws on stabilising the Russian financial market began to operate, according to which the funds of the National Wealth Fund can be placed with Vnesheconombank for deposits up to 31 December 2019 inclusive, for a total amount of not more than 450 billion rubles at a rate of 7% per annum [1].

5 NWF Management

The management objectives of the National Wealth Fund are to ensure the safety of the Fund and a stable level of income from its placement in the long term. The management of its funds for these purposes allows the possibility of obtaining negative financial results in the short term [2].

The National Welfare Fund is managed by the Ministry of Finance of the Russian Federation in the manner established by the Government of the Russian Federation. Separate powers to manage the National Wealth Fund may be exercised by the Central Bank of the Russian Federation. The National Welfare Fund can be managed in the following ways (both individually and simultaneously):

(1) by acquiring foreign currency and placing it on accounts of the Central Bank of the Russian Federation. For the use of funds in these accounts, the Central Bank of the Russian Federation pays interest established by the bank account agreement; and
(2) by placing funds in foreign currency and financial assets denominated in foreign currency, the list of which is determined by the legislation of the Russian Federation.

Currently, the Ministry of Finance of the Russian Federation manages the National Wealth Fund according to the second method, that is, by placing funds in foreign currency accounts of the Central Bank of the Russian Federation. At the same time,

Table 1 Allowed financial assets defined by the Budget Code of the Russian Federation

Allowed financial assets defined by the Budget Code of the Russian Federation	Limit shares established by the Government of the Russian Federation (%)	Regulatory shares approved by the Ministry of Finance of Russia (%)
Debt obligations of foreign states	50–100	80
Debt obligations of foreign state agencies and central banks	0–30	15
Debt obligations of international financial organizations, including those executed by securities	0–15	5
Deposits in foreign banks and credit organizations	0–30	0
Debt obligations of legal entities	not defined	not approved
Shares of legal entities	not defined	not approved
Shares (stakes) of investment funds	not defined	not approved

Source NWF [4]

the Bank of Russia pays interest on balances on these accounts equivalent to the yield of indices formed from debt obligations of foreign states, foreign state agencies and central banks, debt obligations of international financial organisations (Table 1).

The Government of the Russian Federation has determined that at present, the National Wealth Fund may be placed in the same financial assets as the Reserve Fund and has established the following requirements for these financial assets [2]:

1. The National Wealth Fund may be placed in debt instruments in the form of securities of foreign states, foreign state agencies and central banks of fourteen countries: Austria, Belgium, Great Britain, Germany, Denmark, Ireland, Spain, Canada, Luxembourg, the Netherlands, the USA, Finland, France and Sweden.
2. Debt obligations must comply with the following requirements:

   - the debt issuer must have a long-term credit rating not lower than "AA−" according to the classification of rating agencies "Fitch-Ratings" or "Standard & Purs's" or not lower than "Aa3" according to the rating classification Moody's Investors Service agency. If these agencies assigned different long-term credit ratings to the issuer of debt obligations, then the lowest assigned rating is taken as the long-term credit rating;
   - the maturity dates of the debt issues are fixed, the terms of the issue and circulation do not provide for the issuer's right to redeem (repay) ahead of schedule and the right of the debt holder to present them early for redemption (repayment) by the issuer;

- the minimum and maximum maturity standards for debt issues established by the Ministry of Finance of the Russian Federation are mandatory;
- the rate of coupon income paid on coupon debt obligations, as well as the face values of debt obligations are fixed;
- the denomination of debt instruments is expressed in US dollars, euros and pounds sterling, payments on debt instruments are made in the face value currency;
- the volume of the issue of debt instruments in circulation is not less than 1 billion US dollars for debt instruments denominated in US dollars, not less than 1 billion euro—for debt instruments denominated in euros, and not less than 0.5 billion pounds sterling—for bonds denominated in pounds sterling;
- Debt issues are not issues intended for private (non-public) placement.

3. The nominal volume of purchased debt obligations of one issue shall not exceed 15% of the nominal volume of this issue.
4. international financial organizations in whose debt obligations the National Wealth Fund may be placed include:
   - Asian Development Bank, ABD;
   - Council of Europe Development Bank, CEB;
   - European Bank for Reconstruction and Development, EBRD;
   - European Investment Bank, EIB;
   - Inter-American Development Bank, IADB;
   - International Finance Corporation, IFC;
   - International Bank for Reconstruction and Development, IBRD;
   - Nordic Investment Bank, NIB.
5. when placing the NWF on deposits with foreign banks and credit organisations, the total amount of the NWF placed on deposits with one foreign bank or credit institution shall not exceed 25% of the total amount of the NWF placed on deposits with foreign banks and credit organisations.
6. The Ministry of Finance of the Russian Federation has the right to establish additional requirements for debt obligations and deposits with foreign banks and credit organisations within the limits established by the Government of the Russian Federation.

The Ministry of Finance of the Russian Federation approved [4]:

1. The regulatory currency structure of the National Welfare Fund funds as follows (Table 2):

Table 2 Regulatory currency structure of the National Welfare Fund

US dollar	–	45%;
Euro	–	45%;
Pound sterling	–	10%.

Source NWF [4]

Table 3 Current terms to repayment of issues of debt instruments of foreign states, debt instruments permitted for placement of the National Wealth Fund for debt denominated in US dollars and euros

Minimum maturity	–	3 months
maximum maturity	–	3 years

Source NWF

Table 4 Current terms to repayment of issues of debt instruments of foreign states, debt instruments permitted for placement of the National Wealth Fund for debt denominated in pounds

Minimum maturity	–	3 months
maximum maturity	–	5 years

Source NWF [4]

2. current terms to repayment of issues of debt instruments of foreign states, debt instruments permitted for placement of the NWF:
   - for debt denominated in US dollars and euros (Table 3):
   - for debt denominated in pounds (Table 4).

The terms indicated above are valid at the time of acquisition of debt instruments at the expense of the NWF or at the time of formation of indices from debt instruments used to calculate interest accrued on cash balances on accounts of the NWF in authorised foreign currencies.

3. The list of foreign state agencies in which the NWF can be placed (as agreed with the Central Bank of the Russian Federation):
   - State Credit Agency, Spain (Instituto de Credito Oficial, ICO);
   - Motorway Financing Agency, Austria (Autobahnen-und Schnellstrassen-Finanzierungs-Aktiengesellschaft, ASFINAG);
   - Group of banks for reconstruction and development, Germany (Kreditanstalt fur Wiederaufbau Bankengruppe);
   - Canadian Agency for Export Development (Export Development Canada, EDC);
   - Municipal Bank of the Netherlands (Bank Nederlandse Gemeenten, BNG);
   - Medium-Term Railway Network Financing Society, United Kingdom (Network Rail MTN Finance CLG (Plc));
   - Agricultural rental bank, Germany (Landwirtschaftliche Rentenbank);
   - The Federal Home Loan Mortgage Corporation, USA (Federal Home Loan Mortgage Corporation, Freddie Mac);
   - Federal National Mortgage Association, USA (Federal National Mortgage Assosiation, Fannie Mae);

- Federal banks lending to housing, USA (Federal Home Loan Banks, FHLBanks);
- Federal Farm Credit Banks, USA (Federal Farm Credit Banks, FFCB);
- Municipal Credit Fund, France (Dexia Group);
- Social Security Debt Services Fund, France (Caisse d'Amortissement de la Dette Sociale, CADES);
- French Mortgage Fund (Credit Foncier de France, CFF);
- Austrian Export-Import Bank (Oesterreichische Kontrollbank Aktiengesellschaft, OKB).

4. The nominal volume of purchased debt obligations of one issue shall not exceed 5% of the nominal volume of this issue.

The NWF can only be used to co-finance voluntary pension savings of Russian citizens and to ensure the balance (deficit) of the budget of the Pension Fund of the Russian Federation. The volume of the NWF allocated for these purposes is established by the federal law on the federal budget for the next year and planning period.

According to the calculations of the authorities, 2018 should be the last year when the budget deficit will be covered from the NWF. In the future, they promise to turn it into a real welfare fund, which will be replenished with the currency purchased by the Ministry of Finance according to the budget rule, and be spent only on co-financing pension savings of Russian citizens.

6 National Welfare Fund Objectives

According to the budget code of the Russian Federation, the completion and augmentation of the means of the NWF is a triple goal:

- co-financing of pension savings of Russians;
- covering the federal budget deficit;
- ensuring a balanced budget of the Russian Pension Fund.

The most acute today are the issues of the NWF management strategy.

The law assigns the management of the NWF funds to the Ministry of Finance of the Russian Federation. He also allows the exercise of certain powers to manage the fund's assets by the Bank of Russia.

According to official figures from the Russian Ministry of Finance, in 2017 the NWF contracted to 3.753 trillion rubles, or $65.15 billion, which amounted to 14% in the national currency and 9.3% in the US currency. In general, the Fund adheres to currency diversification and has four currencies—about a third of the assets in dollars and euros, a quarter in rubles and a balance (less than 10%) in pounds sterling.

As of January 1, 2018, the Fund's money (liquid part, approximately 60% of the volume) was placed [4]:

- on separate accounts of the Central Bank of the Russian Federation—$15.65 billion;
- on deposits of Vnesheconombank—222.47 billion rubles and $6.25 billion;
- on deposits of VTB and Gazprombank, intended to finance self-sustaining investment projects approved by the Cabinet of Ministers—164.43 billion rubles;
- in the securities of Russian issuers relating to the implementation of "self-sustaining infrastructure projects"—112.63 billion rubles. and $4.11 billion;
- in preferred shares of banking organisations—278.99 billion Russian rubles.

Big questions were raised by the decision to invest the Fund in infrastructure projects. Experts say that investing in the high-speed Moscow-Kazan road, the new ring bypass road in the Moscow Region (TsKAD), as well as the reconstruction of the railroads—the Trans-Siberian Railway and the Baikal-Amur Railway—will not pay for itself. The management of the Fund definitely needs to be improved, reducing raw material dependence and the share of risky projects, since the liquid part of deposits only restrains, but does not overtake, inflation. According to the budget rule in force since 2017, the NWF forwards revenues from oil sales in excess of the price of $40USD per barrel (with an annual indexation of the cut-off price by 2%) [1].

Available funds of the NWF are placed in accounts with the Central Bank and form part of the international reserves managed by the Central Bank. According to the Central Bank's annual report for 2018, 18% of all reserves were invested in gold in Russia, 14% of assets were placed in France (presumably in French government bonds), 14% in Germany, about 10% in the USA (in US government securities Treasuries), 7.5% in Japan, 6.6% in the UK. The volume of the NWF as of September 1 is 8.2 trillion rubles, which is equivalent to $123 billion USD. The liquid part of the NWF (funds in the accounts of the Central Bank) is $98 billion USD, or 6% of the forecast GDP of 2019. The rest has already been invested mainly in infrastructure projects. It is expected that in 2020 the liquid part of the NWF will overcome the legislative threshold of 7%, which will allow the government to start investing the accumulated funds. The question of how to use the accumulated oil and gas revenues is currently being discussed in the government. One of the options—the "Norwegian model"—involves investing part of the Fund in a wider range of assets [4].

The NWF investment strategy is conservative, focused on investments in the most reliable assets, especially sovereign bonds. This scheme is significantly different from the investment policy of the sovereign fund of Norway, which actively invests in shares of foreign companies and bonds of developing countries. Application of the "Norwegian model" can increase the return on investment of the NWF. The average annual yield of the NWF in recent years has been about 1.5 compared to 5.9% per year on average since 1998 for the Norwegian fund. Further conservation of the National Wealth Fund will not ensure the stability of the Russian economy, but will only contribute to its sliding into a recession. Other possible options for using NWF funds include issuing export loans to foreign buyers of Russian non-oil products and investing in infrastructure projects in Russia.

Earlier, the International Monetary Fund recommended that Russia stop extrabudgetary investments from the NWF in the domestic economy, which in the past were aimed at capitalising banks, infrastructure projects and subsidising loans to small-sized and medium-sized businesses. The IMF also advised the Russian government to continue investing NWF funds in high-quality foreign assets—government bonds of Western countries (France, Germany, the USA, etc.) with a high credit rating. The discussion about the possibility of using the funds of the National Welfare Fund escalates as its volumes approach the level of 7% of GDP. Exceeding this level gives the government the opportunity to use savings in various investment projects. The liquid part of the NWF with a volume of 7% of GDP is only a guarantee of relatively stable fulfilment of budgetary obligations in a crisis; a fund of this size cannot be considered a sustainable tool for maintaining and transferring income from exhausted resources to future generations. Opponents of the NWF spending argue the insufficiency of this level for insurance of budgetary obligations, as well as the danger of inflationary pressures as a result of increased costs.

A strict budget rule and active accumulation of reserves were at the centre of macroeconomic policy in recent years and allowed to significantly increase the fundamental stability of the Russian economy to external shocks, return investment ratings and reduce the dependence of the Russian currency on the oil market. The transition to less active savings means a departure from the mechanisms that have been successfully operating in recent years. Details of the use of additional oil and gas revenues (the form of their use, the composition of the projects, the share of funds allocated for their financing, the mechanism for converting the Fund's currency into rubles) are still unknown. Their discussion will be at the centre of the macroeconomic agenda in 2020. Despite the fact that additional infrastructure costs can accelerate the Russian economy, we see significant risks from a modification of the approach to using SWF funds.

The threshold of 7% of GDP, above which additional options for using the NWF funds arise, is established by law, but in fact there is no fundamental reasoned justification for the adequacy of a fund of this size. As the experience of the crises of 2008–2009 and 2014 showed, a fund of approximately this size was sufficient to make up for budget revenues, which had fallen due to falling oil prices for 3 years. A similar conclusion follows from the stress testing of the federal budget prepared by the Ministry of Finance. According to the agency, while maintaining the price of Urals oil at $25USD per barrel within three years, budget shortfalls will amount to 7.6% of GDP. Thus, the NWF of 7% of GDP is only a guarantee of the stable fulfilment of budgetary obligations in the conditions of one next crisis. A fund of this size cannot be considered a sustainable tool for the conservation and transfer of income from depleted resources to future generations. If we assume that in the next 10 years the price of oil will correspond to the current trajectory of the futures curve, that is, it will remain close to $60USD per barrel, then maintaining the current savings strategy would allow us to increase the size of the National Wealth Fund to $400 billion USD by 2028 (17.6% GDP). This size of the Fund allows us to talk about maintaining a very impressive airbag even in the event of a prolonged decline in oil prices. Halving the pace of savings doubles the size of the Fund to $290 billion USD (12.5% of GDP) by 2028, while maintaining a stable oil environment [1].

Naturally, a stable oil market is far from guaranteed; therefore, the actual growth rate of the NWF may be lower. In addition to the risks of deteriorating oil conditions, there are internal risks associated with the depletion of developed reserves. In the budget forecast until 2036, the Ministry of Finance indicates a likely decrease in the effective tax rate for the oil and gas sector due to the expected increase in the share of production at new fields to which preferential tax rates are applied. In such a scenario, the choice in favour of spending rather than saving resource incomes can negate the possibility of actively increasing the size of the Fund.

The official goal of the NWF, defined by the Budget Code of the Russian Federation, is to co-finance voluntary pension savings of citizens and to ensure a balanced budget for the Pension Fund. In our opinion, in order to achieve this goal, given the country's dependence on oil exports, the model of the Fund should perform a saving, and not just a stabilising function. One of the most common approaches to managing resource income is to save it in accordance with the concept of permanent income. According to this concept, it is necessary to appeal to the total present value of the country's natural wealth and to distribute these wealth between current and future generations. That is why the State should save part of the income so that future generations can use the savings accumulated in the past when resources are depleted or structural changes occur that would not allow to benefit from ownership of these resources. Many sovereign funds in other countries (Norway, Saudi Arabia, Kuwait, etc.) follow the same concept, where their size exceeds or approaches 100% of GDP. One of the advantages of this approach is also that as the Fund grows, the absolute size of investment income increases, which can be used while keeping the body of the Fund unchanged. Thus, the use of the NWF is a necessary and even inevitable condition for maintaining economic growth. First of all, fund resources should be directed to infrastructure projects. Further, it is necessary to expand investments in the production of machinery and equipment—such investments are least dependent on inflation, since they involve the purchase of investment goods abroad.

But it is necessary to strictly define the criteria for "self-sufficiency of infrastructure projects" and use estimates of direct return to investors. Indirect evaluations of project performance, such as aggregate additional budget revenues, the creation of additional jobs, and ultimately the multiplier effect on economic growth, are also important. Today, it is obvious that we have a balance between stability and growth that is significantly biased towards stability. Tight monetary policy and extremely tight fiscal policy determine the attenuation of growth, even in favourable external conditions. Further conservation of the NWF will not ensure stability, since in itself it only contributes to the sliding of the Russian economy into a recession, within which the use of accumulated funds is supposed.

It is important to note that the current system provides immunity of funds from foreclosure or seizure by judicial claims of foreign organizations to the Russian Federation. Information on the use of funds by the Central Bank of the Russian Federation is not in the public domain, which makes it impossible to determine the real effectiveness of managing these funds. It is also necessary to increase the transparency of the management mechanism and try to avoid an indirect strategy for managing funds, which exists because financial assets with a high level of risk (debt instruments and shares of legal entities and units of investment funds) are not included in the list of assets allowed to be held with them operations by the Bank of Russia.

7 NWF Management Strategy

It seems that the objective of the NWF is not to cover the Pension Fund of the Russian Federation (PFR) budget deficit in the coming years (currently this is being done at the expense of other current federal budget revenues), but to achieve a significant amount in order to ensure the financial stability of the PFR in the future. An ideal option would be to create such a volume of NWF funds, in which in order to cover the PFR budget deficit in the future, income from managing the NWF funds would be sufficient.

Factors ensuring the safety of the NWF include:

- elimination of the use of Sovereign Wealth Funds (SWFs) without urgent need until the desired amount of SWFs is achieved. Co-financing of additional insurance contributions of citizens for the funded part of the retirement pension and covering the budget deficit of the Pension Fund should be carried out exclusively from the income from the management of the NWF;
- ensuring high reliability of investments.

Factors providing an increase in the assets of the NWF:

- transfer of part of the oil and gas revenues of the federal budget;
- maximization of income from the management of the NWF (in this case, the strategy of the NWF cannot be very conservative, since the income of the NWF should be at least comparable to inflation in the long run).

In addition, it seems appropriate to provide for the management of the NWF the possibility of investing part of the funds in reliable financial instruments in the Russian financial market, such as debt instruments and shares of legal entities, as well as shares (stakes) of investment funds. Investing the Fund in these instruments will ensure the influx of long-term investment resources into the economy of the Russian Federation (in contrast to bank deposits and bank balances with Russian banks and credit organizations, which are sources of short-term and medium-term resources). The unfolding discussion will have to answer two questions—where to invest (in Russia or abroad) and in which assets, so as not to interfere with the fulfilment of the three key tasks of the NWF, which at the same time serves as an airbag for the budget, a cure for the "Dutch disease" and foundation for future generations.

The crises of 2008 and 2014 clearly demonstrated that in order to protect the Russian budget system from falling oil prices, the NWF needs to invest exclusively in highly liquid foreign assets. The only question is whether there is enough "pillow" in this 7% of GDP. In 2013, when the budget was balanced only at a price of more than $100USD per barrel, the answer was obviously negative. However, the transition to a floating ruble exchange rate, coupled with a fiscal rule binding to the average annual oil price, made such a threshold fiscally acceptable. Despite the volatility of oil prices, the probability of their long-term decline below the cut-off price under the budget rule ($41.6USD per barrel) is not high. The main driver of the oil supply today—the American shale oil and gas sector—is experiencing financial and infrastructural constraints. The oil market is unlikely to be able to balance with a sharp, for example twofold, fall in prices. Therefore, it is not worth increasing the threshold for investing in the NWF, for example, up to 10% of GDP. It is better to think about investing the fund in assets whose interest will be higher than those of the same American treasury securities.

Domestic investment will exacerbate the "Dutch disease": rising budget spending will accelerate inflation, the Central Bank will tighten credit conditions in response, and real ruble appreciation will weaken the competitiveness of non-oil sectors. Of course, it is not worth exaggerating macroeconomic risks, but even a slight increase can erase the benefits of investing in Russia. As a compromise, the government could invest the Fund's resources in export projects, whether it be the construction of nuclear power plants abroad or the construction of toll roads. The key condition is the return on investment, coupled with the predominance of private investment over government, including the assets of the NWF.

True, the use of such an approach can enhance the raw material bias of the Russian economy, as can be seen in projects that received NWF funds under individual government decisions—before a threshold of 7% of GDP was set. In particular, it is the world's first Arctic liquefied natural gas (LNG) plant Yamal LNG and the largest petrochemical complex in Russia, Zapsibneftekhim, whose construction assistance was issued in 2015 as a purchase of bonds. To date, three of the four planned lines have been commissioned at Yamal LNG, from which gas is completely shipped abroad. Most of the products of Zapsibneftekhim, the launch of which is scheduled for the end of 2019, will also be exported.

Support for these projects in practice was an export subsidy, indicative of the ratio of private and public investments: in the structure of costs for Zapsibneftekhim ($9.5 billion USD), the share of NWF funds amounted to 19%, and for Yamal LNG ($27 billion USD)—amounted to 9%. However, despite the return on investment and the predominance of private investment, support for these projects only strengthens the raw material specialisation of the Russian economy, making the "Dutch disease" chronic [4].

Overcoming it, as well as taking care of future generations, is possible through diversification of the economy, which is easier to achieve not by supporting environmental business, but by cutting taxes. The best option is to reduce insurance premiums while compensating for the lost profits of the PFR from the NWF, which would fully comply with the mandate of the Fund, as enshrined in the Budget Code. This option is also relevant given that, according to PwC, Russia is among the ten countries with the highest share of labour taxes in the gross profit of a model company (36.4% in 2017), significantly lower than neighbouring Kazakhstan in this indicator (11.3%) [5]. Reducing this share will help primarily non-resource sectors of the economy, and therefore its diversification and avoiding oil dependence, which is one of the tasks of the NWF.

It is clear that due to the exhaustion of the NWF, a reduction in insurance premiums can be planned only for a certain period—for example, five years, during which the total transfer from the NWF will reach 2% of GDP. And by 2024, thanks to the completion of national projects, the government will be able to free up funds to make this option permanent. Moreover, it will suit both the Ministry of Finance, which will receive additional non-oil and gas revenues due to the acceleration of economic growth, and the Central Bank, since a transfer of less than 0.5% of GDP will not affect inflation in any way, and the remaining NWF assets will be invested in highly liquid foreign assets.

Finally, another decision well known to economists is to invest in stocks and bonds in proportion to their capitalisation, as the Norwegian sovereign fund does, whose income is used to pay pensions. A time-tested idea, the implementation of which would insure against the mistakes of "effective" managers, whose investments over the long-term horizon almost always lose to a simple investment in the stock index. True, this measure has limitations—in the form of a relatively small volume of SWFs, due to which the increase in pension will reach only several thousand rubles a year and a low share of Russia in the global financial market, because of which only a small part of the Fund will be invested in Russian valuable paper.

8 Benefits and Costs

Investing NWF assets in projects within the country, of course, is able to give additional acceleration to economic growth. The magnitude of the stimulus, however, does not look impressive. If the price of oil remains near $60USD per barrel, and the government saves only half of the oil and gas revenues generated when the price of oil is above the cut-off price ($40USD per barrel in 2017 and already $50USD per barrel in 2028), and the rest should be invested in infrastructure projects, then additional investments will average $12.5 billion USD per year, which is less than 1% of GDP.

At the same time, the rejection of an active savings strategy may lead to the fact that in the situation of a new crisis, the entire NWF airbag will be used up. There are more short-term negative consequences:

1. The Bank of Russia considers the possible use of NWF funds within the country as an inflationary risk, which is fraught with a slower rate of decline in the key rate. At the same time, only companies associated with this project receive the impetus for growth from financing investment projects, while the stimulating effect of lower rates extends to the economy as a whole.
2. Continuation of the active accumulation of reserves under the budget rule could lead to a further sustainable decrease in the Russian country risk premium, which would reduce the cost of both sovereign and corporate borrowing in foreign

currency. With a slowdown in the rate of accumulation, the likelihood of such a scenario decreases.

Relatively high oil prices in 2018 allowed Russia to replenish the NWF up to 4 trillion rubles. This is a huge amount—about a third of all planned revenues of the country. In 2019, the volume of the NWF promises to grow by another 3 trillion and make up 7% of GDP [4]. Significant amounts were also received by other producing powers, which, like Russia, are putting oil dollars into similar reserve funds. Unlike our country, which so far is only accumulating reserves, most of them have long outlined promising projects that will allow not only to preserve, but also to increase the acquired capital. Russia is in no hurry to take advantage of the investment experience of other producing countries, preferring to save money to cover the federal budget deficit in difficult times for the economy.

9 Conclusion

The need for reserve funds is controversial. A number of economists and politicians believe that it is more efficient not to keep money in reserve, but to use it for import purchases working for the future of the country: for example, to buy patents and equipment, to pay for students to study abroad, etc. Such tactics can avoid the negative consequences of favourable conditions, without resorting to the actual freezing of funds in the Reserve Fund.

Until 2008, there was a single Stabilization Fund in Russia, which was subsequently divided into the Reserve Fund and the National Welfare Fund, each of which had an independent goal. The Reserve Fund was designed to ensure the implementation of budget expenditures in the event of a decrease in oil and gas revenues. In other words, the Fund served as an airbag in a crisis situation. The National Welfare Fund was conceived as part of a sustainable long-term pension mechanism. In 2008–2010 and 2014–2016, the Reserve Fund was actively spending amid falling oil prices. By the end of 2017, the Fund was completely used up and since February 2018 has ceased to exist. From this moment, the savings of oil and gas revenues are accumulated in the NWF. A rather conservative fiscal rule in force since 2017 (saving oil and gas revenues from oil prices in excess of $ 40USD per barrel, followed by annual indexation of this level by 2%), made it possible to ensure a fairly rapid rate of accumulation of funds in the NWF. On 1 August 2019, the Fund exceeded 7% of GDP ($124 billion USD). It is expected that in 2020 the 7% threshold will also exceed the liquid part of the NWF (that is, funds excluding those that were allocated in 2014 to finance infrastructure projects) [1].

Russia is a country rich in gas and oil resources that it successfully exports, and part of the federal budget revenues from the sale of hydrocarbons goes to the National Welfare Fund of Russia. Sovereign wealth funds may in the near future become a leading force in the global economy. Sovereign funds are the newest leading force in the global financial market, which is replacing hedge funds and private investment funds. They also replaced central banks as the largest borrower. Thus, the goal set by the Stabilization Fund—increasing fiscal sustainability—was generally achieved. During the period of the Fund's existence, it was not used to make up for the loss of oil revenues, but it was actively used to pay off Russia's external debt.

The Reserve Fund was also used to ensure fiscal sustainability: financing the federal budget deficit during the crisis in 2009 and 2010. In fact, the Fund was used to compensate for the shortfall in non-oil and gas budget revenues, as well as to finance additional costs to support the economy. The crisis showed that the negative impact of falling oil prices is not limited to a decrease in oil and gas budget revenues. Indirect effects of falling prices and foreign economic instability cause an economic recession, which, in turn, causes a fall in non-oil and gas budget revenues. That is why, due to the Reserve Fund, it was necessary to compensate for the shortfall of non-oil and gas revenues.

During the decline in oil prices and the crisis, the National Wealth Fund has become one of the main sources of ensuring quasi-fiscal measures to support the financial system. Thus, the NWF was actually used to maintain macroeconomic stability and served as an additional reserve for the federal budget. In general, it can be argued that Russia's budget reserves are really necessary, since the degree of dependence of the Russian economy on the external economic situation is still high.

References

1. Official website of the Ministry of Finance of the Russian Federation. Available at: http://www.minfin.ru
2. National Welfare Fund. Available at: https://web.archive.org/web/20081221234448/http:/www1.minfin.ru/ru/nationalwealthfund/
3. Reserve Fund. Available at: http://minfn.ru/ru/perfomance/reservefund/statistics/volume/
4. National Welfare Fund. Statistics. Available at: http://minfin.ru/ru/perfomance/nationalwealthfund/statistics/
5. PwC outlook : Sovereign Investors 2020. A growing force, 2016. Available at: http://www.pwc.com/sovereignwealthfunds

Alaska's Petroleum Industry, Institutions and Sovereign Wealth Fund

Douglas B. Reynolds

I hold it true, whate'er befall; I feel it, when I sorrow most;
'Tis better to have loved and lost; than never to have loved at all.
—Lord Tennyson *In Memoriam*: 27
It is better to have loaned and lost; than never to have loaned at all
—*The Economist*
It is better to have explored and produced oil and lost;
than never to have explored and produced oil at all.
—Logic

Abstract Alaska is a unique petroleum-producing region in that it is a state within the USA and therefore both an independent oil-producing and gas-producing region with an independent State government as well as a region dependant on, and intertwined with, a more powerful developed U.S. economy and its U.S. Federal Government. As such, it is possible to better observe political characteristics in the formation of petroleum policies and how they relate to the larger governmental institutions since some policies are based on State government policies and therefore the State's citizens' desires. Moreover, some policies are driven by the more powerful U.S. Federal Government policies and its American citizens' desires. The chapter looks at Alaska's oil and gas fiscal system, the State's petroleum industry, its petroleum institutions and its sovereign wealth fund and how all those items work together within the broader government functioning.

D. B. Reynolds (✉)
Petroleum and Energy Economics, Department of Economics, School of Management, University of Alaska Fairbanks, Fairbanks, USA
e-mail: dbreynolds@alaska.edu

E. G. Pereira et al. (eds.), *Sovereign Wealth Funds, Local Content Policies and CSR*, CSR, Sustainability, Ethics & Governance,
https://doi.org/10.1007/978-3-030-56092-8_9

1 Introduction

Every petroleum region in the world from Kazakhstan to Saudi Arabia to Venezuela to Alaska has its own petroleum industry with its own set of institutions for petroleum extraction and, more often than not, its own sovereign wealth fund to help save money for the ups and downs of the petroleum industry. One way in which these institutions work is to define which property rights the industry enjoys and which responsibilities the government undertakes in order to maximize the benefits the oil province citizens receive from the industry. The institutions include how oil taxes and other fiscal systems are setup, how the industry is regulated, and how the government and industry try to influence each other. The purpose of such institutions is for the oil province to gain as much value as possible from its petroleum resources, including reducing pollution, developing the local economy and providing social welfare to its citizens.

Theoretically though, these institutions can vary from complete government ownership and control to having totally free markets. When government control is complete, the producing companies and mineral rights are government run. Under totally free markets, producing companies and mineral rights are privately run (other than a court system to adjudicate property rights). For example, a country can theoretically benefit from a resource industry by simply allowing the value of the petroleum resources to be earned directly by privately owned, international oil companies (IOCs) or smaller independent oil companies and their workers, who then indirectly, through the so-called trickle down effect of secondary spending, help the local economy; a small tax on infrastructure might be imposed.

Alternatively, by having the government own, control and produce all the petroleum with its own State-owned company then, theoretically, all of the oil profits can be captured and used to provide roads, schools, subsidies and other social welfare to the general population, and where a high level of government services can enhance other businesses either ancillary to the petroleum industry or alternative to the petroleum industry within the country or region. The downside of total government control though is that there is normally a lack of incentives to find and produce more oil or to reduce the costs of production so that government revenues are enhanced. Totally free markets tend to enhance income inequality. Therefore, most provinces have a mixed system with at least some government control and taxation and at least some private industry and free-market competition. Alaska is no different.

The way in which Alaska's institutions work for petroleum extraction, also called crude oil and natural gas production, is that the State of Alaska or the U.S. Federal Government owns all underground mineral rights within the region of Alaska, but where the mineral rights are leased out to major IOCs or minor independent oil and gas companies who in turn explore for, develop and produce the oil and gas resources for a profit. The State and Federal governments, then, tax those profits using a complex set of taxes and royalties. Each mineral lease is for a specific underground area and for a specific mineral, such as oil and gas, and where the lease lasts as long as the company that owns the lease is able to develop and produce that particular mineral resource. Therefore, Alaska has an institutional set up that is somewhere between a wholly government directed and controlled system and a totally free market, privately owned system, but which is closer to the free-market side of the spectrum.

Another aspect of oil and natural resource production is that most oil-producing countries have what Ross (2013) calls *The Oil Curse*, which is how a regional economy can be adversely affected by having a dominant petroleum industry and where the petroleum-producing government is a rentiér state, i.e. where the government gets most if not all of its revenues from the rent of petroleum mineral rights and where the country often only has that one single petroleum industry to carry its economy. Indeed, many oil producers similar to Alaska are unable to develop a large robust set of alternative industries inside their regions. One interesting aspect of the Oil Curse is where a local economy bubbles up high during periods of high oil prices or high oil production levels and then collapses during periods of low oil prices or low oil production levels and where such cyclical ups and downs are severe and create social distress.

The way to mitigate such economic cycles of commodity price, or production level, increases and decreases is to have a sovereign wealth fund, i.e. a savings account, which the government can tap into during low cycle busts and build up during high cycle booms. Running such a fund appropriately can act like a modern, Keynesian economic fiscal or monetary policy, which is designed to reduce recessions during low cycle troughs and inhibit booms during high cycle expansions. However, most governments, and Alaska in particular, are ineffective in being able to run such spending and taxing policies effectively because it is so hard to forecast those ups and downs and therefore there is political resistance to operating any such countercyclical policies properly. Indeed, since it is the Alaska State Legislature that controls the State's Sovereign Wealth Fund, and not a set of independent economists similar to how the U.S. Federal Reserve runs, then it is no wonder that the Oil Curse occurs in Alaska.

Interestingly, the Oil Curse cycles and the institutions of petroleum-producing regions often work in tandem with each other to enhance the curse rather than diminish it. According to new institutional economic theory, it is even possible that oil prices themselves can affect how institutions change. For example, Acemoglu et al. (2001) suggest that institutions are endogenous and as much caused by general market factors as they are able to shape those markets. However, Alaska is complex. For example, Alaska is driven by U.S. Federal Government institutions and traditions that are distinctly American, such as using free markets, having less government control of its mineral industry and only using privately owned companies to produce oil. In Alaska, those traditions and institutions are not easily changed. Furthermore, the State of Alaska depends on U.S. Federal Government direction and the U.S. economy in a way that most oil-producing countries cannot, such as U.S. provided health care, U.S. welfare programs and the U.S. constitution and court system. Nevertheless, Alaska has managed to tweak its institutions in a uniquely alternative way as a result of changes in oil market factors, particularly the price and production level of oil. Unfortunately, Alaska's government policies are shaped by oil market forces, more so than being able to effectively counteract them. That is, Alaska does show signs of the Oil Curse characteristics of a boom and bust economy even though there is a larger U.S. Federal Government to backstop the Alaskan economy, and even though Alaska does have policies to help stabilize its economy.

In this chapter, we will look at the fiscal system of Alaska, i.e. its oil and gas taxes, which have changed over the years. We will consider Alaska's version of the Oil Curse and its Permanent Fund, which is Alaska's sovereign wealth fund, and how that fund has interacted with the Alaska economy. And we will look at the way in which Alaska has been able to assuage its own Oil Curse by being able to leverage the wider U.S. economy to help diversify its own economy. Note, in American English the word "petroleum" is separated into the terms "oil", which includes hydrocarbons from ethane (C2) and upwards to as high as (C55) in carbon content and "gas", which includes natural gas based on methane (C1) and where the low-weight liquids like propane (C3), butane (C4) and sometimes pentane (C5) are often called liquid petroleum gas (LPG). These distinctions are important for American tax law practices when considering (petroleum) oil and gas extraction, and they are important in Alaska where vast reserves of northern natural gas have yet to be developed due to a lack of a northern natural gas pipeline.

2 Alaska's Petroleum Fiscal System

A fiscal system of a petroleum government can be taxes or other forms of government revenue from its oil and gas industry, which is often called the "Government Take" by the industry. On both U.S. Federal Government land (which includes 220 million acres [90 million hectares] within the State of Alaska or about 60% of the State) and State of Alaska land (which includes a little more than 100 million acres [40 million hectares] or about 30% of the State), oil production by producing companies, often called "producers", is subject to a royalty which is defined as the presiding government usually recovering 12.5%, or one-eighth, of all oil and gas produced irrespective of costs or quantities. The royalty is a guaranteed income for the government owner. On top of the royalty, both the U.S. Federal Government and the State of Alaska have corporate income taxes that take a per cent of all oil company's net positive income. Interestingly, though, for Alaska, if the oil company is based outside of the state, then the income tax is based on a prorated income according to how much of that income was derived from within the State of Alaska.

Crucially, the State of Alaska has one more tax in addition to royalties and the income tax, and that the U.S. Federal Government often does not have, which is a production tax (also known as a severance tax or a profits tax) and which is based on the actual amount of oil produced and other factors. It is this production tax that over the years has changed due to either concerns for government revenues or concerns for new oil and gas exploration and development investments. Often oil-producing and gas-producing governments around the world raise government taxes or other forms of "government take" when oil prices rise and they lower the "government take" when oil prices decline; the "lower tax during lower oil price eras" strategy is meant to lead to added oil production and therefore added oil and gas tax revenue later. These kinds of cyclical petroleum government fiscal systems are what Reynolds (2016) refers to as the Wal-Mart (low oil price) and the Neiman Marcus (high oil price) strategies, and Alaska is no exception. What this means is that government taxes or other forms of "government take" change in reaction to the price of oil, but more to the point, the government institutions change, even if only slightly, parallel and along with the tax changes. High oil prices induce a higher "government take" and often a change in institutions so that the government has more control over the petroleum industry, whereas low oil prices induce a lower government take and typically less control over the petroleum industry. For example, in the 1970s and early 1980s, Alaska was raising its production taxes and some regulations when the price of oil and also Alaska's oil production was rising in order to increase government revenues, but at other times when oil production or oil prices were declining, Alaska actually lowered taxes and even some regulation in order to induce more investment so that production could increase.

However, on top of those tax changes, the State has one more complicating factor which is the tariffs on the Trans-Alaska Pipeline System (TAPS), which includes the 800-mile-long oil pipeline (but not natural gas) that delivers the North Slope oil to the southern ice-free port of Valdez, Alaska. TAPS also includes the oil-processing facilities on the North Slope and the oil tanker port on the Southern Coast in Valdez. When TAPS tariffs are high, the wellhead value of oil is low and consequently the State's government tax revenues are low. Therefore, a TAPS tariff law suit was started in 1977, just as oil prices were rising, and lasted until an out-of-court settlement was reached in 1985. The settlement occurred as oil prices were heading down from $38USD per barrel to $26USD per barrel and soon to plunge to $11USD. The settlement then was a State compromise towards the petroleum producers' favour, ostensibly to try to induce more oil production, and to this day some Alaskan oil observers feel the settlement went too much towards the producer's favour. But the declining oil prices may have induced a quicker settlement.

One interesting aspect of the early Alaska tax system was the Economic Limit Factor (ELF), which was actually an ingenious method to help account for the costs of oil production. One of the problems in oil tax accounting is to be able to declare that a specific cost is a legitimate cost for one or another oil field or alternatively a cost that is associated with exploration for new oil and gas resources or a cost associated with other industrial endeavours, such as downstream development in oil refineries. Therefore, in order to simplify the accounting and force producers to reduce costs and keep taxable profits higher, the State used ELF to tie a simple cost per oil well to its tax system. The State tweaked the ELF system to the benefit of the State in order to raise taxes in 1989 when the price of oil was just starting to gyrate up again, although some see the Exxon Valdez incident, where a large oil tanker accident occurred on Alaska's coast line near Valdez that released about one quarter of a million barrels of oil off the coast of Alaska into coastal fishing waters, as a call to push the oil producers to be more responsible and, in addition, to pay their fair share of taxes. At another low point in oil prices in 2002 and 2003, the State of Alaska gave additional tax credits that were in essence monetary incentives to explore for and produce more oil.

Eventually, the price of oil went up again starting in 2004, and so in 2005, a new rule was put into place to aggregate exploration and production into higher tax categories within the ELF system in order for the State to increase its revenues, although the actual State's oil and gas tax rates remained unchanged. However, in 2006, with ever-higher oil prices, and with costs of oil wells declining over the years due to better technology, the ELF taxation method allowed too little taxes to the State and it had to be changed. Therefore, a mechanism of a profits tax whereby all of the costs and revenues were explicitly measured and taxed was passed in addition to the normal corporate profits tax. So, the old tax system that gave high deductions to the detriment of tax revenues, including the old ELF tax cost deductions, was changed to a new petroleum production tax that was put in place, although it was changed again in 2007 as oil prices rose even more to give even higher tax revenues to the State, although the 2007 change did give some tax credits to help spur production.

In 2013, with oil production in Alaska falling, a new tax rate and tax deduction regime called SB 21 (for Senate Bill #21), and also called MAPA (for More Alaskan Production Act), was put in place to help spur development, in particular development by smaller independent oil producers rather than the major IOCs. One impetus that drove this change was that oil in Alaska has to go through the 800-mile TAPS pipeline, which is in essence a monopoly pipeline and where the two major oil producers on the North Slope own a majority of TAPS and where smaller new producers in Alaska's competitive market are forced to pay high fees and the high tariffs in order to use TAPS. The end result is the high TAPS fees and tariffs often caused independents to sell out and transfer ownership of their new found oil fields to the IOCs, which then gives greater control of the North Slope oil to the two major IOCs, which at that time were BP and Conoco Philips. MAPA was meant to induce new independent producers to start searching for oil more aggressively than what the majors had done.

Most of the pipeline is above ground due to permafrost and seismic issues, and therefore, if the oil throughput within the Trans-Alaska Pipeline were to become too low, the pace of pumping the oil would slow down enough that the oil inside the pipe could freeze-up and gel, especially due to the very cold ambient temperature surrounding the pipeline. Such a freeze could catastrophically reduce Alaskan oil production to near zero for months and cause irreparable damage to TAPS as well as the North Slope oil fields, which is why Alaska wanted to incentivize more oil to go through TAPS.

While some believe the new SB 21 tax law was meant to help increase TAPS throughput, others thought it to be a giveaway by Alaska to the producers and so a vote on the 2013 tax by Alaskan residents occurred in November of 2014. The vote occurred right when oil prices were just in decline, which may have scared people enough to vote for keeping the lower tax in place in order to get more oil flowing, although shrewd marketing may also have changed the vote. What this shows is that the State has typically reacted to oil prices and production levels with either a low oil price Wal-Mart strategy during petroleum downturns (less "government take" during eras of low oil prices in order to increase production and therefore future revenue streams) or a high oil price Neman Marcus strategy during petroleum upturns (more "government take" during eras of high prices in order to increase current revenue streams). The end result is Alaska's oil-based economy is highly volatile.

3 The Oil Curse and Government Spending

According to Ross (2012), petroleum-dominant economies have neither above average long-run growth rates nor below average long-run growth rates, even though their economic growth per person should be higher than average owing to the early intense resource extraction and economic stimulus when there is an early oil boom. Nevertheless, there is still the possibility for an Oil Curse due to the volatile nature of a petroleum-based economy. For example, when a region first discovers oil there is an early boom in the petroleum sector that can easily crowd out other business sectors such that when the petroleum industry finally goes into decline, those other business sectors are not available to take the slack. Such an economy is characterized by a much more severe business cycle where the economy cycles acutely high during periods of high oil prices and high production levels and cycles uncommonly low during periods of low oil prices or low production levels and Alaska is no exception. In general, those oil-based business cycles can cause extreme volatility within an economy and wreak havoc on normal long-run business and governmental planning, and oil cycles can be much more severe than the average developed economy economic cycle.

In Alaska's case, when oil production and oil prices are high, the democratically elected State government is inundated with revenue, so much so, that its legislators are politically pressured by public opinion to do three things: (1) They are pressured to spend money on infrastructure and government programs whether such spending is needed or not, (2) they are pressured to reduce Alaska's separate State corporate income tax, State individual income tax and other State taxes to nothing in order to further stimulate the local economy, where the tax reduction is outside of the petroleum sector and is intended to stimulate non-oil related industries but which ends up being a government giveaway to citizens and therefore causes a further stimulus to an already booming economy, and (3) they are pressured to use the earnings on the State's savings account, its sovereign wealth fund (Alaska's Permanent Fund), to give money directly to citizens, i.e. the fund gives money directly to Alaskan citizens, which is meant to subsidize the high cost of living in Alaska.

For example, in 1980, when oil prices were high and rising, the Alaska State legislature voted to rescind the individual income tax as there were plenty of oil revenues at that time. Then in 1986, Alaska had a major recession owing to the collapse of oil prices, but it was thought that during such a recession, that that was not the time to increase taxes since keeping such taxes low would help spur new businesses. Such programs as the Alaska ferry system, the University of Alaska system and a project to create a huge hydroelectric dam expanded during the oil boom, but during the oil recession in 1986, those programs received less State money or were cut entirely.

Later in the early 2000s, State revenues went up again and programs expanded again. Then when revenues declined in 2015, Alaska cut government spending again and many agencies were cut, university funding was cut, funding for infrastructure was cut, and funding for schools in expensive per student rural districts was cut. However, during the government contraction, no new non-oil taxes were implemented, and no new Statewide taxes, such as a State personal income tax, were implemented, the idea being that keeping taxes low would induce new businesses to come to Alaska. This exasperated the economic decline.

Still, in order to understand better how the government was ineffective in reducing the Oil Curse in terms of the volatile oil cycles, consider what would have happened if Alaska were to have tried to develop alternative industries simultaneously during the first oil boom. During those days of the early oil boom, if Alaska had tried to help develop alternative industries to oil, it would have taken a lot of government support to do that, just when there was an oil boom, an impossible practicality. During the oil boom, the government already had its hands full with taking care of the existing economic expansion, with regulating the new petroleum industry, with revenue collection from that industry, and with setting up health and welfare safety nets and other programs for the many new people who arrived daily during the boom. There was no time, no ability, nor available government employees to focus simultaneously on the chaotic oil industry let alone on developing an alternative industry.

More problematic during the oil boom years were the costs for labour, housing and transportation services, which were very high due to the influx of many people moving into the distant remote U.S. State. So, even if the State government had wanted to try to develop additional industries on top of the oil industry, it would have meant that those alternative industries would experience very high costs of labour, capital and transportation, and therefore, those industries would not have been able to compete against similar companies located in the Lower 48 states. It is a comparative advantage question. Alaska has a comparative advantage in oil and California has a comparative advantage in the high tech industry while Michigan and Ohio have comparative advantages in manufacturing that Alaska could not compete with. So, ideally, the strategy should be to develop the oil when the oil industry is booming and then wait until there is an oil bust to start developing alternative industries as they would be more cost competitive during an era with an oil bust. Unfortunately, during an oil bust, there is no money to help develop alternative industries as the State has a deficit maintaining existing services.

This raises another aspect of the Oil Curse, which is that during the oil industry's decline in 2015 to 2019 when State revenue declined, suddenly it was thought that the size of government was too big and inefficient. Many Alaskan citizens wanted to cut government spending and did not want to impose any new general taxes, again, because it was thought that new general taxes would discourage developing new alternative industries, just when the oil industry was in decline. So not only did the oil industry rise during the oil booms and decline during oil busts but so did State government spending rise and fall in tandem, exacerbating the boom-bust cycles. Rather than the State government trying to use a Keynesian strategy of smoothing the cycles, the State strategy resulted in amplifying the cycles. However, there was always one possibility of at least some mitigation of the Oil Curse in Alaska and that was through developing a sovereign wealth fund, which Alaska did manage to do.

4 The Sovereign Wealth Fund

Governments worldwide want a smooth, reliable, steady source of government revenue from which the government can plan its programs of education, development and other services. Likewise, a government wants smooth and reliable personal or corporate tax rates so that industry and labour can make long term plans for investment and business. In addition, if a government has extractive industries where the price of the commodity of that natural resource can vary substantially, then the government normally wants to give that industry a reliable fiscal system, or system of "government take", that the extraction industry can plan around to do its investment with. In this way, governments following the macroeconomic prescriptions of John Maynard Keynes try to smooth variances in an economy by actively intervening to mitigate booms and reduce busts such that the economy can be made smoother in order to have less volatility in the lives of residents. Although since Alaska has the prerogative to depend on U.S. Federal Government programs for individual welfare help, the State can therefore afford more readily not to embrace Keynesian economic principles.

Still, Alaska's economy has been at times very volatile owing to a lack of State government intervention. Since State government revenues from extractive industries can vary substantially, and since government programs and a desire to stabilize the economy means that government spending should theoretically vary counter-cyclically to petroleum booms and busts (i.e. the government should spend more during a recession and spend less during a boom), then it follows that a government needs a sovereign wealth fund. Such a fund can be used to fill in the gaps between spending and revenues whereby government programs can remain steady in the face of declines in petroleum extraction taxes and revenues. This would further allow non-petroleum companies' taxes or individual taxes to remain steady during booms and busts and to allow government services to remain steady during booms and busts or even allow for the government to spend money counter-cyclically compared to the petroleum industry. This is where Alaska's own sovereign wealth fund comes into play.

Over the years, the State of Alaska has reaped huge benefits from its oil industry. However, due to the concern that oil prices could decline or that Alaska's oil resources could run out, causing the State government to encounter an Oil Curse where the State would be left with no available industry from which to obtain tax revenue from in order to continue to run State government with, and where the U.S. Federal Government does not fund local and State government functions, then the State needed a sovereign wealth fund. The State of Alaska established a savings account, which is its own sovereign wealth fund, affectionately called the Permanent Fund

(PF) in 1976. The fund was set up to use only a small portion of the government's petroleum take, typically 10% of the total State government's petroleum revenue from oil production taxes and royalties based on a set formula, to put into the fund. The Permanent Fund has since become a separate entity within the State government with a separate board of directors and a separate functioning.

The Alaska Sovereign Wealth Fund, the Permanent Fund, is coded in Alaska's Constitution and cannot be withdrawn, and so it is, therefore, "permanent". However, the earnings from the fund can be withdrawn, and usually, some of the earnings are returned back to the Permanent Fund for what is called inflation proofing, i.e. to keep increasing the size of the fund in order to overcome inflation. Note, also a U.S. Federal Government judge could conceivably if the State of Alaska were to default on any of its bond or legal obligations, use federal law to take over the fund even if Alaska's Constitution does not allow for that. U.S. Federal debts or obligations may trump any State laws or State constitutions should the Alaskan government debt be deemed to be a federal obligation.

The interesting thing about the Permanent Fund, though, is that theoretically the State of Alaska can invest in alternative economic growth-inducing industries such as the high tech industry, manufacturing, real estate, and transportation. But instead of developing those industries inside of Alaska, Alaska through its PF can develop those industries outside of the State by investing in them in other State's or regions. Thus, Alaska gets to keep the earnings from the non-petroleum industries but where those industries can develop outside of the State wherever it is best for them to develop. It is as if Alaska developed new alternative industries for itself where Alaska gets to keep the earnings from those industries without the hassle of forcing them to be developed physically inside the State. This means Alaska did not have to restructure its own State agencies, education system or its infrastructure in order to attract those industries that may or may not have come to the State. Rather, Alaska created alternative industrial diversification for itself indirectly, reduced the boom and bust cycles of the petroleum industry indirectly and created a balanced State funding source indirectly, all through the use of the Permanent Fund. This is the power that a sovereign wealth fund can give.

Plus, the Permanent Fund has over the years increased in value, not so much because of the oil royalties funding it, but rather because the money was invested in stocks, bonds and real estate that have all increased in value over the years. Its value has increased so much so that it is now the portfolio side of the fund, rather than the government's per cent of oil income side, which determines the fund's size and growth. The Alaska oil tax and royalty inputs into the fund are now a small drop in the bucket compared to the portfolio earnings of the fund.

5 The People's Petroleum Dividend: The PFD

The original purpose of the Permanent Fund was to have a rainy day savings account for the State's government in order to effectively carry out a Keynesian economic policy. However, it took a sizeable political effort just to create the fund. This raises another peculiar circumstance of Alaska, which is that it has a very small population and a relatively transparent political democracy where the people wanted the oil money not just for schools, roads and infrastructure but also to reduce State and local taxes and to subsidize energy costs. Once the Permanent Fund existed, though, many thought it would too easily be spent and be drained to nothing. Therefore, in order to have received enough political support and political will to keep the Permanent Fund growing, and to have the fund permanently inscribed as part of the State's Constitution, it was decided that about one-third of the earnings of the fund per year, based on a five-year averaging formula, would be given directly to every man, woman and child in the State in the form of what every Alaskan euphemistically calls the PFD (Permanent Fund Dividend) cheque, which is a cash give-a-way to its citizens. That is, where most petroleum-producing governments give substantive subsidies for fuel, education, roads, food and housing, Alaska simply gives money directly to its citizens and only subsidizes a few things like education, roads and some rural electric power among other things. Although most Alaskans consider the PFD a return to the people for the people's oil, where most of the gas has yet to be developed, it was really meant as a savings account for the government. Interestingly though, the PFD can act as an incentive for having larger families.

The PFD is a cheque to every man, woman and child in Alaska who is a State resident by legal definition. The PFD is designed to give Alaskans a stake in the oil industry, but it is in fact completely divorced from that industry as 95% or more of all the earnings are based on stock, bond and real estate value growth in the Lower 48 and the world and not on oil prices or oil production in Alaska. This is because the fund is so big now, the fund's internal growth dwarfs any miniscule amount of oil royalty additions it receives nowadays. So while most Alaskans think of the fund as their stake in the oil industry, it is in fact free money from the government's stock and bond Permanent Fund savings account.

Still, the earnings from the Permanent Fund can be used for funding State government services in case the rest of the oil taxes decline in value such that the State could no longer provide all of its obligations to build and maintain roads or create industry or subsidize schooling. Such an emergency happened during the oil price collapse of 2015 to 2019, as well as in 1986, where suddenly the State's revenues and other non-Permanent Fund alternative savings accounts declined even as the budget could not be cut very easily. This resulted in a State fiscal crisis of deficit spending requiring the use of the Permanent Fund's earnings for State operations and thus a PFD reduction to every citizen. Such reductions in PFD's are not happily accepted by the public though.

The interesting thing about the State's recessions was that by reducing the PFD cheque to each Alaskan, the State was being funded by a highly regressive effective personal income tax rate. Normally, if a government needs money, it has a progressive tax that in percentage terms, takes relatively more money from the rich and less from the poor, but by reducing the PFD, Alaska's State government became a highly regressive revenue collector. The poor in Alaska paid more to the government, in terms of reduced PFD cheques, than did rich Alaskans.

Consequently, what ends up happening with the Permanent Fund is that during periods of natural resource industry growth, there is pressure on the state government to allow earnings on the Fund to be given directly to the people since the government does not need the earnings and everyone believes they should have a "piece of the natural resource action". Then, when resource revenues decline, there is also a general recession with the State where both the resource industry and government programs have to be cut. However, in that case, it only makes sense to give people money during a recession and so again there is a policy to give each resident the PFD. But then the Permanent Fund is not big enough to help both the State's finances and give money to the people at the same time. Plus, it is not allowed to spend the principle of the fund so only the earnings of the PF were used to fund State government, i.e. the PFDs were cut although State services were also cut. So, the end result is that the State's government programs rise during the resource booms and are cut during the resource declines, and in addition, the PFD cheques end up rising during booms and being cut during recessions. Thus, the State exacerbates the rises and falls of the State's economy, completely counter to Keynesian economics.

Also, the U.S. Federal Government does still give some welfare help to Alaska. However, even with U.S. Federal Government spending as a backup for State government services, nevertheless, the State was left with no available source to obtain tax revenues to run the State government with and it had to cut both PFDs and State services.

6 The Lower 48 Relationship

One interesting aspect of the State of Alaska is that it is under the U.S. Federal Government, so when Alaska had an oil industry-induced recession in the 2015 to 2019 period, and in 1986, where both State government and private industry were forced to cut back, many of the private industry and government workers left Alaska and moved to the Lower 48 for jobs in other States in other government employment or in other industries. Therefore, one of the ways that Alaska takes care of its own Oil Curse of oil booms and busts is to have an overarching U.S. economy where unemployed Alaskan workers can quickly move outside of Alaska but inside of the USA in order to find other jobs in other states. That is, the labour mobility within the U.S. as a whole keeps Alaska from having a full-blown Oil Curse that other countries might encounter.

In fact, the U.S. Federal Government also provides social services to Alaskan citizens that include subsidized health care, student loans, food and other federal safety net programs. Accordingly, even if the State of Alaska falters in its responsibility to take care of Alaskans while it reduces its State spending, the U.S. Federal Government is still there to provide help, and Alaska's representation in the U.S. Federal Government is quite high owing to it having two U.S. senators for a State of only 700,000 residents as compared to the State of California, which has 40 million residents and has only two U.S. Senators in Washington D.C. However, California does have a large number of U.S. congressional house representatives.

Another interesting aspect of Alaska is that whenever a State worker or State legislator or even an industry worker is in some way tempted to accept bribes, launder money or embezzle funds, the U.S. Federal Government institutions are there to find out about it and prosecute such activity as opposed to State institutions, which are often too small or have too little funding or are too politically entrenched to be able to do such an investigation. For example, in 2006 when the State legislature was trying to determine an appropriate petroleum tax cut for a potential natural gas industry, which included a large natural gas pipeline, a prominent business person in Alaska, Bill Allen who headed VECO, an oil field service company, was convicted of bribing legislators to have them pass the tax break. In that case, it was the U.S. Federal Government not Alaska's State government that discovered and prosecuted the misdeed; otherwise, Alaska would have been under the thumb of corporate interests. Hence, by Alaska being under U.S. Federal institutions, this is another way in which Alaska is able to ward off the Oil Curse. However, it cannot totally free itself from the Oil Curse's grip due to the pro-cyclical government spending.

Another aspect of the Oil Curse is the way local currencies can fluctuate. While Alaska's currency is stable due to the U.S. Federal Government's control of the dollar through the Federal Reserve, nevertheless, Alaska's state fiscal policy rises and falls as the State's petroleum industry rises and falls. When the petroleum industry goes up and down, even though the currency is stable, nevertheless, the State's government spending is still volatile and the State's economy goes up and down. Still, the one saving grace for Alaskan workers is that they can move down to the Lower 48 U.S. States and find jobs there. Plus, Alaskan citizens can still receive U.S. government health and welfare benefits including jobless benefits.

One last help for Alaska's volatile petroleum industry though, is that there is the mining industry of non-petroleum minerals such as gold, copper or coal to help create jobs. While the mining industry has been developed to a degree, it is not mature enough as an industry to effectively create a counter-cyclical alternative non-oil industry. Therefore, whenever the oil industry has gone down, the entire State has had all of its human capital, all of its State agencies and all of its infrastructure, which was set up for oil and not for new major mining operations or for new high tech industries or for manufacturing, decline. Hence, when the oil industry went down in 2015 the first and only strategy by the State government, and for that matter, the only policy advocated by public demand, was to build a large natural gas pipeline in order to develop a natural gas industry that would mimic an oil industry. In other words, even when the hydrocarbon industry goes down, the only strategy put forth by politicians and embraced by the general public is to get the petroleum industry to come back up again by using an alternative (natural gas) hydrocarbon industry to create jobs, State government revenue and economic growth. Therefore, even when a region such as Alaska is in the worst of the Oil Curse, it has very little opportunity to escape the curse, even if it wants to.

7 Conclusion

The State of Alaska is a major oil-producing and gas-producing state with a large developed economy, the USA, backing it. This situation has caused Alaska to have a certain set of institutions in its petroleum industry, such as no Alaska State-owned and operated petroleum company. Therefore, Alaska has tended to be a mostly free-market system rather than a government-controlled system. Alaska did manage to create a sovereign wealth fund to at least create some alternative industrial development (outside of the state) and which could be used as a backup source of funding for its government.

Alaska's democratically elected government has tended to increase the Oil Curse rather than to mitigate it by, for example, having more spending during oil booms and decreases in government spending during oil busts, creating a larger boom and bust cycle for its economy. When oil prices and production are rising, there is a political push to increase government spending. Not only were government services and infrastructure investment increased, but the Alaskan sovereign wealth fund earnings were given away directly to the people. Once all this stimulus is going full blast, then suddenly oil prices and oil production can decline, but (1) all the government programs and infrastructure still need to be maintained, and continue to require budgeting, (2) the legislature does not want to raise any taxes as it would cause a further cost to new businesses and industry just when the State economy needs to induce those industries, and (3) the State cannot easily retract the citizen give-away programs just when citizens need extra money during a recession without tremendous political turmoil.

The end result is that the State of Alaska's spending and government fiscal system is highly pro-cyclical in nature and in conjunction with the price of oil and to some degree the State's rate of production of oil. Thus, Alaska has the Oil Curse in spades, although as stated at the beginning of the chapter, it is far better to have produced oil and suffered an economic loss than to never have produced oil at all. The only saving grace for Alaska is the wider U.S. Federal Government and the other States where workers can flee to and where the U.S. Federal Government provides some backup welfare to Alaskans and can effectively police and prosecute corruption inside of Alaska. Thus, it is the U.S. Federal Government that looks into illegal activities by public or private employees so that Alaska's corruption is kept to a minimum and where the currency is held steady. Therefore, even though Alaska exhibits the Oil Curse, it is in some way muted by the U.S. Federal Government. And finally, when there is a recession, many workers can migrate to other states for better opportunities.

References

Acemoglu, D., Johnson, S., & Robinson, J. (2001). The Colonial Origins of Comparative Development: An Empirical Investigation. *The American Economic Review, 95*(5), 1369–1401.

Thomas, C. S., Savatgy, l., Nakazawa, A. T., & Klimovich, K. (Eds.) (2016). *Alaska politics and public policy: The dynamics of beliefs, institutions, personalities and power*. University of Chicago Press and University of Alaska Press.

Ross, M. L. (2013). *The oil curse: How petroleum wealth shapes the development of nations*. Princeton University Press.

Reynolds, D. B. (2016). *Cold war energy: The rise and fall of the Soviet Union*. Alaska Chena.

Non-renewable Natural Resource Wealth Management and Distribution in Canada: National, British Columbia, Northwest Territories, Quebec

Andrew Bauer and Sarah Daitch

Abstract Canada possesses some of the world's largest and most valuable non-renewable resource deposits. As of 2017, Canada was the world's fourth largest natural gas producer, fifth largest crude oil producer and a significant producer of gold, copper, coal, potash and iron ore. Management of these resources is complex, but largely resides with subnational governments, whether provinces or territories. As such, fiscal regimes, environmental and social regulations, distribution of resource revenues, and management of resource revenues vary across the country. This chapter summarizes the management of oil, gas and mining resources and revenues in Canada, with a focus on subnational sovereign wealth fund governance. Among our conclusions, we highlight low average effective tax rates by global standards, limits to the benefits that can be captured by the territories and a tendency towards discretionary use of sovereign wealth funds. We present three subnational sovereign wealth fund case studies from British Columbia, the Northwest Territories and Quebec, since these funds are explicitly meant to be financed in part by resource revenues. The Alberta Heritage Savings Trust Fund is covered in a separate chapter, and the Manitoba Stabilization Fund is not explicitly financed by natural resource revenues nor resides in a resource-dependent province. While some funds, namely the Quebec fund, have incorporated many good global practices in sovereign wealth fund management, the case studies underscore the need for withdrawal rules that help governments smooth fiscal expenditures and promote intergenerational equity, along with a need for greater public oversight.

Keywords Canada · Fiscal terms · Sovereign wealth fund · British Columbia · Northwest Territories · Quebec · Natural resource · Revenue management · Oil · Mining

---

A. Bauer (✉)
Public Finance Consultant, Montreal, Canada
e-mail: drewbauer@gmail.com

S. Daitch
Daitch & Associates, Montreal, Canada
e-mail: sarah@daitchassociates.ca

E. G. Pereira et al. (eds.), *Sovereign Wealth Funds, Local Content Policies and CSR*, CSR, Sustainability, Ethics & Governance,
https://doi.org/10.1007/978-3-030-56092-8_10

1 Who Owns Extractive Resources in Canada?

Canada is a federal state with ten provincial and three territorial governments (Northwest Territories, Nunavut and the Yukon). Ownership of subsoil assets is complex and rooted in a history of colonization. In brief, 89% of Canadian onshore territory is 'Crown land', owned by federal or provincial governments. Of this amount, approximately 46% of Crown land is owned by the federal government—the three territories as well as national parks, reserves and military bases—while 54% is owned by the provincial governments. Offshore resources are fully owned by the federal government (Government of Alberta 2019).

The remaining 11% of Canadian subsoil resources are owned by private entities. 'Freehold land' is a relic of the seventeenth century to the early twentieth century when settlers, the Hudson's Bay Company and the railroad companies were granted rights to encourage colonization of the western provinces, develop the fur trade and build the transcontinental railway (Martin 1973). While these companies have retained some land, a portion has been sold to private individuals or corporations. In the province of Manitoba, approximately 75% of mineral rights are privately owned (Government of Alberta 2018b).

Subsoil ownership must be distinguished from rights to manage subsoil resources. Under the *Constitution Act* (1867, amended 1982), minerals, oil, gas and other natural resources are managed by the government of the province in which they are located. The federal government manages resources on federal lands, in offshore waters, and on the continental shelf (Government of Canada 2016). In order to mitigate any conflict between different levels of government, offshore oil and gas production is regulated by boards with representation from both federal and provincial governments, despite management being the legal responsibility of the federal government. Each jurisdiction in Canada (province, territory and federal government) has its own mining, petroleum, environmental and occupational health and safety legislation.

Direct federal involvement in the regulation of onshore oil and mining operations in the provinces is limited. For example, it includes some responsibility for safe handling of uranium, a nuclear fuel, and shares responsibilities for environmental protection with the provinces. Provincial governments are responsible for mining, oil and gas, including the exploration for, and the development and extraction of, mineral resources, and the construction, management, reclamation and close-out of mine sites within their respective jurisdictions. Provinces have the exclusive right to levy royalties and mineral taxes, though the federal government may collect corporate income tax from extractive companies (Government of Canada 2016). The one exception is on freehold land, where private landowners may lease mineral rights and levy royalties and taxes (Government of Alberta 2018b).

The three territories do not enjoy independent constitutional status (Baldwin and Fipke 2010). However, in recent years, lawmaking and regulatory responsibilities in the mining, oil and gas sectors have been delegated to the territories through a process called "devolution" (Irlbacher-Fox 2012). Prior to devolution, the federal government, through the Department of Indian Affairs and Northern Development, was directly responsible for extractive activities in the three territories North of the 60th parallel, including mineral exploration and extraction, the development, management and reclamation of mine sites, and the collection of resource revenues and royalties in the Yukon and Northwest Territories (NWT). Through devolution, the federal government has reduced its role by devolving provincial-type responsibilities to the territorial governments.

The argument in favour of devolving these responsibilities is that it enhances "made in the North decision-making" concerning the development of mineral resources, allowing the territories to keep a portion of resource revenues (Government of Canada 2016). However, there is a strong argument that some Devolution Agreements have not been accompanied by adequate funds or administrative support to guarantee that the territorial governments' new administrative responsibilities are effectively carried out, and that the share of resource revenues that can be retained without other transfers from the federal government being clawed back is too low to incentivize an equitable arrangement in the resource sector (Irlbacher-Fox 2012; Irlbacher-Fox and Mills 2007).

Currently, the three territories have responsibilities in the areas of land-use planning, environmental assessment, water resources, and royalty and tax collection (Government of Canada 2016). The resource sector is regulated by co-management boards consisting of representatives from Indigenous groups, the federal government and the territorial governments. In addition, Indigenous governments in Canada have particular rights and title to lands, protected by the Constitution of Canada. For Indigenous governments who have signed modern land claim agreements, these agreements often include rights for the Indigenous government to manage a portion of subsoil resources on their settlement lands. While the percentage of settlement lands on which Indigenous governments have rights to subsoil resources varies by agreement, in many cases this portion makes up about 15% of the Indigenous governments' total settlement land (Baldwin and Fipke 2010). Additionally, under Canadian law, the Government of Canada has a duty to consult Indigenous nations and governments regarding extractive and other natural resource developments on their traditional lands (Baldwin and Fipke 2010).

2 What Are the Main Extractive Resources in Canada?

Although it has a diversified economy—the extractive sector generates approximately 8% of GDP on average—crude oil, natural gas and raw minerals represented more than 30% of exports in 2017 (MIT Observatory of Economic Complexity 2019). While most Canadian governments do not depend on natural resource revenues to finance their budgets, the governments of the provinces of Alberta, Newfoundland and Labrador and Saskatchewan are resource-dependent; around 20% of their fiscal revenues in recent years have been collected from oil, gas or mining companies. The Northwest Territories and Nunavut, both territories rather than provinces, are resource-dependent as measured by GDP, though not in terms of fiscal revenues (Statistics Canada 2019b).

Canada produces more than 60 types of minerals. Natural Resources Canada, the federal ministry mandated to promote the natural resource sector, valued Canadian non-oil mineral production at $43.9 billion in 2017, an increase of 11.3% from the 2016 total of $39.4 billion (Government of Canada 2016). Canada's top five mineral products by value in 2017 were gold ($8.7 billion), coal ($6.2 billion), copper ($4.7 billion), potash ($4.6 billion) and iron ore ($3.8 billion). British Columbia, Ontario, Saskatchewan and Quebec were the largest producers. All commodity groups recorded gains in value, but results for individual commodities varied, with coal recording the largest year-on-year increase at 55.6% or $2.2 billion, a notable development in light of Canada's commitment to the Paris Agreement and striving towards climate change targets.

Canada's oil and gas sector is even larger than the mining sector, generating CAD 118.5 billion in 2017 (Statistics Canada 2018a). In the same year, Canada was the world's 5th largest crude oil producer and the world's 4th largest natural gas producer (British Petroleum 2018). The country holds 10% of the world's proven oil reserves, more than any other country except Saudi Arabia and Venezuela (British Petroleum 2018). Canada was also the largest foreign supplier of crude oil to the USA, accounting for 43% of total US crude oil imports (Natural Resources Canada 2018b). 99% of all Canadian crude oil and equivalent exports go to the USA. The oil sands of Alberta and Saskatchewan supply the vast majority of Canadian production. In 2017, oil sands production was 2.7 million barrels per day compared to 1.5 million barrels per day of conventional oil (Natural Resources Canada 2018b).

Given the capital-intensive nature of the industry, the oil, gas and mining sector has generated approximately 8% of GDP on average in recent years but employed fewer than 200,000 workers, or 1.2% of the labour force, in 2017. However, these jobs offer some of the highest wages in the country; the average salary in the extractive sector is more than CAD 100,000 per year (approximately USD 76,000) (Statistics Canada 2019a).

Furthermore, economic multipliers are among the smallest of any economic sector in terms of labour income, jobs and output.[1] Studies from Alberta, British Columbia, the Northwest Territories and the federal government have each shown that the non-renewable resource sector produces the fewest jobs per million dollars in output (e.g. 1.3 jobs in the extractive sector relative to 23.4 jobs in forestry and logging in the Northwest Territories) and wage multipliers are among the smallest of any industry (Northwest Territories Bureau of Statistics 2012; Ryser et al. 2008; Statistics Canada 2019). The upstream petroleum sector has the smallest GDP multiplier of any industry (Government of Alberta 2018a; Statistics Canada 2018c).

[1]Multipliers are the change in one variable relative to the change in another variable. For example, in this context, the wage multiplier is the percentage change in overall wages from a given increase in extractive sector wages. Similarly, the GDP multiplier is the percentage change in overall GDP from a given increase in the extractive sector domestic production.

In theory, additional benefits could be generated from secondary and tertiary processing, for example petrochemicals, copper roofing or jewellery. However, processing from the extractive sector represents less than 1% of provincial GDP in each of the resource-dependent provinces. After all, producing raw materials usually does not give a country a competitive advantage in downstream processing of those materials (Toledano and Maennling 2018). Only Ontario has significant downstream mineral activity representing approximately 2% of provincial GDP (Alexander et al. 2018; Statistics Canada 2019b).

3 What Is the Fiscal Regime in Canada?

Canada's fiscal regime for oil, gas and minerals is complex by global standards. The federal government collects taxes and fees from extractive companies, such as federal corporate income tax. However, since natural resource governance is provincial jurisdiction under the Canadian constitution, each of Canada's ten provinces has the right to set its own natural resource-specific taxes, royalties and fees. As mentioned, the federal government has recently begun devolving responsibility over natural resource management, including fiscal regimes, to its three territories as well.

In general, Canada's tax environment can be described as favourable, even charitable, to oil, gas and mining companies. Among the most generous aspects of the various fiscal regimes in Canada are: (1) relatively low income taxes; (2) no royalty or royalty with cost deduction in most jurisdictions; (3) accelerated depreciation of capital assets; (4) tax credits for intangible assets that can be carried forward up to 20 years; (5) operating loss carry forward for up to 20 years; and (6) generous withholding tax limits or exemptions through tax treaties (KPMG 2016). A single federal tax incentive known as 'flow-through shares' that allows investors in exploration companies to deduct their costs from their own income taxes on unrelated activities cost Canadian taxpayers CAD 440 million dollars a year on average between 2007 and 2012, without any proven impact on productive exploration activity (Jog 2016). Table 1 provides a simplified summary of Canadian fiscal regimes in selected jurisdictions for mining for illustrative purposes.

The average effective tax rate, or 'government take', on a given mine depends on many variables, including commodity prices, costs of exploration and production, lifespan of the mine, and administrative capacity by governments to minimize tax avoidance. Therefore, it is impossible to attribute a single figure to the portion of oil, gas and mining rents being captured by the government. Ultimately, each field and mine is subject to a different 'government take'. That said, Canadian governments collect among the smallest shares of oil, gas and mining profits anywhere in the world. In 2012, Natural Resources Canada modelled an average-sized metallic mine using cost and price assumptions consistent with that time period. The study found that the average effective tax rate in Canada ranged from 20 to 30%, depending on the province, compared to 35-80% in Australia, Indonesia, Peru and Tanzania (Natural Resources Canada 2012). The one exception is Saskatchewan potash whose

Table 1 Simplified metallic mineral fiscal regime in Canadian jurisdictions (PwC 2016)

Jurisdiction	Income tax (%)	Royalty	Mining tax	Notable incentives	Mineral production (CDN billion) (Natural Resources Canada 2018a)
Canada (federal)	15			Generous deductions; investment tax credits; flow-through shares	43.9
Alberta	12	Greater of 1% gross revenue or 12% of net revenue		No cap on processing allowance	2.44
British Columbia	11		2% of net "proceeds" plus 13% of net revenue	Indefinite carry forward on most expenses; no cap on processing allowance	8.84
Manitoba	12		10–17% of net profits	Tax holiday until full costs are recovered, subject to govt approval	1.66
New Brunswick	12		2% of net revenue plus 16% of net profits		0.39
Newfoundland and Labrador	14		15% of net profit plus 20% of profit minus mineral tax paid		2.93
Northwest Territories	11.5	0–14% minus costs		Generous deductions from royalties	2.07
Nova Scotia	16		Greater of 2% of net revenue or 15% of net income	Accelerated depreciation	0.24

(continued)

Table 1 (continued)

Jurisdiction	Income tax (%)	Royalty	Mining tax	Notable incentives	Mineral production (CDN billion) (Natural Resources Canada 2018a)
Nunavut	12	0–14% minus costs		Generous deductions from royalties	0.84
Ontario	11.5		5–10% of net profits	Generous deductions; 36 month tax exemption for new mines or expansions	9.86
Prince Edward Island	16				0.00
Quebec	11.9		5% of net earnings plus 16–22.9% of profit, depending on profitability	Generous deductions	8.61
Saskatchewan	12	Dependent on mineral (e.g., 2.1–4.5% on potash; 5–15% on uranium); 3% resource surcharge charged on sales	Various taxes dependent on commodity	10-year tax holiday; costs are deducted from royalties; accelerated depreciation; no cap on processing allowance	5.72
Yukon	15	0–12% minus some costs			0.30

'government take' on large, viable projects is aligned with international standards at between 45 and 90% depending on prices (Keen et al. 2014).

'Government take' is largely in line with US rates, which are also among the lowest in the world. These percentages do not take into account base erosion, profit shifting and other tax avoidance measures, implying that the effective 'government take' is lower in each country.

Currently, the mining industry is taxed more lightly than other industries in Canada and provides significant benefits to investors. Depending on the province, the marginal effective tax rate on metallic minerals ranges from −9 to 21%. In comparison, the oil and gas sector's marginal effective tax rate ranges from 13 to 40%, and the rate for non-resource industries ranges from 2 to 25% (Chen and Mintz 2013).

The mining sector's low, even negative, rates are mostly attributable to the generous rules around expensing of assets and low royalty rates or mineral taxes.

The Canadian petroleum sector is also characterized by low 'government take' by global standards. The IMF's standard fiscal model demonstrates that, on paper, the average effective tax rate for conventional oil in Alberta and Saskatchewan would be approximately 55%, making a number of assumptions, including an oil price of USD 50 per barrel. The same field would garner an average effective tax rate above 60% in the UK, North Dakota, Oklahoma and Texas, and more than 80% in Algeria and China. The gap between Canada and other countries is similar for shale oil (Daniel et al. 2017).

The World Fiscal Rating of Oil Terms published by Van Meurs Energy provides a more sophisticated assessment of 'government take' in the oil sector, though one that is difficult to communicate. The rating is based on modelling of different field size, cost and price scenarios. Canada's overall rating, which is an average of ratings across provinces and territories, ranks its fiscal regime in the 25th percentile globally, meaning that 'government take' in the oil sector is lower than in 75% of countries (Van Meurs Energy 2019).

It is more difficult to assess the fairness of fiscal terms on oil sands—which represent approximately two-thirds of Canadian oil production—since few countries outside of Canada and the USA produce such high-cost unconventional oil. However, one study by the Government of Alberta royalty review panel found that the government collects 60–100% of the "super-rents" generated by oil sand companies (Royalty Review Advisory Panel 2016).[2] The World Fiscal Rating of Oil Terms confirms that the fiscal regime for Alberta oil sands provides a higher 'government take' than for oil production in other Canadian jurisdictions. Still, if Alberta were a country, 'government take' would be lower than in 62% of the world (Van Meurs Energy 2019). On the other hand, the high cost of production on the oil sands and high transport costs means that fields are only viable when oil prices are relatively high, above USD 44 per barrel as of 2018 (CERI 2018).

Natural Resources Canada and the Canadian oil and mining industries view these low rates as a virtue, encouraging investment in the extractive sector. However, this perspective fails to recognize that, unlike automotive factories or software companies, oil fields and mines are location-specific, generate large economic rents and are assets owned by governments. As such, the fiscal regime can be adjusted to maximize rent collection without harming investment on viable fields and mines (Hogan and Goldsworthy 2010; Mintz and Chen 2012). Moreover, governments have a responsibility to their citizens to maximize revenue from sales of their assets, whether physical buildings or minerals under the ground.

The Canadian extractive sector is already appealing to global investors due to Canada's comparative advantages. These include a skilled and experienced workforce, well-established supplier networks, political and regulatory stability, relatively

[2]"Super rent" figures are based on rents above a 10% 'hurdle rate'.

low electricity and water costs, and good infrastructure, especially non-pipeline transport. Fiscal incentives are therefore needed less in Canada than in most countries to attract investment.

The IMF suggests that 'government take' in the oil sector can be as high as 90%—as has been the case in Angola, Kazakhstan and Norway in certain years—without harming investment on major projects (Daniel et al. 2017; Goldsworthy and Zakharova 2010). 'Government take' in the mining sector is generally lower than in the oil sector; however, the average effective tax rate can be raised to between 60 and 80% without diminishing investment on viable mines, provided that the fiscal regime is designed to be progressive and there is political and social stability, as is the case in intermediate and high price scenarios on medium-sized copper mines in Chile, Indonesia and Western Australia (Manley 2017; Keen et al. 2014).

The impacts of existing incentives are substantial exploration activity, development of marginal mines and oil fields, and shifting significant economic rents from Canadian governments to shareholders of large mining companies. Since the majority of shares are held by foreigners—for example, oil sands production is approximately 70% owned by foreigners, even if the majority of operators and their staff are based in Canada—the fiscal regime represents an enormous transfer of wealth to foreigners at the expense of Canadian governments (Financial Post 2012). The loss to governments totals many billions of dollars annually. As a result, less money is available for hospitals, schools, roads and other public services.

4 Revenue Collection and Distribution in Canada

As mentioned, the federal government collects corporate income tax from oil, gas and mining companies. It also collects value added tax (the goods and services tax), customs duties and withholding taxes, though these are largely offset by deductions or exemptions. Approximately one-third of resource sector payments in Canada are made to the federal government in an average year.

Provincial governments collect provincial corporate income tax, royalties and mineral taxes (when they are levied), value added taxes and various fees. Since the signing of Devolution Agreements with Canada's three territories (the Northwest Territories, Nunavut and the Yukon) between 2003 and 2016, royalties have been collected by the territorial governments and transferred to the federal government. Property taxes, fuel taxes, and fees for water, land and road use are collected and retained by the territories. Corporate income taxes are collected and audited by the Canada Revenue Agency on behalf of the territories and transferred to them.

Some Indigenous governments also levy taxes, royalties or fees, often negotiated directly with companies. Payments are usually delineated in Impact Benefit Agreements or in land claim agreements. For instance, the Dehcho First Nations are entitled to 2.45–12.25% of royalties collected on mineral production in the Mackenzie Valley, depending on the royalty amount (Deh Cho First Nations—Government of Canada 2003).

In 2017, Canadian governments collected at least USD 8.87 billion from publicly traded oil, gas and mining companies, representing just under 6% of the gross value of production (Natural Resources Canada 2012; Resource Projects 2019; Statistics Canada 2018d). Of this amount, approximately USD 2.5 billion was allocated to the federal government (Resource Projects 2019). In comparison, Norway's government collected approximately 20% of the gross value of oil and gas production in 2017, though, in fairness, costs of production were lower in Norway than in Canada (EITI 2018; Norwegian Petroleum 2019).

The federal government makes transfers to the provinces and territories through various channels. The two most important for this discussion are equalization payments and the territorial formula financing (TFF), though the federal government also makes large health and social transfers. Equalization transfers are calculated based on as a standard amount each of Canada's 10 provinces should need to cover expenses ('the standard') minus the amount collected ('fiscal capacity'). Importantly, only half of natural resource revenues are included in the calculation, implying that natural resource revenue generation is penalized to a smaller degree than revenue generation from other sectors (Feehan 2014). The largest oil-producing provinces have regularly sought to exclude all natural resource revenues from the formula since they are the legal 'owners' of these resources and therefore feel they should retain all fiscal revenues generated by them. On the other hand, the recipient provinces support inclusion of all natural resource revenues since inclusion would increase the federal government's overall equalization payments pool (Béland et al. 2017).

Equalization payments totalled USD 14.3 billion in FY 2017–18, with Quebec the largest recipient. The most natural resource-dependent provinces—Alberta, Newfoundland and Labrador, and Saskatchewan—did not receive any payments (Department of Finance Canada 2017).

Under Territorial Formula Financing (TFF), the formula that determines the annual unconditional transfer from the Government of Canada to the territories, for each dollar a territory raises itself in taxes, approximately 70 cents are removed from the federal transfer. In other words, even if corporate income taxes from the resource sector rose significantly, much of the revenue would be clawed back by the federal government. Each territory is subject to its own Devolution Agreement, which sets special rules around distribution of natural resource revenues. The Northwest Territories Devolution Agreement, for example, allows the territorial government to retain the lesser of: 50% of mineral, oil, gas and water-related revenues (not including corporate income tax); or 5% of an amount called the 'Gross Expenditure Base', calculated at between CAD 70–100 million per year over the coming decade.[3] Therefore, for the first CAD 100 million in resource revenues, which consist in largest part by royalties, the Government of the Northwest Territories would retain a maximum of CAD 50 million (Bauer 2017).

[3]The definition of resource revenues can be found on p. 13–14 of the Devolution Agreement, Government of the Northwest Territories (2013) available online at http://devolution.gov.nt.ca/wp-content/uploads/2013/09/Final-Devolution-Agreement.pdf.

This provision generates a massive disincentive to raise additional revenue and expand mineral production. As a result, the federal government loses out on the corporate income tax, sales taxes and other sources of revenue it would collect from the sector, not to mention any economic activity mining would generate in terms of spillovers. The transfer system also reinforces the Northwest Territories' dependence on the federal government since it creates a disincentive for the territory to raise its own fiscal revenues, further costing the Government of Canada in terms of fiscal transfers.

Fiscal transfers are also made to some Indigenous governments. For example, the Northwest Territories Devolution Agreement requires that 25% of the territory's resource revenues be transferred to the 9 of 12 Indigenous governments that have signed the resource revenue sharing agreement.[4] Once other Indigenous governments sign the agreement, they too will be eligible for a share of the transfers. The money flowing to Indigenous governments cannot be spent on operational expenditures; they must be spent on capital investments or used for debt repayment.

In 2017, transfers to Indigenous governments in the NWT totalled CAD 10–15 million, distributed based on cost of living and population indicators. Despite not hosting active mines, the Gwich'in Tribal Council, Inuvialuit Regional Corporation and Sahtu Secretariat Incorporated receive approximately two-thirds of the transfers due to high cost of living in their territories and the size of their populations (Bauer 2017).

At the federal level, resource revenues are pooled with general fiscal revenues and spent according to legislation and annual budgetary allocations. At the provincial and territorial level, several governments have established special funds to manage a portion of their natural resource revenues or the fiscal surpluses engendered by resource production. For example, the Alberta Heritage Savings Trust Fund was established in 1976. The history and details of this fund are covered in Chap. 11 in this book. Quebec created the Generations Fund in 2006 and the Stabilization Reserve Fund, which is an account rather than a sovereign wealth fund (SWF), in 2009. In 2012, the Northwest Territories Heritage Fund was established. In 2015, the British Columbia Prosperity Fund was established, though it remains an embryonic SWF and may be discontinued.

5 Sovereign Wealth Funds

According to the International Forum on Sovereign Wealth Funds, a sovereign wealth fund is defined as a government-owned entity, established for a macroeconomic purpose, which does not have liabilities and invests at least partly in foreign assets (IFSWF 2019). As of 2019, there were approximately 60 SWFs financed by oil, gas or

[4]At present, there are nine signatories, including the Gwich'in Tribal Council, Inuvialuit Regional Corporation, Northwest Territories Métis Nation, Sahtu Secretariat Incorporated and Tłı̨chǫ Government. Three groups have not signed the agreement.

mineral revenues or by fiscal surpluses in countries dependent on natural resources. Canada is host to at least four such funds, all at the subnational level.

There are several government-owned pension funds that meet the definition of a SWF, as well as mineral-financed funds owned or co-owned by Indigenous governments, such as the Raglan Trust. Canadian governments have also created a panoply of oil and mineral funds to finance resource exploration or promote the industry. However, in this chapter, we will focus purely on the funds listed above. For each case, we will discuss the fund's: (1) history and objectives; (2) deposits and withdrawals; (3) investments; (4) institutional structure; (5) transparency and oversight; and (6) performance and political context.

5.1 British Columbia's Prosperity Fund

The British Columbia (B.C.) Prosperity Fund was announced in 2013 with the aim of channelling the province's liquified natural gas revenues to: (1) debt reduction (50% of the fund); (2) specific public investments (25%); and (3) savings for future generations (25%) (Government of British Columbia 2016). There is no legislation governing the fund, meaning all investments as well as transfers to and from the fund are discretionary.

Notwithstanding the fact that large natural gas projects have been cancelled, the government made two large deposits into the fund, a CAD 100 million deposit in FY 2015/16 and a CAD 400 million deposit in FY 2016/17 (see Fig. 1). As of February

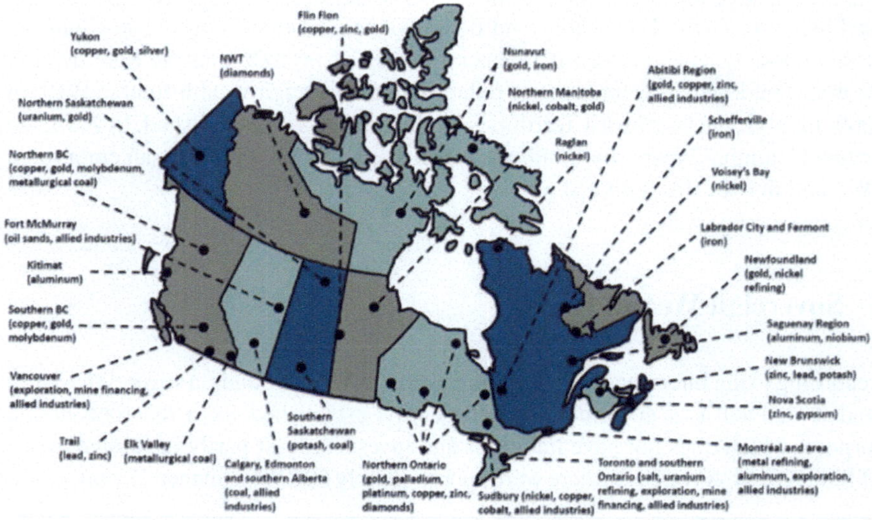

Fig. 1 Canada's mineral production by province and territory. *Source* Mining Association of Canada (2017)

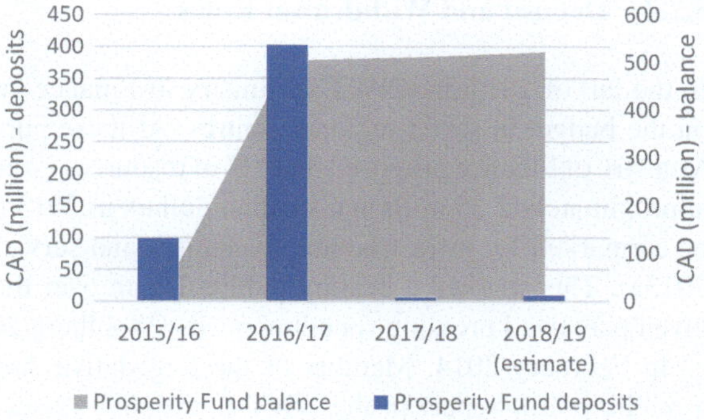

Fig. 2 B.C. Prosperity Fund deposits and balance. *Source* Annual budget documents, British Columbia Ministry of Finance

2019, the fund is still operational, though the only deposits over the last two years have consisted of interest earned on the existing balance.

The fund's asset allocation has not been made public. Nevertheless, we can assume a low-risk mandate; the fund earned a mere 1.4% return in FY 2017/18 and 2.1% in FY 2018/19 (Government of British Columbia 2019). Given that the Government of British Columbia 10-year bonds yielded 2.6% as of February 2019, should the fund's investment strategy remain unchanged, it would make financial sense for the balance to be used to pay down public debt. The Prosperity Fund has not published any annual reports or financial statements, nor is it subject to independent external audits except by the British Columbia Auditor General (Fig. 2).

5.2 Northwest Territories Heritage Fund

5.2.1 History and Objectives

In preparation for new resource royalties flowing from Devolution, the Government of the Northwest Territories (GNWT) passed legislation establishing a NWT Heritage Fund in 2012 (GNWT 2013b). The Northwest Territories Heritage Fund Act is vague on the fund's objectives. According to the Act, the fund's purpose is "to ensure that future generations of people of the Northwest Territories benefit from on-going economic development, including the development of non-renewable resources" (2012: 3). Financial statements from 2016 to 2017 state the Heritage Fund's single objective is to "maximize long-term growth" of the money in the fund while "avoiding undue risk" (Beers 2018). The fund's balance sits at 17.1 million Canadian dollars, which does not keep pace with current levels of inflation, as of July 2018 (GNWT 2018).

5.2.2 Deposit and Withdrawal Rules

In the fall of 2013, the NWT's Ministry of Finance undertook public consultations on the budget in seven regional centres. At these public consultations, the NWT's Ministry of Finance proposed that 5% of resource revenues be placed into the fund, or approximately 2.25 million Canadian dollars in 2013 (Wohlberg 2013). 95% would be earmarked for infrastructure investment and servicing the GNWT debt (GNWT 2013a). This sparked a healthy public debate over the appropriate deposit amount given perceived pressing spending needs (Wohlberg 2013).

In February 2014, Member of the Legislative Assembly Wendy Bisaro tabled a public policy report in the NWT Legislative Assembly pressing the Minister of Finance to commit more than 5% of revenues to the fund, and to introduce legislation to administer it. Following the debate, the Minister of Finance announced that 25% of GNWT extractive royalties would be allocated to the new Heritage Fund. The remaining 75% of resource revenues would be allocated for two other expenditure items proposed by the NWTs Ministry of Finance: debt repayment and infrastructure (GNWT 2013c). As of January 2019, this deposit commitment from the Minister of Finance, though on the public record, remains an informal policy statement (GNWT 2014, Personal communication, GNWT Ministry of Finance staff January 23, 2019). Neither the fund's purpose, nor many of the rules that generally govern sovereign wealth funds, such as deposit amounts, have been clarified in legislation or regulation.

GNWT's legislation introduced a 20-year period during which withdrawals from the fund are not permitted. Once the legislated twenty-year term has ended, the NWT faces a number of options for how much should be withdrawn from the fund and on what the money should be spent. One approach is to withdraw a five-year average of the interest earned (less inflation) while leaving the principal entirely in the fund, thereby establishing a 'permanent fund', and spending the interest via the annual budget process since territorial government spending through the budget is independently audited (Daitch et al. 2014).

5.2.3 Investment Rules

The current approach to investing the NWT Heritage Fund is very conservative, even compared to other funds that the NWT legislature oversees. Investing for the Heritage Fund allows only low-risk investments, including in government and bank bonds. In Fiscal Year 2016/17, the fund yielded a return of 1.05%, less than rate of inflation (1.525%). As a result, the fund posted a real return of −0.45%, equivalent to a loss of CAD 50,000 given the CAD 10.6 million balance in that year (Beers 2018). The mandate of the GNWTs 18th Legislative Assembly laid out that a review of the Heritage Fund Act was to take place, but this has not occurred to date. The fund continues to lose money against inflation and does not have a defined revenue stream (GNWT 2019).

The fund is currently administered by the Financial Management Board's Secretary, who is appointed by the Minister of Finance. The Financial Management Board overseeing the fund is made up of a committee of cabinet members. An independent assessment suggested that the fund's low returns are the product of investment rules poorly suited to the NWT context as well as inadequate asset management capacity (Daitch et al. 2014). In February 2018, during a Legislative Assembly debate on this topic, government representatives indicated that there would be recommendations put forward to the Minister of Finance to allow for external fund management, proposing that the additional cost of external management would be offset by better fund performance. Subsequently, steps have been taken to secure external fund management, though it is not yet confirmed to be in place as of January 2019 (GNWT 2018; Personal Communication, GNWT Department of Finance, January 23, 2019).

5.2.4 Institutional Structure

The NWT Heritage Fund is currently managed by the GNWT Department of Finance. GNWT Legislation states that the Financial Management Board is authorized to act as trustee of the Fund (GNWT 2012). The Financial Management Board, composed of Cabinet Ministers and Chaired by the Minister of Finance, is responsible for monitoring the performance of the Heritage Fund and, on an annual basis, for directing and supervising the Secretary of the Financial Management Board. The Secretary, a member of the public service, is responsible for carrying the administration and maintenance of the Heritage Fund as directed by the Board. The fund does not currently use external asset management services (Fig. 3).

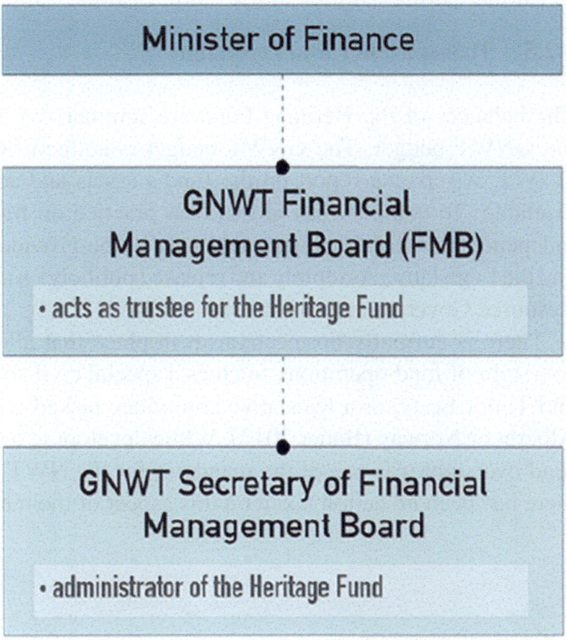

Fig. 3 NWT's current fund governance model. *Source* Daitch et al. (2014); GWNT 2012c

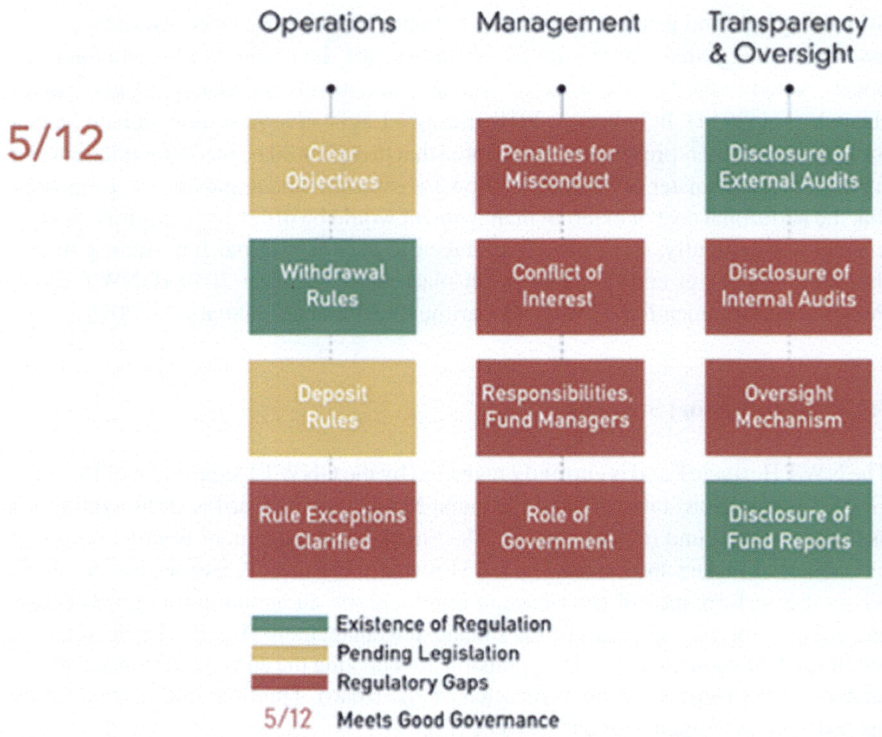

Fig. 4 NWT Heritage Fund: good governance and gaps in regulation. *Source* Daitch et al. (2014)

5.2.5 Transparency and Oversight

The balances of the Heritage Fund are summarized in a separate notes section of the GNWT budget. The GNWT budget is audited by the Auditor General of the GNWT. A separate report on the fund's assets and activities is not made publicly available. To comply with global best practice on transparency, both internal and independent external audits would need to be provided to the Ministry of Finance and the Legislative Assembly and released publicly on a government website (Natural Resource Governance Institute 2017; IFSWF 2008).

There is currently no mechanism in place that allows for independent external oversight of fund operations, such as a special civil society committee, as in Ghana and Timor-Leste, or a legislative committee tasked with overseeing the fund, as in Alberta or Norway (Bauer 2013). While developing a stronger role for the public in fund oversight was one of the mandates for the NWT's 18th Legislative Assembly, there has been no action taken on this aspect of the mandate to date (GNWT 2018).

5.2.6 Performance and Political Considerations

As non-renewable resources are discovered and developed in the territory, one of the government's priorities will continue to be to foster the well-being of future generations. Yet, despite large scale extractive projects operating in the territory since 1933, high poverty rates persist in the NWT's Indigenous communities (Irlbacher Fox 2012). Quality education, healthcare and nutritious foods remain inaccessible to many citizens.

On April 1, 2014, through devolution, NWT citizens gained greater control over their lands and resources for the first time since Canada's confederation. A Heritage Fund is a relatively new concept to most residents of the NWT that could, in theory, support the territory's devolution process as well as economic development over the long-run. Although the GNWT has held public consultations on the Heritage Fund in 2013, public education and engagement will be important in order to promote public awareness and help better define the fund's mandate, and further action has not been taken to develop this critical element (GNWT 2013e; GNWT 2018; GNWT 2019).

In addition, the fund's overall potential is stymied by very low resource royalties. NWT Member of the Legislative Assembly, Kevin O'Reilly, pointed out in a 2019 Legislative debate that the NWT will raise as much money from tobacco taxes and liquor revenues (about $40 million) as the projected $47 million in resource revenues for 2019–2020. A review of resource royalties, through re-examining the territorial formula financing and other aspects, has been pushed to the next legislative assembly; however, there is no guarantee this review will be done publicly (GNWT 2019). Meanwhile, extractive revenues flow out of the territory (Bauer 2018).

Future efforts to engage the public in a stronger role can help equip residents with the skills and knowledge to act as independent overseers to benefit the fund and its future. The more the public understands and supports the long-term objectives of the fund, the more it will hold current and future governments to account to protect the integrity of its original purpose. For example, Norway's fund enjoys broad public support as a point of national pride (MacKinnon 2013). If governance improves, the Heritage Fund could become a similar symbol of good resource stewardship.

5.3 Quebec's Generations Fund and Stabilization Reserve 'Fund'

5.3.1 History and Objectives

Quebec has one sovereign wealth fund and an account that serves as a budget balancing mechanism: The Generations Fund and the Stabilization Reserve Fund. The Generations Fund was established as a long-term savings fund in 2006 to address high and rising public sector indebtedness in the context of an ageing population. In that year, the share of net debt to provincial GDP was approximately 40% and rising quickly; today it stands at 42%. The consequence was an annual interest rate of 4.7% for 10 year bonds and more than 12.5% of budget expenditure being allocated to debt servicing annually (Government of Quebec 2007).

In response, the major political parties agreed to establish a fund to reduce the province's debt burden. The Generations Fund would accumulate money, which would be invested in financial markets, earning a higher rate of return than the interest paid on public debt. In this way, public assets would increase so that net debt figures would improve. The book value of the Generations Fund was CAD 13.8 billion as of the end of 2017 (Caisse de dépôt et placement du Québec 2018). However, its balance was expected to drop to CAD 7.7 billion in 2019 as funds will be used to pay down the public debt (Finances Québec 2018a).

The Stabilization Reserve Fund was established in 2009 to capture a portion of fiscal surpluses in order to balance the budget during economic downturns. However, the fund is not actually a fund but rather an account used to comply with the government's balanced budget rule. Its balances therefore represent the fiscal space available to the government to meet its medium-term balanced budget target. As of March 2018, the Stabilization Reserve Fund 'balance' was CAD 7.2 billion (Finances Québec 2018b).

5.3.2 Deposits and Withdrawals

Under the *Balanced Budget Act* (1996, amended), the Quebec government may not run a budget deficit except in moments of economic or social crisis. Any surplus collected above expected revenue is used to repay short-medium to medium-term public debt. At the same time, the surplus is credited to the Stabilization Reserve Fund. The fund's balance therefore represents the fiscal deficit that is permissible in subsequent years without breaking the balanced budget rule. As such, the Stabilization Reserve Fund 'balances' are considered a part of revenues in budget calculations. In practice, Stabilization Reserve Fund 'balances' have been used to cover fiscal deficits even in boom years without circumventing the balanced budget rule. The fund's long-run average 'balance' for precautionary purposes is expected to be CAD 2.4 billion.

The *Act to Reduce the Debt and Establish the Generations Fund* (2006, amended) sets debt reduction objectives, which are elaborated in budget documents. By FY 2025–26, the gross debt must not exceed 45% of GDP and the "debt representing accumulated deficits" must not exceed 17% of GDP (Government of Quebec 2018a). The Generations Fund is meant to help achieve these objectives.

The Generations Fund was originally financed by hydropower royalties, earnings on hydropower outside of Quebec, sales of government assets, and earnings on fund investments. Later, two new sources of financing were introduced: All mining revenues and CAD 500 million annually from taxes on alcoholic beverages. In the four years since mining revenues have been added to the list of the fund's revenue streams, CAD 602 million in mining proceeds have been added. Mining revenues

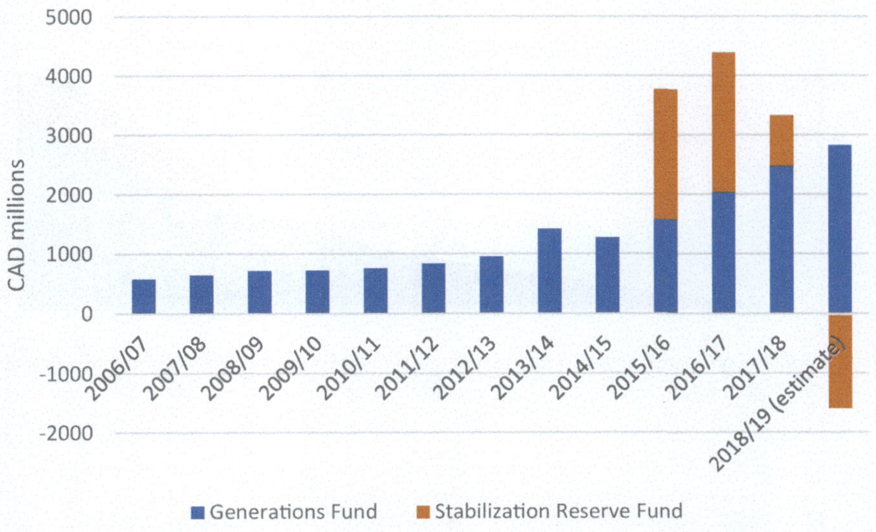

Fig. 5 Annual deposits and withdrawals from Quebec funds. *Source* Quebec Ministry of Finance budget documents. *Note* A Stabilization Reserve Fund deposit implies repaying public debt, while a withdrawal implies short-term to medium-term borrowing

have never represented more than 14% of new deposits into the fund (Government of Quebec 2017).

The government may also transfer money from the Stabilization Reserve Fund to the Generations Fund and has done so twice since the fund's inception. In practice, this 'transfer' implies borrowing money to transfer to the Generations Fund. The government may draw any sum from the Generations Fund at any time, but only to repay the gross debt (Figs. 5, 6).

5.3.3 Investments

The Generations Fund investment policy is determined by the Ministry of Finance in collaboration with the fund's day-to-day operational manager, the Caisse de dépôt et placement du Québec (the 'Caisse'). The Caisse is a publicly owned institutional investor that manages Quebec's more than CAD 300 billion in public pensions and insurance plans.

The Generations Fund is not subject to legislated investment rules. However, its target asset allocation is printed in Government of Quebec budget documents and risk management framework is disclosed on the Caisse website.

The Fund's risk appetite is elevated relative to most SWFs—62.5% of the fund is invested in equities and alternative assets—though it is invested in slightly less risky assets than the Caisse's pension holdings (see Fig. 4). The Fund is invested in

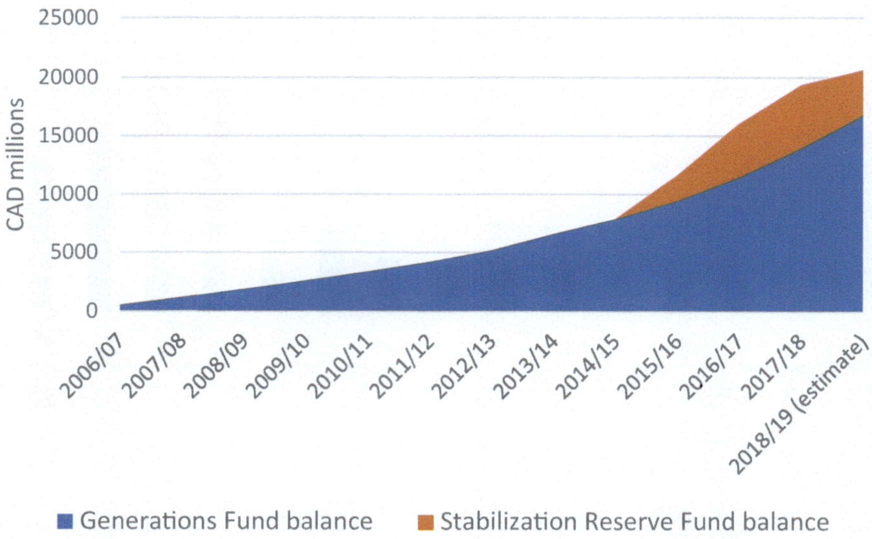

Fig. 6 Quebec fund cumulative balances, based on deposits (not book value). *Source* Quebec Ministry of Finance budget documents. *Note* Stabilization Reserve Fund balance represents fiscal space while complying with the balanced budget rule

even the riskiest asset classes, including private equity, real estate, infrastructure and derivatives for currency hedging purposes.

This risk-level has led to healthy returns; the average annual rate of return on the fund has been 6.1% over the first 11 years of its existence. The Generations Fund's financial performance ranks it among the best performing SWFs in the world, similar to the Alaska and Texas funds and with higher returns than in Chile, Norway, Trinidad & Tobago or Wyoming (Bauer 2018). Returns remain strong due to a robust risk management framework. This includes a list of eligible assets, performance targets, benchmarks, risk limits, and independent oversight, including external audits and Board reviews (Caisse de dépôt et placement du Québec 2019).

On the other hand, an elevated risk appetite has also led to significant volatility in returns. In 2008, the fund lost more than 22% of its value, though balances were less than USD 1 billion at the time. In contrast, returns were 12.3% in 2010. The fund has earned CAD 3.4 billion in interest for Quebec citizens since inception (Government of Quebec 2017).

Since the Stabilization Reserve Fund is only a balancing account, it does not manage any assets (Fig. 7).

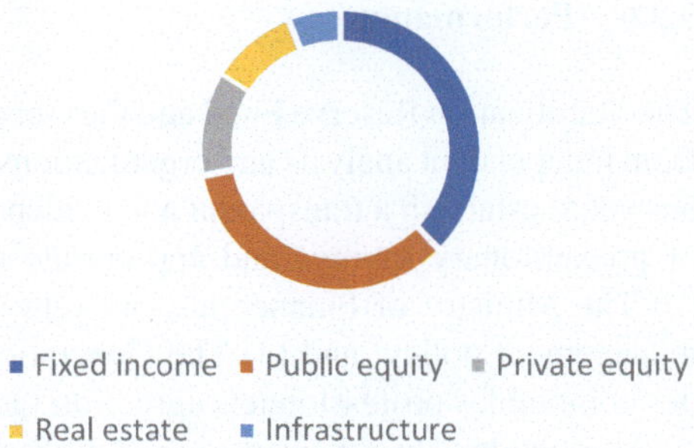

Fig. 7 Asset allocation for Generations Fund investments. *Source* Government of Quebec (2018b)

5.3.4 Institutional Structure

The ultimate owner of the Generations Fund is the Government of Quebec; however, the Fund is controlled and managed by the Quebec Ministry of Finance. The day-to-day operational management has been outsourced to the Caisse. A small portion of the Caisse's holdings have been outsourced to external investment managers. While deposits into the fund are subscribed in legislation, withdrawals to pay down public debt must be approved by the National Assembly through the annual budget process.

As with the Generations Fund, the Stabilization Reserve Fund is controlled and managed by the Ministry of Finance. While deposits into the Stabilization Reserve Fund are determined by the size of the budget surplus, withdrawals are relatively arbitrary. Therefore, the National Assembly has full power to determine both deposits and withdrawals. In practice, since the government and the majority in the National Assembly are usually one and the same under Quebec's Westminster system of government, cabinet decisions are often rubber stamped by the National Assembly.

5.3.5 Transparency and Oversight

According to legislation, the Minister of Finance must report to the National Assembly, in the Budget Speech, on the evolution of both the debt representing the accumulated deficits and the gross debt, on the sums credited to the Generations Fund and on any sums used to repay the gross debt. Deposit and withdrawal amounts for the Stabilization Reserve Fund are published in annual budget documents. Additionally, the Ministry of Finance has published extensive details on the Generations Fund's operations and finances on its website.

All funds placed at the Caisse are subject to an annual independent external audit as well as oversight by the Board of Directors. In its annual report, the Caisse provides details of its board and staff members, asset management strategy as well as every asset it holds.

The National Assembly and the Auditor General of Quebec examine both funds' finances as part of the annual budget process. Neither has performed a performance audit or investigation since their inception dates.

5.3.6 Performance

The Stabilization Reserve Fund and the Generations Fund have come under criticism from independent analysts and organizations along several lines: (1) The government has yet to establish a transparent and evidence-based framework for the optimal size of precautionary savings and appropriate use of the Stabilization Reserve Fund; (2) The Ministry of Finance has not released the Generations Fund's investment management policy; and (3) The Generations Fund delays debt repayment (Ordre des comptables professionnels agréés du Québec 2018; Laurin 2018).

As mentioned by others, the use of the Stabilization Reserve Fund is discretionary. This is a weakness in legislation that allows for pro-cyclical fiscal policy as a result of the balanced budget rule, an issue that could be addressed by replacing the existing rule with a more formal structural balanced budget rule or an expenditure growth rule (Bauer 2013).

We agree with others that Generations Fund governance could be improved if it released more detail on its investment management policy, for instance target return and volatility tolerance, as well as the specific assets held by the fund. However, given the Caisse's stellar performance history, we are less concerned about this issue.

With regard to the issue of financial savings and debt repayment, the math speaks for itself. As of February 2019, Quebec government 10-year bonds yield 2.7%, slightly lower than most provinces, and the government maintains a high grade credit rating.[5] Quebec is no longer Canada's most indebted province, though 9.3% of provincial expenditure in FY 2017/18 was still spent on debt servicing, the government's third largest expenditure item after healthcare and education (Government of Quebec 2017).

By comparison, the fund has generated an annual average return of 6.1% since inception. As a result, in the current environment, reducing the debt by USD 1 billion at the expense of the Generations Fund would lead to an opportunity cost of approximately USD 34 million. As can be seen from this example, what matters more than gross debt is *net debt*.

Should the gap between the average return and provincial bond yields shrink significantly, a greater portion of fiscal surpluses ought to be allocated to debt reduction. However, the government's savings policy has been sound. Of concern, Quebec's new government, elected in 2018, plans to use CAD 10 billion of the Generations Fund to pay down the public debt. This would lower debt payments but, given the spread between sovereign debt interest rates and the rate of return on the fund, would lead to a loss of hundreds of millions of dollars to future generations of Quebecers.

[5]The drop in borrowing costs is mainly a product of lower interest rates across Canada, not an improved macroeconomic context in Quebec.

6 Conclusion

Canada is often lauded as a model of good economic and public sector governance. Its citizens benefit from relatively high per capita income, broad and fair access to quality healthcare and education, low crime rates, political freedoms and environmental protections (Social Progress Imperative 2018; OECD 2017). Yet these strengths and successes often mask specific failings, such as leading the world in per capita greenhouse gas emissions and chronic poverty and weak social indicators among Indigenous groups (Boothe and Boudreault 2016; Eisler 2018).

Similarly, the success of Canada's oil, gas and mining companies' in expanding production in Canada and abroad—along with Canada's well-earned reputation as a global leader in exploration, extractive technologies, and extractive sector project management—mask serious weaknesses in the sector's governance framework at home. This chapter has highlighted three challenges: low 'government take' by global standards, limits to the benefits that can be captured by the territories, and a tendency towards discretionary use of sovereign wealth funds. To these we can add challenges that fall outside the scope of this chapter, such as an antiquated licensing regime—for instance, in some jurisdictions, companies can physically "stake" and explore land without consulting nearby residents—and weaknesses in environmental oversight (Bauer 2017). Canada had one of the highest rates of known tailings accidents in the world from 2007 to 2017, second only to China (Roche et al. 2017). And, as of 2013, the Alberta government intervened in less than 1% of cases of alleged contravention of environmental regulations in the oil sands (Canadian Press 2013).

Canada's resource governance ought to be measured relative to its potential rather than in sheer production figures. Given the country's vast natural resource wealth, paying for social services and infrastructure should be easy for Canadian provinces and territories. Standard of living for citizens should be comparable to Iceland, Norway, Qatar or the UAE, where poverty has been all but eliminated among citizens and infrastructure is cutting-edge. Instead, several resource-rich provinces or territories—namely Alberta, Newfoundland and Labrador, Saskatchewan and the three Northern Territories—still struggle with poverty, especially in Indigenous communities living close to extractive sites; Austria and Oman are Canada's nearest comparables in terms of general standard of living.

The resource sector can fill the financing gap. While fiscal regimes in Canada incentivize exploration and development of marginal deposits, they generate some of the lowest government revenue per unit extracted anywhere in the world. Relative to fiscal regimes in most other countries, this system shifts billions of dollars in economic rents each year from Canadian taxpayers to shareholders of oil and mining companies. Furthermore, since many if not most of the shares of extractive companies operating in Canada are owned by foreigners, the system represents an implicit transfer of wealth from Canada to foreign countries. As has been shown, the employment benefits and economic multipliers do not justify such low tax rates.

A technical review of Canadian fiscal regimes for oil, gas and mining is warranted, as is an evidence-based public discussion on the net benefits of extraction in Canada.

At present, resource taxation policy discussions are usually focused on increasing production rather than increasing the benefits from extraction. Notably, the strength of industry sector interest groups—the Mining Association of Canada, the Prospectors and Developers Association of Canada, the Canadian Association of Petroleum Producers, provincial industry groups and public relations departments of the largest companies—in capturing policymakers and academic institutions makes broadening the discussion to include net benefits and 'government take' exceedingly difficult.

A proper review of distribution of resource revenues to the territories may also increase the net benefits accruing to Canadians. At present, Territorial Formula Financing and the Devolution Agreements, particularly in the Northwest Territories, discourage extractive activities and severely limit the benefits that can accrue to Northerners. Reforming these systems could help transform lives in the North, particularly among Indigenous governments who are entitled to a share of resource revenues and whose education and healthcare systems are chronically underserved.

Finally, Canadian governments, especially in resource-dependent provinces and territories, may wish to review aspects of how they manage their resource revenues. The key to benefiting from non-renewable resource wealth is investing the proceeds in financial assets (e.g. savings funds), physical assets (e.g. infrastructure) or human capital (e.g. universities) rather than consuming them. At the same time, mitigating the negative effects of resource revenue volatility on budget expenditures often requires a stabilization mechanism, whether a fund or through counter-cyclical debt management (Collier et al. 2009).

Historically, Canadian provincial governments have often made the implicit choice to prioritize consumption over investment of resource revenues. This emphasis on consumption shows up in some net debt figures. For example, while Alberta and Saskatchewan managed to keep debt levels low until recently, Newfoundland and Labrador's net debt per capita nearly doubled in the decade from 1997 to 2007, the years of peak oil production. Today, the province has the highest rate of net debt per capita in Canada (Government of Newfoundland and Labrador 2018). Only the governments of Alberta and Quebec have significant financial savings in sovereign wealth funds, though in Alberta, the savings are much less than economic models suggest are optimal (Van der Bremer and Van der Ploeg 2014).

In some regions, the choice to consume rather than invest has taken the form of low taxes, leading to greater discretionary income for households at the expense of public investment. Alberta, British Columbia, Ontario, Saskatchewan, and the Territories, several of the most resource-dependent regions, have some of the lowest personal income tax rates in the OECD (EY 2017). These same jurisdictions plus Manitoba minus Ontario have some of the lowest value added tax rates (OECD 2018). In a different subset of jurisdictions, governments have underinvested in public services and infrastructure, as evidenced by Canada's more than CAD 180 billion infrastructure gap (Government of Canada 2018). On an optimistic note, resource-dependent provinces and territories seem to have learned lessons from the past; today, they are the ones invested most heavily in modern infrastructure and education (Statistics Canada 2016; Statistics Canada 2018b).

What's more, most resource-rich provinces have historically not countered the negative effects of commodity price volatility on government budgets, either because it was unnecessary due to low resource revenues or as a result of a lack of appropriate counter-cyclical fiscal rules (Atkinson et al. 2016). However, over the last three decades, several provinces have taken steps to introduce mechanisms to smooth fiscal expenditures. For example, Manitoba established a stabilization fund in 1989 (current balance less than CAD 200 million) (Government of Manitoba 2018; Tapp 2013).[6] Quebec established the Stabilization Reserve Fund in 2009 to support its cyclically adjusted balanced budget rule. The British Columbia Prosperity Fund was established in 2013, partly to stabilize the budget. And Alberta created the Contingency Fund in 2013 to mitigate the effects of oil revenue volatility (NRGI-CCSI 2013). Interestingly, Saskatchewan established a Fiscal Stabilization Fund in 2000, which was replaced by the Growth and Financial Security Fund in 2008. The government abandoned efforts at counter-cyclical fiscal policy when it wound down the fund in 2016 (Graham 2016).

Foreign governments could draw on some of Canada's more successful experiences to improve their natural resource revenue management. With respect to revenue distribution, Canada's equalization formula and some Territories' revenue sharing formulas with Indigenous governments are both useful models that could be adapted to different contexts. With respect to fiscal stability and sustainability, Alberta's *Fiscal Management Act* (2013) and Quebec's *Act to Reduce the Debt and Establish the Generations Fund* (2006) both represent practical examples of how to manage large, volatile and finite revenue flows. It should be noted that, similar to other global experiences of resource revenue management, in each of these cases, governments mismanaged their public finances for decades before enacting more responsible and broadly accepted fiscal management legislation.

Canadian governments may wish to build on the success stories within Canada and draw lessons from countries like Chile, Norway and the UAE on how to smooth fiscal expenditures, promote fiscal sustainability, and invest in projects that will help diversify their economies. Regarding the funds captured in this chapter, we have specific recommendations.

In British Columbia, the disparity between the fund's earnings and the cost of capital to the provincial government suggests that the fund's balances would best be used to pay down public debt or be allocated to finance public services. Furthermore, the province is not resource-dependent nor is it expected to be in the foreseeable future, meaning that existence of a subnational sovereign wealth fund to manage resource revenues may not be justified. On the other hand, the government could convert the fund into a contingency or stabilization fund that would serve during economic downturns.

The Northwest Territories Heritage Fund is a nascent fund and, as such, may require significant changes to its governing legislation and regulations to meet international good practices in sovereign wealth fund governance. Among our recommendations are: (1) clarifying objectives, as the investment strategy and deposit and

[6]New Brunswick also had a short-lived fiscal stabilization fund.

withdrawal rules each flow from clear fund objectives; (2) formalizing the deposit amount; (3) clarifying how withdrawals will be used once the 20-year no-withdrawal period has ended; (4) drawing on independent expertise to revise the investment mandate and hire external managers; (5) requiring independent audits once the fund reaches a critical size; and (6) publishing annual reports online that cover, among other information, balances, returns, assets, fund managers, significant fund activities and transactions. Since fiscal revenues are smoothed through federal transfers to the Northwest Territories, we would recommend that the fund serves as a future Generations Fund and that the returns be earmarked to underfunded expenditure items, such as renewable energy or education.

In many ways, Quebec's Generations Fund is a model of good governance. Its deposits and investment strategy are rules-based and consistent with the fund's objectives, it is well-managed by a professional entity, it is transparent and audited annually, and the fund serves a logical macroeconomic purpose. Its main weakness is the ability of the government to make arbitrary and discretionary withdrawals. Withdrawals ought to be a function of the spread between sovereign debt rates and the long-term return on the fund; the rule should be codified in legislation. Similarly, the use of the Stabilization Reserve Fund should be clarified by formalizing a counter-cyclical fiscal rule in legislation. Part of the challenge may be that the respective funds' roles in promoting intergenerational equity and stabilizing fiscal expenditures are poorly understood, not only by the public but also by Quebec policymakers.

Finally, we would recommend that both Newfoundland and Labrador and Saskatchewan consider enacting counter-cyclical fiscal rules that would smooth fiscal expenditures and better balance spending today with the needs of future generations. This may imply establishing sovereign wealth funds, certainly in the case of Saskatchewan given low public debt levels, provided that fund design is aligned with global best practices such as the Santiago Principles. In 2017, Saskatchewan collected at least CAD 1.4 billion (USD 1.1 billion) in revenue mainly from oil and potash, representing more than 20% of fiscal revenues. Newfoundland and Labrador collected at least CAD 989 million (USD 761 million) in the same year, mainly from offshore oil, representing just under 20% of provincial fiscal revenues (Resource Projects 2019; Government of Newfoundland and Labrador 2017; Government of Saskatchewan 2018). Both provinces are resource-dependent, and government spending tends to oscillate with the ups and downs of commodity prices, amplifying boom-bust cycles that harm growth and lead to poor public investment choices (RBC 2018). Along with other parts of Canada, Saskatchewan and Newfoundland and Labrador could benefit from a more evidence-based and longer-term vision of resource revenue management.

References

Act to Reduce the Debt and Establish the Generations Fund. Available at: http://legisquebec.gouv.qc.ca/en/pdf/cs/R-2.2.0.1.pdf.

Alexander, I. et al (2018) *Provincial and territorial natural resource indicators, 2009 to 2016.* Statistics Canada. [online] Available at: https://www150.statcan.gc.ca/n1/pub/13-604-m/13-604-m2018088-eng.htm [Accessed 3 Mar. 2019]

Atkinson, M. M. et al. (2016). Fiscal rules in Canadian provinces: Abject failure or qualified success? *Canadian Public Administration, 59*(4), 495–515. http://www.pbo-dpb.gc.ca/web/default/files/files/files/Publications/Fiscal_Rules_Oct_2010.pdf.

Balanced Budget Act 2001, c. 56, s. 10. Available at: http://legisquebec.gouv.qc.ca/en/ShowDoc/cs/E-12.00001.

Baldwin, C., & Fipke, J. (2010). Canadian Mining Law. In Paper presented at a Rocky Mountain Mineral Law Foundation Mining Law Short Course. Lawson Lundell.

Bauer, A. (Ed.). (2013). *Managing the public trust: How to make natural resource funds work for citizens.* Columbia Center on Sustainable Investment—Natural Resource Governance Institute. [online] Available at: https://academiccommons.columbia.edu/doi/10.7916/D80P10KX.

Bauer, A. (2017). *Northwest territories mineral sector review and benchmarking.* Government of the Northwest Territories. [online] Available at: https://www.engage-iti.ca/3808/documents/7981.

Bauer, A. (2018). *How good are sovereign wealth funds at investing money made from natural resources?* Natural Resource Governance Institute. [online] Available at: https://resourcegovernance.org/blog/how-good-are-sovereign-wealth-funds-investing-money-made-natural-resources.

Beers, R. (2018). N.W.T. Heritage Fund not keeping up with inflation, gov't borrows money to feed it. CBC News. [online] Available at: https://www.cbc.ca/news/canada/north/nwt-heritage-fund-inflation-1.4734937.

Béland, D., Lecours, A., Marchildon, G. P., Mou, H., & Olfert, M. R. (2017). *Fiscal federalism and equalization policy in Canada: Political and economic dimensions.* Toronto: University of Toronto Press.

Boothe, P., & Boudreault, F. -A. (2016). *By the numbers: Canadian GHG emissions.* Lawrence National Centre for Policy and Management, Ivey Business School at Western University. [online] Available at: https://www.ivey.uwo.ca/cmsmedia/2112500/4462-ghg-emissions-report-v03f.pdf.

British Petroleum. (2018). *bp statistical review of world energy*, June 2018. [online] Available at: https://www.bp.com/content/dam/bp/business-sites/en/global/corporate/pdfs/energy-economics/statistical-review/bp-stats-review-2018-full-report.pdf [Accessed 20 Feb. 2019].

Caisse de dépôt et placement du Québec. (2018). *Annual report additional information.* [online] Available at: https://www.cdpq.com/sites/default/files/medias/pdf/en/ra/ra2017_renseignements_add_en.pdf.

Caisse de dépôt et placement du Québec. (2019). *Risk management.* [online] Available at: https://www.cdpq.com/en/investments/risk-management. [Accessed 2 Mar. 2019].

Canadian Press. (2013). Enforcement lacking in oilsands infractions, study finds. Canadian Broadcasting Corporation. [online] Available at: https://www.cbc.ca/news/canada/edmonton/enforcement-lacking-in-oilsands-infractions-study-finds-1.1383760.

CERI. (2018). *Canadian oil sands supply costs and development projects (2018–2038).* [online] Available at: https://ceri.ca/studies/canadian-oil-sands-supply-costs-and-development-projects-2018–2038 [Accessed 25 Feb. 2019].

Chen, D., & Mintz, J. (2013). *Repairing Canada's mining-tax system to be less distorting and complex.* University of Calgary School of Public Policy. [online] Available at: https://www.policyschool.ca/wp-content/uploads/2016/03/mining-tax-chen-mintz.pdf [Accessed 22 Feb. 2019].

Collier, P., Van der Ploeg, F., & Venables, A. J. (2009). *Managing resource revenues for developing economies.* Oxcarre—CSAE, University of Oxford. [online] Available at: http://citeseerx.ist.psu.edu/viewdoc/download?doi=10.1.1.554.9061&rep=rep1&type=pdf.

Daitch, S., Schwann, A., Bauer, A., Dias, A., & Fan Li., J. (2014). *A question of future prosperity: Transforming natural resource wealth into citizen well being through the northwest territories heritage fund.* Arctic Yearbook. [online] Available at: https://arcticyearbook.com/images/yearbook/2014/Scholarly_Papers/10.Daitch_et_al.pdf [Accessed 3 Mar. 2019].

Daniel, P., et al. (2017). *How should shale gas extraction be taxed?* IMF Working Paper 17/254. [online] Available at: https://www.imf.org/~/media/Files/Publications/WP/2017/wp17254.ashx [Accessed 22 Feb. 2019].

Deh Cho First Nations—Government of Canada. (2003). *Deh Cho first nations interim resource development agreement*. [online] Available at: https://dehcho.org/docs/Interim-Resource-Development-Agreement.pdf.

Department of Finance Canada. (2017). *Federal support to provinces and territories*. Government of Canada. [online] Available at: https://www.fin.gc.ca/fedprov/mtp-eng.asp [Accessed 3 Mar. 2019].

Eisler, D. (2018). *The grim reality of Canada's biggest policy failure*. University of Saskatchewan School of Public Policy. [online] Available at: https://www.schoolofpublicpolicy.sk.ca/research/publications/policy-brief/the-grim-reality-of-canadas-biggest-policy-failure.php.

EITI. (2018). *Norway*. [online] Available at: https://eiti.org/norway [Accessed 3 Mar. 2019].

EY. (2017). *2017 personal income tax calculator*. [online] Available at: https://www.ey.com/ca/en/services/tax/tax-calculators-2017-personal-tax.

Feehan, J. (2014). *Canada's equalization formula: Peering inside the black box... and beyond*. University of Calgary School of Public Policy. [online] Available at: https://www.policyschool.ca/wp-content/uploads/2016/03/feehan-equalization.pdf [Accessed 20 Feb. 2019].

Finances Québec. (2018a). *The generations fund*. Government of Quebec. [online] Available at: http://www.budget.finances.gouv.qc.ca/fondsdesgenerations/index_en.asp.

Finances Québec. (2018b). *Quebec's Debt*. Government of Quebec. [online] Available at: http://www.finances.gouv.qc.ca/en/Quebecs_debt325.asp.

Financial Post. (2012). Majority of oil sands ownership and profits are foreign, says analysis. The Financial Post, May 10. [online] Available at: https://business.financialpost.com/news/majority-of-oil-sands-ownership-and-profits-are-foreign-says-analysis.

Goldsworthy, B., & Zakharova, D. (2010). *Evaluation of oil fiscal regime in russia and proposals for reform*. IMF Working Paper 10/33. [online] Available at: https://www.imf.org/external/pubs/ft/wp/2010/wp1033.pdf [Accessed 22 Feb. 2019].

Government of Alberta. (2018a). *Alberta economic multipliers 2013*. [online] Available at: https://open.alberta.ca/publications/alberta-economic-multipliers [Access 20 Feb. 2019].

Government of Alberta. (2018b). *Freehold mineral rights in Alberta*. [online] Available at: https://www1.agric.gov.ab.ca/$department/deptdocs.nsf/all/ofa16605/$file/freeholdnew.pdf?OpenElement [Accessed 3 Mar. 2019].

Government of Alberta. (2019). *Mineral ownership*. [online] Available at: https://www.alberta.ca/mineral-ownership.aspx [Accessed 3 Mar. 2019].

Government of British Columbia. (2016). *Budget and fiscal plan 2016/17-2018/19*. [online] Available at: https://www.bcbudget.gov.bc.ca/2016/bfp/2016_Budget_and_Fiscal_Plan.pdf [Accessed 2 Mar. 2019].

Government of British Columbia. (2019). *Estimates: Fiscal year ending March 31, 2020*. [online] Available at: https://www.bcbudget.gov.bc.ca/2019/pdf/2019_Estimates.pdf [Accessed 2 Mar. 2019].

Government of Canada. (2016). *Exploration and mining in Canada: An investor's brief*. Ottawa, ON: Natural Resources Canada.

Government of Canada. (2018). *Investing in Canada—Canada's long-term infrastructure plan*. [online] Available at: https://www.infrastructure.gc.ca/site/alt-format/pdf/plan/icp-pic/IC-InvestingInCanadaPlan-ENG.pdf.

Government of Manitoba. (2018). *Manitoba budget 2018*. [online] Available at: https://www.gov.mb.ca/finance/budget18/papers/Summary_Budget_r.pdf.

Government of Newfoundland and Labrador. (2017). *Estimates of the program expenditure and revenue of the consolidated revenue fund 2017–18*. [online] Available at: https://www.budget.gov.nl.ca/budget2017/estimates/estimates.pdf.

Government of Newfoundland and Labrador. (2018). *Economic and project analysis*. [online] Available at: https://www.economics.gov.nl.ca/EB-Oil.asp [Access 3 Mar. 2019].

Government of Quebec. (2007). *Public accounts 2006–2007.* [online] Available at: http://www.finances.gouv.qc.ca/documents/Comptespublics/en/CPTEN_vol1-2006-2007.pdf.

Government of Quebec. (2017). *Generations fund—A renewed commitment to fostering intergenerational equity.* [online] Available at: http://www.budget.finances.gouv.qc.ca/budget/2017-2018/en/documents/Budget1718_GenerationsFund.pdf.

Government of Quebec. (2018a). *The generations fund: Quebec is repaying its debt.* [online] Available at: http://www.budget.finances.gouv.qc.ca/budget/2018-2019/en/documents/GenerationsFund_1819.pdf.

Government of Quebec. (2018b). *The Quebec economic plan.* [online] Available at: http://www.budget.finances.gouv.qc.ca/budget/2018-2019/en/documents/EconomicPlan_1819.pdf#page=499.

Government of Saskatchewan. (2018). *Saskatchewan provincial budget 18–19: On track.* [online] Available at: http://publications.gov.sk.ca/documents/15/106322-2018-19%20Budget%20for%20WEB.pdf.

Government of the Northwest Territories. (2012). *Northwest territories heritage fund act.* [online] Available at: http://www.justice.gov.nt.ca/collections/legislation/acts/NWT%20Heritage%20Fund%20Act.pdf.

Government of the Northwest Territories. (2013). *Public Accounts 2012–2013, consolidated financial statements and government indicators.* [online] Available at: http://www.fin.gov.nt.ca//public-accounts/index.htm.

Government of the Northwest Territories. (2013). *Northwest territories lands and resources devolution agreement.* [online] Available at: http://devolution.gov.nt.ca/wp-content/uploads/2013/09/Final-Devolution-Agreement.pdf.

Government of the Northwest Territories. (2013). R*esourcing our legacy.* [online] Available at: http://www.fin.gov.nt.ca/Dialogue%202013/documents/ResourcingourLegacy.pdf.

Government of the Northwest Territories. (February 11, 2019). Northwest territories legislative assembly hansard. [online] Available at: https://hansard.opennwt.ca/debates/2019/2/11/kevin-oreilly-6/.

Government of the Northwest Territories. (February 23, 2018). Northwest territories legislative assembly hansard. [online] Available at: https://hansard.opennwt.ca/dbates/2018/02/23/mr-stewart-9/.

Government of the Northwest Territories. (June 5, 2013). Northwest territories legislative assembly hansard. [online] Available at: http://www.assembly.gov.nt.ca/sites/default/files/hn130605.pdf.

Government of the Northwest Territories. (October 6, 2013). Public consultation report. [online] Available at: http://www.fin.gov.nt.ca/Dialogue%202013/Dialogue2013.htm.

Graham, J. (2016). Saskatchewan repeals 'balanced budget' law". The Canadian Press. [online] Available at: https://globalnews.ca/news/2745116/saskatchewan-repeals-balanced-budget-law/.

Hogan, L., & Goldsworthy, B. (2010). International mineral taxation. In P. Daniel, M. Keen & C. McPherson (Eds.) *The taxation of petroleum and minerals: Principles, problems and practice.* Washington: IMF.

IFSWF. (2008). *Sovereign wealth funds: Generally accepted principles and practices "Santiago Principles".* [online] Available at: https://www.ifswf.org/santiago-principles-landing/santiago-principles.

IFSWF. (2019). *International forum of sovereign wealth funds.* [online] Available at: http://www.ifswf.org/ [Accessed 3 Mar. 2019].

Irlbacher-Fox, S. (2012). *Gahcho Kué economic impacts and NWT devolution,* 4 May 2012 [Web log message]. Retrieved from: http://www.northernpublicaffairs.ca/index/irlbacher-fox-gahcho-kue-economic-impacts-and-nwt-devolution/.

Irlbacher-Fox, S. & Mills, S. J. (2007). *Devolution and resource revenue sharing in the Canadian North: Achieving fairness across generations.* Walter and Duncan Gordon Foundation.

Jog, V. (2016). *Rates of return on flow-through shares: Investors and governments beware.* SPP Research Paper No. 9–4. [online] Available at: https://papers.ssrn.com/sol3/papers.cfm?abstract_id=2741764.

Keen, M., et al. (2014). *Israel: Reviewing the fiscal regime for mining*. IMF Country Report No. 14/125. [online] Available at: https://www.imf.org/external/pubs/ft/scr/2014/cr14125.pdf.
KPMG. (2016). *A guide to Canadian mining taxation*. [online] Available at: https://assets.kpmg/content/dam/kpmg/pdf/2016/04/KPMG-Mining-Taxation-Guide-2016.pdf [Accessed 12 Feb. 2019].
Laurin, A. (2018). *The generations fund fails the test of fiscal transparency for Quebec's dudget*. C.D. Howe Institute. [online] Available at: https://www.cdhowe.org/intelligence-memos/alexandre-laurin-generations-fund-fails-test-fiscal-transparency-quebec%E2%80%99s-budget.
Manley, D. (2017) *Ninth time lucky: Is Zambia's mining tax the best approach to an uncertain future?* Natural Resource Governance Institute. [online] Available at: https://resourcegovernance.org/sites/default/files/documents/ninth-time-lucky.pdf.
Martin, C. (1973). *"Dominion Lands" policy*. Toronto: McClelland and Stewart Ltd.
Mining Association of Canada. (2017). *Facts and figures 2017: Facts and figures of the canadian mining industry*. Toronto: Mining Association of Canada. [online] Available at: http://mining.ca/sites/default/files/documents/Facts-and-Figures-2017.pdf.
Mintz, J., & Chen, D. (2012) *Capturing economic rents from resources through royalties and taxes*. University of Calgary School of Public Policy. [online] Available at: https://www.policyschool.ca/wp-content/uploads/2016/03/economic-rents-mintz-chen.pdf [Accessed 25 Feb. 2019].
MIT observatory of economic complexity. (2019). *OEC-Canada (CAN) exports, imports, and trade partners*. [online] Available at: https://atlas.media.mit.edu/en/profile/country/can/ [Accessed 3 Mar. 2019].
Natural Resource Governance Institute. (2017). *Resource governance index 2017*. [online] Available at: https://www.resourcegovernanceindex.org/.
Natural Resources Canada. (2012). *Taxation of mineral income 2012–How Canada compares*. [online] Available at: https://www.nrcan.gc.ca/mining-materials/markets/8358 [Accessed 22 Feb. 2019].
Natural Resources Canada. (2018a). *Canadian mineral production*. [online] Available at: https://www.nrcan.gc.ca/mining-materials/publications/17722 [Accessed 12 Feb. 2019].
Natural Resources Canada. (2018b). *Crude oil facts*. Government of Canada. [online] Available at: https://www.nrcan.gc.ca/energy/facts/crude-oil/20064#L1.
Northwest Territories Bureau of Statistics. (2012). *NWT economic multipliers—Overview and results*. [online]. Available at: https://www.statsnwt.ca/economy/multipliers/Multiplier%20Report-2012.pdf [Accessed 12 Feb. 2019].
Norwegian Petroleum. (2019). *Exports of Norwegian oil and gas-Norwegianpetroleum.no*. [online] Available at: https://www.norskpetroleum.no/en/production-and-exports/exports-of-oil-and-gas/ [Accessed 3 Mar. 2019].
NRGI-CCSI. (2013). *Alberta heritage savings trust fund and contingency account*. Natural Resource Governance Institute—Columbia Center on Sustainable Investment. [online] Available at: https://resourcegovernance.org/sites/default/files/NRF_Alberta_September2013.pdf.
OECD. (2017). *OECD better life index: Canada*. [online] Available at: http://www.oecdbetterlifeindex.org/countries/canada/.
OECD. (2018). *OECD tax database*. [online] Available at: http://www.oecd.org/tax/consumption/tax-database.htm#VATTables.
Ordre des comptables professionnels agréés du Québec. (2018). *Réflexions sur les priorités budgétaires due Québec en 2018–2019*. [online] Available at: https://cpaquebec.ca/~/media/docs/salle-de-presse/memoires/2018-03-06-priorites-budgetaires-quebec-2018-2019.pdf.
PwC. (2016). *Canadian mining taxation*. [online] Available at: https://www.pwc.com/ca/en/mining/publications/pwc-canadian-mining-taxation-2016-11-en.pdf [Accessed 12 Feb. 2019].
RBC. (2018). *Canadian federal and provincial fiscal tables*. [online] Available at: http://www.rbc.com/economics/economic-reports/pdf/provincial-forecasts/prov_fiscal.pdf.
Resource Projects. (2019). *Resource projects: An open-source data repository on oil, gas and mining projects*. [online] Available at: https://resourceprojects.org/country/Canada [Accessed 3 Mar. 2019].

Roche, C., et al. (Eds.) (2017) *Mine tailings storage: Safety is no accident*. United Nations Environment Programme and GRID-Arendal. [online] Available at: https://www.grida.no/publications/383.

Royalty Review Advisory Panel. (2016). *Alberta at a crossroads: Royalty review advisory panel report*. [online] Available at: https://open.alberta.ca/dataset/01b3af57-350c-43ec-8c73-d70fe74523cb/resource/94fca310-354d-4604-bccb-d897c2f68b91/download/2016-royalty-framework-report.pdf [Access 22 Feb. 2019].

Ryser, L., et al. (2008). *Extracting the value*. Community Development Institute, University of Northern British Columbia. [online] Available at: https://www.unbc.ca/sites/default/files/assets/community_development_institute/extracting_the_value_draft_1.pdf [Accessed 21 Feb. 2019].

Social Progress Imperative. (2018). *2018 social progress index*. [online] Available at: https://www.weforum.org/agenda/2016/07/these-countries-have-the-highest-quality-of-life.

Statistics Canada. (2016). *Consolidated per capita spending by selected Canadian classification of functions of government, 2016*. [online] Available at: https://www150.statcan.gc.ca/n1/daily-quotidien/180111/t001b-eng.htm.

Statistics Canada. (2018a). *The daily—Oil and gas extraction, 2017*. [online] Available at: https://www150.statcan.gc.ca/n1/daily-quotidien/180924/dq180924d-eng.htm [Accessed 3 Mar. 2019].

Statistics Canada. (2018b). *Infrastructure economic account, 2017*. [online] Available at: https://www150.statcan.gc.ca/n1/daily-quotidien/180913/dq180913b-eng.htm.

Statistics Canada. (2018c). *National and provincial multipliers, 2014*. [online] Available at: https://www150.statcan.gc.ca/n1/en/catalogue/15F0046X [Accessed 21 Feb. 2019].

Statistics Canada. (2018d). *Oil and gas extraction, 2017*. [online] Available at: https://www150.statcan.gc.ca/n1/daily-quotidien/180924/dq180924d-eng.htm [Accessed 20 Feb. 2019].

Statistics Canada. (2019a). *Employment by industry, annual*. [online] Available at: https://www150.statcan.gc.ca/t1/tbl1/en/tv.action?pid=1410020201 [Accessed 21 Feb. 2019].

Statistics Canada. (2019b). *Provincial and territorial natural resource indicators, 2009 to 2016*. [online] Available at: https://www150.statcan.gc.ca/n1/pub/13-604-m/13-604-m2018088-eng.htm [Accessed 3 Mar. 2019].

Tapp, S. (2013). The use and effectiveness of fiscal rules in Canadian Provinces. *Canadian Public Policy, 39*(1), 45–70.

Toledano, P., & Maennling, N. (2018). *Local content policies in the mining sector: Fostering downstream linkages*. IISD and IGF. [online] Available at: https://www.iisd.org/sites/default/files/publications/local-content-policies-mining-downstream-linkages.pdf. [Accessed 3 Mar. 2019].

Van den Bremer, T., & Van der Ploeg, R. (2014). *Digging deep for the heritage fund: Why the right fund for Alberta pays dividends long after oil is gone*. University of Calgary School of Public Policy. [online] Available at: https://www.policyschool.ca/wp-content/uploads/2016/03/digging-deep-bremer-ploeg.pdf.

Van Meurs Energy. (2019). *Fiscal rating of oil terms*. [online] Available at: https://vanmeursenergy.com/World-Fiscal-Rating [Accessed 3 Mar. 2019].

Non-Renewable Resource Revenue Savings and Distribution in Canada: Alberta

Shantel S. Jordison and Niloo Hojjati

Abstract The Alberta Heritage Savings Trust Fund is Canada's first and largest sovereign wealth fund. Like Norway's oil fund, the Heritage Fund is fuelled by oil wealth. As one of the first funds of its kind, in 1976 it inspired many other countries to start funds of their own but unfortunately, it has experienced persistent challenges from the start. It has had different, and often conflicting objectives, low public buy-in, and inconsistent and discretionary deposit and withdrawal rules. With the current balance of the Heritage Fund standing at a mere $17.5 billion (CBC News 2004) (CDN), many argue that the Heritage Fund has largely failed to achieve its savings mandate, especially when compared to other funds around the world in countries with similar populations and resource profiles. However, while the Heritage Fund is much smaller than its Alaskan and Norwegian peers, Alberta has made key policy decisions that have set it on a unique path. It has chosen to maintain historically low tax rates (with no provincial value added tax), a competitive royalty framework, and historically has spent the most dollars per capita on public services for its citizens. This chapter explores the history, governance, oversight, investment policy, operational and deposit/withdrawal rules of the Heritage Fund. It also discusses how wealth is transferred from Alberta to other parts of Canada and offers some critical observations about the Fund's successes and failures.

Keywords Natural resource management · Non-renewable resource revenue savings · Alberta heritage savings trust fund · Canada sovereign wealth fund · Ralph Bucks

---

S. S. Jordison (✉)
Alinea International, Calgary, Canada
e-mail: shanteljordison@gmail.com

N. Hojjati
Extractive Resource Governance Program, School of Public Policy, University of Calgary, Calgary, Canada
e-mail: niloo.hojjati@gmail.com

E. G. Pereira et al. (eds.), *Sovereign Wealth Funds, Local Content Policies and CSR*, CSR, Sustainability, Ethics & Governance,
https://doi.org/10.1007/978-3-030-56092-8_11

1 Introduction

The Alberta Heritage Savings Trust Fund is Canada's first and largest sovereign wealth fund. As the name suggests, the Fund was established in the Western Canadian province of Alberta as a long-term savings fund to collect and save a portion of non-renewable resource revenue. With a current balance of $17.5 billion CAD (approx. $13.2 billion USD)[1] the Heritage Fund alone is larger than the balance of all of Canada's other sovereign wealth funds put together. Like Norway's oil fund, but unlike other Canadian sovereign wealth funds, the Heritage Fund is fuelled by oil wealth. Canada has the world's third largest proven petroleum reserves next to Venezuela and Saudi Arabia, with Alberta producing more than eighty per cent of the country's total crude oil. Current figures from the Government of Alberta Ministry of Energy place provincial crude oil production at 2.5 million barrels per day (bpd) and natural gas reserves stand at 33 trillion cubic feet (tcf) (Jordison 2018, p. 79) (Fig. 1).

Established in 1976, nearly fifteen years before the onset of Norway's savings program, the Heritage Fund was born out of an era of rising oil prices and economic growth. As one of the first funds of its kind, it inspired many countries around the world to develop similar savings funds. But unfortunately, the Heritage Fund has experienced persistent challenges from the start. It has had different, and often conflicting objectives, low public buy-in, and inconsistent and discretionary deposit and withdrawal rules. Since its establishment, $43 billion has been withdrawn from the Alberta Heritage Fund to support government spending in healthcare, education, infrastructure, debt reduction and social programs (Government of Alberta; Alberta Treasury Board and Finance 2018). And with the current balance of the Heritage Fund standing at a mere $17.5 billion (CDN), many argue that the Heritage Fund has largely failed to achieve its savings mandate, especially when compared to other funds around the world in countries with similar populations and resource profiles. By way of comparison, the Heritage Fund's current balance constitutes only 5.3% of Alberta's total GDP[2] (Government of Alberta 2019) whereas Norway's fund is valued at 2.5 times its annual GDP.

Interestingly, in 2015 the Calgary Chamber of Commerce calculated that Alberta's sovereign wealth fund would be worth $40.9 billion if it followed Alaska's model of taxation and $163.7 billion in the case of Norway (Wit 2017). But unfortunately these comparisons do not take into account important differences between Alberta, Alaska, and Norway, and have been widely criticized as being unfair. While the Heritage Fund is much smaller than Norway's oil fund, Alberta has made key policy decisions that have set it on a different path. Alberta has chosen to maintain low tax rates (with no provincial value added tax), maintain a competitive royalty framework, and historically has spent the most dollars per capita on public services for its citizens.

[1] Current market rate at February 23, 2019.

[2] As of September 30, 2018.

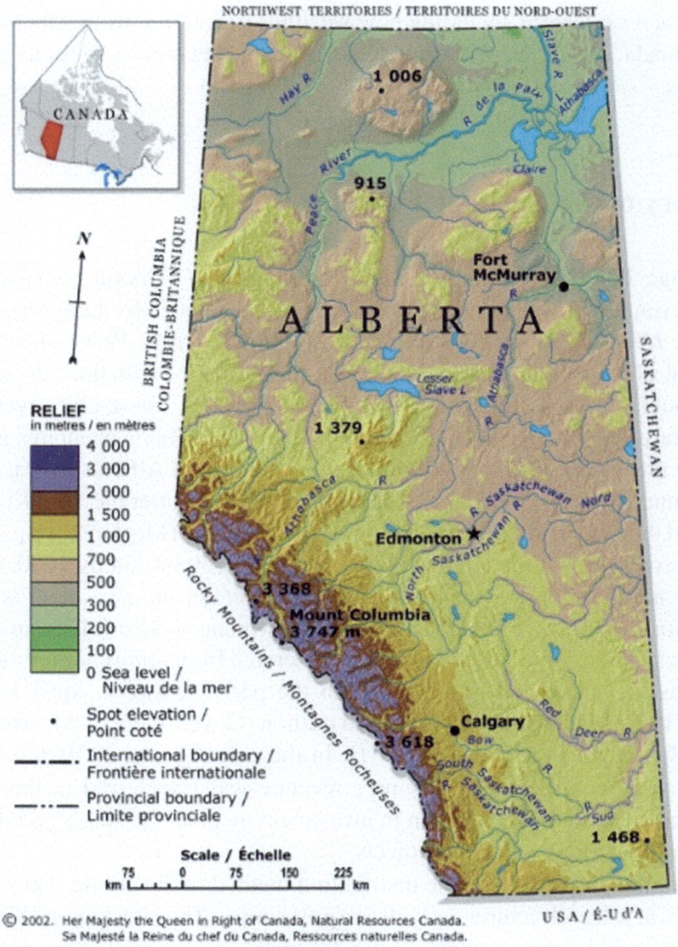

Fig. 1 Province of Alberta, Canada. *Source* Natural Resources Canada (2017)

Undisputedly, Albertans experience a very high quality of life. Indeed, according to the Economist Intelligence Unit's 2018 Global Liveability Index, the thriving oil and gas metropolis of Calgary was ranked as the fourth most liveable city in the world (The Economist 2018).

Thanks to the work of our other Canadian colleagues (Bauer et al.), who provided an overview of Canada's fiscal landscape and smaller sovereign wealth funds in the previous chapter, we focus exclusively on Alberta in this chapter. We will explore the history, evolution, and current status of the Heritage Fund and also explain its governance, oversight, investment policy, operational and deposit/withdrawal rules. We will conclude by offering some information about how the Heritage Fund is distributed across Alberta, including how wealth is transferred from Alberta to other parts of Canada. Lastly, we will offer some critical observations about its successes and failures.

2 History of the Heritage Fund

The Heritage Fund is the main vehicle for savings of Alberta's non-renewable resource revenues. It was established by Alberta Premier Peter Lougheed through the *Alberta Heritage Savings Trust Fund Act (AHSTF Act)* in 1976, when the price of oil was at a historical high. The Fund was initially created with three different, and as time would tell—sometimes incompatible—objectives. These objectives were to save for future generations, strengthen and diversify the Alberta economy, and lastly to improve the quality of life of Albertans (Government of Alberta 2019a, b, c).

At the time, the price of oil increased dramatically, first after the Yom Kippur War in 1973 and then again after the Iranian revolution in 1979 (Morton 2018 p. 7). Early estimates projected that the Heritage Fund could top $50 billion by 2000; however, this would never become a reality. The *AHSTF Act* set out two main sources of financing for the Heritage Fund: an initial endowment of $1.5 billion in cash and other financial assets; plus 30% of revenues generated from non-renewable resources collected by the provincial government (in the period between April 1, 1976 to March 31, 1977), which amounted to $620 million (Government of Alberta; Alberta Treasury Board and Finance, 2018, p. 1). In the period between 1976 and 1983,[3] $9.69 billion in non-renewable resource revenue was transferred to the Heritage Fund, which generated $4.13 billion in investment income. Of this, $1.6 billion was used to fund spending on capital projects.

These capital projects were one distribution method to the public. They included parks, medical research centres, as well as a golf course. The now famous Kananaskis Golf Course included summer and winter destinations where Albertans could experience the beauty of the Canadian Rockies outside of the National Parks (Kananaskis Country Golf Course 2019). Kan-Alta Golf Management Ltd. still exists today and has managed the facility since it first opened in 1983. These projects were not without criticism. At the time of their development, they were harshly criticized for being wasteful and needlessly luxurious. One example was the import of products to the golf course, such as white sand from the Caribbean. According to Nelson (2015) "The Kananaskis course was controversial from the very start. Many questioned why $25 million from the Heritage Fund's energy royalties—supposed to diversify the economy or be saved for when the oil runs out—was going instead to build a luxury golf course during a time when Alberta was bleeding from the effects of the National Energy Program" (Nelson 2015).

By 1983, the Heritage Fund had accumulated assets with a net value of $8.3 billion (Government of Alberta; Alberta Treasury Board and Finance 2018, p. 21). Around

[3]Note: fiscal year.

this time, $1.9 billion from the Heritage Fund was also loaned by the province of Alberta to six borrowing Canadian provinces (Government of Alberta 2019a, b, c, p. 2). These provinces included: Manitoba, Quebec, Newfoundland, New Brunswick, Nova Scotia, and Prince Edward Island. These loans offered the opportunity for the borrowing provinces to use Alberta's credit rating to their advantage (Government of Alberta 2019, p. 2) and represented an indirect distribution of wealth from resource-rich Alberta to the rest of Canada, which was struggling at the time.

Unfortunately due to the volatile nature of oil prices, by the mid-1980s, Alberta was experiencing significant economic turmoil due to falling oil prices, leading to a sharp drop in its resource revenues. With a looming election in 1982, the Heritage Fund's architect, Premier Lougheed, reduced deposits from 30 to 15% into the Fund in order to help maintain government spending during the campaign period. This proved to be a troubling move that other politicians would emulate in the future, eventually eroding the deposit rule and removing it completely (Morton 2018, p. 7). Oil prices continued to drop into the late 80s, and by 1987 deposits of resource revenues into the Heritage Fund were suspended altogether, reducing the deposit rate to zero. Thus, all investment income from the Heritage Fund was transferred to the provincial government's general revenue to finance its operating and capital expenditures (including the financing of capital projects), while the principal remained untouched. Alberta has always relied heavily on oil and gas royalties to fund its public spending, rather than relying on other sources of revenue (such as a value added tax for instance) as is the case in other Canadian provinces.

As a result of this, by March of 1996, the Heritage Fund shrank to 11.8% of Alberta's GDP, a substantial decrease of 7.5% since 1983 (see Fig. 2.) According to critics, "the variety of objectives in its first 20 years made it difficult to pursue any one effectively. An even worse design flaw of the Heritage Fund was the decision

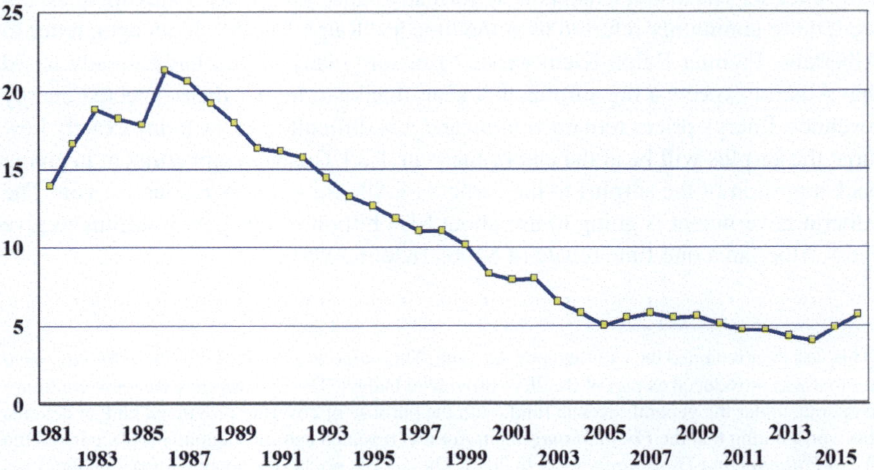

Fig. 2 Alberta heritage savings trust fund, as a % of provincial GDP (1981–2015). *Source* Dahlby (2017)

to make contributions voluntary and subject to political approval, and to give those same politicians access to the investment income it generated" (Fawcett, 2016).

In 1995, the Alberta government conducted a public survey to determine the perspective of Albertans on the future of the Heritage Fund. The results of the survey indicated that a vast majority of Albertans believed that the primary focus of the Heritage Fund should be centred on savings for future generations and for long-term investments. Consequently, in 1996, an amendment was made to the *AHSTF Act* to reflect this and the Heritage Fund was reconstructed so that it could no longer be used for direct social investment or economic development purposes (Government of Alberta 2019a, b, c, p. 3). To ensure the fulfilment of the Heritage Fund's new objectives for long-term savings, a Legislative Standing Committee (Province of Alberta 2019, p. section 6), separate from the provincial government, was created in 1997 to review and approve the Fund's new business plans (Government of Alberta 2019a, b, c, p. 3).

The 2000s brought with them increasing oil prices, and consequent improvements in Alberta's resource-dependent economy. In 2003 Alberta added a new fund, the Alberta Sustainability Fund[4] (Fiscal Planning and Transparency Act, 2015). It was established as an account within the General Revenue Fund with the aim of providing short-term fiscal stabilization. Often this fund has been referred to as the "Rainy Day" Fund (Wood, 2015). On 12 July 2004, Alberta proudly became the only debt-free province in the country, with then-Premier Ralph Klein stating in the news: "I've been dreaming about this day for quite some time now. Today, I'm very, very proud to announce that Alberta has slain its debt" (CBC News 2004). This was followed by a $7.4 billion surplus in 2005, and a ground breaking decision to share Alberta's wealth in an unprecedented manner. The Premier of the day Ralph Klein, commonly referred to as "King Ralph"—as a reference to his political longevity and his management style—made the decision to give each resident in Alberta and their children, a tax-free $400 cheque. The disbursements were formally called prosperity bonus cheques, but were more commonly referred to at the time as "Ralph Bucks". In an open letter to Albertans, Premier Ralph Klein wrote, "I'm sure many of you have already heard that Alberta expects a big surplus this year, thanks to higher-than-expected energy revenues. Energy prices remain volatile and it is difficult to know today exactly how large the surplus will be at the end of the year. [...] This year, government is giving back a portion of the surplus to the owners of Alberta's energy resources: you. The Alberta government is going to use about $1.4 billion of this year's surplus to give every Albertan a one-time rebate of $400" (Klein 2005).

[4]This was later renamed the Contingency Account. This came as a result of the *Fiscal Management Act* that was introduced as part of the 2013 provincial budget. The Contingency Account continues to operate under the general revenue fund, with the purpose of covering provincial budget deficits. It is worth noting that the *Fiscal Management Act* was repealed and subsequently replaced with the *Fiscal Planning and Transparency Act* in 2015. The *Act* did not have a notable impact on the flows in/out of the Fund, but rather has been used to cover the provincial budget deficits. As of March 2018, the balance of the Contingency Account is $1.7 billion, a decrease of $638 million from the prior year (Government of Alberta; Alberta Treasury Board and Finance, 2019, p. 8).

The total cost to government of the program was $1.4 billion, representing 20% of that year's $6.8 billion budget surplus. While Premier Klein said that there might be more cheques in the future for Albertans, the program was a one-hit-wonder, and no future cheques were issued. Largely, the program was heralded as a success by the public, but it also faced harsh criticism. Former Premier Lougheed, the architect of the Heritage Fund, did not approve of the prosperity bonus cheques. In an open letter in the Calgary Herald on 16 February 2006, he wrote: "If we do not save a sufficient portion of these oil and gas revenues, history proves that much of it will be dissipated on non-essential expenditures and we will not have much to show for it 10 years or so from now" (Fawcett 2016).

Later in 2005, non-renewable resource revenues peaked at over $12.5 billion, a new historical high for the province (Morton 2018, p. 8). And later in the year, the Heritage Fund began retaining a portion of its investment income in order to counteract the effects of inflation. An estimated $226 million was retained in the Fund to protect it against inflation (Government of Alberta 2019a, b, c, p. 9), while *The Act* was amended to ensure that the inflation proofing became a statutory requirement (Province of Alberta 2019, p. section 11).

The next big change came in 2007, when the Alberta Investment Management Corporation (AIMCo) was established through an act of legislation (Province of Alberta 2007 Current as of 12 June 2013). AIMCo would come to serve as the external managers of the Heritage Fund's investments. Prior to this, the Ministry of Finance carried out the daily management of the Fund and its investment portfolio. AIMCo's establishment signified an important milestone in the Heritage Fund's history. It enabled the Ministry to utilize the extensive technical expertise of a new, arms-length body to make investment decisions, and also provided a new level of transparency and credibility to the Heritage Fund's management.

In 2007, an advisory committee chaired by Canada's foremost tax policy expert, Jack Mintz, set an ambitious target for the Heritage Fund: $100-billion by 2030. They recommended "a fiscal adjustment" to make that possible. "Alberta's non-renewable resources should provide significant benefits not just to Albertans today, but also for our children and grandchildren. When our stock of non-renewable resources dwindles, Alberta's economy will need to rely only on its people—not its natural resources—to create wealth. Alberta should not look like a ghost town in the next century when the resources are depleted" (Fawcett 2016).

Between 2010 to present day, there have been some modest changes to the Heritage Fund; however, the most drastic changes occurred in the Fund's earlier days. Today, the Alberta Investment Management Corporation (AIMCo) invests the Fund in a globally diversified portfolio across many assets classes, such as stocks, bonds, and real estate. Throughout its history, the Heritage Fund's mandate has changed numerous times, but the current objective is to: "provide prudent stewardship of the savings from Alberta's non-renewable resources by providing the greatest financial returns on those savings for current and future generations of Albertans" (Province of Alberta 2019, p. preamble) (Fig. 3).

Since its inception in 1973, the Heritage Fund has contributed $43 billion to the benefit of Albertans through financing provincial priorities such as health care,

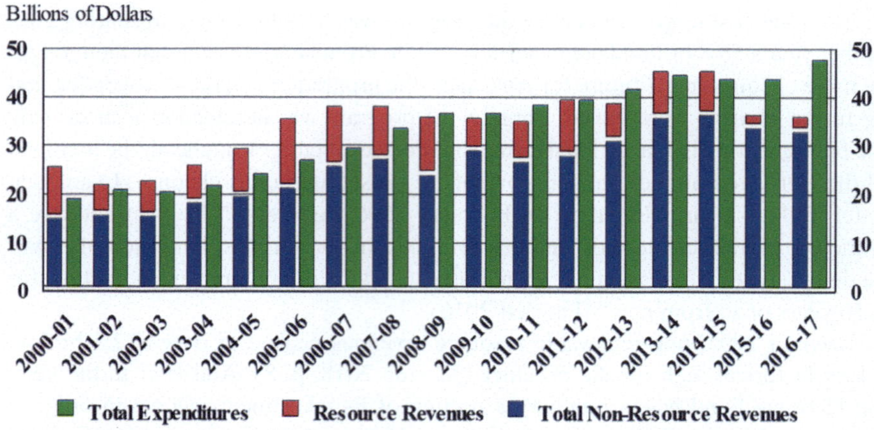

Fig. 3 Alberta government total expenditures, resource revenues and non-resource revenues. *Source* Kneebone (2016)

education, infrastructure, and other social programs. Since the 80s, there has only been three ad hoc deposits made into the Fund tracking against higher oil prices later in the 2000s (Dahlby 2014, p. 1). Fast forward to 2017/2018, the Heritage Fund earned $1.787 billion in net income, $230 million of which was retained in the Fund for inflation proofing and $1.557 billion was transferred to the province's general budget to support government spending (Government of Alberta 2019a, b, c, p. 4).

3 Heritage Fund Governance

Now that we have explored the history of the Heritage Fund, we will dive into some of the distinguishing characteristics of the Fund. The Heritage Fund operates under the legislative authority as established through the *AHSTF Act,* which provides a prudent governance structure with clear roles and responsibilities for the management and oversight of the Fund. The following section outlines the bodies that hold responsible roles in the management of the Heritage Fund and its investments, illustrated in Fig. 4.

Under *The AHSTF Act,* the President of Treasury Board, Minister of Finance, holds ultimate responsibility for the management of the Heritage Fund and its investments (Province of Alberta 2019). The Minister is supported in the investment management of the Heritage Fund by two main groups which includes: (i) the Ministry of Treasury Board and Finance; and (ii) the Alberta Investment Management Corporation (AIMCo) (Government of Alberta 2011, p. 6). The Ministry of Treasury Board and Finance determines the Statement of Investment Policies and Goals (SIP&G), which sets out the Fund's long-term investment strategy and policy through ongoing research and analysis. The Ministry is also responsible for the risk management and financial reporting of the Heritage Fund, which occurs on a monthly, quarterly, and

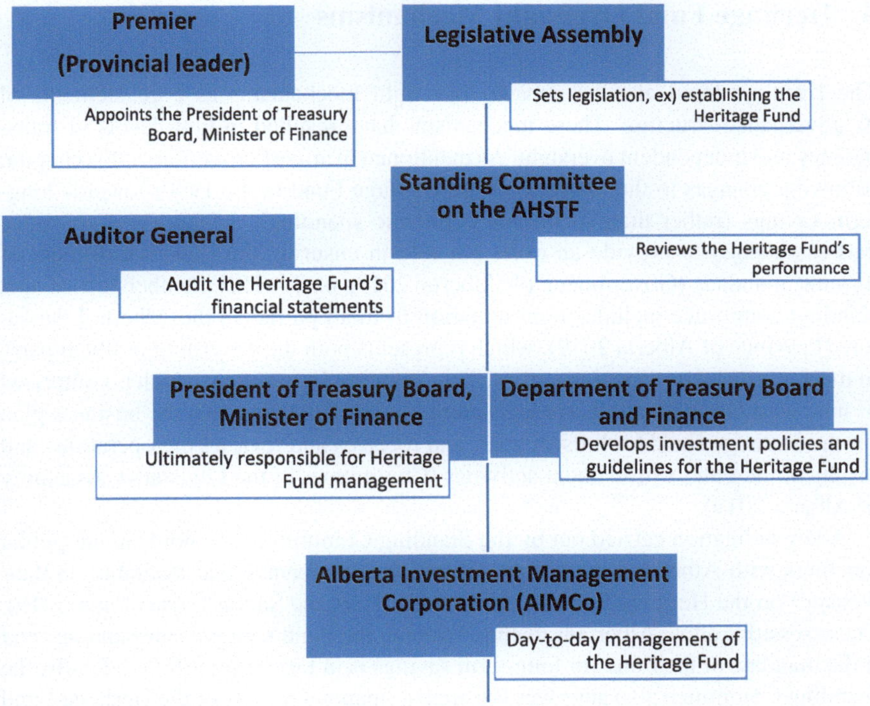

Fig. 4 Governance of the Alberta Heritage Fund

annual basis (Government of Alberta 2011, p. 6). The second body, which provides support to the President of Treasury Board, Minister of Finance in the Fund's investment management is AIMCo. AIMCo, a provincial Crown Corporation[5] (Province of Alberta 2019), is an external body that is responsible for the day-to-day management of the Fund and its investments in accordance to the guidelines in the SIP&G, as developed by the Ministry.

AIMCo manages the purchase of the Heritage Fund's investment portfolio that includes a range of investments in public and private companies, real estate, bonds, mortgages, and infrastructure developments[6] using internal AIMCo staff as well as external managers (Government of Alberta 2011, p. 7). AIMCo also supports the Ministry in its preparation of the Heritage Fund's publicly available financial reporting documents, by providing reports on the performance of investments and their associated costs. In addition, to ensure that the investment portfolio is reflective of changing market conditions, AIMCo conducts ongoing research to provide recommendations to the Ministry in regards to potential improvements to the SIP&G (Government of Alberta 2011, p. 7).

[5]A crown corporation refers to a corporation that is established and regulated by the Canadian government.

[6]Further details on the Fund's investment assets will be discussed in the next section of this chapter.

4 Heritage Fund Oversight Mechanisms

The Heritage Fund also has several oversight mechanisms as core elements of its governance structure. These mechanisms have resulted in high levels of transparency and independent oversight. As mentioned in an earlier section of this chapter, following changes to the mandate of the Heritage Fund in the 1990s towards long-term savings (rather than social and economic spending), a Standing Committee was established to provide an oversight role in ensuring the Fund's fulfilment of its new mandate (Government of Alberta 2019, p. 3). The membership of this standing committee includes representation from all parties in the Alberta Legislature (Province of Alberta 2019), which represents both the governing political party of the day, as well as opposing parties. This Standing Committee, which is comprised of nine members of the Alberta legislature, reviews and approves the business plan for the Heritage Fund each fiscal year; and receives and reviews the operations and results of the Fund's investment activities (Committees of the Legislative Assembly of Alberta 2019).

A key obligation carried out by the Standing Committee is to hold annual public meetings with Albertans (recordings of which are streamed and available on their website) on the Heritage Fund's performance (Heritage Savings Trust Fund 2019). These sessions allow Albertans to enquire about the Fund's various investments, and if the mandate of the Fund for long-term savings is in fact being fulfilled. Lastly, the Standing Committee also approves the annual financial reports of the Heritage Fund prepared by the Ministry of Treasury Board and Finance and reports on the status of the Fund to the Alberta Legislature (Province of Alberta 2019). The final element of oversight in the Heritage Fund's management is an independent Auditor General, which is responsible for performing an external audit of the Fund's annual financial report (Province of Alberta 2019). These annual external audits are performed to ensure that the reports prepared by the Ministry are accurate, fair, and meet Canadian accounting standards. The Auditor General's findings are summarized in a financial statement, which is included in the Heritage Fund's publicly available annual financial reports (Government of Alberta; Alberta Treasury Board and Finance 2018, p. 24).

5 Heritage Fund Investment Policy and Operational Rules

As described in the previous section, the investment authority of the Heritage Fund lies with three bodies: the President of Treasury Board, Minister of Finance, Ministry of Treasury Board and Finance, as well as AIMCo. The investment policy of the Heritage Fund is developed by the Ministry (and approved by the Minister) through the SIP&G, as previously described. The SIP&G determines the Fund's target policy portfolio, risk profiles, and expected rate of returns; all of which governs the parameters available to AIMCo for investment decisions.

Fig. 5 Heritage Fund long-term target policy asset mix. *Source* Government of Alberta; Alberta treasury board and finance (2018, p. 7)

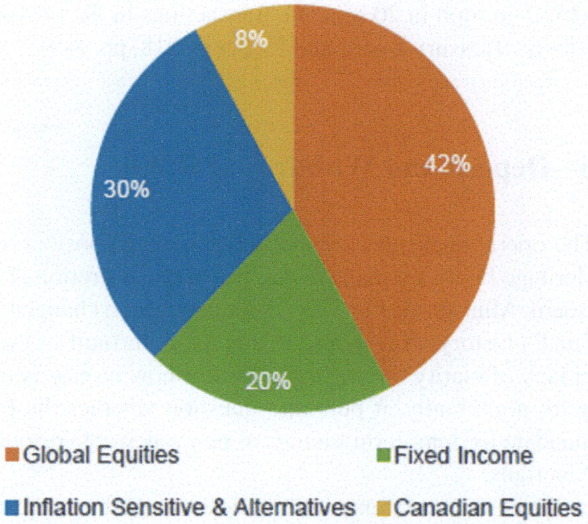

The investment policy of the Heritage Fund is based upon two vital concepts: a long-term planning approach and a diverse portfolio to manage and mitigate risk levels. The Heritage Fund is invested globally and across three broad asset classes consisting of: 50% in equities (42% global equities and 8% Canadian equities); 20% in money markets and fixed income; and 30% in inflation sensitive and alternative investments (such as real estate and infrastructure) (Government of Alberta 2011, p. 9). Through these mix of assets, which is illustrated in Fig. 5, the Heritage Fund's investment goal is to earn a target rate of return of 4.5% above the rate of inflation (as measured by the Canadian Consumer Price Index) over a 5-year period. Furthermore, it is expected that AIMCo will utilize its expertise to make investment decisions, which earn an additional 1% on top of the 4.5% rate of return (Government of Alberta 2011, p. 8).

The Heritage Fund's investment policy can be reviewed or revised at any time; however, the Ministry must formally review the policy at least once each calendar year (Government of Alberta 2011, p. 7). A significant recent revision to the Fund's investment policy occurred in November of 2015, when the provincial government introduced a new investment mandate for AIMCo aimed towards encouraging direct investments within Alberta. This new mandate, aptly named the "Alberta Growth Mandate", requires AIMCo to invest up to three per cent of the Heritage Fund (approximately $500 million) in investments, which create jobs in Alberta, build new infrastructure in the province, diversify the economy and support growth, connect Alberta businesses to export markets, and lastly, develop subject matter expertise within the province (Alberta Investment Management Corporation 2015).

Prior to this announcement, the Heritage Fund did not have clear mandate to invest directly in the Alberta market (Government of Alberta; Alberta Treasury Board and Finance 2018, p. 20). Since the announcement of this mandate, AIMCo has invested

$362.1 million in 20 separate transactions in the province (Government of Alberta; Alberta Treasury Board and Finance 2018, pp. 8-9).

6 Deposit and Withdrawal Rules

The operational rules around the amounts deposited into, and withdrawn from the Heritage Fund, are made in practice at the discretion of the President of the Treasury Board, Minister of Finance. These rules have changed several times throughout the Fund's history, in response to economic turmoil in the province. This has resulted in lack of clarity on the limitations of how money is transferred into the Fund, but more importantly, it puts into question whether the Fund has in fact achieved its mandate for long-term savings of non-renewable resources for future generations of Albertans.

While the variations in the rules surrounding deposits and withdrawals into the Heritage Fund have been highlighted in earlier sections of this chapter, it is worthwhile to revisit these changes in the context of the impact on the growth of the Heritage Fund over the years (Fig. 6).

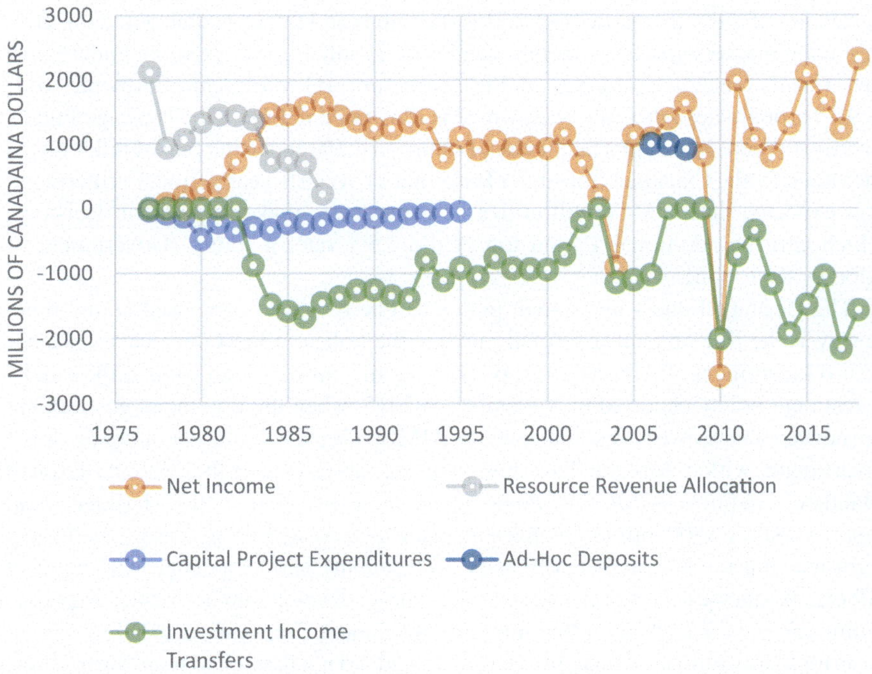

Fig. 6 Flows to and from the Heritage Fund. *Source* Government of Alberta; Alberta treasury board and finance (2018)

Upon the creation of the Heritage Fund in 1976, there was a statutory requirement to deposit 30% of non-renewable resource revenues into the Heritage Fund. However, this percentage was reduced to 15% in 1982 in response to economic downturn in the province, and by 1987, deposits into the Fund were suspended all together (Government of Alberta 2019, p. 2). From 1987 to present day, there is no longer a statutory requirement for a particular percentage to be deposited into the Fund. Since 1987, the provincial government has only deposited into the Heritage Fund on two occasions (in 2006 and 2008) for a total amount of $3.9 billion due to provincial budget surpluses (Government of Alberta; Alberta Treasury Board and Finance 2018, p. 4).

As it relates to the rules around withdrawals from the Heritage Fund, the principle amount in the Heritage Fund cannot be withdrawn. However, all investment income earned each year through AIMCo's efforts minus the amount retained for inflation proofing (as required by legislation) is transferred to the provincial government's general budget (Province of Alberta 2019, p. section 8(2)). In 2017–18, the Heritage Fund earned $1.787 billion in net income, $230 million of which was retained in the Fund for inflation proofing, while $1.557 billion was transferred to the province's general budget to support government spending (Government of Alberta; Alberta Treasury Board and Finance 2018, p. 4).

7 Distribution Mechanisms

While some intra-provincial wealth distribution mechanisms have already been discussed above with regard to State fund and natural resource revenue, we have not yet discussed Canada's most important wealth distribution mechanism: equalization. As already mentioned, Canada's system of federal–provincial relationships is complex, including wealth distribution mechanisms between provinces: "there are areas of joint or overlapping federal and provincial jurisdictions, grey areas of jurisdiction, complex tax harmonization and co-ordination arrangements, shared-cost and joint programs, and substantial transfers from the federal to provincial governments and territories. The pillars of those federal transfers are the Canada Health Transfer (CHT), the Canada Social Transfer (CST) and equalization payments" (Feehan 2014, p. 2).

Simply, equalization is a national Canadian program that distributes wealth across the country from rich to poor; it affects all provinces and territories. There is a formula applied on an annual basis whereby provinces with stronger economies and higher income households and businesses (referred to as "have" provinces) distribute their wealth via transfer payments, to provinces with weaker economies (referred to as "have-not" provinces). Fifty per cent of natural resource revenues, including royalties from oil and gas count towards this calculation. Ultimately, this means that there is a wealth transfer from resource-rich provinces like Alberta, to other provinces with

Table 1 Critical observations

Successes	Failures
• High level of transparency • Public disclosure of annual reports and audits • Independent oversight • High level of technical expertise of fund manager • Low tax rates; high quality of life; high quality public services; competitive royalty structure	• Lack of coherency in Fund objective • No deposit rules since 1987 • Lack of coherency in government policy • No clear withdrawal rules • Lack of government commitment to savings over time • Legislators frequently used fund for public expenditures • Lack of public understanding and commitment to Fund

weaker economies. In the history of equalization, Alberta has always been a "have" province, meaning that it has never received equalization payments.[7]

8 Critical Observations

While it is clear that higher deposits and fewer withdrawals from the Heritage Fund in its 43 years of existence would have yielded a higher balance than what exists today, there are other important lessons to be learned from the Alberta experience (Table 1).

9 Conclusion

As one of the first funds of its kind, the Heritage Fund holds interesting and historically relevant lessons to be learned regarding public wealth management and natural resource revenue. Chief among its successes are that it has achieved a high level of transparency, with public disclosure of annual reports and audits. It also boasts independent oversight with a high level of technical expertise/fund management. Alberta has chosen the path of levying low taxes with a highly competitive royalty structure and has historically offered high-quality public services contributing to a high quality of life for its citizens. It undeniably holds a special place among the world's earliest sovereign wealth funds, and with a current balance of $17.5 billion CAD (approx. $13.2 billion USD),[8] the Heritage Fund is larger than the sum of all other Canadians sovereign wealth funds put together. However, its savings are a far cry from what could have been if the province had been able to maintain consistent contributions and had high levels of public spending using investment income.

[7]Please see chapter in this book from Bauer et al. for more information on equalization.

[8]Current market rate at 2019-02-23.

Among its primary failures are that the Fund lacks coherency in its objective, which has changed several times since its inception; absence of clear deposit rules since 1987; and lack of clear withdrawal rules. Perhaps the most significant failure of the Fund has been the lack of an overall government commitment to accruing savings over time. Historically, the Fund has been used to fund public expenditures, and there has been an overarching lack of public understanding and commitment to the Fund and its mandate over time. Because revenue from oil and gas extraction is temporary, and revenues end when resources become depleted or extraction becomes too costly, it remains very important to save resource revenue (van den Bremer, 2016, p. 2), and so, while there have been some successes along the way, there remains room for much improvement in the future of the Heritage Fund.

References

Alberta Investment Management Corporation. (2015). *Alberta Growth Mandate,* Edmonton: s.n.

CBC News (2004). *Alberta debt free, Klein says,* Calgary: s.n.

Committees of the Legislative Assembly of Alberta (2019). *Standing Committee on the Alberta Heritage Savings Trust Fund,* Edmonton: s.n.

Dahlby, B. (2017). *The Alberta Experience.* Calgary, s.n.: Alberta Heritage Savings Trust Fund Graph.

Dahlby, B. a. N. H. (2014). The Challenge of Saving Natural Resources Revenues: An Overview of the Alberta Heritage Savings Trust Fund. *ITIC Special Report 19th Annual Edition,*July.

Fawcett, M. (2016). Heritage Trust Fund: Better Late Than Never for Saving. *Alberta Views,* 1 March.

Feehan, J. (2014). Canada's Equalization Formla: Peering Inside the Black Box…And Beyond. *The School of Public Policy SPP Research Papers,* September. 7(24).

Fiscal Planning and Transparency Act (2015).

Government of Alberta; Alberta Treasury Board and Finance. (2018). *Alberta Heritage Savings Trust Fund 2017-2018 Report,* Edmonton: s.n.

Government of Alberta; Alberta Treasury Board and Finance (2019). *Alberta Treasury Board and Finance, 2019-19 Third Quarter Fiscal Update and Economic Statement,* Edmonton: s.n.

Government of Alberta. (2011). *Alberta Heritage Savings Trust Fund Statement of Investment Policy and Goals: Heritage Fund,* Edmonton: s.n.

Government of Alberta. (2019a). *Alberta Heritage Savings Trust Fund Historical Timeline,* Edmonton: s.n.

Government of Alberta. (2019b). *Heritage Savings Trust Fund.* (Online) Available at: https://www.alberta.ca/heritage-savings-trust-fund.aspx#toc-2. Accessed 01 03 2019.

Government of Alberta. (2019c). Alberta Treasury Board and Finance, 2018–19 *Third Quarter Fiscal Update and Economic Statement* Available at: https://open.alberta.ca/dataset/9c81a5a7-cdf1-49ad-a923-d1ecb42944e4/resource/a0f104f4-75a6-4d7e-8e52-dbad812c3941/download/2018-19-3rd-quarter-report-fiscal-update.pdf. Accessed 13 Sept 2019.

Heritage Savings Trust Fund. (2019). *Assembly Online Website: Standing Committee on the Alberta Heritage Savings Trust Fund,* Edmonton: s.n.

Jordison, S. (2018). Canada. In I. Overland (Ed.), *Public Brainpower: Civil Society and Natural Resource Management* (pp. 75–90). s.l.: Palgrave Macmillan.

Kananaskis Country Golf Course (2019). *The Alberta Advantage.* (Online) Available at: https://www.kananaskisgolf.com/blog/110-title-of-blog-3. Accessed 23 Feb 2019.

Klein, P. R. (2005). *An Open Letter to Albertans,* Edmonton : s.n.

Kneebone, R. a. M. W. (2016). Canadian provincial government budget data 1980/81 to 2013/14 (updated annually). *Canadian Public Policy,* 42(1).

Morton, F. (2018). Why Alberta needs a fiscal constitution. *The School of Public Policy Publications SPP Research Paper,* pp. 1–24.

Natural Resources Canada (2017). *Reference Maps; Alberta Relief.* (Online) Available at: http://geogratis.gc.ca/pub/nrcan_rncan/raster/atlas_6_ed/reference/bilingual/alberta_relief.jpg. Accessed 23 Feb 2019.

Nelson, C. (2015). *Calgary Herald: Kananaskis golf course has taxpayers in the rough.* (Online) Available at: https://calgaryherald.com/opinion/columnists/nelson-kananakis-golf-course-has-taxpayers-in-the-rough. Accessed 23 Feb 2019.

Province of Alberta. (2007). Current as of June 12, 2013. *Alberta Investment Management Corporate Act,* Edmonton: s.n.

Province of Alberta (2019). *Queen's Printer: Laws Online/Catalogue.* (Online) Available at: http://www.qp.alberta.ca/1266.cfm?page=A23.cfm&leg_type=Acts&isbncln=9780779726646&display=html. Accessed 01 03 2019.

The Economist (2018). *Global Liveability Index 2018,* s.l.: s.n.

Van den Bremer, T. S. a. F. c. d. P. (2016). Saving Alberta's resource revenues: Role of intergenerational and liquidity funds. *Energy Policy*, pp. 1–16.

Wit, J. (2017). *Norway's Oil Savings Just Hit $1 Trillion. Alberta Has $17 Billion. What Gives?,* s.l.: The Narwhal.

Wood, J. (2015). *Alberta's Rainy Day Fund to Be Liquidated Next Year*. Calgary: Calgary Herald.

Sovereign Wealth Funds and Impact Investing in Australia

Rochelle Spencer, Eduardo G. Pereira, and Fadzai Matambanadzo

Abstract Sovereign wealth funds (SWFs) have gained traction in recent years as effective capital pools created by governments to invest surplus funds in markets, both internationally and domestically. This chapter looks at the role that SWFs can play in Australia's next resources boom, and whether SWFs can provide a sustainable development pathway through an alternative and innovative investment structure such as social impact investing. We shed light on the dynamics and role of SWFs in promoting sustainable development in Australia and the role of impact investing for accelerating sustainable development.

Keywords Extractives industry · Sovereign wealth funds (SWFs) · Sustainable development · Social impact investing

1 Introduction

The increasing global demand for commodities due to population growth, ongoing processes of industrialisation and rising urbanisation necessitates governments to embrace sustainable development and meet their commitments to the Paris Agreement (UNFCCC 2020); in turn, this will affect the nature of commodity demand

R. Spencer (✉)
Centre for Responsible Citizenship and Sustainability, Murdoch University, Perth, Australia
e-mail: rochelle.spencer@murdoch.edu.au

E. G. Pereira
Siberian Federal University, Krasnoyarsk, Russia
e-mail: egp.portugal@gmail.com

The University of West Indies, St. Augustine Campus, St. Augustine, Trinidad and Tobago

University of São Paulo, São Paulo, Brazil

F. Matambanadzo
Socio-Economic Development and Impact Strategist, Perth, Australia
e-mail: fadzai@theafricandreamgroup.com

E. G. Pereira et al. (eds.), *Sovereign Wealth Funds, Local Content Policies and CSR*, CSR, Sustainability, Ethics & Governance,
https://doi.org/10.1007/978-3-030-56092-8_12

as we see nations becoming wealthier. In Australia, a resource-rich country, gold rushes were pivotal in the early development of the nation state, marking mining as a cornerstone of the Australian economy. Australia has experienced a number of mining booms through the 1960s, 1980s and early 2000s, with the states of Western Australia and Queensland being the most resource-rich areas of the country. However, Australia has arguably not made the best use of harnessing the wealth from these resource booms to sustainably finance a new era of sustainable development as we see in other nations that have cultivated sovereign wealth success. Some have argued that Australia failed to capitalise on the mining booms because of political short-sightedness for votes and powerful mining lobby interests (Fernyhough 2015; Szatow 2020). Instead of putting the increased corporate tax revenue during the mining boom into a sovereign wealth fund, the government instituted personal tax cuts. At the same time, the mining lobby spent more than $20 million AUD on an anti-tax advertising campaign to sway public opinion and warn the Federal government not to implement further tax increases on the extractives sector; it was successful. Some critics have compared Australia's squandered iron ore boom with Norway's successful oil-funded SWF, urging Australia not to squander the next boom–electrification, and to ensure that a SWF benefits all Australians (Szatow 2020).

Sovereign wealth funds (SWFs) constitute just one of many sovereign investment vehicles[1] and are heterogeneous in nature, making them dynamic, complex and multifaceted investment instruments. Defining the key characteristics of SWFs beyond being state-owned investment vehicles can therefore be challenging. Most SWFs have been established in countries that are rich in natural resources, with oil and gas-related SWFs being the most common and largest group (Bortolotti et al. 2015). Such funds have their origins in either commodity-based surplus such as oil, gas and minerals, or non-commodity trade-based surpluses such as currency reserves. This chapter focusses on the former; the state funds that are created in Australia as a result of commodity exports and are either privately owned assets and taxed, or state-owned assets. Moreover, we delve into how SWFs can be a central function in sustainable development through 'impact investing' in Australia.

2 Defining Sovereign Wealth Funds

SWFs have gained traction in recent decades as effective capital pools set up by the state for macroeconomic purposes to invest surplus funds in foreign countries or domestically. They are increasingly being seen as effective state wealth redistribution vehicles, particularly in resource-rich countries such as Australia. The World Economic Forum (2017) identifies over 40 SWFs created since 2005 with a value of over $8.4 trillion (SWFI 2020). Indeed, the recent expeditious growth of SWFs has

[1]Types of sovereign investment vehicles include SWFs, Pension Funds, State-Owned Enterprises, and Sovereign Wealth Enterprises (Sovereign Wealth Fund Institute, 2008-2018).

led to the rapid evolution of the role of the state in recent years. While many governments began privatising state-owned assets since the 1970s, over the 2001–2012 period governments globally acquired more assets through stock purchases (US$1.52 trillion) than they sold through share issue privatisations and direct sales (US$1.48 trillion) (Bortolotti et al. 2015). The world has thus witnessed two powerful, simultaneous and apparently contradictory economic phenomena over recent years: continuing sales of state-owned assets and enterprises to private investors by some governments, coupled with increasingly large purchases of private, often listed, corporate equity by other governments (Bortolotti et al. 2015).

The International Monetary and Financial Committee (IMFC)[2] has recognised SWFs as well-established institutional investors and important participants in the international monetary and financial system (IWG 2008). This phenomenon can be called the rise of the fiduciary state, and SWFs are the single most important expression of this; we have seen their total assets grow to exceed those of hedge funds and private equity combined (Bortolotti et al. 2015). Within their host countries, particularly countries that are high exporters of natural resources, SWFs play an instrumental role in improving the management of state surpluses and creating an environment that promotes sustainable investment practices, both of state funds and of the entities in which they invest (Karametaxas 2017). The overall structure of SWFs means they are well placed to substantially shift the dial on socio-economic and environmental outcomes in their host countries and the regions the funds directly invest in, particularly in emerging economies. As demonstrated during the global financial crisis, SWFs can have a catalyst effect on global financial markets (Karametaxas 2017). They would equally have the same transformative effect on sustainable development in emerging economies through the rapid emergence of 'impact investing', discussed below in the Australian context.

While it is difficult to broadly define the dynamic nature of sovereign wealth funds, Rozanov (2005) first coined the term describing it as the accumulation of national budget surpluses from long-term budget planning and spending restraint, with the aim of building a nest egg for future generations to spend on social and economic development initiatives. Perhaps the most cited definition comes from the International Working Group of Sovereign Wealth Funds (IWF-SWF) convened by the International Monetary Fund (IMF) commonly known as the Santiago principles (2008: 7–9) that defined SWFs as:

> special-purpose investment funds or arrangements that are owned and created by the general government for macroeconomic purposes; which hold, manage, or administer assets to achieve financial objectives, and employ a set of investment strategies that include investing in foreign financial assets. The SWFs are commonly established out of balance of payments surpluses, official foreign currency operations, the proceeds of privatisations, fiscal surpluses, and/or receipts resulting from commodity exports.

The above definition is broad and offers little insight to the meaning of 'macroeconomic purposes' and 'financial objectives'. It also only places emphasis on

[2]The IMFC is a committee of the Board of Governors of the International Monetary Fund (IMF), comprising representatives of all 185 IMF member countries..

investments being foreign assets. The Organisation for Economic Cooperation and Development (OECD) has loosely defined SWFs as:

> government investment vehicles funded by foreign exchange assets that manage those assets separately from official reserves. Some are funded directly through commodity exports with an explicit decision to transfer wealth to future generations, while others result from trying to make the best from official reserves accumulation. Countries running current account and budget surpluses for macroeconomic stabilisation purposes, as well as those accumulating reserves to maintain a pegged exchange rate, have found themselves with amounts of currency well above levels deemed necessary to face external shocks. The creation of SWFs is then seen as an attempt to increase the return of this accumulated wealth (OECD 2007: 42).

This definition has a narrower view on SWFs than previous definitions, tracing their establishment to excess foreign exchange reserves, and their main objective as macroeconomic stabilisation strategies and general management of foreign currency reserves. But as we have seen throughout the world, SWFs are established using various state capital sources, and their objective and investment strategies vary greatly. The above definitions do not offer enough insight as to what role SWFs play in the context of other alternative investments such as environmental, social, and corporate governance (ESG) and sustainable development investments. These definitions attempt to define SWFs purely as traditional institutional investment vehicles without defining their dynamic characteristics that result from them being extensions of the state and consequently a reflection of the state's broad and diverse mandate of medium- to long-term social and economic development across generations. The strategic objectives of SWFs vary; therefore, defining SWFs in the context of their broad strategic objectives not only contextualises the SWF, but also gives greater clarity. Therefore, with the strategic objective of sustainable development in mind, we consolidate these definitions to offer a more refined vision for strategic development SWFs that we would like to see emerge in the Australian context, where arguably, Australia wasted SWF opportunities from the previous mining booms:

> Government owned, controlled and managed investment vehicles established as a result of balance of payments surpluses, official foreign currency operations, the proceeds of privatisations, proceeds of equity holdings in national assets, fiscal surpluses, and/or receipts resulting from commodity exports. These funds make strategic risk adjusted domestic and cross border investments in traditional assets, ESG assets and/or large-scale social or environmental projects that generate a mixture of financial and social or environmental returns for the host or recipient country.

Despite the variation of definitions, it is important to note that as alternative SWFs continue to grow in size and importance, so will their potential impact on diverse markets. Furthermore, while not nearly as homogeneous as central banks, public pension funds or traditional private investment funds, SWFs do have a number of unique characteristics, which make them a distinct and potentially valuable tool for achieving certain public policy and macroeconomic goals (Rozanov 2005). Their overall structure and dynamic characteristics could make them a transformative driver of sustainable development (particularly in emerging economies) when coupled with impact investing models.

3 Australian Sovereign Wealth Funds

Despite a number of mining booms, Australia came to the SWF party rather late, finally establishing a SWF, The Future Fund, in 2006 via The Future Fund Act 2006 (the Act). The Act established the Future Fund Special Account, the Future Fund Board of Guardians and the Future Fund Management Agency (FFMA), and collectively, they are referred to as the Future Fund. The FFMA has a mandate to invest the assets of five special purpose public asset funds: The Future Fund, the DisabilityCare Australia Fund,[3] the Medical Research Future Fund[4] and three Nation-Building Funds.[5] The FFMA has no role in determining how drawdowns will be distributed out of the various funds. The Fund received contributions from a combination of budget surpluses, proceeds from the sale of the state holding of the national telecommunications company Telstra and the transfer of remaining Telstra shares. Each of the Funds in Australia is invested across public and private markets domestically and internationally. Within each category, they develop an investment strategy for each Fund that is consistent with its investment objectives and approach to the total portfolio construction. The Future Fund provides for unfunded superannuation liabilities that will become payable when the ageing population of Australia is likely to place significant pressure on the Commonwealth's finances with the drawdown from 2026 to 2027.

In addition to the above Future Funds, The Western Australian Future Fund (WA Future Fund) was established by the Western Australian Future Fund Bill 2012 (Australian Parliament 2012) and is the only natural resource fund in Australia. It was launched with an initial investment of $300 million by the state of Western Australia, and distributions towards infrastructure will begin in 2032. The objective of the WA Future Fund is to set aside and accumulate a portion of the revenue from the state's finite mineral resources for the benefit of future generations of Western Australians (Parliament of Western Australia 2012).

The WA Future Fund was established with seed capital from the Royalties for Regions Fund totalling an estimated $1.04 billion AUD between 2012 and 2016.

---

[3]The DisabilityCare Australia Fund was established in 2014 as a result of the DisabilityCare Australian Funds Act 2013, to help the National Disability Insurance Scheme (NDIS), with the goal to help support Australians with significant and permanent disabilities and their families and carers. The Fund reimburses States, Territories and the Commonwealth for expenditure incurred in relation to the National Disability Insurance Scheme Act 2013 and to fund initial implementation of this Act.

[4]The Medical Research Future Fund was established by the Medical Research Future Fund Act 2015 to improve the health and wellbeing of Australians by providing grants of financial assistance to support medical research and medical innovation.

[5]The Nation-Building Funds were established in 2008 as a result of the *Nation-Building Funds Act 2008*, namely the Building Australia Fund (BAF) and the Education Investment Fund (EIF). The BAF enhances the Government's ability to make payments in relation to the development of transport, communications, energy and water infrastructure and in relation to eligible national broadband matters. The EIF enhances the State's ability to make payments in relation to the development of higher education infrastructure, research infrastructure, vocational education and training infrastructure. Health and Hospitals Fund: to support capital investment in health infrastructure.

Since 2008, the Royalties for Regions Fund has invested over $6 billion AUD of the mining and onshore petroleum royalties in Western Australia to nearly 4000 projects and programmes including transfers to the WA Future Fund. Since 2016–17 onwards, the Fund has been credited each and every year with a minimum 1% of the state's annual royalty income. For the first 20-year accumulation period (until 2032), earnings on the WA Future Fund balance are retained and reinvested in the Fund whereby drawdowns from the Fund are prohibited. The balance of the WA Future Fund at 30 June 2032 will be maintained into perpetuity, and the income earned on the balance of the Future Fund after that date may be applied for the purpose of providing public works and other public infrastructure in the metropolitan and regional areas of the state of Western Australia (Australian Parliament 2012).

Western Australia covers a land area over 2.6 m km^2 (to put that in some perspective, it is four times the size of Texas and twelve times the size of the UK). Therefore, the division of the Fund's income between the metropolitan area and the regions must be comprehensively negotiated between the state's Treasurer and the Minister for Regional Development. The Minister for Regional Development also consults with the Western Australian Regional Development Trust in considering the application of the Future Fund in the regions across the state.

In legal parlay, the Australian commonwealth and state governments own Australia's mineral and petroleum resources and share the responsibility in the taxing of projects. Irrespective, Australia is still grappling with recognition of the custodianship over land and sea that Aboriginal and Torres Strait Islander people have held for tens of thousands of years. At the time of writing this chapter, the extractives sector is backing a referendum for enshrining the voice of Australia's First Nations people in the Constitution. However, some critics ostensibly question the support of the big mining giants, saying:

> When mining companies are able to control Indigenous lands, extract from them and prevent them from being lovingly cared for by the peoples who are its custodians, then their profits keep incurring *sovereign debt*. They prosper from the exploitation of land that they have never had to pay for. This perhaps is so obvious that it is easily forgotten; or worse, dismissed. And yet in any other profit-making context it would be unintelligible for an industry to be able to access its raw materials, the basis for its profit and prosperity, for free (Giannacopoulos 2019).

While Australia has made some important headway in terms of Native Title, it does not equal land rights, particularly when faced with the prosperity of mining interests. Giannacopoulos (2019) argues that the extractives sector may well support a constitutional reform for First Nations peoples in Australia, 'but this is just to side-step the real question of self-determination … it underscores the fact that mining wealth creation actually heightens the austere conditions Aboriginals face in their own country'. Understanding this contested background provides the basis for the need to think about SWFs as instruments for impact investment in sustainable development in Australia.

In the Australian context, the state levies company taxes and royalties on mining projects, and these are the main fiscal instruments used by governments to collect revenues from the extractives industry. Despite the mining sector in Australia paying almost a quarter of all company tax nationally, it is a fiercely debated arena where many lobby for the government to introduce a stricter fiscal regime for the extractives sector in Australia. The existing model sees most of the economic value obtained from minerals and petroleum concessions derived largely from royalties and corporate income tax. The corporate tax rate is 30%, which is relatively high among other OECD countries (PWC 2018: 10). However, statistics from the Australian Tax Office reveal the amount of tax credits increased from $282 billion AUD to $324 billion AUD highlighting that Australia has generous tax concessions available on spending that is greater than a company's assessable tax receipts, which can be carried over from year to year. This means that the largest global oil and gas companies accumulate hundreds of billions in tax credits in Australia and pay much lower amounts or even close to nothing. Lobbies are calling for a 10% royalty to replace the existing petroleum resource rent tax (PRRT) established in the late 1980s to apply to new offshore projects (Centre for International Corporate Tax Accountability and Research 2019). The PRRT was revised in 2012 at the peak of the most recent mining boom to apply to all oil and gas production including coal seam gas and shale oil and gas. Where previously large resources companies, such as Shell and Chevron, failed to pay PRRT in Australia despite earning billions of dollars, the 2010 Henry Tax Review reinforced that the PRRT revenues are declining and the system fails to capture an appropriate share of economic resource rents that major gas projects deliver to their owners (Treasury Department 2010). When compared to Qatar's similar LNG export volumes, which collects over $25 billion in royalties alone, Australia will collect nothing in PRRT from its booming offshore gas industry. A review of the PRRT by The Australia Institute (2017: 4) called for Australia to learn from the example of Norway where 'the total tax on profits can approach 90% without deterring investment'. They propose that the majority of the profits should be returned via the PRRT to the people who own the resource with enough left for the project to operate efficiently. Interestingly, The Australia Institute (2017: 6) recommends that the current 40% PRRT should be increased to 70% on projects that have earned double their costs.

Despite the above critiques, mining contributed 6–7% of GDP in 2016–17 making it the fourth largest contributor to the national economy, and according to the Minerals Council of Australia, the mining equipment, technology and services sector brings this share of GDP to over 15% (Minerals Council of Australia n.d.). Mining is the largest source of export revenue, representing 60% of total exports (Heath 2019). Australia is the world's largest exporter of iron ore, accounting for 53% of world trade in 2018 (Department of Industry 2019), the largest exporter of metallurgical (54%) coal (ibid) and lithium (44%) (Mayyas 2018) and second largest exporter of thermal coal (20%) in the world. Significant investment has been made in mining of minerals used in modern technologies such as electronics, renewable energy and electric vehicles, particularly in Western Australia where minerals projects are increasingly value-adding through further downstream processing. Taxes and royalties from mining are a relentlessly debated topic in Australia, precisely because they provide a major source of income for the national, state and territory governments in the contested context of Native Title. Approximately $31 billion AUD in company taxes and royalties was received in the 2017–2018 financial year (Minerals Council of Australia n.d.b). A previous study from Deloitte Access Economics released in January 2018 showed mining companies alone paid an effective tax rate of 51% in 2015–16 in company tax and royalties (ibid).

Australia has a strong performing economy and is the only major economy in the world to be entering its 28th year of continuous economic growth with no recession as at 2019 (ATIC 2020). According to the IMF (2018) between 2019 and 2023, the Australian economy is predicted to outperform every other major advanced economy. Despite the size of the extractives and agriculture sectors, their exposure to global commodity price volatility and the global financial crisis of 2007–2008, growth has been inclusive and broad based. The Australian economy has proved to be a resilient and diversified economy, highlighting that Australia is well placed to responsibly and ethically manage its resource base for sustainable development outcomes that all Australians will benefit from.

4 Sustainable Development Challenges and Opportunities

The broad and overarching goal of sustainable development is to achieve both intra-generational and inter-generational equity (Hanley et al. 2007). This goal evokes Brundtland's now infamous definition of sustainable development from the World Commission on Environment and Development report: to fulfil the needs of the present generations, while ensuring future generations have the opportunity to fulfil their needs. This is particularly relevant in resource rich countries such as Australia where a large proportion of export receipts (whether directly or indirectly) are derived from non-renewable resources often exposed to the volatility of commodity prices. As further elaborated by Ashiem (2013; 2017), sustainable development is a requirement of our generation to manage the resource base such that the quality of life we ensure ourselves can potentially be shared by all future generations. This chapter recognises sustainable development as fundamentally an equity imperative rather than an efficiency issue, as pointed out by Howarth and Norgaard (1993). This does not suggest that efficiency is not a fundamental requisite to sustainable development, but it is merely not a core focus for the purposes of this chapter.

In 2015, a gathering of world leaders convened at the United Nations in New York to endorse the Sustainable Development Goals (SDGs) with the Secretary General exclaiming that the SDGs and Agenda 2030 signal a 'paradigm shift for people and the planet' (UN 2014: para. 24). This new agenda differs from the preceding Millennium Development Goals in numerous ways; there is greater integration among SDGs than the MDGs (Le Blanc 2015) with all countries being responsible for progressing a pathway forward considering different national realities, capacities and levels of development and respecting national policies and priorities. As well, all sectors of society are responsible including businesses, governments and civil society. The extractives sector arguably has particular strengths that can be brought to bear on delivering the SDGs (Lucci 2012; Porter and Kramer 2011) whereby cutting-edge technologies, big data and specialised skills underpin the competition required to drive innovation and efficiency (Kramer 2014). The question for Australia is whether the extractives sector will make a meaningful contribution to achieving the SDGs or continue with 'business as usual' where we will see social, economic and environmental winners and losers *vis-à-vis* the extractives sector.

The extractives industry is at the centre of modern industrial and post-industrial societies. Since the mid-1990s, global mining companies have actively engaged in shaping sustainability discourses, particularly in the lead up to the 2002 Johannesburg Summit (Danielson 2006). Typically, the extractives sector has been championed as leading investment, creating job opportunities and developing new economic prospects. Now with the introduction of the SDGs, it is incumbent on the extractives sector to be a 'consciously engaged agent of development' (Blowfield 2012: 415) rather than simply contributing to economic growth.

Arguably, host governments face enormous economic pressure, both internationally and nationally, to overexploit their resource base and keep up with growing global consumer demands. Paradoxically many of the SDGs would not be reached without the contribution of minerals, metals, oil and gas, which fuel manufacturing, create jobs along supply chains (Mancinia and Sala 2018) and generate much needed capital to drive sustainable development projects. Despite this, the extraction and processing of raw minerals, metals, oil and gas generates social, economic and environmental impacts that are often contradictory to the SGDs (Hickel 2019), particularly for host communities in close proximity to the projects, and specifically for women in mining and artisanal and small scale miners in developing countries. Despite the extractives sector representing a large portion of exports and current accounts, the majority of positive economic impacts from the extraction of these resources do not flow equitably in host countries and/or communities who experience the lion's share of negative impacts.

The extractives sector is thus often criticised for its role in perpetuating inequality, corruption and even violence (Spencer 2018), with critics 'highlighting patterns of continuity and the enduring legacy of extraction, exploitation and empire building' (Gilberthorpe and Rajak 2017: 188). But, in the context of Agenda 2030, the extractives industry is increasingly recognised as important because of the capacity to 'draw on their global resources for community-based 'sustainable' development projects' (Spencer 2018: 75). Exercising some caution obliges us to monitor that resource companies do not simply adopt the discourse of sustainable development to signify economic growth (where they bear no responsibility for development outcomes), rather than to signify social equality and poverty reduction. Agenda 2030 explicitly shifts the logic from resource companies simply doing business in emerging economies, to being far more intentional and instrumental, to doing development; not just development, but sustainable development (Blowfield and Dolan 2014). Sustainable development within the context of the extractives industry in Australia can work to ensure economic and social benefits from the sector are broad-based and inclusive, and environmental impact to the host communities and First Nations people are minimised to their full extent.

In the context of planetary boundaries and tipping points (Rockström et al. 2009; Steffen et al. 2015), the extractives sector unquestionably contributes to the spectre of climate change impacts (Intergovernmental Panel on Climate Change 2018) through the extraction and use of fossil fuels, which are increasingly conflictive in an era of SDGs (Klein 2014; Luke 2017; Ritter 2018). We increasingly see sector specific codes adopted by extractives companies to ensure mining activities have reduced impact on environment and communities and lasting value for society, demonstrating that the sector is a vocal proponent of sustainability in extractives (see the International Council on Mining and Metals 2001, 2015). Indeed, in Australia the extractives sector has made some progress towards improving environmental impact protocols (Morrison-Saunders et al. 2015) and mine site rehabilitation (Gardner and Bell 2007), and without doubt, the extractives sector could play a key role in energy transition and climate initiatives like carbon sequestration, cleaner fuel, eliminating/reducing flaring, etc. Nevertheless, with the SDG framework mandating the private sector to contribute to achieving Agenda 2030, a key issue for extractives companies in Australia is to advocate for the fair and equitable distribution of wealth emanating from the SWFs in order to mitigate the resource curse.

5 Sovereign Wealth Funds as Impact Investors

The objective of sustainable development and the integrated nature of the global environment and development challenges pose problems for institutions that were established on the basis of narrow profit preoccupations. However, funding models can be transformed or improved to sustainably finance a new era of sustainable development. SWFs offer governments and ethical investors concerned with sustainable development an opportunity to reframe solutions with regards to financing sustainable development. SWFs can have a transformative impact on the development of Australia in terms of how the wealth is redistributed equitably (World Bank 2015/2016). Like SWFs, social impact investments are rapidly emerging and gaining traction globally with an increasing recognition of the power of investment capital to address social and environmental challenges. The Global Impact Investing Network's *Sizing the Impact Investing Market* 2019 report provides an in-depth analysis of the current size and composition of the global impact investing market estimating its size at \$502 billion USD at the end of 2018. Meeting the SDGs imperative requires trillions of dollars, and already we see a quarter of professionally managed assets now embracing sustainability principles (USSIF 2018). Indeed there is great potential for impact investing to intentionally progress sustainable social and environmental development.

There are lessons to be gleaned from the SWFs of other countries, which could have a bearing on the intentional social impact investing of Australia's SWF. Norway, for example, used its SWF to invest outside the country thereby protecting its currency and market against so-called Dutch disease. Alternatively, SWFs could be invested within Australia to develop key strategic areas of social development or environmental conservation. In any case, regardless of whether SWFs are invested abroad or domestically, impact investing requires that the SWF supports sustainable development projects. In the Australian context, where the large portion of the SWFs is derived from the extractives sector, there is a push from lobby groups for the government to prioritise energy transition. When the Western Australia Environmental Protection Authority recently recommended a carbon neutral proposal for new or expanding projects to offset their emissions, oil and gas companies strongly resisted attempts to enforce them to pay to limit the impact of their pollution. It's a familiar story where the Australian government claims it is serious about a climate policy that accords with the Paris Agreement while vastly increasing its fossil fuel sales to developing countries. This is where utilising Australia's SWFs for impact investing could lead to sustainable development outcomes from the extractives sector. The World Bank investigated the relationship between SWF investments, economic growth and long-term financing (Diallo et al. 2016) finding that the average GDP growth rate positively correlates with SWF investments (Diallo et al. 2016: 8). This suggests that countries receiving SWF investments grow faster, and these investments positively affect economic growth, illustrating the transformative potential of SWFs in sustainable development.

An emerging trend that responds to multiple SDGs is social impact investing.[6] This is the notion that investors can pursue financial returns while also intentionally addressing social and environmental challenges (Bugg-Levine and Emerson 2011). Impact investing has also been described as actively placing capital in enterprises that generate social or environmental goods, services or ancillary benefits with expected financial returns ranging from the highly concessionary to above market (Brest and Born 2013). The overall simplicity of the definition can often seem ominous to some mainstream investors who see social impact and sustainable development as purely the role of government, aid and philanthropy. The intention is to make social/environmental impact the primary qualifying criterion and motivation is the presiding factor; investments that unintentionally result in social good are not regarded as impact investments (Monitor Institute 2009). The idea that capital can create blended value[7] may seem like a dichotomy. All organisations, whether they are for-profit or non-profit, create value that has economic, social or environmental significance; however, failing to manage for blended value or developing an intentional impact strategy, means business managers and investors miss opportunities to capture their total value potential (Bugg-Levine and Emerson 2011).

SWFs have a dual mandate. On the front-end, they must invest surplus funds for future generations and achieve the best rate of return relative to their strategic objective. On the back-end, they must distribute drawdowns equitably in a manner that has the maximum socio-economic or environmental impact. Impact investing can address both of these roles–capital can be used to secure financial and social/environmental

[6] Social impact implies the inclusion of social, environmental and economic impact in its meaning.

[7] Blended value is the recognition that capital, community, and commerce can create more together than the sum of the three independently. See Bugg-Levine & Emerson (2011).

returns simultaneously at the front-end, while being used to distribute funds equitably at the back-end. For the Australian context, we focus here on the front-end and discuss how SWFs can structure strategic development funds *as* impact investment vehicles to seek out both financial and social impact dividends. It is not to say the distribution of funds is any less important; however, given the maturity stage of SWFs and impact investing in Australia, the investment mandate is our main focus (Monitor Institute 2009: 12).

Impact investors place capital across all asset classes, including private equity, debt, working capital lines of credit, soft loans, loan guarantees, fixed income and real estate. In addition to this, impact investing has various financing innovations, namely social impact bonds (or development bonds), and pays for entrepreneurship (World Bank 2015/2016). The generally accepted rate of return varies among investors; however, market-rate return is the expectation. Growth in impact investing has predominantly been fuelled by:

- prominent family offices looking for large-scale innovative opportunities to social problems,
- clients of private banks needing more investment options than traditional investments or philanthropy,
- private equity funds looking to provide growth capital to business with social environmental solutions,
- mutual funds who dedicate a portion of their funds to emerging companies with social and environmental solutions,
- pension funds and sovereign wealth funds wanting to diversify their funds,
- corporations who want to materially improve people's lives, and governments investing in funds that support economic development (Monitor Institute 2009).

SWFs have a significant role to play in the growth of the impact investment market globally and in particular in Australia. Despite being at an embryonic stage in Australia, impact investing is increasingly being used by countries to deliver aid programmes and achieve development policy outcomes, like reducing poverty (Commonwealth of Australia 2017). The overall investment mandate for SWFs is closely aligned to the mandate of impact investing. As investors of social and environmental progress, strategic capital partnerships between the private institutional impact investors and SWFs, could be transformative.

Thus far, both private and public investors involved in impact investing agree that financial and sustainable development returns can go hand-in-hand (OECD 2019) and that impact investing offers a unique opportunity to leverage private capital at a large scale to achieve sustainable development. Despite the fact that impact investors have a common vision of balancing financial with social/environmental returns, their rate-of-return expectations can still conflict and therefore these investors are categorised into two broad groups. The Monitor Institute, in their *Investing for Social and Environmental Impact* report, defines these two broad groups as: financial first investors who seek to optimise financial returns with a floor for social/environmental impact. This group tends to consist of commercial investors who search for investment vehicles that offer market-rate returns while yielding some social/environmental good.

The second group are defined as impact first investors who seek to optimise social or environmental returns with a financial floor. This group uses social/environmental good as a primary objective and may accept a range of returns, from return of principal to market rate. This group is willing to accept a lower than market rate of return on investments that may be perceived as higher risk in order to help reach social/environmental goals that cannot be achieved in combination with market rates of financial return.

There are several ways to invest capital as an impact investor (e.g. traditional investments like public equities and hedge funds or social impact investments like special bonds and aid development). However, given the maturity stage of the impact investment market and SWFs, social impact bonds (SIBs) or development bonds (DBs), are an asset class that could potentially play a strategic role in driving the impact investment market in economies of the Global South. SIBs or DBs are pay-for-success instruments (OECD 2019) that not only allow investors and governments to share risk, but the leverage lender (which is usually government, but not always) only pays for successful outcomes and investors receive a coupon at a premium, usually above market rate[8] when outcomes are met by the operator or service provider. SIBs or DBs allow governments to test innovative solutions for sustainable development while sharing the risk with private investors. In the context of the SWF as an investor (and not the leverage lender),[9] the investment model for SIBs/DBs where the SWF is a sole or co-investor means SWFs can play a more active and even riskier role as an investor rather than just a leverage lender. However, in this case, a SWF must consider not only the risks associated for its 'shareholders' but also the SIBs/DBs goals, thereby providing a more balanced approach than a traditional investor seeking profits as their main goal.

Underpinning the 17 SDGs is the need for sustainable financing to drive each SDG. Goal 17 (revitalise the global partnership for sustainable development) addresses the need to improve financial resources available to drive sustainable development. In particular, the SDG targets 17.1, 17.3, 17.6 and 17.17 (UN n.d.). call for a need to strengthen domestic resource mobilisation, the mobilisation of additional financial resources from multiple sources, global multi-stakeholder partnerships for sustainable development that mobilise financial resources to support the achievement of the SDGs in all countries, and encourage effective public-private and civil society partnerships (UN n.d.b). Impact investing is one pathway to bridge this gap, particularly in relation to enhancing public–private partnerships in the mobilisation of a diverse suite of financial resources. By leveraging the private sector, these investments can provide sustainable development solutions at a scale that philanthropy, aid or traditional business interventions usually cannot reach (OECD 2019).

In Australia, impact investing is at a critical phase in its market adoption. It has transitioned from uncoordinated innovation to market building (Monitor Institute

[8]Coupon rates are determined by considering the potential savings that government would receive as a result of the successful outcomes.

[9]SWFs could also be leverage lenders (or the outcome funder), at which point they will no longer be investing, but distributing funds.

2009). During this market building phase, impact investors have shifted from activist investors to mainstream investors, social businesses have improved their business models, and there has been a general improvement in impact measurement practices. In Australia, the impact investment market has somewhat matured, and there have been five SIBs launched in several states (Western Australia, New South Wales and Victoria) (SWFI 2020b). These SIBs were launched without the involvement of any SWFs but have been a collaboration between state governments and institutional investors such as private banks, major philanthropists and corporations. Integration of SWFs as major investors and drivers of market maturity is critical. Increasingly, investors are looking for the best ways to achieve financial return and impact and are eager to source deals in diverse settings.

General practice is that SWFs establish several vehicles that have distinct strategies and objectives. In order to integrate impact investing into the investment strategies of Australia's SWFs, a specific impact investing development fund would need to be established. Depending on the goals and objective of that fund, the SWF investor could seek financing partners from other impact investors. Each partnership would decide the investment motivation; profit first or impact first. Ultimately, motivation will play a role in determining the types of investments each investor will consider regardless of the asset class or geography in which they choose to invest. Given the market stage of impact investing globally and in Australia, the government needs to initially see itself as a key market enabler and stimulator and invest in building and developing a coordinated impact investment ecosystem.

The Australian Government highlights the main principles that it considers essential to implement social impact investments (Treasury Department n.d.):

1. Government as market enabler and developer
2. Value for money
3. Robust outcomes-based measurement and evaluation
4. Fair sharing of risk and return
5. Outcomes that align with the Australian government's policy priorities
6. Co-design.

These principles are somewhat self-explanatory; however, what's important is that they should not be seen in isolation. Rather these principles form a coordinated impact investment ecosystem in which the government enables and facilitates a holistic approach between diverse governmental authorities, private sector investors and civil society in order to achieve such principles. Questions concerning these principles might include what are the government priorities? How to fairly distribute wealth from the investments across different regions of such a geographically large country? What are the criteria to determine the value for money? Is there enough human capital to attend to demand for the desired areas of investments? Are goals short-, mid- or long-term? The answers to these questions might be complex but necessitate a transparent and open debate, otherwise it will be difficult to provide accountability to verify whether the SWF performance is consistent with its goals and the prescribed Australian government principles.

While SIBs are suggested as a potential strategic entry point for SWFs into impact investing, the Australian Government's Treasury Department report on social impact investing noted that social impact investing is not suitable for funding every service or programme; it merely provides an opportunity to address problems where existing policy interventions and service delivery are not achieving the desired outcomes (Commonwealth of Australia 2017). Determining where these opportunities exist is a fundamental action towards deciding which areas of development social impact investing might have the most leverage to deliver better sustainable development outcomes for Australia's SWFs. This section of the chapter has delved into social impact investment as a central function for how Australia's SWFs manage their resource wealth for sustainable development outcomes of such investments.

6 Conclusion

Sovereign wealth funds can play an essential role in assisting resource-rich countries to protect against volatility of commodity prices as well as save valuable resources for future generations. Australia provides an example of a resource-rich nation that has historically lacked vision and planning on how to manage its natural resources wealth. Although Australia has extracted natural resources (especially via mining) for over a century, it took decades to develop a SWF. However, it is never too late to review such decisions, and fortunately, Australia and one of its key resource states (Western Australia) have developed such financial mechanisms as they move into what may be another resources boom.

Australia faces serious challenges with its extractives sector if it is to strike a balance between its climate commitments to the Paris Agreement, its recognition of its Indigenous people via Native Title and in the Constitution, and managing its sovereign wealth funds responsibly and equitably. We have suggested one pathway, social impact investing, for how Australia might harness the wealth from its natural resources to sustainably finance a new era of sustainable development as seen in other nations that have cultivated sovereign wealth success.

References

Asheim, G.B. (2017). *Sustainable Growth. Social Choice Welfare, 49,* 825–848.

Asheim, G.B. (2013). Hartwick's rule. In J. Shogren (Ed.), *Encyclopedia of Energy, Natural Resource and Environmental Economics* (Vol. 2, pp. 314–320). Amsterdam: Elsevier.

Australia Institute, The. (2017). *Review of the Petroleum Resource Rent Tax*, Canberra. https://treasury.gov.au/sites/default/files/2019-03/R2016-001_The-Australia-Institute.pdf. Accessed 3 March 2020.

Australian Parliament. (2012). Western Australian Future Fund Bill 2012: Explanatory Memorandum. Online at http://www.parliament.wa.gov.au/parliament/bills.nsf/6B335AD31398DA1748257A5B00181391/%24File/EM%2B300-2.pdf. Accessed August 2019.

ATIC. (2020). "Why Australia." Australian Trade and Investment Commission. Online at: https://www.austrade.gov.au/International/Invest/Why-Australia/robust-economy. Accessed 26 March 2020.

Blowfield, M. (2012). Business and development: making sense of business as a development agent. *Corporate Governance, 12*(4), 414–426.

Blowfield, M., & Dolan, C. (2014). Business as a Development Agent: Evidence of possibility and improbability. *Third World Quarterly, 35*(1), 22–42.

Bortolotti, B., Fotak, V., & Megginson, W. L. (2015). The Rise of Sovereign Wealth Funds: Definition, Organization, and Governance. In Stefano Caselli, Guido Corbetta, & Veronica Vecchi (Eds.), *Public Private Partnerships for Infrastructure and Business Development* (pp. 296–318). New York: Palgrave MacMillan.

Brest, P., & Born, K. (2013). *Unpacking the Impact in Impact Investing, Stanford Social Innovation Review, Fall 2013: Leland Stanford Jr.* California: University.

Bugg-Levine, A., & Emerson, J. (2011). Impact Investing: Transforming How We Make Money while Making a Difference, *Innovations: Technology. Governance, Globalization, 6*(3), 9–18.

Centre for International Corporate (2019) Tax Accountability and Research, Submission of the Centre for International Corporate Tax Accountability & Research (CICTAR) on the *Treasury Laws Amendment (Petroleum Resource Rent Tax Reforms No. 1) Bill 2019* https://treasury.gov.au/sites/default/files/2019-04/t344421-centre_for_international_corproate.pdf.

Commonwealth of Australia. (2017). *Social Impact Investing: Discussion Paper*. Australian Government: Canberra.

Danielson, L. (2006). *Architecture for Change: an Account of the Mining*. Berlin: Minerals and Sustainable Development Project, Global Public Policy Institute.

Department of Industry. (2019). Australian Government, Department of Industry, Innovation and Science. Office of the Chief Economist. *Resources and Energy Quarterly*, June 2019. Online at https://publications.industry.gov.au/publications/resourcesandenergyquarterlyjune2019/documents/Resources-and-Energy-Quarterly-June-2019-Iron-Ore.pdf.

Diallo, B., Tchana Tchana, F., & Zeufack, A.G. (2016). *Sovereign wealth funds and long-term investments in Sub-Saharan Africa* (English). Policy Research Working Paper; no. WPS 7903. Washington, D.C.: World Bank Group.

Fernyhough, J. (2015). *How Australia Squandered the mining boom*. Online at https://thenewdaily.com.au/finance/finance-news/2015/01/15/australia-squandered-mining-boom/.

Gardner, J. H., & Bell, D. T. (2007). Bauxite mining restoration by Alcoa World Alumina Australia in Western Australia: Social, Political, historical, and environmental contexts. *Restoration Ecology, 15*(s4), S3–S10.

Giannacopoulos, M. (2019). Blood money: Why mining giants are backing an Indigenous voice to parliament', Opinion, ABC Religion & Ethics, Posted Fri 15 Feb 2019, 10:17am Updated Thu 30 May 2019, 8:50am. https://www.abc.net.au/religion/why-mining-giants-are-backing-an-indigenous-voice-to-parliament/10815008. Accessed 7 Mar 2020.

Gilberthorpe, E. & Rajak, D. (2017). The Anthropology of Extraction: Critical Perspectives on the Resource Curse, *The Journal of Development Studies, 53*(2), 186–204.

Hanley, N., Shogren, J. F., & White, B. (2007). *Environmental Economics in Theory and Practice*. UK: Red Globe Press.

Heath, A. (2019). *Address to the association of mining and exploration companies*, Perth. 5 June. Online at https://www.rba.gov.au/speeches/2019/pdf/sp-so-2019-06-05.pdf.

Hickel, J. (2019). The contradiction of the sustainable development goals: Growth versus ecology on a finite planet. *Sustainable Development*, pp. 1–12.

Howarth, R.B., & Norgaard, R.B. (1993). Intergenerational transfers and the social discount rate. *Environ Resource Econ, 3*, 337–358.

Intergovernmental Panel on Climate Change. (2018). *Global warming of 1.5°C. An IPCC Special Report on the impacts of global warming of 1.5°C above pre-industrial levels and related global greenhouse gas emission pathways, in the context of strengthening the global response to the*

threat of climate change, sustainable development, and efforts to eradicate poverty. Geneva, Switzerland: IPCC.
International Council on Mining and Metals. (2001). *Sustainable Development Framework: ICMM Principles*. London: ICMM.
International Council on Mining and Metals. (2015). *Sustainable Development Framework: ICMM Principles*. London: ICMM.
International Monetary Fund. (2018). *World Economic Outlook April 2018: Cyclical Upswing*. Structural Change, Washington: International Monetary Fund, Publication Services.
IWG (International Working Group of Sovereign Wealth Funds). (2008). *Sovereign Wealth Funds: Generally Accepted Principles and Practices—"Santiago Principles"* (Washington). http://www.iwg-swf.org/pubs/eng/santiagoprinciples.pdf. Accessed 4 Mar 2020.
Karametaxas, X. (2017). Sovereign wealth funds as socially responsible investors. In: Adinolfi G., Baetens F., Caiado J., Lupone A., Micara A. (Eds.) *International Economic Law*, Springer. pp. 271–288.
Klein, N. (2014). *This Changes Everything*. New York: Simon & Schuster.
Kramer, M. (2014). Profit: *The new corporate motivation to do good*. https://sharedvalue.org/groups/profit-new-corporate-motivation-do-good. Accessed 4 Mar 2020.
Le Blanc, D. (2015). Towards integration at last? The Sustainable Development Goals as a network of targets. *Sustainable Development, 23*, 176–187.
Lucci, P. (2012). *Post-2015 MDGs: What Role for Business?*. London: ODI.
Luke, H. (2017). Social resistance to coal seam gas development in the Northern Rivers region of Eastern Australia: Proposing a diamond model of social license to operate. *Land Use Policy, 69*, 266–280.
Mancinia, L., & Sala, S. (2018). Social impact assessment in the mining sector: Review and comparison of indicators frameworks. *Resources Policy, 57*, 98–111.
Mayyas, A. (2018). Are there enough materials to cover li-ion batteries?. Clean Energy Manufacturing Analysis Center August 15. Online at https://www.manufacturingcleanenergy.org/blog-20180815.html. Accessed 26 Mar 2020.
Minerals Council of Australia. (n.d.). *Driving Prosperity*. Online at https://minerals.org.au/driving-prosperity. Accessed 26 March 2020.
Minerals Council of Australia. (n.d.). *Taxation*. Online at https://minerals.org.au/taxation. Accessed 26 Mar 2020.
Institute, Monitor. (2009). *Investing for social and environmental impact: A design for catalysing and emerging industry*. USA: Global Impact Investing Network.
Morrison-Saunders, A., Bond, A., Pope, J., & Retief, F. (2015). Demonstrating the benefits of impact assessment for proponents. *Impact Assessment and Project Appraisal, 33*(2), 108–115.
OECD. (2007). *OECD Economic Outlook*, 82, December.
OECD. (2019). *Social impact investment 2019: The impact imperative for sustainable development*. Paris: OECD Publishing.
Porter, M.E., & Kramer M.R. (2011). The big idea: creating shared value. how to reinvent capitalism—and unleash a wave of innovation and growth. *Harvard Business Review, 89*(1/2), 62–78.
PWC. (2018). PWC Australia Africa practice report 'battle of the taxes who comes out on top: Australia and Africa compared. Online at https://www.pwc.com.au/industry/energy-utilities-mining/africa-practice/assets/battle-of-the-taxes-australia-and-africa-compared-2018.pdf.
Ritter, D. (2018). *The coal truth: The fight to stop Adani, defeat the big polluters and reclaim our democracy*. Perth: University of Western Australia Press.
Rockström, J., Steffen, W., Noone, K., et al. (2009). Planetary boundaries: Exploring the safe operating space for humanity. *Ecology and Society, 14*(2), 32.
Rozanov, A. (2005). Who holds the wealth of nations? *International Journal of Central Banking, 15*(4), 52–57.

Sovereign Wealth Fund Institute. (2020a). *Top 89 largest sovereign wealth fund rankings by total assets*. https://www.swfinstitute.org/fund-rankings/sovereign-wealth-fund. Accessed 5 Mar 2020.

Sovereign Wealth Fund Institute. (2020b). List of sovereign wealth fund profiles in Australia and Pacific. Online at: https://www.swfinstitute.org/profiles/sovereign-wealth-fund/australia-and-pacific. Accessed 24 Mar 2020.

Spencer, R. (2018). Corporate social responsibility as sustainable development: a critical anthropology of development perspective. In M. Brueckner, R. Spencer, & M. Paull (Eds.), *Disciplining the Undisciplined: perspectives on responsible citizenship, corporate social responsibility and sustainability*. UK: Springer.

Steffen, W., Richardson, K., & Rockström, J. (2015). Planetary boundaries: guiding human development on a changing planet. *Science, 347*, 6223.

Szatow T. (2020). Australia needs a sovereign wealth fund like Norway for the next boom—electrification. Online at https://www.michaelwest.com.au/australia-needs-a-sovereign-wealth-fund-like-norway-for-the-next-boom-electrification/. Accessed 6 Mar 2020.

Treasury Department. (2010). Australia's future tax system review final report (2010). Online at https://treasury.gov.au/review/the-australias-future-tax-system-review/final-report. Accessed 9 Mar 2020.

Treasury Department. (n.d.). Australian Government principles for social impact investing. Online at https://treasury.gov.au/programs-initiatives-consumers-community/social-impact-investing/australian-government-principles-for-social-impact-investing. Accessed 24 Mar 2020.

United Nations (UN). (2014). *The Road to Dignity by 2030: Ending Poverty, Transforming All Lives and Protecting the Planet. Synthesis Report of the Secretary-General on the Post-2015 Agenda*. New York. http://www.un.org/ga/search/view_doc.asp?symbol=A/69/700&Lang=E. Accessed 4 Mar 2020.

UN. (n.d.). *Goal 17: Revitalize the global partnership for sustainable development. Sustainable Development Goals*. Online at https://www.un.org/sustainabledevelopment/globalpartnerships/. Accessed 24 Mar 2020.

UN. (n.d.). #Envision2030 Goal 17: Partnerships for the Goals. UN Department of Economic and Social Affairs Disability. Online at https://www.un.org/development/desa/disabilities/envision2030-goal17.html. Accessed 21 Mar 2020.

UNFCC. (2020). *Paris Agreement*. Online at https://unfccc.int/process-and-meetings/the-paris-agreement/the-paris-agreement. Accessed 16 March 2020.

United States Social Impact Foundation (USSIF). (2018). *Report on US Sustainable, Responsible and Impact Investing Trends 2018*, Washington. https://www.ussif.org/files/Trends/Trends%202018%20executive%20summary%20FINAL.pdf. Accessed 23 March 2020.

World Bank. (2015/2016). *Long-Term Finance. Global Financial Development Report*, Online at: http://documents.worldbank.org/curated/en/955811467986333727/pdf/99100-PUB-REVISED-Box393195B-OUO-9-until-9-14-15.pdf. Accessed 24 Mar 2020.

World Economic Forum. (2017). *What is a sovereign wealth fund*. https://www.weforum.org/agenda/2017/10/what-you-need-to-know-about-sovereign-wealth-funds/. Accessed 4 Mar 2020.

Norway's Sovereign Wealth Fund

Jonathon W. Moses

Abstract Norway's sovereign wealth fund (SWF), the Government Pension Fund Global (GPFG), has become the largest in the world. But size is not the most unique or interesting feature of Norway's petroleum fund. This contribution describes how the Norwegian authorities took their time before committing to a SWF, deciding instead to spend the country's first oil revenues on developing its welfare state and paying down its debts. Once established, the GPFG developed several unique characteristics. What is perhaps most remarkable about the GPFG is the way that its funds can (and cannot) be accessed by political officials. In this way, the GPFG plays an important role in protecting the Norwegian economy from some of the curses we often associate with resource wealth. Finally, this contribution examines the unique way in which Norway manages the GPFG, by including strong political and ethical components when determining where it will place its investments.

Keywords Norway · Sovereign wealth funds (SWF) · Government Pension Fund Global (GPFG) · Petroleum fund

Norway's sovereign wealth fund (SWF), The Government Pension Fund Global (GPFG), is the largest in the world. In 2017, the value of the GPFG tipped over the $1 trillion USD threshold, and it shows little sign of slowing down. Given Norway's relatively small size (population, ca. 5.1 million people), the GPFG corresponds to roughly $200,000 USD for every Norwegian citizen.

The success of this fund has won it admirers from around the world, and it has become increasingly common for young oil states to establish a petroleum fund, in hopes of following Norway's example. This example is often misunderstood, as Norway itself did not rush to establish a petroleum fund, and that the most remarkable

This contribution draws heavily from my book with Bjørn Letnes. See Moses and Letnes (2017a).

J. W. Moses (✉)
Department of Sociology and Political Science, Norwegian University of Science and Technology, Trondheim, Norway
e-mail: jonathon.moses@ntnu.no

E. G. Pereira et al. (eds.), *Sovereign Wealth Funds, Local Content Policies and CSR*, CSR, Sustainability, Ethics & Governance,
https://doi.org/10.1007/978-3-030-56092-8_13

component of the GPFG is not its size, nor its make-up, but the way it can (and cannot) be accessed by elected officials.

For these reasons, this contribution aims to provide some nuance to the Norwegian example. It does so by describing three of the more unique and admired components of the Norwegian fund: how Norway postponed the creation of its petroleum fund, spending the money instead on pressing political needs; how the GPFG is used to buffer the national economy from volatile prices and the threat of Dutch Disease; and how the fund provides political leadership to the global investment community by being managed in an ethical and socially responsible manner.

1 The Birth of a Petroleum Fund

Norway first discovered oil, in its impressive Ekofisk field, at the end of 1969. Production began soon thereafter, in 1971. Throughout the 1970s, the level of oil (and eventually gas) production increased gradually, and this increase continued over subsequent decades, as seen in Fig. 1. Today Norway is the third largest exporter of natural gas in the world and supplies about 25% of the EU's demand (NPD 2019a).

It is not by accident that Norway was developing its productive petroleum capacity at the very time the global price of oil shot through the roof, in a decade characterized by "oil crises". The North Sea was developed because it offered political stability, when the Middle East was in crisis (even if the weather and production conditions were anything but stable). Consequently, Norway's oil activities generated substantial economic rewards throughout this period. With all this oil money pouring into the country, it is noteworthy that Norway did not immediately begin to invest its money in a sovereign wealth fund. For the first 25 years of Norwegian oil production, Norway's

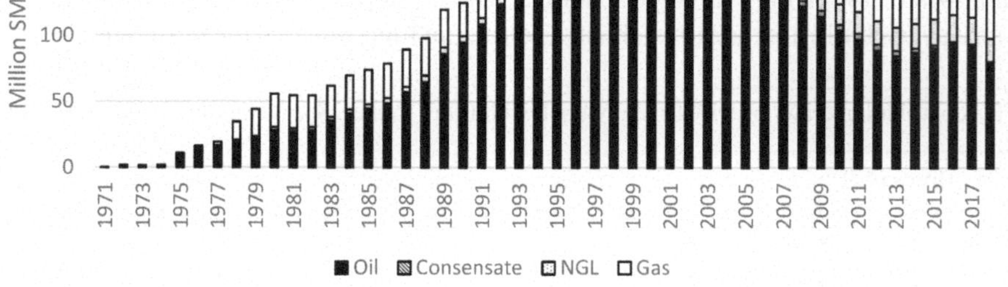

Fig. 1 Norwegian production figures, 1971–2018. *Source* NPD (2019b)

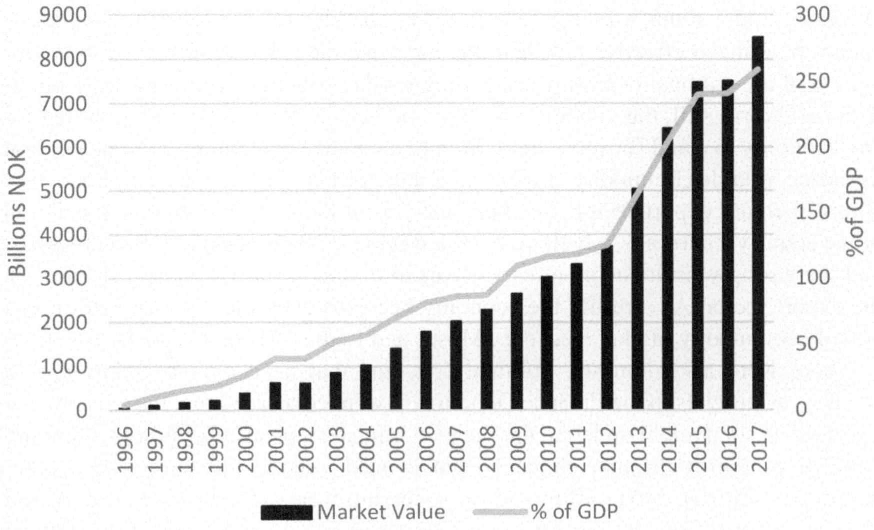

Fig. 2 GPFG, 1996–2017. *Source* NPD (2018)

oil income was being consumed just as rapidly as it was being produced. As seen in Fig. 2, the first instalments to what we now call the GPFG began in 1996, nearly three decades after the initial North Sea discoveries. Once established, however, the size of the Fund increased steadily and significantly over the subsequent years, along with the production figures seen in Fig. 1, to become eventually the global leader that it is today.

This temporal gap—between the wealth coming in, and the Norwegian government's willingness to invest it in a petroleum fund—can be explained by Norway's longstanding and growing debt at the time. Recognizing the enormous economic promise latent in its proven petroleum reserves, the Norwegian government borrowed substantially throughout the 1970s. These loans were used to prepare the country for its subsequent management of that resource (e.g. developing professional and technical competencies) and to ensure that the Norwegian population would benefit from this increased wealth (by investing in welfare and infrastructure). In other words, Norway used its early oil money to develop the national economy: to pay off its existing debts, encourage local development and growth, and to improve the basic welfare of its population. It was only after these basic necessities were secured, and when the country began to worry more about the inflationary consequences of increased spending, that a petroleum fund was introduced.

This is not to say that Norwegian officials were unaware of the macroeconomic threat from its new-found oil. Developing an oil-based economy is as much a curse as it is a blessing (Karl 1997; Ross 1999). Already in the early 1970s, there was a growing awareness of the threat that oil revenues could have on the competitiveness of Norway's tradeables sector. Today we refer to this threat with reference to Dutch

Disease[1]: that too much money, entering too quickly into the domestic economy, cannot be utilized effectively. When this happens, the result is inflation, an appreciation of the national currency, and falling levels of international (price) competitiveness. Worse still, the volatility of global oil prices makes oil a very poor leg for any economy to stand (or rely) upon. Should a country come to rely too heavily on this price-volatile commodity, it will find it quite difficult to sustain an alternative or complementary export sector. An abundance of oil money can fuel an appreciation of the country's currency, which makes it tempting (and cheaper) to rely on imports, and increasingly difficult (more expensive) to maintain (price) competitiveness in the export sector. As a result, the economy becomes even more dependent on this volatile commodity market (see, e.g., Moses and Letnes 2017a: Chap. 7).

There were also domestic political concerns that needed to be addressed in Norway, as officials in the Finance Department were concerned that the existence of a petroleum fund might undercut the Finance Minister's position of power—funding political measures through channels that existed outside the normal budgetary routines (Lie 2010: 345). All too often, sovereign wealth funds offer an easy and alternative source of funding for those looking to avoid constitutional or political constraints on political authority. This additional pile of money constitutes a problem in that it can reduce government transparency and accountability. Norway's unique institutional response to these concerns is described below.

In 1973, a Norwegian government white paper explicitly recognized that controlling the cost and pace of adjustment to the new oil wealth was one of the most important issues facing the country (*St. meld. nr.25* (1973–74), pp. 6–7). To address that challenge, the government established what became known as the "Pace Committee", under the leadership of Hermod Skånland—subsequently governor of Norway's central bank (NOU 1983: 27). The name of this committee is derived from its conclusion: the committee recognized that the easiest and most realistic way to avoid the inflationary consequences of increased oil revenues was to decrease the pace of petroleum activities in Norway.

The Pace Committee entertained the possibility of creating a petroleum fund (what they called a "buffer fund"), invested off shore, but they were (rightly) sceptical that politicians would be able to keep their hands off this fund.[2] Although it made economic sense, the committee doubted the political realism of a fund. Indeed, countries with a significant need for capital investment or those that suffer from corruption might do more harm than good in establishing an offshore investment fund. It is for these reasons that the Pace Committee concluded it was better to maintain international competitiveness by reducing the country's pace of resource extraction, rather than developing a sovereign wealth fund. Reducing the pace of production was still

[1] *The Economist* news magazine first coined this term in 1977. See *Economist* (1977).

[2] It should be noted that this was not the first time that the idea of a fund was publicly entertained in Norway. Already in the mid 1960s, when Norway was establishing its sovereign claim of the Norwegian continental shelf, the Norwegian Prime Minister, Einar Gerhardsen, spoke of the potential for a Norwegian oil fund (NBIM n.d.).

possible at a time when the oil industry remained nascent in Norway.[3] Over time, as the industry grew in size and political power, it has become increasingly difficult (and politically unrealistic) to argue for a decreased pace of extraction.

It was not before 1990, almost a decade later, that the Norwegian government adopted its Act on the Government Petroleum Fund (Act of 22 June 1990, No. 36). This Act created "The Government Petroleum Fund", but its name was changed in 2006 to the "Government Pension Fund Global" (GPFG).[4] Despite the subsequent name change, the GPFG is not (and has never been) formally linked to the Norwegian pension system, nor does it include any pension liabilities. The initial motivation for the fund was to establish a pile of savings that could help Norway transition to a post-oil economy once the petroleum ran dry. To date, there has been no explicit political decision as to how the money in the fund might be used in the future; the reference to pensions was simply a convenient way for politicians to secure broad political support for the fund.

Although it was created in 1990, the first transfer of funds to the Government Petroleum Fund did not occur until 1996 (as we saw in Fig. 2). Since that first, somewhat tardy, deposit, the GPFG has grown beyond the wildest dreams of its original creators. Today, as shown in Table 1, the GPFG is the largest sovereign wealth fund in the world, with investments spread around the world and across a varied portfolio that includes equity (roughly 67%), fixed-income (roughly 30%) and real estate investments (roughly 3%) (NBIM 2018). The manner by which these investment decisions are made is described in more detail below. But the end result is that the GPFG owns shares in more than 9000 companies, invested in 72 different countries around the globe, on behalf of the Norwegian people (NBIM 2017a). Calculated in another way, Norwegians—who constitute less than one-thousandth of the world's population—own roughly 1.3% of the world's listed companies (Moses and Letnes 2017a: 135).

The Norwegian experience offers a remarkable Cinderella story. Like many young oil nations, Norway faced a long list of pressing economic and social needs when it first found oil. It is not surprising, then, that the country decided to spend its early oil money to develop Norwegian society and to pay off its sundry debts. It was only later, when these basic needs were satisfied, and increased petroleum revenues risked inflation and appreciation, that Norway diverted this money to an offshore fund.

This should be the lesson learnt from the Norwegian experience: it is not the existence or size of its sovereign wealth fund that matters, but that it was introduced

---

[3]Other countries could benefit from following Norway's example, but they will find it difficult to ignore intense political pressures to develop the resource as quickly as possible.

[4]It is important to recognize that the Norwegian government has another fund, with a similar name: The Government Pension Fund Norway (GPFN), which should not be confused with the GPFG. Established in 1967, the GPFN is much smaller and functions as a sort of national insurance fund. The GPFN is managed separately from the GPFG, by *Folketrygfondet* (the National Insurance Scheme Fund), and its investments are limited to domestic and Scandinavian investments—making it a key stock holder in many large Norwegian companies, predominantly via the Oslo Stock Exchange. By the middle of 2018 (Q2), the market value of the GPFN was 250bn NOK, while the GPFG (by contrast) was worth 8337 bn NOK. See Norwegian Government (2018a, b).

Table 1 Largest SWFs by assets under management

	Country	Name	Assets	Origin
1	Norway	Government pension fund, global	1074.6	Oil
2	China	China investment corporation	941.4	Non-Commodity
3	UAE-Abu Dhabi	Abu Dhabi investment authority	683	Oil
4	Kuwait	Kuwait investment authority	592	Oil
5	China-Hong Kong	Hong Kong monetary authority investment portfolio	522.6	Non-Commodity
6	Saudi Arabia	SAMA foreign holdings	515.6	Oil
7	China	SAFE investment company	441	Non-Commodity
8	Singapore	Government of Singapore investment corporation	390	Non-Commodity
9	Singapore	Temasek holdings	375	Non-Commodity
10	Saudi Arabia	Public investment fund	360	Oil

Source SWFI (2018)
Note Assets listed in Billions USD

after basic needs were satisfied, and in order to protect its economy from becoming over-reliant on a very volatile commodity market. Developing countries should also be allowed to use their oil wealth to grow their national capacities, before investing the money offshore; but these domestic investments need to be done in a way that does not undermine the country's international competitiveness. In short, much of the magic of Norway's petroleum fund does not lie in the GPFG itself, or in its impressive size, but in how the Fund is accessed by elected officials. It is to this component that we now turn.

2 Accessing the Fund

One reason for global interest in the GPFG is its very size. Another reason, perhaps a more important one, is the success that the fund has brought to political authorities in protecting the Norwegian economy from Dutch Disease. While the Norwegian economy remains dependent on oil and gas, the government works hard to ensure that the Norwegian price level and alternative export sectors (such as fish) remain internationally competitive. Norway's capacity to avoid Dutch Disease has less to do with the existence of a fund, and more with the way the GPFG is integrated into Norway's system of public financing.

The money found in most sovereign wealth funds is the result of government budget surpluses. This means that the government receives its "oil money", decides how to spend it, and then invests whatever remains in the sovereign wealth fund. Not only does this approach facilitate wasteful spending, as it is tempting to use more of this money than may be prudent, but the approach can be destabilizing as well. As the price of oil will vary substantially from year to year, letting it flood into the government's budget introduces a great deal of uncertainty and volatility in an important source of government financing.

In Norway, the temptation for the government to spend its oil money is reduced by the fact that its "oil money" goes directly to the GPFG (offshore), and the government budget is only granted an annual share of the expected return from that Fund. This allows Norwegian governments a steady income stream over time (de-linked from the volatile price of oil), and it ensures that the wealth generated from oil is not allowed to flood the domestic economy, generating inflation, an appreciation of the currency, and threatening the international competitiveness of the country's exposed sector. In effect, the government follows something akin to Hartwick's (1977) Rule, where the resource owner (here the Norwegian State) is encouraged to invest (immediately) the resource rents in reproducible capital, goods or assets, and to spend only the returns from these investments to sustain consumption.

2.1 A Rule of Thumb

This need to limit access to the Fund was already recognized by the 1982 Pace Committee, introduced in the previous section. When discussing their ("buffer") fund option, the committee explicitly recognized the need for a budgetary rule, or rule of thumb (*handlingsregel*), to ensure that the rate at which the oil money entered the Norwegian economy would be both consistent (stable) and economically prudent. While the caveat was clear, the particular details of this rule of thumb were not established until 2001, when the Finance Ministry published a white paper (*St. meld. nr. 29* (2000–2001)).

In short, the Norwegian budgetary rule has two main components, each of which is informally constituted, largely by way of Norway's consensual and democratic norms. The first component concerns how the GPFG's money is to be spent. When the Norwegian parliament agreed to this rule of thumb, it emphasized that the money generated should be aimed at measures that could increase productivity, and with it the rate of economic growth, in the rest of the economy. In practice, however, it has proven impossible to trace how this money is being spent, in that the money from the GPFG (via the budgetary rule) is not earmarked for particular activities: it is (in effect) lost when mixed together with the rest of the general budget.

The second, and perhaps the most important, component concerns the amount of money that can be channelled back into the government's budget. The government's white paper noted that this return should be determined by the expected real rate of return from the fund (*St. meld. nr. 29* (2000–2001), p. 9)—but what this actually means has been a matter of varying political interpretation. When selling the idea to the broader public, subsequent politicians have referred to alternative but more familiar principles of resource management in Norway, such as silviculture (i.e. forest management techniques), where the rate of harvest should not exceed the forest's expected rate of growth (see, e.g., Reinertsen 2009). The government in 2001, with broad political support in parliament, employed a 4% threshold, believing that this would be corresponded with the long-term return on the Fund's investments. In 2017, another government reduced that threshold to 3%, in light of global economic conditions, arguing that the long-term return of the Fund was now likely to be lower.

This (3%) level is not established by law, but reflects an informal agreement among the leadership in all of Norway's political parties. A future government could, in principle, increase the share of the returns being repatriated, but it can expect significant political pushback. The informal nature of the agreement allows the government a good deal of flexibility, as is often required with a wildly fluctuating price for oil, or in response to global economic crises. In practice, what this arrangement means is that there is an implicit recognition that the money saved in the Fund reflects actual budget surpluses, while the government aims to balance its (non-oil) revenues and expenditures. It then estimates what a 3% return on the Fund is expected to look like, and readjusts its budget balance accordingly (see Moses and Letnes 2017a: 132f for more details).

This arrangement varies from that in many other States, in that Norway's oil money is not cycled through the government budget. Instead, Norway's oil wealth is directed offshore immediately into the GPFG, and then the government is allowed access to a stable and predictable return on the investment from that Fund. Each year, more and more oil money is poured into this (offshore) Fund, and the returns on the Fund's investment are (mostly) reinvested—providing another source of revenue for the Fund. (I say mostly, because 3% of these returns are going to fill the government's coffers.) The relatively modest size of this transfer (from the Fund to the government's budget) is not sufficient to inflate the domestic economy, or undermine the country's international competitiveness. All the while, the remainder of the money is allowed to grow, offshore, to be repatriated later—when needed.

In this way, Norway's oil money can enter into the economy at a steady and predictable rate (in contrast to the volatile price of oil), allowing governments to plan accordingly. At the same time, the system is flexible enough to allow authorities access to more money in times of economic crisis (See Moses and Letnes 2017b). Hence, the GPFG acts like both a savings and a stabilization fund: it saves money for future needs, but the Fund is also used to stabilize the broader Norwegian economy from the volatility associated with global oil prices, and the threat that the resulting wealth brings in the form of Dutch Disease.

As members of the 1982 Pace Committee noted: this is a remarkable (and rather unlikely) political feat. As an immigrant to Norway, I marvel at the willingness and ability of Norwegian elected officials to keep their hands out of the cookie jar. After all, every country—even a rich country like Norway—has a long list of things that it hopes to secure, and politicians will always be tempted to promise more than they can deliver. There can always be better roads, shorter hospital queues, bigger defence and educational budgets… As Norway's elected officials sit on a huge pile of money, in the form of the GPFG, it must be incredibly tempting to access this money to pay for important and popular objectives. The budgetary rule is the economy's first line of defence that keeps this from happening.

3 Managing the Fund

We have now seen the immense size of the GPFG, and how it is designed to stabilize the Norwegian economy from the threat of Dutch Disease and volatility in the global price of oil. This final section examines the particular means by which the Norwegian government manages its Fund. Like the other two aforementioned components, the way that Norway manages its GPFG is quite unique.

Originally, and before 1998, the Fund was managed by Norway's central bank (Norges Bank), using the same strategy that the central bank uses to manage its foreign reserves. In other words, the investment strategy had a very conservative (low-risk, low-return) focus on government and government-guaranteed bonds, invested offshore.

After 1998, Norwegian policymakers decided to increase the risk exposure of the Fund, and with it the potential return. To do this, management of the Fund was transferred over to the Norges Bank Investment Management (NBIM), which manages the Fund on behalf of the Norwegian people and their representative, the Norwegian Ministry of Finance (MoF). When the management reins were handed over to NBIM, one of the first things it did was to convert about 40 percent of the Fund's bond portfolio into equities (NBIM n.d.: at 1998). In particular, the MoF decides upon the investment strategy for the Fund, on advice provided from both the NBIM and the people's representatives in parliament. Today, the government is more ambitious, in that it hopes:

> to achieve the highest possible return with an acceptable level of risk. Management shall be transparent, responsible long-term and cost effective. There is a broad political consensus that the Fund shall not be used as a foreign policy or climate policy instrument (*Meld. St. 13* (2017–2018), p.5).

The formal framework for managing the Fund is laid down by the Norwegian parliament, in the form of the same Government Petroleum Fund Act, as mentioned above. While the Ministry of Finance has the overall responsibility for the Fund's management, it issues guidelines through its Management Mandate for the Government Pension Fund Global. Formally, the central bank (Norges Bank) remains responsible for managing the Fund, but its Executive Board has delegated the operational management of the Fund to the NBIM. In 2017, a government commission (NOU 2017: 13) recommended that the GPFG be given more autonomy from the central bank—but (to date) this has not happened.

Over time, this investment portfolio was diversified with an eye at maximizing the highest possible risk-adjusted return, combing equity, fixed-income, and (eventually) real-estate holdings. What has remained constant is a requirement that the Fund cannot invest in Norwegian holdings, and that it is not allowed to buy more than 10% of the voting shares in any individual company (real estate companies are exempted from this rule) (Ministry of Finance 2010: Sect. 3.4, §10). The likely reason for this ceiling is an earlier, politically contentious, attempt by Swedish Social Democrats to establish "wage earner funds" in the 1980s. The Norwegian authorities are afraid of upsetting international markets (thus reducing their returns) by appearing to use their ownership position to secure explicitly political objectives. This obviously limits the Fund's ability to actively influence the firms in which it invests, so the Fund has developed other, less direct, means of influence. These include the use of ethical guidelines (and the threat of exclusion), explicit voting strategies (given their limited holdings in any given company), and simply talking to the companies in which they invest.

3.1 Ethical Guidelines

In 2004, the Petroleum Fund's Advisory Council on Ethics was established, and since that time, the Fund has been required to follow strict ethical guidelines (with new ethical measures being added subsequently, see https://etikkradet.no/en/). Today, the Council on Ethics is an independent body made up of five members (with a staff of 8) that makes (public) recommendations, on the basis of guidelines determined by the MoF to Norges Bank (see Ministry of Finance 2017). These guidelines help to establish whether a given company should be excluded from the GPFG, or placed under observation.

These guidelines include product-based as well as conduct-based exclusion criteria. The product-based criteria can be used to exclude firms that produce tobacco, coal, or certain types of weapons; while the conduct-based criteria search out gross corruption, human rights violations, environmental damage, and/or unacceptably high greenhouse gas emissions (Council on Ethics 2017: 7; see also Council on Ethics 2019). Much political attention has been directed to the fact that a petroleum-generated GPFG is now prohibited from investing in coal-based industries.[5] Given the size of the GPFG, and the fact that the Council's recommendations are made public (Council of Ethics n.d.), prospective firms and other investors pay attention to the Council's recommendations. Then, when the NBIM makes its decision against the backdrop of these recommendations, the list of excluded companies is made public (see NBIM 2019a). In 2019, for example, three companies were held out for special mention: Texwinca Holdings Co was excluded for serious or systematic human rights violations; while Evergy Inc. and Washington H. Soul Pattinson & Co Ltd. were excluded based on an assessment of the product-based coal criterion (NBIM 2019b). A full list of the excluded companies can be found here: https://www.nbim.no/en/the-fund/responsible-investment/exclusion-of-companies/.

---

[5]The original Standing Committee on Finance and Economic Affairs was tasked to look into excluding both petroleum and coal companies (Recommendation 290 S (2014-2015)), but the resulting parliamentary resolution focused only on coal (see https://www.stortinget.no/en/In-English/About-the-Storting/News-archive/Front-page-news/2014-2015/hj9/). In particular, the resulting exclusion criterion holds that mining companies and power producers can be excluded from the fund if they themselves (or through entities they control) derive 30% or more of their revenue from thermal coal, or base 30% or more of their operations on thermal coal.

In addition to the ethical guidelines, the NBIM uses its ownership rights in different companies to promote long-term value creation. It does this in two ways: by voting its share and by joining in dialogue with company boards.

3.2 Ownership Strategy

By exercising its voting rights, the NBIM seeks to strengthen governance, improve performance, and promote sustainable practices in the firms in which it invests (NBIM 2017b:13). Obviously, the impact of this voting behaviour is limited by the fact that the NBIM can only secure 10% of the voting shares in any given company. In practice, this means that the NBIM votes in a way to support long-term value creation, sustainable business practices, board accountability, shareholder rights, equal treatment of shareholders, and transparent company communications. In particular, their public voting guidelines (see NBIM 2016a) are used to motivate their voting decisions, and a public record of how they have voted helps to send the message home (NBIM 2016b). For example, in 2017, the NBIM voted at 11,084 shareholder meetings (NBIM 2017b: 28).

In their dialogues with companies, the NBIM asks the companies in which they invest how they intend to address global challenges related to child labour, water management, climate change, and (more recently) human rights (NBIM 2017b: 6). In 2017, the NBIM held 3252 meetings with 1380 companies, on a number of strategic topics, including issues concerning the environment, social issues, shareholder rights, CEO remuneration, corruption, etc. (NBIM 2017b: 38–45).

4 Conclusion

It is not generally advisable to follow slavishly the economic model of another country. Countries and contexts change over time, making any country's lesson difficult to transfer. This is also true with regard to the use of particular policy instruments, such as a sovereign wealth fund. It is important to consider the local conditions on the ground before carefully deciding whether a SWF is useful for a country, and how it should be designed. But the key lessons of Norway's SWF can be fruitfully transmitted to other countries; its approach is transparent, democratic, and fiscally conservative.

This contribution has focused attention on three of the most unique and often overlooked characteristics of Norway's successful SWF: the GPFG. Unlike the SWFs in many other countries, Norway's Fund was relatively late in coming, and when it was first established, it was managed in a very conservative manner. More importantly, the GPFG was set up in a way that keeps temptation at bay: it was created to keep the money offshore (away from the domestic economy) and to ensure only a (steady) fraction of the money gets repatriated. This is key to Norway's macroeconomic success and its ability to (mostly) avoid the Dutch Disease. Finally, by using ethical and socially responsible investment guidelines, the Norwegian authorities are able to influence international investment trends in a positive direction, without appearing to be intervening, politically, in the market.

The design and make-up of Norway's GPFG, and the NBIM that manages it, draw a significant amount of domestic political attention to the Fund. This attention is of two types: in terms of the ethical guidelines—of how Norway aims to invest for the future—but also in terms of how the money should be spent; e.g. when (or if) it is appropriate to dip into the Fund in light of pressing and/or special circumstances. This is very much the case at the time of this writing, when a conservative coalition government is hinting that it will take money out of the Fund to replace a naval frigate that sunk in a recent collision (among other unforeseen expenses). The opposition, in return, declares this is the top of a very slippery slope—making it easier to draw down the Fund in the future, rather than make difficult decisions about political priorities. Hence, the political consensus that lies at the centre of Norway's SWF continues to be challenged and defended.

Such is the nature of an active and effective sovereign wealth fund: it *should* draw attention to, and provoke disagreement about, how a country's collective savings are to be saved (and spent).

References

Council on Ethics. (n.d.). Recommendations. Online at: https://etikkradet.no/recommendations/.

Council on Ethics. (2019). Council's Activities. Council on Ethics for the Norwegian Government Pension Fund Global. Online at https://etikkradet.no/councils-activities/.

Council on Ethics. (2017). Annual Report 2017. Council on Ethics for the Norwegian Government Pension Fund Global. Online at https://nettsteder.regjeringen.no/etikkradet3/files/2018/03/Etikkradet_arsmelding_2017_eng_UU.pdf.

Economist, The (1977). The Dutch Disease. November 26.

Hartwick, J. M. (1977). Intergenerational equity and the investing of rents from exhaustible resources. *The American Economic Review, 67*(5), 972–974.

Karl, T. L. (1997). *The Paradox of Plenty: Oil Booms and Petro-States* (Berkeley: University of California Press).

Lie, E. (2010). *Over Evne: Finansdepartementet 1965-1992* (Oslo: Pax).

Meld. St.13 (2017–2018). The Government Pension Fund 2018. (Oslo: Norwegian Ministry of Finance).

Ministry of Finance. (2017). Guidelines for observation and exclusion of companies from the Government Pension Fund Global. 17 February draft. Online at https://www.regjeringen.no/globalassets/upload/fin/statens-pensjonsfond/formelt-grunnlag/guidelines-for-observation-and-exclusion-from-the-gpfg—17.2.2017.pdf.

Ministry of Finance, Norway. (2010). Management mandate for the Government Pension Fund Global. Adopted 8 November. Online at https://www.nbim.no/en/the-fund/governance-model/management-mandate/.

Moses, J., & Letnes, B. (2017a). *Managing Resource Abundance and Wealth: The Norwegian Experience* (New York: Oxford University Press).

Moses, J., & Letnes, B. (2017b). Breaking brent: Norway's response to the recent oil price shock. *Journal of World Energy Law and Business, 10*(2017), 103–116.

NBIM. (n.d.). The History. Online at: https://www.nbim.no/en/the-fund/the-history/.
NBIM. (2019a). Observation and exclusion of companies. 17 January draft. Online at https://www.nbim.no/en/responsibility/exclusion-of-companies/.
NBIM. (2019b). Decisions to exclude companies from the government pension fund global. 17 January draft. Online at: https://www.nbim.no/en/the-fund/news-list/2019/decisions-to-exclude-companies-from-the-government-pension-fund-global/.
NBIM. (2018). Market Value. 26 October. Online at: https://www.nbim.no/en/the-fund/market-value/.
NBIM. (2017a). Holdings. December 31. Online at: https://www.nbim.no/en/the-fund/holdings/holdings-as-at-31.12.2017/?fullsize=true.
NBIM. (2017b). Responsible investment. Global pension fund global. No. 4 (Oslo: Norges Bank Investment Management).
NBIM. (2016a). Voting Guidelines. 22 March draft. Online at: https://www.nbim.no/en/responsibility/our-voting-records/voting-guidelines2/.
NBIM. (2016b). Our Voting Records. 4 February draft. Online at: https://www.nbim.no/en/responsibility/our-voting-records/.
Norwegian Government. (2018a). Market value: government pension fund global (GPFG). 6 November. Online at https://www.regjeringen.no/en/topics/the-economy/the-government-pension-fund/government-pension-fund-global-gpfg/market-value-and-capital-inflow/id696852/.
Norwegian Government. (2018b). Market value: government pension fund Norway (GPFN). 6 November. Online at https://www.regjeringen.no/en/topics/the-economy/the-government-pension-fund/government-pension-fund-norway-gpfn/market-value-gpfn/id710461/.
NOU 2017:13. *New Central Bank Act* (Oslo: Norges Offentlige Utredninger).
NOU 1983:27. *Petroleumsvirksomhetens framtid: det framtidige omfanget av petroleumsvirksomheten på norsk sokkel* (Oslo: Norges Offentlige Utredninger).
NPD [Norwegian Petroleum Directorate]. (2019a). Exports of Oil and Gas. *Norwegian Petroleum.* Online at https://www.norskpetroleum.no/en/production-and-exports/exports-of-oil-and-gas/.
NPD [Norwegian Petroleum Directorate]. (2019b). Annual Production. *Norwegian Petroleum.* Updated 10 Jan 2019. Online at: https://www.norskpetroleum.no/en/facts/production/#annually.
NPD [Norwegian Petroleum Directorate]. (2018). The market value to the government pension fund global, 1996–2017. *Norwegian Petroleum.* 17 Feb 2018 draft. Online at: https://www.norskpetroleum.no/en/economy/management-of-revenues/.
Reinertsen, M. (2009). Oljefondets utspring. *Morgenbladet*, 22 May.
Ross, M. L. (1999). The political economy of the resource curse. *World Politics, 51*(2), 297–322.
St. meld. nr. 25. (1973–74). Petroleumsvirksomhetens plass i det norske samfunnet (Oslo: Finansdepartementet).
St. meld. nr. 29. (2000–2001). Retningslinjer for den økonomiske politikken (Oslo: Finansdepartementet).
SWFI. (2018). Sovereign wealth fund rankings. November 2018 update. Online at: https://www.swfinstitute.org/sovereign-wealth-fund-rankings/.

LCP

Local Content in the Extractive Resource Industry in Nigeria

Chilenye Nwapi

Abstract Most of Nigeria's application of local content in the extractive industry is focused on the oil and gas sector, the mining sector receiving only marginal attention, without a well-established policy or legislation. In the oil and gas sector, local content is pursued through three key vehicles: state participation, the Oil and Gas Industry Local Content Development Act, 2010, and the marginal fields policy. These vehicles express the Nigerian government's understanding of local content in the oil and gas sector, which is not limited to value addition. In its broadest sense, value addition is about what the industry as a whole can contribute to the wider economy, for instance, the contributions of oil and gas companies not only to the domestic oil and gas industry but also to the growth of other industries. The key vehicle being utilised to achieve value addition is the afore-mentioned Act. Nigeria's local content policy, however, embraces the somewhat wider concept of 'local participation'. Local participation aims to ensure that the citizens take charge of the development of the industry. The extent to which the content of the industry is local is therefore determined not only by the value brought to the local economy by the industry, but also by the extent to which the participants in the industry are 'local'. The key vehicles being utilised to achieve local participation are state participation and the marginal fields policy. This chapter discusses each of the three vehicles, highlighting their successes and challenges.

Keywords Local content · Value addition · Local participation · State participation · Marginal fields

C. Nwapi (✉)
Canadian Institute of Resources Law, Calgary, Canada
e-mail: enyenwapi@yahoo.com

E. G. Pereira et al. (eds.), *Sovereign Wealth Funds, Local Content Policies and CSR*, CSR, Sustainability, Ethics & Governance,
https://doi.org/10.1007/978-3-030-56092-8_14

1 Introduction

The need for local content promotion has been recognised in the Nigerian extractive industry, particularly the oil and gas sector, for a long time. While International Oil Companies (IOCs) dominated the oil and gas sector during the first decades of commercial oil discovery in Nigeria,[1] during the building of the first refinery by Shell in 1965, the Nigerian government and Shell agreed to 'give earnest consideration to the possibility of enabling' the government or the Nigerian public to invest in the refinery 'provided a minimum response [was] anticipated' (Nwokeji 2007). This is said to be the first deliberate opportunity offered to the Nigerian public to invest in the oil and gas sector (Nwokeji 2007). Thirteen years after the official discovery of oil in commercial quantities, the Petroleum Act, 1969, was enacted, drastically reshaping the investment atmosphere in three key relevant ways, namely by (1) mandating that oil and gas concessions could only be granted to companies incorporated in Nigeria[2]; (2) reserving for the Nigerian government the option to acquire participation rights in oil and gas concessions[3]; and (3) mandating every holder of an oil mining lease to ensure that within 10 years of the grant of its lease, the number of Nigerian citizens in their employ reaches a specified percentage.[4] While the first requirement was immediately implemented, the government did not immediately exercise the option contained in the second requirement (Nwokeji 2007). The third requirement was incorporated in contractual arrangements entered into between the Nigerian government and IOCs.

While the local content provisions of the Petroleum Act were meagre, the Petroleum (Drilling and Production) Regulations 1969 enacted pursuant to the Act contained more elaborate provisions. Regulation 26 of the Regulations mandated every holder of an Oil Prospecting Licence to submit for the Minister's approval within 12 months of the grant of its licence a detailed programme for the recruitment and training of Nigerians in all phases of oil and gas operations. A similar obligation was placed by the same regulation on every holder of an oil mining lease upon the grant of its lease. Scholarship schemes proposed by a licencee or lessee must be approved by the Minister of Petroleum and might not be varied without the Minister's permission.

Much of the focus on local content in the extractive industry in Nigeria is the oil and gas sector. This is unsurprising given the dominance of oil and gas development in Nigeria's economy. The mining sector has far from gathered as much attention. In fact, local content is being implemented in the mining sector without a well-established official policy or legislation. The Nigerian Mines and Minerals Act, 2007—the principal law regulating mineral activities in the country—contains no local content provisions other than the standard provision that only Nigerian citizens and companies incorporated in Nigeria can be granted mining licences. The key

[1]Oil was discovered in commercial quantities in Nigeria in 1956.

[2]Petroleum Act, 1969, s 2(2).

[3]Petroleum Act, 1969, First Schedule, para 35(a).

[4]Petroleum Act, 1969, First Schedule, para 38.

reason for the lack of a well-established local content policy for the mining sector is, arguably, the decades-old neglect of the sector by successive governments. The sector is significantly backward not only when compared to the oil sector but also when compared with the mining sector in other African countries. To revamp the sector, the government is currently focusing on the attraction of investors (Ministry of Mines and Steel Development 2016). Introducing binding local content obligations at this time would therefore likely be counterproductive.

Another reason for the lack of official local content policy for the mining sector is the fact that most of the mining activities that have been going on for decades are Artisanal and Small-scale Mining (ASM), a sub-sector that is dominated by local actors rather than by foreign multinationals. Since the content is already local, there is no need to introduce a local content policy. Instead, attention is focused on formalising ASM (which for years was operating informally and without accountability) and in jump-starting investments in large-scale mining. The latter requires rebuilding market confidence through, *inter alia*, the establishment of a clear policy direction and assuring investors that the government will be consistent in its policy implementation (Ministry of Mines and Steel Development 2016). That is not to say that the government does not recognise the importance of local content in the sector. As stated in its 2016 Roadmap for the Growth and Development of the Nigerian Mining Industry, one of the government's objectives is stated as '*promoting formal small scale operators* through expanding access to funding and supporting knowledge development to drive local content' (Ministry of Mines and Steel Development 2016). However, this policy objective has not been elaborated in any policy document or operationalised in legislation.

This is in clear contrast to the oil and gas sector where, apart from the Petroleum Act, the government has adopted free-standing legislation (the Oil and Gas Industry Local Content Development Act 2010) to promote local content and regulate its implementation. The enactment of this law marked a watershed in the evolution of the local content policy in the extractive industry in Nigeria. Given the limited application of the local content policy in the mining sector in Nigeria, this chapter focuses on the oil and gas sector.

Local content is discussed in the literature mainly in terms of value-added. However, a key point made in this chapter is that Nigeria has adopted a broad understanding of local content in the oil and gas sector that is not limited to value-added. Value-added is a narrow concept that speaks to the value the operations of a company or industry bring to a country. It is thus primarily about obligations placed on companies, e.g. to use local raw materials, to employ the citizens of a country, and to use local suppliers in the procurement of goods and services. In its broadest sense, value addition is also about what the industry as a whole can contribute to the wider economy, for instance, the contributions of the oil and gas industry to the growth of other industries. Nigeria's local content policy embraces the wider concept of local participation. This is well captured by the local content policy objective established in the National Petroleum Policy 2017: 'Building Nigerian human resources, so that Nigerians can take a full role in managing our own resources, is a key objective of the Nigerian petroleum policy'. Local participation aims to ensure that citizens take over the development of the industry. The extent to which the content of the industry is local is therefore determined not only by the value brought to the local economy by the industry (value-added), but also by the extent to which the participants in the industry are local. Participants including not only employees and suppliers of goods and services, but also the primary holders of the concession rights.

Accordingly, Nigeria's local content policy in the oil and gas sector is implemented through three key vehicles: (1) direct state participation, (2) the Oil and Gas Industry Local Content Development Act of 2010, and (3) the marginal fields policy. This chapter discusses the nature of local content obligations created in the oil and gas sector in Nigeria under these key vehicles. It also considers the successes Nigeria has achieved so far as well as the main challenges to the implementation of those obligations.

2 Direct State Participation

During the first decade of commercial oil operations in Nigeria, there was little state participation, as the government's involvement was limited to regulation and revenue collection from IOCs. In 1962, an agreement between the Nigerian government and Italian oil giant Agip (now ENI) gave the government the right to acquire 33.3% ownership in Agip's Nigerian subsidiary (the Nigerian Agip Oil Company) (Frynas 2000; Pearson 1970). However, the Nigerian government did not exercise this option until 1971, meaning that the government continued to limit its involvement in oil and gas operations to regulation and revenue collection (Frynas 2000). The civil war of 1967–1969 changed the government's perception of the significance of oil in international politics and the need for the government to have tighter control of the industry (Frynas 2000). With the industry looking very promising globally, the government enacted the Petroleum Act 1969, *inter alia* officially introducing state participation into the oil and gas sector by giving the Nigerian government the option to acquire part-ownership in oil and gas concessions. In 1971, it joined the Organisation of Petroleum Exporting Countries and also established the Nigerian National Oil Corporation (NNOC) as a state-owned company to implement its indigenisation policy. Due to lack of capital and technical capacity, however, the NNOC could not operate on its own, compelling it to enter into joint venture agreements with the IOCs (Intergovernmental Forum on Mining, Minerals, Metals and Sustainable Development 2018). But institutional rivalry between the NNOC and the Ministry of Petroleum prevented the government from taking advantage of the oil price boom in the early 1970s. Efforts to address the problem led to the merger of the NNOC and the Ministry to form the Nigerian National Petroleum Corporation (NNPC) in 1977 (Intergovernmental Forum on Mining, Minerals, Metals and Sustainable Development 2018).

The NNPC embarked on extensive contractual arrangements with IOCs aimed at establishing the government's interest in the industry. It used two principal participation arrangements: joint venture agreements and production sharing contracts and represents the government in all the arrangements. But the NNPC's role was not limited to the acquisition of interests in the IOCs. It also used those contractual arrangements to implement the local content provisions of the Petroleum Act and the Petroleum (Drilling and Production) Regulations 1969 enacted pursuant to the Petroleum Act, which contains more elaborate local content provisions. Its model service contract obliges the contractor to:

> make the maximum use if available indigenous Nigerian manpower in the conduct of petroleum operations under this contract. The contractor shall within six months after the effective date and after consultation with the NNPC submit for NNPC's approval a detailed recruitment programme, and within 12 months, submit for NNPC's approval, a training programme for all Nigerians employed by the contractor in the conduct of all petroleum operations (see Onyi-Ogelle 2016:138).

3 Local Content as Value-Added: The Oil and Gas Industry Local Content Development Act 2010

Before the introduction of the Local Content Development Act in 2010, the Petroleum Act 1969 and the Petroleum (Drilling and Production) Regulations had already established local content obligations. However, these did not represent a focused approach to local content development. Such an approach began with a workshop organised by the National Petroleum Investment Management (NAPIMS) in 2001, during which the participants produced a communique recommending the creation of a National Committee on Local Content Development (NCLCD) (Falola and Achberger 2013). A report produced by the NCLCD in 2002 proposed a local content target of 40% in 2005 and 60% in 2010 and recommended the enactment of local content legislation to address local content development in a more targeted manner.

A 2003 report of the Department of Petroleum Resources that was commissioned by the Norwegian Agency for Development Cooperation and the Norwegian Ministry for Petroleum and Energy '[t]o assess the capabilities of the Nigerian Supply and Service Industry and propose measures to enhance Nigerian private sector development based on Petroleum Activities' (Heum et al. 2003:1) strongly influenced the drafting of the Local Content Development Act 2010. The report defined local content as 'value addition activities taking place in Nigeria' and linked value addition 'to the magnitude of manufacturing and service production that is taking place in Nigeria' (Heum et al. 2003:2). According to the report, this calls for the development of indigenous companies and the encouragement of foreign investment in the country, as value addition would occur through the interaction between indigenous companies and foreign investors. This reflects a supply chain perspective to local content and the report's explanation is significant:

> The oil industry … consists of a few major oil companies … and a large number of mainly nationally based-oil companies. They have numerous suppliers, offering a wide variety of goods and services, from the very sophisticated to more standardized products. Some are global players as oil majors; others are more locally based. Contracts between the oil companies and their suppliers frequently involve a hierarchy, or a chain, of subcontractors. This

> supply chain may be regard as a linear sequence of activities organised around the flow of materials from source of supply to finished products, after-sales services and often also recycling. It is in this context that the Nigerian industry has to perform and prove useful, if the ambition of private sector development based on petroleum activities is to be met. In this supply chain perspective, **activities are only justified when they add value to the overall process** (Heum et al. 2003:5).

The report strongly decries the 'dominance of foreign over local companies' in both the oil industry and the service industry, leading to most jobs being executed by foreign contractors (Heum et al. 2003:24). The report therefore recommended that 'the ultimate goal of a viable local content policy should be to create jobs by enhancing sustainable industrial growth and national wealth' (Heum et al. 2003:63).

The Local Content Development Act (section 106) defines local content as 'the quantum of composite value added to or created in the Nigerian economy by a systematic development of capacity and capabilities through the deliberate utilisation of Nigerian human, material resources and services in the Nigerian oil and gas industry'. At the core of this definition is, thus, value addition to the country. This is to be achieved by facilitating the participation of Nigerians in the oil and gas industry and ensuring that all relevant economic activities in the oil and gas industry take place in Nigeria. To this end, the Act prescribes several actionable steps that would operationalise the local content policy. They include:

- Nigerians shall be given 'first consideration' before other nationals in matters of employment and training.[5] The rationale for this is not merely to help address the unemployment problem in the country, but also to help to build the technical skills of Nigerian nationals in the industry, which, in turn, would enhance their full participation in the industry.
- In matters of procurement, raw materials available in Nigeria, services provided by Nigerians, and goods manufactured in Nigeria shall receive first consideration.[6] The rationale for this is to concentrate economic activities in the industry across the entire spectrum of the value chain within Nigeria. By prohibiting the importation of raw materials that are available in Nigeria, for instance, the Act ensures that indigenous assets are utilised, which in turn would increase the value of those assets, thus boosting backward linkages. The utilisation of those assets would also invariably involve the participation of indigenous suppliers.
- In the award of oil blocks, oil field licences, oil lifting licences, and in all projects for which contract is to be awarded, indigenous independent operators shall receive first consideration, subject only to conditions to be specified by the Minister.[7] This prescription is addressed to the licencing authority rather than to operators, since it is the government that awards licences. The provision itself is intended to enhance indigenous participation and is not focused solely on value addition.

---

[5] Nigerian Oil and Gas Industry Content Development Act, 2010, s 28(1).

[6] Nigerian Oil and Gas Industry Content Development Act, 2010, s 10(1).

[7] Nigerian Oil and Gas Industry Content Development Act, 2010, s 3(1).

- Indigenous service companies with evidence of 'ownership of equipment, Nigerian personnel' and requisite capacity shall receive 'exclusive consideration' for contracts and services listed in the Schedule to the Act.[8] 'Exclusive consideration' means that no other companies are to be considered for such contracts. The rationale is to ensure that no Nigerian company that has all the requisite qualifications to execute a contract is rejected in preference for a non-Nigerian company.
- Operators are to carry out a programme for the promotion of technology transfer to Nigeria.[9] The essence of technology transfer is to ensure that Nigeria retains the technical skills brought by the IOCs. However, it is one thing to invite technology transfer; it is another to absorb the transferred technology (see Olawuyi 2018).

Furthermore, the Act provides targets for a progressive increase in local content, from the pre-existing 2007 target of 45–70% in 2010 and 80% by 2020 (Ihua et al. 2011). It establishes minimum targets for Nigerian participation in specified categories of oil services, which include: engineering, fabrication, materials and procurement, finance, research and development, shipping and logistics and other categories (Ovadia 2013).

Prospective operators are required to submit a Nigerian Content Plan with their bids, setting out in detail how they will fulfil their local content obligations.[10] To specifically ensure the employment and training of Nigerians, the Act requires operators to submit an Employment and Training Plan detailing their employment and training needs, expected local skills shortage, project-specific training requirements, etc.[11] It establishes a Local Content Development Fund into which operators are required to contribute 1% of the value of their contracts to support local training and business support services (Ramdoo 2015).

There is empirical evidence that since the introduction of the Local Content Development Act, Nigeria has slowly begun to localise its oil and gas industry and that backward linkages are deepening (McCulloch et al. 2017). There has been a noticeable growth in the number and range of local firms providing services to IOCs (Ovadia 2013). In 2013, the Local Content Board reported that Nigeria's local content policy led to the investment inflow of $5 billion into the economy and the creation of about 38,000 jobs (see Aoun et al. 2015). The IOCs have also reported significant increases in the share of local procurement and the number of local employees in their companies (see Aoun et al. 2015). Through greater participation of local firms, the local content policy has had positive impacts on local value creation (Adedeji et al. 2016).

Still, the overall impacts of the Local Content Development Act have in fact been mixed. Using a structural equation modelling technique to analyse data from a survey of over 200 local firms, Adedeji et al. (2016) find that the local content policy has had a positive and significant impact on value creation in Nigeria, but that local value

[8]Nigerian Oil and Gas Industry Content Development Act, 2010, s 3(2).

[9]Nigerian Oil and Gas Industry Content Development Act, 2010, s 43.

[10]Nigerian Oil and Gas Industry Content Development Act, 2010, s 7.

[11]Nigerian Oil and Gas Industry Content Development Act, 2010, s 29.

creation traceable to the policy is lower than expected. In another study that considered the impact of local content policy on Small and Medium Enterprises (SMEs) in Nigeria, Ihua (2010) finds that the policy has led to increased award of contracts to local SMEs, but that this does not mean increased local SME participation, as the contracts are awarded to the existing SMEs while new SMEs find it difficult to break into the market. That is to say, the number of SMEs participating in the industry as a result of the local content policy is largely stagnant. In Sect. 4, the chapter identifies some of the major weaknesses of the Act that have contributed to its lack of greater impact on local content development in Nigeria.

4 The Marginal Fields Policy

Marginal fields are fields allocated to IOCs, but which have remained undeveloped because, for a number of factors, they are considered unprofitable by the IOCs. The National Petroleum Policy 2017 defines such fields as 'any field that has producible reserves booked of less than 10,000 bbls/d and [that] has remained unproduced for over 10 years'. The reasons that the IOCs have regarded such fields as unprofitable include: the size of the oil reserves, availability of needed infrastructure in the reserve area, excessive costs of development, technological constraints, environmental and political concerns, and the price of the produced oil or gas (Dentons 2014). As stated by the Ministry of Petroleum Resources, marginal fields must have some or all of the following characteristics:

- Fields not considered by licence holders for development because of assumed marginal economics under prevailing fiscal and market terms.
- Field with at least one exploration well drilled and have been reported as oil and or gas discovery for more than 10 years with no follow up appraisal or development effort.
- Fields with crude oil characteristics different from current streams (such as crude with very high viscosity and low API gravity), which cannot be produced through conventional methods or current technology.
- Fields with high gas and low oil reserves.
- Fields that have been abandoned by the leaseholders for upwards of three years for economic or operational reasons.
- Fields that the present leaseholders may consider for farm-out as part of portfolio rationalisation programmes (Ministry of Petroleum Resources 2013).

Section 17 of the Petroleum Act provides the legal framework for the development of such fields by authorising the President to 'farm-out' marginal fields if they have been left unattended to for at least 10 years 'from the date of the first discovery of the marginal field' and when it is in the public interest to farm them out. It is the policy of the government, which dates back to the mid-1990s,[12] to farm-out these fields exclusively to indigenous companies as a way of promoting indigenous participation (by giving local companies greater access to oil fields), generating employment for Nigerians and the enhancement of the exploitation of oil and gas reserves (Atsegbua 2005; Intergovernmental Forum on Mining, Minerals, Metals and Sustainable Development 2018). The general consideration was thus to find an indigenous company with the potential to be successful in the development of the fields.

Since the marginal fields policy was aimed to promote indigenous participation in the petroleum sector, the eligibility of participating operators is determined based on the definition of a Nigerian company for the purposes of local content, i.e. under the Local Content Development Act 2010. Thus, eligible operators are companies formed and registered in Nigeria and in which Nigerians hold at least 51% of equity shares.[13] Guidelines for Farmout and Operation of Marginal Fields published by the Ministry of Petroleum Resources in 2013 contained provisions encouraging states and local communities to register companies to bid for contracts in marginal fields bid rounds (Ministry of Petroleum Resources 2013).

The award of marginal fields to indigenous companies became the official policy of the government in 2001. There are said to be about 183 marginal fields under concession in Nigeria (Ezeani and Nwuke 2017). Twenty-four marginal fields were awarded in 2003 to 31 indigenous operators. Thirty-seven marginal fields were announced for award during a 2014 bid round and as at 30 May 2016 had been awarded (Ezeani and Nwuke 2017). Plans to hold marginal field bid rounds in 2013 and 2017 were not followed through due in part to the stalled efforts to enact new petroleum legislation, as public pressure mounted on the government to enact a new legal framework before any bid rounds were to be conducted.

5 Weaknesses and Challenges in the Implementation of Local Content in Nigeria

5.1 State Participation

The implementation of state participation in Nigeria has produced mixed results. While the acquisition of equity interests in the operations of the IOCs has resulted in an increase in the amount of revenues the government generates from the oil and gas industry, the effective control of the operations has remained in the hands of the IOCs which see to the day-to-day running of the operations. This means that the

[12]The idea of marginal fields was introduced by the Petroleum (Amendment) Decree No 23 f 1996 which amended the Petroleum Act.

[13]Nigerian Oil and Gas Industry Content Development Act 2010, section 106.

IOCs remain major players in the industry while the NNPC, which operates through several subsidiaries, remains a relatively minor player.

State participation has had a negative impact on the capacity of the industry to contribute to broad-based economic development. This is mainly because the NNPC, which represents the government in the participation arrangements, has, for the most part of its existence, been embroiled in political struggles over who is to control its affairs (Nwokeji 2007). The main reason for its establishment was to advance the government's interests in the industry. However, its ability to adequately advance those interests was strongly undermined from the very beginning by the fact that it was established to serve as both industry participant and regulator—two incompatible roles. This has made it difficult for it to perform either role well, as it creates a conflict of interests in the corporation's discharge of its functions (see Sayne et al. 2012). Calls to restructure the corporation by divesting it of its regulatory powers so that it would focus only on its commercial responsibilities have not been adequately implemented due to lobbying by special interests and a lack of political will on the part of the federal government. In theory, regulatory functions belong to the Department of Petroleum Resources (DPR), an arm of the Ministry of Petroleum Resources. However, the centrality of the NNPC in Nigeria's political economy has made it very difficult for the DPR or even the Petroleum Minister directly to control it, and incidents of power clash between the corporation and the Minister are not uncommon (Centurion Lawyers and Advisors 2017). While it is not unprecedented for a government to form partnerships with entities it regulates, it is important for the government to establish independent regulators to oversee the operations of those entities to avoid a conflict of interest.

On revenue generation to the country, NNPC's oil sales constitute roughly half of the country's total oil production (Sayne et al. 2015). However, its approach to oil sale fails to maximise returns to the government. Proceeds of its oil revenues have not always been transferred to the government coffers (KPMG 2010). A 2015 forensic audit of its remittances to the government found that the NNPC spent 46% of its domestic oil revenues on operations and subsidies, a situation the study decried as unsustainable (Pricewaterhouse Coopers 2015). The corporation's record of corruption is one of the worst among state-owned companies in the world (see Nwapi 2014). A study by the Natural Resource Governance Institute identified political interference as the root of the problem (Sayne et al. 2015). In addition, despite having been in existence for more than 40 years, the NNPC has failed to develop its own commercial or operational capabilities (Sayne et al. 2015), as a result of which it is unable to compete favourably with IOCs.

These problems undermine the capacity of state participation in the oil and gas industry to yield the desired result. The 2017 National Petroleum Policy proposed a radical restructuring of the oil sector, including the NNPC, with a view to vesting the regulatory, commercial and policy-making functions in separate institutions. State participation will be operationalised through the establishment of a new state-owned company to replace the NNPC, and which will focus solely on the commercial interests of the government. The bill intended to operationalise aspects of the policy dealing with institutional restructuring was passed by the federal legislature in mid-2018 but the President declined to sign it into law due to concerns over the reduction of his control over the oil and gas sector. This act (or omission) by the President reflects the political power struggles that have undermine state participation in the oil and gas sector and which have generally torpedoed meaningful restructuring of the sector.

5.2 *The Local Content Development Act*

While there is empirical evidence that the Local Content Development Act has led to significant increases in the level of local procurement and job creation in the Nigerian oil and gas industry, including deepening backward linkages, the implementation of local content in Nigeria still faces significant challenges. One of the major challenges is oil price volatility. This is a challenge because when prices are low, companies are not only unwilling to make new investments, they cut jobs and sometimes divest. This situation is inimical to local capacity development and has been blamed for the low level of local content in the oil and gas industry in Nigeria (Ajayi 2017).

The Local Content Development Fund established under the Act has not been adequately implemented. Local companies have found it difficult to access the fund due mainly to inefficiency in the implementation of the fund (Ajayi 2017). The Nigerian Content Development and Monitoring Board has noted the lack of a transparent system for the disbursement of the fund and the lack of adequate mechanism to ensure that recipients of the fund pay back what they have been given (Ajayi 2017).

Another major challenge is the monitoring and evaluation of local content implementation. The World Bank (2015) has noted that Nigeria lags behind in monitoring and evaluation of local content implementation in the oil and gas sector and noted specifically a lack of transparency in the monitoring system. \query{Please check whether the sentence 'Adequate monitoring is critical to meaningful implementation because...' conveys the intended meaning.}Adequate monitoring is critical to meaningful implementation because it provides an avenue for the government to understand weaknesses in the law and how they might be strengthened.

Another point of concern is the potential of the Local Content Development Act to facilitate corruption. Section 92 of the Act provides broad discretionary powers to the Content Development Board responsible for approving local content plans submitted by companies. The Board is authorised to accept gifts of various kinds from the public, without any restriction with regard to the person offering the gift. In other words, the Board can accept gifts even from the very entities that it regulates. The only restriction is that the conditions attached to the gift shall not be incompatible with the Board's functions. Since it is the person offering the gift that determines its conditions or purposes, the restriction has little or no practical value. This is because a Board that accepts gifts from a company is unlikely to be able to evaluate a local content plan submitted by such a company in a fair and independent manner. Such gifts could therefore serve as conduits for facilitation payments. Thus, notwithstanding the restriction, the provision creates a potential conflict of interest with regard to the Board's ability to discharge its functions (Nwapi 2015:94). In addition, the discretionary powers of the Board to determine compliance with the Act are broad and aggrieved parties do not have the right to challenge the Board's decisions. This may lead to uneven application of the Act. Also, the Minister of Petroleum Resources has the power to grant waivers with regard to compliance with the Act, based on criteria that are not known to the public. Such a lack of transparency opens also the gate to uneven and corrupt application of the Act (Martini 2014; World Trade Institute Advisors 2013).

Relatedly, there is also the problem of capture, which has been facilitated by opacity in the oil and gas licencing process (Nwapi 2014). Although public officials are bound by the Code of Conduct for Public Officers under the Nigerian Constitution, which prohibits them from using their positions to secure financial gains for themselves, there is no legal requirement for beneficial ownership disclosures by companies participating in oil and gas licencing. Civil society investigations reveal the existence of mystery investors in the oil and gas industry in Nigeria, who seek to take unfair advantage of the Local Content Development Act (Global Witness 2012).

There are also structural barriers in the investment climate that militate against the effective implementation of the Act. The barriers include poor infrastructure (particularly energy and roads), high cost of business finance, and security (especially in the oil region). These barriers increase the cost of doing business for both IOCs and local firms, preventing them from effectively engaging in local value addition activities. For instance, lack of finance for local firms makes it difficult for IOCs to find local firms with the financial capacity to execute supply contracts. Poor energy and road infrastructure affects the quality of local inputs (McCulloch et al. 2017).

Thus, while progress has been made to enhance the benefits of the local content policy in the oil and gas industry in Nigeria, there still exist significant barriers to meaningful local content development in the country.

5.3 The Marginal Fields Policy

The implementation of the marginal fields policy cannot so far be described as successful as only about 30% of the allocated fields have reached commercial production—a disappointing outcome given the fanfare that greeted the policy when it was officially announced in 2001. The implementation weaknesses are highlighted by the government's inconsistency in the conduct of the bid rounds. For instance, plans to hold bid rounds in 2013 and 2017 were not followed through. For the 2013 rounds, the government had published a bid timetable and had conducted road shows to clarify the guidelines and the bid process to the public. However, without any explanation to the public, the bid rounds were delayed and no revised timetable was published. Nothing further was heard about the rounds, which ultimately did not take place. Sources had it that the rounds failed due, at least in part, to disputes between the government and some IOCs that held the original licences over the fields regarding which fields should be included in the round (Dentons 2014). It appeared that due to technological advances since the fields were abandoned, some fields that had been considered unprofitable had become potentially profitable, with the result that the original licence holders had once more become interested in developing them and wanted to have them back (Dentons 2014).

Another major obstacle to the implementation of the policy is lack of access to finance by indigenous operators awarded the marginal fields. They lack the ability to provide adequate collateral for loans (Okonkwo 2011). Also, their lack of industry experience, including managerial skills, denied them the technical capability to operate the fields (Humphrey and Dosunmu 2017). Other obstacles include lack of adequate geological data and lack of adequate infrastructure around the marginal fields (Machunga-Disu and Sayne 2014).

Furthermore, while the provision in the 2013 Guidelines allowing communities and states to register companies to acquire marginal fields is commendable, its practical application is very difficult. While it is not clear whether any communities or states have registered any indigenous companies for the purpose of participating in the marginal fields programme, any such newly registered company would lack the industry experience—and most likely also the financial resources—to meet the requirements for a successful marginal fields development. As a result, such a company would be unable to launch a competitive bid. One way it can overcome this obstacle is to either acquire an existing indigenous company that has the requisite experience—if it has the resources to do so—or co-bid with such an existing company.

6 Conclusion

The local content policy in the oil and gas industry in Nigeria is pursued through three key vehicles: state participation, the Local Content Development Act, and the marginal fields policy. These vehicles represent the Nigerian government's understanding of local content in the oil and gas sector, which is not limited to value addition. In its broadest sense, value addition is about what the industry as a whole can contribute to the wider economy, for instance, the contributions of oil and gas companies not only to the oil and gas industry in the country but also to the growth of other industries. The key vehicle being utilised to achieve value addition is the Local Content Development Act. Nigeria's local content policy, however, embraces the wider concept of 'local participation'. Local participation aims to ensure that the citizens take control over the development of the industry. The extent to which the content of the industry is local is therefore determined not only by the value brought to the local economy by the industry, but also by the extent to which the participants in the industry are 'local'. The key vehicles being utilised to achieve local participation are state participation and the marginal fields policy.

This chapter has considered the nature of the three key vehicles utilised by the Nigerian government as well as key challenges and weaknesses related to each of them. While state participation has enabled Nigeria to earn substantial revenue from oil and gas resources, the benefits of state participation have been significantly below expectation due to a combination of factors, including mainly the lack of a clearly streamlined role for the state oil company (the NNPC), lack of transparency and downright corruption. The Local Content Development Act marked a watershed in the evolution of the local content policy in the oil and gas industry. However, the ability of the country to derive optimal benefit from it has been marred by its ineffective implementation but also by downturns in oil prices (which slowed development activities and resulted in job losses). Lastly, the marginal fields policy has achieved very limited success due mainly to lack of available indigenous companies with the requisite technical and financial capacity to undertake effective development of marginal fields. Despite these barriers to local content development in Nigeria, the country has shown determination to advance its efforts to increase the benefits of the policy. More work, however, needs to be done particularly in the areas of financing local companies, implementation monitoring, and corruption.

References

Adedeji, A. N., Sidique, S. F., Rahman, A. A., & Law, S. H. (2016). The role of local content policy in local value creation in Nigeria's oil industry: A structural equation modelling (SEM) approach. *Resources Policy, 49*, 61–73.

Aoun, M.-C., Mathieu, C., & IFRI Centre for Energy. (2015). *Local content strategies in the oil and gas sector: how to maximise benefits to host communities*. Institut Français des Relations Internationales, 2012–2015 Triennium Work Reports. https://www.clingendaelenergy.com/inc/upload/files/IGU-2015_Local_Content_TF3_IGU_Final_May_2015.pdf.

Ajayi, O. (2017). *Why local content in Nigeria's oil, gas industry is low—Executive Secretary*. NCDMB, Nigerian Content Development and Monitoring Board. https://ncdmb.gov.ng/2017/05/why-local-content-in-nigerias-oil-gas-industry-is-low-executive-secretary-ncdmb/.

Atsegbua, L. (2005). Issues in the Development of Marginal Oilfields in Nigeria. *Journal of Energy and Natural Resources Law, 23*(3), 323–336.

Bloomfield Law Practice. (2017). The proposed 2017 marginal fields bid round in Nigeria—An investor's guide. *Bloomfield Law Practice*. https://bloomfield-law.com/Publications/BLP_NEWSLETTER%20_THE_PROPOSED_2017_MARGINAL_FIELDS_BID_ROUND_IN_NIGERIA_AN_INVESTOR'S_GUIDE.PDF.

Centurion Lawyers and Advisors. (2017). *The Nigerian National Petroleum Corporation v the Ministry of Petroleum Resources*. https://centurionlg.com/2017/11/20/the-nigerian-national-petroleum-corporation-v-the-ministry-of-petroleum-resources/.

Dentons. (2014). *Marginal fields in Nigeria: Recent developments*.

Ezeani, E. C., & Nwuke, C. (2017). Local content and the marginal fields programme: Challenges for indigenous participation in the Nigerian oil industry. *Oil, Gas and Energy Law Intelligence, 15*(1). https://openair.rgu.ac.uk/bitstream/handle/10059/2225/EZEANI%202017%20Local%20content%20and%20the%20Marginal%20Fields.pdf?sequence=3&isAllowed=y.

Falola, T., & Achberger, J. (Eds). (2013). *The political economy of development and underdevelopment in Africa*. Taylor and Francis.

Frynas, J. G. (2000). *Oil in Nigeria: Conflict and litigation between oil companies and village communities*. Lit Verlag Munster

Global Witness. (2012). *Rigged? The scramble for Africa's oil, gas and minerals—A report by Global Witness*.

Heum, P. et al. (2003). *Enhancement of local content in the upstream oil and gas industry in Nigeria—A comprehensive and viable policy approach.* SNF Report No. 25/03 (2003). https://pdfs.semanticscholar.org/3532/457b23ddd81b6f9112b037fc4eb7cb86e52a.pdf.

Humphrey, O., & Dosunmu, A. (2017). The critical success factors for marginal oil field development in Nigeria. *Journal of Business and Management Sciences, 5*(1), 1–10.

Ihua, U. B. (2010). Local content policy and SMEs sector promotion: The Nigerian oil industry experience. *International Journal of Business and Management, 5*(5), 3–13.

Inter-Governmental Forum on Mining, Minerals, Metals and Sustainable Development. (2018). *Nigeria: National Capacity—Building Capacity in the Oil Sector through Indigenization Policies (Case Study), IGF Guidance for Governments: Leveraging Local Content Decisions for Sustainable Development.* IISD, Winnipeg. https://www.iisd.org/sites/default/files/publications/case-study-nigeria-national-capacity.pdf.

KPMG. (2010). *Interim report on the process and forensic review of NNPC.* https://www.premiumtimesng.com/docs_download/Consolidated%20Detailed%20Findings-v1.pdf.

Machunga-Disu, L. L., & Sayne, A. (2014). *Maximizing value from Nigeria's marginal oil fields: A survey of past awards.*

Martini, M. (2014). *Local content policies and corruption in the oil and gas industry.* U4 Anti-Corruption Resource Centre

Ministry of Petroleum Resources. (2013). *Guidelines for Farmout and Operation Marginal Fields.* https://africaoilgasreport.com/wp-content/uploads/2013/12/GUIDANCE-NOTES-FOR-NIGERIAN-2013-MARGINAL-FIELDS-BID-ROUND.pdf.

McCulloch, N., et al. (2017). *Local content policies and backward integration in Nigeria.*

Nwapi, C. (2015). Corruption vulnerabilities in local content policies in the extractive sector: An examination of Nigerian oil and gas industry content development act, 2010. *Resources Policy, 46,* 92–96.

Nwapi, C. (2014). Enhancing the effectiveness of transparency in extractive resource governance: A Nigerian case study. *Law and Development Review, 7*(1), 23–47.

Nwokeji, U. (2007). *The Nigerian National Petroleum Corporation and the Development of the Nigerian Oil and Gas Industry: History, Strategies and Current Directions.* James Baker Institute for Public Policy and Japan Petroleum Energy Center, Rice University. https://www.bakerinstitute.org/media/files/page/9b067dc6/noc_nnpc_ugo.pdf.

Okonkwo, O. (2011). *Inter-partner learning in asymmetric alliances between foreign and indigenous companies in the Nigerian oil industry.* European universities series publisher. Peter Lang International academic publishers.

Olawuyi, D. S. (2018). From technology transfer to technology absorption: Addressing climate technology gaps in Africa. *Journal of Energy and Natural Resources Law, 36*(1), 61–84.

Onyi-Ogelle, H. O. (2016). National participation in the Nigeria oil and gas industry: Prospects and challenges. *Journal of Emerging Trends in Educational Research and Policy Studies, 7*(2), 134–144.

Ovadia, J. (2013). *Measurement and implementation of local content in Nigeria—a framework for working with stakeholders to increase the effectiveness of local content monitoring and development.* Facility for Oil Sector Transparency in Nigeria. https://cpparesearch.org/wp-content/uploads/2014/12/FOSTER-Measurement-and-Implementation-of-Local-Content.pdf.

Pearson, S. R. (1970). *Petroleum and the Nigerian economy.* Stanford University Press.

Pricewaterhouse Coopers. (2015). *Investigative forensic audit into the allegations of unremitted funds into the federation accounts by the NNPC.* https://www.wgei.org/wp-content/uploads/2015/10/Full-report-20billion-dollars-missing-oil-money_SAI-Nigeria.pdf.

Ramdoo, I. (2015). *Resource-based industrialisation in Africa: Optimising linkages and value chains in the extractive sector.* European Centre for Development Policy Discussion Paper No 179. https://www.extractiveshub.org/servefile/getFile/id/7232.

Sayne, A., Gillies, A., & Katsouris, C. (2015). *Inside NNPC oil sales: A case for reform in Nigeria.* Natural Resource Governance Institute. https://resourcegovernance.org/sites/default/files/NRGI_InsideNNPCOilSales_CompleteReport.pdf.

Sayne, A., et al. (2012). *The petroleum industry bill and the future of NNPC*. Revenue Watch Institute. https://resourcegovernance.org/sites/default/files/rwi_bp_nnpc_synth_rev2.pdf.
World Bank. (2015). *A practical guide to increasing mining local procurement in West Africa*. Report No AUS6324. https://documents.worldbank.org/curated/en/266701468119062079/pdf/AUS632 40WP0P130I0Guide0Eng0Feb02015.pdf.
World Trade Institute Advisors. (2013). *Local content requirements and the green economy, Ad Hoc Expert Group Meeting on domestic requirements and support measures in green sectors: Economic and environmental effectiveness and implications for trade*, June, 13–14 2013, Salle XXI, Palais des Nations, Geneva.

Complexities of Local Content in Kenya's Extractive Sector: An Appraisal of Policy, Legal and Institutional Frameworks and Practice

Sarah Nduku Muyonga

Abstract The debate on local content has taken center stage since the discovery of commercial quantities of oil in 2012 by Tullow Oil that could generate up to $1.2 billion in revenue for the Kenyan government. This chapter follows the evolution of local content development in Kenya's extractive sector, describing the existing legal, institutional and policy frameworks promoting local content in the industry. Defining the 'local' in local content has been a major source of conflict in the Kenyan context. From the existing legislation, it is apparent that 'local' is synonymous with 'national content,' meaning that inputs and skills can be sourced from anywhere in the host country as opposed to local sourcing from the resource rich regions. Development of favorable policies and legislation in the recent past have embolden the industry though it is not clear that they would automatically resolve these issues and address other glaring gaps. The chapter concludes that there is a need to prioritize the development of an overarching Local Content Development Policy in Kenya that will define the focus of local content, which should be based on the country's priorities, needs, and specific contexts. It should also address the question of 'community content' verses 'local content.' The policy would take into consideration the actual state of local industry, noting that local content development is a long-term goal that needs to be phased to adapt to the demands of each stage of development of the resources.

Keywords National content · Kenya local content policy · Community content · Local equity participation

1 Introduction

The debate on local content has taken center stage since the discovery of commercial quantities of oil in 2012 by Tullow Oil that could generate up to $1.2 billion in revenue for the Kenyan government. The discoveries were made in the remote,

S. N. Muyonga (✉)
Natural Resource Governance Institute, New York, USA
e-mail: nduku3@gmail.com

E. G. Pereira et al. (eds.), *Sovereign Wealth Funds, Local Content Policies and CSR*, CSR, Sustainability, Ethics & Governance,
https://doi.org/10.1007/978-3-030-56092-8_15

underdeveloped and historically marginalized Turkana region of Kenya. As expected, there have been great expectations on the oil revenues and the potential benefits for the local communities of Turkana and the country as a whole, well before actual drilling actually commenced. These expectations have inevitably caused numerous challenges, especially at the community level, characterized by disputes between the locals, their leaders and the oil companies. They have also contributed to delays in the earlier anticipated production start date of 2020.

Local content is an evolving concept with slightly different deviations as different countries with natural resource endowments have taken different approaches based on contextual considerations. The issue of what constitutes local content is the subject of different interpretations but the evolution of the debate can be linked to efforts in addressing the so-called *resource curse.* The resource curse (also known as the paradox of plenty) refers to the failure of many resource-rich countries to benefit fully from their natural resource wealth, and for governments in these countries to respond effectively to public welfare needs (NRGI 2015). Some argue that the so-called natural resource curse and Dutch disease are outcomes of bad economic decisions, while other arguments have advanced the notion that the limited linkages between the extractive sector and the rest of the economy via sector's value chain propagates the problem (Torres et al. 2013a). The latter provides the basis of the development and evolution of local content policies and strategies, as a response to the tackling the challenge.

There is no universal definition of local content but it can be generally defined as the ways countries benefit from extractives projects beyond the government revenue accrued. These benefits include employment, skills, knowledge, technology transfer, business for local companies and creation of local supply chains and markets to enhance linkages between the oil, gas and mining sector and other sectors of the economy. The World Bank defines local content as 'the *extent to which the output of the sector generates further benefits to the economy beyond the direct contribution to its value add, through its links to other sectors*' (Torres et al. 2013b). The structures and definition of local content as mentioned vary in different contexts, the common thread in all is the 'value addition'. In this case, the concept of value addition focuses on the optimization of the economic value derived from the resource. The critical understanding and application of considerations that relate to specific contexts determine the success of policy design and implementation (Youssef 2016).

Kenya is one of the new producer countries currently seeking to develop local content polices to increase local participation along the value chain of the oil, gas and mining sector. The challenge for Kenya is on how to maximize the potential benefits from the resources to avoid the under-developmental pathway that has been the trajectory of major oil producing countries in Africa such as Nigeria and Angola while taking into consideration the various complexities of the country's context.

This chapter will discuss the evolution of local content development in Kenya's extractive sector, describing the existing legal and policy frameworks promoting local content in the industry. It will further discuss some of the gaps in these frameworks, the main challenges facing the development of local content in the extractive sector and highlight some good practices. It is fair to note that the focus and evolving discussions on local content in Kenya have been largely centered around the oil sector with the discovery of oil and in anticipation of its potential economic benefits. The mining sector has received less attention given its negligible contribution to the economy at present, accounting for less than 1% of gross domestic product (KPMG 2016). This is however not to say that there have not been concerted efforts by government to include local content development in the mining sector and its legislation.

2 Institutional Frameworks, Legal and Policy Instruments; Local Content Obligations in Kenya's Extractive Sector

The growing importance of the petroleum sector has led to the recent overhaul of the legal and regulatory framework for the oil and gas industry. This has been achieved through the promulgation of the Constitution of Kenya in 2010 as well as the subsequent enactment of the Petroleum Act 2019 and the development of the Local Content Bill 2018 to create the necessary legal framework for local content in the country.

Kenya however lacks a set overarching policy that underpins or informs any existing local content legislation in the extractive sector. The existing framework is contained in various legislations related to the oil, gas and mining industry but it is hard to identify the vision and policy objectives for these legal tools. The development of the Local Content Bill, 2018 has proceeded in the absence of this kind of policy. The government recently announced, through the Minister of Petroleum and Mining, that it is in the process of developing such a policy that '*will promote a national development agenda by nurturing local participation and international competitiveness of Kenyan nationals and firms at community, county and national levels*' (Business Today 2019).

2.1 The Constitution of Kenya

The Kenyan Constitution promulgated in 2010 creates an excellent starting point for local content development. It lays a foundation for the management of natural resources and has enhanced protection and enforcement of fundamental rights among other gains spelt out in the Bills of Rights. The document provides for a two-tier structure of government, i.e., the National and the County Governments. It then distributes functions and powers between these two levels of government, allowing for local level government participation in governance processes. To this end, it provides the locals, through the county governments, the legitimacy to participate in decisions concerning management of extractive resources.

Article 66(2) of the Constitution requires parliament to enact legislation that ensures investments in property shall benefit the local communities and their economies. In addition, Section 69 (1) (a) provides that the state shall '*ensure sustainable exploitation, utilization, management and conservation of the environment and natural resources, and ensure the equitable sharing of the accruing benefits.*' These provisions are viewed as the bedrock of local content development in Kenya and provide the legal basis of local content policies and have equally led to the enactment of legislation for the extractive industry that include the recently enacted Mining Act 2016, the Energy Act 2019 and the Petroleum Act 2019. Additionally, with the discovery of commercial quantities of oil, the 2nd Medium Term Plan (MTP) of Kenya's Vision 2030 included oil and other minerals in the economic pillars and recognized the sector's potential to contribute '*to increased export earnings, higher GDP growth, broader social development, and a major spur for infrastructure development and job creation*' (Second Medium Term Plan 2013–2017). The plan highlights areas of intervention to include enhancement of local expertise and skills development in petroleum exploration and production.[1]

The Oil and Other Mineral Resources sector was identified as an additional priority sector under the Economic Pillar of Kenya Vision 2030. Even though the sector then accounted for only one percent of GDP and three percent of total export earnings, discoveries of oil, gas and other mineral resources pointed to an increasing importance of this sector to contribute to increased export earnings, higher GDP growth, broader social development, and a major spur for infrastructure development and job creation.

2.2 *The Oil and Gas Sector*

2.2.1 The Petroleum Act, 2019

In the oil sector, prior to the enactment of the Petroleum Act 2019, the legal framework in existence was not sufficient to support the implementation of local content. The only existing laws were contained in the Petroleum (Exploration and Production) Act, chapter 308 and its model Product Sharing Agreement (PSA). The Petroleum (Exploration and Production) Act 1986 requirements for local content were broad statements that become subject to different interpretations. The Act included an obligation for contractors to give preference to locally available goods and services and to prioritize Kenyans in employment and training.

This Act had some important limitations. Perhaps most importantly, the drafters did not provide for targets outlining how much local representation is needed. It is unclear whether this was a deliberate approach, as there are some arguments against the setting of targets—for example, that overly optimistic or rigid targets such as those defining the percentage of local ownership in any joint ventures tend to create rent-seeking behavior and market abuse that actually deny locals the very opportunities the legislation is seeking to create (Esteves et al. 2013). The Act also did not provide for monitoring or reporting. The lack of clarity meant that local content creation was

[1] Ibid p. 67.

essentially left for negotiations between the government and the oil companies. No standard requirements were set and different contracts were likely to have different provisions guaranteeing no uniformity in practice. Local communities continued to feel marginalized, as neither are they party to the negotiations or privy to the agreements as contracts are not yet in the public domain.

The Petroleum Act, 2019 and its model Product Sharing Contract (PSC) schedule 18, also provide for local content development in the oil and gas sector. The 2019 Act defines local content as '*the added value brought to the Kenyan economy from petroleum-related activities through systemic development of national capacity and capabilities and investment in developing and procuring locally available workforce; resources and supplies; for the sharing of accruing benefits*'.[2] The Act also allows the Cabinet Secretary to make regulations in respect to local content development. Other provisions require that priority is given to services provided and good manufactured in Kenya, provided that they meet the specifications of the industry as set by Kenya Bureau of Standards or, in the absence of this, any other international standards provided by the authority created to regulate the sector. The Act further requires that priority is given for the employment of or engagement of qualified and skilled Kenyans at all levels of the value chain. The Act creates a Local Content Training Fund and all moneys raised by contractors will be used for the purpose of training Kenyan nationals in the upstream petroleum operations. Contractors are required to submit a local content plan to the Energy Petroleum Regulatory Authority (EPRA), the new sector regulator, before commencing any petroleum operations and each consecutive year, thereafter, together with corresponding work programs. All contractors are also required to evaluate their existing contractual and procurement arrangements to ensure compliance with the local content requirement.[3]

2.3 National Energy Policy 2018 and Energy Act 2019

The Kenyan government published the National Energy Policy in 2018. This document highlighted, among other things, the need for government to develop and implement a local content policy; to develop and implement education frameworks for human capital development to build knowledge and technical capacity in the sector; to develop and implement legislative framework to prioritize utilization of locally available goods and services and human resource and ensure compliance by investors and contractors to local content requirements. The Energy Act 2019, which was enacted shortly after and also came into force at the same time as the Petroleum Act 2019, is intended to among other functions, regulates midstream and downstream petroleum activities. The Act has a similar definition of local content as to that in the Petroleum Act, imposing a requirement on contractors to comply with local content requirements and to submit to the Energy Petroleum Regulatory Authority annual

[2]The Kenya Petroleum Act, 2019.

[3]Clause 50, *The Petroleum Act*, 2019.

local content plans for approval. The Act provides that the local content plan should ensure that priority is given to services provided and goods manufactured in Kenya. The Act requires the government to establish the Local Content Development and Monitoring Unit. It also provides that skilled and qualified Kenyans should be given first priority in respect to employment at all levels of the value chain. In this regard, the provisions on local content in the Energy Act are in harmony with those of the Petroleum Act 2019.

2.4 *Mining Policies, Laws and Regulations*

As mentioned above, most debates on local content have centered around the petroleum sector, though the government has recently introduced some new mining policies and legislation that target increased local participation as part of broader efforts to improve the sector's performance. The first was the Mining and Minerals Policy Sessional Policy No. 7 of 2016. This policy calls for local content requirements for goods and services for the mining sector and in supporting industries. The policy's focus includes local equity participation in investment in the mining sector as one of its objectives. It also provides for maximization of mining benefits through use of local goods and services by promoting horizontal/lateral and vertical/backward and forward linkages in the mining industry. Alongside these efforts, the government passed the Mining Act, 2016. Its key highlights include provisions for preference in local procurement and the employment of Kenyans. The Mining Act also provides local content requirements for capacity building, equity participation, research and development, and recognizes the local community in areas where the mining industry is located. The government subsequently developed supporting regulations through the Mining (use of local goods and services) Regulation, 2016 and Mining (Employment and Training) Regulations 2017 (Employment Regulations) 2016.

2.5 *Draft Local Content Bill, 2018*

In 2018, Parliament debated the Local Content Bill introduced as a Senators Bill in the upper house. The bill is yet to be considered by the lower house. Some of its key provision spell out the roles of the national and county governments in maximizing development and implementation of local content, emphasizing that the national government would ensure consultation with country governments in undertaking its functions. Although it supposedly covers the oil, gas and mining sectors, the text focuses only on oil and gas. It defines local content '*as the added value brought to the Kenyan economy from extractive industry through systemic development of national capacity and capabilities and investment in developing and procuring locally available workforce, services and suppliers, for the sharing of accruing benefits.*' The bill also provides for the establishment of The Local Content Development Committee, the proposed structure and roles of which are highlighted further below.

3 Institutional Frameworks

3.1 Energy Petroleum Regulatory Authority (EPRA)

The Energy Act, 2019 established the Energy Petroleum Regulatory Authority (EPRA), which the Petroleum Act, 2019 empowers to do a range of important things with respect to local content. The EPRA is an independent government agency established as the successor to the Energy Regulatory Commission (ERC) with an expanded mandate of inter alia regulation of upstream petroleum and coal. The functions of the Authority also include the enforcement of local content requirements. Specifically, Section 51 of The Petroleum Act 2019 mandates the EPRA to oversee, co-ordinate and manage the development of local content and prepare guidelines in setting targets and formats for local content plans and reporting; make appropriate recommendations to the Cabinet Secretary for the formulation of local content regulations; set minimum requirements for local content and undertake audit, monitoring and enforcement.

3.2 Proposed Local Content Development Committee (LDC)

The Local Content Bill as stated above, proposes to establish the *Local Content Development Committee (LCDC).* The Bill spells out the function of the committee that include overall oversight and coordination of the management of local content development in Kenya. The committee is tasked with making recommendations and advising the Cabinet Secretary on policy and strategies for local content development and implementation. The committee will appraise and approve annual local content plans from operators and consult with county governments to ensure compliance. The committee will also keep a register of all equipment and services required to effectively service the industry. The chairperson of the committee with be appointed by the Cabinet Secretary.

4 Challenges in Implementing Local Content Development in Kenya

The first and most common challenge arising in local content development in any country is in defining the 'local' in local content within the applicable policy and legal frameworks. Some jurisdictions, for example, have made distinctions between 'local content' and 'local participation'. In these cases, local content is defined as the quantum or fraction of locally produced materials, personnel, goods and services rendered to the oil, gas and mining industries, all of which can be measured in monetary terms. Local participation, by contrast, is defined as the level of equity ownership local citizens hold in these business or companies servicing the extractive industry. This distinction has led to the emergence of numerous definitions for a local company in different contexts (Esteves et al. 2013).

Defining the 'local' in local content has been problematic and a major source of conflict in the Kenyan context. From the existing Kenyan legislation and the draft Local Content Bill, it is apparent that 'local' is synonymous with 'national content,' meaning that inputs and skills can be sourced from anywhere in the host country as opposed to local sourcing from the resource-rich regions. The existing legislation, i.e., the Petroleum Act 2019 and Energy Act, 2019, expressly speak of systematic development of national capacity while also defining local community as '*people living in a sub-county within which a petroleum resource under this Act is situated and are affected by the exploitation of that petroleum resource.*' However, neither Act in its respective local content provisions apply or make reference to 'local community' in describing the requirements or obligations of operators in the sector.

In Turkana, there have been reported cases of conflicts involving local communities, government and oil companies in recent times. The conflicts have been linked to issues around land compensation, the influx of people from areas outside Turkana, oil exploration's possible impacts on livestock grazing zones and water sources, displacement of households and preferential recruitment of outsiders (Nanok and Onyango 2017). A common factor in these disputes is dissatisfaction among local communities, many of which already claim to feel disenfranchised and strongly believe that they should be benefiting more from the oil produced from their lands. They have expressed their concerns in the use of 'local' to mean 'national' appreciating that their current capabilities would present limited opportunities in meeting the product and employment standards met by the industry that will favor people or businesses outside their community.[4] This is definitely expected as Turkana is Kenya's third-poorest region. Official data indicates that more than half the local economy relies on pastoral farming and the per capita income is less than one-fifth of levels seen in Kenya's capital. According to a Kenya National Bureau of Statistics, around 55% of Turkana residents live below the poverty line.[5]

Despite these challenges, it is not clear that local content policies and legislations would automatically resolve these issues and address the glaring gaps. Tullow Oil has been at the receiving end of most of the conflict and has deployed many community engagement strategies and programs to mitigate the challenges. In early November 2013, for example, Tullow's operations in Turkana suffered a two-week shutdown due to demonstrations by the local community, which was protesting against exclusion

[4]Fourth Workshop Report of the Extractive Sector Forum: Local Content Development at the County Level—Turkana (November 2016) Meeting Report.

[5]Kenya National Bureau of Statistics, *Kenya Integrated Household Budget Survey Report* 2015/16.

from employment and supply of goods and services (Financial Times 2013). Tullow has taken up the task of responding to community demands and has also embraced voluntary public disclosures on its activities. To this end, they have published annual reports including data and figures related to local content expenditure, employment and social investment and other payments to government.

However, this approach outside clear and concerted efforts by government in deploying appropriate strategies of addressing the glaring challenges seems futile. Overseas Development Institute (ODI) introduces the concept of 'community content' defining it as 'the *interface of community investment programs with local content*.' ODI suggests that community content can be seen as 'merit good,' meaning they are not necessarily specific in nature but are targeted at those affected by oil and gas development and this to some degree are exclusionary. They therefore differ from local content programs that are less exclusionary and a 'public good' (Warner 2007). Such arguments suggest that community content should be pursued as a distinct policy from local content, even as community content would still be treated as local content. Achieving community content would however require concerted and conscious efforts in building the capacity of local skills to access the opportunities in extractive sector, and strategies for community content should be designed differently from those of local content development.[6] The Kenyan case needs to assess the viability of this approach in dealing with some of the community-specific challenges. Partly in recognition of this, the County Government of Turkana developed its Policy Framework for Extractive Industries in Turkana in 2018. If successfully implemented, this document could help ensure the local government prioritizes in its development plans, support in developing skills and business capacity of the local community in becoming competitive to participate in the sector.

Another major challenge in implementing local content development in Kenya is the lack of an overarching National Local Content Policy. Kenya seems to have adopted a local content approach that consists of setting legally binding targets. Although the existing laws have stipulated the relevant institutions under legislation would set those the operators are obligated to comply. It is argued that setting these kinds of targets implies that the government has sufficient human capacity in skills, goods and services that the foreign operators require through the value chain (Youssef 2016). However, this is not the case as there is a limited pool of expertise and skills in the sector both at the national and local levels.

5 Conclusion

Kenya has made some meaningful strides in the development of local content in the extractive sector. In adopting a broader approach to the issue, it has adopted national legislation for the same. The existing laws make a good attempt at harmonizing the definition of local content as seen in The Petroleum Act, 2019 and its model

[6]Ibid p. 6.

PSC, The Energy Act, 2016, the Mining and Mineral Policy, the Mining Act and the Proposed Local Content Bill, 2018. All these policies and legislation offer some guidelines for developing and enforcing measures that will ensure a competitive local workforce, facilitate knowledge and skill transfer and promote the use of local goods and services.

The regulatory frameworks have also adopted some good practices on local content. Examples on these include the provision for the establishment of an Energy Petroleum Regulatory Authority (EPRA) in the Energy Act, 2019 and the Petroleum Act and its model Product Sharing Agreement in the Petroleum Act, 2019. The EPRA is the independent upstream industry regulator tasked with enforcement of local content requirements among other regulatory duties. The Act also introduces a Training Fund to train Kenyan nationals in the upstream mainly guaranteeing a funding for skills development.

At the same time, some serious challenges remain. Foremost among these are the multiplicity of regulations and regulatory institutions, lack of clear distribution of power and responsibilities between governments at national and county levels, enormous development setbacks in extractives host communities, and the gaps in the expectations of locals. In addition, a lack of requisite skills and competence continues to limit employment opportunities and participation by host communities. The case of Turkana, highlighted above, illustrates some of the conflicts in the region, which has been a result of limited opportunities due to these gaps. It is therefore important that government and company interventions on local content should not only tighten the legislative and regulatory frameworks but also inquire into the social dynamics as well as into governance and accountability structures at the community level, paying special attention to community content.

One unique development is the proposed Local Content Bill that seeks to establish a Local Content Development Committee that would oversee, co-ordinate and manage the development of local content in the country. While it is anticipated that passage of the bill will be a step forward for local content development, it may not necessarily address all the complexities. For instance, if the bill does not clarify the functions of the Local Content Development Committee compared to those of the already existing EPRA, it may create clashes of interest and overlapping roles. In particular, special attention should be paid when developing the roles and functions and appointments of this committee, to avoid too much bureaucracy and numerous reporting requirements for companies that would create bottlenecks in contracting and subcontracting and subsequently in the smooth running of the sector. The final draft of the bill also needs to say more about its applicability to the mining sector, as the current draft reads more like a Petroleum Sector Local Content Bill.

The country also needs to prioritize the development of an overarching, non-sector-specific Local Content Development Policy that will define the focuses and goals for local content in line with the country's priorities, needs and specific contexts. The policy needs to take into consideration the actual state of local industry, noting that local content development is a long term goal that needs to be phased to adapt to the demands of each stage of development of the resources. It should also reflect a common understanding among all the stakeholders of what constitutes local content, who is responsible for all actions and implementation—e.g., the national versus county governments, companies, etc. The policy should state clear yet realistic targets and goals that will provide a basis for Local Content Laws and Regulations and in the case of Kenya, ensure a harmonized approach and timelines for implementation. Consultations with wider stakeholders, including the community in the development of the policy should ensure that local content is understood as part of a broader policy approach and directly linked to broader economic development policies such as Vision 2030. It should in essence Act as a springboard for additional economic activities and facilitate a strong multiplier effect to achieve this vision. Clear monitoring and evaluation mechanisms to ensure compliance have to be established in this context while assigning responsibilities for implementation and monitoring of the policy.

Other important steps in the development of local content include prioritization of physical development, improving the enabling environment for businesses and targeting efforts at human capital development. The national and local governments need to conduct an objective and comprehensive analysis of local capacities available and develop a detailed plan, including the right institutions and funding support to enhance the same. This process should involve a demand side value chain analysis that will diagnose the company standards, operational requirements at each stage of projects and identify opportunities for local skills, good and services. Equally, the process should assess the supply side, i.e., the local skills and commodity capabilities and conduct a baseline to establish the gaps. This will enable interventions to be targeted at areas in ways that respond to existing demand and supply dynamics. Funding arrangements and loan plans should be provided for Small and Medium Enterprises to aid growth and competitiveness of locals. Moreover, targeted efforts should be made in the Counties Integrated Development Plans to reduce capacity gaps among locals in the resource-rich areas. Periodic social and economic assessments of communities should be prioritized to understand local needs and aspirations. For all these reasons, the government should prioritize data collection and analysis to measure progress and inform development of future local content measures. In addition, a robust conflict management strategy should be developed to help facilitate better relations between locals, investors and the government. And critically, for the success of all these efforts, the government should resist the urge to create processes that will encourage patronage, corruption and rent seeking for those in power and focus more on the growth of legitimate businesses and human capacities in the extractive sector.

References

Business Today, May 8, 2019.

Esteves, A. M., Coyne, B., & Moreno, A. (2013). *Local content initiatives: Enhancing the subnational benefits of the oil, gas and mining sector* (p. 6). NRGI

Financial Times, October 18, 2013.

KPMG (2016). *Analysis of mining act 2016* (p. 1).
Nanok, J. K., & Onyango, C. O. (2017). A socioeconomic and environmental analysis of the effects of oil exploration on the local community in Lokichar, Turkana County, Kenya. *International Journal of Management, Economics and Social Sciences (IJMESS), 6*(3), 144–156. ISSN 2304-1366, IJMESS International Publishers, Jersey City, NJ.
NRGI. (2015). *The resource curse: The political and economic challenges of natural resource wealth*
Second Medium Term Plan. (2013–2017). *Transforming Kenya: Pathway to devolution, socio-economic development, equity and national unity* (p. 67). www.vision2030.go.ke.
Tordo, S., Warner, M., Manzano, O. E., Anouti, Y. (2013). *Local content policies in the oil and gas sectors* (pp. 2–3). Washington DC: The World Bank., Grossman, G. M. (1981). The theory of domestic content protection and content preference. *Quarterly Journal of Economics, 96*(4), 583.
Torres, N., Afonso, O., & Soures, I. (2013a). *A survey of literature on the resource curse: Critical analysis of the main explanations, empirical tests and resource proxies* (No 1302). Universidade do Porto, Falcudade de Economia do Porto
Warner, M. (2007).*Community content: The interface of community investment programmes with local content practices in the oil and gas development sector* (p. 5). Overseas Development Institute (ODI) Policy Brief Note 9.
Youssef, E. H. (2016). *Local content in the extractive sector: Exploring policy issues in new producing countries'*. Policy Centre for the New South, Blog posted November 11, 2016.

Local Content Policy in Indonesia Oil and Gas Industry

Eko A. Prasetio and Elisabeth D. Kumalasari

Abstract As a net oil importer, the Government of Indonesia (GoI) put a lot of effort into attracting new investment for oil and gas exploration and exploitation. The gross split scheme was introduced in 2017 as an alternative to cost recovery. Meanwhile, the GoI has been setting up various regulations and a road map to increase local content to stimulate local manufacturers in the oil and gas supporting industries. How this significant policy change affects the local content requirement and implementation becomes the central focus of this chapter. In the cost recovery scheme, local content requirement was enforced through various regulations with close supervision from the government's executive task force in upstream oil and gas business activities; meanwhile, the gross split scheme uses an incentive approach by incorporating local content as a part of variable split. The impact of new policy started to emerge in the form of new investments in the gross split scheme as well as the corresponding local content procurement. Challenges in the local content implementation include concerns over the quality and price of locally produced goods. Some cases suggested that locally procured goods are not yet up to the standard and specifications required by the contractors of cooperation contract, and yet, the price is higher than imported ones. As these challenges could hinder the target of local content realization, government, local manufacturers of oil and gas supporting industries and the contractors need to work together to overcome those.

Keywords Local content · Oil and gas · Cost recovery · Gross split · Indonesia extractive resource

E. A. Prasetio (✉)
School of Business and Management, Bandung Institute of Technology, Bandung, Indonesia
e-mail: eko.prasetio@sbm-itb.ac.id

E. D. Kumalasari
Ministry of Energy and Mineral Resources, Jakarta, Republic of Indonesia
e-mail: elisabeth.kumalasari@esdm.go.id

E. G. Pereira et al. (eds.), *Sovereign Wealth Funds, Local Content Policies and CSR*, CSR, Sustainability, Ethics & Governance,
https://doi.org/10.1007/978-3-030-56092-8_16

1 Introduction

This chapter discusses the local content policy in Indonesia's oil and gas sector as part of the extractive resource industry. The overview and short history of the Indonesian oil and gas industry are described, including the change in the scheme of profit sharing, i.e., cost recovery vs. gross split.

The elaboration on how the change of sharing schemes affects the local content requirements is presented. The goals and implementation of local content policy are described and analyzed. The analyses will also include the challenges in the implementation of local content requirement as well as the recommendation measures required to overcome those challenges.

2 Background and Overview of Indonesia Oil and Gas Industry

Indonesia is a net oil importer since 2004, due to decreasing oil production, which is not balanced by the increasing domestic oil consumption. The production of crude oil of 786 thousand barrels per day in 2015 is only 70% of its 2005 daily production. The decline is caused by the natural maturation of producing oil fields as well as the slow reserve replacement rate.

Consequently, the contribution of the oil and gas sector to state revenue decreases significantly along with the decrease in production and reserve. In 2016, the oil and gas sector contribution fell to 3% compared to 15% in the year 2010 (Fig. 1).

The Government of Indonesia declared several new upstream oil and gas projects to boost production, e.g., Jangkrik field development, Tangguh Train-3, the Indonesia Deepwater Development Project and Genting's Kasuari block *(source: Oil and Gas in Indonesia, PwC,* 2017*)*.

High risk and uncertainty, huge amounts of investment and the requirement of sophisticated technology characterize the upstream oil and gas business activities. A production sharing contract (PSC) has been applied since 1966 in the upstream oil and gas business activities, when PERMINA signed a PSC with Independence Indonesian American Oil Company (IIAPCO).

As a metaphor in agriculture, we might think of the PSC model as the business model between the owner of the paddy field and the farmer who manage the field. The government acts as the owner of the field and the oil and gas company as the farmer who provides all the investment and production equipment. The investment by the "farmer" needs to be approved by the "field owner" since this investment will be returned eventually at harvest time. This is how the cost recovery scheme works in the oil and gas business. This scheme was considered the best model applied in Indonesia, since this sharing model ensured the ownership and power of the country upon its oil and gas resources and protected the country from the excessive risk and uncertainty in the upstream oil and gas business activity.

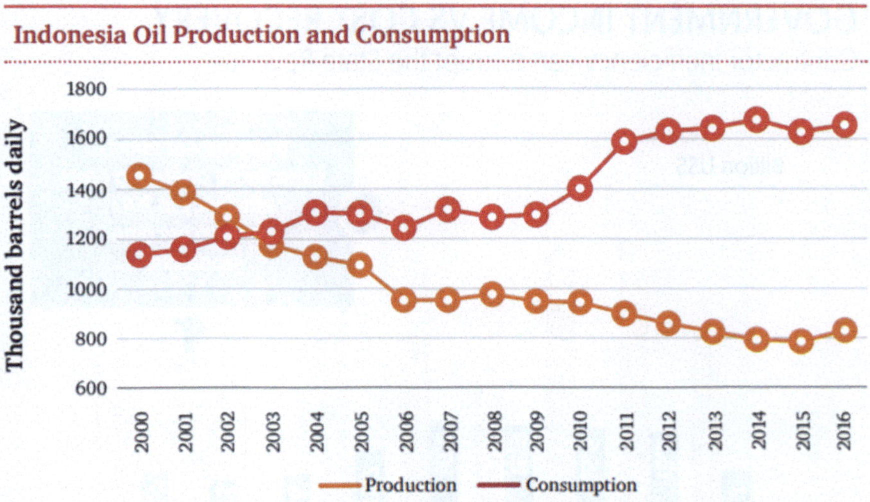

Fig. 1 Indonesia oil production and consumption. *Source* Oil and Gas in Indonesia: PwC Report (2017)

In 2015, however, the value of cost recovery was unexpectedly higher than the state revenue coming from the oil and gas lifting. This of course became the controversy (Fig. 2). This occurred because there was no obligation for the contractors of cooperation contract to be efficient in their operational activities.

(https://money.kompas.com/read/2014/11/12/161158726/Mengenal.Kontrak.Hulu.Migas.Indonesia).

The Government of Indonesia eventually issued the gross split scheme in the beginning of 2017. The scheme is expected to encourage the contractor of cooperation contract to be able to maintain the efficiency of operational expenses while continuously improving the technology management, human resources and the system. Unlike the cost recovery, the contractor of cooperation contract covers the overall expense in the gross split scheme (Fig. 3).

The other difference between cost recovery and gross split lies in the share of split. On one hand, in the cost recovery, after tax, the split is fixed with the government obtaining 85% while the contractor obtains 15% for oil; whereas it is 70 versus 30% for gas. On the other hand, the gross split scheme is dynamic where the split depends on the variable split (Fig. 4).

The criteria of profit sharing in the gross split scheme include:

1. Reservoir depth;
2. Field location, whether the field is onshore, offshore or remote;
3. Reservoir condition;
4. Level of difficulty based on the geological condition;
5. Reservoir characteristics, i.e., whether the reservoir is categorized as conventional or non-conventional oil and gas block, as well as the technology utilized by the contractor.

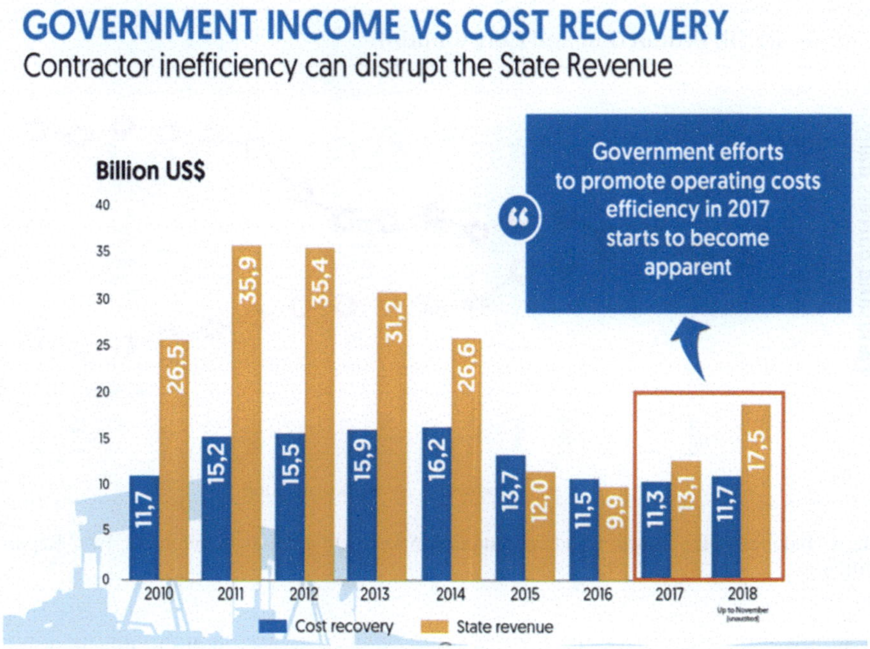

Fig. 2 Government income versus cost recovery

This policy started to show an impact as shown in Figs. 5 and 6. In 2019, 40 oil and gas blocks have been using the gross split scheme. After the new scheme was introduced, 5 and 9 blocks have been successfully bid in 2017 and 2018, respectively.

3 Target and Implementation Road Map of Local Content Requirement

Indonesia uses local content requirement *(Tingkat Komponen Dalam Negeri* or *TKDN* in Bahasa Indonesia*)* as part of its industrial policy in the oil and gas industry for various reasons. According to Regulation No. 15/2013, the reason for using local products and services for upstream oil and gas business activities includes: the multiplier effect for national economy, developing local innovation and technology, and producing goods and services effectively and efficiently. Local content requirement is also expected to ease the current account deficit due to imports of goods and services.

The government believes that local content requirement policy would be able to reduce import for basic industrial needs, including in the oil and gas sector, up to USD 20 billion (*source: Vice Minister of Finance, GOI, CNBC Indonesia interview on 24th of July* 2018*)*. Therefore, the government established a road map and target

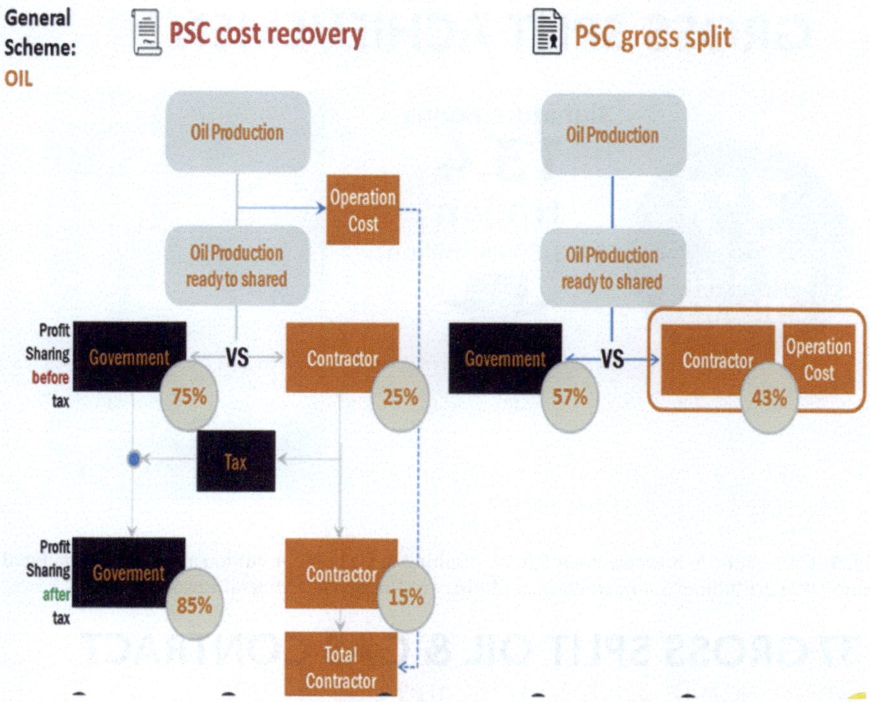

Fig. 3 PSC cost recovery versus the PSC gross split

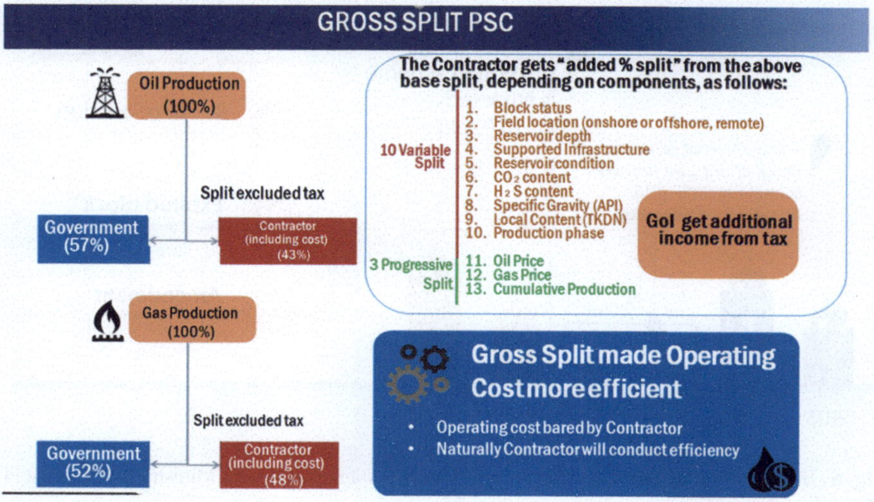

Fig. 4 Gross split scheme and the local content as part of variable split

Fig. 5 Gross Split Achievement until 2019, resulting in USD 895.4 million and Exploration Fund worth USD 2.1 billion. *Source* Indonesia Ministry of Energy and Mineral Resources

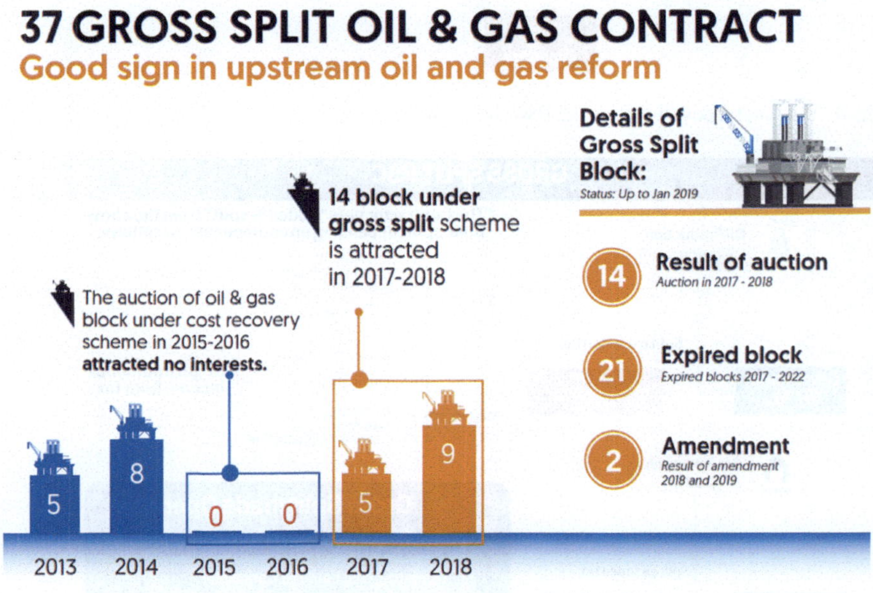

Fig. 6 Impact of newly introduced gross split scheme. *Source* Indonesia Ministry of Energy and Mineral Resources

Table 1 Road map of local content of goods in upstream oil and gas business activities

No.	Commodities	Target of local contents (%)		
		Short term	Medium term	Long term
		2013–2016	2017–2020	2021–2025
1	Fuel	60	75	95
2	Lubricants	50	60	70
3	Drilling pipe			
	– Hi grade	25	40	55
	– Low grade	15	25	40
4	Linepipe	50	65	80
5	Drilling mud, cement and chemical	40	55	70
6	Electrical submersible pump	15	25	35
7	Pumping unit	40	55	70
8	Machinery and equipment	20	30	40
9	Wellhead and X Mas Tree			
	– Onshore	40	55	70
	– Offshore	15	30	40
10	Other goods	15	25	40

Source Annual Report of Energy and Mineral Resources Ministry of Republic of Indonesia 2013

of local content for short, medium and long term as shown in the table, as stated in the Minister of Energy and Mineral Resources Regulation Number 15/2013 (Tables 1 and 2).

MoEMR issued a new regulation in the spirit to make Indonesia become more investment-friendly. For example, MoEMR Regulation No. 14/2018 regarding Oil and Natural Gas Supporting Business Activities simplifies the bureaucracy for oil and gas supporting businesses in obtaining Supporting Business Capacity Certificate (Surat Kemampuan Usaha Penunjang, or SKUP). The SKUP is classified into oil and gas construction services, oil and gas non-construction services, and oil and gas supporting industries. By issuing Reg 14/2018, the previous Regulation No. 27/2008 that requires the oil and gas supporting industries to acquire Registration Certificate on top of SKUP is abolished and the issuance of SKUP that previously took 10 days is shortened to 3 days.

4 Local Content Policy in the Indonesian Oil and Natural Gas Sector

Indonesian local content requirements are spread across various laws, regulations and decrees. The policy is based on the Constitution of the Republic of Indonesia 1945 and requires natural resources to be directly controlled by the state. The Law

Table 2 Road map of local content of services in upstream oil and gas business activities

Services				
No.	Commodities	Target of local contents (%)		
		Short term	Medium term	Long term
		2013–2016	2017–2020	2021–2025
1	Chartering service EPCI			
	– Onshore	50	70	90
	– Offshore	35	45	55
2	Drilling services			
	– Onshore	55	70	90
	– Offshore	35	45	55
3	Shipping services	75	80	85
4	Airline services	80	90	95
5	Services survey, seismic and geological studies			
	– Onshore	60	75	90
	– Offshore	15	25	35
6	FEED services	60	70	80
7	Other services	40	55	75

Source Annual Report of Energy and Mineral Resources Ministry of Republic of Indonesia 2013

No. 22/2001 on Petroleum and Natural Gas ("Law 22") is the primary oil law. It is amended before the legislature following a 2012 Constitutional Court Ruling that found parts of the Law unconstitutional. Under Law 22, private companies could carry out exploration, development and production as contractors to the government under the profit-sharing contract (PSC) scheme.

The Regulation No. 35/2004 is on upstream oil and gas business activities (known as "Reg. 35"). This regulation was last amended by Reg. No. 55 of 2009. Under this regulation, companies must give preference to qualified Indonesian personnel and train such personnel for staff positions including in administration and executive management (Reg. 35, Art. 82). According to Law 22, Art. 40(4) and Reg. 35, Art. 79(2), business entities are required to prioritize local goods, services and technology, as well as Indonesian design and engineering capabilities so long as they are of comparable quality, price and availability. Goods, services, technology, and design and engineering capabilities can be imported if they are not produced domestically (Reg. 35, Art. 80).

SKK Migas, or Executive Task Force for Upstream Oil and Natural Gas Business Activities, issued the Working Guidelines of the Implementing Body for Oil and Gas Business Activities No. 007 Revision-IV/PTK/I/2011 or "PTK 007." It is followed by Regulation No. 15/2013 or "Reg. 15" that codifies Pedoman Tata Kerja or PTK (Working Guideline) 007, abbreviated as PTK 007, which states a 5% pricing preference for Indonesian companies, defined as companies with 51% of voting shares and two-third of Board of Directors (BOD) seats held by Indonesian citizens *(source: SSEK Indonesian Legal Consultant, "Oil and Gas Procurement Amendment to PTK 007, Indonesian Insight, 16 May, 2013).*

The Government of Indonesia established the regulations for oil and gas to increase the domestic product competitiveness. Operational material and equipment is defined as all the material and equipment required for upstream oil and natural gas operation activities, including field processing activities, transportation, storage and sales of production results. Contractor of cooperation contract may use the local or imported operational material and equipment.

Ministry of Energy and Mineral Resources issued the Regulation No. 15/2013 on local content requirement in upstream oil and natural gas business activities with goals in mind, i.e., encouraging the local products and services to be able to support the upstream oil and natural gas business activities and providing value added for the economy. The regulation also has a goal of creating jobs for locals, as well as giving opportunity for local products and services to be able to compete at the national, regional and international levels. The regulation of local content requirement is intended to support and to develop innovation and technology of local products.

By applying this regulation, it is expected that the increase in the local content of oil and gas upstream business activities maintains the effectiveness and efficiency principle, as well as maintains the good governance of the implementation.

Meanwhile, government had to monitor the material and equipment that will be used by the contractor of cooperation contract. It has to do with the exemption of custom and import tax of material and equipment that will be required by the contractors. The contractor of cooperation contract needs to acquire Import Plan Document (Rencana Impor Barang, or RIB, in Bahasa Indonesia) before they are allowed to import. According to Ministerial Regulation No. 17/2018 on the import of material and equipment for the contractor of cooperation contract, usage of local product is one of the important aspects for a contractor of cooperation contract to obtain the RIB document. Contractors, under the cost recovery scheme, who import goods without the Import Plan Document will bear the consequence, i.e., unable to recover the costs borne by importing.

Indonesian MoEMR continually updates the list of products and services that can be locally produced in a Master List of Local Goods and Services. This Master List acts as a reference for the contractor of cooperation contract in the search for required goods and services for their operational activities. In formulating operational equipment and materials, the contractor of cooperation contract has to prioritize locally produced goods, services, technology and engineering capabilities in a transparent way, referring to the Master List of Local Goods and Services (or *Buku APDN in Bahasa Indonesia*). (http://skkmigas.mic.ads2.kompas.com/post/30/mengenal.kontrak.hulu.migas.indonesia).

In the gross split scheme, the government considers local content as one of the profit-sharing variables. In principle, the higher the local content used by the contractor of cooperation contract in the upstream business activities, the more the contractor has the opportunity to get the bigger split.

5 The Elaboration of the Difference of Local Content Requirement Under Cost Recovery and Gross Split

In the cost recovery scheme, the procurement of goods and services of the contractors of cooperation contract refers to PTK 007 issued by SKK Migas (Executive Task Force of Upstream Oil and Gas Business Activities), wherein the guidelines regulate and supervise the prioritization of domestic or locally produced goods and services to be maximized. PTK 007 Revision 04 and its operational guidelines govern the provision of stronger use of domestic goods and services. In this revision, it is stipulated that the use of domestically produced goods is an "obligation," rather than the previously used term "prioritize," in the upstream oil and gas business activities.

Greater preference is given to the use of ships and domestic production drilling rigs. This regulation also strengthens the multiplier effect in the area of oil and gas operations because the implementation of tenders with a value of up to Rp. 10 billion can only be followed by goods/service providers domiciled in the province where the contractor operates.

While in Article 18 of the ESDM Ministerial Regulation No. 08 of 2017 concerning gross split profit-sharing contracts, it is stated that procurement of goods and services is carried out independently by the contractor of cooperation contract. The variable of local content in the gross split scheme is expected to be able to encourage contractor of cooperation contract to better utilize domestic products.

Moreover, with the existence of this variable local content, the domestic oil and gas supporting industries are capable of further developing their products. For example, the way to increasing local content in the pipe industry in Indonesia is by producing pipes from billets. Previously, the local pipeline manufacturer only carried out heat treatment and threading facilities with a maximum local content of 25–30%. However, by adding the process of making green pipes from billets, the local content could rise to more than 40%. Optimizing the use of domestic products (goods and services) by the contractor of cooperation contract is expected to be able to support the development of domestic industries and provide optimal share split for contractor of cooperation contract. (Source: https://finance.detik.com/energi/d-3585578/ini-aturan-baru-tender-barang-dan-jasa-hulu-migas.)

6 Implementation of Local Content Policy Under Cost Recovery and Gross Split Schemes

Efforts taken by the Ministry of Energy and Mineral Resources are showing positive results. In 2006, the total expenditure of all contractor of cooperation contract amounted to USD 8523.9 million. The local content of all goods and services procurement was 43%. However, in 2013, the total expenditure of goods and services was USD 12,553 million with USD 7030 million domestically sourced, or equivalent to

56%. The more recent local content value of procurement of goods and services in the upstream oil and gas business is shown in Fig. 7.

The Ministry of Energy and Mineral Resources (MoEMR) delegates the oversight of the upstream oil and gas business activities to the Executive Task Force for Upstream Oil and Natural Gas Business Activities (SKK Migas), including local content compliance. Until 2012, that job was delegated to Executive Agency for Upstream Oil and Gas Business Activities (BP Migas).

In terms of monitoring compliance with local content requirements, the responsibility lay with the companies themselves since, under PTK 007, the government has limited supervision. Reg. 35 increased the government's monitoring and enforcement authority including the ability to witness production processes, making verification of local content levels compulsory and giving the Executive Task Force of Upstream Oil and Gas Business Activities *(SKK Migas)* the ability to set the local content level a company needs to achieve in its procurement plan.

In 2011, MoEMR issued the Ministerial Regulation No. 0982/2011 to form a team of Enhancement of Local Production Use (Peningkatan Penggunaan Produksi Dalam Negeri or P3DN) in the upstream oil and gas business activities. The team consists of different stakeholders, e.g., government (directorate general of oil and gas, MoEMR and Executive Task Force of Upstream Oil and Gas Business Activities), business association of oil and gas, and Director of Metal Industry, Ministry of Industry. However, this P3DN team is inactive at the moment although it has a very important role in the implementation of local content policy in upstream oil and gas business activities.

In the gross split scheme, the government considers local content as one of the profit-sharing variables. According to the Regulation of Minister of Energy and

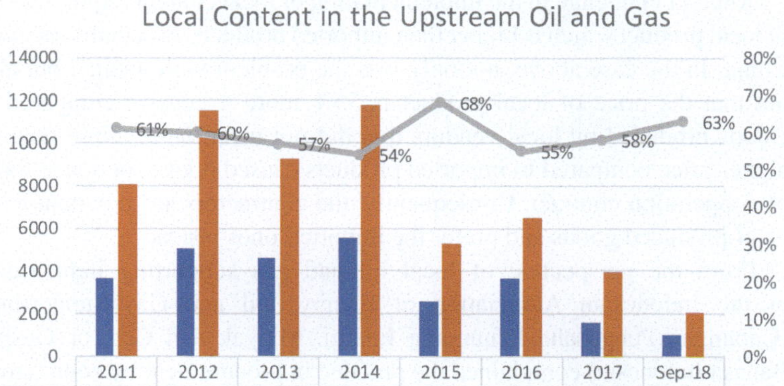

	2011	2012	2013	2014	2015	2016	2017	Sep-18
Goods (mn US$)	3706	5082	4616	5548	2590	3699	1637	1034
Services (mn US$)	8109	11531	9304	11807	5319	6496	3999	2121
% Local Content	61%	60%	57%	54%	68%	55%	58%	63%

Fig. 7 Local content value for procurement of goods and services in the upstream oil and gas business. *Source* Indonesia Special Task Force for Oil and Gas Upstream Business Activities (SKK Migas) Report, 2018)

Mineral Resources No. 52 Year 2017, for a contractor that achieves local content of 30% or more up to less than 50%, the contractor will receive 2% more of the split from gross revenue; from 50% or more up to less than 70%, the contractor will receive 3% more; and from 70% or more up to less than 100%, the contractor will receive 4%, respectively.

7 Opportunities and Challenges in the Implementation of Local Content Policy in Indonesia

Local content requirements give an opportunity for the Indonesian oil and gas supporting industry to develop. However, the challenges are plenty. The capability of the domestic oil and gas supporting industries is still limited due to the low level of research and development efforts. The other challenge is the lack of raw material in the upstream industry, such as green pipe, seamless pipe, round bar, stainless steel and steel plate. Also, the capacity of the domestic industry is still low for products such as valve, stud bolt, pressure gauge, forging and mechanical seal (source: *Annual Report of Energy and Mineral Resources Ministry of Republic of Indonesia,* 2013).

There is a lack of quality for some local products. Some cases in point are as follows. A case related to the lack of quality of domestic-sourced goods occurred in 2018 in the contractor of cooperation contract (Kontraktor Kontrak Kerja Sama, or KKKS, in Bahasa Indonesia), which has a working area in the eastern part of Java island. The products that have been domestically ordered were not produced in accordance with the specifications in the purchase order. This eventually caused damage to operating equipment in the contractor's field.

Another challenge in the implementation of local content requirement is the price of local product which is higher than imported products, especially the products from China. In the case above, not only was the problem with quality, but also it turned out that the price of local product is 35% more expensive compared to imports. Goods produced by local vendors that did not meet the required quality and have higher price compared to imported products caused poor experience for contractors of cooperation contract. Consequently, the contractors are reluctant to procure the local-produced goods and prefer the imported ones instead.

From the perspective of local oil and gas supporting industries, the chief of the Indonesian Association of Energy, Oil and Gas Supporting Business (Gabungan Pengusaha Penunjang Energi, Minyak dan Gas, or Guspenmigas in Bahasa Indonesia) complained the practice in procurement by contractor of cooperation contract who deliberately avoiding using local products or services. The chief also complained about the dominance of expatriates in the decision-making process in the contractor of cooperation contract, which makes the effort to further increase local content ineffective.

(Source: https://pelakubisnis.com/2018/05/tkdn-migas-meningkat-tapi-tetap-kompetitif/).

8 Policy Measures to Overcome the Challenges in the Implementation of Local Content Requirements

Contractors of cooperation contract hesitate to procure locally manufactured products since there are still problems in quality and delivery. The local products fail to match the required specification. In addition, there can be lateness in product delivery that has a negative impact on the completion of the contractor's project. Furthermore, considering the fluctuation of oil and gas prices, the gross split contract demands for local players to be more efficient and competitive. Recommendations for policy measures to address these challenges include, the government:

1. Coaching and support for local manufacturers that involve the contractor of cooperation contract in matching the quality requirement of operational needs. An initiative such as a joint assessment program—between the Directorate General of Oil and Natural Gas, MoEMR, Executive Task Force for Upstream Oil and Gas Business Activities (SKK Migas) and contractors of cooperation contract that has taken place—needs to be maintained and even strengthened. Investment is often an obstacle for local manufacturers. The input and advice given on the results of the joint assessment often cannot be carried out due to constraints on the amount of investment that must be spent, and there is no guarantee that their products will be used by the contractor.
2. Plans to formulate the list of the needs of operational goods and equipment for the next 5 years. In so doing, local participation in the provision of operational goods and equipment can be well prepared. Local manufacturers might be able to anticipate the needs of operational goods and equipment well ahead and plan for a timely product development, production and delivery.
3. Has to provide information transparency regarding the price and availability of supply of locally manufactured products.
4. Should encourage and build positive communication between contractor of cooperation contract and local companies by giving opportunity for qualified local companies to promote their products and achievements to the contractors.
5. Needs to consider changing the formulation of local content. The current formula, which is based on "Cost-to-Make," is considered problematic from several aspects, i.e., the calculation, the verification, the application as well as the supervision aspect. A new formula that is simpler and based on the depth of upstream industry might be preferable. Since the authority to determine the new and simpler formula is in the hands of the Ministry of Industry, the smooth coordination between MoEMR and the Ministry of Industry requires no further explanation.
6. In this case MoEMR, should reactivate the Team of Enhancement of Local Production Use (Peningkatan Penggunaan Produksi Dalam Negeri or P3DN) to further coordinate, monitor and evaluate the implementation of the enhancement of local production use in the upstream oil and gas business activities.

9 Conclusion

Local content in the oil and gas industry is regulated through various regulations and amendments. The recent introduction of the gross split scheme in production sharing contract, as an alternative to cost recovery, changes how local content is promoted and enforced. Unlike the cost recovery scheme, in the gross split scheme, local content becomes part of variable split, which means the higher the local content, the bigger the split for the contractor.

The government set up a target and road map of local content for short-, medium- and long-term periods for local manufacturers. However, the implementation of the local content policy faces several challenges including the quality of locally manufactured products, the delivery of locally manufactured products, and the high price compared with imported products. Since these challenges might hinder the target of local content, the government needs to work together with local companies in the oil and gas supporting industries as well as the contractors of cooperation contract.

In this chapter, the discussion is focused on the local content policy change according to different PSC schemes employed by the Indonesian Government. Under the previous cost recovery scheme, local content requirement was enforced through regulation; however under the new gross split scheme, local content becomes an incentive for contractors as it is part of variable split. The lessons learned of local content policy formulation and implementation in Indonesia are as follows:

(1) New local content policy under the gross split scheme accommodates local content increase in the Indonesian oil and gas sector.
(2) In an effort to match the locally manufactured products with contractors' operational needs and to improve the quality and availability of locally manufactured products, MoEMR of Indonesia is cooperating with contractors of cooperation contract and the Executive Task Force for Upstream Oil and Gas Business Activities *(SKK Migas)* in a joint assessment program.
(3) Despite those encouraging efforts discussed above, in reality, contractors of gross split cooperation contract have difficulty achieving even 30% of local content. The reasons vary from inability of local manufacturers in providing high-tech products according to contractors' operational needs to the insufficiency of local product quality and/or uncompetitive price.
(4) The results of the joint assessment program in point (2) regarding the suggestions for improving product quality are seldom followed up by local manufacturers due to required high investment.

References

Annual Report of Energy and Mineral Resources Ministry of Republic of Indonesia. (2013).

Minister of Energy and Mineral Resources Regulation No. 0982/2011 concerning the National Team of Enhancement of Local Production Use in oil and gas business activities.

Minister of Energy and Mineral Resources Regulation No. 15/2013 concerning the Use of Domestic Products in Upstream Oil and Gas Business Activities.

Minister of Energy and Mineral Resources Regulation No. 17/2018 concerning Import of Operational Goods for Upstream Oil and Gas Business Activities.

Nurdin, Y. T. (2018). *TKDN Migas meningkat Tapi tetap Kompetitif*. https://pelakubisnis.com/2018/05/tkdn-migas-meningkat-tapi-tetap-kompetitif/. Accessed May 2018.

Oil and Gas Guide in Indonesia, PwC Report. (2017).

Vice Minister of Finance, GOI, CNBC Indonesia interview on 24th of July, 2018. https://www.cnbcindonesia.com/market/20180724154249–17-25124/ini-strategi-pemerintah-tekan-impor-barang-migas.

Local Content and Extractive Industry in Brazil

Israel Lacerda de Araújo and Hirdan Katarina Medeiro da Costa

Abstract Local content clause is the obligation of hiring a share of national goods, employment, as well as services in the exploration, development and production processes. If companies do not comply with this clause, they will incur a fine applied by government agents. The Brazilian oil and gas industry has used the local content clause since 2003. The clause and regulation is long, and the details allow the contractor to know each parameter before the bidding round, to be prepared, and to be aware of the penalty in case of disobeying. This chapter answers the following questions: (i) What are the main obligations related to local content in Brazil within the extractives resources industry? Are there any peculiarities? (ii) What are the main challenges that extractives resources are facing with regards to local content practices and local involvement? (iii) What are the best and worst experiences with local content requirements in the extractive resources sector in Brazil?

Keywords Local content · Oil industry · Market failure · Effective protection

1 Introduction

During the second half of the last century, how developing countries, mainly those rich in natural resources, have performed economically (income per capita) worse than the other country group, those countries without much resources. Auty (2001) wrote about these patterns, well known as the "resource curse" phenomenon, where the relationship between a nation's natural resources and its long-term economic growth is negative. The limited spillover effect of natural resource sector and the rest of the economy that occur when legal regimes fail to determine fair rules to share the benefits of their resources sector explains weak economic growth in that

I. L. de Araújo · H. K. M. da Costa (✉)
IEE, University of São Paulo, São Paulo, Brazil
e-mail: hirdan@usp.br

I. L. de Araújo
e-mail: israel@senado.leg.br

E. G. Pereira et al. (eds.), *Sovereign Wealth Funds, Local Content Policies and CSR*, CSR, Sustainability, Ethics & Governance,
https://doi.org/10.1007/978-3-030-56092-8_17

period (Torres et al. 2013). The extant literature conveys different causes for the resource curse, ranging from structural factors such as a lack of productivity growth in the extractives sector, governance issues such as weak property rights in resource-rich economies, or else macroeconomic factors such as resource price volatility that impact fiscal stability (Kakanov et al. 2018).

In order to avoid the resource curse, many resource-rich nations, especially in oil, started to implement local content policies as a way of promoting national industrialisation (technological transfers) for sustainable economic growth, to maintain sectorial employment rates and qualified workers in the country, as well as transferring economic demand through supply chain as a spillover effect of natural resource activity (Malcolm 2013; Costa and Santos 2013; Lucas et al. 2015; Araujo et al. 2018a, b; Costa 2018).

In general, mineral and oil sectors involve capital-intensive activities (Costa and Santos 2013), which require governments to adopt policies to increase national participation beyond the "government take" (royalties and taxes) (Pascal 2013; Araujo et al. 2018a, b). Hufbauer et al. (2013) provide examples and suggest it is possible for governments to provide not only training for the workforce within the natural resources sector, but also build infrastructure unrelated to the main activity (hospitals, schools and roads).

Brazil historically is a natural resource-rich country but with resource-poor characteristics. After the 1950s, the country experienced a huge growth on mining sector (particularly in iron and manganese) and began its oil exploration programme of onshore sedimentary basins. However, since 1930, the national economy has developed in strategic sectors and remained as a world-class agricultural producer (Furtado 2007; Bresser-Pereira 1985; Araujo et al. 2018a, b). In other words, despite the growth of its natural resources sector, the country maintained other industrial and agricultural sectors as a way to counterbalance its mining and oil growth. This resulted in a economy being less exposed to the resource curse compared with other developing countries with a high dependency on commodities exportation.

During this period, local content was not an explicit local content policy. Nonetheless, a large number of private natural resources companies were created downstream supply chains including petrochemical, fertilisers, petroleum refining and the respective sale to the domestic market and exportation (Araujo et al. 2018a; b). More recently, since 1995, the government has promoted structural reforms that included extractives industry and changed patterns of national development policy made through local content requirements.

In respect to this, our goal in this chapter is to answer the following questions:

- What are the main obligations related to local content in Brazil within the extractives resources industry? Are there any peculiarities?
- What are the main challenges that the extractives resources sector faces with regards to local content policies and local uptake?
- What are the best and worst experiences with local content requirements in the extractives resources sector in Brazil?

- As an introduction, we may think that the mechanisms of local content undertaken by the Brazilian government sought the development of the national economy; however, they do not resemble those made by other countries highly dependent on extractives. Thus, this chapter sets out to explore these issues.

2 Energy Law and Local Content Policy

Heffron and Talus (2016a: 189) point out that "energy law and policy plays a vital role in the energy sector in the twenty-first century. It aims to ensure that societies meet their energy targets whether that is about the provision of increased energy security and/or economic benefits, and/or environmental goals." In addition, we highlight the need to reflect on distributive justice, an important issue for the energy sector, in order to assure sustainable development within resource-rich nations (Heffron and Talus 2016b; Heffron 2018; Heffron and Mccauley 2017). This chapter explores the extractives industry, mainly the new energy law field, as well as its set of guiding principles (Heffron et al. 2018) presented in Table 1 and includes an overview of local content policy in Brazil.

According to Heffron and Talus (2016a: 191) "energy law concerns the management of energy resources. This is a simple definition, and disguises that it is arguably one of the more complex areas of law. It demands that a scholar in the area engage with other disciplines to some degree, such as politics, economics, geography, environmental sciences and engineering." The pros and cons of energy law can be used as argument by decision-makers regarding development strategies and for local content policies (Krugman et al. 2015). The most important is the infant industry case, which advocates that new industries are not able to be competitive at the same level as mature industries. Looking for solutions, countries create trade barriers through a myriad of instruments well known in the academic literature, such as import quotas and subsidies. The main point about infant industries is the protection they need to improve income and growth given the initial profits are low.

Table 1 Principles of Energy Law

Principles of energy law
1. The principle of natural resource sovereignty
2. The principle of access to modern energy services
3. The principle of energy justice
4. The principle of prudent, rational and sustainable use of natural resources
5. The principle of the protection of environment, human health & combatting climate change
6. Energy security and reliability principle
7. Principle of resilience

Source Heffron et al. (2018)

Energy policy, national development strategy and local content policies commonly appear intertwined in discussions concerning the role of the State in certain sectors of the economy, especially in the extractives sector. Energy policy is aimed at the rational use of available energy resources, in order to guarantee supplies nationally, attract investments (national or international), increase the competitiveness of the domestic industry against foreign counterparts and promote free competition (Brazil 1997). National development, in a broad sense, includes actions, guidelines and policies aimed at building the necessary infrastructure for private enterprise to generate employment, income and taxes, while observing the dictates of social justice (Malard 2006). It also seeks to increase the industrial and technological capacity, directly or indirectly, of companies under national jurisdiction.

Obligations of local content can be understood as a mechanism for national development as well as a way to assure distributive justice. In Brazil, this is referred to as local content clause; it is the obligation of hiring a share of national goods, employment, as well as services in the exploration, development and production processes of the extractives sector. If companies do not comply with this clause, they incur government-imposed fines. The Brazilian oil and gas sector has used the local content clause since the first concession under the Energy Policy, but the turning point was 2003, when detailed clauses were imposed by the State in new concessions. The clause is lengthy in order to provide comprehensive details to allow the contractor to understand each parameter before the bidding round and to be aware of the penalty in the case of non-compliance.

3 Local Content Obligations Related to the Extractives Resources Sector in Brazil

Local content requirements (LCR) necessitate that in the manufacture of a specific product, a certain amount of the inputs, parts or raw materials used are produced within the country. Examples are the use of local labour, the development of managerial and operational skills, the acquisition of goods and services provided by companies established in the host country, the obligation to form compulsory partnerships with local enterprises and the increase of local technological capacity (IPIECA 2011; Franklin 2013). This requirement may be a prerequisite for the marketing of the respective goods or services, or a condition for producers' access to some advantage or benefit, whether of a fiscal, credit or other nature. Sometimes, the minimal local content is quantified as a percentage of the product or service's value, sometimes by reference to physical units. As a rule, taxes are not considered as local content due to the variations in time and space, which can result in distortions.

Among the various forms used to benefit those who observe the local content rules, we can mention

(i) Granting of tax incentives: legal rules that exempt certain sectors from the payment of taxes or reduce their tax burden, depending on compliance with local content requirements established by law or contract. A recent example is the Inovar-Auto programme, which reduced the amount to be paid as an industrialised product tax (IPI) for cars whose production had a minimum nationalisation rate. These rules had to be adapted to conform to trade agreements of which Brazil is a signatory, such as Mercosur, but they generated international protests for violating World Trade Organisation (WTO) agreements.

(ii) Allocation of financing lines with local content clauses: the Brazilian Bank for Economic and Social Development (BNDES), which since its inception has been earmarking funds that, although without due legislative discussion, contain implicit subsidies. On the other hand, the beneficiary companies generally have to comply with local content obligatory clauses.

(iii) Preferences in concessions or government purchases: one of the Executive Branch's actions is the determination of local content obligation in costing and investment expenses under oil contracts, such as concession and production sharing agreements, or the determination to contract with a national company overpaid to the detriment of an international competitor.

An alternative to stimulate production of certain goods or services within the country is the creation of State-Owned Enterprises (SOE). For sectors in which direct intervention is desired, companies whose shareholding control is held by the State were created. Consequently, some of these companies gain relevant positions in certain sectors, sometimes reaching the position of monopolies. This mechanism can be used when the private sector, at an initial moment or due to market failure, has no interest in the development of the activity. In addition, they can act in the transfer of technologies of interest for the development of the national industry, contracting national goods and supply chain's companies, which, indirectly, is a way of stimulating local content.

The Brazilian extractive industry has used SOEs since the industrial period, over the last century. PETROBRAS has been the major SOE on exploration and production of oil and natural gas since its foundation in 1953 and the company compel downstream supply chain's to develop technology and to improve local content. In contrast, Vale do Rio Doce Company (VALE), an SOE in the mining sector and a key stakeholder in regional development and building the infrastructure needed for the iron ore sector, such as the mine-railroad-port complex, does not have a local content policy.

Among such instruments, three are worth highlighting for the purposes of this chapter: subsidised financing lines with clauses of local content, the use of SOE to direct demand to domestic companies and the determination of local content obligation through concession contracts. The main financial agent has been the Brazilian Investment Bank (BNDES), through a specific budget directed to capital costs (machines and equipment) called FINAME. BNDES was responsible for providing credit to the extractives industry from the second half of the last century until financial scarcity reached Brazil during the 1980s. Second, SOEs may, by virtue of their relevant market power, direct contractors to execute national contracting policies rather than cheaper imported ones. If the resale price of the product does not change, the higher cost ends up being transferred to the shareholder, who transfers it to the taxpayer. If the price of the product is increased due to the higher costs resulting from LCR, consumers will be penalised. Finally, some contractual obligations related to public service concessions impose high fines for non-compliance with national content clauses or LCR. This is the case of the Brazilian oil sector.

Industrialised countries have adopted some type of policy in order to allow the new segment of industry to grow. Auty (1994, 2011) highlights nascent industrial protection policies implemented not only in Brazil, but also in Mexico, China, India, Taiwan and South Korea that can be linked to huge industrial facilities in their territories. It is possible to list some tools available to benefit those who meet the local content requirements. There are states that have been using, for example, tax incentives mechanism as a way of reducing the necessary investment costs for the development of a particular economic sector (Mullins 2010). In Brazil, Law nº 13,586, of 28 December 2017, introduced a huge fiscal incentive applied just to the oil sector in order to reduce this sector's costs.

The establishment of LCR is a form of protectionism. It is one of the instruments that, along with import and export taxes (also called tariffs), quotas, voluntary export restrictions (voluntary export restraint) and sanitary barriers represent a means of trying to protect the national economy from external competition, aiming to encourage local production, internal tax generation and jobs. This does not necessarily mean that the imposition of any of these instruments represents an effective protection of the economy as this will depend on the interaction between the measures taken. In addition to verifying whether, in practice, the adoption of some specific protectionist measure has the power to produce the initially intended goals, it would be necessary to evaluate whether these goals, generally sectoral, bring advantages for the economy as a whole. In other words, it would be necessary to weigh if the adopted measures actually protect and, second, whether protection confers some advantage to the national economy.

4 Challenges Faced by the Extractives Sector Regarding Local Content Practices and Local Participation

The adoption of protectionist measures is usually discouraged in economic literature based on arguments such as the distortions it causes in international trade and in the international division of production, which results in companies in disagreement with the principle of comparative advantage, since unilateral protectionist measures tend to stimulate retaliation, which in the aggregate does not create more jobs but only an inefficient redistribution of employment across the planet. Protectionism, when motivated by macroeconomic reasons, is classified as a "neighbour-impoverishment policy" or beggar-thy-neighbour policy (Varsano et al. 2015), and it is reasonable to suppose that this neighbour will react to the protectionist initiative. For these reasons, protectionist measures have also been discouraged within the framework of the World Trade Organisation (WTO) and by regional integration initiatives.

However, according to the new theories of international trade developed around the 1980s and the Dutch disease phenomenon, the use of protectionist measures is advisable in some specific and often temporary situations:

(i) Learning curve: The first reason to encourage the protection of a given sector is the possibility of creating development conditions through "learning". Because certain comparative advantages are innate and others are acquired, a productive sector with a high average production cost may, under certain circumstances, benefit from an average cost decline over time, provided that the companies are given the time needed to develop, absorb or adapt production techniques and technologies. In other words, provided they have the opportunity to learn how to produce more cheaply, which is why it is necessary that these companies receive some protection, without which they would not resist external competition. However, this protection, known as the infant industry argument, is necessarily temporary and must be extinguished as soon as its effects are felt or as soon as its unfeasibility has been diagnosed.

(ii) Economies of scale: The second reason that would support the protection of a sector is the possibility of a fall in production costs over time due to simple production increases. It matters little if this increase in production occurs within each company (the so-called internal economies) or by the expansion of the number of companies (through the external economies); the protection applied to industry should be temporary.

(iii) Abundance of natural resources: The third reason to justify the use of some form of protectionism is the episodes of sudden discovery of natural resources or of sudden appreciation of already known reserves that gave rise to Dutch disease. The most well-known cases are those involving the discovery of hydrocarbon deposits. When the volume of deposits or the rise in price induces a massive inflow of foreign currency, either through export revenue or capital attraction, there is a tendency towards exchange appreciation and therefore the loss of competitiveness of goods subject to international competition, especially those of industrial nature. In these cases, the literature supports the adoption of trade defence measures, not necessarily for a limited time.

Dutch disease is often associated with institutional deterioration and education quality loss. The second effect would be job loss vacancies to most qualified professions, due to the expansion of extractives activity. The first would be the reflection of the expansion of activities linked to the mere dispute over the surplus value created by natural wealth (rent-seeking behaviour). There is no consensus among researchers if in fact the Brazilian economy suffers from Dutch disease (Bresser-Pereira 2017, 2013; Strack and Azevedo 2012). It is remarkable that the use of protectionist instruments also creates opportunities for the development of "business counters" and the growth of corrupt practices, as a way of influencing political decisions in favour of specific interest groups.

Any protectionist solution creates a tension between sectoral development and national development. Protection implies the sacrifice of cheaper and more efficient sources of production, which represents a cost to the country. This cost can be temporary or permanent, depending on the duration of the measures. Moreover, protectionism has different effects along the production chain. If the protection covers the final stage of the chain, the trend will be the growth of this final step, with or without the induction of the previous stages (since the raw materials, parts and components can be imported). If however an initial or intermediate stage of the production chain is protected, the advantage it will receive will have a higher production cost in subsequent stages, including the final. Protection of the initial and intermediate steps tends to deprotect the following steps.

If the final product is not under international competition, the result of protecting the national element will be the increasing on merchandise prices, thus, transferring production costs to consumers. However, if the industry is subject to competition from abroad, either because the import of the final product competes in the domestic market, or because at least part of the production is exported, the increase of the national product as a result of the protection probably cannot be passed on to the market, and the product will lose competitiveness.

This happens regardless of the protection instrument used. In the specific case of LCR, the increase in cost is a function of the difference between the prices of imported goods and the prices of goods produced in the country and the fraction of the total value that should be reserved for local production. This tension between each of the sectors along the production chain of a product subject to LCR, which means the protection granted to one phase implies a greater lack of protection of the following phases, is the reason why it could be used as an indicator of effective protection. Granting protection to certain stages of production may result in the deprotection of other steps. Therefore, it is important to assess the final effect of protection measures along the production chain.

It is possible that LCRs in the Brazilian oil and gas industry have created an effective deprotection situation, as oil is a commodity traded in the international market, and Brazilian production does not have the power to affect international prices. The fall of the oil price in the international market, insofar as it reduced the margin of the extractives industry and thus the space for absorbing higher costs, may have exposed the difficulties created by the local content policy. In that case, it would be advisable to withdraw these requirements as a way to encourage or boost investments in the oil and gas sector.

There are reasons to believe that assessments favouring local content creation do not apply to Brazil. First, because currently the Brazilian economy is not suffering from Dutch disease and it does not seem likely that this problem will arise in the foreseeable future. The oil and gas sector is not relevant in terms of balance of payments to affect other sectors. The loss of industry space, for example, is more easily explained by the expanding effects of the Chinese economy, a phenomenon that has affected the industrial sector worldwide. Second, it is necessary to check carefully whether the application of protection strategies to an infant industry would be justified. It does not appear that LCR protection responds to studies that identify potential gains with local capacity development or economies of scale.

Before concluding this section, we will deal briefly with two questions. The first concerns the establishment of local content rules in which the percentage of local content can be satisfied by acquisitions and expenditures with any good or service (LC rule). When the rule is not individualised, that is, if it does not refer to minimum percentages in specific sectors, the objective of protecting the domestic industry, even if it has no prospects of productivity gains, is apparent. It would make more sense that a policy aimed at harnessing these potential gains would be based on the individualised indication of the most promising sectors. As it is modelled, the current policy seems to be an attempt to prevent Dutch disease.

The second is the dilemma between eliminating the previous LC rules (i.e. contracts already signed) or just from now on (i.e. only contracts to be signed). This is a real dilemma, because eliminating or smoothing LCR retrospectively involves rewriting bidding rules, which had specific winners and losers that by the revised rules could have been victorious at the time; and because eliminating the requirements just for future contracts places the various companies under different competitive conditions. It must be recognised that these LCRs were explicitly accepted by the companies who voluntarily undertook to comply with them, obtaining on the other hand, a reduction of the amount to be paid as a signature bonus.

5 Lessons Learned from Local Content Requirements in the Extractives Sector in Brazil

Local content has been the object of public policy since industrialisation processes over the last century. Both the mining and oil sectors were the centre of national development, despite only the latter showing prominent LCR. Mining directive was linked to rational use of limited natural resources in order to provide foreign exchange reserves and budgetary resources to the country. On the other hand, local content policy, even in the mining SOEs, was not among the priorities of the mining sector.

Regarding the oil sector, since the establishment of PETROBRAS as monopoly, LCRs have been the object of concern for decision-makers. As a leader, before the break of the monopoly of oil and gas exploration and production granted to PETROBRAS, industry development was stimulated by financing lines offered for State-Owned Banks through tax incentives. This was to substitute similar goods produced in Brazil and by the use of the state-owned company's purchasing power to direct the sector's demand to companies under national capital stock control. Both strategies were widely used by other countries during their industrial maturation periods. The results, however, were not favourable to all those who did, as was the case in Brazil.

Amendment 9 to the Brazilian Federal Constitution allowed private companies to conduct research and development of hydrocarbons. Law nº 9,478 of 6 August

1997 regulated the new constitutional provision. Companies with legal, technical and economic qualification became able to carry out the activities, under the trusteeship of a concession agreement preceded by a specific bidding for this purpose. However, the Act does not regulate local content obligations. The closest of this would be, at the time, the objectives of the National Energy Policy contained in Article 1, such as promoting development and increasing the country's competitiveness in the international market (items I and XI). However, both objectives may conflict with that of promoting free competition (item IX).

LCR appeared in the bidding documents and in the concession contract for successive rounds of exploratory block bidding. Local content criteria were used since the first rounds as an object of classification of bidders for a particular block, despite just a small percentage of the note corresponded (15%, with 3% referring to exploration phase and 12% to the development of production, in case of discovery of a new oil field), the signing bonus—value offered and paid at the time of signing the concession agreement, with a greater weight.

Prior to the year 2000, there was no need to prove compliance with local content clauses, since the reporting was purely declaratory. This failure is due to the adaptation period of the Petroleum, Natural Gas and Biofuels National Agency (ANP) to its legal attributions. Until 2002, the concessionaires had to classify the expenses by origin (national or foreign) in quarterly reports. The improvement was progressive. In 2003, it was necessary to require a minimum percentage of local content, which varied depending on the location of the block (land, shallow water and deep water) and gave greater weight to this item for classification purposes in the bidding process. In that period, the grade was weighted according to three criteria: signature bonus (30%), minimum exploratory programme (30%) and local content (40%). As of 2005, the weight of the criteria was redefined, with local content having a weight of 20% to the final note of the bid.

This system, however, began to incorporate several sub-items that made up the global obligation of local content. Oil companies were operationally obliged to have specific sectors to monitor and perform activities related to the compliance with contractual clauses of local content: this is called transaction cost. For instance, oil companies had to demonstrate by certificates emitted by third parties whose reliance to inspect and certify local content adherence was conceded by ANP, according to published regulations on inspection and certification methods. Those independent entities verified whether enterprises accomplish the minimum of national expenditure as local content in each sub-item, item and global obligation, and the process was repeated in each concession contract.

In 2013, the rules aforementioned were simplified to make it easier to understand and apply after the oil industry complained about regulations which made it difficult to deliver assets within the required quality into the budget and when they needed to.

Although the requirements were minimal, companies began participating in block auctions by considering the availability of cash and the prospect of high local content compliance. As a result, successful bids were those with heavy contractual local content obligations that would only have significant impact years after the contract was signed at the production stage (the one with the largest financial outlay). This behaviour demonstrates at least two different explanations. It may be the effect of calculating the present value of the additional costs that LCR would imply, from a high discount rate, due to the risk involved, which would make immediate expenditure more relevant in financial terms. Or it may be the effect of a perceived temporal inconsistency of government, which as the likely bidders believed at some future time would give up LCR. This expectation would justify the acceptance of high local content rates, which would be eliminated in the future as a means of lowering immediate financial expenses. This factor can be observed today. The concessionaires are probably not willing to comply with the local content commitment offered but would show willingness to comply with the minimum bidding requirements.

At the same time that concession contracts started to be changed, in terms of LCR, the State has established a specific institution for better implementation of public policy. The National Program for Oil and Natural Gas Industry Mobilisation (PROMINP) was coordinated by the Minister of Mines and Energy and promotes Brazilian industry of goods and services, in competitive and sustainable terms, not only in Brazil, but in the global oil sector. Through an Executive Order, also called Presidential decree no. 4.925 (Brazil 2003), it was proposed an institutional framework to coordinate and promote projects and actions to maximise the spillover effect in the national economy and was attended by public sector representatives, oil industry associations and mainly PETROBRAS.

PROMINP was the focal point and the main arena for discussion and bargain between the State, PETROBRAS, suppliers and others associations involved, and it was focused on diagnosing and proposing solutions to prevent a lack of economic resources for growth of the national oil sector. Into this programme, instruments were developed to offer low interest funds (more competitive than those offered to the general Brazilian economy) for the supply chain of the oil industry, and mechanisms were established with major banks to allow using Goods and Services Purchase Agreements (GSPA) between PETROBRAS and suppliers as guarantee of payment (BNDES 2009).

Labour training was one of the most important components of PROMINP with investment in R&D resources (an obligation linked to concession contracts) paying for technical education scholarships to provide qualified human resources for the whole oil sector, from exploration to downstream. Over ten years, the programme has trained approximately 100,000 people. Despite this, less than 18% were employed in the oil industry (BNDES 2009; FIRJAN 2015).

This experience brings important lessons to bear on LCR. The main observation was how capable agents were to organise themselves in order to attain and improve the common good from the extractives sector. In general, the result was positive; however, some odds can be pointed out. First, PROMINP, despite a ministerial coordination, PETROBRAS was responsible for operational coordination, and the enterprise placed its interest above those of Brazil when conflicts arose between them. Second, as an SOE and the most important company in Brazil, it used its influence to superimpose its interests over the other agents of the oil industry. Finally, industrial policy was commonly divergent to energy policy, and the decision-making body was under the aegis of the Ministry of Mines and Energy.

It should be remembered, however, that the whole rule was established on an under-registered basis and under concession contracts with stability clauses in the Federal Constitution of 1988 (article 5, item XXXVI). Discussions about possible changes to the local content clause have started. In terms of private sector participation in the oil and natural gas industry, private investors typically see the local content clause as a barrier because of its difficult fulfilment. Even PETROBRAS faces difficulties in this issue: the company was fined more than R$200 million in 2015. After that, policy makers started to discuss mechanisms to solve energy-industrial policy conflicts caused by local content issues and created another governmental framework. Presidential decree no. 8.637 instituted a programme to stimulate competitiveness of the productive chain and development and improvement of oil industry suppliers, called PEDEFOR (Brazil 2016). The Executive Order changed part of the local content rules and added investments in research and development and innovation to meet the demands of concession contracts and production sharing agreements. The new Directive Committee was responsible to propose bonus for local content interchangeable between concessions and to offset strategic technological gains for the oil industry, but of small value compared to expenses related to the whole concession contract. Moreover, PEDEFOR determined that the main coordination had to be transferred to the minister responsible for the industrial sector.

There are even directions for the application of research resources, but without interfering in the regulation promoted by governmental regulatory office. This decree turned fine into investments, which is an improvement of the policy considering, the purpose of local content other investment results, as the export of Brazilian items and technology development, for example.[1]

According to the decree, companies can use a committee to consider local content crediting, named local content units (LCU). Those LCUs may be obtained when the company facilitates the installation of new suppliers in the country, promoting the expansion of production capacity and technological innovation process from suppliers, purchase goods and systems here to export pioneers and acquire lots of goods and systems developed in the country (OMS 2016).

In 2016, an ANP's Resolution focused on the inclusion of deduction of the amounts of national plots of items classified as materials, of the imported parcels values when these are incorporated into goods and systems of foreign origin manufactured in Brazil under the special customs regime for export and import of goods intended for research activities and exploitation of petroleum and natural gas (REPETRO). These items are included in ANP Directive 19 of 14 June 2013, which deals with local content certification.

In the short term, this easing of the requirement of local content clause brings economic benefits for oil companies because they will only buy a product or hire a service when they have better proposals. In addition, with those changes, the companies have greater contractual freedom, including reducing transaction costs with the

[1] According News from R7 (2016), authorities informed that the local content attendance was not be used as a metric in order to evaluate proposals in the coming bidding rounds for petroleum exploration areas.

possibility of incorporating major economic gains in their budget. For instance, new rules for local content may benefit the relationship between PETROBRAS and Sete Brasil, with a reduction in the loss of State investment in the company created to enable the construction of the pre-salt probes—especially in shipyards dedicated exclusively to Sete, as the Paraguaçu, Bahia, whose shareholders are Odebrecht, OAS, UTC and Kawasaki. If the oil operator facilitates the installation of new shipyards or encourages the export of components for Brazil, for example, it can earn credits and make a smaller acquisition of local content in other sectors. PETROBRAS would have at least part of the injury compensated because it would have to rent these probes at prices above the international market. Prices of probes belonging to Sete did not follow the decrease of general costs of the oil industry in the world, resulting from decrease in the price per barrel in recent months.

Foreign operators in Brazil may have advantages with this decree, as it aims to stimulate multinationals to bring suppliers to Brazil, which are located in other parts of the world and provide not only for operations in the country but also as a global platform of production.

The criteria were used in granting under concession only. For those areas in the pre-salt tendered under a production sharing modality, the auctions will count only one item to determine the winner: the surplus in oil for the Union. This modification is recent and delegates to the National Council of Energy Policy (CNPE) the assignment to determine the local content indexes for each round under the sharing mode. Although not ideal, it is more doable than the model previously applied.

There are favourable arguments to these obligations. At the time of the opening of the oil sector, it was expected that the international companies of the sector would contract goods and services in their host countries. Several countries have set up their own companies and, since their maturation in their territories, started to develop activities abroad, but directing the demand for goods and services of their contracts for the exploration and production of hydrocarbons to subsidiaries or affiliated companies. This is the case of Equinor, for instance, which started to contract mainly Norwegian suppliers in those activities where they had the capacity to export, regardless of availability on competitive bases in the host territory of the contract.

These content clauses aim to prevent such behaviour from companies. However, using such an instrument as a way of developing the industry has a significant deleterious factor: the clauses may fall on a link that does not have the expertise to perform the role of industry inductor. On one hand, the Public Power demands investments and attraction of technological poles. On the other hand, suppliers of goods and services want to obtain market reserve so that they can make gains beyond what would be economically achieved under free market logic. In the middle of both, the oil companies end up being burdened, having to play the role of industrial policy enforcer, when their locus of action is to search for hydrocarbons and to tillage them. The balanced solution for these clauses is one that would not allow foreign state-owned companies to direct their oil sector demands to their suppliers of the same flag without economic criteria advantageous to both sides, nor would affect them to the point of sterilising the activities of the petroleum sector with impacts on the Gross Domestic Product (GDP).

Looking to the positive lessons, it is possible to highlight the certification system by third party. Neither government nor oil companies could evaluate without trends how much local content was achieved in each agreement. In order to improve confidence in the institutional framework, specialised companies authorised by a regulatory agency started to perform as local content certifiers avoiding being captured by both parties involved.

Therefore, Brazil's best experience was building normative and institutional frameworks that allowed attempts to upgrade LCR in future bidding rounds in order to protect partially internal demand to local supply chains. On the opposite side, mistakes made during past decades are not easily repaired because of the inflexibility of LCR regulation and rules under concession contracts, which could sterilise the whole oil industry.

6 Conclusion and Policy Implications

In Brazil, the extractives sector along with agriculture and livestock is an important component of the trade balance and regional development linked to local content. The mining sector has not employed local content policy so far, although it may in the long term. Thus, the Brazilian experience we report here is with the oil sector.

The question remains, is local content policy an effective protection measure or market reservation? It is possible to affirm that LCR are a form of protection of the local industry and, as such, is a practice generally condemned in the economic literature. However, there are three situations in which protectionism is defensible: the learning curve, the economies of scale (internal to the company or external to it) and to fight against Dutch disease. Of these three, learning curve and economy of scale are justified only on a temporary basis. Their implementation should be based on rigorous diagnoses as to the possibilities of productivity gains and give space to negotiations whose results are not always aligned with the collective or national interest. There are no studies that point out the technical and economic viability of the implantation of supplier industries for the oil exploration sector in the country in similar conditions to the external ones, which suggests that this is not a classic case of protection to infant industries.

The third case, in turn, may last as long as the effects of the extractives industry are felt on the trade balance, the exchange rate and other sectors of the economy. It does not seem to be the case, least in the current situation, that the Brazilian economy might suffer from Dutch disease, either because the Brazilian economy is large and diversified, or because the oil markets and their substitutes do not point to a situation where Brazilian exports will have such a profound influence on the exchange rate and the economy.

It seems that the lack of protection created by LCRs, coupled with falling international oil prices, appears to have contributed to reduced investment in the sector, one of the reasons contributing to the current recession and, indirectly, to the fiscal crisis.

Moreover, the main instrument applied in the Brazilian oil sector was oil concession contracts and production sharing agreements to impose obligations linked to LCRs, but without State coordination capable to deal with the complexity of interests related to oil companies, industries, suppliers for whole oil supply chain, and also the State. Therefore, there were many instruments of rent seeking into local content policy. Given that the main protagonist was an SOE, PETROBRAS, outcomes became dependent on its performance, and board decisions influenced significantly how the national sector has behaved.

Finally, looking to the future, with the increasing number of major operators beyond PETROBRAS doing exploration in the pre-salt oil province, policymakers will be less dependent on national oil company's decisions, and they will have more freedom to implement realistic local content policies.

Acknowledgements We gratefully acknowledge support of the RCGI—Research Centre for Gas Innovation, hosted by the University of São Paulo (USP) and sponsored by FAPESP—São Paulo Research Foundation (2014/50279-4) and Shell Brasil, and the strategic importance of the support given by ANP (Brazil's National Oil, Natural Gas and Biofuels Agency) through the R&D levy regulation.

References

Araujo, I. L., de Medeiros, H. K. C., & Musarra, R. M. L. M. (2018a). As cláusulas de conteúdo local e seus efeitos em contratos na área do pré-sal: Precisa-se de um Novo Marco Legal? In *Proceedings of Rio Oil & Gas 2018, 2018, Rio de Janeiro. Rio Oil & Gas 2018* (pp. 1–10). Rio de Janeiro: Instituto Brasileiro de Petróleo—IBP.

Araujo, I. L., Costa, H. K. M, & Musarra, R. M. L. M. (2018b). Breve análise do marco regulatório das Cláusulas de Conteúdo Local. Rev do Direito da Energia.

Auty, R. M. (1994). Industrial Policy reform in six large newly industrializing countries: The resource curse thesis. *World Develepment, 22*, 11–26.

Auty, R. M. (2001). *Introduction in resource abundance and economic growth*. Oxford: Oxford University Press.

BNDES—Banco Nacional de Desenvolvimento Econômico e Social. (2009). *Desenvolvimento da cadeia produtiva de petróleo e gás e investimentos em E&P*. São Paulo: Bain & Company/Tozzini Freire Advogados.

Brazil, (1997). Lei nº 9.478, de 6 de agosto de 1997, e alterações. Publicada no Diário Oficial da União de 7.7.1997. Available at www.planalto.gov.br/ccivil_03/leis/L9478.htm. Accessed June 22, 2019.

Brazil, (2003). Decreto nº 4.925, de 19 de dezembro de 2003. Publicada no Diário Oficial da União de 22.12.2003. Available at https://www.planalto.gov.br/ccivil_03/decreto/2003/D4925.htm. Accessed June 22, 2019.

Brazil. (2016). Decreto nº 8.637, de 15 de janeiro de 2016. Publicada no Diário Oficial da União de 18.1.2016. Available at https://www.planalto.gov.br/ccivil_03/_Ato2015-2018/2016/Decreto/D8637.htm. Accessed June 22, 2019.

Bresser-Pereira, L. C. (1985). *Desenvolvimento e crise no Brasil 1930–1983*. São Paulo: Brasiliense S.A.

Bresser-Pereira, L. C. (2013, July–September). The value of the exchange rate and the Dutch disease. *Brazilian Journal of Political Economy, 33*(132), 371–387.

Bresser-Pereira, L. C. (2017). *How to neutralize the Dutch disease notwithstanding the natural resources curse. Texto para Discussão nº 452. Escola de Economia de São Paulo da Fundação Getulio Vargas Brasília*. São Paulo.

Costa, H. K. M. (2018). *Royalties de Petróleo, Justiça e Sustentabili-dade* (1 ed.). Synergia, Rio de Janeiro.

Costa, H. K. M., & Moutinho, S. E. (2013). Institutional analysis and the-resource curse-in developing countries. *Energy Policy, 63*, 788–795.

FIRJAN. (2015). *A Política Brasileira de Conteúdo Local para o Setor Petróleo e Gás Análise e Sugestões de Aperfeiçoamento*. Rio de Janeiro: Federação das Indústrias do Estado do Rio De Janeiro.

Franklin. B. (2013). Types of local content policies. In *Local content policies in the oil and gas sector (Chap. 3)*. Washington D.C.: International Bank for Reconstruction and Development (The World Bank).

Furtado, C. (2007). *Formação econômica do Brasil* (34th edn). São Paulo: Companhia das Letras. (1st ed. 1959).

Heffron, R. J. (2018). The application of distributive justice to energy taxation utilising sovereign wealth funds. *Energy Policy, 122*, 649–654.

Heffron, R. J., & Mccauley, D. (2017). The concept of energy justice across the disciplines. *Energy Policy, 105*, 658–667.

Heffron, R. J., & Talus, K. (2016a). The evolution of energy law and energy jurisprudence: Insights for energy analysts and researchers. *Energy Research and Social Science, 19*, 1–10.

Heffron, R. J., & Talus, K. (2016b). The development of energy law in the 21st century: A paradigm shift? *Journal of World Energy Law and Business 9*(3),189–202.

Heffron, R. J., Ronne, A., Bradbrook, A., Tomain, J. P., & Talus, K. (2018). A Treatise for Energy Law. *Journal World Energy Law Business, 11*(1), 34–48.

Hufbauer, G. C., Schott, J. J., Cimino, C., Vieiro, M., & Wada, E. (2013). Local content requirements: A global problem. Peterson Institute for International Economics.

IPIECA. (2011). Local content strategy: A guidance document for the oil and gas industry. London.

Kakanov, E., Blöchliger, H., & Demmou. L. (2018). Resource curse in oil exporting countries: Economics department working papers no. 1511. OECD Working Paper. Available at https://www.oecd.org/officialdocuments/publicdisplaydocumentpdf/?cote=ECO/WKP(2018)59&docLanguage=En. Accessed February 2020.

Krugman, P., Obstfeld, M., & Melitz, M. (2015). *Economia Internacional: Teoria e Política – 10ª edição* (pp, 198–221). São Paulo: Pearson.

Lucas, L. J. R., Costa, H. K. M., & Santos, E. M. (2015). Engenharia de Projetos e aumento da participação do Conteúdo Local. *Revista Petroquímica, Petróleo, Gás & Química, 1*, 52–54.

Malard, N. D. (2006). *O desenvolvimento nacional: objetivo do Estado nacional* (Vol. 3, pp 255–282). Direito, Políticas Públicas e Mundialização, Prismas.

Malcolm, F. (2013). The case for (and against) local content policies in the petroleum sector. In *Local content policies in the oil and gas sector (Chap. 2)*. Washington D.C.: International Bank for Reconstruction and Development (The World Bank).

Mullins, P. (2010). International tax issues for the resources sector. In *The taxation of petroleum and minerals: Principles, problems and practice (Chap. 13)*. New York, NY: International Monetary Fund (IMF).

Oms, C. (2016). Valor Econômico. Decreto do governo facilita regra de conteúdo local de petróleo. 18/01/2016. Available at https://www.valor.com.br/empresas/4396662/decreto-do-governo-facilita-regra-de-conteudo-local-de-petroleo. Accessed October 2016.

Pascal, B. (2013). Defining local content. In *Local content policies in the oil and gas sector (Chap. 1)*. Washington D.C.: International Bank for Reconstruction and Development (The World Bank).

R7. (2016). Notícias. Governo aliviará regra para conteúdo local antes do leilão de áreas de petróleo. Available at https://noticias.r7.com/brasil/governo-aliviara-regra-para-conteudo-local-antes-do-leilao-de-areas-de-petroleo-17102016. Acessed October 2016.

Strack, D., & Azevedo, A. F. Z. (2012). A doença holandesa no Brasil: Sintomas e efeitos. *Revista Economia E Desenvolvimento, 24*(2), 2012.

Torres, N., Afonso, O., & Soares, I. (2013). *A Survey of Literature on the Resource Curse: Critical analysis of the main explanations, empirical tests and resource proxies* (p. 2013). Faculdade de Economia do Porto: Universidade do Porto.

Varsano, R., Ferreira, S. G., & Afonso, J. R. (2015). Fiscal competition: A bird's eye view. Discussion Paper nº 114. Brasilia-DF.: Instituto de Pesquisa Econômica Aplicada (IPEA).

Local Content Requirements in Iran

Zoha Abdolalizadeh

1 Introduction

With the fourth largest reserves of oil and second largest reserves of gas in the world, Iran is one of the resource-rich Gulf countries in the Middle East. The oil and gas industries dominate and overshadow all its other resources. This chapter surveys and examines how the country is increasingly introducing local content requirements (LCRs) into its legal system. This is being done through various means, such as legislation, regulations, policies, guidelines and industry contracts.

The study of legal, economic, political and social aspects of Iran's extractive resources requires the examination of a complicated system, which contains different beneficiaries, a huge number of laws and regulations, parallel policies and contradictory practices. Socio-political issues as well as international relations need to be considered, and to get a clear picture, an explanation of Iran's extractive resources is also necessary. The country's dependency on the oil industry, as well as its international geopolitical situation, makes oil its most important asset. Increasing its capacity to produce crude oil, as well as natural gas, directly affects Iran's role in international relations and consequently, its foreign exchange earnings.

Part One discusses the most important organization in Iran's oil industry, the National Iranian Oil Company. This study will be followed by scrutinizing its financial structure. The second part of Chapter Two looks into the structural framework of oil contracts. Until the government nationalized the oil industry, there had been several mandatory contractual frameworks. This section explores the most recent and common agreements: Buy-Back contracts and Iran Petroleum contracts (IPCs). This

Z. Abdolalizadeh (✉)
Senior Legal Specialist, Dana Energy Company, Tehran, Iran
e-mail: zoha.abd@gmail.com

E. G. Pereira et al. (eds.), *Sovereign Wealth Funds, Local Content Policies and CSR*, CSR, Sustainability, Ethics & Governance,
https://doi.org/10.1007/978-3-030-56092-8_18

analysis shows there are various limitations in providing a precise understanding of the country's key resource industries.

Part Three then reviews the Sixth Development Plan (2016–2021) pertinent to the country's oil industry. Iran's legal system allows different authorities to pass legislation and regulations which brings the power-based hierarchical arrangements and assigned duties lead to a mélange of precedents, and various interpretations of the regulations. In addition, some of the main challenges—such as the incomplete process of privatization and underdeveloped management in extractive resources—will be investigated.

In conclusion, it will be clear that Iran's integration of local content requirements into legal practices around extractive resources has great potential; however, regressive policies, paradoxical laws and a lack of rule of law, together with mismanagement, have moved the country away from effectively valued LCR.

2 Part One: Main Obligations Related to LCR in Iran Within the Extractive Resource Industry: The National Iranian Oil Company (NIOC)

The control of domestic oil has been a significant feature in Iran's national, economic and political independence.[1] In April 1951, the newly elected Prime Minister, Mohammad Mosaddegh, led a movement in the country's parliament that resulted in the nationalization of oil. Until then, the industry had been run by private companies, mostly controlled by foreign interests. The Anglo-Iranian Oil Company (AIOC) was founded in 1908 and mainly extracted petroleum from the south of the country. With nationalization, it was displaced and renamed the National Iranian Oil Company (NIOC). It became the sole governmental national oil and gas producer, distributed under the supervision of the Ministry of Petroleum. Since its very beginning, the NIOC has been in the spotlight as the sole organization in charge of the development of Iran's two key industries. It manages all activities regarding exploration and production of crude oil, natural gas and condensate, as well as the drilling and development of hydrocarbon reservoirs.[2]

Revenues from oil exploration first entered Iran's economy in 1908. From then till now, the importance of these revenues has grown rapidly, with the country's economy becoming increasingly dependent on these earnings. Almost 70% of Iran's budget in 2006 came from the proceeds of oil, and all Development Plans have been passed and executed on the basis of these revenues. Iran's dependency on the oil industry instituted an economic structure with specific limitations that impacted socio-political decision-making and policy implementation, as well as creating an organizational framework for the private sector. As "oil export earnings are high,

[1]Noreng (1994).

[2]For more information, see the document "Nationalization of the Oil Industry of Iran." 1951. *Middle East Journal* 5(3): 353–354.

totaling roughly $60 billion in 2007 and constituting 35% of GDP, it's unavoidable that many aspects of the relationship between oil and Iran's macro economy are classic examples of the symptoms displayed by the so-called resource curse or Dutch disease."[3]

Iran is among 15 countries with the world's most resource reserves and is estimated to have 7% of the planet's extractive resources. Aside from the main energy sources, oil and gas, the country holds 70 types of minerals including copper, coal, aluminum, alumina, iron ore, zinc and crude steel. The country's citizens make up almost 1.08% of the global population.

NIOC's production totals 4% of the world's total crude oil production, and 5% of total gas production.[4] According to Petroleum Intelligence, the company is the second largest globally[5] after Saudi Arabia's state-owned ARAMCO, but there is a significant gap. Although the NIOC holds almost 9.3% of the world's liquid hydrocarbon reserves and 18% percent of gas reserves, there are a number of reasons to explain the big difference, besides the size of its reservoirs:

(a) Laws and regulations related to the NIOC have not changed significantly since its establishment. There is a dire need to amend and evolve regulations and practices affecting the performance of the company.
(b) The Islamic Republic of Iran puts an undue burden on the symbolic sovereignty of the NIOC by giving it sovereign immunity. This gives the company considerable privileges, special facilities and exemptions by artificially reducing the risk exposure. Granting such exemptions results in a lack of transparency in actual performance, and an increase in its monopoly power, thereby reducing the accuracy of assessing its profitability and performance. It is therefore necessary that the government seriously revises and implements limits on these privileges, such as tax and customs exemptions.
(c) Financing is another factor that explains the huge gap between ARAMCO and the NIOC. In recent years, discussions around funding NIOC's projects have gained attention. It is obvious that any delay or shortfall in the required investments primarily leads to the reduction of production capacity of oil, natural gas and other derivatives. Sanctions by the United States play a huge role in discouraging and limiting foreign investments in Iran, as well as impeding the country's ability to sell oil. At the same time, sanctions caused Iran to stay out of international competition with Saudi's Aramco and the Russian state-run company, Gazprom.

Laws and regulations in Iran on extractive resources are one of the most perplexing and troublesome areas. First of all, it should be noted that legal points of extractive resources are reflected in different kinds of laws, which include the Constitution, the civil code and specific laws. Here, the discussion will begin with contradictory articles in the Constitution and the civil code related to the matter of local content.

[3] Green and Wehrey (2008).

[4] Ebrahim-Fathabadi (2014).

[5] https://www.energyintel.com/pages/top100-2018.aspx.

The issue of local content and making use of existing domestic capacities have been always concerned Iranian officials. Because, after a hundred years of commercial exploration of Iran's oilfields, the most part of a project involving oil technology and equipment still relies on IOCs. And the improvement of domestic companies in the field is negligible, and in most cases, operations are costly and take more time. In addition, following the revolution and the withdrawal of foreign specialists and experts, and subsequently the imminent war and serious damage to oil and gas equipment, efforts were made to improve local content. But for reasons such as the priority of oil production and the exploitation of wells, maintaining the level of production, the lack of technical and technological know-how and the lack of sufficient capital to finance, these efforts have not been successful. This trend continued until previous sanctions and problems arose, therefore, some national actions have been taken, most notably the construction of ten priority groups of commodity families and equipment for the oil industry.

Apart from the general laws and constitution, the Petroleum Law (1987–amended 2011) is considered as the most important existing legal text on the oil and gas industry that explains the principles governing it. One of these principles is the requirement of local content stating in article 10 as "The Ministry of Petroleum is committed to continually training and equipping human resources and accessing advanced technology, growth and development in the various fields of the petroleum industry…. by designing training programs and training centers and establishing research and laboratory complexes. Ministry shall continually strive to take effective measures to increase the level of knowledge and scientific and practical information of staff and experts and to create an environment conducive to the recruitment and encouragement of efficient and committed experts[6]".

In addition, three laws have been passed after revolution to strengthen domestic industries including oil and gas industry. The first one was "The law of maximum utilization of technical, engineering, production, industrial and executive power of the country in the implementation of projects and creation of facilities for exporting services" passed in 1996. It took six years to formulate its executive directives by Management and Planning Organization, though subsequently, the law was suspended due to the merger of the Economic Council with the State Economic Commission, the dissolution of the Management and Planning Organization, and the establishment of the Vice President of Strategic Planning and Supervision, which had a structural difference with the Management and Planning Organization. Later, in 2012, another law was passed called "Law on the maximum use of production and service capacity to meet the needs of the country and strengthen the production and service capacity in exports and amend article 104 of the direct tax Law" which repealed the previous one. The last one was ratified in 2019 titled "Law on maximum Use of production and service capacity to meet country's needs and strengthen the production and service capacity in exports".

The main indicator that underpins all three laws is the move in the direction of the "domestic production", with at least a 51% share of domestic production in the

[6]Petroleum Law, ratified 1987-amended 2011.

total project value. The latest law adds this point that 51% share should be calculated regardless of the value of land, buildings and general utilities. The "capacity" is not defined in any of these laws; however, some other concepts such as "work referring", "domestic production", "Iranian company" and "working inside the country" which are related to the concept of capacity are defined and explained. The concept of "capacity" is defined in appendix N of Buy-Back Contract (2008 model) titled "Maximum Utilization of Iranian Content" stating that "Iranian content means the total value of work performed and services provided by Iranian natural persons, Iranian companies and JVCs/Iranian partnerships to carry out a part of the project that will be fully calculated as Iranian content". In this regard, the methods of employing Iranian manpower, engineering capability, manufacturing capability, equipping capability, materials and services are all separately identified as some of the key constituents of local content.[7]

In this regard, Buy-Back contract model is silent about mandatory application or wavier of the requirement for minimum use of local content; however, its appendix indicates that the contractor shall be obliged to comply with the requirement for the use of local content when the required goods, equipment and services of the project are sufficiently available at a reasonable price and quality in the domestic market. Otherwise, the contractor may import items from abroad in the case that they are not available, or of poor quality or not competitively priced with their foreign similar goods in the Iranian market. It is noteworthy that in practice, it has not adopted a uniform procedure regarding the conditions of use of local content and different criteria have been applied in different contracts. For example, the contract for the development of Phase 12 of the South Pars gas field specifies some requirements for using domestic potentials including quality, quantity, proportion of the project objectives and the delivery period, while does not mention the price criterion.

Following the initiating IPC framework, specific terms and conditions regarding local content have been determined including:

1. Presentation of technology transfer and development plan by foreign contractor,
2. IPC contracts are subject to "Law on the maximum use of production and service capacity to meet the needs of the country and strengthen the production and service capacity in exports",
3. rotation of executive management positions in production period,
4. establishing a joint operation company for the development and operation of the fields (contractor is required to be a joint venture (incorporated or unincorporated) with an Iranian partner approved by NIOC),
5. 51% of the value of the contract be awarded to Iranian entities,
6. training obligation for foreign contractor.[8]

A critical issue is the legal approach to the distribution of wealth that originates from natural resources. At the end of Article 45 of the Constitution, the government

[7]Ebrahimi and Ghsemi Moghaddam (2018).

[8]Mohammadi et al. (2016).

is obligated to treat the mentioned resources "in accordance with the public interest". The said public interest has never been clarified by officials. In recent years, many interpret it as a social justice issue that includes equal opportunities for development, such as access to decent jobs, affordable and trustable health systems and environmental protection. These matters are raised because the cities and provinces, such as Khouzestan—where major and valuable reserves are located—are severely marginalized and underdeveloped. They also suffer from environmental degradation, and in recent years, the communities have faced air, water and soil pollution.[9] In December 2019, for instance, the air pollution indicator rose to 400 in the city of Ahvaz in Khouzestan province.

As a matter of fact, Article 48 of the Constitution originated from justice-based ideologies which prevailed among revolutionists. This article states that "there should be no discrimination among various provinces and regions in the country in extracting natural resources and using national incomes, and in allocating economic activities to them. Each area according to its own needs and aptitude for growth should have access to the necessary capital and provisions."

The related Act obligates the government to undertake specialized studies and assessments to implement relevant policies that were passed in 1982. The government must undertake spatial planning studies to establish which measures are required, including the elimination of all kinds of discrimination in the exploitation and use of resources, and to provide proper infrastructure for development in all regions and provinces according to their needs. It must also appropriately distribute financial resources and economic activities and find opportunities and comparative advantages in each region to optimize development. The Government Obligation to Indemnify Underdeveloped Regions Act was passed in 2001, though it does not mention equitable distribution of revenues from oil and gas production.

Later, the government made efforts to amend the current precedent and to improve conditions in underdeveloped provinces. The special budget under the Fifth and Sixth Development Plan (2011–2021) was allocated to such provinces with oil and gas reservoirs, and the Fifth Development Plan indicated that 2% of revenues from crude oil and gas sales should be proportionally disbursed in constructive plans to them. This trend continued in the Sixth Development Plan, in which the Petroleum Ministry was obligated to allocate 3% of its revenues from oil and gas contracts to improve infrastructure, reduce environmental damage and compensate the regions around the oil and gas fields.

Given LCR, and despite these plans and regulations, the situation in these cities is getting worse. They do not receive direct revenues from resources, neither do other regions with significant reserves. A clear instance is Zanjan province in which 472 mines are located, including the Angooran mine, Iran's biggest reserve of zinc and lead. According to a recent study, Zanjan province is the most underdeveloped region in Iran, confirmed by social and economic inequality indicators.[10]

[9]Khavarian-Garmsir et al. (2018).

[10]Nazmfar et al. (2018).

The Research Center of the Iranian Parliament reports that current financing undertaken by the Ministry of Petroleum is mainly limited to a few sources: the internal resources of the Ministry of Petroleum, the National Development Fund, oil contracts and corporate bonds.[11] The current financing strategies are different from those in previous annual budgets and in the Five Year Development Plans. The 2018 budget required banks to issue corporation bonds. Banks were also required to grant different kinds of facilities to the Ministry of Petroleum as well as investors in private, cooperative and non-governmental entities for upstream oil and gas development plans. The budget emphasized only the provision of banking facilities to investors in the petroleum industry, and the issuance of bonds; no other financing revenue was discussed. Private, cooperative and public companies were allowed to establish oil and gas industrial units, including refineries and petrochemical plants. They could also participate in the exploration, production and operation (not ownership) of oil and gas fields—especially joint fields. Although these measures are not nearly sufficient to secure and ensure the structure of the industry's financing, they are important steps in shaping the financing structure of the oil industry outside the domestic financial market.[12]

In June 2015, the general policies of the Sixth Development Plan, set out in 80 paragraphs, were put forward by President Rouhani. The Sixth Development Plan offered some similar policies to those in the budget. It also provided a changed notion on oil and gas to develop the energy market, and the supply of crude oil and petroleum products on the energy stock exchange.

In 2018, NIOC reported "the first consignment of Iranian light crude was presented at the International Energy Bourse in November 2018", and it consisted of "oil of eight shipments of 35,000 bpd of crude oil on the same day, at $74.85 per barrel, through three brokerage firms".[13]

3 Part Two: Contractual Framework of Oil Industry

3.1 Historical Context

Since its establishment in 1951, the National Iranian Oil Company has experienced five main periods. The first was when Mossaddegh, the Shah's political adversary, became prime minister, and the nine-points law was enacted by parliament, passed by

[11] Report by The Research Center of Parliament "Requirements for Upstream Oil and Gas Financing Under Sanctions", 2018, No. 16163.

[12] Article 12 of The Sixth Development Plan https://www.rrk.ir/Files/Laws/%D9%82%D8%A7%D9%86%D9%88%D9%86%20%D8%A8%D8%B1%D9%86%D8%A7%D9%85%D9%87%20%D9%BE%D9%86%D8%AC%D8%B3%D8%A7%D9%84%D9%87%20%D8%B4%D8%B4%D9%85%20%D8%AA%D9%88%D8%B3%D8%B9%D9%87.pdf.

[13] "A look at the National Iranian Oil Company's report in 2018", Edited by NIOC Public Relations, Accessible at: https://www.nioc.ir/portal/file/?321174/NIOC-En.pdf, p 22.

the senate and received royal assent. The nine-points law covered the implementation of nationalization and the necessary arrangements for continuing to sell oil to the former customers of AIOC.[14]

The Churchill–Truman Proposals shaped the second period of the NIOC which fully acknowledged that the management and control of oil industry in Iran would be held by Iranians, and the settlement offers did not include foreign management and control of oil operations in Iran. Although the US government and Mosaddegh's initial response to these proposals in February 1953 was positive, he rejected them in March 1953 and failed to make nationalization beneficial for Iran.

In 1954, we entered the third period in the history of the NIOC as a result of the 1954 Consortium Oil Agreement. A memorandum was provided for the establishment of a consortium in which the shares would be as follows: 40% for AIOC; 14% for Royal-Dutch-Shell; 8% for Standard Oil; 8% for Socony, 8% for Socal, 8% for Texas and Gulf and 6% for Compagnie Francaise des Pétroles (CFP). The consortium was incorporated in England where the headquarters were based at the time. It effectively controlled Iranian oil operations and shared the profits equally with the Iranian government.[15] Mosaddegh wanted to put an end to British control over Iran's oil. However, the supporters of nationalization lacked a plan or strategy on how the industry would be managed.[16] This disorganization within the NIOC provided the Shah with an opportunity to accumulate huge personal wealth from the oil reserves and to generate a new kind of authoritarianism, mixed with his developmental plans.

The fourth period began in early 1973 when, the NIOC presented the consortium with an ultimatum: to hand over all management and control over Iran's oil, or to give up all privileged access to the country's oil, and be treated like any other customer. External rights were passed by law in the same year. The 1974 Petroleum Act and Risk Service Contracts stated:

… [T]he Petroleum resources and the Petroleum industry of Iran belong to the Nation. The experience of sovereignty right of Iranian Nation over the Petroleum resources of Iran with respect to the exploration, development, production, exploitation and distribution of Petroleum throughout the country and its continental shelf is entrusted exclusively to the National Iranian Oil Company who shall act thereupon directly, or through its agents and contractors.[17]

In late 1978, the consortium and other foreign companies left Iran because of the violent turn of the revolution against Shah. At that time, the NIOC possessed four offshore joint ventures (JVs) in the golf producing oil. This action prompted foreign companies to sue the government and resulted in a halt on exploration activities a year later. In 1980, the NIOC established the National Iranian Drilling Company (NIDC), whose main task was to explore new sites. Eight years later, Asaluye became one of the NIDC's discoveries, alongside Darkhovin and Azadegan.

[14]Elm (1992).

[15]Heiss (1994).

[16]Minutes of Persia Committee Meeting February 2, 1954.

[17]Yong (2013).

The correlation between politics and oil industry in Iran became more complicated after the Islamic Revolution in 1979, when any relations between oil and foreigners were interpreted as dependency. As a result, Iran reduced its oil production in 1980 and 1981. As the most principal body in power and in a failed attempt to avoid this dependency by reducing its revenues from foreigners, the Revolutionary Council terminated most of external oil-related agreements. All these revolutionary decisions inflicted heavy losses on the country and created confusion in the relationship between oil and Iran's political structure.

3.2 *Buy-Back and IPC Models:*

The five periods mentioned briefly examined the historical context of petroleum agreements and will be discussed in greater detail in the next section. The pattern of oil contracts in Iran began with Darcy's concession contracts and evolved from consortium to corporate and service contracts. The main two current approaches were Buy-Back contracts and Iran Petroleum contracts (IPCs).

The Buy-Back approach was one of the contractual models for oil implementation projects.[18] Despite the fact that it is legal for foreigners to have properties in Iran, establishing a company by a foreigner is prohibited by Article 81 of Constitution. However, if the company is established outside of Iran, it is allowed to have a legal representative or to register a branch office in Iran, after obtaining relevant authorizations according Registration of Companies Act 1931. There is an essential condition to obtain these approvals indicating that the foreign company should "have a contract with the Iranian government or one of its state-controlled institutions".[19] In fact, the foreign company is allowed to establish a branch or have a representative to perform the contract with Iranian party. Furthermore, Foreign Investment Promotion and Protection Act was ratified in 2002 in which it is asserted that foreign investors are treated the same as domestic investors. Thus, IOCs do not face any obstacle to invest in Iran, while the specification of oil and gas contracts shall be met by them.

The Constitution of the Islamic Republic does not allow the granting of concession which is why the government came up with a Buy-Back formula, where foreign companies that develop oil or gas fields are repaid their costs and given a rate of return on the initial production. It also includes extended access to hydrocarbons. Buy-Back contracts are "a kind of service contract with unique features" struck between the NIOC and an international oil company. "Therefore it is sometimes regarded as a separate kind of agreement. Buy-Back contracts might be concerned only with the development of discovered oil fields, or with both their exploration and development." With these contracts, foreign investors, as contractors, conduct oil and gas field development operations with their own capital and financing over a specified period of time. With the commencement of production, the contractor transfers all

[18]Bahmaei and Zere (2018).

[19]Kakhki (2008).

field operations to the NIOC and receives up to 60% of the project's product output as depreciation of cost and reimbursement for a specified period of time, such as five to seven years. It can be concluded that the focus of the Buy-Back oil contracts is the exploration and description of the oil and gas fields, the development of these fields and the transfer of related services. Once these operations have been completed, the contractor agrees to be paid the price in full or in part, by purchasing the product produced from the oil and gas fields. Eleven field projects were offered in 1989, but the response was not as promising as planned, so in 1995, the rate of return was increased to attract more companies. In 1997, during Mohammad Khatami's presidency, total signed a $2 billion deal for South Pars Phases 2 and 3. The Buy-Back formula was used for a second term in 1998, when the NIOC presented four categories of projects: exploration; offshore fields for further development; onshore fields for urther development and three projects for the Abadan oil refinery expansion.[20] Although the NIOC is responsible for the development of oil and gas industry in Iran, it lacks the capacity and sufficient financial resources to develop all energy resources without cooperating with international oil companies (IOCs).

3.3 Iranian Petroleum Contract Model

Until 2015, Iran practiced only Buy-Back contracts that presented several risks for IOCs. These included the effects of the fluctuation in the international price of oil and gas; higher capital expenditure requirement than agreed upon; short-term contracts and short trial periods for project delivery; inadequate distribution over the construction period and lack of advanced technology. Moreover, under Buy-Back contracts, IOCs are involved only for exploration and development, which basically reduces their role to technology service companies,[21] as when full production is achieved, the NIOC takes over the operation.

The evidence of these shortcomings suggests that Buy-Back contracts did not convince foreign investors to participate in the upstream sector. Also, the impacts of sanctions, alongside a reduction in oil prices, cutbacks on loans and bank lending, and other strict international regulations related to banking and climate change, have led to a decrease in the tendency of foreign and domestic oil companies to invest in Iran's oil projects.

The government felt the need to amend the contractual framework through which it might be able to dispel doubts investors may harbor. In 2012, the Parliament passed the "Rights and Duties of Petroleum Ministry Act", by which the ministry enters into production—sharing contracts. Pursuant to this Act, the Cabinet adopted a regulation titled General Conditions, Structure and Pattern of Upstream Oil and Gas Contracts, which ultimately outlined the overall framework of these new contracts.

---

[20] Shammas (2001).

[21] Mazraati and Groenedaal (2006).

The Iranian Petroleum Contract model was drafted as a result of problems associated with Buy-Back method. In 2017, government passed a resolution called "General conditions, structure and pattern of upstream oil and gas contracts" in which the requirements of new model of contracts named Iranian Petroleum Contract (IPC) have been specified. The client is defined as NIOC or its subsidiaries on behalf of NIOC and contractor is defined as qualified oil companies that have contracted for each exploration, characterization, development and operation and implementation of well optimization or recovery projects, or all of them continuously, together with all necessary funding, through the legal process. The following principles apply to all contracts made pursuant to this resolution:

1. Exercising sovereignty and public ownership over the country's oil and natural gas resources and reserves through the Ministry of Petroleum on behalf of the Islamic Republic of Iran.
2. All risks and costs shall be borne by the contractor if no commercial field or commercial oil tank is discovered.
3. Risks of failure to meet the intended contractual objectives or the inadequacy of the field or reservoir product for the depreciation of financial liabilities is incurred by the contractor.

Three key elements of IPCs are as follows: an emphasis on collaboration through incorporated or unincorporated joint ventures to facilitate the transfer and upgrading of national technology in the field of upstream business, empowering national companies, establishing and strengthening oil companies; a focus on technology transfer and the dissemination of the technology-related knowledge into oil and gas industry; and prioritizing the enhancement of oil recovery for developed and undeveloped fields.[22] IPCs can be defined as a combination of production sharing and Buy-Back contracts.[23] Like production sharing contracts, they are "prepared for long-term use and cover a significant portion of a field's lifespan",[24] and also retain the exact ownership feature of the Buy-Back model.

In this new arrangement, Iranian exploration and production companies deemed qualified by the Ministry of Petroleum act as a partner to a reputable foreign oil company. This requirement is set to fulfill the need for technology transfer. First, the foreign contracting party must annually submit a technology transfer and a development plan. Second, operational measures and enforcement policies are attached to the contract as a "contract technology attachment". Given that the annual operational financial plan must be approved by the Iranian party—either the NIOC or its subsidiaries—it has the authority and opportunity to incorporate the principles and procedures of the technical knowledge transfer into the contracts. Unlike the Buy-Back model, the contractor is permitted to be part of the production and operational stages.

[22] For more information see: James et al. (2020). Tagliapietra (2016).

[23] Sahebonar et al. (2016).

[24] Kohan Hoosh Nejad et al. (2018).

4 Part Three: Sixth Development Plan (2016–2021)

After the 1979 revolution, the NIOC experimented with several budget plans. Three main financial systems have been implemented throughout the 40 years of the Islamic Republic. The first period lasted for 20 years. It started when the Ministry of Petroleum was established according to the Act passed by The Council of the Islamic Revolution in 1980. Under this Act, three important companies, the National Iranian Oil Company, the National Iranian Gas Company and the National Iranian Petrochemical Company became subsidiaries to the newly established Ministry of Petroleum, which procured their budgets from the public sector treasury.

The budget plan of 1980 ratified that all oil agreements should be signed by the NIOC on behalf of Iranian government, and its revenues deposited into the treasury. The Central Bank was obliged to sell as much foreign currency to the NIOC as it required. It meant that all of the NIOC's revenues were considered part of the public budget, which would also provide it with funds to invest in resource exploration. Meanwhile, the Iran-Iraq war broke out, and public budget was unable to provide funds to invest in the oil and gas fields anymore.

Budgeting for the NIOC in the Third Development Plan (2000–2005) was the preface of the second period. From 2000 onwards, the Iranian government tried to provide infrastructure in order to improve the mining industries and increase a share of production. It passed a law to establish the Iranian Mines and Mining Industries Development and Renovation Organization (IMIDRO) as a holding company with several subsidiaries that included some of the key businesses working to extract resources, such as the Al-Mahdi Aluminum Company, the Mobarakeh Steel Company and the Esfahan Steel Company. The goal was to optimize the country's economic system through the mining sector and related industries, as well as to supervise the use of mineral reserves and their proper and effective exploitation.[25] Yet it took five years after its establishment before this organization could participate in decisions that involved mining development plans. According to Article 120 of the Third Development Plan, all revenues from oil and gas were considered as a means to develop the oil and gas fields and downstream sections. The failure of this system soon became apparent as the NIOC needed revenues from oil and gas product sales to invest in upstream and downstream sections, and it could not change the existing refinery patterns.[26] It was necessary to reduce furnace oil production and increase production of distillery products; however, such changes resulted in a fall in NIOC's revenue, and it was unable to replace the funds immediately. Also, under this system, increasing crude oil production had no impact on NIOC's revenues. Therefore, the NIOC lacked any incentives to make an amendment or reform its products.

---

[25] https://www.tejaratefarda.com/%D8%A8%D8%AE%D8%B4-%D8%A7%DB%8C%D8%B1%D8%A7%D9%86-%D8%B5%D9%86%D8%B9%D8%AA-16/22360-%D8%B3%D8%A7%D9%84-%D8%B1%D8%A6%DB%8C%D8%B3. Accessed on 2nd of November 2019.

[26] "A Review of the Financial System of National Iranian Oil Company and its Impacts on Oil and Gas Production" 2009, *Economy and Energy*, No. 116–117: 46–49.

The Fourth Development Plan (2005–2010) ratified that the NIOC could receive a share from crude oil and gas production which heralded the beginning of the third period. It also ratified that domestic refineries and their products had to purchase crude oil at the international price (Persian Gulf FOB price), and the government would pay the price difference as a subsidy. These changes had significant effects: the production of light petroleum and distillery products with higher values resulted in an increase in total revenue, as well as boosting efficiency and proficiency.[27] In addition, the rise in crude oil production boosted the NIOC's revenue, as it benefited from oil and gas production. All these variations led to increased investment in both upstream and downstream sections.

The Fifth Development Plan (2011–2015), which was delayed by a year, had not performed well during the tough international conditions and hardship of country's economy (particularly in the oil and gas sector). This was the most unsuccessful of all the Plans, though its failing was not entirely due to the difficult international conditions and sanctions. Proper policies and measurements could have prevented a considerable portion of its shortcomings. From the outset,[28] it could have been predicted that the Plan's goals were overly ambitious. Added to this, specialized and non-specialized duties were imposed on the oil and gas sector, plus tasks mismatched to the country's technical and national capacity. Targeting the $2 billion-dollar foreign investment in the upstream sector of the oil industry was clearly at odds with government policies at the time.

The Fifth Plan had 15 articles on the oil and gas industry. Article 125 indicated that the goals were set to achieve a daily production capacity of 5.1 million barrels of crude oil and 1470 million cubic meters of natural gas. These figures should be compared to daily production of 4 million barrels of crude oil and 620 million cubic meters of natural gas in 2010 (the last year of the Fourth Plan). Article 126 emphasized the need for Buy-Back contracts to explore and develop new fields in all parts of the country. However, due to international restrictions (sanctions), development projects were suspended. International oil companies announced their withdrawal, leaving Iran with the remaining projects focused on exploration in the border areas. Article 129 focused on the issue of exploration licenses, and the development and production of oil and gas fields. The performance indicated the issuance of more than 80 such licenses, but invitations to the private sector and promoting competition and productivity did not take place. Five years after the Fifth Plan was implemented, a total of $70 billion was invested in the petroleum industry, with the NIOC contributing $55 billion. However, this fell short of the predicted $226 billion investment in the upstream oil and gas sector. Also, 75% of this actual amount came from domestic sources, which, given the limited resources, put additional pressure on the oil industry.

The Sixth Plan aimed to increase the NIOC's daily production capacity to 4.7 million barrels, its daily gas production capacity to 1.3 billion cubic meters and its liquids and condensate production capacity to 1.1 million barrels per day. The Plan

[27] Modarresi (2009).

[28] Mostaghel (2016).

aims to achieve new technologies, localize and commercialize these technologies in the oil and gas industry and support the NIOC's corporate social responsibility endeavors. The program also aims to increase oil export potential, as well as the amount of gas and water injected into the fields.[29] This Plan was revised with Iran's Nuclear Deal in mind, and estimates were made under the assumption that foreign investors would be eager to take part in Iran's oil industry.

Four years after this Plan, domestic mismanagement and corruption were not the only impediments to production and development; sanctions imposed by Trump after withdrawing from the Nuclear Deal forced foreign investment to pull out of Iran and prevented the country from engaging in oil and gas, or other business transactions. As a result, Iran experienced its lowest production rate in the past 20 years. The country is located in an oil-rich region, which means it has "11% of proven global oil reserves and 16% of the world's natural gas resources, amounting to 133 billion barrels of oil (17 billion tons) and 27 trillion cubic meters of gas, which totals to $4000 billion by current price of oil and gas".[30] Even though these reserves rank fourth among oil-rich countries, and first among gas producers, its energy production was severely reduced, and foreign pressure also deprived Iran from achieving development. This has hit its economy, which includes an impact on the mining industry and the export of raw materials.

As with its predecessors, the Sixth Plan has not been successful in fulfilling the potential of oil industry and achieving the desired prosperity. The Plan Sixth was unrealistic, and its anticipated goals were not in line with existing capabilities and capacities. These ambitious and unrealistic plans were turned into useless and irrelevant documents that were seriously affected when they coincided with several frequently occurring international crises and conflicts.

5 Conclusion

This chapter discussed the main challenges that Iran's extractive resources face with regard to both local content practices and involvement. As the most important gas and oil organization, the NIOC's history is intrinsically rooted in the strict connection between the energy resource industry and Iranian politics. The establishment is responsible for sovereign as well as operational duties. In addition, complex international relations and an authoritarian political system have an extreme effect on the integration of local content resources on extractive resources at every stage of its processing, from discovery (where foreign investment and sufficient funding is inaccessible), to the final stages that results in the sale of raw and processed material.

[29]Daneshjafari and Karimi (2016).

[30]2nd Iran Mines and Mining Industries Summit, IMIDRO Projects, 2016, https://imidro.gov.ir//parameters/imidro/modules/cdk/upload/content/general_content/216/14844805897047ber9qj6malrg2rb07lnbenom7.pdf.

Lack of access to high-tech equipment and up-to-date technical knowledge makes precise reporting on LCR near impossible.

Another challenge for integrating LCR is that the NIOC is a state-owned entity whose budgeting and development plan is designed by government and has to be officially ratified by parliament. The complexity and ambiguity of the gas and oil industry's legal documents create several contradictions between laws that have different beneficiaries and can cause confusion. Iran also faces the problem of an accumulation of laws and regulations passed by different competent authorities that increases rather than resolves the chaos. Another challenge is that most laws are not applied as predicted, and even if they may be efficient, appear useless. This also applies to those regarding the development of regions that are rich in reserves but remain marginalized, and where the local communities have not been considered or involved.

Iran has practiced different kinds of contracts since it began oil production. Apart from concessions and consortiums that were in operation till the mid-1900s, Iran has insisted on unrevokable ownership of its oil and gas reserves. This approach strengthened after the revolution when the government designed a definite contractual structure based on service agreements. The first were Buy-Back contracts, which remained in operation for 20 years. This kind of service contract was only used for the exploration and development of oil fields and brought bounded revenues. It could not meet the demands of the oil industry. To overcome its limitations, the the Iran Petroleum contract (IPC) was introduced, a new model that combined features of the Production Sharing contracts (PSC) with some characteristics of the Buy-Back contracts. The unveiling of IPCs coincided with a new round of US sanctions that resulted in discouraging foreign company investments. As a result, the strengths and weaknesses of the IPC model have yet to be examined, especially with regard to LCR.

All in all, local content resources have had to confront many challenges, both as a result of domestic inefficiencies as well as international events. To overcome these problems, Iran has to find ways to finance its projects and to provide a platform on which it can expand the integration of LCR into its gas and oil industry. If principle rules and structures remain unchanged, the 40 years of experience under the current regime indicate a worse outcome for the future.

References

Bahmaei, M., & Zere, S. D. (2018). Iran's new oil order and its reflection in designing contractual patterns of upstream oil industry projects. *Legal Studies, 10*(3), 29–65.

Daneshjafari, D., & Karimi, S. (2016). Oil, sixth socio-economic development plan and resilient economy. *Strategic and Macroeconomic Policies Quarterly, 8*(2), 1–35.

Ebrahim-Fathabadi, H. (2014). A look at the international position of the National Iranian Oil Company. *Oil and Gas Exploration and Production Journal, 129*, 14–20.

Ebrahimi, N., & Ghsemi Moghaddam, S. (2018). Investigating the legal requirements for the use of local content in the upstream oil and gas industry. *Journal of Public Law Research, 20*, 59, 63–95.

Elm, M. (1992). *Oil, power, and principle: Iran's oil nationalization and its aftermath*. Syracuse University Press.

Green. J. D., & Wehrey, F. (2008). Wolf, Charles "Understanding Iran". RAND Corporation.

Heiss, M. A. (1994). The United States, Great Britain, and the creation of the Iranian oil consortium, 1953–1954. *The International History Review, 16*(3), 511–535.

James, D., Black, A., Wood, C., Christopher M.-O., Hariri, R., Humphrey, D., Nadir, G., Cheung, A. (2016, February 6). New Model Iranian Petroleum Contract. *Mondegh*. https://www.mondaq.com/Energy-and-Natural-Resources/481064/New-Model-Iranian-Petroleum-Contract.

Khavarian-Garmsir, A. R., Pourahmad, A., Hataminejad, H., & Farhoudi, R. (2018). A comparative assessment of economic and physical inequality between shrinking and growing cities: A case study of Khuzestan province, Iran. *International Journal of Urban Sciences, 22*(1), 104–122.

Kohan Hoosh Nejad, R., Manzoor, D., & Amani, M. (2018). Economic evaluation of fiscal regime of buy back contracts in comparison with production sharing contracts (case study: Azadegan oil field). *Iranian Economic Review, 22*(2), 579–598.

Mazraati, M., & Groenedaal, W. (2006). A Critical Review of Iran's Buyback Contract. *Energy-Policy., 34*(18), 3709–3718.

Modarresi, M. (2009). Budgeting at the National Iranian Oil Company and its effects on oil and gas production. *Exploration and Production Journal, 58*, 18–21.

Mohammadi, S., Asghari Agh Mashhadi, F., & Haji Hosseini, M. J. (2016). Investigating the specific rules of upstream oil contracts with emphasis on the new IPC oil contract. *Administrative Law Quarterly, 3*, 9, 113–135.

Mostaghel, B. (2016). Sixth five year socio-economic development plan from the perspective of oil industry. *Oil and Gas Exploration and Production Monthly*, No. 124, Introduction.

Nazmfar, H., Mohammadi H. S., & Akbari, M. (2018). The inequality of development in the provinces in order to achieve sustainable development. *25*(10), 297–340.

Noreng, Ø. (1994). National Oil Companies And Their Government Owners: The Politics Of Interaction And Control. *The Journal of Energy and Development, 19*(2), 197–226.

Sahebonar, H., Taherifard, A, Farimani, F. (2016, August 28–31). Economic analysis of New Iranian Petroleum Contract (IPC): the case study of caspian sea fields. Presented at: Energy Economics Emerging from the Caspian Region: Challenges and Opportunities, Baku Azerbaijan.

Shammas, P. (2001). Iran: Review of petroleum developments and assessments of the oil and gas fields. *Energy Exploration & Exploitation, 19*(2/3), 207–260.

Tagliapietra, S. (2016). *Energy Relations in the Euro-Mediterranean: A Political Economy Perspective*. Springer.

Yong, W. (2013). *NIOC and the state*. Oxford Institute for Energy Studies.

The Development and Implementation of Local Content in the Extractive Industries in Trinidad and Tobago and Guyana

Alicia Elias-Roberts

Abstract This chapter discusses the main features of the legal framework regulating local content in Trinidad and Tobago and Guyana within the extractive industries. It also discusses the role that local content can play in the development of the two countries under review. The problem of illegal quarrying in Trinidad and Tobago and the incident where a major oil company in Trinidad and Tobago decided to withdraw a major project are discussed to demonstrate some of the problems associated with local content. A case study from Guyana concerning the dissatisfaction with the draft local content policy is reviewed as well as general pitfalls that Guyana needs to avoid in its implementation of local content policies and laws. Further, the potential for local content policies and law to conflict with international and regional trade obligations are briefly discussed.

Keywords Local content · Extractive industries · Sustainable development · Oil and gas law · Production sharing agreement

1 Introduction

The development and implementation of local content in the extractive industries in Guyana and Trinidad and Tobago is a topic which is often bantered about by politicians in these two jurisdictions, but the discussions often lack depth and critical analysis. This chapter offers a comprehensive comparative analysis of the legal and regulatory framework of local content development in both countries. Guyana and Trinidad and Tobago, two countries in the Commonwealth Caribbean, make for an interesting comparative analysis because the two jurisdictions have very different experiences in their extractive industries' history. Trinidad and Tobago has over 100 years of experience in oil and gas exploration and production, and from that perspective, we can review lessons learnt in that jurisdiction. On the other hand,

A. Elias-Roberts (✉)
Faculty of Law, University of the West Indies, St. Augustine, Trinidad and Tobago
e-mail: aeliasroberts@gmail.com

E. G. Pereira et al. (eds.), *Sovereign Wealth Funds, Local Content Policies and CSR*, CSR, Sustainability, Ethics & Governance,
https://doi.org/10.1007/978-3-030-56092-8_19

Guyana is a new frontier with the first major announcement of a significant discovery of oil in 2015 and therefore, we can discuss lessons to be learned and pitfalls to avoid in the context of Guyana.[1] Interestingly, both countries have over 60 years of experience in other extractive industries, with Guyana[2] having experience in bauxite, stone, sand, gold and diamond mining and Trinidad and Tobago[3] having experience in quarry operations in sand and gravel, blue limestone, yellow limestone, clay, porcellanite and sand.

This chapter will begin by briefly discussing the main legislative and regulatory framework related to local content in Trinidad and Tobago and Guyana within the extractive industry. It will also discuss the role that local content can play in the development of the two countries under review. Next, some experiences and practices with local content requirements in the extractive industries in Trinidad and Tobago and Guyana will be discussed. Before concluding the problem of illegal quarrying in Trinidad and Tobago and the incident where a major oil company in Trinidad and Tobago decided to withdraw a major project, from that, jurisdiction will be discussed as two case studies. A case study from Guyana concerning the dissatisfaction with the draft local content policy in that country will be discussed as well as general pitfalls that Guyana, a new frontier in the petroleum industry, needs to avoid in its implementation of local content policies and laws. Also, in both jurisdictions, the obligations that each State has under several bilateral investment treaties[4] and the CARICOM regional agreement[5] will be highlighted to discuss how these obligations may conflict with or relate to local content laws and policies. The final section will conclude and make a few suggestions about how Trinidad and Tobago and Guyana can implement and maximise the benefits from local content policies and laws.

[1] 'ExxonMobil Announces Significant Oil Discovery Offshore Guyana,' press release, May 20, 2015 at https://news.exxonmobil.com/press-release/exxonmobil-announces-significant-oil-discovery-offshore-guyana accessed Feb. 19, 2019.

[2] Guyana Extractive Industries Transparency Initiative (GYEITI) report for 2017 Fiscal Year (April 2019).

[3] See Government of the Republic of Trinidad and Tobago Ministry of Energy and Energy Industries, 'White Paper on National Minerals Policy,' June 2015 https://www.energy.gov.tt/wp-content/uploads/2014/01/White-Paper-on-National-Minerals-Policy-June-2015.pdf accessed 4 July 2019; See also Trinidad and Tobago Extractive Industries Transparency Initiative Report 2016 https://www.tteiti.org.tt/wp-content/uploads/TTEITI-Report-2016.pdf accessed 4 July 2019.

[4] For instance, the Government of Guyana has entered into bilateral investment treaties with China, Cuba, Germany, Indonesia, Korea, Kuwait, Switzerland and the United Kingdom https://investmentpolicy.unctad.org/international-investment-agreements/countries/89/guyana. Accessed 11 July 2019.

[5] See Chap. 3 on the Establishment, Services, Capital and Movement of Community Nationals in the *Revised Treaty of Chaguaramas establishing the Caribbean Community including the CARICOM Single Market and Economy*, 2259 UNTS 293 (Nassau, Bahamas 05/07/2001).

2 The Main Obligations Related to Local Content in Trinidad and Tobago and Guyana Within the Extractive Industry.

2.1 Trinidad and Tobago

Trinidad and Tobago is currently the largest oil and natural gas producer in the Caribbean, and since the early 1960s, its economy has been characterised by its heavy dependence on the production and export of oil and gas.[6] Between 1973 and 1982, high oil windfall afforded rises in income, expansion of public sector employment and improvement in physical infrastructure and living conditions. However, a large share of expenditure was not sustainable, including subsidies and high public sector salaries.[7] When international oil prices declined in the 1980s, the economy was deeply affected. Again in 2014, when oil prices declined, Trinidad and Tobago went into a recession soon after. The country still faces a wide range of socio-economic challenges, including sustainable development beyond the petroleum sector.

The petroleum sector in Trinidad and Tobago is dominated by subsidiaries of large international oil companies. Exploration and production activities are carried out under the terms of production sharing agreements or contracts (PSAs or PSCs). The Petroleum Company of Trinidad and Tobago Limited (Petrotrin), wholly owned by the government, was incorporated in 1993 and was mandated to engage in petroleum activities along the sector value chain. For several years, Petrotrin owned and operated the only refinery in the country, but in late 2018, the government made a decision to close it down due to financial and other reasons.[8] The National Gas Company of Trinidad and Tobago (NGC) was established in 1975 and is engaged in the purchase, transportation and distribution of natural gas to industrial users. NGC is still in operation. In addition, it participates in the Atlantic LNG company. Approximately 50% of natural gas production is used locally, mainly by the petrochemical industry and for power generation.

Even though Trinidad and Tobago's first commercial oil production in south-west Trinidad started in 1908, it was only in the 1960s that the island's energy sector took on a national identity. In 1963, a commission of enquiry was set up by the government to review the oil industry and to recommend a legal framework for the industry among other things. Today, there are several laws, regulatory instruments, contracts and policy documents that channel the government's local content policy in Trinidad and Tobago. Under the recent *Public Procurement and Disposal of Public Property Act* (2015), local content is defined as: "the local value added to goods, works or services measured as the amount of money or percentage of each dollar of expenditure

[6]See, The Oil & Gas Year Report (2019), Pawan G Patil, John Virdin, Sylvia Michele Diez, Julian Roberts, Asha Singh, 'Toward A Blue Economy: A See also, Promise for Sustainable Growth in the Caribbean; An Overview' (Report No: AUS16344 The World Bank 2016) 70.

[7]Silvana Tordo et al. (2013) *supra* n. 6 at 116.

[8]See Ghourlal (2018), Sant (2018).

remaining in Trinidad and Tobago after the production of the good or the performance of the work or service".[9] The outdated *Petroleum Act* of 1969 and implementing regulations from 1970, while not expressly using the term "local content", makes reference to local content indirectly.[10] Regulations 42 (2) (f) provides that a licensee shall perform such of the general obligations specified in sub-regulation (2) in his licence and a licensee shall:

> minimise the employment of foreign personnel, ensure that such employees are engaged only in positions for which the operator cannot, after reasonable advertisement in at least one daily newspaper circulating in Trinidad and Tobago, find available nationals of Trinidad and Tobago having the necessary qualifications and experience; determine the rules of employment including salary scales in such manner as to ensure that all employees in the same category enjoy equal conditions irrespective of nationality....[11]

Investment incentives for local content are granted by the Government of Trinidad and Tobago and are coordinated through the Industrial Development Division of the Tourism and Industrial Development Company of Trinidad and Tobago Limited (TIDCO). These may include training subsidies for developing new skills, export credit insurance and exemption from value-added tax on inputs for companies exporting 80% of production. The *Fiscal Incentives Act*[12] allows for the granting of a tax holiday (or partial holiday) for periods up to ten years for the manufacture of approved products by approved enterprises. These fall into separate classifications including the highly capital intensive enterprises investing in excess of TT$50 million (US equivalent $7.38 million)[13]; export enclaves, where products are manufactured exclusively for export; and enterprises using a significant portion of local inputs.[14] These concessions are discretionary and require applications to the Ministry of Trade and Industry via TIDCO.[15]

Other legislation related to local content includes the *Unemployment Levy Act*[16] that provides for training and relief employment for the unemployed and the *Petroleum Production Levy and Subsidy Act*[17] used to fund subsidies on petroleum products. The *Income Tax Act*[18] provides for several special classes of companies that are entitled to a tax credit of 15% of their chargeable income for seven years.

---

[9] *Public Procurement and Disposal of Public Property Act*, (No. 1 of 2015), Laws of Trinidad and Tobago, sec. 4.

[10] *Petroleum Act* (no. 46 of 1969), Cap. 62:01, Laws of Trinidad and Tobago; see also the *Petroleum Regulations* (Legal notice 5 of 1970), made pursuant to section 29 of the Petroleum Act, Laws of Trinidad and Tobago.

[11] Ibid., regulations 42(2)(f).

[12] *Fiscal Incentives Act*, Cap. 85:01, Laws of Trinidad and Tobago, as amended, which grants fiscal incentives to qualified companies in areas that are considered important to economic development.

[13] Currency conversion done at xe.com.

[14] *Fiscal Incentives Act, supra* n. 14, sections 2, 5 and 9.

[15] See Government of the Republic of Trinidad and Tobago Ministry of Finance (2019).

[16] *Unemployment Levy Act* (no. 16 of 1970), Cap. 75:03, Laws of Trinidad and Tobago.

[17] *Petroleum Production Levy and Subsidy Act* (no. 14 of 1974), Cap. 62:02, Laws of Trinidad and Tobago.

[18] *Income Tax Act,* Cap. 75:01, Laws of Trinidad and Tobago.

This reduces their effective tax rate to 20% from the statutory rate of 35%. These special classes of companies include approved small companies, approved companies trading in a regional development area and approved activity companies. Special companies must be locally incorporated and owned.[19]

In addition to the various laws quoted above, the *Model Production Sharing Agreement* (2012)[20] provides for local content and stipulates that the contractor shall provide for the maximum utilisation of services and facilities available from Local Enterprises. The model PSA also provides that the contractor shall also employ the following obligations in all aspects of Petroleum Operations:

> 39.1 Contractor shall comply with the Government's Local Content Policy in force and as modified from time to time.
>
> 39.2 Contractor shall maximise to the satisfaction of Minister the level of usage of Local Goods and Services, businesses, financing and the employment of nationals of the Republic of Trinidad and Tobago...
>
> 39.8 Contractor shall ensure the development of people by imparting to nationals technology and business expertise in all areas of energy sector activity including but not limited to:
>
> 39.9 Contractor shall ensure that nationals are selected and trained consistent with Contractor's performance standards in relation to activities referred at 39.8....[21]

As shown above, the government has formulated policies to increase the number of national workers, where priority will be given to the employment of national workers. The PSAs mandate the contractor to provide training to national workers in line with the contractor's performance standards and contractors must also develop and implement training programmes that enable nationals to replace foreign workers. Production sharing agreements require the contractor to maximise the use of local goods, services and financing. Interesting, the PSAs stipulate that seismic processing work shall be undertaken in-country, and in the award of subcontracts, reasonable preference will be given to qualified local contractors or suppliers who meet the quality, cost and schedule requirements. Note, however, that there is no explicit margin of domestic preference.

Turning to institutional responsibilities, the current *Local Content and Local Participation Policy and Framework* was launched in 2004, and by 2006, the government appointed the Permanent Committee on Local Content to "monitor the various activities in the energy sector to ensure that, as far as possible, all energy projects include opportunities for the development of the expertise of nationals and maximise the level of local content and local participation".[22] The committee is tasked with developing policies and strategies to ensure the transfer of knowledge and technology that improve local capabilities, in addition to business and the capital market; update

---

[19]Ibid.

[20]Government of the Republic of Trinidad and Tobago Ministry of Energy and Energy Industries, *Model Production Sharing Contract* (2012), Article 39.

[21]Government of the Republic of Trinidad and Tobago Ministry of Energy and Energy Industries, *Model Production Sharing Agreement* (2010).

[22]Government of the Republic of Trinidad and Tobago Ministry of Energy and Energy Industries (2004).

local content and participation policies, as required; and ensure compliance with the set policies.

The goal of the policy framework was not to simply capture value through fiscal measures in the form of taxation and the direct participation by Government but to incorporate other measures such as the maximisation of the depth and breadth of local control and financing, maximisation of the usage of local goods and services and local capability development. The Trinidad Offshore Fabrication Company (TOFCO) Fabrication Yard, as well as the significant level of local input into the Atlantic LNG project stands out as major achievements coming out of the *Local Content and Local Participation Policy and Framework*. It was reported by the government that domestic investment in these projects came from both the government and local private sectors through the building of the fabrication yard at La Brea, the development of training institutions, training of the nationals in Trinidad and procurement by local businessmen of equipment, technology, facilities and know how.[23] Despite the significant levels of domestic and foreign investments, however, the model failed to create a sustainable local service industry, notwithstanding Trinidad and Tobago's globally competitive local capabilities. More recently, the Permanent Local Content Committee was re-established with a detailed mandate.[24] However, according to a former Minister of Energy in Trinidad and Tobago, in recent times, in spite of obligations in Production Sharing Contracts and under the existing Petroleum Act and Regulations, such signature successes have not been pursued or replicated, and the Permanent Local Content Committee has in fact become inactive and non-functional.[25]

3 Guyana

Local content is critical to the development of Guyana's extractive industry and its oil and gas economy in particular. It must be highlighted that even before the announcement of oil in 2015, Guyana had laws concerning local content in the extractive industries, and Guyana has several laws that make reference to local content implicitly and allows foreign investors to engage in extractive activities, such as mining. However, like many other new oil and gas producing countries, Guyana sought to develop a local content policy specifically for the petroleum sector after the first announcement of oil in 2015. In May 2018, the Government of Guyana published a second draft of the *Local Content and Value Addition Policy Framework*.[26] The new

[23]See Speech by the Honourable Nicole Olivierre, Minister Of Energy And Energy Industries at the Energy Chamber's 3rd Local Content Form, Trinidad and Tobago, November 18, 2015 https://www.energy.gov.tt/wp-content/uploads/2015/11/Speech-Local-Content-by-Minister-Olivierre-16-Nov-2015.pdf, accessed August 23, 2019.

[24]Ibid.

[25]Ibid.

[26]Local Content and Value Addition Policy Framework https://doe.gov.gy/published/document/5af7293bdc677720ccdc33ba accessed 11 August 2019.

policy stated that it aims at delivering benefits beyond the payment of royalties and taxes and will provide a full definition of the local content. The policy also delineates the framework by which local content will be understood, developed, measured and secured.

The *Guyana Geology and Mines Commission Act* (1979),[27] as amended in 1987[28] created the Guyana Geology and Mines Commission (GGMC), and it sets out its functions and roles. GGMC provides effective stewardship of all mineral resources by ensuring that opportunities for mineral resources development (exploration and extraction) increase, by promoting and supporting investment in the mining sector. The *Forest Act*, Cap 67:01, which was repealed by *Forest Act* (no. 6 of 2009) s.83 also provides for exploratory permits, and section 6(2)(c) provides for proposals for the employment and training of Guyanese.[29]

The *Mining Act,* Cap. 65:01 (1989)[30] is the principal Act regulating the mining sector in Guyana. It provides the legal basis under which mining exploration, development and production are to be conducted. It defines the rules for granting exploration licences or permits. The *Mining Act* (1989) provides for the grant of prospecting licences under section 30(2)(c) and mining licences under section 46(1)(d) and specifies obligations concerning the employment and training of Guyanese nationals as well as the procurement of goods and services that can be procured locally. This Act provides that all subsurface mineral rights in Guyana are owned by the State and authorises the Guyana Geology and Mines Commission (GGMC) to manage these resources. Section 30 of the Act concerning the granting of prospecting licence provides as follows:

> "30. (1) Subject to this Act, on application duly made, the Commission may, with the approval of the Minister, grant on such conditions as it determines, or refuse to grant, a prospecting licence in respect of any parcel or parcels.
>
> (2) No prospecting licence with respect to any mineral shall be granted to an applicant unless the Commission is satisfied that…
>
> (c) his proposals for the employment and training of citizens of Guyana are satisfactory, or, if the Commission is not so satisfied, in the opinion of the Commission there are special circumstances which justify the granting of the prospecting licence and the Minister approves of the grant thereof to the applicant".

With regard to the grant or refusal of mining licence, the *Mining Act* also places certain provisions relating to local content. Section 46 provides as follows:

---

[27] *Guyana Geology and Mines Commission Act* (1979), Cap 65:09, Laws of Guyana, subsequently amended by the Guyana Geology and Mines Commission (Amendment) Act no. 3 of 1987, available at https://www.guyaneselawyer.com/lawsofguyana/Laws/cap6509.pdf accessed 11 July 2019.

[28] *Guyana Geology and Mines Commission (Amendment) Act* no. 3 of 1987, Laws of Guyana, available at https://parliament.gov.gy/documents/acts/8163-act_no._3_of_1987_guyana_geology_and_mines_commission_(amendment)_act_1987.pdf accessed 11 July 2019.

[29] *Forest Act*, Cap 67:01 Laws of Guyana, subsequently repealed by Forest Act no. 6 of 2009 available at https://www.doe.gov.gy/published/document/5ae8f275b4d000153ca57a90 accessed 11 July 2019.

[30] *Mining Act,* (no. 20 of 1989). Cap. 65:01, Laws of Guyana, available at https://parliament.gov.gy/documents/acts/8532-act_20_of_1989_mining.pdf accessed 11 July 2019.

> 46 (l) A mining licence in respect of any mineral shall not be granted to an applicant therefor unless the Commission is satisfied that—
>
> … (d) the applicant's proposals for the employment and training of citizens of Guyana are satisfactory;
>
> (e) the applicant's proposals with respect to the procurement of goods and services obtainable within Guyana are satisfactory….

Turning to the *Petroleum (Exploration and Production) Act* (1986),[31] which is similar to the *Petroleum Act* of Trinidad and Tobago, it too is generally outdated. Similar to the *Mining Act*, the *Petroleum (Exploration and Production) Act* includes general obligations concerning the employment and training of Guyanese nationals as well as the procurement of goods and services that can be procured locally. Note that the *Petroleum (Exploration and Production) Act* and the *Petroleum (Exploration and Production) Regulations* 1986[32] do not mention the term "local content". Section 36. (1) of the Act provides as follows:

> A petroleum production licence shall not be granted to an applicant therefor unless—…
>
> (iv) the applicant's proposals for the employment and training of citizens of Guyana are satisfactory;
>
> (v) the applicant's proposals with respect to the procurement of goods and services obtainable within Guyana are satisfactory….

4 The Main Challenges that the Extractive Resources Are Facing with Regards to Local Content Practices and Local Involvement

4.1 Case Study 1—The Incident Where BPTT Decided not to Have a Major Platform Fabricated in Trinidad and Tobago

Media in Trinidad and Tobago reported that protests by the Oilfields Workers' Trade Union (OWTU) and delays in the negotiation of a new BP/National Gas Company (NGC) contract were to blame for BP's decision not to build the Angelin platform in La Brea.[33] That was included in a statement made by former Minister of Energy, Kevin Ramnarine, who also said that the country and the community of La Brea had

[31] *The Petroleum (Exploration and Production) Act* (no. 3 of 1986), Cap. 65:10, Laws of Guyana available at https://parliament.gov.gy/documents/acts/8170-act_no._3_of_1986_petroleum_(exploration_and_production)_act_1986.pdf accessed 11 July 2019.

[32] The *Petroleum Regulations* (Legal notice 5 of 1986), made pursuant to section 70 of the Petroleum (Exploration and Production) Act, (no. 3 of 1986), Cap. 65:10, Laws of Guyana.

[33] "Union protests, delay in agreement causes of BP's decision on Angelin,' *Loop, Trinidad and Tobago* (April 6, 2017) https://www.looptt.com/content/union-protests-delay-agreement-causes-bps-decision-angelin accessed August 23, 2019.

lost out, along with many families, contractors and suppliers. Ramnarine said that while the decision taken by BP is disappointing, it should come as no surprise.

The former Minister noted that there are two issues that led to the current situation. "Firstly, the energy sector's ability to deliver projects has been severely impacted by industrial relations unrest and increasing red tape. Angelin's predecessor the BP Juniper platform endured no end of stoppages and protests".[34] He stated that the OWTU played a major role in these events, which led to schedule slippage. "As a result of the loss of time on Juniper, BP decided to move to jacket component to Texas in mid-2015. This was done to preserve the project timeline. The topside was completed in La Brea in January 2017".[35] The second issue, Ramnarine said, lies in the length of time taken to negotiate a new BP/NGC contract, which is reportedly nearing completion. He explained that these negotiations should have been completed in December 2016 if the target of first gas from Angelin was to be achieved (January 2019), as it takes two years to develop a project like Angelin. To account for the loss of time due to the late agreement, BP will now have to compress its timelines which means building the platform where it can be completed faster, Ramnarine noted.[36]

4.2 *Case Study 2—The Problem of Illegal Quarrying in the Extractive Industry in Trinidad and Tobago*

The problem of illegal mining (quarrying) has long plagued the Government of Trinidad and Tobago.[37] Despite the fact that in the past several persons were brought before the courts in Trinidad and Tobago charged with offences related to illegal quarrying the practice continues. These illegal quarrying activities have led to destruction of the environment and depletion of the nation's natural resources. Large areas of forests have been destroyed without any provision for restoration or rehabilitation. Illegal quarrying can result in detrimental impacts on citizens of the country, both directly or indirectly. Quarrying produces raw materials that are utilised in all facets of life, especially in the construction industry, communities and in various businesses in Trinidad and Tobago. Quarrying operations have significant impact on various sectors of the economy and play a major role in sustaining the livelihood of a large percentage of persons in rural and semi-rural areas, such as Sangre Grande and its environs. It is therefore important that this sector is properly regulated, so that high quality and returns are derived from extracted resources, and the country benefits from the best possible earnings. The reasons illegal quarrying occurred include the

[34] Ibid.

[35] Ibid.

[36] See also 'BPTT confirms Angelin platform will not be fabricated in T&T' *Loop, Trinidad and Tobago* (April 6, 2017) https://www.looptt.com/content/bptt-confirms-angelin-platform-will-not-be-fabricated-tt, accessed August 23, 2019, where BPTT's spokesperson said in a statement that though the decision was difficult, it was necessary to preserve Angelin's project schedule.

[37] See, Alicia (2016).

fact that it is lucrative and there are deficiencies in national enforcement mechanisms, as well as the regulatory and legislative mechanisms. The conclusion of a *White Paper on the National Minerals Policy* of 2015[38] stated that "given the current trends and challenges, especially illegal quarrying, in the minerals sector of Trinidad and Tobago, it is imperative that legislative amendments be made which will allow the proper functioning of the *Minerals Act* and its Regulations".[39]

As such, there is recognition that there is the need for the legislative framework for mining to be amended. However, like many areas of the law, while there is a general recognition of a need for law reform, there is delay in effecting the necessary changes. Chief among the pieces of legislation that must be amended is the Minerals Act. The amendment must provide for the effective prosecution of offences relating to illegal quarrying, powers of arrest without warrant, powers of forfeiture of illegally quarried minerals, as well as attaching liability to the directors of mining companies. There should also be a requirement for rehabilitation to be undertaken by quarrying operators.

Currently, the Environmental Management Authority (EMA) in collaboration with the Ministry of Energy and Energy Industries (MEEI) along with the assistance of police officers work towards eradicating illegal quarrying. The EMA also collaborates with the Minerals Advisory Committee (MAC), especially in streaming and reinforcing the Certificate of Environmental Clearance (CEC) decision-making process. There are other stakeholders that are also involved because a multidisciplinary, multi-stakeholder approach is necessary to eliminate illegal quarrying.

Both the *Minerals Act*,[40] and the *State Lands Act*[41] make quarrying without a licence illegal. The fine and imprisonment for a person who conducts quarrying without a licence were increased. Previously, the penalty for illegal quarrying under section 45(1) of the Minerals Act was a fine of TT$200,000.00 and imprisonment for a term of 2 years. Now, the fine is TT$500,000 and 5 year terms of imprisonment upon first conviction. For repeat offenders, upon a subsequent conviction, previously the fine was TT$300,000.00 and a term of imprisonment 3 years, and now, the fine is TT$700,000.00 and a term of 7 years imprisonment for this category of repeat offenders.

Offences under the *State Lands Act*, Chapter 57:01 were also amended by the *Finance Act* of 2014.[42] Under the amended section 25 of the *State Lands Act* a person who digs, wins or removes material from any State Lands without a licence is liable to a fine of TT$300,000.00 and imprisonment for 3 years upon a first conviction and a fine of TT$500,000.00 and imprisonment for 5 years upon a subsequent conviction where the material is asphalt. Where the material is other than asphalt, the fine and imprisonment are lesser. Upon a first conviction, the fine is TT$120,000.00, and upon a subsequent conviction, the fine is TT$300,000.00 and imprisonment for 3 years.

---

[38] See White Paper on the National Minerals Policy of 2015, *supra* n. 3.

[39] *Minerals Act,* Cap. 61:03, Laws of Trinidad and Tobago.

[40] Ibid.

[41] *State Lands Act,* Cap. 57:01, Laws of Trinidad and Tobago.

[42] *Finance Act*, (no. 4 of 2014), Laws of Trinidad and Tobago, sec. 6.

4.3 Case Study 3—Dissatisfaction with Draft Local Content Policy in Guyana

As stated above, in April 2017, a draft local content policy to regulate the new oil and gas industry, was prepared by the Ministry of Natural Resources, and released by the Government of Guyana.[43] This policy intends to guide or set the stage for the quantum of locally produced materials, personnel, financing, goods and services that are to be rendered to the oil industry. Unfortunately, however, the policy was greeted by severe criticisms from the local business community. The Georgetown Chamber of Commerce (GCCI), a non-governmental organisation in Guyana, is of the opinion that the Local Content Policy leaves much to be desired.[44] The GCCI is of the view that the draft local content policy makes little provision for local business owners and Guyanese in general. There are several areas where the GCCI found the draft policy to be lacking. GCCI stated that its discontent with the draft policy is because it appears there will be no regulations for local content and little provision is made for jobs for locals, as well as it not having enough objectives.

The Chamber said that the objectives should include scalable development in priority areas in order to have local personnel and goods and services supplied to operators by a local business. The Chamber thinks that the policy should also set out to ensure operators and contractors from abroad have partnering systems to enable local companies and investors to learn about oil and gas industry and its supply needs. In response to the criticisms, the government responded that the policy will be developed over time, and in tandem with the growth of the industry and said: "At present, regulations will not be promulgated, but may become necessary as the industry unfolds and expands, and impetus is needed to steer the process or to solidify gains".[45] In contrast to the government's position, the GCCI thinks that regulation on local content should be promulgated earlier in the life cycle of oil and gas capacity development.

The GCCI said that the language used in the policy tapers expectation and felt that it failed to reflect the job opportunities, both directly and indirectly that can be obtained from the new petroleum sector. Further, the GCCI is of the opinion that the policy's entire implementation strategy needs to be re-worked. "It does not spell out priority areas, near term initiatives, medium-term GAP development areas, long-term structural improvement of the state and programs to enable private sector to function as a core network of the suppliers to serve the industry". The GCCI, in its response to the Ministry of Natural Resources noted that the private sector companies either in supply of goods and services, wharfing onshore facilities, etc., can be beneficiaries in the early stages of seismic testing, drilling, well development and pre-production set up. It said too that after production starts, the government benefits from royalties.

[43] Ministry of Natural Resources (2017).

[44] Abena (2017).

[45] Ibid., statement by the Hon. Minister Raphael Trotman, Minister of Natural Resources.

The government can benefit from tax revenues once routing of supplies and services are done via Guyana and not anywhere else.

4.4 Regional and Bilateral Obligations that May Conflict with Local Content Laws and Regulations

As mentioned in the introduction, both Guyana and Trinidad and Tobago have entered into several bilateral investment treaties that create obligations, which may conflict with local content laws and policies. Local content policies are also at odds with the CARICOM treaty[46] that requires the inclusion of CARICOM nationals in local content policies. Both Guyana and Trinidad and Tobago are also WTO members and that treaty regime grants most-favoured-nation treatment to all its trading partners.

Trinidad and Tobago and the United States are party to a bilateral agreement on the Encouragement and Reciprocal Protection of Investment that the Ministry of Energy and Energy Affairs[47] acknowledged would need to be partially renegotiated to be consistent with the government's local content policy. Under the bilateral investment treaty between the *Government of the United Kingdom and Guyana* Article 3 on the National Treatment and Most-favoured-nation Provisions and provides as follows:

> (1) Neither Contracting Party shall in its territory subject investments or returns of nationals or companies of the other Contracting Party to treatment less favourable than that which it accords to investments or returns of its own nationals or companies or to investments or returns of nationals or companies of any third State.
>
> ...
>
> (3) The provisions of this Article relative to the grant of treatment not less favourable than that accorded to nationals Of Companies of either Contracting Party shall not prevent either Contracting Party from according special incentives to its own nationals or companies in order to stimulate the creation of local industries, provided those incentives are considered carefully on each occasion by the Contracting Party concerned, which shall in particular, ensure that they do not significantly affect the activities or investments of nationals or companies of the other Contracting Party.[48]

As the above highlights, States need to be aware that they can confront significant problems posed by local content laws and regulations under private international law. As part of a strategy to attract investment, in the formative years of industrialisation by invitation, many developing countries entered into bilateral investment treaties with developed countries. Also, the economic might of multinational corporations

---

[46]See Chap. 3 on the Establishment, Services, Capital and Movement of Community Nationals in the *Revised Treaty of Chaguaramas supra* n. 5.

[47]*Treaty between the Government of the United States of America and the Government of the Republic of Trinidad concerning the Encouragement and Reciprocal Protection of Investment,* https://www.sice.oas.org/bits/triusa_e.asp, accessed August 23, 2019.

[48]*Agreement between the Government of the United Kingdom of Great Britain and Northern Ireland and the Government of the Co-operative Republic of Guyana for the Promotion and Protection of Investments* (London, 27 October, 1989) entered into force 1990.

led to the inclusion of many of these clauses in contracts with host governments and producer states, particularly in the oil industry.

Additionally, a common feature of the contractual and legal relationship between host governmental and multinational investment companies is resort to international arbitration to protect the foreign investor. The International Centre for the Settlement of Investment Disputes ("ICSID"), which was established in 1966 in Washington DC under the auspices of the World Bank, is such an international arbitral tribunal. It seeks to promote an atmosphere of mutual confidence between States and foreign investors conducive to increasing the flow of private international investment for development. The ICSID is established under the *Convention of the Settlement of Investment disputes between States and Nationals of Other States*, which entered into force on October 14, 1966. Eight Caribbean states have ratified the conventions, including Guyana and Trinidad and Tobago[49] and two others have signed.[50]

5 Conclusion

Currently the laws regulating local content in Trinidad and Tobago and Guyana can be considered outdated. The main *Petroleum Act* in Trinidad and Tobago was drafted in the 1960s and the Guyana Act, which has less details than the Trinidadian Act, was enacted in 1986. As mentioned above, both the Guyana *Petroleum (Exploration and Production) Act* 1986 and the Guyana *Petroleum (Exploration and Production) Regulations* 1986 do not mention the term "local content", but the laws and regulations imply local content and make general reference to the employment and training of nationals as well as the procurement of goods and services that can be procured locally. We saw that in both jurisdictions, they have several pieces of legislation and various policy documents, which fill the gaps and help to augment the legislative and regulatory framework governing the local content policies in the various extractive industries.

While there are various tools that a State can adopt to implement local content such as quotas, training programs, employing local work force and fiscal incentives, it is posited that there are lessons to be learnt from various jurisdictions that have legislated very specific local content requirement and others that have very vague local content policies. For instance, there are very specific local content requirements in the *Indonesian Bill on Mineral and Coal Mining* of 2008 which requires all companies to process and refine mining products in Indonesia and the *Nigerian Oil and Gas Content Development Act* of 2010 has provisions to enhance local participation in all aspects of oil operations, including the following: 65% of divers in offshore energy

[49]The Convention entered into force for the following countries on the dates indicated: Bahamas, November 18, 1995; Barbados, December 1, 1983; Grenada, June 23.1991; Guyana, August 10, 1969; Jamaica, October 14, 1966; St Kitts & Nevis, September 3, 1995; St Lucia, July 4, 1984′ Trinidad and Tobago, February 2, 1967.

[50]The Convention was signed by the following countries on the dates indicated: Belize, December 19, 1986; St. Vincent and the Grenadines, August 7, 2001.

projects must be Nigerian; 60% of steel ropes used in projects must be made locally; all contracts awarded in excess of $100 million must include a "labour clause", mandating the use of minimum percentage of Nigerian labour or the use of indigenous companies of a minimum size, etc. On the other hand, some jurisdictions have more relaxed provisions, which give more discretion to the contractors. The *Afghan Amu Daya Basin contract* requires that the "contractor agrees to as far as possible train and employ qualified Afghan nationals…and…will undertake the schooling and training… The contractor will require the contractors and subcontractors to do the same", and the *Timor-Leste's production sharing contract* for Area A stipulates that "the Contractors shall draw to the attention of suppliers based in Timor-Leste, in such a manner as the Ministry agrees, all opportunities for provision of good and services in petroleum operations".

The legal and regulatory framework in Trinidad and Tobago and Guyana fall somewhere in the middle of the two examples cited above. They have laws that are very general, and then, the contractual arrangements and various policy documents tend to lay out more details. While this is understandable, the problem with having very specific legislation is that the countries may not have the capacity to fulfil the requirements, which can lead to delays in projects and might deter investments. The Trinidad and Tobago's Vision 2020 Energy Subcommittee Report[51] recognises the importance of local content and outlines specific objectives and initiatives to achieve local content goals. Nevertheless, it is recommended that governments must have clear communication with investors and the public regarding the bodies responsible for implementing and undertaking various obligations in this area. Further, with regard to law reform and policy reform in these areas, it is recommended that the implementation of the new local content laws and policies must not be left mostly to the private sector. It is submitted that local content policy and law must be based on certain core principles and has to be linked to the wider national economic development goals of the country. It is also submitted that governments must seek to establish policies, which are easily adaptable and procurement strategies must be integrated early in the development and planning stages of extractive projects.

From the first case study concerning the incident where BPTT decided not to have a major platform fabricated in Trinidad and Tobago amid protest and delays, many lessons can be learnt. It must be highlighted while the Government in some jurisdictions may pass laws or issue regulations on the local content requirements for major construction and infrastructural development relating to oil and gas exploration projects to help improve domestic production; however, the government must also ensure that the necessary skilled persons and workforce are available. In these types of situations, the government must also ensure that there is the supply of various goods to ensure that projects are completed in a timely manner in accordance with international standards.

The second case study on illegal quarrying demonstrates that apart from the energy sector, other extractive industries are in need of legislative reform and support from

[51] Vision 2020 Energy Sub-committee (2004) *Vision 2020 Energy Report*, Trinidad and Tobago.

the government to ensure that laws are implemented to safeguard sustainable management practices. Indeed, the time is ripe for amendments to be made to the *Minerals Act* and regulations to help to effectively eradicate illegal quarrying in Trinidad and Tobago. This will also support the proper management and regulation of the mining sector and help to promote sustainable development of the nation while ensuring that the State receives its revenue due from the exploitation of its natural resources.

The third case study concerning the reaction to local content policy in Guyana illustrates that clear communication with the public is critical. The energy sector and other extractive industries does not have the capacity to create the volume of jobs required to make significant changes to employment demographics in Trinidad and Tobago and Guyana. Hence, public expectations must be carefully managed. The expectations of the GCCI are unrealistic. While they have general dissatisfaction with the draft policy and regulatory framework for local content development, the reality is that many of the extractive industries are not a great *direct* employer. The oil and gas sector in particular is not a great *direct* employer. For instance, in Nigeria, employment in the sector is less than 1% of direct employment, in Namibia the mining sector is about 1% and in Trinidad & Tobago it is about 3.3% of the workforce, before the closure of the national oil refinery company (Petrotrin).[52]

A final important factor which must be highlighted is that the local content laws, policies or regulations that the government decides to adopt and implement must not be developed in isolation but must be streamlined to enhance the countries' ease of doing business and develop its international portfolio. From the point of view of international investors, Guyana is ranked 134, and Trinidad and Tobago is ranked 105 out of 190 countries surveyed and ranked on the World Bank's Ease of doing business ranking.[53] The report compares the business regulation for domestic firms in 190 economies against some international benchmarking standards. Having clear local content laws and policies as well as relevant data on local content implementation publicly available could enhance a country's raking in this survey.

References

Abena, R.-C. (2017, June 14). Oil and gas development…Draft local content policy offers little or nothing to Guyanese—GCCI. *Kaieteur News*. https://www.kaieteurnewsonline.com/2017/06/14/oil-and-gas-developmentdraft-local-content-policy-offers-little-or-nothing-to-guyanese-gcci/. Accessed February 19, 2019.

Alicia E.-R. (2016, July 5). Illegal quarrying. *The Trinidad Express*. https://www.trinidadexpress.com/news/local/illegal-quarrying/article_c4ae11d5-f431-50d9-b8db-5207af4786eb.html. Accessed June 14, 2019.

[52]These figures were reported before the collapse of the national oil company, Petrotrin, in Trinidad and Tobago. See Ramdass (2018).

[53]*Doing Business 2019—Training for Reform,* A World Bank Group Flagship Report, 16th edition (2019); See also Stedman and Green (2018).

Ghourlal, D. (2018, November 30) Petrotrin closed; $2.7 billion paid out to workers. *Loop, Trinidad and Tobago*. https://www.looptt.com/content/petrotrin-closed-27-billion-paid-out-workers. Accessed August 23, 2019.

Government of the Republic of Trinidad and Tobago Ministry of Energy and Energy Industries. (2004). *Local content and local participation policy and framework*' (2004). https://www.energy.gov.tt/local-content-local-participation-framework/. Accessed August 23, 2019.

Government of the Republic of Trinidad and Tobago Ministry of Finance. (2019). Major Investments Incentives in Trinidad and Tobago. https://www.finance.gov.tt/wp-content/uploads/2017/08/MOF-Investment-Incentives-in-Trinidad-and-Tobago-web.pdf. Accessed August 23, 2019.

Ministry of Natural Resource. (2017, April). *Making the most of our oil and natural gas—Maximising benefits and value retention from guyana's petroleum resources through capacity development, local content and value addition: A policy framework—Working draft for discussion and input*. https://www.doe.gov.gy/published/document/5af7293bdc677720ccdc33ba. Accessed October 11, 2019.

Ramdass, N. (2018, 29 August). Petrotrin's refinery shutdown explained (and why it was done). *Loop, Trinidad and Tobago*. https://www.looptt.com/content/petrotrins-refinery-shutdown-explained-and-why-it-was-done. Accessed July 11, 2019.

Sant, R. (2018, September 28). Petrotrin to shut down November 30. *Trinidad and Tobago Guardian Newspapers*. https://www.guardian.co.tt/news/petrotrin-to-shut-down-november-30-6.2.678617.781afbbf9e. Accessed August 23, 2019.

Stedman, A., & Green, K. P. (2018). *Fraser Institute Global Petroleum Survey*. Available at https://www.fraserinstitute.org/sites/default/files/global-petroleum-survey-2018.pdf. Accessed July 11, 2019.

The Oil & Gas Year Report. (2016). Trinidad and Tobago 2019: The oil & gas year: Information is power. At https://www.theoilandgasyear.com/market/trinidad-and-tobago-2/. Accessed February 19, 2019.

Local Content Within Extractive Resources Industry in the Russian Federation

Olga S. Kirillova

1 Introduction

The Russian Federation (the "RF") is considered to be a non-OPEC oil producer, which actually influences the oil market significantly due to high oil production volumes. It is possible to conclude that the oil and gas industry plays a great role in the Russian Federation economy. It is a well-known fact that the RF is rich in natural resources and the federal state budget relies heavily on the sale of oil and gas. According to the RF Prime Minister Dmitri Medvedev (2017), the Russian budget has reduced its reliance on the sale of oil and gas, and its revenues from other sectors now account for more than half. Nevertheless, obviously, with such a strong dependency on the tax earnings from the export of raw materials significantly influences state policy and places a special role on the companies operating in the oil and gas industry sector.

Taking this into account, it is no surprise to conclude that the largest Russian companies operating in the energy sector are: Rosneft Oil Company, Gazprom JSC, Novatek JSC, Transneft JSC, Lukoil JSC, Zarubezhneft JSC, Surgutneftegas JSC, SIBUR JSC. According to a report from Skolkovo Energy Center (2018), the RF oil and gas companies are very dependent on foreign technologies and therefore sanctions applied by the USA and EU during 2014–2018 significantly influenced the development of the Russian energy market. Nevertheless, despite the sanctions, international cooperation between the Russian energy companies and international corporations is still happening, but not within the area of sanctioned projects (i.e. shale, deep water and Arctic projects).

O. S. Kirillova (✉)
ZN Vostok, Ltd, Lukinskaya Street, 8, 119, Moscow, Russia
e-mail: olga.s.kirillova@gmail.com

E. G. Pereira et al. (eds.), *Sovereign Wealth Funds, Local Content Policies and CSR*, CSR, Sustainability, Ethics & Governance,
https://doi.org/10.1007/978-3-030-56092-8_20

Nevertheless, despite all the above-mentioned restrictions imposed by the USA, EU and even the Russian state authorities, joint projects are still considered to be one of the most attractive ways for cooperation between the major Russian oil and gas companies and their foreign counterparts. The main reasons for Russian companies to enter into joint projects with foreign partners include: technology transfer, experience sharing, new difficult project execution, local staff development and education, production localisation, capital intensity, risk sharing (Ernst and Young report 2017). Hereinafter, the chapter reviews the RF peculiarities regarding the local content requirements ("LCRs") from theoretical and practical perspectives as they significantly influence foreign investor operations in the RF extractive resources industry.

2 Local Content Requirements Within the Extractive Resources Industry in the Russian Federation

The Russian energy sector consists of the following subsectors: oil industry, gas industry coal industry and electric power industry (atomic, hydropower). The main legal framework for the extractive resources centre on the following documents:

1. Civil Code of the RF;
2. RF Federal Law N 69-FZ dated 31.03.1999 "On gas supply in the RF";
3. RF Federal Law N 147-FZ dated 17.08.1995 "On natural monopolies in the RF";
4. RF Federal Law N 135-FZ dated 26.07.2006 "On the protection of competition";
5. RF law N 2395-1 dated 21.02.1992 "On subsoil";
6. RF Federal law N 225 dated 30.12.1995 "On production-sharing agreement".

There is no specific legislation regarding the LCRs in the Russian Federation. Nevertheless, usually the LCRs mean the requirements expressed in the percentage of the local goods, local employees, local raw materials to be used in the manufacturing process for the purpose of creating value addition to the local economy.

It is necessary to note that sometimes the LCRs are mixed with the import substitution industrialisation ("ISI" or localisation) model. Due to the absence of the specific legislation and normative acts regarding the LCRs in the oil and gas industry, it is possible to assume that localisation can also be considered part of LCRs.

It is interesting to note that the value and importance of the LCRs (including localisation) can be illustrated by the experiences of Rosneft Oil Company and Gazprom. Since 2014, each annual report of Rosneft Oil Company contains a chapter devoted to equipment and technology localisation. Gazprom has also developed a set of measures, which are intended to replace foreign procurement of goods, works and services. This reflects the willingness of oil companies, on the one hand, and growing opportunities of domestic engineering enterprises, on the other hand, to build a long-term relationship (Semykina 2017).

Analysis of the latest joint projects indicates that the majority of the onshore projects have being implemented in the area of tight oil exploration. Actually, Russian oil and gas companies are able to develop the usual (light oil) fields by themselves on the bases of the corresponding licences and using available resources and technologies. On the contrary, possibility of exploration and production of the difficult-to-extract hydrocarbons projects depends on specific technologies and know-how that are used by international oil and gas companies worldwide but not available to the Russian market.

As it was stated by the US Energy Information Administration (2015), the majority of Russia's current oil production (nearly two thirds) comes from the Western Siberia field and Volga-Urals region. Nevertheless, due to the fact that these oldest fields have been producing since 1940s, presently they are declining even with the new technologies and focus on secondary recovery and hydrofracturing. Therefore, Russian oil companies are very interesting in the drilling and completion techniques used in the USA for the purpose of development of their unconventional oil and gas resources.

The RF unconventional oil reservoirs are huge and still not developed due to the lack of experience and technologies. For example, such technological solutions as hydrofracturing and coiled tubing are necessary for tight oil exploration but well-known only to international companies (i.e. Weatherford, Baker Hughes, Halliburton, Schlumberger). Russian companies do not have enough experience in such operations (Skolkovo Energy Center 2018). For example, at Vankor field (one of the largest oil fields in West Siberia), five years ago most components and equipment used in the project came from abroad: top drive drilling by Canrig or Varco (USA), chisels by Hughes Christensen (USA), solutions for horizontal wells of MI-SWACO (USA), logging technology MWD and LWD (USA) (Semykina 2017). Horizontal drilling is performed by Schlumberger Company, which has its own repair base for the maintenance and restoration of drilling systems on Vankor. Local procurement mostly comes from the so-called industrial upstream including building materials, metalwork, electric poles, sawmill products, cement and so on (Semykina 2015).

Therefore, it is logical that Russian companies are interested in cooperation with international corporations with respect to joint project implementation in the area of tight oil exploration. As it was stated by the Russian Council on International Affairs (2018), the foreign service companies have about 25% of the corresponding Russian market of oilfield services (e.g. they have significant market share in such critical areas as fracture stimulation (90%), seismic interpretation (50%) and horizontal drilling (25%)).

Taking into account all above-mentioned, it is possible to summarise that the main areas of cooperation regarding the local content usage by foreign investors are creation of the joint ventures with the local partners and localisation process (R&D and production of goods).

It is also necessary to mention the production-sharing contract (hereinafter—the "PSA") as a way of cooperation between the foreign investors, the RF and Russian companies. There is a special law "On production-sharing agreements" as of 1995. According to this law, for the purpose of PSAs implementation the Russian companies have the prevailing rights to be the contractors, Russian citizens have prevailing rights to be hired as employees and locally manufactured goods shall be used (article 7 of the mentioned law). Nevertheless, PSAs as a form of cooperation with the foreign investors are not very popular in the RF (the most famous PSAs implemented in the RF are the following: "Sakhalin-1" signed in June 1995, "Sakhalin-2" signed in June 1994 and "Kharyaga" signed in December 1995). According to the PSAs, the specific tax regimes shall be applied for the participants. For example, participants of "Sakhalin-1" shall pay profit tax in the amount of 35%. Nevertheless, the RF Ministry of Finance does not consider the PSAs as beneficial for the Russian economy (Kozlov and Zanina 2018).

3 Experience of Local Content Requirements in the Russian Federation

3.1 Localisation Aspects in the RF

At present, the RF Government has implemented a strict policy for the purpose of encouraging foreign investors to localise R&D and production of their goods in the RF and for the purpose of stimulation of Russian companies to develop their competences. For example:

(i) The Ministry of Industry and Trade of the Russian Federation has adopted special rules regarding recognition of the goods produced in the RF;
(ii) The special Centre of the competences for the technological development of the import substitution was established by the Ministry of Industry and Trade of the Russian Federation in December 2018.
(iii) The special Fund of Industrial Development was established for the purpose of soft financing. For example, the RF Chamber of Commerce together with the Fund of Industrial Development supports municipal and regional investors (i.e. 351 projects were financed). The loans for such projects are provided for 7 years under 2–5% annual interest with the minimum initial financing required from the municipal and regional investors.

There are also several special programmes developed for the purpose of cooperation between foreign investors and local companies [i.e. preferential taxation, single source procurement for the purpose of special investment contracts ("SICs")]. The most active and positive dynamic in the area of localisation is in the automobile industry in the RF. About 85% of cars are produced in the RF. The power plant industry also has several examples of localisation (i.e. the manufactory for the purpose of turbine blade repair based in Siemens technology opened in 2018).

The SICs can be offered to foreign investors for the purpose of localisation of the production of their goods. According to the available information, about 25 SICs have been signed for the purpose of localisation stimulation (6—in pharmacy industry, 4—in automobile and chemical industries, 3—in oil-and-gas machine building, 2—in agricultural machinery industry and metallurgy, 2—in power engineering industry, machine tool manufacture, aircraft construction and pump equipment manufacturing). It is expected that realisation of the above mentioned SICs will gain about 410 billion RF Rubles in the budget, creating about 9900 new working places and the volume of the produced goods exceeding 4 trillion and 177 billion of the RF Rubles.

A special plan regarding localisation purposes was developed by the RF Government and according to this plan the dependence from the foreign goods and equipment in the industrial area shall decrease to 43% by 2020. According to the information available from the RF Ministry of industrial trading (2018), the level of dependence from the foreign technologies has already decreased from 56 to 52% in industrial manufacturing. It was planning to achieve the zero level of the dependency from the foreign technologies in the oil and gas industry by 2020, but this goal was unrealistic.

According to the opinion of the RF expert Arkhipov (2018), the purpose of localisation is to push foreign investors to incorporate R&D centres in the RF, to launch local production of goods and equipment, to implement the technology transfer and experience sharing; only such an approach could guarantee localisation rather than just assembly plants. Therefore, on the basis of the above mentioned, it is possible to illustrate the following successful samples of the localisation process in the RF:

1. Localisation samples in Upstream areas

    1.1. *Creation of the joint R&D centres and local manufacturers*

It is possible to mention the "AETC Sapphire" Limited Liability Company (LLC) and "Advanced Research and Technology Centre" LLC, the joint ventures created by Rosneft Oil Company and General Electric for the purpose of conduction research and development operations and local production in the RF:

(i) On 21 June 2013, Rosneft Oil Company and General Electric entered into a Strategic Cooperation Agreement pursuant to which the parties agreed, inter alia, to pursue a multifaceted, strategic cooperation relationship for the purposes of leveraging existing technologies of the General Electric Company and its affiliates to help Rosneft improve efficiency and yields in its assets, and exploring opportunities to jointly develop, manufacture, test, assemble, sell and market equipment for use in the oil and gas industry in the Russian Federation;
(ii) the joint ventures and its subsidiaries shall conduct, among other things, certain business activities and initiatives in the Russian Federation in any areas that may be agreed by the shareholders of the Joint Venture through a process to be agreed by the parties. Specifically, the parties intend to mutually develop a project that contemplates the production of wellhead equipment and Christmas trees (flowing wellhead equipment) for specific exploitation conditions in the territory of the Russian Federation;
(iii) Parties shall consider and explore the possibility of investing in a project that would comprise construction of a plant for production and/or assembling

of wellhead equipment and Christmas trees on the territory of the Russian Federation;
(iv) The Parties intend to reach the level of the equipment production with a target of localisation seventy per cent (70%) by 2025.

1.2. *Local production of the machineries*

The special road map was developed by the RF Ministry of Foreign Trade according to which the level of dependency from the foreign machines has to decrease significantly in the nearest years. Since 2015 until 2019, about 90 projects in the area of localisation of equipment production were supported by the RF Government for the total amount of approximately 14 billion RF Rubles (Uchenov 2018).

1.3. *Local producing of proppant*

One of the examples of the localisation of the production process is manufacturing of the proppant for the completion operations in the well. It is interesting to note that this did not exist in the RF 10 years ago; all proppant was bought from the foreign companies. Nevertheless, for now due to Gazproneft's efforts, proppant has been successfully produced in the RF and used during the completion process in the RF and even abroad.

1.4. *Development of local software*

Usage of the high-quality software is critical in the Upstream area. For many years, Russian companies had to use foreign software due to absence of the levant local one. Nevertheless, as of today, there is available local Russian software for 3D modelling of the hydro-fracking operations that is actively being used by Russian companies. The RF Security Council stated in 2018 that development of the local software and decreasing dependency on foreign companies are extremely important for the RF and even influence the country's security.

2. Localisation samples in the refinery area

Local production of the catalysts and additions can be considered as successful examples of the localisation of the goods' manufacturing in the refinery area. Earlier most of the catalysts and additions were bought from foreign suppliers. As for today—Rosneft Oil Company has been successfully producing its own catalysts at Angarsk refinery plant. Gazpromneft has been also launching its own catalysts production at Omsk refinery plant. As for the additions production, Transneft is planning to start their production in Alabuga.

3.2 Joint Ventures Specific in the RF

On the basis of the information available, it is possible to say that joint projects can be considered a very attractive way of developing business relationships in Upstream projects (i.e. projects related to the oil and gas exploration and production). Many international oil and gas companies are interested in international communication and business development. There are many joint ventures (JVs) in the RF created by the major participants of the international oil and gas market, i.e. ENI (Italian oil company), Exon Mobile (USA oil company), Repsol (Spain oil company), Equinor (Norwegian oil company, ex Statoil), British Petroleum (English oil company), with their local partners, i.e. Rosneft Oil Company (Russian oil company), Zarubezhneft (Russian oil company), Gazpromneft (Russian oil company).

It is possible to indicate the following benefits of the joint projects in comparison with the other forms of international business cooperation:

(i) *Capital intensity*

Oil and gas projects are considered very expensive, therefore, possibility of foreign investment guarantees the additional funding and resources to a project. It is interesting to note that according to the latest trends, some of the JVs are originally being funded only by the foreign partner under condition of returning the money later during the operational phase in case of a project's success.

For example, most of the Russian Federation JVs between Rosneft and its foreign partners (i.e. ENI, Equinor, BP) are originally funded by the foreign partners according to the shareholders and operating agreements. Given drilling operations are very expensive (i.e. cost of one exploration well could cost up to 20 million USD), such partnerships in JVs are strategically important for the foreign partners.

(ii) *Risk mitigation*

It is necessary to note that many JVs are created for the purpose of developing exploration projects in the oil and gas sector. From a practical perspective, such projects are considered to be very risky and only 30% of the projects are successful (Ernst and Young report 2017). It is quite often, that a single company does not want to invest solely in an unproven technology or region. In such cases, it is more reasonable to share risks and costs correspondingly.

Most of the JV projects in the RF are structured in a way that foreign partners invest funds in the first stage of the project implementation and later the local partner will compensate such costs proportionally in case of success according to the agreed terms and conditions.

(iii) *Technology transfer*

Usually, the main motivation for a JVs with a foreign partner is the opportunity for technology transfer. The level of oil and gas technologies development is very different around the world, and therefore, local oil and gas companies in developing countries are very interested in cooperation with foreign partners who have modern and effective technologies and equipment.

This international cooperation is relevant to the projects in the area of difficult-to-extract oil because a lot of contemporary technologies are required for the successful implementation of such projects (i.e. coil tubing, frack equipment, etc.). For example, project agreements between Rosneft and Equinor have special conditions regarding experience sharing, technology transfer, local staff education.

(iv) *Access to resources*

It is a common practice that the JVs shareholders delegate the staff to the project on the basis of the corresponding staff services agreements and/or general services agreements. Because it is extremely important for the JVs to have an experienced and qualified staff, such an approach is quite useful for the projects.

For example, usually it is stated within the project agreement that each participant shall provide educated and experienced staff according to the agreed organisational chart and allocation of management positions. It is necessary to note that there are special requirements in the RF regarding hiring of foreign employees. The general rule is that the RF Government approves the quota for foreign employees who can work in the RF. Nevertheless, if the foreign employee is considered as highly qualified based on income, education and experience, that candidate can be hired on special conditions from the quota approved by the RF Government. For local partners, it is also very valuable to develop its staff through cooperation with the representatives of the foreign partners and via mutual educational programs and trainings.

(v) *Supply chain optimisation, market positioning and scale*

Taking into account that the projects are usually implemented in very tight timely conditions, it is quite important for the JV to be able to use its shareholder privileges and benefits during negotiations with potential contractors. Such negotiations could help to decrease the costs and production time significantly. For example, it is a common practice to attract the local shareholder representatives to the procurement process implemented by the JVs. This approach could help to secure better commercial offers from the tender participants and to decrease price significantly. Usually, the local partner is a strong local market player, has reliable business reputation and relies on its local network, experience and knowledge of the local market. In case of efficient strategy JVs can develop a strong market position for the purpose of its potential business development. For example, while planning the procurement process and project implementation, JVs rely significantly on the information provided by the local partner, which is aware of the local country context and can share important information.

(vi) *Awareness of the local norms and requirements*

It is necessary to say that all countries have their local regulations regarding the possibility of the foreign company to take part in the local oil and gas projects. In the case of local partner availability, it is possible to get the qualified assistance and advice regarding this local requirement. Usually, while structuring the project it is very important to develop the right legal scheme and order of cooperation.

For example:

- three offshore projects between Rosneft and Equinor were structured in a way when the JVs was a foreign legal entity with a branch in the Russian Federation,
- one onshore project was structured in a way when the JVs were established in the Russian Federation as a Russian legal entity.

All these constructions are possible according to Russian legislation.

(vii) *Political sensitivity*

All international projects in the oil and gas sector are very political. Many decisions regarding a projects' implementation and relevant partners are taken at the top level. At the same time, changes in political forces in the host country, such as new sanctions, could significantly influence the projects' implementation. Unfortunately, many of the JVs were put on hold due to sanctions imposed in 2014 by the USA and EU.

Usually, the JVs communication is considered to be more palatable to government, labour groups and communities. According to a report by Skolkovo Energy Center (2018), the latest international joint projects were implemented in the areas of onshore tight oil exploration and offshore deep water. Table 1 presents the latest main international joint projects implemented in the RF.

On the basis of the abovementioned, it is possible to conclude that the main international oil and gas companies that take part in the joint projects are Exxon Mobil, ENI, Equinor (ex. Statoil) and BP. The main areas of international cooperation are Upstream projects (offshore and onshore difficult-to-extract hydrocarbons), where foreign technologies are extremely important for the successful implementation of projects. According to the forecasts of Gazprom (2018), if the new technologies are successfully implemented, the share of the Russian tight oil production is expected to be 2.5 million barrels per year by 2025.

4 Conclusions

Taking into account all above mentioned, it is possible to summarise the following:

(i) There is no specific legislation regarding the local content requirements within the extractive resources in the RF. The general understanding under this term is the following: usage of the local goods and products, technology transfer, employment of the local staff, new working places' creation, and localisation of R&D and production.

(ii) Joint ventures between Russian companies and foreign companies are considered very attractive for the development of localisation.

(iii) The consequences of the LCRs in the Russian Federation are not so unequivocal. Advantages include the growth of production and gaining benefits; nevertheless, reduction of foreign investments, increasing costs for buyers can be considered as disadvantages.

Table 1 Main international joint projects implemented in the RF

Project	Parties	Status
Offshore projects		
Well "Universitetskaya-1"	JV between Rosneft (51%) and Exxon Mobil (49%)	Delayed due to sanctions
Vostochno-Prinovozemelskiy-1,2,3; Severo-Karsky; Ust-Olenekskiy; Ust-Lenskiy; Aninsko-Novosibirskiy; Severo-Vrangelevskiy – 1,2,3; Uzhno-Chukotskiy Tuapsinskiy arch	JV between Rosneft (67%) and Exxon Mobil (33%)	Exxon left projects due to sanctions Rosneft decided to implement by itself
Barents Sea and Black sea	JV between Rosneft (67%) and ENI (33%)	Delayed due to sanctions
Onshore projects		
Bazhenov and Achimov formations in the West Siberia	JV for Pilot Phase between Rosneft (51%) and Exxon Mobil (49%)	Delayed due to sanctions
Domanik sediments in Orienburg	JV for Pilot Phase between Rosneft (51%) and BP (49%)	Delayed due to sanctions
Development of Bazhenov formation in Khanty-Mansiyski Autonomy district	JV between Lukoil and Total	Total transferred its share to Lukoil
Development of Bazhenov formation in Khanty-Mansiyski Autonomy district	JV between Shell and Gazpromneft	Shell stopped operations
Domanik sediments in Samara region	JV for Pilot Phase between Rosneft (51%) and Equinor (49%)	Pilot phase is going on
North-Komsomolskiy field	JV for Operational Phase between Rosneft (77%) and Equinor (33%) ("Sevkomneftegaz" LLC)	Operational Phase started in 2017

(iv) LCRs and localisation within the extractive resources sector are considered very important by the RF Government, which supports them by way of different types of the state measures.

(v) Local content in the RF was developed due to the sanctions because business had to find alternative options to progress without foreign support. Different joint ventures have been incorporated in the RF between the Russian companies and their foreign partners for the purpose of R&D and local production of goods.

References

British Petroleum. (2018). *Energy outlook* [Online]. Available from https://www.bp.com.

Deloitte. (2018). *2018 outlook on oil and gas* [Online]. Available from https://www2.deloitte.com.
Ernst & Young. (2014a). *Spotlight on oil and gas megaprojects* [Online]. Available from https://www.ey.com.
Ernst & Young. (2014b). *Navigating geopolitics in oil and gas* [Online]. Available from https://www.ey.com.
Ernst & Young. (2017a). *Joint ventures for oil and gas megaprojects* [Online]. Available from https://www.ey.com.
Ernst & Young. (2017b). *Global oil and gas transactions review* [Online]. Available from https://www.ey.com.
Ernst & Young. (2018). *Global oil and gas transactions review* [Online]. Available from https://www.ey.com.
Kazakova, M. V. (2010). *Dependence of Russian budget from raw materials export*. Gaidar Institute for Economic Policy [Online]. Available from https://www.iep.ru/en/dependence-of-russian-budget-from-raw-materials-export.html.
Kozlov, D., & Zanina, A. (2018). *Sakhalin-1 gets new life*. Commersant [Online]. Available from https://www.kommersant.ru.
Pertsova, V. (2018). *Oil under sanctions*. Forbes [Online]. Available from https://forbes.ru.
PWC. (2017). *Making energy industry joint ventures work toward improved governance and decision making* [Online]. Available from www.pwc.com. Accessed August 30, 2018.
Semykina, I. (2015). Opportunities and benefits of local content requirement policy: case of Eastern Siberian oil and gas industry. In *55th Congress of the European Regional Science Association: "World Renaissance: Changing roles for people and places"*, Lisbon, Portugal, August 25–28, 2015. Louvain-la-Neuve: European Regional Science Association (ERSA) [Online]. Available from http://hdl.handle.net.
Semykina, I. O. (2017). Managing regional economic development through local content requirements in oil and gas industry. *Ekonomikaregiona [Economy of Region], 13*(2), 457–464 [Online]. Available from https://www.econstor.eu.
Simola, H., & Laure, S. (2017). *Overview of Russia's oil and gas sector*. Bank of Finland, BOFIT Institute for Economies in Transition [Online]. Available from https://helda.helsinki.fi.
Skolkovo Energy Center. (2018). *Perspectives of the RF oil and gas industry under sanctions* [Online]. Available from https://energy.skolkovo.ru.
Uchenov, A. (2018). *Participation in localization projects helps development of the Russian companies* [Online]. Available from https://rg.ru/2018.

Alaska's Tug-of-War on Land Rights

Douglas B. Reynolds

Abstract Alaska's petroleum industry is run by national and international oil companies (IOCs) coming to Alaska with their workers and expertise to develop and produce oil. However, a more pertinent issue surrounding the petroleum industry in Alaska is who owns and who controls the land of Alaska including its mineral rights. Alaska's land encompasses about 360 million acres (146 million hectares) or about the size of Germany, France, Italy, and the United Kingdom combined, but with a population of only 700,000 people. After subsequent treaties and laws, the land has become split between Native Alaskan land, U.S. Federal Government land, and Alaska State land where by much of the splitting only occurred after the discovery of the great Prudhoe Bay oil field on Alaska's North Slope in late 1968. The story of how the land was split and how much of Alaska's land has been made into nature preserves, which have been saved from mineral rights development, is quite a story. The different interests of Americans, Native Alaskans, and average Alaskan citizens have conflicted with each other every step of the way and have affected land rights and even some business practices including how Alaska's native companies work.

Alaska became, "the greatest conservation fight in history."
Morris Udall, October 8, 1982

"The most fragile thing up at Prudhoe Bay is not the environment,
It is man!
It is tough up there … (and) environmentalist do not understand."
Governor of Alaska, Wally Hickel, 1969

D. B. Reynolds (✉)
Department of Economics, School of Management, University of Alaska, Fairbanks, USA
e-mail: dbreynolds@alaska.edu

E. G. Pereira et al. (eds.), *Sovereign Wealth Funds, Local Content Policies and CSR*, CSR, Sustainability, Ethics & Governance,
https://doi.org/10.1007/978-3-030-56092-8_21

1 Introduction

There are three prevalent social groups in Alaska all with different needs, different desires for Alaska, and different ideas about how Alaskan lands should be used. Those groups that comprise Alaskan society include: (1) First, there are the Alaskans who live in the bigger Alaskan towns and cities, the population of which comprise only about 600,000 people. (2) Then, there are many first nation or native culture peoples of Alaska, called Native Alaskans, many of whom live in small villages in and around Alaska's vast bush (frontier) regions, although many of whom also leave the villages and live in the cities, and who still identify with the villages that they left behind. Indeed, they often go back to visit their home villages. Native Alaskans comprise about 100,000 Alaskans depending on how their ethnicity is counted. (3) Finally, the last group of people, who are one of the least understood groups, and yet who are unusually heavily involved in Alaska, and who do not even reside in Alaska, are the so-called Lower-48ers.[1] who are the bulk of the United States' 325 million citizens. Lower-48 Americans are not often thought about when you think of Alaska since these Americans live in the contiguous United States of America, often called the Lower-48 states, and where Lower-48 Americans are often called Lower-48ers for short. These Lower-48 Americans have a lot of opinions about how Alaska should or should not be run, especially in regard to environmental standards, wildlife management, and economic development.

While many first nation or native peoples of Alaska look at Alaska's regions as ancestral native lands that rightfully belong to them, but which through treatise, laws, and decrees now belong to the U.S. Federal Government or the State of Alaska's Government (Alaska), or to some Native Alaskan villages, nevertheless, they do not have a common view of Alaska. Native Alaskans have varying views on land usage such as whether saving a certain wildlife population is important or not. Some tribes may want to protect wildlife populations of certain animal species while others will want more economic development of lands that could detrimentally affect certain wildlife populations even as the development creates a lot of opportunity for their tribal members. Sometimes a mineral resource development project that helps one tribe can be the same development project that hurts another tribe, like, for example, creating jobs for some Native Alaskans, but hurting salmon fishing for other Native Alaskans. And so even among native groups or among Alaskans themselves, or for that matter among Lower-48 Americans, there are wide differences of opinion about what constitutes good land use policy for Alaska.

In general, the typical Alaskan encompasses a Eurocentric ethnicity and culture. Most Alaskans live in and round Anchorage (population 500,000 in the metro area) or Fairbanks (population 100,000 in the Fairbanks borough), or Juneau (population 50,000 in the metro area). Then, there are number of smaller towns, especially in the Kenai Peninsula area (population 50,000 and over 15 million acres [6 million hectares]). These urban dwelling Alaskans also interact with the vast Alaskan frontier

[1] The "Lower-48," are the fifty United States of America (USA) that do not include Alaska or Hawaii; they are south of Alaska and thus "lower" on the map.

lands where they hunt, fish and ski even though most Alaskans stay in their urban areas. It is these typical, city dwelling Alaskans, who tend to want more development particularly during economic recessions.

The end result is there is a kind of tug-of-war between all three of these groups in Alaska, where Lower-48 Americans are mostly represented by the U.S. Federal Government; Alaskans are mostly represented by Alaska's State Government, and Native Alaskans are represented by villages, regions, the state, or the U.S. Federal Government. The Lower-48 Americans are politically strong since they influence the U.S. Federal Government regulations of Alaska the most. And, while it is hard for non-Alaskans to believe that Alaska is under the cultural influence of the Lower-48, it is not hard for Alaskans to believe it. Most Lower-48 Americans would like to keep much of Alaska's Federal land as undeveloped land in order to preserve natural environments, but where many Alaskans would like to make much of Alaska into developed land and would like to see the many U.S. Federal lands used for the express purpose of developing mines. Native Alaskans are split on the issue of development depending on if their village or region will benefit of not.

Interestingly, some Native Alaskans can use the Lower-48ers to their advantage. That is, there is a kind of leveraging that takes place between Native Alaskans and Lower-48ers. For example, if one tribe wants to open up an area for exploration and development, say in the Arctic National Wildlife Refuge (ANWR), and another tribe does not want it to open; then the latter tribe can send an elder to the U.S. Congress and proclaim that all of the native tribes do not want to open up such an area, and viola, the Lower-48ers believe that all Native Alaskans want this closing up of Alaskan lands. Thus, there is a smaller tug-of-war between Native Alaskans, between those who want to close up Alaskan lands and those who want to open up Alaskan lands to such things as oil or mining development. Indeed, the economic benefits of, and the very responsible environmental methods used in, drilling and producing oil and mining make them very advantageous to many first nation peoples economically, especially where development is close to Native Alaskans' villages.[2]

During the Alaska land distribution process, the U.S. Federal Government (Lower-48 Americans) went ahead and designated many of Alaska's federal lands as permanently withdrawn from consideration of being Alaskan State lands and withdrawn from consideration of any oil and gas or other mineral development. The U.S. Federal Government did this by making many different designations for its federal land such as the Arctic National Wildlife Refuge (ANWR), 20 million acres, the Nation Petroleum Reserve Alaska (NPRA), 24 million acres, and Gates of the Arctic National Park, 8 million acres, among other lands that are now restricted in, or closed off completely from, development potential. While many of the U.S. Federal lands in Alaska are still open to mineral development; Alaskans would like more of these lands to be open since development of federal land tends to help Alaska's economy but tends to have little effect on the Lower-48's economy.

Native Alaskans were also not given the right to take large sections of land, other than the roughly one eighth of the state that was given, but they were given

[2]See Strohmeyer (1993) for more on Native rights.

an alternative to land, which is something called Native Corporations. The Alaska Native Claims Settlement Act (ANCSA) is the law that settled the native claims to all of Alaska lands. The ANCSA law allowed that Native Alaskans be given a small portion of land (44 million acres or roughly 12% of Alaska), but where they still were allowed to hunt, fish and whale along wide swaths of land and sea not owned by them. Often, after the hunting, whaling or fishing was done, and the native peoples would go back to live in small villages. On the one hand, if each native tribe were only given a small reservation based on their village, then they would lose out on the hunting and fishing outside theirs and other's villages. On the other hand, all of the valuable oil, mineral, and forest resource lands were taken by the U.S. Federal Government or the State Government that Native Alaskans lost out on but which they did receive some compensation over (roughly $1 billion in 1970 dollars) and where they could now invest that money in businesses using the concept of Native Corporations.

The way the Native Corporations work is that Native Alaskans can invest in businesses in or outside of their village in order to make money, and they receive tax exemptions. The Native Corporations, then, distribute that money to villages or native residents identified with that particular Native Corporation. This was a compromise between the Native Alaskans not receiving valuable land for mineral development but being allowed to make economic gains for themselves. The Native Corporations do, however, help in a lot of mineral development and profit from it, so at least the Native Alaskans do benefit economically from the mineral development of their ancestral lands, although Wall Street tends to be more lucrative for their investments.

2 Early State History

In the early history of Alaska, Alaska was governed under various U.S. appointed governors, who were mostly interested in caretaking Alaska rather than in developing it. While there was fishing and mining development, much of it was done to the benefit of the mining and fishing companies, which resided on the Lower-48 West Coast, and little of the value of mining and fishing went into the development of the State's infrastructure that could improve the business environment and help develop other industries or other mining and fishing. This meant that much of Alaska's politics was under the thumb of outside business interests and not to the advantage of most Alaskans. Finally, this caused enough Alaskans to push for Alaska becoming a state, which it did in 1959. Since Alaska was entirely owned by the U.S. Federal Government, though, there was a need to have state-owned lands, rather than all federal land, so as to allow the State of Alaska to develop. So, the Alaska Statehood Act included allowing that U.S. Federal lands should be transferred from the U.S. Federal Government to the State of Alaska in the amount of a little more than 100 million acres (40 million hectares).

Nevertheless, there is a dichotomy about what Alaska is. To many Americans, Alaska represents undisturbed land meant for having environmental services like unbroken wilderness areas, lots of wildlife, and vast untouched ocean coasts, but to the people living in Alaska it is a harsh cold environment with huge landmasses that are distant to get to, but which need to be developed in order to give economic opportunity to Alaskans. Ernst Gruening's "Freedom from Colonialization" speech in 1959 just before Alaska became a state, makes the point clear. Gruening said, "Conservation of natural resources cannot be separated from the issues of conservation of human resources." That is you need jobs and a clean environment, but you have to develop the land in order to get those jobs; although the development doesn't have be done in an environmentally detrimental manner. You have to build expansive roads that are costly to maintain to get any industry going. So infrastructure costs more, wages are higher due to difficult conditions; energy is expensive due to the long distances which it has to be carried. And so this makes business in general difficult to develop. However, the way to overcome these difficulties is to try to open up more resources to development such that you can obtain more government revenues to pay for and maintain more infrastructures that can in turn increase development further.

Then, on top of the high costs of normal day to day business, due to expensive infrastructure and due to long distances and due to the requirement of higher wages, there are also natural disasters that can be very prevalent in Alaska such as earthquakes in South Central Alaska, flooding rivers that happened in Interior Alaska, and arctic storm surges on the coasts. When these natural disasters happen, there is a certain amount of help from the U.S. Federal Government, but the more enduring method for overcoming them is to create a more vibrant economy that can provide jobs and give tax revenues that can help to overcome the high costs of disasters in Alaska.

For example, when the great earthquake of 1964 happened in South Central Alaska, it forced the state government to change how it was going to proceed with land development, especially in terms of leases for oil exploration. The state government knew that a more vibrant economy could help people pay for new housing or to fix old homes and to build more expensive homes that are resilient to earthquakes. Also in 1967, Fairbanks had a great flood that inundated the entire town and ruined lots of buildings and houses there. While a dam was built that could help keep future Fairbanks' floods away, nevertheless, the damage was done and the Interior of Alaska, just like South Central Alaska, needed an expansion of the Alaskan economy in order to pay for the damages and allow people to work to make money to pay for their losses.

Both of these disasters caused the State of Alaska to accelerate its policy on taking land from the U.S. Federal Government that had been going slow until then and speeding up the process. In particular, the State declared the central North Slope mineral rights land for its own State land in order to jump start new oil development there. This was in stark contrast to the U.S. Lower-48, which had a booming economy in the 1960s and which therefore wanted to see more natural resources kept safe from development, especially in Alaska even though mining and oil exploration and production can be done in a very environmentally clean manner when done appropriately.

3 Alaska State Lands

When Texas became a state in the United States of America (USA), the U.S. Federal Government did not own any land in Texas. All of the land was owned privately and any land that the U.S. Federal Government wanted, they would have to purchase. In diametric contrast to Texas, Alaska was completely owned by the U.S. Federal Government, so once the U.S. purchased Alaska from Russia in 1867, there was no privately held land. At first, the State of Alaska was a territory with an appointed governor from Washington D.C., who would run the territory. And while some lands were sold to settlers and some mineral rights were leased and coastal fishing rights were given, the vast bulk of Alaska lay undeveloped.

Then, when Alaska became a State, it was negotiated that the U.S. Federal Government would keep 60% or roughly 200 million acres (81 million hectares) of Alaska land and that the State of Alaska would receive 30% or roughly 100 million acres (40 million hectares) through a process of transfers. The transfers had Alaska taking mostly land in and round its major cities and towns or along road ways or in areas proven to have lucrative mineral rights, or potential mining lands. And Alaska has, as of 1994, claimed its share of federal land, however, the final adjudication needs to be completed with land surveys. Still, much of Alaska's mining development potential was and is in the hands of the U.S. Federal Government rather than in the hands of Alaska's State Government, although, Alaska has claimed the central North Slope area where Prudhoe Bay exists, and the state benefited greatly from its development. Indeed, it was issues surrounding the Prudhoe Bay oil field that instigated a lot of the land transfers.

Later on Native Alaskans would be given a right to 12% or roughly 44 million acres (18 million hectares) of Alaska land. On top of that, the U.S. Federal Government, i.e., Lower-48 Americans through their representatives, has permanently taken away many of the best oil and gas producing and mineral producing lands from consideration for Alaskan land such as much of the North Slope and in and around the Brooks range mountains on the north of the state where many of the most potentially lucrative oil and mineral production exists. Americans often view Alaska as one of the few remaining natural environments that they want to protect, whereas Alaskans and even many Native Alaskans think of Alaska as a vast frontier land that is hardly used other than for wildlife hunting and fishing, and so they want to use that land for potential mining and petroleum production. What this meant was that, completely different from Texas, the Federal or the state government owned and controlled all mineral rights for mining and petroleum production in Alaska, rather than private individuals, which reduced incentives for exploration in comparison to Texas. By contrast, in Texas, oil companies buy leases from private landowners for the right to extract oil and gas and pay a royalty to those landowners, which is why in Texas not only are there private oil companies but private mineral lease owners as well, and which may be why Texas is more robustly developed economically than Alaska.

In Alaska, all the value of the mineral rights goes to the federal, the state, or a few native governments, and it is those governments that have to approve the lease sales, which normally means that the lease sale must be much more lucrative before the state or federal governments will engage in the sale of mineral rights, compared to privately held land where the land owner will more readily sale their mineral rights. Nevertheless, there was petroleum exploration in Alaska in the 1960s, usually on U.S. Federal Government land because Alaska was slow to declare which lands it would take from the U.S. Federal Government. Although, the U.S. Federal Government set up onerous rules for lease sales on the North Slope that required petroleum exploration companies to only be allowed to claim small parcels of land at a time, and where no one company could own a lot of parcels together.

The reason the U.S. Federal Government wanted many different companies to own leases was so that no one company would control all of the mineral rights in one general area such as all of the Prudhoe Bay area. The Federal policy was designed to reduce a monopoly power (or a monopsony) over all the mineral rights in one area that would make it hard for workers to get jobs. Unfortunately, if one company had a single parcel of land and if they found oil, then all the nearby parcels would also have a right to that oil, and so other companies would gain profit at the expense of the first company who invested all the money and work and who took all the risk in exploring for and finding the oil in the first place. That is, the first company would pay all the costs to look for oil, only to have others receive all the reward. So what companies wanted to do to mitigate those costs was to own many parcels all next to each other so that if they take all the risks of exploration and finding oil, then they will receive all (or at least most) of the reward.

Since the U.S. Federal Government was not going to change its rules and its way of doing business, it was determined that if the State of Alaska took those central North Slope lands and made them into State lands, then Alaska could sale a lot of parcels together creating large lease holding areas so that a single company could reduce their exploration risks of not getting back all their costs of exploration. So the oil companies asked Alaska to take those Federal lands. Alaska hesitated to do that owing to a lack of state regulators to run such leases, but after the great Alaska earthquake of 1964 that devastated Anchorage and South Central Alaska, Alaska decided that one way to help overcome the economic consequences of the earthquake was to finally take those U.S. Federal lands and organize them into leases with larger parcels of mineral rights in order to develop the oil industry faster. Then, eventually in early 1968 the stupendous Prudhoe Bay oil field was found on the newly acquired state lands using the large parcels of lease sales, and it was soon followed by other discoveries.

What the Prudhoe Bay example shows is that the tug-of-war over land rights can delay the exploration and development of oil production. Most Alaskans view this as a needless delay since they are interested in developing mineral rights in order to gain economic development. Whereas most Americans feel this delay is a good way to preserve the natural environment. Clearly Alaska is unique in that so many people, like the 325 million or so Lower-48 Americans, feel that they have a say in Alaska even though each American in his own state feels that only his state residents should be in charge. An example of this out-of-staters opinion on Alaska was when the late great senator John McCain of Arizona stated that, "I wouldn't drill in the

Grand Canyon unless the people in Arizona wanted to." But then he further stated that drilling in ANWR is different because it is a wildlife refuge, and therefore, he voted against opening up ANWR to oil development, even though he was a republican and did vote for offshore oil development when many other Senators did not.[3]

It is as if an Illinois senator were to say that Wall Street is sacred to all of America and that therefore the trading on Wall Street in New York City should be limited to preserve New York's historic trading buildings, while the trading in Chicago should be encouraged, say with lower taxes to make its financial markets more vibrant. While such a claim may or may not be true, many New Yorkers would find it odd that an Illinois senator finds it compelling that the U.S. should treat New York and Chicago trading differently. Likewise, if an Alaskan U.S. senator were to say that drilling in ANWR is okay, but drilling in the Grand Canyon has nothing whatsoever to do with Arizonians, then that would seem a bit odd.

4 The U.S. Federal Government Versus Alaska and Alaska Native Land Claims

While the U.S. Federal Government lands in Alaska have not been completely developed, they have been developed to some degree. However, when Alaska finally took some of its allotment of Federal land by taking the Central North Slope oil lands and some of the North Slope coastal plains regions and sold oil leases there, which helped in the discovery of Prudhoe bay; there was still an important role for the U.S. Federal Government to play before Prudhoe Bay that could be fully developed because the U.S. Federal Government had not settled much of the Native Alaskan land claims from previous treatise and laws. This is why the above mentioned ANCSA was finally settled, so that a large corridor of land for the right-of-way of the Trans-Alaska Pipeline System (TAPS), which transports North Slope oil to market, could be permitted. That is why, the first thing the U.S. Federal Government had to do after Prudhoe Bay was discovered was to settle all the native land claims, and it did so with ANCSA.

However, one other interesting aspect about ANCSA was that since much of Alaska's land mass held oil and other minerals and many Native Alaskans would hunt and fish on wide swaths of land, then it was thought that instead of a reservation system, like that in the Lower-48 Indian country, Native Alaskans within the ANCSA act would have a different kind of settlement. The Lower-48 Indian reservation system was where large swaths of land was given to individual native tribes, but this same

[3]We should note one anomaly in Senator McCain's position. Unlike many of his Republican counterparts in the Senate, McCain has long been opposed to drilling in Alaska's Arctic National Wildlife Refuge, even though most Alaska residents support it. In Houston, McCain explained why ANWR is a special case.

"Quite rightly, I believe, we confer a special status on some areas of our country that are best left undisturbed," McCain said. "When America set aside the Arctic National Wildlife Refuge, we called it a "refuge" for a reason."

idea was thought to not be workable in Alaska because: (1) The Alaskan Native villages had very small populations that could not control large reservation lands. (2) No matter how large reservation lands could be; they would not include all traditional hunting, fishing, and whaling areas that Native Alaskans were used to using, and so the reservation type system would constrain their subsistence hunting rights. And (3) there were valuable mineral rights lands that the State and the U.S. Federal Government wanted and valuable whaling and hunting areas that the State and the U.S. Federal Governments wanted to protect, and so they did not want to hand those lands over to a native reservation land system. When these issues came out in U.S. congressional hearings on ANCSA, and with some Native Alaskans wanting to keep their hunting and fishing areas protected as natural land preserves rather than having those lands designated as mineral development land, and with many Lower-48 Americans also wanting to preserve lands, then, the idea quickly came to fruition that Lower-48 Americans can jump into the discussion of Alaskan lands settlement on the side of at least some Native Alaskans who wanted less development. That is Lower-48 Americans joined forces with a number of Alaskan Natives to emphasize natural preservation of Alaskan lands as opposed to mineral right development of those lands. After all, as stated, some Native Alaskans did want mineral and mining development depending on how close to the development they lived, but others did not.

Americans, i.e., Lower-48 Americans who outnumbered Alaskan Americans by about 600 to one had, as of 1968 when Prudhoe Bay was discovered, gone through a metamorphosis in regards to environmental consciousness. In 1962, Rachel Carson published a book, *Silent Spring*, about how pollution, particularly insecticide used by agriculture and city dwellers, was inadvertently killing birds and then reducing the sound of their singing (tweeting), especially in the spring time, i.e., the springs were more "silent" with less birds singing. This emboldened the environmental movement causing Americans to suddenly grow more conscious of how a number of beautiful lands within America, including Alaskan lands, was being mined, built over by cities, and in general developed. So, these Americans saw the issue of native land claims in Alaska as a way to attach a number of environmental concerns on to an Alaska settlement, particularly land preservation concerns, where such lands can be permanently kept undeveloped and left in their natural state. Since Alaska had more undisturbed land than anywhere else in America, then Alaska became a focal point of preservation, and ANSCA was the perfect legislative tool to use to get more land preservation happening.

With the ANCSA law process, congress eventually ruled that some Native land will be given back to Native Alaskans but that Native Alaskans will also receive some financial compensation for their land claims that were not returned. But the Native Alaskans will still be able to hunt, fish and whale in traditional ways and over their traditional territory even if they do not own the lands. Also in ANSCA, Native Alaskans would be able to set up the aforementioned "Native Corporations" to help in their economic development.

However, on top of the native issues and due to the large environmental movement in the 1960s and 1970s, one issue that was taken up within ANSCA, and added on, was the so-called D-2 provision, whereby the U.S. Federal Government would permanently preserve large areas of Alaska into national parks, wilderness, wildlife refuges, and other designations. The D-2 provision stated that in addition to Alaska native lands, of about 44 million acres given to Native Alaskans, there should also be 83 million acres (33 million Hectares) of Alaska, later to be increased to over 100 million acres (40 million hectares) that Alaska will never be able to take as their own state lands and that, in addition, will be permanently a part of U.S. Federally protected lands such as national parks, wildlife refuges, and wilderness. But there was not enough time to choose those lands, so the issue was to be decided over the course of the next three years after ANCSA went into law; and then, the U.S. Federal Government would choose which lands to preserve or not. However, after the three years were up, the issue was still not decided so congress voted to delay the final decisions over the D-2 lands. Then, finally with newly elected President Ronald Reagan about to take office in early 1981 and congress about to change to a more republican controlled body, President Jimmy Carter and the democratically controlled U.S. Congress put together a list of Alaskan lands that would be permanently preserved. This was signed into law as the Alaska National Interest Lands Conservation Act (ANILCA).

What ANILCA essentially did was create a compromise between the pro-mining mineral development advocates, mostly Alaskans, who wanted to develop Alaska and the nature preserving land, environmentally conservation minded Americans, which are mostly Lower-48 Americans, who wanted to keep the land undeveloped. To this day, both the debate in America and the debate in Alaska put a majority of Americans, who are nominally environmentalists, and a majority of Alaskans, who are nominally pro-development including mining development, at odds with each other. Certainly, Alaska wants to develop the land in an environmentally responsible manner, but develop it nonetheless. The compromise in ANILCA was put in force by the U.S. Federal Government, and so it is not easily changed, but the way Lower-48 Americans forced Alaskans to have less development is still contentious.

5 The U.S. and the Trans-Alaska Pipeline System (TAPS)

Even though ANCSA was decided in 1970 in order to pave the way forward for Prudhoe Bay's development, there was still one last land development problem with being able to get a pipeline permitted on U.S. Federal land from the North Slope of Alaska all the way to Valdez Alaska in the south. The problem with getting the Trans-Alaska Pipeline System (TAPS) permitted was that most of the right-of-way for the pipeline was on U.S. Federal land. However, there was a law from the late 1800s that said any land access to a mine or mineral right on Federal land could only use a 100 foot wide (30 m) corridor. Well 100 feet was good enough for an old road to a mine in the 1800s, but to put in place TAPS, a 48 in. pipeline with all the equipment needed for its construction; there was going to be a need for a wider corridor. So, U.S. environmentalists sued the U.S. Federal Government to stop permitting the pipeline because of the 1800s law, and indeed, it was found in the courts that TAPS could not be put in place due to that law.

Therefore, it took an act of the U.S. Federal Government, i.e., an act of congress, called the Trans-Alaska Pipeline Authorization (TAPA) Act, to get the pipeline permitting approved. That is, Washington D.C. (the capital of the U.S. Federal Government) was more important than Juneau, Alaska (the capital of the State of Alaska) in getting the North Slope oil developed. However, the TAPA Act not only permitted the pipeline it also outlined specific requirements for certain environmental protections such as TAPS being earthquake proof, so that it would not leak during an earthquake, and TAPS having to be built above ground, so it would not destroy the permafrost. So TAPs was put above ground on slider supports so it could withstand earthquakes, where the pipe would be able to slide along its supports, but not melt the permafrost. Although, TAPS did have some below ground lengths of pipeline in certain areas so caribou could get through to the other side of the pipe. Also the pipeline has a zigzag design so that it can have some give and take to it should a large earthquake hit. Interestingly while Alaska does have more earthquakes than the Lower-48, those earthquakes are mostly in the southern region of the State such that much of the design of the TAPS is overkill and not particularly needed.

A cheap option for the pipeline would have been to bury the pipeline underground with a lot of insulation, and then, every time an earthquake would erupt or the permafrost melt and cause a sink hole; you just shut down the pipeline oil flow, for a few days, empty the pipe out at that area of breakage, and repair the pipe and the surrounding ground support and then open it up again to restart the flow. Nevertheless, after an extensive environmental impact statement (EIA), the pipeline was built mostly above ground and made able to withstand a sizable earthquake on the order of 8.0 on the Richter scale.

One of the interesting aspects of TAPS is that since it is mostly above ground, then it can succumb to cold weather. The cold can cause the oil inside the pipe, even with insulation, to cool and harden as the crude oil throughput declines. The oil goes more slowly through the pipeline as oil production declines giving it a chance to cool more and more as it proceeds through the 800 mile long pipe. One idea to be able to allow less oil to get through TAPS was to simply heat up the oil as it goes through, and one way to do that is to have coal-fired steam driven heat pumps at each pump station where the excess heat of the steam engine can be used to heat up the oil inside the pipe to keep it from cooling down to a hardened state.

One of the reasons for not using coal fired heat pump system is pressure from the Lower-48 regarding environmental standards such as carbon emissions, although labor costs and control system issues were important too. So the oil companies who own the pipeline get pressure by the general public of the Lower-48, leading them to use electric pumps, which probably uses more energy and probably emits more carbon due the way the electricity is generated than if they used simple steam turbines with the excess heat being used to heat up the oil. Such a system could allow for lower throughputs to keep the pipeline open longer.

Still, with oil production lower on the North Slope, the lower production means less throughput in the pipeline, which allows the oil companies to claim that they need lower taxes from the State of Alaska to find more oil in order to keep the pipeline fuller. All of this is due to the fear of the oil pumping more slowly through TAPS and cooling to a hardened state and shutting down.

Granted you would need more personnel and more electronic controls to run the coal fired pumping system, and yet much of these systems can be run remotely using radio signals. There would possibly be a higher tariff to run the equipment but given the technologies available; it should be possible to run TAPS at lower throughputs. Nevertheless, there is an interesting technical-political interface with the oil pipelines, and where public opinion from the Lower-48 seems to drive how the TAPS pipeline is designed.

6 Alaska Native Land Issues

One last issue with Native Alaskan lands is how the different regions have different incentives for development. One of the biggest examples is the Pebble Mine, a gold, copper, and molybdenum mine in South West Alaska with roughly $100 billion worth of minerals below the surface. The easiest way to mine those minerals is with an open-cast mine where the over burden is put into a holding pond. The pond would be highly toxic with acids and chemicals from the underground minerals and from some of the mining chemicals used, but where the holding pond would slowly evaporate leaving a hardened land fill behind, which would be covered by gravel and soil and replanted. The resulting area would not be any more of an environmental hazard than if it had not been mined. The problem is that once the mine is running, the pond would be held in place by a dam, and since the area is earthquake prone, the dam could give way under the power of a 7.0 or 8.0 earthquake, which is not unheard of in that area. Then, if the dam is not built strong enough to take such a quake, the pond would dump chemicals into rivers and streams and affect salmon fisheries down river and on the coasts of Alaska.

Interestingly, the Native Alaskans who are nearer the mine and who would benefit most from the mine are different than the Native Alaskans, usually of a different tribe that use the salmon fishery downstream. So two different Native Alaskan groups can often have opposing views to each other. The Native Alaskans close to a mine will want it developed and the Native Alaskans far away from a mine, but close to the salmon fishery that would be affected by the mine if there was a release of the holding pond, are against the Pebble Mine.

This is in stark contrast to many Alaskans and Lower-48 Americans who believe that all Native Alaskans have a similar view of Alaskan development in that they do not want any at all. Nevertheless, many Native Alaskans, as is the case in the Lower-48 Indian country and on Indian reservation land, believe that economic development, including mining, gives jobs to native peoples and helps their education funding and provides welfare benefits for the tribal peoples.

ANCSA allowed the U.S. Federal Government and the State to own most of the mineral resources of Alaska and sale mineral right leases for development. Although,

Native Corporations could buy land or put together their own oil companies to lease oil minerals, it is often the big, private mining or petroleum companies that develop mineral rights, but those big companies often involve the Native Corporations such as the Arctic Slope Region Corporation (ASRC) to help in the mineral developments such as having petroleum service companies. Indeed, native corporations set up many service companies that do a lot of the work for the major oil producing companies. These service companies would invest in capital equipment, labor, and management that would conduct the work, and then, the native corporations receive a profit share from that work, which goes to the owners of the native corporations, who are the native tribe members based on living in certain regions and being in certain tribes.

The native corporations are involve in many businesses inside and outside Alaska, too, so they tend to not to be heavily involved in the petroleum industry exclusively. Rather the petroleum industry tends to have a few large international corporations with many smaller contracting companies and where many of the contractors contract offices are in Anchorage but where most of the worker's, executives, and engineers live outside of the state in sunnier, nicer climates. It's rather like having all of the industrial aspects of the industry in one remote location, on the North Slope, with a few small offices in one city, Anchorage, and with the workers living all across the Lower-48. While there are a few refineries in Alaska, they are relatively small so that even the downstream industrial aspects of the petroleum industry are outside of the state.

However, since the petroleum industry provides most of the state's tax revenue, it makes the state's economy very dependent on the price of oil rather than on the jobs within the industry. Alaska is therefore heavily government dependent economy rather than a heavily petroleum industry job dependent economy, and with only 700,000 residences, many of whom live in remote villages or towns and where a cold climate, sparse infrastructure, and high costs keep people and industry away; then, there is little opportunity to expand industrial jobs, although many residents like it that way.

7 Conclusion

Alaskans have always wanted to develop Alaska more, while Americans in the Lower-48 have always wanted to preserve Alaska with nature preserves so that less of Alaska gets developed. The battle for land development was implicit in Alaskan native land rights. The chain of land rights events started with the U.S. Federal Government owning all of Alaska and Alaskans not being able to develop the land very effectively under a U.S. designated government. Then, Alaska became a State but was only given one third of the land, and Native Alaskans were given one eighth of the land. However, the land that Alaska would take was determined by how oil development would proceed. Alaska managed to get land where the great Prudhoe Bay oil field was discovered, but further native land claims and pipeline legislation was needed to be able to build TAPS and develop Prudhoe Bay and the surrounding Central North

Slope oil lands. These U.S. Federal Government acts included ANCSA the TAPA Act and eventually ANILCA. The gist of all these interactions was that Alaska got its land and its oil, but Alaska lost a lot of lands that are permanently preserved and not available for mineral development. Native Alaskans both won and lost in the land tug-of-war depending on if they were close to or far away from mineral development.

References

Alaska national interest lands conservation act ANILCA 104 million acres taken from possible Alaskan use. Signed on December 2 1980, by President Jimmy Carter.

D-2 Provision of ANCSA: allotted 83.5 million acres of wilderness, national parks, wild rivers, wildlife refuges to be determined by conferees by the end of 1978.

Monday, August 4, 2008. https://www.politifact.com/. Accessed February 7, 2019. On Senator John McCain.

Strohmeyer, J. (1993). *Extreme conditions: Big oil and the transformation of Alaska.* Anchorage, Alaska: Cascade Press.

Local Content Policies in the Extractive Industry in Canada

Chilenye Nwapi

1 Introduction

Canada is not recognized as a 'go-to' nation in local content policies (LCPs) in the extractive industry. Its LCP reputation stems from renewable energy development. Even then, it is an unenviable reputation that is connected with one of its provinces' (Ontario's) feed-in tariff programme for wind and solar energy launched in 2009. The programme required electricity generators in Ontario to procure a minimum share of goods and services required for wind and solar energy projects from local Ontario manufacturers and service providers. A minimum of 25% was established for wind energy (which would increase to 50% in 2012) whilst 50% was established for solar (which would increase to 60% in 2012). Compliance with this minimum requirement was a precondition for the receipt of subsidies and other governmental support from Ontario. Joined by the European Union, Japan brought a complaint against Canada in the World Trade Organization (WTO), arguing that the Ontario programme violated certain provisions of the General Agreement on Tariffs and Trade (GATT) and certain provisions of the Agreement on Trade-Related Investment Measures (TRIMs Agreement).[1] Upholding the decision of the WTO panel, the WTO Appellate Body ruled that the programme violated Article III of GATT and Article 2.1 of the TRIMS Agreement and directed Canada to ensure that the Ontario scheme complied with them (see Kuntze and Moerenhout 2013; Kiragu 2015). This decision—the first WTO ruling, at least in recent times, on the legality

[1] See Canada-Measures affecting the Renewable Energy Generating Sector, and Canada-Measures relating to the Feed-In tariff Programme, Reports of the Appellate Body (WT/DS412/AB/R: WT/DS426/AB/R, May 6, 2013).

C. Nwapi (✉)
Canadian Institute of Resources Law, Calgary, Canada
e-mail: enyenwapi@yahoo.com

E. G. Pereira et al. (eds.), *Sovereign Wealth Funds, Local Content Policies and CSR*, CSR, Sustainability, Ethics & Governance,
https://doi.org/10.1007/978-3-030-56092-8_22

of LCPs—delivered a fatal blow on Ontario's feed-in tariff programme, spurring global academic debates on the international validity of LCPs generally (see Nwapi 2015; Ramdoo 2015, 2016) and reinvigorating debates on the use of industrial policy by developing countries to accelerate economic development (see Nwapi and Lee 2016). Outside this case, not much is known about LCPs in Canada. Yet, LCPs have a long history in Canada, including in the extractive sector, even if the Canadian government does not openly describe its policies as LCPs.

The extractive industry is of critical importance to the Canadian economy. The mining industry alone contributes more than CDN$40 billion annually to the Canadian economy (Prospectors and Developers Association of Canada 2019), with a total of about 426,000 workers directly and another 208,000 indirectly employed in the various stages of the mining value chain, bringing it to a total of 634,000 (Prospectors and Developers Association of Canada 2019). The industry is estimated to be the largest private sector employer of Canada's First Nations and other indigenous peoples (Mining Association of Canada 2018). Globally, about 75% of mining companies trace their home to Canada. Around 2013, it was estimated that about 70% of global equity capital is raised on the Toronto Stock Exchange and the Toronto Venture Exchange (Nwapi 2014). However, a recent report by the Prospectors and Developers Association of Canada (2019) shows an 'extreme drop in equity financing' in the first half of 2019 in Canada.

With respect to oil and gas, Canada is the fourth largest oil producer in the world, with about 170 million barrels in proven reserve (Natural Resources Canada 2018; Canadian Association of Petroleum Producers 2018a). More than 95% of the oil reserves are, however, held in the Alberta oil sands (Natural Resources Canada 2018). In 2017, the oil and gas industry contributed $101 billion to Canada's GDP notwithstanding the oil price downturn, and the industry is estimated to contribute about $1.8 trillion between 2017 and 2027 (not including expected contributions to Alberta) (Canadian Association of Petroleum Producers 2018b). Employmentwise, the industry is responsible for 528,000 direct and indirect jobs (Canadian Association of Petroleum Producers 2018b).

Local content development in Canada's extractive industry is strongly influenced by Canada's constitutional framework. The Constitution Act, 1867 grants ownership of non-renewable natural resources, such as minerals, oil and gas except uranium, to the provinces where the resources are located (Section 92A(1)). Accordingly, regulatory authority over extractive resource development resides with provincial governments. It is up to each provincial government to decide the regulatory instruments to utilize to ensure optimal benefit of the resources to the people of the province. As a result, there is no uniform local content policy across Canada. On the other hand, federal interest in extractive resource development is mainly indirect. Section 91(24) of the Constitution Act, 1867 gives the federal parliament authority over matters relating to 'Indians, and lands reserved for Indians'. Under this authority, the federal government has limited involvement in decisions regarding extractive resource development in order to ensure that Aboriginal interests are protected. Efforts to protect Aboriginal interests in extractive resource development have evolved considerably through the application of the duty to consult Aboriginal peoples enshrined in the

Constitution Act, 1982.[2] The duty to consult is that of the Crown and not of companies but companies play a critical role in making the duty meaningful through direct consultation with Aboriginal communities. Lastly, as will be seen later in this chapter, whilst the duty was not conceived as a local content policy, the measures by which extractive companies have sought to make the duty meaningful in their relationship with Aboriginal peoples have involved measures typically associated with LCPs.

LCPs in the extractive industry in Canada can best be understood under three separate analytical contexts: (1) under the federal foreign investment policy; (2) under benefits plans established under federal oil and gas legislation and under federal-provincial agreements; and (3) under the constitutional duty to consult Aboriginal peoples before natural resource development projects are undertaken. These are discussed in turn and followed by some concluding remarks.

The discussion makes some key finds. (1) The Canadian federal government does not have an official LCP as we know it, as there is no policy document that makes explicit reference to local content. But this does not mean that the policy is unknown in Canada. In fact, unbeknownst to many, the federal government has incorporated the policy in some form under the federal foreign investment policy since at least 1973. (2) Local content is fully embedded in requirements for the submission of benefits plans under accords negotiated between the federal government and the Atlantic provinces of Newfoundland and Labrador and Nova Scotia to enable the provinces to benefit from offshore oil and gas development in Canada's continental shelf appurtenant to those provinces. (3) Unlike in most other countries, local content in Canada is not characterized by the setting of stringent local content targets that companies must meet but instead adopts a more flexible approach. (4) The understanding of 'local' under Canada's local content regime embraces three types of 'local': Canadian nationals (and permanent residents) as a whole, provincial residents and Aboriginal people. Each of these locals is the focus of separate local content programmes. This approach shows how the LCP can be strategically harnessed to address the socio-economic situation of different segments of the national population—an approach that contains policy lessons for other countries.

2 LCPs Under the Federal Foreign Investment Policy

2.1 The Rise of Economic Nationalism

Although economic nationalism has a long history in Canada—dating back at least to Prime Minister John MacDonald's launching of the National Policy of 1878—local

[2]Canada operates a mix of written and unwritten constitutions, with the written component consisting of two separate documents: the Constitution Act, 1867 and the Constitution Act, 1982. Whilst the former established Canada as a federation, the latter established enforceable human rights and ended British involvement in constitutional amendments in Canada. The unwritten component is a relic of its British colonial heritage.

content policies in their current iteration are relatively new. MacDonald's National Policy sought to promote the growth of Canadian industries by curtailing the importation of foreign goods into Canada through the imposition of high tariffs (Avery et al. 2013; Wesson 2007).[3] His goal was to restore struggling industries, including mining, agriculture and manufacturing (Wesson 2007). But the policy did not directly regulate foreign investment. In fact, foreign investors, especially American investors, with strong interest in the Canadian market began to incorporate subsidiaries in Canada in order to avoid the high tariffs (Olson 1980). Yet, the policy represented a new thinking in Canada regarding the impact of foreign investments on Canadian economic development.

Especially after World War II, Canada witnessed a huge influx of foreign investments, which included foreign acquisitions of Canadian businesses (Olson 1980). In the beginning, the Canadian government encouraged such investments because they resulted in an increased domestic production of Canadian goods. As the investments grew, however, Canadians began to question their long-term value to Canada. One of the main issues of concern was that both the management staff of the investors and the technology required to run the investments were supplied by the foreign parents (Avery et al. 2013; Olson 1980). As Canadians were not involved in technology development, Canada's technological growth suffered, with likely repercussions on the international competitiveness of Canadian companies. Another major problem was that the Canadian subsidiaries participated only in the supply of raw materials to their foreign parents and did not participate in production activities. The raw materials were processed outside Canada by the foreign parents and imported into Canada as finished products (O'Sullivan 1980).

Canadians were also concerned that the influx of foreign investors wrestled decision-making power over the direction of Canada's economic development from Canadians to the management of the foreign parent corporations which were making important decisions for their Canadian subsidiaries (Rostein 2015; O'Sullivan 1980). There were also concerns about the extraterritorial effect of US laws, specifically the Trading with the Enemy Act, 1917, which prohibited US companies, including their foreign subsidiaries, from trading with countries regarded as US enemies. The implication was that the Canadian subsidiaries could not trade with such countries, which could otherwise have been profitable (O'Sullivan 1980).

Canadian commentators have noted that these concerns are still present today and are reflected in the types of undertakings, which the Canadian government requires of foreign investors as a precondition for the approval of their investments (Avery et al. 2013). The Canadian government responded to the concerns by setting up several task forces to investigate the impact of foreign investments. One of the task forces, headed by Walter Gordon (1957), confirmed the rise of foreign direct investments in Canada, and observed that 'legitimate Canadian interests' were being sacrificed in the process, and recommended to the government to mandate at least part-ownership

[3]High import tariffs had been imposed in the past under the leadership of Prime Minister Alexander Mackenzie, but that was only for the purpose of raising revenues.

of the foreign-owned subsidiaries by Canadians (Rostein 2015). Of note also is the Watkins Report of 1968, which even took a less moderate tone:

> The major deficiency in Canadian policy has been not its liberality toward foreign investment per se but the absence of an integrated set of policies, partly with respect to both foreign and domestic firms, partly with respect only to foreign firms, to ensure higher benefits and smaller costs for Canadians from the operations of multinational corporations (Watkins 1968:392).

Other recommendations of the Watkins Report included the establishment of 'a special' agency to coordinate Canadian government policies with respect to foreign corporations to enable the government to better supervise their actions (Watkins 1968:395; Bishop 2016). Another report (the Wahn Report) published in 1970 recommended 51% Canadian ownership of foreign corporations (House of Commons 1970). Yet another report with remarkable consequence was the Gray Report of 1972, which stressed the need for a foreign investment review process 'as an economically rational instrument' (Gray 1972:454). Gray recommended that foreign firms wanting to establish in Canada should be required to show (1) that their products were needed in Canada and would not merely duplicate products already available in Canada; (2) the nature of the technology to be employed in comparison with technology available in Canada; (3) the type of employment opportunities for Canadians; (4) their plans for local purchase of materials, components and services; and (5) their plans for research and development as well as their product innovation in Canada (see Rostein 2015). These recommendations speak clearly to local content as we know it and set the tone for the later adoption of the LCP in Canada.

2.2 The Establishment of the Foreign Investment Review Agency

The foreign investment reports, but particularly the Gray Report, led to the enactment of the Foreign Investment Review Act (FIR Act) in 1973, with the explicit purpose of enabling 'the ability of Canadians to maintain effective control over their economic environment' (Section 2(1)). The FIR Act established the Foreign Investment Review Agency (FIRA) with the task of reviewing proposals for foreign acquisitions of Canadian businesses as well as proposals for the establishment of new foreign businesses in Canada (Hilmer 2013). The standard for approval of an investment by FIRA was that it would result in 'significant benefits' to the Canadian economy. 'Significant benefits' were assessed based on five factors, namely (1) contribution to job creation; (2) the participation of Canadians in the management of the investment; (3) 'the effect… on productivity, industrial efficiency, technological development, product innovation and product variety in Canada'; (4) the effect on competition with existing industries in Canada; and (5) compatibility with federal and provincial industrial and economic policies (FIR Act, Section 2(2)).

Criticisms trailed the FIR Act and its implementation. First, the review process was considered too long and uncertain, with added costs for investors, often leading to outcomes considered 'commercially unreasonable' (Bishop 2016:8; Rose 1986; Globerman 1984). There were no legal provisions for challenging FIRA's determinations. Observers note that although following the establishment of FIRA, there was a noticeable increase in the Canadian acquisition of Canadian firms and a noticeable decline in foreign acquisitions, they argue that the Canadian acquisitions were anti-competitive (Globerman 1984). Also, in 1983, a GATT dispute settlement panel found that the FIR Act's undertakings for domestic sourcing were inconsistent with Canada's obligation to ensure the equal treatment of domestic and imported products under Article III.4 of GATT (Rose 1986).

On the other hand, studies show that the government oftentimes negotiated with investors and obtained their written commitments to accept to perform certain local content-related activities, such as employment of Canadians (United Nations Commission on Trade and Development (UNCTAD 2011). The implementation of the commitments was monitored through requirements for the submission of regular progress reports (UNCTAD 2011) although the government showed reluctance in enforcing the commitments (Hoffman 2010). However, criticisms of FIRA's application of the 'significant benefits' test, coupled with 'a domestic recession' that began in 1982 led to a policy shift that saw the relaxation of the review process to attract foreign investments (Hoffman 2010; Bishop 2016).

2.3 *The Establishment of the National Energy Programme*

Another major aftermath of the foreign investment reports was the establishment of the National Energy Programme (NEP) in 1980, tasked with protecting the security of Canadian energy supply to end dependence on foreign suppliers amidst global concerns around the future of global oil supply and rising oil prices. NEP was also tasked to device measures to increase Canadian ownership of the oil and gas industry in Canada. By then, foreign control of the energy industry exceeded 90%, three-fourth of which was held by US investors (O'Sullivan 1980). NEP made three key recommendations:

> (i) at least 50% Canadian ownership oil and gas production by 1990; (ii) Canadian control of a significant number of the larger oil and gas firms; and (iii) an early increase in the share of the oil and gas sector owned by the government of Canada (Sheppard and Hardwicke-Brown 1992).

NEP resulted in increased Canadian ownership of the oil and gas industry (Sheppard and Hardwicke-Brown 1992). However, it did not find popularity in Alberta where most of Canada's oil and gas resources were located as the provincial government saw it as an avenue for the federal government to take control of Alberta's oil (Bregha 2016). Some of the objectives of NEP, specifically that of increasing Canadian ownership of the oil and gas industry, aligns closely with the objective of local participation adopted as part of LCPs in many countries.

2.4 *The Establishment of the Investment Canada Agency*

In 1985, FIRA was abolished through the enactment of the Investment Canada Act (IC Act), which established the Investment Canada Agency (ICA) and substantially limited the categories of foreign direct investments that were reviewable. The IC Act's stated purpose is:

> to encourage investment in Canada by Canadians and non-Canadians that contribute to economic growth and employment opportunities and to provide for the review of significant investments in Canada by non-Canadians in order to ensure such benefit to Canada (Section 2).

The new Act replaced the 'significant benefit' test with a 'net benefit' test. Thus, foreign investments or acquisitions in Canada must undergo a ministerial review process to determine if they are 'likely to be of net benefit to Canada' (Section 16(1)).

The IC Act does not define 'net benefit'. However, it establishes that the new Agency's decisions are to be guided by the following six considerations:

(a) the effect of the investment on the level and nature of economic activity in Canada, including, without limiting the generality of the foregoing, the effect on employment, on resource processing, on the utilization of parts, components and services produced in Canada and on exports from Canada;
(b) the degree and significance of participation by Canadians in the Canadian business or new Canadian business and in any industry or industries in Canada of which the Canadian business or new Canadian business forms or would form a part;
(c) the effect of the investment on productivity, industrial efficiency, technological development, product innovation and product variety in Canada;
(d) the effect of the investment on competition within any industry or industries in Canada;
(e) the compatibility of the investment with national industrial, economic and cultural policies, taking into consideration industrial, economic and cultural policy objectives enunciated by the government or legislature of any province likely to be significantly affected by the investment; and
(f) the contribution of the investment to Canada's ability to compete in world markets.

These guidelines are identical with the guidelines established under the FIR Act, except for the sixth factor, which was absent. The ICA assesses these factors based on the information, representations and undertakings provided by the investor. It has been argued that the above criteria were intended to be subjective in order to allow the Canadian government to 'exercise considerable latitude in the review process' (Sheppard and Hardwicke-Brown 1992). The factors have however been described as 'protectionist' and 'economically incoherent' and that the Act itself is based on false 'assumptions that foreign firms act differently than Canadian firms' (Bishop 2016:1).

The factors set out in the IC Act for determining net benefit to Canada align clearly with the objectives of LCPs. In fact, the foregoing review shows that the Canadian government has the same concerns as other countries that have more explicitly adopted LCPs. Even though there are no explicit mandatory requirements for employment and training of Canadians or minimum thresholds for the procurement of Canadian goods and services, a potential foreign investor would, in practical terms, need to make undertakings in these regards in order to demonstrate net benefit to Canada. A report published by Industry Canada in 2010 providing guidance on the assessment of net benefit states that the type of undertakings potential investors can provide varies from transaction to transaction but usually relate to employment of Canadians, capital spending, Canadians' participation in the business, and investments in research and development particularly in industries driven by research and technology (Industry Canada 2010).

These undertakings are not taken lightly by the Canadian government. The IC Act itself contains clear enforcement provisions for non-compliance with undertakings. For instance, the government can issue a demand letter to require an investor to remedy a compliance default. It can seek an order of court directing an investor to comply with an undertaking. An erring investor can be subject to $10,000 fine for each day of the default. The government may also seek an order of divesture against the investor (IC Act, Sections 39–40). The government has stepped up enforcement of such undertakings. Its first formal attempt to enforce undertakings was in 2009 when it brought judicial proceedings against US Steel for non-compliance with undertakings relating to local production and employment of Canadians following the company's 2007 acquisition of Canadian steel company Stelco. The government brought the proceedings notwithstanding the price fall in the steel industry after the 2008 global financial crisis. After an unsuccessful challenge of the validity of certain provisions of the IC Act, US Steel negotiated a settlement with the government (Sheppard and Hardwicke-Brown 1992). Thereafter, the government developed guidelines for the settlement of such disputes through mediation.

Observers have noted that the oil and gas industry performed better than other industries under FIRA because the industry 'tended to provide benefits relating to increased employment and capital injection into the Canadian economy, as well as the increased use of Canadian parts and services to support increased levels of resource processing' (Sheppard and Hardwicke-Brown 1992:349). Thus, the more local content there is in a foreign investor's proposal the more likely the proposal would pass the net benefit test; and the more the investor complies with such undertakings the more the goals of the legislation are realized.

However, Canada's approach seems to be more flexible than those of most LCP countries and is not characterized by the setting of stringent local content targets. The ICA has wide discretion to determine what is of net benefit to Canada whereas most other countries establish less flexible and more mechanical or technocratic standards for determining compliance with local content requirements. In addition, not all foreign investment transactions in Canada under the IC Act require local content or other undertakings. It depends on the importance of the transaction to the Canadian economy and, specifically in the case of acquisitions, on the health of the Canadian business that is being acquired (Industry Canada 2010). This is a most defining characteristic of the application of the LCP under Canada's foreign investment policy. However, there appears to be no transparency in the review process for determining net benefits. Also, the flexible and uncertain nature of the review process has rendered it vulnerable to political manipulations, as the decision whether a transaction meets the net benefit test has, compared to other regulatory approval processes in Canada, been considerably influenced by the political interests of the ruling party (Sheppard and Hardwicke-Brown 1992; Bishop 2016).

3 Local Content Under Benefits Plans

3.1 The Canada Oil and Gas Operations Act, 1985

Although Section 92A(1) of the Constitution Act, 1867 grants ownership of non-renewable natural resources, such as minerals, oil and gas, to the provinces where the resources are located, the federal government still retains some regulatory authority over oil and gas operations in defined areas or circumstances, such as oil and gas development within Canada's continental shelf, or where inter-provincial or international oil and gas pipelines are to be built. Based on this power, the federal government enacted the Canada Oil and Gas Operations Act, 1985 (COGOA), which applies to oil and gas development in that part of the onshore that is under the administration of a federal minister and oil and gas exploration and exploitation in Nunavut, Sable Island, and in 'the continental shelf of Canada and the waters superjacent to the seabed of that continental shelf', etc. (Section 3).

COGOA prohibits the approval of any plan for oil and gas development as well as the issuance of an authorization of any work or activity unless the project proponent or applicant has submitted a 'benefits plan' to the Minister for approval and which the Minister has approved, unless the Minister has waived the requirement to submit a benefits plan (COGOA, Section 5.2(2)).[4] The Act defines a benefits plan as:

> a plan for the employment of Canadians and for providing Canadian manufacturers, consultants, contractors and service companies with a full and fair opportunity to participate on a competitive basis in the supply of goods and services used in any proposed work or activity referred to in the benefits plan (COGOA, Section 5.2(1)).

What distinguishes the above provision from most local content requirements is that the provision does not require that Canadians be provided employment or opportunities for the supply of goods and services on a preferential basis in relation to non-Canadians. But this is a minor distinction that is of little or no consequence, because once a benefits plan providing, e.g., for the employment of Canadians, has been submitted, its effective implementation would require preferential consideration of Canadians, otherwise it would be very difficult, if not impossible, for a company

[4]This provision is adopted by the Canada Petroleum Resources Act, 1986, Section 21.

to uphold the terms of the plan. Moreover, the plan will be implemented within the broader context of Canada's employment policy, which requires all companies operating in Canada, regardless of the sector, to demonstrate that there is no Canadian or permanent resident of Canada who is qualified and available for a job before they can hire a foreigner to take up a job. As will be seen later, however, benefits plans required in Newfoundland and Labrador (NL) and Nova Scotia explicitly incorporate the first consideration principle.

COGOA also contains an affirmative action provision allowing the Minister to require project proponents to include in their benefits plans provisions that specifically provide 'disadvantaged individuals or groups' access to opportunities for training and employment and to enable such persons and businesses owned and operated by them to participate in the procurement of goods and services (Section 5.2(3)).

COGOA does not provide guidance for the operationalization of the benefits plans provisions, but it provides the National Energy Board (NEB) the authority to establish guidelines and interpretation notes for the application and administration of the benefits plans (Section 5.3). It is not clear whether any such guidelines have been established by the NEB.

3.2 The Atlantic Accords Benefits Plan

The most active adoption of the traditional language of LCPs in Canada is found in the Canada-Newfoundland Atlantic Accord Implementation Act, 1987 (C-NAAI Act) and the related Canada-Nova Scotia Atlantic Accord Implementation Act, 1987. These Acts were enacted to fully formalize and make binding two 1985 memoranda of understanding (MoUs) signed separately between the Government of Canada and the Provincial Governments of NL and Nova Scotia on offshore oil and gas resource management and revenue sharing.[5] The MoUs were the product of political settlements between the federal government and the two provinces to assuage the latter's bitterness over the decision of the Supreme Court of Canada in the *Hibernia Reference* in which the court ruled that rights over the Canadian continental shelf rested with the federal government rather than the government of the province bordering the shelf (Clarke 2004; Macdonald and Thompson 1985). At the time, the two provinces were faced with high levels of unemployment (Barber 2018). The agreements thus represent the federal government's recognition that the provinces where natural resources are found ought to be the primary beneficiaries of the resources.

The C-NAAI Act provides that before any oil and gas development plan can be approved, a 'benefits plan', called the Canada-Newfoundland and Labrador Benefits

[5]The two provincial governments enacted their own legislation of the memorandum, the Canada-Newfoundland Atlantic Accord Implementation (Newfoundland) Act; and Canada-No Scotia Atlantic Accord Implementation (Nova Scotia) Act. References in this chapter are made to the federal version of the legislation between Canada and Newfoundland and Labrador.

Plan, must be submitted and approved by the Canada-Newfoundland and Labrador Offshore Petroleum Board (C-NLOPB) established under the Act. After a benefits plan is approved, the C-NLOPB will consider the development plan in its entirety. Its approval of a development plan is regarded as a fundamental decision requiring a subsequent approval by the relevant Ministers of Canada and NL. However, the approval of a benefits plan is not regarded as a fundamental decision and so ends with the Board (C-NLOPB 2006).

A benefits plan is defined as:

> a plan for the employment of Canadians and, in particular, members of the labour force of the Province … for providing manufacturers, consultants, contractors and service companies in the Province and other parts of Canada with a full and fair opportunity to participate on a competitive basis in the supply of goods and services used in any proposed work or activity referred to in the benefits plan (C-NAAI Act, Section 45(1)).

A benefits plan must include the following provisions:

- a plan to establish an office in NL where appropriate levels of corporate decisions are to be made;
- a plan for the employment of Canadians and for giving residents of NL 'first consideration' in matters of employment and training;
- a plan of expenditures to be made on research and development and education and training to be carried out in NL; and
- a plan for giving 'first consideration' to goods manufactured in NL and to services provided from within the province 'where those services and goods are competitive in terms of fair market price, quality and delivery' (C-NAAI Act, Section 45(3)).

The C-NAAI Act creates an affirmative action provision requiring benefits plans to include plans to ensure the effective participation of disadvantaged individuals or groups in training and employment opportunities and for businesses owned by such persons to participate in the supply of goods and services (C-NAAI Act, Section 45(4)). Of significant importance is a requirement in the Act that collective agreements between a corporation and its employees shall not undermine the right of access of residents of the province to training and employment opportunities on a first consideration basis (C-NAAI Act, Section 45(3)(b)).

The overall intent of the benefits provisions of the C-NAAI Act is to ensure that oil and gas development in NL makes a long-term contribution to the sustainable development of the province. A key vehicle for achieving this is to provide residents of the province avenues to participate in the economic opportunities that the development of the resources brings. The resource development approval process is therefore made dependent on the quality and quantity of such opportunities proposed by a prospective investor.

Rather than itself provide detailed guidance on the preparation of benefits plans, the C-NAAI Act allows the C-NLOPB to issue guidelines and interpretation notes for the preparation of benefits plans to assist project proponents to comply with the intendments of the benefits plans provisions. In February 2006, the C-NLOPB issued the *Canada-Newfoundland and Labrador Benefits Plan Guidelines* (C-NLOPB 2006). New Draft Guidelines were published in January 2016 but it is unclear whether they were finally adopted, as a publicly available version of the Guidelines is described as 'Draft' (C-NLOPB 2016). The difference between the two Guidelines is not significant at least for the purposes of this discussion and the difference relates more to the way certain guidelines are framed than to the actual substance of the guidelines themselves.

Under the Guidelines, a benefits plan is required to be 'as detailed and specific as possible', clearly specifying the manner in which each of the elements of a benefits plan will be addressed. Project proponents are further required to 'describe the consultative, monitoring and reporting procedures' they intend to utilize to achieve the objectives of the benefits plan (C-NLOPB 2006:3–4). Benefits plans are subject to amendment, especially when an operator seeks to amend its development plan. In fact, the operator is required to amend the benefits plan unless it can provide a rationale that an amendment of the benefits plan is not warranted in the circumstances (C-NLOPB 2006).

Benefits plans are subject to a high level of scrutiny to ensure that they promote the objectives of the Act. For instance, project proponents and operators are not only required to establish procurement procedures that do not unfairly put local suppliers at a disadvantage, they must equally demonstrate that they have taken all reasonable measures to ensure that local suppliers are provided an opportunity to participate in the procurement process on a competitive basis. As explained in the 2016 Draft Guidelines, '[s]electing suppliers on a competitive basis' means using the 'fair market price, quality and delivery' of a good or service to determine which supplier is to be selected (C-NLOPB 2016:6). It thus means that operators cannot determine the outcome of the procurement 'solely on the basis of lowest price' (C-NLOPB 2006:7). A public hearing may be conducted to determine the compliance of an operator's benefits plan with the intendments of the Act.

Providing provincial residents opportunities to participate in procurements involves investments in local supplier development. This includes assessing provincial and other Canadian supplier capabilities with a view to identifying their strengths and constraints, establishing policies and programmes to enable provincial residents and other Canadians to participate in the operator's national and international activities, and technology transfer to Canadians (C-NLOPB 2006). It also includes providing timely information to provincial and Canadian supplies about project opportunities and requirements. The establishment of management offices in the province is intended to facilitate the involvement of provincial residents in management responsibilities. For matters of employment, an operator is required to identify training requirements that are needed to facilitate the participation of provincial residents and other Canadians in its labour force and the impact of the project demand on existing educational facilities of the province (C-NLOPB 2006).

To facilitate the giving of first consideration to residents of NL on matters of employment and procurement of goods and services, the Guidelines define who a 'resident' means, namely a Canadian or permanent resident of Canada who has resided in NL for the last six months. Thus, the person need not be a Canadian citizen, but holders of work permits and other visas who reside in the province do not qualify as residents. The reason for the adoption of the first consideration principle, as has been explained, was to address the high unemployment levels in NL to support families and communities, a policy that was 'expedient for the provincial government' (Barber 2018).

The Guidelines explain in detail what 'first consideration' means. Operators are free to stipulate the required qualifications, expertise and experience that candidates for an advertised position must possess. Once these have been established, operators must first look to NL candidates and give the job to any one of them meeting those qualifications, expertise and experience. Furthermore, if there are positions in the operator not initially held by residents of NL, the operator must establish a succession plan for qualifying residents to take over the positions (C-NLOPB 2006).

The 'first consideration' principle applies also to goods and services procurement. Local suppliers of goods and services that are competitive based on fair market price, quality and delivery are to be given first consideration. In the C-NLOPB's view, the price quoted by a local supplier should not outweigh other factors, so that where the local supplier's bid is not relatively low in relation to price, he should still receive first consideration if his is otherwise 'competitive in relation to the low bidder, and meets or is comparable to other evaluated factors (e.g. technical, HSE, benefits, etc.)' (C-NLOPB 2016:7). Furthermore, first consideration allows an operator 'to limit bids only from providers from the Province, should there be competitive capacity in the Province' and an operator might choose to source from a sole provider if the Province has only one supplier with the requisite competitive capacity (C-NLOPB 2016:7; C-NLOPB 2006:8). In other words, as long as there is even only one qualified provider, an operator cannot go outside the province. However, the process of ascertaining the existence of a qualified provider in the province will usually warrant a Canada-wide search unless qualified provincial providers have become well known in the industry.

The Guidelines do not establish minimum employment or procurement thresholds for companies to meet. Neither do they establish targets that the province must meet within any defined timescale. However, the Guidelines require project proponents to provide a high level of detail and specificity regarding their local employment and procurement levels. For instance, project proponents are to specify the proportion of work and employment that can be performed by provincial and Canadian businesses and workers (C-NLOPB 2016). Also, they are to establish an estimate of the goods and services required annually to support their production operations and provide, for each major cost category identified for each production system, an assessment of the available capacity in the province and in Canada to perform the work associated with each category, including any constraints thereto (C-NLOPB 2016).

To ensure that operators follow the principles of the benefits plan as well as meet their commitments, the Guidelines provide for monitoring and reporting of operators' implementation of their benefits plans. Operators are thus required to submit quarterly and annual reports, which will be shared with the public. At the expiry of their project approvals, they are to provide an assessment of the results of the application of the benefits plan and the likelihood that further benefits might be realized (C-NLOPB 2016).

In addition to the statutorily mandated benefits plans, benefits agreements are also negotiated between NL and project operators for all offshore petroleum projects. The Board has no statutory authority with regard to negotiated agreements, whether in terms of the negotiation process or in terms of the enforcement and monitoring process. However, negotiated agreements typically include provisions that grant monitoring and oversight functions to the C-NLOPB. In such cases, the C-NLOPB's role is usually limited to monitoring the implementation of benefits agreements to ascertain whether operators are complying with the terms of the agreements and to notify the government of NL accordingly. It is the responsibility of the government of NL, and not that of the C-NLOPB, to resolve any disputes that may arise between it and operators regarding compliance with a negotiated benefits plan (Wood Mackenzie 2018).

Negotiated agreements have the advantage of flexibility over mandated benefits plans. As such, they can be more easily tailored to meet the needs of the government. In addition, as the negotiating process will normally involve considerable dialogue between the government and the project proponent, negotiated agreements are an important instrument for building relationships. However, mandated benefits plans have established guidelines for their development, which can contribute to their effectiveness.

3.3 Benefits Plans in Practice

There is not much academic discussion of the Atlantic Accords' benefits plan, as a result of which not much is known of the plan's performance. The little available information shows that as a result of the benefits plans, the offshore petroleum industry in NL has delivered substantial and sustainable economic development to the province, as there has been significant growth in employment and local supplier capability, in addition to growth in the contribution of the sector to provincial GDP. The province has also, at least partly as a result of the benefits plans, witnessed a significant growth in the number of university graduates with skills relevant to the oil and gas sector, a small but flourishing research and development community, 'an increasingly diverse and cosmopolitan urban culture, and improved external transportation links' (Shrimpton 2012).

Details of benefits plans are made public and operators submit quarterly and annual implementation reports, which can be found in the public domain (see, e.g., Iron Ore Company of Canada 2017; Hibernia Management and Development Company 2016; Husky Energy 2016; Wood Mackenzie 2018). A 2018 Annual Report submitted by Suncor Energy in connection with the Terra Nova development project showed that as of 31 December of that year, a total of 1032 persons were directly employed in the project, approximately 88% of which were residents of NL, 9% other Canadian residents and 3% non-Canadian residents. Of the 1032 employees (which were disaggregated according to discipline, e.g., management, engineering, professional, technicians, etc.), 910 were male whilst 122 were female. The report also showed that during the reporting period (1 January 2018 to 31 December 2018), a total of 4,712 purchase orders with a cumulative value of about CDN$58 million were awarded, of which about 62% were awarded to NL vendors, 26% to other Canadian vendors, and 12% to non-Canadian vendors. Also, about CDN$2.3 million was spent on research and development during the same period whilst over $1 million was spent on education and training (Suncor Energy 2018).

The 2017–2018 Annual Report of the C-NLOPB shows that as of 31 December 2017, there were 4,915 NL and other Canadian residents in direct employment in the offshore petroleum industry. Production activities accounted for annual expenditures of $1.7 billion, of which approximately 63% occurred in NL and a further 18% occurred in other parts of Canada. Expenditures on research and development, together with education and training, accounted for CDN$23 million, spent on such areas as science, technology, engineering and math programs; employee health and safety; enhanced oil recovery; seismic exploration technologies; and educational scholarships (C-NLOPB 2018).

The publication of the performance reports can facilitate policy learning, i.e., the ability and willingness of an agency to sit back and measure its progress and review its interventions to identify where changes are needed (see Vogel 2017). The LCP Guidance for Governments published by the Intergovernmental Forum on Mining, Minerals, Metals and Sustainable Development (2018) stresses the importance of performance measurement in the success of LCPs. Furthermore, the availability of benefits plans and annual implementation reports in the public domain speaks to the transparency of the benefits regime, a contrast to the way in which, as will be seen later, impact benefit agreements with Aboriginal communities are treated. Transparency enlists the public, especially civil society organizations, into the local content project and empowers them to provide important oversight that would help to improve processes and enhance the benefits of the programmes. It helps the government to obtain more information about the performance of the programme and increases its ability to analyse and target its resources to where they are most needed. It also encourages firms to review their supply chains to identify where alternative local suppliers are available (Geipel and Hetherington 2018).

It is remarkable that the reports track the gender diversity of the companies' employees. Furthermore, a look at the benefits plans—both statutorily mandated plans and negotiated agreements—shows the existence of strong local content commitments as well as 'best-endeavours' clauses. For instance, a negotiated benefits agreement between the Government of NL and ExxonMobil in connection with the Hebron Oil Field project—the province's fourth offshore development project operated by ExxonMobil under a joint venture with Chevron Canada Resources, Suncor Energy, Statoil Canada and the Energy Corporation of Newfoundland and Labrador—includes a project partner's commitment of CDN$1 million to enhance skills training in the College of North Atlantic and Memorial University of Newfoundland (a firm commitment) and a commitment to a comprehensive gender equality programme (a best-endeavours commitment).

A 2018 study by Wood Mackenzie noted that the requirement of benefits plans has rendered the development permitting process for oil and gas development in NL relatively less attractive (Wood Mackenzie 2018). The study also notes that whilst both provincial policy-makers and industry players share the broad goals of LCPs in the offshore oil and gas sector, industry players have raised concerns regarding the need for provincial suppliers and service providers to strive to become internationally competitive (Wood Mackenzie 2018).

Other commentators have noted that project operators face difficulties in determining what is 'full and fair opportunity' to Canadians and provincial residents as well as in applying the 'first consideration' criterion. For instance, an operator may be presented with multiple bids having competing degrees of local content, but which otherwise are equal in terms of price and quality. The operator is expected to select the bidder with the highest degree of local content. However, whether bids are equal may depend on the eyes of the beholder and, '[i]f pressed, an operator can usually identify some "material" difference, be it technical, commercial or otherwise' (Taylor and Dickey 2001:79).

On monitoring, it has been noted, albeit with regard to Nova Scotia, that whilst the Board monitors the content of employment and supply contracts, it is mostly the commitments made by project proponents that are monitored and that the Board but does not specifically require project proponents to maintain any level of local content (Clarke 2004). In addition, there has been little evidence that failure by oil companies to maintain adequate levels of local content in their projects will attract sanctions (Clarke 2004).

Disputes have sometimes arisen over the implementation of benefits plans. One such dispute relates to the Hebron oil field. The Government of NL signed an MoU with the Hebron project partners to provide benefits commitments to the province. The MoU required the project to provide to the province's residents first consideration in matters of employment and in the supply of goods and services and to construct their machinery within the province. It emerged that a key equipment module was constructed in South Korea instead of in NL. The parties settled the dispute with ExxonMobil paying CDN$150 million to the government (Wood Mackenzie 2018).

The benefits Guidelines published by the C-NLOPB have also come under direct legal challenge. In 2012, Mobil Investments Canada Incorporated, joined by its project partner Murphy Oil Corporation, brought proceedings against the Canadian Government before an arbitral tribunal at the International Centre for the Settlement of Investment Disputes challenging the Guidelines for research and development expenditures, which required investors to spend a fixed percentage of their revenues from offshore petroleum products within NL. They argued that the fixed expenditures established under the Guidelines had a significant financial impact on them and that they contravened Article 1106(1)(c) of the North American Free Trade Agreement, which prohibits a Trading Party from requiring an investor from another Trading Party 'to purchase, use or accord a preference to goods produced or services provided in its territory, or to purchase goods or services from persons in its territory'. The arbitral tribunal found that Canada violated the said provisions of NAFTA and order Canada to pay approximately CDN$19 million to the investors.[6]

Based on the foregoing, it can be concluded that benefits plans have made significant contributions to the economic development of NL and Canada as a whole. However, the performance of benefits plans as revealed in this discussion should be taken with caution, for other than through annual and quarterly reports, which were published by companies and the C-NLOPB, not much is known about the successes or limitations of benefits plans. Detailed independent studies are needed in order to properly and fairly assess the performance of benefits plans.

4 Local Content Under Impact Benefit Agreements with Aboriginal Communities

Development policy in Canada has had 'a chequered history' with regard to Aboriginal peoples (Young 1995). Filled with colonial mentality, Canadian development policy for a long time showed an unwillingness to recognize Aboriginal voices regarding both the process and the goals of development (Young 1995). This generated enormous acrimony in the relationship between the government and Aboriginal peoples. However, it has since been recognized that reconciliation can be achieved only through careful consultation between the government and Aboriginal peoples, conducted within a framework that incorporates Aboriginal social and economic values (Young 1995). To this end, Canadian courts have determined that the Crown has a duty to consult Aboriginal peoples whenever the Crown intends to act in a manner that may have an adverse impact on the enjoyment of Aboriginal or treaty rights guaranteed under Section 35 of the Constitution Act, 1982.[7]

The Crown's duty to consult is rooted in the 'Honour of the Crown' and is legally binding and enforceable.[8] The duty is not owed only by the federal Crown but also applies to provincial Crowns and unless the Crown can obtain the consent of Aboriginal peoples, any infringement of Aboriginal or treaty title must be fully justified.[9] Furthermore, the Crown cannot delegate the duty to consult and accommodate to third parties (Newman 2014). By implication, project developers cannot legally assume the duty to consult. However, it has become 'a non-optional practice', connected with the need to obtain a social licence to operate, for project developers to engage in consultations with Aboriginal peoples before the commencement of projects (Newman 2014:80). Whilst this is not as a matter of legal obligation, it is

---

[6]See Mobil Investments Canada Inc & Murphy Oil Corporation v Canada (2012), ICSID Case No. ARB(AF)/07/4; Attorney General of Canada, v Mobil Investments Canada Inc. and Murphy Oil Corporation, 2016 ONSC 790, https://www.italaw.com/sites/default/files/case-documents/italaw7160.pdf.

[7]Haida Nation v. British Columbia (Minister of Forests), 2004 SCC 73, 3 S.C.R. 511.

[8]Ibid.

[9]Tsilhqot'in Nation v British Columbia, 2014 SCC 44.

arguable that consultation by industry evolved out of the entrenched nature of the Crown's constitutional duty to consult coupled with pressures exerted by Aboriginal peoples who threatened to obstruct the take-off of resource development projects and demanded control over natural resources on their lands.

An important offshoot of the evolution of the Crown's duty to consult is the development of Impact Benefit Agreements (IBAs) between project developers and Aboriginal communities. Broadly speaking, IBAs are contractual arrangements negotiated between project developers and Aboriginal communities with the principal aims of accommodating Aboriginal interests and building relationships. Historically, the negotiation of IBAs in Canada began in the mid-1970s as agreements between the government and industry, whereby the government acted on behalf of Aboriginal communities to provide socio-economic benefits for them (Public Policy Forum 2006; Gilmour and Mellett 2013). Since the 1990s, however, IBAs have been generally negotiated directly by Aboriginal people and project developers. Empirical research shows that the negotiation of IBAs is part of industry's 'good neighbour' policy intended to ensure business stability by fending off opposition and securing the support of Aboriginal communities (Public Policy Forum 2006; Gilmour and Mellett 2013). The direct involvement of Aboriginal communities in the negotiation of IBAs has been largely attributed to the evolution of the recognition of Aboriginal rights by the courts as well as through treaties and land claim agreements (Public Policy Forum 2006).

IBAs are generally not legally mandated (Gilmour and Mellett 2013; Lacasse 2005).[10] In fact, they occupy a regulatory void, for there are no regulations governing their negotiation or implementation. Typically, an Aboriginal community that signs an IBA commits to support the resource development project to which the IBA relates in return for a 'package of measures' that include both economic benefits and the minimization of the adverse socio-environmental impacts of the resource development on the Aboriginal community (Imai 2017; O'Faircheallaigh 2003).

There is little guidance on what should be the contents of an IBA, but IBAs have become an important local content instrument in Canada even though the Canadian government does not see them to be so (OECD 2017). Typically, IBAs often provide for the employment of members of Aboriginal communities who are parties to the agreements and may provide for preferential consideration for the community members before other candidates are considered. Often, IBAs set targets or quotas for the employment of Aboriginal people on a project (Sosa and Keenan 2001). To facilitate the employment of Aboriginal people, some IBAs require that positions be advertised in Aboriginal newspapers, which are published by Aboriginal organizations primarily for the readership of their members (McKinley and Huebner 2018). Some IBAs give priority to Aboriginal employees over seniority in times of lay-off (Sosa and Keenan 2001).

Associated with employment is the provision of training and apprenticeship programmes as well as educational, including scholarship, programmes to enable

---

[10]Some initial IBAs were mandated under land claim settlements. See Gilmour and Mellett (2013); Lacasse (2005).

Aboriginal people to acquire relevant skills. Some IBAs extend these obligations of project developers to their contractors and subcontractors even though these are not parties to the agreements (Sosa and Keenan 2001). Cash payments by project developers to Aboriginal communities can also form part of IBA provisions (Kennet 1999). Other typical elements of IBAs include clauses that prohibit Aboriginal communities from objecting to projects, or from engaging in specified kinds of resistance (Gilmour and Mellett 2013). From the perspective of project developers, such clauses are the key element in IBAs.

Unlike benefits plans and other local content measures in other countries, IBAs are ad hoc in nature and are negotiated with specific Aboriginal communities rather than with the government. There is little government oversight regarding their content and implementation. Also, unlike benefits plans, they target specific social groups within the population rather than all locals. From a local content perspective, therefore, local is understood in the context of IBAs as only the Aboriginal community to which the IBA relates, who have faced historical marginalization in Canada (Grice 2018). This approach is different from the notion of 'community content' (also referred to as 'local–local content') found is some jurisdictions, such as the Philippines.[11] Whilst community content is location-specific, Aboriginal content is social group-specific. However, both community content and Aboriginal content spring from the same intuition, for, like Aboriginal content, community content is rooted in the fact of marginalization that communities located near extractive resource projects have historically faced, especially in developing countries (Nwapi 2015; Marcel et al. 2016; Esteves et al. 2013). A major advantage of IBAs over legally mandated local content requirements is that being a product of dialogue and negotiation between project developers and Aboriginal communities, IBAs serve as an important instrument for relationship building and conflict prevention and resolution.

IBAs are not without their problems. Scholars have highlighted the power imbalance in the negotiation of IBAs (Caine and Krogman 2010). Such power imbalance influences the actual contents of an IBA, in terms of the type, quality and quantity of benefits project developers are to provide to Aboriginal communities. The ad hoc, flexible nature of IBAs has also meant that the quality and quantity of benefits to Aboriginal communities depends largely on their negotiation strategy and might in some cases imply weak monitoring of the implementation of IBAs (OECD 2017). The power imbalance has been blamed in part on the absence of Crown involvement in most IBAs although Crown non-involvement is on the other hand viewed as a recognition of the capacity of Aboriginal people to determine their priorities and to speak for themselves (Fidler and Hitch 2007).

There is no designated agency to supervise and monitor the design and implementation of IBAs. This means that the prospects for policy learning are very limited. There is also a problem of transparency. Unlike benefits plans under the Atlantic

[11] Section 136(d) of the Revised Rules and Regulations of the Philippines Mining Act of 1995 stipulates that priority should be given to Filipinos living in 'local' or neighbouring communities or the province where a mine is located.

Accords, IBAs incorporate confidentiality clauses that prohibit their public disclosure (Gilmour and Mellett 2013). As such, the actual contents of specific IBAs are shrouded in secrecy. Furthermore, where Aboriginal land claims have not been settled or where the impacts of a project extend from the boundaries of one land claim into another, the negotiation of IBAs is complicated. This is because in such situations, there is often a proliferation of claims and demands for IBAs and 'there is no principled basis for determining eligibility to negotiate IBAs' in such situations (Kennet 1999:35). Project developers have often cited Aboriginal people's 'lack of basic education, skills, business experience and capital' as the main reasons why many mining projects do not generate significant employment and business opportunities for Aboriginal people. And efforts by project developers to prepare Aboriginal people to take advantage of the opportunities often takes a long time to produce results (Kennet 1999). As a result, most of the available jobs for Aboriginal people are unskilled jobs whilst non-Aboriginal people disproportionately occupy the skilled and higher paying positions (Gibson et al. 2005).

The use of 'fixed quotas and guaranteed jobs or contracts'—a common element in IBAs—is widely considered by Canadian commentators to be ill-suited to enable Aboriginal people to become economically independent (Kennet 1999). Likewise, unskilled jobs hardly provide long-term benefits to Aboriginal people, for when, for instance, a mine shuts down, unskilled employees are hardly able to find new jobs due to their lack of skill (Kennet 1999). Similarly, the long-term economic value of cash payments to Aboriginal peoples has been questioned due to the absence of accountability mechanisms to ensure that cash payments are responsibly utilized to promote economic development (Brock and Migone 2018; Blue and O'Faircheallaigh 2018; Kennet 1999). It has also been found that Aboriginal organizations that have been awarded contracts by project developers often subcontract them to non-Aboriginal companies or form joint ventures in which their 'operational role' is very minimal (Kennet 1999).

As a local content instrument, the greatest weakness of IBAs is perhaps the lack of any guidance established by an independent body regarding the negotiation process, content and implementation monitoring of IBAs. This has tilted the scale heavily to one side. It is also a serious shortcoming that IBAs are kept confidential. Being confidential, there is no way the public can credibly assess claims made by project developers regarding their commitment to provide benefits to Aboriginal communities impacted by their projects.

5 Conclusion

Local content in Canada's extractive industry is pursued through three vehicles: the federal foreign investment policy, benefits plans under federal oil and gas legislation and under federal accords with the Atlantic provinces of Newfoundland and Labrador and Nova Scotia, and under IBAs negotiated between project developers and Aboriginal communities. These vehicles also incorporate the understanding of the different types of 'local' in LCPs, namely national content, provincial content (which roughly aligns with the notion of local-local content) and Aboriginal content. The idea of Aboriginal content is unique in that it focuses on a particular social group, rather than a geographical location, which has faced historical marginalization within the Canadian population. The Canadian approach thus shows how the LCP can be utilized to address the economic situation of other marginalized groups in other countries. It must be warned, however, that the status of Aboriginal peoples in Canada in relation to extractive resource development is unique, as recognized under the Canadian constitution, and may not directly be transposed to other marginalized groups around the world. However, it may be possible for other jurisdictions to adapt the LCP to their own situation to address the challenges faced by marginalized groups within their territory.

Of the three local content vehicles, benefits plans represent the strongest incorporation of local content requirements and align with the practice in most other jurisdictions. This is because of the explicit requirements on local employment and procurement, the giving of first consideration to residents of the provinces (NL and Nova Scotia) near the offshore oil and gas resources and the establishment of a Board to supervise the design and implementation of benefits plans. However, there is very limited independent assessment of the successes and challenges of benefits plans available, as available sources emanate from the C-NLOPB and companies. On the other hand, whilst IBAs with Aboriginal communities facilitate relationship building vital for the success of projects, they are tormented by lack of guidance on their contents, lack of a watchdog agency to monitor their implementation, and lack of transparency about their actual contents. And whilst federal foreign investment policies provide a useful instrument to promote local content through the assessment of net benefits, their success has been marred by political considerations.

The statistics show that benefits plans have led to a large volume of goods and services procurement being localized. Massive local employment has also been created and significant investments in training and in research and development have equally been made. Experience with the benefits plans shows that where the government has a clearly defined goal, it is easier to measure progress. The background to the emergence of benefits plans in NL (and in Nova Scotia) shows that the government's primary goal was to address the massive unemployment problem in the province as well as to ensure that other benefits of offshore oil and gas development are retained in the province. Whilst the federal government understood the importance of spreading the benefits across the country, it recognized that the province bordering the continental shelf ought to be primary beneficiary of the resources. Accordingly, it conceded to a benefits plan regime that gives first consideration to the subnational locality likely to be impacted the most by the development of the resources rather than adopting a broader national content strategy. This is an important lesson for other LCP countries. Every country is unique and should establish its goals accordingly, as there is not a one-size-fits-all approach.

References

Attorney General of Canada v Mobil Investments Canada Inc. and Murphy Oil Corporation, 2016 ONSC 790. https://www.italaw.com/sites/default/files/case-documents/italaw7160.pdf. Accessed November 1, 2019.

Avery, A., Glossop, P., & Olexiuk, P. (2013). Foreign investment in Canada's oil and gas sector: New and emerging challenges. *Alberta Law Review, 51*(2), 343–384.

Barber, L. (2018). Inside the mobilities regime of Newfoundland and Labrador's construction megaprojects: Experiences of rotational work close to home. *Journal of Canadian Labour Studies, 81,* 187–212.

Bishop, G. (2016). A rule of reason for inward FDI: Integrating Canadian foreign investment review and competition policy. *The School of Public Policy Research Papers, 9*(34). University of Calgary. https://www.policyschool.ca/wp-content/uploads/2016/10/Foreign-Investment-Bishop-Final.pdf. Accessed November 9, 2019.

Blue, L., & O'Faircheallaigh, C. (2018). Indigenous autonomy and financial decision-making in communities. *Financial Planning Research Journal, 4*(2), 39–50.

Bregha, F. (2016). *National energy programme.* https://www.thecanadianencyclopedia.ca/article/national-energy-program. Accessed October 20, 2019.

Brock, K., & Migone, A. (2018). Financial capacity of First Nations in Canada. *Canadian Public Administration, 61*(4), 479–498.

Caine, K. J., & Krogman, N. (2010). Powerful or just plain power-full? A power analysis of impact and benefit agreements in Canada's north. *Organization & Environment, 23*(1), 76–98.

Canadian Association of Petroleum Producers. (2018a, July). *Canada's oil sands.*

Canadian Association of Petroleum Producers. (2018b). *Economic report series: leveraging opportunities: Diversifying Canada's oil and natural gas markets.*

Canada—Measures affecting the Renewable Energy Generating Sector, and Canada—Measures relating to the Feed-In tariff Programme. Reports of the Appellate Body (WT/DS412/AB/R: WT/DS426/AB/R, May 6, 2013).

Canada-Newfoundland Atlantic Accord Implementation (Newfoundland) Act.

Canada-Nova Scotia Atlantic Accord Implementation (Nova Scotia) Act.

Canada-Newfoundland and Labrador Offshore Petroleum Board. (2006). The *Canada-Newfoundland and Labrador Benefits Plan Guidelines.*

Canada-Newfoundland and Labrador Offshore Petroleum Board. (2016). *Canada-Newfoundland and Labrador benefits plan guidelines (Draft) 2016.* https://www.cnlopb.ca/wp-content/uploads/cnlopbsn/bpguide.pdf. Accessed October 15, 2019.

Canada-Newfoundland and Labrador Offshore Petroleum Board. (2018). *2017–2018 annual report.* https://www.cnlopb.ca/wp-content/uploads/ar2018e.pdf. Accessed October 15, 2019.

Clarke, M. (2004). Primary beneficiaries: Newfoundland and Nova Scotia's struggle to achieve the promise of petroleum wealth. *Dalhousie Journal of Legal Studies, 13*(2), 1–30.

Esteves, A. M., Coyne, B., & Moreno, A. (2013). *Local content initiatives: Enhancing the subnational benefits of the oil, gas and mining sectors.* Natural Resource Governance Institute Briefing Note. https://resourcegovernance.org/sites/default/files/Sub_Enhance_Benefits_20151125.pdf. Accessed November 11, 2019.

Fidler, C., & Hitch, M. (2007). Impact and benefit agreements: A contentious issue for environmental and aboriginal justice. *Environments Journal, 35*(2), 45–69.

Geipel, J., & Hetherington, D. (2018). Local content policy: What works, what doesn't work. https://assets.publishing.service.gov.uk/media/5c7d425aed915d4063af98df/BERF-Zimbabwe-What-Works-in-Local-Content-Policy-30Oct18.pdf. Accessed November 11, 2019.

Gilmour, B., & Mellett, B. (2013). The role of impact and benefit agreements in the resolution of project issues with First Nations. *Alberta Law Review, 51*(2), 385–400.

Globerman, S. (1984). Canada's Foreign Investment Review Agency and the direct investment process in Canada. *Canadian Public Administration, 27,* 313.

Gordon, W. (1957). *Royal Commission on Canada's Economic Prospects. Final report*. Ottawa: Queen's Printer.

Gray, H. (1972). *Foreign direct investment in Canada*. Ottawa: Information Canada.

Grice, T. (2018). Local content policies in the mining sector: Stimulating direct local employment. In *Intergovernmental Forum on Mining, Minerals, Metals and Sustainable Development and the International Institute for Sustainable Development*. <https://www.iisd.org/sites/default/files/publications/local-content-policies-mining-direct-local-employment.pdf. Accessed October 16, 2019.

Haida Nation v. British Columbia (Minister of Forests), 2004 SCC 73, 3 S.C.R. 511.

Hibernia Management and Development Company Ltd. (2016). *Canada—Newfoundland and Labrador benefits report*. https://www.hibernia.ca/2016AnnualReportingofIndustrialBenefits.pdf. Accessed November 1, 2019.

Hilmer, N. (2013). Foreign Investment Review Agency. The Canadian encyclopedia. https://www.thecanadianencyclopedia.ca/article/foreign-investment-review-agency. Accessed November 9, 2019.

Hoffman, A. (2010). *Broken promises mark foreign mining deals*. https://www.theglobeandmail.com/globe-investor/broken-promises-mark-foreign-mining-deals/article574587/. Accessed November 9, 2019.

House of Commons—Standing Committee on External Affairs and National Defence. (1970). *Special Committee respecting Canada–US relations: Report to the House*.

Husky Energy. (2016). *Canada-Newfoundland and Labrador benefits annual report*. Husky Energy Ref. No. WR-ADM-RP-0044. https://huskyenergy.com/downloads/areasofoperations/eastcoast/CNLOPB_2016_Annual_Report_Website.pdf. Accessed November 4, 2019.

Imai, S. (2017). Consult, consent and veto: International norms and Canadian treaties. In M. Coyle, & J. Borrows (Eds.), *The right relationship: Reimagining the implementation of historical treaties* (p. 405). Toronto: University of Toronto Press.

Industry Canada. (2010). *Investment Canada Act: annual report 2009–2010*. https://www.ic.gc.ca/eic/site/ica-lic.nsf/vwapj/ica-report-2010-eng.pdf/$file/ica-report-2010-eng.pdf. Accessed November 10, 2019.

Intergovernmental Forum on Mining, Minerals, Metals and Sustainable Development (IGF). (2018). *IGF guidance for governments: Local content policies*. Winnipeg: IISD. https://www.iisd.org/sites/default/files/publications/igf-guidance-for-governments-local-content.pdf. Accessed November 11, 2019.

Iron Ore Company of Canada. (2017). *IOC Wabush 3 open pit mine project industrial and employment benefits plan 2017*. http://www.ironore.ca/CLIENTS/1-iocv4/docs/upload/sys_docs/IOC_Wabush_3_Benefits_Plan__January_27_2017.pdf. Accessed November 7, 2019.

Kennet, S. (1999). *Issues and options for a policy on impact and benefits agreements*. Paper prepared for the Mineral Resources Directorate, Department of Indian Affairs and Northern Development, Canadian Institute of Resources Law. http://www.cbern.ca/content/uploads/sites/4/2016/03/Issues-and-Options-for-a-Policy-on-Impact-and-Benefits-Agreements.pdf. Accessed November 9, 2019.

Kiragu, E. (2015). *Transition into a Green Economy: Are there limits to government intervention?* World Trade Institute Working Paper No. 05/2015. <https://www.wti.org/media/filer_public/5b/dd/5bddb3d9-5ed8-448a-8d38-ff3325c4cd97/wti_seco_wp_05_2015.pdf. Accessed October 13, 2019.

Kuntze, J.-C., & Moerenhout, T. (2013). *Local content requirements and the renewable energy industry—A good match?* International Centre for Trade and Sustainable Development, Ad hoc Expert Group Meeting on Domestic Requirements and Support Measures in Green Sectors: Economic and Environmental Effectiveness and Implications for Trade Salle XXI, Palais des Nations, Geneva, June 13–14, 2013. https://unctad.org/meetings/en/Contribution/DITC_TED_13062013_Study_ICTSD.pdf. Accessed November 1, 2019.

Lacasse, J.-P. (2005). Impacts and benefits agreements on aboriginal title lands. In D. Dorey, & J. Magnet (Eds.), *Legal aspects of aboriginal business development* (p. 311 at 321). Markham: Butterworths.
MacDonald, C. P., & Thompson, R. S. G. (1985). The Atlantic accord: The politics of compromise. *Alberta Law Review, 24*(1), 6.
Marcel et al. (2016).
McKinley, T., & Huebner, C. (2018). *Extractive industries and consumption linkages to enhance industrialisation.* Paper prepared for the Consultancy for the United Nations Development Programme Regional Service Centre for Africa 'Extractive Industries for Sustainable Development' Programme, January 2018.
Mining Association of Canada. (2018). *Facts and figures of the Canadian mining industry.* https://mining.ca/wp-content/uploads/2019/03/Facts-and-Figures-English-Web_0.pdf. Accessed November 11, 2019.
Mobil Investments Canada Inc. & Murphy Oil Corporation v Canada. (2012). ICSID Case No. ARB(AF)/07/4.
Natural Resources Canada. (2018). Crude oil facts. https://www.nrcan.gc.ca/science-data/data-analysis/energy-data-analysis/energy-facts/crude-oil-facts/20064. Accessed November 9, 2019.
Newman, D. (2014). *Revisiting the duty to consult Aboriginal peoples in Canada, revised edition.* Saskatoon: Purich Publishing.
Nwapi, C. (2015). Defining the 'local' in local content requirements in the oil and gas and mining sectors in developing countries. *Law & Development Review, 8*(1), 187–216.
Nwapi, C. (2014). Resource extraction on the courtroom: The significance of Choc v Hudbay Minerals Inc. for transnational justice in Canada. *Asper Review of International Business and Trade Law, 14,* 121–160.
Nwapi, C., & Lee, Y. S. (2016). Trade and development in Africa. In Y. S. Lee (Ed.), *Reclaiming development in the world trading system* (2nd ed., pp. 334–360). New York: Cambridge University Press.
O'Faircheallaigh, C. (2003). *Accountability in the negotiation of indigenous mineral development agreements.* Brisbane, Queensland, Australia: Centre for Governance and Public Policy, Griffith University.
Olson, T. H. (1980). Foreign investment restrictions on Canadian energy resources. *International Lawyer*, 579–608.
O'Sullivan, B. J. (1980). Canada's Foreign Investment Review Act revisited. *Fordham International Law Journal, 4*(1), 174–198.
Prospectors and Developers Association of Canada. (2019). *Industry news.* CORE. https://issuu.com/pdac-web/docs/core__fall_2019_?e=36081262/72726050. Accessed November 8, 2019.
Public Policy Forum. (2006). *Sharing in the benefits of resource developments: A study of First Nations-Industry impact benefits agreements.* https://ppforum.ca/wp-content/uploads/2018/05/Sharing-in-the-Benefits-of-Resource-Development-PPF-report.pdf. Accessed November 9, 2019.
Ramdoo, I. (2016). *Local content, trade and investment: Is there policy space left for linkages development in resource-rich countries?* ECDPM Discussion Paper No. 205. https://ecdpm.org/publications/local-content-trade-investment/. Accessed November 9, 2019.
Ramdoo, I. (2015). Unpacking local content requirements in the extractive sector: What implications for the global trade and investment frameworks? *E15Initiative.* Geneva: International Centre for Trade and Sustainable Development (ICTSD) and World Economic Forum. https://ecdpm.org/wp-content/uploads/Unpacking-Local-Content-Requirements-Isabelle-Ramdoo-September-2015-ECDPM.pdf. Accessed November 9, 2019.
Reference re Newfoundland Continental Shelf, [1984] 1 SCR 86.
Revised Rules and Regulations of the Philippines Mining Act of 1995.
Rose, W. B. (1986). Foreign investment in Canada: The New Investment Canada Act. *International Lawyer, 20,* 19–44.

Rostein, A. (2015, 29 June). *Economic nationalism.* The Canadian Encyclopedia. https://www.thecanadianencyclopedia.ca/en/article/economic-nationalism. Accessed October 30, 2019.

Sheppard, M. J., & Hardwicke-Brown, M. (1992). Overview of the Investment Canada Act with a focus on investments in Canada's upstream oil and gas industry. *Alberta Law Review, 30*(1), 4–56.

Shrimpton, M. (2012). Delivering benefits: Approaches, initiatives and outcomes. In *OTC Arctic Technology Conference*, Houston, TX. https://www.onepetro.org/conference-paper/OTC-23706-MS. Accessed 6 November 6, 2019.

Suncor Energy. (2018). *Terra Nova development: Canada-Newfoundland and Labrador benefits public annual report.*

Taylor, A., & Dickey, J. (2001). Regulatory regime: Canada-Newfoundland/Nova Scotia Offshore Petroleum Board Issues. *Dalhousie Law Journal, 24,* 51–86.

Tsilhqot'in Nation v British Columbia, 2014 SCC 44.

United Nations Conference on Trade and Development. (2011). How to attract and benefit from FDI in mining: Lessons from Canada and Chile. Best practices in investment for development case studies in FDI. New York and Geneva. https://unctad.org/en/Docs/diaepcb2010d11_en.pdf. Accessed November 9, 2019.

Vogel, D. (2017). The role of policy learning and regulation in regulatory excellence. In C. Coglianese (Ed.), *Achieving regulatory excellence* (pp. 207–224). Washington, DC: Brookings Institution.

Watkins, M. (1968). Task Force on the Structure of Canadian Industry. *Foreign ownership and the structure of Canadian industry—Report of the Task Force on the Structure of Canadian Industry.* Ottawa: Queen's Printer.

Wesson, T. (2007). The sectoral structure of Canada's economy. In Tom Wesson (Ed.), *Canada and the new world economic order: Strategic briefings for Canadian enterprise* (3rd ed., pp. 68–82). Concord: Captus Press.

Wood Mackenzie (2018). *Newfoundland & Labrador competitiveness in oil & gas investment.* https://www.nr.gov.nl.ca/nr/publications/energy/Competitiveness_Oil_Gas_Investment.pdf. Accessed October 29, 2019.

Young, E. (1995). *Third world in the first: Development and indigenous peoples.* London: Routledge.

Local Content Policies for Regional Economic Development in Western Australia

Simon White

Abstract This chapter examines the efforts of the Western Australian Government to promote new economic, business and employment opportunities for local firms and residents in its non-metropolitan regions. Specific attention is given to the use of public policies, institutions and programmes to realise these opportunities based on increasing investments in the extractives sector.

1 Introduction

Like many territories engaged in resource extraction, mining sites in the Australian State of Western Australia (WA) are largely located in areas of low population. While this creates logistical challenges for the extraction, processing and transportation of minerals, based on distance and infrastructure, it also creates a challenge for the development of sustainable non-metropolitan regional communities. While governments often welcome the significant investments mining brings, they are often eager to ensure these benefits extend beyond the payment of mineral royalties by contributing to new jobs, improved infrastructure and increased business for local firms. Local content policies have been used by governments to enhance these broader benefits, often with mixed results.

The WA economy is dominated by mining, and while this has led to significant investments into new jobs and better infrastructure, the sustainability of these outcomes is not assured. Thus, it is important to use these investments to transform the economy in ways that deepen their industrial base, improving local skill and firm capabilities. This chapter examines the case of WA and the attempts by the Western Australian Government to increase local content provisions for local workers and businesses.

S. White (✉)
Publicus Pty. Ltd., Subiaco, Australia
e-mail: simon@simonwhite.com.au

E. G. Pereira et al. (eds.), *Sovereign Wealth Funds, Local Content Policies and CSR*, CSR, Sustainability, Ethics & Governance,
https://doi.org/10.1007/978-3-030-56092-8_23

2 Mining and the WA Regional Economy

The development of the State's non-metropolitan regions has been driven by agriculture and minerals. Late in the nineteenth century, when WA was still a colony, some 32% of its population lived in non-metropolitan settlements. Many of these communities were transient, based on the economic opportunities of the time. Some fifty years after Federation, a series of State resource booms shifted the economy towards minerals (Wilson et al. 2004). Indeed, the sectoral contribution to the WA economy has changed significantly over time. While the service sector now dominates the State economy, mining accounted for 30% of Gross State Product (GSP) in 2017–2018, while agriculture (around A$79 billion) forestry and fishing accounted for only two per cent. Indeed, Western Australia accounted for 52% of Australia's mining gross value added in 2017–2018. Iron ore, from the Pilbara Region, accounted for 50% of the State's minerals and petroleum sales in 2018, followed by liquefied natural gas at 21%. Other key contributors, albeit at a smaller scale are gold (9%), alumina and bauxite (6%), crude oil (2%), and lithium, tin and tantalum (1%) (Government of Western Australia 2019).

Both the Federal and State governments have supported regional development with varying objectives. At the Federal level, government influences development in regions in six main ways. First, it redistributes income from the more populous to the less populous States, primarily through direct payments to Local and State governments. Second, it contributes to regional development through policies dealing with 'adjustment failures' within those regions with persistently high unemployment and declining economic activity. Third, it has policies that seek to promote selected regional 'winners' by strengthening their comparative advantages and industry linkages. Fourth, it has introduced general decentralisation measures to spread people and industry beyond the capital cities. Fifth, it influences regional development through industry policy and the provision and pricing of public infrastructure, notably in relation to roads, transport and communications. Finally, it provides various types of assistance to promote regional growth directly (Department of Tourism and Department of Industrial Relations 1992).

State governments have supported development in the regions through a more direct involvement with rural regions, albeit with fewer funds, while struggling to balance urban policies with rural ones. Since Federation, State governments have been involved in opening up new land for development. This largely focused on the provision of essential infrastructure. However, in the past 50 years, State governments have also had to deal with some of the social and economic consequences of rural decline. Most States have established a network of regional development organisations with various organisational and funding arrangements. The WA network is discussed further below.

The current WA *Regional Development Strategy: 2016–2025*, prepared under the former Liberal-National Government (2008–2017), describes how the State government would 'direct greater resources towards establishing the socio-economic foundations for development while also driving growth and new investment in the regions'. This includes fostering the growth of emerging industries and the attraction of new private investment in the form of 'capital and capability (Government of Western Australia 2016).

A constant challenge for Federal and State governments has been to find ways to broaden the benefits of mining investments and build a broader set of industrial capabilities. In the 1980s, the Australian Parliament commissioned an investigation into Australian industry participation in major projects. The resulting report, *A sea of indifference: Australian industry participation in the North West Shelf Project,* found that government had failed to 'play a significant role in promoting and fostering local industry in major projects' and is 'indifferent to the real opportunities for domestic industry growth' (Chairperson's Foreword). As a result, the inquiry argued the case for local content provisions and the development of cross-sectoral linkages (House of Representatives Standing Committee on Industry Science and Technology 1998).

Comparing Australia with Norway, Hunter (2014) argues that the Australian government's failure to use policy to diversify its industrial base in the petroleum resources sector has meant that domestic industry diversification in Australia has faltered, with minimal economic diversification occurring and few cross-sectoral linkages established and a failure to promote local suppliers'. Australia's primary exports are dominated by minerals and agricultural products, while Norway exhibits a more diversified portfolio. Norway has used revenues from oil and gas to create a sovereign wealth fund with over one trillion US dollars. In addition to petroleum and gas, Norway's economy includes machinery and equipment, metals, chemicals, ships and fish products.

More recently, the 2018 report of the Australian Parliament's Committee on how the mining sector can support businesses in regional economies, *Keep it in the regions*, describes a range of initiatives resource companies have employed to promote local content. The Committee notes, however, that despite many large mining companies having local procurement programmes, 'regional communities are missing out on these opportunities' (House of Representatives Standing Committee on Industry Innovation Science and Resources 2018: 43). Indeed, small and medium enterprises (SME's) were found to face a number of barriers to their participation in mining contracts, including the centralisation of mining company offices outside of the regions in which the company operates, complex tendering processes, loose definitions of 'local' used by mining companies, poor regional infrastructure and a lack of local industry capacity.

Thus, government efforts to build an industrial base for increasing value addition have proved to be relatively shallow and inadequate. It is in this context that the efforts of the WA Government to support regional economic development through local content mechanisms are examined. This incorporates the institutional and policy framework for regional development, including support for the economic and industrial development of the State's non-metropolitan regions. More specifically, this review includes the major policies, laws and programmes that attempt to integrate mining into the regional economies in which they operate.

3 Institutions for Regional Development

As with most other States, WA has established a network of regional institutions designed to coordinate and support development in the nine demarcated non-metropolitan regions. Each region contains a Regional Development Commission (RDC) to oversee the State government's support for development. These agencies coordinate government service provision, conduct studies of social and economic infrastructure requirements, provide information to new and existing firms, and identify regional growth opportunities and impediments to development with the aim of attracting investment to their respective regions. RDCs are legislated by the *Regional Development Commissions Act of 1993*. State Government funding varies significantly between regions, both in absolute and per capita terms.

The RDCs develop and implement regional development policies, strategies and plans. While these typically cover a wide range of issues, including government services, social and community development, and infrastructure, they also have a particular emphasis on economic strategies to enhance growth and job creation. The economic component of these regional development plans typically identifies strategic industry sectors, both current and emerging, and describes the ways RDCs, in concert with other State and Local government entities, and in some cases Federal Government instrumentalities will work with the private sector and civil society to achieve desired outcomes. Within this context, several regional communities in Western Australia have formulated regional 'blueprints'. Regional blueprints assist in future regional development and set a firm foundation for regional action. They are prepared by the RDC and are designed to strengthen regional policy leadership and horizontal policy integration. Blueprints incorporate land use planning, while identifying future needs for a better and broader range of government services (Government of Western Australia 2010).

As statutory authorities with the status of a State government agency, RDCs play a pivotal role in planning, in collaboration with the peak planning agency, the Western Australian Planning Commission. The Western Australian Planning Commission is the decision-making body responsible for guiding the future development of the State. It plays a major coordinating role across all aspects of the State Government's planning process and operates as a partnership between the community, business and all levels and sectors of government. It has a broad range of strategic responsibilities, including the preparation and implementation of the State Planning Strategy that proposes a vision for the future development of Western Australia. Within this mandate, the commission has responsibilities for metropolitan and non-metropolitan matters.

The State government has invested heavily into regional planning, requiring a wide range of State and Local government agencies to prepare and submit plans on a regular basis. In 2014, the Planning Commission released the *State Planning Strategy 2050*. This document sought to reach 'beyond traditional land-use planning' by placing 'a priority on economic and population growth as the key drivers of land use and land development' (pp. 7). This would 'help to align and deliver regional development programmes and services' (Western Australian Planning Commission 2014:8). The strategy also identified a hierarchy of 27 regional centres and sub-regional centres, also referred to as 'Super Towns'. State government has supported the preparation of Blueprints for the development of selected regional centres and towns and has also supported regional centres in the preparation of economic growth plans (Department of Regional Development and Lands 2011).

Another State planning structure, the Regional Development Council is the peak advisory body to the Minister for Regional Development. Established by the Regional Development Commissions Act 1993, the Council consists of the chairs of the nine RDCs and advises the Minister for Regional Development on all regional development issues. This includes promoting the development in the regions, developing policy proposals, facilitating liaison between commissions and relevant government agencies, and liaising between Local, State and Federal Government bodies with respect to regional issues (Government of Western Australia 2017). In addition, the Regional Development Trust was established in 2009 as an independent statutory advisory body under the Royalties for Regions Act. The Trust provides independent and impartial advice and recommendations on the allocation of funds from the Royalties for Regions Fund.

Paül and McKenzie (2015) describe how regions in Australia have evolved from an 'undefined and sometimes confused meaning' to become 'particular and bounded territories with specific attributions'. WA regions have evolved from being 'ambiguous rural areas to constitute certain territorial shapes with a distinct unit in the spatial structure of the society. Despite the top-down, State-driven approach to regional development, WA's regions are unique in Australia for their process of regionalisation and regional consistency. The RDCs have played a significant role in this regard, providing sustained institutions for planning.

Appointed by the Minister for Regional Development, RDC members typically include private sector and community representatives from the region. Despite this, the level of private sector participation in economic planning and development initiatives is somewhat limited.

4 State Agreements and Public–private Partnerships

The State Government established a range of initiatives to engage large mining projects in an effort to maximise the economic and social benefits of mining and mitigate environmental risks. Sensitive to the dangers of scaring off international investors, these initiatives have sought to encourage, rather than regulate for greater integration by local firms and workers into supply chains, while promoting some level of corporate social responsibly. The obligations placed upon firms vary from project to project.

One of the State Government's earliest and most long-lasting attempts to regulate for local content in major resource projects was through the use of State Agreements.

These are legal contracts between the Western Australian Government and a proponent of a major project. Under such agreements, proponents take or share responsibility with the State for developing infrastructure specific to the project (Department of Jobs Tourism Science and Innovation 2019). Thus, a State Agreement is a 'contract containing financial and non-financial concessions granted by the State in return for project obligations accepted by private companies' (Horsley 2013).

Since 1952, these agreements had their genesis in assisting the establishment of a fledgling iron ore mining project for domestic usage only, due to the then Federal Government's embargo on the export of iron ore. This project contemplated mining of iron ore deposits at Yampi Sound in the State's West Kimberley region. In fact, the first State Agreement was formulated in 1947, principally to reduce the amount of land that could be held under a mining lease due to a concern that limited strategic resources should not be controlled by a single party. However, first State Agreement in the form it is understood today was signed by the State government and BHP to support the development of an oil refinery in Western Australia, the Oil Refinery (Kwinana) Agreement Act 1952. This initiated the subsequent widespread use of State Agreements to facilitate development of major resource-based projects commenced with the development of the State's iron ore industry in the Pilbara.

Over time, local content provisions have been incorporated into State Agreements. One example of this is the agreement entered into as part of the Gorgon Project, which requires the use of local labour, suppliers and professional services where it is reasonable and economically practicable to do so. The proponents are also required to provide local suppliers with a fair opportunity to tender or quote during design and the tendering and letting of contracts.

Horsley (2013) describes how the 1960s State Government vision for an integrated industrial complex in the desert was written into the series of State Agreements entered into for the development of the Pilbara region. These agreements all contained clauses calling on the companies to maximise local content and proceed through secondary processing to an integrated iron and steel industry. However, it became evident that the forward and backward industrial linkages were not emerging and could not be left to the market alone. As a result, the government attempted to take a more proactive approach through regional industrial development policies.

5 Royalties for Regions

The Royalties for Regions programme was the centrepiece of the State government's regional development agenda at the height of the mining boom in the late 2000s. Shortly after the election of the Liberal-National alliance Government in 2008, the Royalties for Regions Act 2009 was passed. This provided the legislative framework for the creation of the Royalties for Regions Fund and the Western Australian Regional Development Trust. The Act provides for the equivalent of 25% of the State's forecast annual mining and onshore petroleum royalties to be directed into the fund, which is separate from the government's Consolidated Account. The Royalties for Regions Fund balance cannot exceed A$1 billion at the end of a financial year with any surplus funds returned to the Consolidated Account. The object of the Act is to promote and facilitate economic, business and social development in regional Western Australia through the operation of the Fund (Royalties for Regions Act 2009). Monies allocated from the Fund are over and above existing or planned normal expenditure by government agencies and used for three purposes: to (1) provide regional infrastructure and services; (2) develop and broaden the economic base of the regions, and (3) maximise regional job creation and career opportunities.

The Act formalised the architecture of the allocation of royalty funds to the overarching Royalties for Regions Fund. The remainder of the programme was established in administrative provisions within the Department of Regional Development (now the Department of Primary Industries and Regional Development). These administrative provisions have changed over time, in line with government priorities and the findings of a series of external reviews into the structure, processes and governance of the programme.

While Royalties for Regions was generally well received in the non-metropolitan regions, over time many questions were raised regarding the allocation of fund resources. Towards the end of the two-term Liberal-National alliance Government, there were concerns the Fund was being misused. In May 2017, the new Labour-led State Government commissioned a special inquiry into the decision–making processes, transparency and financial consequences of 31 projects commissioned between 2008 and March 2017. This included a number of projects funded under the Royalties for Regions programme. In February 2018, Former Under Treasurer John Langoulant AO released the 900-page inquiry report, which contained 107 recommendations to the State Government. The inquiry found a 'significant deficit in the rigour applied to project selection and poor targeting of funding towards projects that would deliver lasting economic and social outcomes to regional Western Australia'. It was common practice for government to establish 'buckets of money' from which future expenditure projects would dip into without adequate business cases was concerning. The so-called ad hoc nature of many housing projects funded by Royalties for Regions in the Pilbara underlined the failure of rigorous planning (State of Western Australia 2018: 142).

In response to the special inquiry report, the State Premier indicated that the State government had already integrated the Royalties for Regions budget into the normal State Budget cycle, but would maintain the programme with a defined strategy, plan and set of priorities. In addition, the State government would develop 'a full suite of economic and social data at a regional level to provide the evidence base to guide decision-making' and develop a formalised programme evaluation framework' (Government of Western Australia 2018b).

The Langoulant inquiry also examined government procurement, which the Department of Finance manages under the State Supply Act 1991. The inquiry found a significant need for improvement. There was a lack of central leadership and confusion around the complex legislative framework governing procurement. This had resulted in a situation in which the framework "did not provide optimal support for the government's commitment to maximising local content, providing opportunities for jobs on government projects, supporting small and medium businesses and Aboriginal businesses" (pp. 19). Generally accepting the recommendation of the inquiry, the Labour-led Government indicated it would undertake further consultations towards the introduction of a single procurement act (State of Western Australia 2018).

The Royalties for Regions programme represents an innovative approach by the State government to provide the regions with resources that were a direct result of the significant levels of private investment into mining and gas extraction. Typically, mining royalties are paid directly into Consolidated Revenue and regions would seek allocation through the normal budgetary processes, which are often not favourable to regions with populations that are substantially smaller than the Perth (the WA State capital city) metropolitan area. The earmarking of royalties for use in the regions was well received and led to investments in a wide range of community and social development and infrastructure projects in the regions. Staden and McKenzie (2019) claim that regional development challenges are framed from a place-based perspective and argue that the Royalties for Regions programme does not view the development challenge as a place-neutral mobility issue. Instead, the programme put money back into regional communities in order to create sustainable communities. The problem for regions is the growth of Perth, at the expense of rural populations.

Ellem and Tonts (2018) also describe the programme positively, suggesting it was designed to disrupt development so that 'wealth and power accumulate in places other than the site of extraction'. Thus, Royalties for Regions sought 'to challenge the marginalisation of remote regions'.

The problem, however, was that very little programme funds were used to stimulate local industries and businesses, either in the mining supply chains or in other regional economic sectors. Indeed, the special inquiry criticised the project selection process as inadequate with 'a lack of clearly defined regional development policy outcomes'. The absence of an overarching strategic framework and appropriate measurement of regional development outcomes made it difficult for government to determine whether funded projects contributed to improved economic and social outcomes in the State (State of Western Australia 2018: 143).

6 The WA Plan for Jobs and Jobs Act

In 2017, the Minister for Jobs in the new Labour State Government introduced the *Western Australian Jobs Act 2017* to maximise local small business participation in the State economy through the use of State Government procurement processes in the supply of goods and services. State Government procurement amounts to around A$27 billion per annum, and the Act puts greater responsibility on State government agencies to provide local industry with full, fair and reasonable opportunity to access and win State Government supply contracts. It also puts a focus on the reporting of economic outcomes of local industry participation.

The Jobs Act is a legislative component of the WA Plan for Jobs, which identifies a number of focus areas, including support for innovation and new industries, tourism marketing, infrastructure and industrial lands development. It also includes a government commitment to local content and jobs on government projects and serviced industrial land developed under the Industrial Lands Authority to foster private sector investment in the regions.

A key component of the Jobs Act is the Western Australian Industry Participation Strategy (WAIPS), which seeks to promote the diversification and growth of the WA economy by opening up supply opportunities for local industry, providing local suppliers with increased access to, and awareness of supply opportunities and building local industry capability. The WAIPS requires prospective suppliers to complete and submit a participation plan as part of their tender bid. This is an outline of how the project proponent will support the principle of full, fair and reasonable opportunity and give ample consideration to the project's local economic impacts. The government requires project proponents to report on its contracting outcomes every six months and at practical completion (Government of Western Australia 2018a). Depending on the value of the supply contract, proponents are required to prepare either a 'core' or 'full' participation plan. A core participation plan is a simple plan that outlines the local economic benefits they will produce should they be awarded the contract. A full participation plan is more detailed, requiring prospective suppliers to demonstrate local economic benefits and provide a full, fair and reasonable opportunity to local industry, should they be awarded the contract. Participation plans will be used as part of the evaluation, award and contracting process (Government of Western Australia 2018c).

A further support to the policy is the Industry Capability Network WA (ICNWA). This is a specialist WA and Australian vendor identification service provided as a joint industry government initiative. ICNWA is part of the Australia-wide ICN network assisting project proponents to ensure they are not paying a premium on imported equipment, spare parts and services that can be obtained at a more competitive cost from local suppliers. The State Government will invest A$1 million per annum to reinvigorate ICN WA to support WA businesses to compete for government contracts.

As indicated above, the Jobs Act focuses on public procurement and not procurement by large private miners. The latter is somewhat covered by the older Building Local Industry Policy, which endeavours to ensure that competitive local industry receives full, fair and reasonable opportunity to participate in the State's major private sector projects. This policy was amended in October 2018 to transfer public procurement processes to the WAIPS, while retaining procurement provisions for the private sector. The Building Local Industry Policy seeks to achieve a cooperative and coherent approach from government, project proponents, suppliers and unions, while assisting more WA businesses to compete for major project work in the private sector and provide improved access to local businesses for major project proponents. A key element of the policy is the requirement for project developers to prepare an Industry Participation Plan (IPP) for all private sector projects 'where the Government makes a significant contribution'. The purpose of the IPP is to show how project proponents plan to include WA industries in their project-wide procurement strategy and demonstrate how local industry will be given full, fair and reasonable opportunity to supply goods and services to the project's development and operational stages.

The Department of Jobs, Tourism, Science and Innovation began implementing the Jobs Act in October 2018. However, in the non-metropolitan regions, the Department of Primary Industries and Regional Development implements the act through a regional network of Local Content Advisers who provide advisory and referral services to businesses in regional areas. Interviews conducted for this study with Local Content Advisers found that their role is often divided between working with public sector procurement managers and local firms who are interested in bidding for government contracts. Interestingly, public procurement managers are often in short supply in regional centres due to centralised procurement processes of many government agencies that are managed by head offices in Perth. In some cases, Local Content Advisers are also working with mining procurement offices and firms interested in participating in mining supplies. However, these interactions are mostly driven by the initiative and networks of the Local Content Advisers.

This is not the first time the government has attempted to promote local content. There have been previous initiatives, such as in 2002 when the State Government announced the Buy Local Policy aimed at recognising the contribution of local businesses to the State economy and confirming the government's 'commitment to buying locally'. The policy promotes the decentralisation of purchasing functions and responsibilities to regional areas in order to stimulate local competition and provide increased opportunities for local businesses. This includes the use of two types of regional price preferences: Regional Business Preferences, which provides regional businesses, located within a prescribed distance from a regional purchase or contract point of delivery, with a price preference that is applied to the total cost of the bid; and Regional Content Preferences, which provides a price preference to all WA businesses that purchase services or materials for use in regional contracts, from regional businesses (Government of Western Australia 2002).

A Regional New Industries Fund of A$4.5 million for regional WA will provide grants to support new and emerging businesses in the regions, the agriculture sector will be encouraged to grow, and more support will be provided for tourism and hospitality in the regions. In addition, the State Government has used Royalties for Regions to fund Regional Economic Development (RED) Grants, which seek to support community-driven projects designed to create jobs and boost economic growth across regional Western Australia.

7 Conclusions and Lessons Learned

This case study of the Government of Western Australia's attempts to promote local content in the extractives sector in order to benefit the residents, firms and workers of non-metropolitan regions highlighting the strengths and limitations of government policy. Over the last fifty years, the State Government has sought to maximise the benefits of mining activity through regionally focused policy instruments and institutions. However, these have typically focused on social and community benefits to regional life and have rarely addressed specific regional industry or economic concerns. While WA's regions are eager to attract investment, with substantial support from the State Government, relatively little success has been achieved in stimulating local industries and their supplier networks.

Formal State Agreements sought to ensure mining firms provided with government financial and non-financial concessions were required to fill a range of local obligations. Over time, these obligations included the need to maximise local content and support secondary processing that would integrate mining into the iron and steel industry. However, these efforts relied heavily on market forces and ultimately did little to stimulate regional industry transformation.

The Royalties for Regions programme was a unique and innovative attempt to channel mining royalties back into the regions. While this led to a number of benefit projects, very few of these focused on strengthening local industries and increasing local content. The more recent policy initiatives to promote local content and employment has largely focused on public procurement. While this is an important first step, the challenge remains as to how to effectively engage and partner with large mining projects.

Mining royalties have largely been used to fund regional development outcomes, rather than stimulate economic diversity and employment growth. Royalties have been an additional, and rather large, source of revenue that has arguably distorted State Government efforts to promote regional development efforts and undermined strategic planning. While the use of royalties to support economic and employment growth in regional areas is to be commended, a more strategic and industry aligned approach to the use of these funds would have greater success in diversifying regional economies and improving labour market dynamics.

References

Department of Jobs Tourism Science and Innovation. (2019). What is a state agreement. Retrieved from https://www.jtsi.wa.gov.au/what-we-do/manage-state-agreements/what-is-a-state-agreement.

Department of Regional Development and Lands. (2011). Regional centres development plan (supertowns) framework. Retrieved from Perth.

Department of Tourism and Department of Industrial Relations. (1992). *Impediments to regional industry adjustment*. Canberra: Australian Government Publishing Service.

Ellem, B., & Tonts, M. (2018). The global commodities boom and the reshaping of regional economies: The Australian experience. *Australian Geographer, 49*(3), 383–395.

Government of Western Australia. (2002). Buy local policy. Retrieved from Perth.

Government of Western Australia. (2010). Structuring regional development for the future; a review of the functions and responsibilities of regional development commissions. Retrieved from Perth.

Government of Western Australia. (2016). regional development strategy 2016–2025; building vibrant regions with strong economies. Retrieved from Perth.

Government of Western Australia. (2017). Regional development council. Retrieved from https://www.drd.wa.gov.au/about/PORTFOLIO/Independent-Partners/Regional-Development-Council/Pages/default.aspx.

Government of Western Australia. (2018a). Buiulding local industry policy. Retrieved from Perth.

Government of Western Australia. (2018b). Media statement: 'McGowan government response to the special inquiry implementation program' [Press release]. Retrieved from https://www.mediastatements.wa.gov.au/Pages/McGowan/2018/02/McGowan-Government-driving-major-changes-across-government.aspx.

Government of Western Australia. (2018c). Western Australian industry participation strategy. Retrieved from Perth.

Government of Western Australia. (2019). Western Australian Economic Profile March 2019. Retrieved from Perth. https://www.jtsi.wa.gov.au/docs/default-source/default-document-library/wa-economic-profile-0319.pdf?sfvrsn=e609731c_4.

Horsley, J. (2013). Conceptualising the state, governance and development in a semi-peripheral resource economy: The evolution of state agreements in Western Australia. *Australian Geographer, 44*(3), 283–303. https://doi.org/10.1080/00049182.2013.817038

House of Representatives Standing Committee on Industry Innovation Science and Resources. (2018). Keep it in the regions: Mining and resources industry support for businesses in regional economies. Retrieved from Canberra.

House of Representatives Standing Committee on Industry Science and Technology. (1998). A sea of indifference: Australian industry participation in the North West shelf project. Retrieved from Canberra.

Hunter, T. (2014). Law and policy frameworks for local content in the development of petroleum resources: Norwegian and Australian perspectives on cross-sectoral linkages and economic diversification. *Mineral Economics, 27*(2), 115–126. https://doi.org/10.1007/s13563-014-0051-y

Paül, V., & McKenzie, F. H. (2015). 'About time the regions were recognised': Interpreting region-building in Western Australia. *Australian Geographer, 46*(3), 363–388.

Staden van, J.-W., & McKenzie, F. H. (2019). Comparing contemporary regional development in Western Australia with international trends. *Regional Studies*. https://doi.org/10.1080/00343404.2019.1584394.

State of Western Australia. (2018). Special inquiry into government programs and projects, final report volume 1. Retrieved from Perth.

Western Australian Planning Commission. (2014). State planning strategy 2050 Retrieved from Perth.

Wilson, E., Layman, B., & Christmas, D. (2004). An economic history of Western Australia since colonial settlement; Research Paper. Retrieved from Perth.

Norwegian Local Content Policy

Jonathon W. Moses

Abstract The Norwegian petroleum industry is today strong and internationally competitive. Of course, this was not always the case: the industry was supported, protected, and nurtured at a time when states enjoyed much greater scope for sovereign autonomy. During a relatively short period of time, from about 1970–1986, the Norwegian authorities employed four main instruments to develop local competence in the sector: their unique licensing system, a strong national oil company (Statoil, now Equinor), a series of Technology Agreements and an autonomous regulatory regime that benefited domestic producers. None of these tools are used today, as Norwegian authorities now embrace a remarkably "hands-off" approach to managing the sector. This chapter describes this brief period of political activism, and the industry it created.

Keywords Norway · Local content policy · National oil company · Sovereignty

Although Norway's oil production only began in the early 1970s, oil and gas have become and remain major contributors to the Norwegian economy. The country's Petroleum Directorate estimates that petroleum activities have contributed to more than 14,000 billion in current NOK to Norway's GDP—a figure that does not include the related service and supply industries (NPD 2019a). In 2018, the petroleum sector's influence on the broader Norwegian economy is formidable, representing 17% of Norwegian GDP, 19% of the country's total investments, 43% of its exports, and providing 21% of the Norwegian state's revenues (ibid).

The influence of petroleum in Norway is both smaller and greater than these macroeconomic figures imply. On the one hand, the petroleum industry is hardly a

This contribution draws heavily from my book, co-authored with Bjørn Letnes (Moses and Letnes 2017a); especially Chap. 8.

J. W. Moses (✉)
Department of Sociology and Political Science, Norwegian University of Science and Technology, Trondheim, Norway
e-mail: jonathon.moses@ntnu.no

E. G. Pereira et al. (eds.), *Sovereign Wealth Funds, Local Content Policies and CSR*, CSR, Sustainability, Ethics & Governance,
https://doi.org/10.1007/978-3-030-56092-8_24

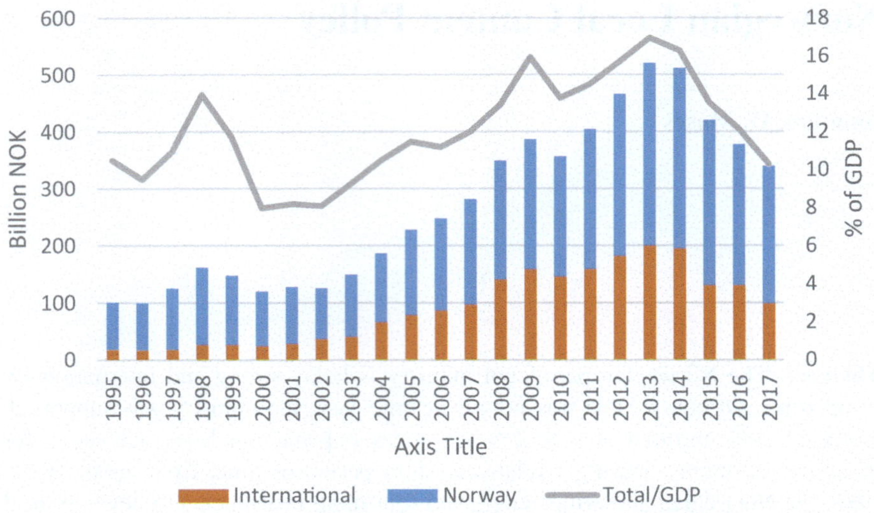

Fig. 1 Norwegian petroleum supply industry. *Sources* Rystad Energy (2018: 8, 19); SSB (2019)

major employer—both in Norway or elsewhere. In 2017, only 6% of the Norwegian labor stock (or a little over 170,000 people) were directly employed in the oil sector (NPD 2018). Petroleum-related employment reached its peak in 2014 (at around 9%), when it fell alongside the price of oil (Moses and Letnes 2017b). On the other hand, the impact of oil extends much farther than the usual economic suspects. While most people associate the Norwegian oil industry with the country's National Oil Company (NOC), Statoil (now Equinor), there is a much more varied collection of firms flying just under the radar. In particular, there are other Norwegian oil companies (both smaller and private[1]), a sizeable research and development sector associated with offshore petroleum production, and a significant oil service and supply industry, which is spread out across the entire country, and which exports throughout the world.[2]

Figure 1 provides an overview of the influential supply industry associated with Norwegian petroleum. Here, we can find a plethora of firms delivering everything from seismic studies and subsea equipment (the latter is the largest sector) to firms that provide transport and logistics. As we can see in Fig. 1, these companies are most active within Norway, but a significant share is also active internationally. Altogether

[1]See NPD (2019b) for examples.

[2]In 2017, the largest Norwegian supply companies, in terms of international presence, were (listed alphabetically): ABB, Akastor, Aker Solution, BW Offshore, Cameron, DeepOcean, DNV GL, DOF, TechnipFMC, Fred Olsen Energy, Kongsberg Gruppen, NOV, Odfjell Drilling, PGS, Rolls-Royce Marine, Siem Offshore, Solstad Farstad, TGS, and Vard and Wärtsilä. This list includes 7 rig- and ship-owners; 12 offshore and maritime equipment and service providers and one shipbuilding company (Rystad Energy 2018: 6).

(domestic and international), these activities captured roughly 10% of Norway's GDP in 2017. Since the 2014 fall in oil prices, it would appear that these industries are having a hard time, both in Norway and internationally, as their numbers are falling quickly.

To understand how this broad and mostly successful industry came to be, we have to go back in time. Students of Local Content Policies (LCPs) will search in vain for any explicit sign of local content legislation or regulation on Norway's books today. Rather, Norway's local petroleum sector was nurtured and protected as an infant industry in the 1970s, then encouraged to grow with a strong and supportive policy mix. This sort of policy support could be done at a time when political intervention in the economy was more widespread and acceptable, and when countries had access to a broader array of tools to encourage local content. Now that these firms are mature and internationally competitive, the Norwegian government has mostly dropped its domestic support; it now works actively to open foreign markets. In the doing, Norwegian policymakers try to limit other states' access to the sort of discriminatory policies from which their own industry once benefited.

In short, Norway developed its oil industry at a time when it was much more common and possible to employ state policies to prioritize local economic activity. A network of international legal obligations, and a changed ideological setting, have made this element of the Norwegian model much more difficult for other states to emulate today. This contribution begins by revisiting the historical context that allowed for this development, and then describing the four main instruments that Norway used to build up its vibrant local petroleum industry.

1 Historical Context

Every country hopes to turn its natural resources into engines of local economic activity. In contrast to the more limited opportunities facing many countries today, Norway had two things working to its advantage. The first was that it developed its oil industry in an international context that facilitated greater democratic control over national investment, production and regulatory regimes. The second was the unique and demanding requirements for producing oil on the Norwegian Continental Shelf (NCS): New tools and thinking was absolutely necessary, and Norway was well positioned to provide both.

1.1 A Changed International Environment

It is difficult to exaggerate the difference that separates our world of economic exchange from that of the late 1960s.[3] While sundry GATT[4] rounds of negotiations were slowly extending the scope of free trade (across a rather narrow set of sectors), national capital markets remained heavily regulated, global capital mobility was minuscule (and mostly linked to direct investment and trade opportunities), and it was not uncommon for states to nationalize industries facing significant challenges. Given the lack of international capital mobility (at the time), states were able to pursue more autonomous monetary, fiscal, regulatory and even procurement policies. Indeed, the international arrangements crafted at Bretton Woods in 1944 were designed to provide states with the tools they needed to steer the national economies in ways that could minimize economic and political turbulence, and to avoid another crisis like that from the interwar period (Ruggie 1982). In other words, in the late 1960s and 1970s, economic management was seen to be part of the legitimate purview of sovereign officials. The only international constraints on policy that mattered were respect for contracts and private property rights—beyond these, states had a much freer hand to do what they wished (and what their people wanted).

This scope for policy autonomy was exploited by the Norwegian authorities at the time: they wielded extensive regional (rural) policies, an active monetary policy, a substantial welfare state fed by a progressive tax system, and strong regulations to protect workers and the environment. The Norwegian state was not averse to letting oil companies make money on the Norwegian continental shelf—even sizeable amounts of money, as it turns out—but it was determined that these rewards would not come at the expense of the Norwegian people or environment.

This sort of freedom no longer exists for contemporary states hoping to create national champions. International organizations (such as the World Trade Organization), along with regional organizations (such as the European Union), lending bodies (such as the World Bank and the International Monetary Fund) and a plethora of bilateral agreements increasingly require signatory states to jettison sovereign control and influence in the name of greater efficiency and a level playing field. Today's states face a myriad of restrictions on their ability to conduct autonomous procurement, preferential, concessions and competition policies. Until recently, political momentum was extending free trade and lending agreements to an ever-broader array of areas and instruments, including the greater liberalization of investments (TRIMs) and intellectual property (TRIPs), trade in services (GATS) and even procurement policies (GPA),[5] while at the same time, including Investor State Dispute Settlement

[3]See Chap. 2 in Moses and Letnes (2017a).

[4]General Agreement on Tariffs and Trade (GATT), subsequently the World Trade Organization (WTO).

[5]There are lots of acronyms in this paragraph. These four are directly associated with programs at the World Trade Organization (WTO): TRIMs: Trade-Related Investment Measures; TRIPs: Trade-Related Aspects of Intellectual Property Rights; GATS: General Agreement on Trade in Services; and GPA: General Procurement Act.

(ISDS) and/or so-called stabilization clauses in a growing number of international agreements and contracts. All of these changes limit a state's ability to influence the level (and type) of domestic activity in the petroleum sector.

1.2 Special Needs

In the more-forgiving international context of the 1970s, Norwegian elected officials enjoyed greater scope to influence domestic economic conditions. This freedom could be leveraged because of the especially challenging conditions where Norway's petroleum resources were located. This is the second advantage Norway enjoyed, relative to young oil states today. When international oil companies (IOCs) first came to Norway, they quickly discovered that the conditions in the North Sea were much more demanding than those that existed in the Gulf of Mexico (where much of the existing equipment and expertise had been developed). The colder climate, deeper depths, rougher weather and higher seas meant that existing, off-the-shelf, technologies could not be employed offshore. The industry needed to replace much of its equipment and expertise with better (more suitable) alternatives.

One does not usually associate Norwegian climate and weather with the concept of advantage, but the challenges associated with Norwegian geography provided the country with a number of unique opportunities. Given Norway's challenging conditions, it was necessary to develop new tools and approaches to access the resource—and Norway's own industries were well positioned to fill this need.

More to the point, Norway already enjoyed technical competence in tangential fields that could be re-tooled to meet the needs of the oil industry. Norwegian shipbuilders could use their knowledge of local conditions to build stronger and more suitable offshore platforms and supply ships. The engineering and materials expertise used to develop Norway's hydroelectric industry could be re-tooled to meet the needs of the oil industry, and its institutions of higher education were revamped accordingly. Given the demand for new solutions to Norway's unique challenges, a broad political consensus in support, and the prior existence of relevant pockets of expertise and protection, all the Norwegian political authorities needed to do was to provide the appropriate incentives to match this new demand with existing local suppliers.

A good example of the new challenges, and the opportunities they offer, can be seen in the first generation of supply boats sent over from the Gulf of Mexico to service Norway's offshore platforms. These boats were simply not built to withstand the conditions in the North Sea, and their inability to work under storm conditions meant that production was frequently delayed. It was clear, almost immediately, that new ships would be needed, and that somebody would have to design and build them. This void was filled by Norwegian contractors. After all, Norwegian shipbuilders were already familiar with North Sea conditions and had been building suitable ships for generations. Norwegian shipbuilders could use their local knowledge and expertise to make necessary improvements to the supply ship fleet: building ships with higher freeboard, more powerful engines, bow-thrusters, etc. The Norwegian government could facilitate local content provision by helping companies transition, and by changing contract requirements in a way that made them more relevant and manageable for domestic producers (e.g. designing platforms to consist of smaller, modular, pieces).

These contextual advantages are important to underscore, because young oil states today do not find themselves in a similar situation. Today's international context does not allow states the same degree of policy autonomy: States are hemmed-in by a number of international agreements and contracts are designed to facilitate free exchange (and protect established interests). In the name of efficiency and a level playing field, today's international regime benefits large international players at the expense of those who would help to procure smaller, local producers. In addition, and more often than not, IOCs find that they can employ cheaper off-the-shelf technologies in developing new fields, providing little room for political officials to leverage local companies and competencies.

2 Tools of the Trade

Norwegian local content policy can be divided into three distinct periods. For most of Norway's oil history, the government has not played an active role in encouraging local content. For example, in the early years of the industry (1963–1970), and in the period after 1986, the government has largely remained on the sidelines, trying to create a level playing field upon which Norwegian firms could compete. It is only in the intervening period (1970–1986) that we find the Norwegian government actively encouraging and facilitating the expansion of local content in the petroleum sector. During this period, we see a strong set of signals being developed to guide Norwegian policymaking toward greater local content provision, and the development of four distinct channels of influence. This section considers each of these components.

2.1 Policy Foundations

Before any oil was actually discovered off the Norwegian coast, policymakers were quite hesitant to involve local firms. The capital requirements, and the level of risk exposure, in the petroleum sector were simply too high for Norway's relatively small and conservative firms (see, e.g. *St. meld. nr. 11* (1968–9), pp.7). For that reason, the first emphasis of Norwegian policymakers was to attract as much foreign expertise and capital as possible and entice them to settle in Norway.[6] This meant providing

[6]The first effort was actually to settle the boundaries of the Norwegian Continental Shelf with neighboring Denmark and the UK. Once these territorial boundaries were established and agreed upon, policymakers could begin to formulate a development strategy.

better (more lucrative, and less interfering) terms than could be found on the other side of the North Sea, where Britain was simultaneously trying to attract IOCs to develop its own resource. Once viable and profitable fields were discovered, however, Norwegian policymakers realized the need to develop a regulatory regime that better reflected Norwegian values.

In the early 1970s, a number of government white papers established a basic framework for Norwegian policymaking, and this framework provided a space for (subsequent) democratic interventions. In particular, a 1971 white paper produced what came to be known as Norway's "10 Oil Commandments," which received unanimous support in parliament and became the de facto political standard, against which subsequent oil policy was gauged (see *St. meld. nr 28* (2010–2011), pp.8). Among the 10 Commandments were many opportunities for potential local content development (See Moses and Letnes 2017a: 74). For example, there was provision for national supervision (1); the need to develop new business activities (3); a requirement that the petroleum be landed in Norway (5); a recognition of the need for active state involvement in the development of a Norwegian oil community (7); and the desirability of a national oil company (8).

But when a formal requirement for local content was introduced by Royal Decree in 1972, it was remarkably weak:

> "Licenses shall use Norwegian goods and services in petroleum operations to the extent that these are competitive in terms of quality, service, delivery time and price."[7]

This Royal Decree did not require the hiring of Norwegian firms; it did not set a floor for the minimum share of Norwegian activity on a license; it did not set a price buffer for domestic producers; it did not elaborate on a precise definition of Norwegian content or domestic firms. This Decree was very different from what we think of today when we consider and compare local content policies.

The initial Norwegian intent was to create a level playing field. Policymakers wanted to provide opportunities for Norwegian firms to compete with foreign firms. It was not their (original) intent to provide a leg-up for domestic firms. At the same time (1972), the government established a Goods and Services Office at the Ministry of Industry to map out the degree to which Norwegian firms were being included (or not) on the various tender lists. But the activities of this Office were, at first, aimed at monitoring and informing international companies of the local possibilities that existed.

It did not take long before local (Norwegian) industries became frustrated by the lack of opportunities available to them. Indeed, two large Norwegian firms (Aker and Kværner) protested the first allocation of licenses on the Norwegian Continental Shelf (NCS) for the relative scarcity of Norwegian firms. While a handful of Norwegian firms were included in some of the allocated licenses, their inclusion appeared as a token or gesture. As large international firms ran away with the cake, Norwegian firms were left with the crumbs.

[7] Article 54 of a Royal Decree from 8 December 1972. See Moses and Letnes (2017a: 155).

For example, the first large production facility on the NCS, the Ekofisk field, was operated by Phillips Petroleum, out of their main office in Bartlesville, Oklahoma. Phillips employed large American engineering firms (with whom they were accustomed); their purchasing office was located in London; and most of their equipment and staff was originally imported from their operations in the Gulf of Mexico. This was (is) how the oil industry worked: they were a relatively small band of international companies, whose employees travelled around like nomads from one production location to another, relying heavily on one another's expertise. The Norwegian share of productive activity in this first major field was miniscule, until a Norwegian contractor won its bid for an enormous new subsea storage tank made of concrete. Even after this massive contract, however, the Norwegian share of Ekofisk was limited to about 17% (Skule and Grytli 1997: 38ff).

Then, by the middle of the decade, we can see a change in tide. The first real efforts at encouraging local content began in the wake of the first oil crisis, in 1973, and the devastating effect this crisis had on the international shipping industry. Many companies were forced to cancel their oil-tanker orders, and the Norwegian fjords were starting to fill up with vacant ships. Norway's coast began to look like a parking lot, and many shipbuilders were forced to shut down production. Norwegian policymakers realized that they needed to spark new activity in Norway's shipbuilding industry, in order to replace the shrinking shipbuilding orders that keep people employed up and down the coast.

By the end of the 1970s, and with the government now in the hands of the Labor Party, Norwegian policymakers began signaling a much more activist policy, alongside the development of new institutions for policy and regulation. Informing them all were four fundamental principles (see *St. meld. nr. 53* (1979–80):

- Maximize Norway's economic gain from the resource;
- Contribute to the country's social and economic development;
- Develop and maintain strong environmental and safety standards;
- Develop the resource in a slow, careful and deliberate manner (Moses and Letnes 2017a: 75).

By the end of the 1970s, the aforementioned Goods and Services Office was not only monitoring the inclusion of local content, but requiring international companies to explain why they were not employing Norwegian suppliers (when these were available). It was at this time that the Norwegian net share of deliveries to petroleum operations in Norway began to take off: from 28% in 1975, to 42% in 1976, to 50% in 1977 and to 62% in 1978 (*St. meld. nr. 53* (1979–80), pp. 27). This remarkable rise of Norwegian local content was driven by four specific instruments of policy: Norway's unique concession or licensing system; its NOC; an array of Technology Agreements; and a national regulatory framework that was alien to the IOCs.

2.2 Licensing System

The first instrument that Norwegian policymakers used to secure local content was the very unique means by which licenses were allocated for offshore production. Starting with its second allocation round, the Norwegian licensing system employed a negotiated process by which political authorities decided the specific make-up of the joint ventures that secured a license. Rather than letting the companies decide by themselves who would join together in bidding for a Norwegian production license, the authorities took responsibility for creating the joint ventures, and then used this responsibility to piece together license groups that could develop local competencies (for details, see Moses and Letnes 2017a: Chap. 5). In practice, this meant pairing up different companies as partners in a license group, and then distributing the voting shares of the license such that the state (and/or its NOC) controlled a majority within each license group. This allowed the state to prioritize its national oil company (Statoil), and require that other, more experienced, firms teach Statoil the tricks of the trade. (Statoil was not the only Norwegian company that benefited from this arrangement—only the most prominent.) Statoil subsequently used its position to train a highly-skilled domestic workforce, develop broad operator experience, and prioritize Norwegian suppliers. By using its licensing agreements to link Norwegian firms with IOCs, the Norwegian authorities could encourage cooperation among individual firms and on specific projects. In so doing, they facilitated a substantial level of technological and knowledge transfer, to the benefit of Norwegian firms.

2.3 Statoil

The rise and dominance of Statoil, as the national oil company, can be seen as the second main instrument that the authorities used to secure local content in Norway. Having benefited substantially from the licensing allocation process described above, Statoil grew faster than anyone could have imagined. Starting with the third allocation round (in 1974), the authorities introduced a number of additional measures that helped secure Statoil a privileged position in the Norwegian market. In particular, we can focus on five different (albeit related) privileges provided to Statoil by the Norwegian authorities: operatorship guarantees; licensing shares; carrying clauses; sliding scales; greater voting rights.

- Operatorship guarantees: Statoil was provided with extraordinary opportunities to attain operator status after a very steep learning curve. For example, in the licensing agreement for the Statfjord field (1974), Statoil was promised an opportunity to take over (from Mobil) the operator status of the license after ten short years, assuming that the company was up to the task. Mobil originally agreed to the terms of the agreement, but probably did not expect that a relatively new company, like Statoil, could be prepared to take over operator responsibility for such a challenging field so quickly after it was formed. Mobil apparently miscalculated: it was tasked with teaching Statoil the job, and then forced to resign its operator responsibility after the 10-year training period. Once it became the license operator, Statoil was able to subcontract out many of the required tasks to local companies.

- Licensing shares: Statoil's second privilege came with Norway's third allocation round (1974–1978), when the authorities introduced a rule, by which Statoil was to be given a minimum 50% share in all new licenses. This resulted in an enormous expansion of Statoil activity (and the wealth associated with that activity). This privilege was subsequently dropped in 1996.
- Carrying clauses: The third allocation round also introduced the use of a carrying clause to minimize Statoil's exposure to risk, maximize its financial gains, while providing the company with the experience it needed to manage and develop the resource. In particular, a carrying clause allows a working interest partner (here Statoil) to pay a disproportionately lower share of its costs and expenses (than its working interest share), during the (risky) exploration or development phase of a contract. For example, even if Statoil enjoyed 50% of a license, it was not expected to pay 50% of the exploration costs. As much exploration ends in failure, this is a very costly and risky part of the oil adventure. Once the exploration results in an economically viable well, however, the NOC (here Statoil) was allowed to "back in" as a member of the license and pay its full share of costs (and receive its full share of reward). This particular privilege was dropped from all new licenses in 1987, and from all production licenses in 1991.
- Sliding scale: Statoil also benefitted from Norway's use of a sliding scale (this device was also introduced in the third allocation round). As the size of the resource rent increases with the size of a field's productive capacity, the government introduced a sliding scale to ensure these rents would remain with the Norwegian people (to be more precise, with Statoil). In other words, Statoil's share of total production was allowed to "slide" along with the size of peak production in any given field. Thus, in a smaller field, Statoil's share might be limited to 50%, but as the size of field increased to reveal a gigantic resource, Statoil's share could eventually increase to as high as 80% (Moses and Letnes 2017a: 158–9). This privilege has also been dropped in recent years.
- Voting shares: Finally, Statoil benefited from the way in which the government could (and did) allocate voting shares when it decided how to distribute its licenses. From the third allocation round until 1985, the government ensured that 50% of all license rights went to Statoil (this was enough to commit the license). In controlling a majority of the voting rights, Statoil was then able to influence the way that field was developed and to prioritize local companies and expertise in the process.

By exploiting these five privileges, Statoil became a very effective tool for securing and developing Norwegian competence in the sector. These benefits help to explain why there are so many successful Norwegian supply companies, and why Statoil seems to be unique as an NOC in playing a role as an integrated contributor to the nation's industrial base and the national economy at large (Victor et al. 2012: 894).

2.4 Technology Agreements

While Statoil plays a central role in any explanation of the strength and breadth of Norwegian local content, a fourth instrument played a vital role in developing Norwegian technological competence: the government's use of Technology Agreements. Norway's fourth allocation round (in 1978) introduced a requirement that 50% of research and development (R&D) efforts related to field development needed to occur in Norway. Hence, license applicants were expected to sign technology (or R&D) agreements with Norwegian institutions if they had any hope of securing a production license. These agreements varied in form, but they aimed to encourage collaboration with Norwegian research institutions and to fund specific R&D projects undertaken at Norwegian institutions.

These Technology Agreements were rooted in Norwegian law and then integrated into the licensing/concession framework. The government argued that such agreements were necessary to develop Norwegian competitiveness, but also to ensure that the resource was developed in a way that was compatible with Norway' strict safety and environmental regulations (*St. meld. nr. 53* 1979–80), pp. 67. In general, there existed two types of Technology Agreements in Norway: *50 percent agreements* required an operator to conduct at least 50% of the research necessary to develop its field in Norway; while *offer agreements* mandated companies to cooperate with Norwegian R&D facilities within defined areas, for explicit amounts, if they were to obtain a license. In addition, Norway encouraged non-binding and voluntary *goodwill agreements*, where companies declared their intent to conduct as much of their petroleum-related R&D in Norway.

As a result of these agreements, Norwegian research institutions and universities enjoyed a substantial boost, propelling them to the technical frontiers of the industry. The competitiveness of Norway's internationally-active supply industry rests heavily on the technological prowess that the agreements secured.

2.5 Regulations and Language Requirements

Finally, when Norwegian authorities were willing to impose Norwegian safety and environmental regulations on the offshore industry, it created an implicit competitive advantage for Norwegian firms that had (earlier) experience with the Norwegian system (and the language it was written in). Foreign (mostly American) firms coming to Norway had a very difficult time understanding the scope and style of Norwegian regulations: it was a completely alien environment for them. Norwegian firms, and their Norwegian employees, were better able to understand and exploit these rules and regulations. Hence, IOCs were forced to hire Norwegian sub-contractors or consultants to help them navigate these regulatory waters, or to hire Norwegian staff that held this form of specialized competence.

Over time, these local regulations and standards have been replaced with international standards and regulations (see Moses and Letnes 2017a: Chap. 9). This is true in Norway, as it is for the industry world-wide. In embracing international standards and regulations, we have (perhaps unwillingly) shifted the competitive advantage from local firms to IOCs. In today's global marketplace, it is more often the case that the international oil company has greater competence and insight into the relevant (international) standards and regulations, than do local firms or political authorities.

This brings us full circle, as we are reminded of how today's international context can limit the opportunities of states to follow the Norwegian experience.

3 Conclusion

The Norwegian petroleum industry is a product of significant political invention in the way that petroleum production licenses were regulated and distributed in Norway. This intervention was possible in the 1970s and early 1980s, when international agreements, standards and practices facilitated more democratic control over domestic economic conditions. In this window of opportunity, Norway encouraged the development of an infant NOC and its supply industry, then allowed it to grow behind a protective and supportive wall of licensing requirements.

The success of these policies created an NOC—Statoil—that became so powerful that Norwegian politicians felt a need to cut it down in size. The Conservative Party (Høyre) was particularly concerned about Statoil becoming a state within the state. This perceived threat was dealt with in two blows. In 1985, the government separated Statoil's operator holdings from the Norwegian state's financial interests in various production licenses—in effect cutting Statoil in half. Then, in 2001, Statoil was privatized (while keeping a majority of shares in the government's hand).

That window for national policy autonomy has now been closed. Once these firms were strong enough to stand on their own and began the search for new markets abroad, the Norwegian political authorities quickly changed tack. Norway, its NOC (now Equinor), and the Norwegian supply industry seek to compete with nascent foreign firms abroad, in their own backyards. In doing so, the Norwegian industry must push to minimize the sort of national support and protective measures that they themselves benefited from at their inception. If other countries are allowed to do what Norway did, then Norwegian firms would not be able to compete with them. This story is akin to the one that Ha-Joon Chang (2005) tells about trade policy in general: Norway and Norwegian firms are now "kicking away the ladder."

In doing so, Norway is following the norms of today's international economic environment: countries find it difficult to encourage local champions and support infant industries. As a result, it is difficult for today's new oil states to pursue the same sort of nourishing and supportive policies that Norway did in the 1970s and early 1980s.

References

Chang, H.-J. (2005). Kicking away the ladder: Development strategy in historical perspective. London: Anthem Press.

Energy, R. (2018). Internasjonal omsetning fra norske oljeserviceselskaper. Rapport til Olje- og energidepartementet. 13 November.

Moses, J., & Letnes, B. (2017a). *Managing resource abundance and wealth: the norwegian experience*. New York: Oxford University Press.

Moses, J., & Letnes, B. (2017b). Breaking Brent: Norway's response to the recent oil price shock. *Journal of World Energy Law and Business, 10,* 103–116.

NPD (Norwegian Petroleum Directorate). (2019a). The Government's Revenues. Norwegian Petroleum. 13 February draft. https://www.norskpetroleum.no/en/economy/governments-revenues/. Accessed August 5, 2019.

NPD. (2019b). Companies. Norwegian Petroleum. 13 February draft. https://www.norskpetroleum.no/en/facts/companies-production-licence/. Accessed August 5, 2019.

NPD. (2018). Employment in the Petroleum Industry. Norwegian Petroleum. 9 October draft. https://www.norskpetroleum.no/en/economy/employment/. Accessed August 5, 2019.

Ruggie, J. G. (1982). International regimes, transactions and change: Embedded liberalism in the postwar economic order. *International Organization, 36*(2), 379–415.

Skule, S., Grytli, T. (1997). Teknologisk utvikling og samfunnsendring. Eksempler fra oljehistorien og bankhistorien. FAFO Report 217. Oslo: FAFO.

SSB (2019) Final expenditure and gross domestic product. Current Prices, NOK millions. 8 February version. https://www.ssb.no/en/nasjonalregnskap-og-konjunkturer/tables/nr-tables. Accessed August 5, 2019.

St. meld. nr. 28 (2010–2011). En næring for framtida—om petroleumsvirksomheten. Oslo: Olje- og energidepartementet.

St. meld. nr. 53 (1979–80). Om virksomheten på den norske kontinentalsokkelen. Oslo: Olje- og energidepartementet.

St. meld. nr. 11 (1968–9). Undersøkelse om boring etter petroleumsforekomster på den norske kontinentalsokkel i Nordsjøen. Oslo: Industridepartementet.

Victor, D. G., Hults, D. R., & Thurber, M. (2012). *Oil and governance: State-Owned enterprise and the world energy supply*. Cambridge: Cambridge University Press.

CSR

Corporate Social Responsibility in the Oil and Gas Industry in Nigeria: The Case for a Legalised Framework

Eghosa O. Ekhator and Ibukun Iyiola-Omisore

Abstract This chapter focuses on the extant corporate social responsibility (CSR) practices in the oil and gas industry in Nigeria. The oil and gas industry has been beset by a lot of problems not limited to violence, kidnappings, eco-terrorism, and maladministration amongst others. One of the strategies of curing or mitigating these inherent problems in the oil and gas sector is the use of CSR initiatives by many oil multinational corporations (MNCs) operating in Nigeria. Notwithstanding that the majority of CSR initiatives in the oil and gas sector in Nigeria are voluntary, this chapter avers that CSR initiatives should be made mandatory by the Nigerian government. Furthermore, Civil Society Organisations (CSOs) should play an integral role in the implementation of any legalised framework on CSR that will be developed in the country. This chapter suggests that a CSR law should be developed specifically for the oil and gas industry to mitigate the negative externalities arising from the activities of oil MNCs in the Niger Delta region of the country.

Keywords Nigeria · CSR · Regulation · Oil industry · Niger delta

1 Introduction

CSR is the idea or theory that companies have a duty towards the society beyond its primary obligations to its shareholders or owners and it is said to be voluntary (Amao 2014). CSR is an increasingly important part of international business. Globalisation of world trade and the rise of powerful companies (MNCs) are mainly responsible for the rise of CSR practices. However, in some instances, the activities of MNCs have had negative consequences in many countries (especially developing countries, including Nigeria). Examples include the Bodo oil spill in Nigeria (amongst many

E. O. Ekhator (✉)
University of Derby, Derby, England
e-mail: eghosaekhator@gmail.com

I. Iyiola-Omisore
University of Leeds, Leeds, UK

E. G. Pereira et al. (eds.), *Sovereign Wealth Funds, Local Content Policies and CSR*, CSR, Sustainability, Ethics & Governance,
https://doi.org/10.1007/978-3-030-56092-8_25

other oil spills in the country), Pfizer's clinical trial of a polio drug that killed 11 children and rendered several others blind and deaf in Nigeria and the Union Carbide disaster in India that killed over 5000 people and caused lifelong health damage to up to 100,000 others (Owoeye 2015; Ekhator 2016a). Thus, Owoeye has argued that 'the list of multinational companies' activities that have inflicted costly damage on populations in the developing world in the unconscionable exercise of the corporate power is indeed endless' (Owoeye 2015: 1; Maiangwa and Agbiboa 2013). Hence, the development of CSR initiatives has encapsulated the various strategies to curb corporate misbehaviour (Adeyeye 2010). These strategies have led to a large number of diverse actors and stakeholders developing soft law initiatives and compulsory rules to enhance corporate behaviour (Adeyeye 2010). Adeyeye (2010: 144) avers that the modern manifestation of CSR is 'generally associated with voluntary non-binding rules which corporations adhere to in an attempt to be socially responsive. CSR encompasses the voluntary codes, principles and initiatives companies adopt in their general desire to confine corporate responsibility to self-regulation'. We will discuss some of these CSR mechanisms in a latter part of this chapter.

Due to the impact of the activities of MNCs on the environment and society, MNCs have a responsibility to mitigate the negative consequences of their actions or activities (Ekhator 2014c). CSR makes corporations responsive to a lot more stakeholders other than just shareholders. These stakeholders could include suppliers, customers, shareholders, environment, and communities amongst others (Ekhator 2014a). Thus, corporations are expected to take responsibility for their actions (if any) on the aforementioned stakeholders. CSR explores issues relating to human rights, labour rights or standards, bribery and corruption amongst others (Amao 2011). In essence, CSR extends beyond the traditional and legal requirements, expected of corporations with regard to its impact on stakeholders. Thus, CSR has been described as a 'business approach for addressing the social and environmental impacts of company activities' (Frynas 2009: 1).

This chapter focuses on the extent CSR policies in the oil and gas sector in Nigeria. This work is divided into six parts including this introduction. The second part of the chapter concentrates on the various definitions of CSR. The third part discusses the extant CSR initiatives in the oil and gas industry in Nigeria. Also, some of the international CSR initiatives will be highlighted in this section. The fourth part of the chapter emphasises the case for a mandatory CSR framework in the oil and gas sector in Nigeria. This is due to the failures and weaknesses of the existing voluntary-styled CSR initiatives in the oil and gas sector. The fifth part of the work discusses some suggestions to enhance any mandatory framework on CSR that will be developed in the country. For example, we recommend that Civil Society Organisations (CSOs) should play an integral role in the implementation of a mandatory/statutory framework on CSR in Nigeria. CSOs are already playing similar roles in different sectors in Nigeria. The sixth part concludes the chapter and offers lessons learnt.

2 Defining CSR

The major problems besetting CSR include the lack of an acceptable definition and the use of various academic/theoretical frameworks to measure its spread and influence (Ekhator 2014a). Notwithstanding the fact that there is no acceptable or universal definition of CSR (Ekhator 2014a; Zhao 2017; Okoye 2009), it is a widely accepted concept in the business world (Zhao 2017). However, there are many definitions of CSR and views on the evolution, history, and concept of CSR.[1] One of the first attempts at defining CSR can be traced back to the definition by Bowen in the 1950s (Bowen 1953). His work emanated from the idea that many large businesses were important centres of power and decision-making and the activities of these corporations touched the lives of citizens at several points (Bowen 1953). Although he raised numerous questions, the most important one was, 'What responsibilities to society may businessmen reasonably be expected to assume' (Carroll 1999: 270).

In defining CSR, Bowen sets out the social responsibilities of businessmen: 'it refers to the obligations of businessmen to pursue those policies, to make those decisions, or to follow those lines of action which are desirable in terms of objectives and values of our society' (Bowen, p 153, in Carroll 1999: 270). Similarly, Davis defines CSR as 'businessmen decisions and actions taken for reasons at least partially beyond the firm's direct economic or technical interest' (Davis 1967 in Carroll 1999: 271). He believes that making responsible business decisions lead to long-term economic gain to the company, which is seen as a reward for its socially responsible outlook (Davis 1967).

The 1970s to 1980s ushered in a new era in the understanding of CSR. According to Carroll (1999), a major writer in this era is Johnson (1971) who offered various views on CSR. Two of Jonhnson's definitions are examined here. First, he presented what he labeled "conventional wisdom"; according to him, 'A socially responsible firm is one whose managerial staff balances a multiplicity of interests. Instead of striving only for larger profits for its stockholders, a responsible enterprise also takes into account employees, suppliers, dealers, local communities, and the nation' (Johnson 1971: 50; Carroll 1999: 273). A significant point here is the insinuation of possible stakeholder approach to CSR, for he includes "multiplicity of interests" and actually mentions the various specific interests (Johnson 1971). Johnson's second view of CSR is that 'Social responsibility states that businesses carry out social programs to add profits to their organisation' (Johnson 1971, p 54; Carroll 1999, p 274). Here, he links CSR with financial performance, this term is known as the business case. Scholars espousing this viewpoint believe businesses should engage in CSR for long-term profit maximisation (Jonker and Marberg 2007). On the other hand, it is argued that the business case negatively affects the academic CSR discussion in that

[1] This chapter will not discuss the various arguments on the definitions, history, and evolution of CSR. For an overview of these views, see generally, Ekhator (2014a) and Okoye (2009). Notwithstanding that CSR and its ideals are seen from Western (USA and Europe) influenced perspectives; many scholars have argued that concepts similar to CSR existed in other cultures in different parts of the word. See, generally, Husted (2015).

it 'largely ignores the integral responsibilities of companies that are associated with impacts on stakeholders' (Waddock 2004: 21). CSR in this framework is not about doing what is right by society, but rather doing what is right for the company's bottom line (Jonker and Marberg 2007).

In 1991, Carroll developed a well-recognised and comprehensive framework on CSR (Carroll 1979, 1991, 1999, 2015, 2016) and he gave a comprehensive four-part definition of CSR. However, the four-part definition was first published by Carroll in 1979 and in 1991, Carroll modified the four-part definition into a CSR pyramid (Carroll 2016). He posits that 'the social responsibility of business encompasses the economic, legal, ethical, and discretionary expectations that society has of organisations at a given point in time (Carroll 1999, p 283, 1979 p 500). He explained further that the economic responsibility of business involves its responsibility to make profits for its investors. At the same time, the society expects businesses to obey the law just as it makes profit, which represents the legal aspect (Carroll 1999). It is the duty of businesses to operate within the framework of legal requirement (Carroll 1999). The ethical responsibility of a company is 'the responsibility to do what is right, just, and fair' (Carroll 1979: 500). This includes society's expectations of business (behaviour and practises) over and above (any) legal requirements (Carroll 1979). Companies must strive to exceed legal duties and obligations in their relationships with members of the society (Carroll 1979). Lastly, discretionary responsibilities signify the voluntary duties that companies assume but for which society does not provide as clear-cut an expectation as it does with ethical responsibility (Carroll 1999). They are left at the discretion of individual managers to perform. These voluntary activities include philanthropic contributions to various sectors for social, educational, recreational or cultural purposes (Amodu 2017). Mallin (2019) interprets this definition further as the business first duty to make profit, then abide by legal requirements, to do what is right and fair and to do what might be expected of businesses in terms of supporting local community and making charitable contributions.

Proponents of neo-liberalism have criticised the CSR definitions discussed above. They believe corporations do not have any responsibility towards other stakeholders; their only responsibility is to maximise profits to shareholders (Mallin 2019). A well-known author associated with this view is Milton Friedman. In his book, Capitalism and Freedom (1962: 133), which was based on a series of lectures he gave in 1956, he said:

> the view has been gaining widespread acceptance that corporate officials and labour leaders have a 'social responsibility' that goes beyond serving the interest of their stockholders or their members. This view shows a fundamental misconception of the character and nature of a free economy. In such an economy, there is one and only one social responsibility of business- to use its resources and engage in activities designed to increase its profits so long as it stays within the rules of the game, which is to say, engages in open and free competition, without deception or fraud.

Friedman (1962) does not believe social responsibility should interfere with the workings of a free market economy, especially in the USA at that time. In recent times, focus has shifted from shareholder theory to more global issues, which include environment, human, and labour rights. Stone criticised Friedman's viewpoint, stating that businesses have broader responsibilities that extend beyond owners and shareholders to include employees, customers, suppliers, and host communities (Stone 2012, p 83). He believes that businesses should be considered as increasingly responsible entities, because they play important roles in society, as distributors, taxpayers, investors, and service providers amongst others (Stone 2012). Mitchell (2001, p 74) similarly states 'that the corporate goal of stockholder wealth maximisation not only destroys the corporation but also destroys our social fabric'.

Proponents of stakeholder theory posit that shareholder theory is less economically profitable than the latter paradigm (Mitchell 2001). As the environment in which companies function is far more complex than the simplified reality assumed by the shareholder theory, a managerial focus wholly on duties to shareholders diverts attention and energy away from other groups like employees and customers amongst others (Mitchell 2001). The involvement of these groups within the company in the creation of value and satisfaction with their treatment by the firm is important to the corporations' success or failure (Mitchell 2001).

It is important to note that CSR is often used interchangeably with business ethics and sustainability (Carroll 2015). It demonstrates numerous ways in which companies may promote responsible behaviour. However, some argue that they are different. A close examination of these concepts seems to refer to the same phenomenon but with a slightly nuanced emphasis (Carroll 2015). The next part of the chapter discusses CSR initiatives in the oil and gas industry in Nigeria.

3 CSR in the Oil and Gas Sector in Nigeria

The Nigerian economy is heavily reliant on the revenues accruing from the oil and gas sector (Ekhator 2014a). The Niger Delta region where oil MNCs maintain a significant presence has become a theatre of incessant violent conflicts or crisis (Ekhator 2014a). Some of the negative consequences of the MNCs in the oil sector in Nigeria include gas flaring, oil spills, environmental pollution, negative social impacts, conflict and violence. Thus, due to these negative impacts arising from the activities of oil MNCs, many oil companies in the country have developed CSR initiatives in an effort to mitigate these negative consequences.

MNCs operating in the oil and gas sector in Nigeria have engaged in a series of initiatives to enhance community development in the Niger Delta (Egbon et al. 2018). This is anchored on the premise that '... if oil MNCs can contribute to community development via CSR, it will help to address local grievances, improve community development and promote positive corporate-community relations' (Egbon et al. 2018: 54). Many oil MNCs (and other companies) are involved in a plethora of CSR initiatives in the Niger Delta and other parts of Nigeria. CSR initiatives in Nigeria may include the building of schools, hospitals, markets, and provision of pipe borne water amongst other initiatives (Amaeshi et al 2006). Nevertheless, 'the extent to which the CSR initiatives have contributed to community development in the region remain contested' (Idemudia and Osayande 2016: 2).

Some scholars including Edoho (2008), Frynas (2009), Akpan (2006), Tuodolo (2007, 2009), and others have contended that the CSR process in Nigeria is not far reaching or deeply entrenched (Ekhator 2014a). Thus, it has been contended that some of these CSR initiatives are not carried out on a coherent basis and not always sustained (Amaeshi et al., 2006). Arguably, despite the adoption of various CSR mechanisms by oil companies in Nigeria, the oil-producing communities 'have received a proportionately low amount of benefit compared to the high social and environmental costs of extractive activities' (Lisk et al. 2013: 20). Notwithstanding the minimal contributions of CSR to oil-producing communities in the Niger Delta, many communities still suffer from various ills including gas flaring, oil spillage, and violence.

On the other hand, Idemudia and Ite (2006), (Ite 2007) and Eweje (2006) support CSR initiatives, arguing that CSR is making tremendous progress in the area of local community initiatives in Nigeria. To further elucidate these assertions, Eweje (2006) illustrates that it is becoming increasingly apparent to oil companies that pollution prevention pays whilst pollution does not and under pressure from stakeholder groups, oil companies now routinely incorporate environmental impact assessments into their corporate strategy. Furthermore, Lompo and Trani (2013) averred that CSR initiatives of oil MNCs have enhanced access to basic capabilities like shelter, water, and electricity but have also weakened human development (Idemudia and Osayande 2016: 2). Also, Renouard and Lado (2012) argued that CSR has impacted positively on some of the people and communities close to oil production sites but inequalities still remained entrenched in such communities (Idemudia and Osayande 2016: 2).

Generally, CSR is voluntary in the oil sector in Nigeria. Many MNCs now utilise the Memorandum of Understanding (MOUs) as a means of realising CSR in the Niger Delta (Egbon et al. 2018). Thus, companies engage in the CSR initiatives to obtain "social licence" to operate in that community or region (Ekhator 2016a; Odumosu-Ayanu 2012; Egbon et al. 2018). MOUs are also referred to as examples of Community Development Agreements (CDAs) (Odumosu-Ayanu 2012). Companies such as Chevron and Shell have variants of the MOUs called Global Memorandum of Understanding (GMOU) (Odumosu-Ayanu 2012; Ekhator and Anyiwe 2016). Hence, CDAs are akin to MOUs and GMOUs that are prevalent in the oil and gas sector in Nigeria (Ekhator and Anyiwe 2016).

'The MOU is a bilateral or multilateral agreement between two or more parties. It expresses a convergence of will between the parties, indicating an intended common line of action. In this case, it is an agreement between the MNC and the community on how to implement a set of CSR programs within a given time frame' (Osemeke et al. 2016: 370). However, it should be emphasised that MOUs and GMOUs models (operating in the oil and gas sector) entered into between MNCs and communities are not legally binding contracts (Ekhator 2016a: 26). Notwithstanding the fact that MOUs and GMOUs are not sacrosanct, Niger Delta communities consider them to be binding and oil MNCs are expected to enforce them (Ekhator 2016a). Thus, many Niger Delta communities and NGOs aver that MOUs should be considered as contracts, for example, as an intention to create binding commitments on the part of oil MNCs towards the inhabitants of oil-producing communities (Ekhator 2016a:

26). Thus, this non-enforceability of MOUs and GMOUs have led to many conflicts and crisis in some communities in the Niger Delta.

3.1 International Initiatives on CSR

In recent years, there have been an explosion of initiatives or mechanisms relating to CSR at the international and local levels. Many MNCs have signed up to these international mechanisms. Examples of these initiatives include OECD Guidelines for Multinational Enterprises, and Global Compact amongst others (Ekhator 2016a). This chapter will briefly discuss the Global Compact and the OECD Guidelines.

According to the former UN Secretary-General Ban Ki-moon '[t]he Global Compact asks companies to embrace universal principles and to partner with the United Nations. It has grown to become a critical platform for the UN to engage effectively with enlightened global businesses'(Ekhator 2016a: 7). The Global Compact is a collaborative model for businesses that are willing to commit their operations to align with the ten principles, especially in areas of labour, the environment, human rights, and anti-corruption (Global Compact website, Ekhator 2016a: 7). Presently, the Global Compact has more than 11,000 participating companies from more than 160 countries and is the largest CSR initiative in the world (Ekhator 2016a). There have been divergent views on the impacts of the Global Compact. For example, 'one school praises the Global Compact for its voluntary approach in the absence of supranational regulatory structures in providing principles for business commitment and public scrutiny, differentiation and social approval competition' (Osuji and Obibuaku 2016: 330). However, the Global Compact has been criticised and has been ineffective for lacking compliance, monitoring, and enforcement provisions, particularly, in relation to the quality and veracity of information that its signatories provide (Osuji and Obibuaku 2016). Hence, notwithstanding the large number of corporate participants in the Global Compact initiative, many companies remain indifferent to it (Mujih 2012).

The next international initiative that will be in focus is the OECD Guidelines on Multinational Enterprises (the Guidelines). They are recommendations on responsible business conduct for MNCs operating in or from the adhering countries (Ekhator 2016a). They are voluntary and non-binding, and paragraph 7 of their General Principles encourages self-regulation by MNCs (Mujih 2012; Ekhator 2016a). The Guidelines are very detailed and intended to be localised into national laws or corporate governance by OECD members (Ekhator 2016a). However, some commentators aver that the Guidelines' non-binding model is not a disincentive (Ekhator 2016a). For example, they are used to promote CSR initiatives in different parts of the world and 'represent a consensus on what constitutes good corporate behaviour in an increasing global economy' (Muchlinski 2001, p 24). Furthermore, the guidelines could evolve into hard and binding international laws if states adhere and continually apply them in their business relationships with MNCs (Mujih 2012). However, it can be contended that the voluntary nature of the international CSR mechanisms is a massive weakness and thus MNCs are able to evade liability in many instances of environmental injustices (Gilberthorpe and Banks 2012, p 185). The next part of the chapter focuses on CSR and the laws applying to it in Nigeria.

3.2 CSR and the Law in Nigeria

Many of the existing CSR initiatives in Nigeria are self-regulatory and voluntary in nature. Pillay (2014: 10) avers that '…contemporary ideas of CSR tend to be firmly premised on a shareholder-oriented model of the corporation as a private enterprise whose directors owe enforceable duties only to shareholders' and Nigeria is no different. This part of the chapter will focus on the Companies and Allied Matters Act 1990 (Cap C20 Laws of the Federation of Nigeria, 2004).

The Companies and Allied Matters Act 1990 (CAMA) is 'the primary corporate law legislation in Nigeria' (Amodu 2017: 115). The CAMA is premised on a shareholder-centric model and it 'adopt[s] the traditional primacy model of corporate governance, whereby corporate responsibility towards stakeholder groups such as employees, creditors, local communities, and suppliers, is very limited' (Amodu 2017: 116). Thus, CSR initiatives are organised on a voluntary basis by the companies operating in the country.

It is important to explore the provisions of CAMA, in order to decide whether any improvement has been made in promoting responsibility of MNCs and CSR initiatives in the country. Section 334 of the CAMA enjoins company directors to prepare a financial statement for each financial year (Amao 2008). The information expected to be disclosed under section 334 (2) CAMA includes the following: (a) statement of the accounting policies, (b) the balance sheet as at the last day of the financial year, (c) a profit and loss account or, in the case of a company not trading for profit, an income and expenditure account for the financial year, (d) notes on the account, (e) the auditors' report, (f) the directors' report, (g) a statement of the source and application of funds, (h) a value-added statement for the financial year (i) a five-year financial summary, and (j) in the case of a holding company, the group financial statement.

According to Amao (2008: 101), a remarkable obligation is the provision in (h) that the statement should include 'value-added statement for the financial year' [section 334(2) CAMA]. This provision is thus linked to financial reporting (Amao 2008). This requirement only focuses on reporting of the financial statement of the company, without reference to non-financial or sustainable reporting. However, the overarching effects of section 279(4) and (9) of CAMA 'shows very limited support for CSR in Nigeria, especially as regards employee rights' (Amodu 2017: 116). Section 279 provides that company directors owe duties to only the company (i.e. shareholders), and thus have no legal responsibility or capacity to embark on any other duty apart from their duty to the company and its shareholders. Section 279 (4) of CAMA seems to enjoin directors to consider and have 'regard for and balance employee-related issues and interests in making corporate decisions' (Amodu 2017: 116). However, Sect. 279(9) makes it clear that—whilst employees may believe that their interests are being taken into account in promoting the success of the company, they are not entitled to sue or make any claim whenever the company in making decisions does not consider their interests (Amodu 2017). Since only the company (that is the shareholders as a whole) can sue if this right is violated or appears to have been violated (Amodu 2017), one begins to question the effectiveness of Sect. 279(4) in the first place.

In other words, this provision has no teeth, as employees are unable to sue or make any claim whenever the company does not consider their interests. From the discussion above, it is clear that the CAMA does not really recognise the stakeholder engagement model, as it is based on the shareholder primacy model of corporate governance (Amodu 2017). There is little or no protection afforded to other constituents such as the employees, environment, suppliers, contractors, local community, and other stakeholders (Amodu 2017). There is a risk that directors are likely to focus on profit maximisation for shareholders, without considering the interests of other members affected by the company's decisions.

The debates so far have shown that Nigerian company law has failed to rise to the potential of domestic company law as an instrument for controlling MNCs and promoting CSR in its provisions (Amao 2011). Whilst in one respect, constraining the ability to hold the parent companies of MNCs liable for the acts of its subsidiary under Sect. 54 of the CAMA, Nigerian company law also fails to meet modern realities in companies' operations (Amao 2011). A significant development that highlights the unresponsiveness of Nigerian corporate law system to these challenges are the gaps in the code of corporate governance initiated in the country (Kajola 2008: 25). Whilst other African countries that introduced the codes of corporate governance in the last few years have followed a comprehensive model that includes other stakeholder issues in varying degrees, and the Nigerian code of corporate governance (2003 version) is a significant exception maintaining the traditional shareholder-centric model of corporate governance (Rossouw 2005: 97; Amao 2008: 102). The recent Corporate Governance Codes are no exceptions and they are premised on a self-regulatory model. For example, the Nigerian Code of Corporate Governance (NCGG) 2018 seeks to localise the 'highest standards of corporate governance best practices in Nigerian companies, especially those companies without industry sector codes or regulations' (Okike 2019: 25). This code is a principle-based rather than a rule-based model, hence, it is self-regulatory in practice (Okike 2019). Furthermore, the recently passed Companies and Allied Matters Amendment Bill 2018 does not provide for a legalised CSR model in its provisions (Onyekwere 2019). However, this bill was assented into law by the President of Nigeria in August 2020. The next part of the chapter focuses on a case for legalised/mandatory CSR framework in the oil and gas industry in Nigeria.

4 The Case for Mandatory CSR in Nigeria

The interrogation of government responsibility for ensuring CSR amongst MNCs appears contradictory, in view of the common perspective that CSR is a voluntary led process driven by companies (Okoye 2012; McWilliams and Siegel 2001). This general perception is increasingly contested due to the protests of several stakeholders within the CSR agenda (Okoye 2012). Arguably, there is a need to adequately define or outline an effective role for CSR as a model for embracing business and society relationships. Some recent developments have led to the re-evaluation of the common perspective; including the position of legal and regulatory scholars who support CSR accountability and legitimacy (Okoye 2012). This emphasises a more visible role for law and regulation (Okoye 2012; McBarnet et al. 2007). McBarnet et al (2007: 55) point out that 'social and legal means need not be seen as alternatives for furthering corporate responsibility, but as complementary controls in a new style of corporate accountability that involves both legal and ethical standards'.

As a result, governments of developing countries (including Nigeria) have sought to control MNCs, to promote development, and curb irresponsible behaviour. This section of the chapter seeks to address the question of how CSR is promoted in Nigeria through the enactment of legislation (and bills) by the government and an analysis of the effectiveness of these laws. This chapter argues for a mandatory variant of CSR in the oil and gas sector in Nigeria.

This chapter aligns with the following position taken by Ihugba (2012: 69): 'Compulsory regulation refers to legislative enactments or judicial judgements prescribing roles and sanctions. The advocacy for compulsory regulation is seen as a sure way to promote transparency and accountability and also regain the trust of the public'. Arguably, when companies are left unchecked or unregulated, they can become irresponsible and oppressive (Ihugba 2012: 69). This is evident in the Niger Delta, wherein the oil MNCs are significantly located in Nigeria.

There have been many justifications for a mandatory/legalised model of CSR to be developed in the oil and gas sector in Nigeria. Some of the reasons include the fact that self-regulatory initiatives by MNCs in the oil and gas sector have been ineffectual in the country (Ekhator 2016a). Hence, CSR initiatives are largely unsuccessful in the Niger Delta region (Ekhator 2014a). MNCs are known for flagrantly disobeying laws and damaging the Nigerian environment (Ekhator 2016b, 2014c). The plight of the Ogoni people is apposite in this analysis. Shell started operations in Ogoniland in the Niger Delta in 1958 but withdrew in 1993 due to socio-political protests and to protect the lives of its employees (Ekhator and Anyiwe 2016). Twenty years after the cessation of oil exploration activities, the Ogoni people are still living with health problems arising from the activities of Shell in that community (UNEP Report 2011). In April 2013, Shell went back to the abandoned site to ascertain the state of its properties and determine how best to decommission them (Ventures Publication 2013; Ekhator and Anyiwe 2016). On arrival, it was learnt that the death toll is still on the increase in Ogoniland after so many years (Ventures Publications 2013). To date,

Shell is still unable to operate in Ogoniland and the environment remains heavily polluted and contaminated.

For the above reasons, this chapter advocates for a legalised or statutory CSR model in the oil and gas sector in Nigeria. Furthermore, the Nigerian government has endorsed mandatory CSR in some of its laws and bills. The next part of this work focuses on laws promoting CSR in Nigeria.

4.1 Laws Promoting CSR in Nigeria

There are some laws promoting CSR in the country. This section highlights some of these laws. The use of formal or regulatory CSR in the mining sector in Nigeria has been a successful experiment (Ekhator and Anyiwe 2016; Akinsulore 2016). A major strength of the Nigerian Minerals and Mining Act of 2007 is that it incorporates the use of Community Development Agreements (CDAs) by mining companies in the country (Odumosu-Ayanu 2012).

Section 116(1) of the Minerals and Mining Act of 2007 states thus:

> Subject to the provisions of this section, the Holder of a Mining Lease, Small Scale Mining Lease or Quarry Lease shall prior to the commencement of any development activity within the lease area, conclude with the host community where the operations are to be conducted an agreement referred to as a Community Development Agreement or other such agreement that will ensure the transfer of social and economic benefits to the community.

CDAs are akin to Memorandum of Understanding (MOU) and General Memorandum of Understanding (GMOU) that are prevalent in the oil and gas sector in Nigeria (Ekhator and Anyiwe 2016). The Minerals and Mining Act 2007 also promotes public participation and consultation of host communities in CDAs. For example, Sect. 117 of the Act states that CDAs shall: '... specify appropriate consultative and monitoring frameworks between the Mineral title holder and the host community, and the means by which the community may participate in the planning, implementation, management, and monitoring of activities carried out under the agreement'. This provision is novel and absent in the previous mutations of the Mining Act and there is no similar provision in the Petroleum Act (which governs the oil and gas industry) (Akinsulore 2016: 98). Thus, the CDAs paradigm in the Mining Act is premised on a CSR approach with emphasis on service provision (Odumosu-Ayanu 2012). Furthermore, the Mining Act gives the right to communities to sue companies that refuse to negotiate MOUs with relevant communities (Akinsulore 2016: 112). This is unlike the position in the oil and gas industry wherein MOUs and GMOUs are not binding contractual agreements (Ekhator 2016a).

The Nigerian Extractive Transparency Initiative (NEITI) Act 2007 is another law that promotes CSR in its provisions (Ihugba 2012). This law provides for the inclusion of members of CSOs in the National Stakeholders Working Group (NSWG)—the governing body of the NEITI to promote transparency and accountability in revenue payments in the oil and gas industry (Ekhator 2014b). Furthermore, the NEITI has always engaged CSOs in its activities as a means of improving transparency and opening the process to the Nigerian public (Ekhator 2014b). For example, 'the NEITI has organised a series of activities and engaged in consultation with CSOs in different parts of the country to determine the roles of CSOs in the NEITI process' (Ekhator 2014b: 49). Furthermore, some recent bills in the country have promoted mandatory CSR in their provisions. The next part of the chapter will highlight some of these recent bills.

4.2 Recent Bills Mandating CSR in Nigeria

Some recent bills in the country have promoted mandatory/legalised CSR in their provisions. In order to provide comprehensive and adequate relief to host communities suffering from the negative impacts of corporate irresponsibility, especially from the MNCs, a corporate social responsibility bill was introduced to the Nigerian National Assembly in 2008 (Amao 2011; Okoye 2012). This Bill was titled 'A Bill for an Act to provide for the Establishment of the Corporate Social Responsibility Commission 2008′—popularly referred to as the 'CSR Bill' (Nwagwu 2016). Although the drafting of the bill is apparently poor and some parts unclear (Amao 2011), it is important to examine the bill and the possible implications of its requirements for MNCs and companies. The Bill sought to establish the Corporate Social Responsibility Commission, a regulatory body that would be responsible for the control and regulation of the activities of corporate organisations in Nigeria (Nwagwu 2016). It was proposed that this body will be responsible for the formulation, implementation, supervision, and provision of policies and reliefs to host communities for the physical, material, environmental, or other forms of degradation suffered due to the activities of entities operating in affected communities (Oserogho and Associates 2014).

Section 1 of the CSR Bill proposed the establishment of Corporate Responsibility Commission and under Sect. 5, the Commission is to create standards on CSR initiatives based on international best practices consistent with international standards such as the OECD guidelines and ILO standards. This takes standard setting away from the voluntary initiatives of companies (Okoye 2012). Osuji (2015: 277) criticised the CSR Bill and he states that the Bill is 'an amalgam of incompatibility seeking in general terms to compel CSR without regard to its instrumental or ethical basis'. Furthermore, Okoye (2012) points out that the provisions of this bill reveal a move towards a stronger regulatory approach for CSR. The CSR Bill mandates annual social and environment impact reporting of direct activities on companies on communities, but puts the responsibility for ensuring this on the Commission (Sect. 5(1)(h)). It is observed that this section does not change the reporting system under CAMA, which provides for only financial reporting (Amao 2011). According to Okoye (2012), the negative reaction (towards the Bill) was provoked by the financial sections contained in Sect. 5(1) (i) of the Bill which mandates corporate philanthropy. The main ground of the widespread opposition to the Bill was the legal compulsion by companies to pay a percentage of its profits and companies viewed this as additional tax burden (Okoye 2012; Anyiwe and Ekhator 2016).

The Bill also attempts to increase the group of stakeholders to which corporations are accountable (Amao 2011). Pursuant to Sect. 5 (1) (k), the CSR Commission will ensure companies are accountable not only to employees and their trade unions, but to investors, consumers, host communities, and the wider environment (Amao 2011; Okoye 2012; Amodu 2017). This indicates the move towards effective stakeholder engagement and signifies a move from the traditional form of shareholder primacy as contained in Sects. 41 and 279 of the principal statute on companies in Nigeria under the CAMA (Amao 2011). Thus, the CSR Bill represents a 'perceived shift from voluntary engagement to mandatory compulsion' (Okoye 2012: 371).

Of important note is the shortcomings of the Bill. It is argued that the poor drafting and ambiguity detract from the noble intents of the bill (Amao 2011). Consistent with the Nigerian National Assembly's preceding legislative enactments; the CSR Bill is a reactive legislation and not proactive law. A proactive law acts in anticipation of future problems, needs, or changes (Haapio 2010). Whilst the reactive law waits for events to occur before the governments takes action. The Bill also faced strong criticism for its archaic conception of CSR as corporate gift or charity (Amao 2011; Amodu 2017). Another flaw in the proposed CSR bill is the establishment of a CSR commission, an external CSR regulator with little knowledge of or access to the internal operations of firms, which may have attendant consequences by way of regulatory inefficiencies (Amodu 2017). Furthermore, the MNCs argue that they have no moral or legal duty to communities or stakeholders beyond the payment of taxes and royalties to government (Ekhator and Anyiwe 2016). Finally, due to the concerns of the multinational companies in Nigeria, the CSR Bill is still in abeyance.

Recently (in 2018), a bill titled: 'An Act to Amend the Financial Reporting Council of Nigeria Act 2011 No. 6 to Prescribe Social Corporate Responsibility Requirement by Companies and for Related Matters' passed its second reading in the House of Representatives in Nigeria. According to the sponsor of the bill, Abubakar Amuda-Kannike 'the bill would promote good governance through the allocation of resources to execute corporate social responsibility of companies' (Olatunji 2018). This bill seeks to 'establish mandatory percentages of corporate social requirements by companies who have earned an average of N50,000.00 and above in profits for three succeeding years' (Olatunji 2018). Arguably, the criticisms of the erstwhile CSR bill can be levelled against this current mutation. Many MNCs and businesses will not accept a national or countrywide mandatory CSR model in Nigeria.

The Petroleum Industry Bill (PIB) is one of the most important bills in Nigeria's history (Ako and Ekhator 2016). The PIB has remained stuck in the National Assembly since 2009. However, in December 2015, the PIB was jettisoned and split into a number of distinct bills (Ako and Ekhator 2016). Section 116 of the PIB provides for 'the establishment of a fund to be known as the Petroleum Host Fund (PHC Fund). PIB provides that each upstream petroleum company shall remit to the PHC Fund on a monthly basis 10 per cent of its net profit to be utilised for the development of the economic and social infrastructure within petroleum producing communities'. Hence, the PIB promotes mandatory CSR in its provisions.

The Petroleum Host and Impacted Communities Development Trust Bill (PHICDB), which is one of the offshoots of the PIB, enjoins all oil companies to establish the Petroleum Host Communities Development Trust (Community Trust) in communities where they operate (Spaces for Change 2018). Thus, this bill promotes CSR in its provisions. However, this provision has been criticised. For example, this 'new obligation not only imposes excessive administrative and financial burdens on operators, but also duplicates the existing community development initiatives a that number of oil companies have already implemented/still implementing under their corporate social responsibility programmes, leading to duplication of efforts and wastages' (Spaces for Change 2018: 3).

Some have argued that the government wants to outsource its functions and responsibilities to MNCs in the guise of CSR (Ekhator 2014a, b, c). This is evidenced by the apparent neglect of the Niger Delta region by the government and the region is rife with poverty and under-development (Ekhator 2013). Also, many social initiatives or programmes developed by the federal government already exist in the country. This is evident in the Niger Delta region, wherein many of the oil MNCs are located. Furthermore, the Nigerian government has set up a plethora of social interventionist agencies in the country dedicated to utilising the revenue from the operations or activities of MNCs and the plight of the people living in the Niger Delta is still precarious (Ekhator 2013). Many companies and scholars are against mandatory CSR (in Nigeria) because it will lead to double taxation in their view (Onyekwere 2015). For example, firms already pay different types of taxes in the country. Examples of such taxes paid by companies includes Education Tax, Pay as You Earn (PAYE), Companies Income Tax, and Nigerian Social Investment Trust Fund amongst others (Umoru 2015: 257). Furthermore, some criminal statutes and policies that promote social services and welfare already exist in the country (Umoru 2015). These include Food and Drugs Act, Standard Organisation of Nigeria, and Consumer Protection Council (recently christened as the Federal Competition and Consumer Protection Commission) amongst others (Umoru 2015). Arguably, these laws and policies have not positively affected the people. Notwithstanding the criticisms of a mandatory/legalised CSR model in Nigeria, this chapter advocates for the legalisation of CSR in the oil and gas sector.

The next part of the chapter suggests recommendations to enhance any legalised CSR model that will be developed in the oil and gas sector in the country.

5 Recommendations

This chapter advocates for some suggestions or recommendations to enhance any legislative model on CSR that will be developed in the oil and gas industry in Nigeria. In defining the contributions companies should make as CSR in the country, the local community and relevant stakeholders should be consulted before such actions are taken (Ako et al. 2009). This should be given statutory flavour in any law that will be enacted. Thus, any proposed CSR law should define who a stakeholder is. Hence, stakeholders (including host communities, CSOs, and other relevant stakeholders) should be garnished with the requisite legal standing to sue MNCs if there are any breaches in the proposed law. Furthermore, Akinsulore (2016: 97) claims that the effect of the Mining Act 2007 is 'to empower the community as an important stakeholder, thereby validating the stakeholder thesis therein'. Also, any new law on CSR in the oil and gas sector, should make CDAs, MOUs, and GMOUs binding and enforceable in courts. This will improve access to environmental justice for communities and other relevant stakeholders affected by the impacts of the non-adherence to the various CSR models in the Niger Delta.

Corporate/company law architecture in the country should be reformed to explicitly provide for a stakeholder variant of corporate governance in Nigeria (Amaeshi, Adi, Ogbechie and Amao 2006: 86). Here, CAMA should be expressly amended to incorporate a CSR provision or any proposed CSR law should amend CAMA in this regard (Amao 2011; Amodu 2017: 126–127). Unfortunately, the recent revised version of the CAMA has maintained the *status quo* and did not explicitly provide for mandatory CSR in its provisions (generally see Halliday and Babalola 2019; Oraegbunam and Ubanyionwu 2019). However, similar to the position in Nepal (Pokhrel 2017), the Nigerian constitution should be amended to explicitly provide for a mandatory CSR in the oil and gas sector in the country.

NGOs or CSOs should play a major role in any mandatory CSR process in Nigeria. This is similar to the roles of civil society in the NEITI framework in Nigeria. Civil society should play a role in the implementation of any new CSR law that will be enacted in the country.

Arguably, the recent CSR bills in Nigeria has led to critical debates about the potential of formalising community development initiatives, stakeholder engagement, sustainability reporting, etc., and including CSR in government's formal strategy (Okoye 2012: 372). These debates are evolving in some developing countries including India, Nepal, Mauritius, and Indonesia (Zhao 2017). Indonesia passed a reformed company law and investment law in 2007 to enshrine CSR in law (Zhao 2017). Similarly, Sect. 135 of India's new Companies Act 2013 and its 2014 Companies Corporate Social Responsibility Policy Rules, 2014 (CRS Rules) provide that companies should spend 2% of average net profit on CSR activities (Gatti et al. 2018). Arguably, this is more realistic in terms of compliance by the companies. The passage of the CSR Bill in Nigeria has been jettisoned due to its contentious provisions. The response of companies to the CSR Bill has ranged from lukewarm to hostile (Amao 2011). On the mandatory nature of CSR in Nigeria, Okoye (2012: 374) suggests that government should drive the development of CSR through legislation as it is their primary responsibility to promote development. However, this should be done in a facilitative manner, as the current attempt to make CSR mandatory through the CSR Bill may be unsuccessful (Okoye 2012). This is partly because the bill lacks sophistication in its use of law and partly because of the persistent opposition by companies (Ekhator and Anyiwe 2016; Okoye 2012). In addition, Amodu (2017) submits that the Nigerian government should take adequate steps towards the reform of the existing laws relating to CSR in order to guarantee a more effective sustainability model and CSR regime in the country.

Furthermore, there is no binding treaty or framework regulating the activities of MNCs in international law (Ekhator 2016a). However, a draft or proposed treaty on business and human rights is in the offing and it is expected to be a binding international treaty (Ekhator 2018). Unfortunately, this proposed treaty focuses on human rights and not CSR. In the global sphere, the preferred approach is soft law, which mainly comprises of the adoption of voluntary guidelines for companies and businesses (Ekhator 2016a). Due to the ineffectual nature of CSR initiatives, some scholars have advocated for the development of a binding treaty in the international sphere (Ekhator 2018). The next part of the chapter is the conclusion.

6 Conclusion

CSR initiatives in the oil and gas sector in Nigeria are mainly anchored on voluntary models. As discussed in the earlier parts of this chapter, the consensus in academic literature is that the impacts of these CSR initiatives have not affected the Niger Delta region positively. Many suggestions have been made to remedy or mitigate these negative consequences. This chapter contends that a mandatory CSR law specific to the oil and gas industry is one of the strategies that can be used to enhance CSR initiatives in the sector. Similar laws have been developed in some developing countries to deepen the effectiveness of CSR initiatives. For example, some developing countries (India, Mauritius, Indonesia and Nepal amongst others) have made CSR compulsory or hard law (Pillay 2014; Zhao 2017). In Mauritius and India, laws have been enacted to compel companies to set aside a percentage of their profits for CSR programmes (Zhao 2017). In Nigeria, a CSR Bill, which was supposed to make CSR compulsory or mandatory in the country, was shot down and never got passed into law. This chapter contends that notwithstanding the criticisms of a legalised/mandatory CSR framework in Nigeria, an oil and gas industry-specific law could be one of the ways to mitigate the negative impacts of oil MNCs in the Niger Delta. Similar to the Nigerian Minerals and Mining Act of 2007, a CSR law should be developed specifically for the oil and gas industry incorporating the recommendations made in the penultimate section of this chapter.

This chapter also suggests that CSOs should play a major role in the development and implementation of any CSR law that will be developed for the oil and gas sector in Nigeria. This chapter aligns with the position of Professor Islam, which states that 'Whilst social movements have led regulators to enact new CSR regulation, regulation on its own may not create much improvement of CSR. This means that continuous monitoring of corporate compliance with CSR legislation by social activists is necessary to achieve the regulatory objective of CSR' (Islam 2018). Finally, this chapter does not suggest that voluntary CSR should be jettisoned; this paper avers that both mandatory and voluntary CSR (in the oil and gas industry in Nigeria) can work in tandem to enhance community development in the Niger Delta region.

Although a one-size-fits-all approach does not apply to all jurisdictions, lessons from Nigeria can serve as an example to represent many problems that are commonly shared by developing countries. Since the various approaches to CSR are intimately connected to national economic structures, political and cultural traditions, a transplantation of one national approach to another country is not easily achieved. However, it might still be possible to learn something from the Nigerian experience through the development of economic models, principles, and best practices, which other countries can emulate. Arguably, other countries (especially developing countries) can learn from the mistakes or weaknesses of the CSR regime in the oil and gas industry in Nigeria. For example, by adopting stand-alone CSR laws to enhance and promote CSR in their countries may be one of the strategies to avoid such mistakes.

References

A Bill for an Act to provide for the Establishment of the Corporate Social Responsibility Commission 2008' - popularly referred to as the 'CSR Bill', cited in Nwagwu, I. 'The CSR bill and doing Business in Nigeria' (Business Day March 24, 2016) https://businessday.ng/analysis/article/the-csr-bill-and-doing-business-in-nigeria/ (Accessed 6 June 2019)

Adeyeye, A. (2010). The limitations of corporate governance in the CSR agenda. *Company Lawyer, 31*(4), 114–118.

Amao, O. O. (2008). Corporate social responsibility, multinational corporations and the law in Nigeria: Controlling multinationals in host states. *Journal of African Law, 52*(1), 89–113.

Amao, O. (2011). *Corporate social responsibility, human rights and the law: Multinational corporations in developing countries*. Routledge.

Amao, O. (2014). Emergent state practice on the creation and practice of standards on corporate social responsibility. *State Practice and International Law Journal, 1*(1), 117–137.

Amaeshi, K., Adi, A. B. C., Ogbechie, C., & Amao, O. O. (2006). Corporate social responsibility in Nigeria: Western mimicry or indigenous influences? *The Journal of Corporate Citizenship, 24*, 83–99.

Akpan, W. (2006). Between responsibility and rhetoric: Some consequences of CSR practice in Nigeria's oil province. *Development Southern Africa, 23*(2), 223–240.

Akinsulore, A. (2016). The effects of legislation on corporate social responsibility in the minerals and mines sector of Nigeria. *Journal of Sustainable Development Law and Policy (The), 7*(1), 97–115.

Ako, R. T., Obokoh, L. O., & Okonmah, P. (2009). Forging peaceful relationships between oil-companies and host-communities in Nigeria's Delta region: A stakeholder's perspective to corporate social responsibility. *Journal of Enterprising Communities: People and Places in the Global Economy, 3*(2), 205–216.

Ako, R., & Ekhator, E. O. (2016). The civil society and the regulation of the extractive industry in Nigeria. *Journal of Sustainable Development Law and Policy (The), 7*(1), 183–203.

Amodu, N. (2017). Regulation and enforcement of corporate social responsibility in corporate Nigeria. *Journal of African Law, 61*(1), 105–130.

Bowen, H. R. (1953). *Social responsibilities of the businessman*. New York: Harper & Row.

Carroll, A. B. (1979). A three-dimensional conceptual model of corporate performance. *Academy of Management Review, 4*(4), 497–505.

Carroll, A. B. (1991). The pyramid of corporate social responsibility: Toward the moral management of organizational stakeholders. *Business Horizons, 34*(4), 39–48.

Carroll, A. B. (1999). Corporate social responsibility: Evolution of a definitional construct. *Business & Society, 38*(3), 268–295.

Carroll, A. B. (2015). Corporate social responsibility: The centerpiece of competing and complementary frameworks. *Organizational Dynamics, 44*(2), 87–96.

Carroll, A. B. (2016). Carroll's pyramid of CSR: Taking another look. *International Journal of Corporate Social Responsibility, 1*(1), 3–8.

Davis, K. (1967). Understanding the social responsibility puzzle. *Business Horizons, 10*(4), 45–50.

Edoho, F. M. (2008). Oil transnational corporations: Corporate social responsibility and environmental sustainability. *Corporate Social Responsibility and Environmental Management, 15*(4), 210–222.

Egbon, O., Idemudia, U., & Amaeshi, K. (2018). Shell Nigeria's global memorandum of understanding and corporate-community accountability relations: A critical appraisal. *Accounting, Auditing & Accountability Journal, 31*(1), 51–74.

Ekhator, E. O. (2013). 'Environmental protection in the oil and gas industry in Nigeria: The roles of governmental agencies. *International Energy Law Review, 5,* 196–203.

Ekhator, E. O. (2014a). Corporate social responsibility and Chinese oil multinationals in the oil and gas industry of Nigeria: An appraisal. *Cadernos De Estudos Africanos, 28,* 119–140.

Ekhator, E. O. (2014b). The roles of civil society organizations in the extractive industries transparency initiative in Nigeria. *Int'l Journal of Not-For-Profit L., 16,* 47–52.

Ekhator, E. O. (2014c). Improving access to environmental justice under the African charter on human and peoples' rights: The roles of NGOs in Nigeria. *African Journal of International and Comparative Law, 22*(1), 63–79.

Ekhator, E. O. (2016a). Regulating the activities of oil multinationals in Nigeria: A case for self-regulation? *Journal of African Law, 60*(1), 1–28.

Ekhator, E. O. (2016b). Public regulation of the oil and gas industry in Nigeria: An evaluation. *Annalen Survey of Int'l & Comparative L., 21,* 43–91.

Ekhator, E. O. (2018). Regulating the activities of multinational corporations in Nigeria: A case for the African union? *International Community Law Review, 20*(1), 30–68.

Ekhator, E. O., & Anyiwe, L. (2016). Foreign direct investment and the law in Nigeria: A legal assessment. *International Journal of Law and Management, 58*(1), 126–146.

Eweje, G. (2006). Environmental costs and responsibilities resulting from oil exploitation in developing countries: The case of the niger delta of Nigeria. *Journal of Business Ethics, 69*(1), 27–56.

Friedman, M. (1962). *Capitalism and freedom.* University of Chicago.

Frynas, J. G. (2009). *Beyond corporate social responsibility: Oil multinationals and social challenges* (Vol. 10). Cambridge: Cambridge University Press.

Gatti, L., Seele, P., Cottier, B., & Vishwanath, B. (2018). Are we moving beyond voluntary CSR? Exploring theoretical and managerial implications of mandatory CSR resulting from the new Indian companies act. *Journal of Business Ethics*, 1–12. https://doi.org/10.1007/s10551-018-3783-8.

Gilberthorpe, E., & Banks, G. (2012). Development on whose terms?: CSR discourse and social realities in Papua New Guinea's extractive industries sector. *Resources Policy, 37*(2), 185–193.

Halliday, C., & Modupe Babalola, M. (2019). Critical analysis of salient features of the companies and allied matters act (Repeal and Re-enactment) bill 2018. *The Journal of Property Law and Contemporary Issues, 10*(1), 111–119.

Haapio, H. (2010). Introduction to proactive law: A business lawyer's view. *Stockholm Institute for Scandinavian Law, 49,* 49–52.

Husted, B. W. (2015). Corporate social responsibility practice from 1800–1914: Past initiatives and current debates. *Business Ethics Quarterly, 25*(1), 125–141.

Idemudia, U., & Ite, U. E. (2006). Corporate–community relations in Nigeria's oil industry: Challenges and imperatives. *Corporate Social Responsibility and Environmental Management, 13*(4), 194–206.

Idemudia, U., & Osayande, N. (2016). Assessing the effect of corporate social responsibility on community development in the Niger Delta: A corporate perspective. *Community Development Journal, 53*(1), 155–172.

Ihugba, B. U. (2012). Compulsory regulation of CSR: A case study of Nigeria. *Journal of Poloniae & L., 5,* 68.

Islam, M. 2018. What is Corporate Social Responsibility—does it work? The Conversation Available from https://theconversation.com/what-is-corporate-social-responsibility-and-does-it-work-89710 [Accessed 5 June 2019)

Ite, U. E. (2007). Changing times and strategies: Shell's contribution to sustainable community development in the Niger Delta. *Nigeria. Sustainable Development, 15*(1), 1–14.

Johnson, H. L. (1971). Business in contemporary society: Framework and issues. Wadsworth Pub. Co.

Jonker, J., & Marberg, A. (2007). Corporate social responsibility quo Vadis? *Journal of Corporate Citizenship, 27,* 107–118.

Kajola, S. O. (2008). Corporate governance and firm performance: The case of Nigerian listed firms. *European Journal of Economics, Finance and Administrative Sciences, 14*(14), 16–28.

Lisk, F., Besada, H., & Martin, P. (2013). Regulating extraction in the global south: Towards a framework for accountability. *Background research paper submitted to the high level panel on the post-2015 development agenda.* https://www.post2015hlp.org/wp-content/uploads/docs/Lisk-Besada-Martin_Regulating-Extraction-in-the-Global-South-Towards-a-Framework-for-Accountability-_FINALFINAL.pdf

Lompo, K., & Trani, J. F. (2013). Does corporate social responsibility contribute to human development in developing countries? evidence from Nigeria. *Journal of Human Development and Capabilities, 14*(2), 241–265.

Maiangwa, B., & Agbiboa, D. E. (2013). Oil multinational corporations, environmental irresponsibility and turbulent peace in the Niger Delta. *Africa Spectrum, 48*(2), 71–83.

Mallin, C. (2019). *Corporate Governance.* Oxford.

McBarnet, D., Voiculescu, A., & Campbell, T. (2007). *The new corporate accountability: Corporate social responsibility and the law.* Cambridge University Press.

McWilliams, A., & Siegel, D. (2001). Corporate social responsibility: A theory of the firm perspective. *Academy of Management Review, 26*(1), 117–127.

Muchlinski, P. (2001). Corporations in international litigation: Problems of jurisdiction and the United Kingdom asbestos cases. *International & Comparative Law Quarterly, 50*(1), 1–25.

Mujih, E. (2012). *Regulating multinationals in developing countries: A conceptual and legal framework for corporate social responsibility.* Gower.

Mitchell, L.E. (2001). *Corporate irresponsibility: America's newest export.* Yale University Press.

Nwagwu, I. (2016) The CSR bill and doing business in Nigeria' (business day March 24, 2016) https://businessday.ng/analysis/article/the-csr-bill-and-doing-business-in-nigeria/. Accessed 6 June 2019.

Odumosu-Ayanu, I. T. (2012). Foreign direct investment catalysts in West Africa: Interactions with local content laws and industry-community agreements. *NC Central L. Reviews, 35,* 65.

Okike, E. (2019). *Corporate governance in commonwealth countries.* International Centre for Research in Accountability and Governance.

Okoye, A. (2009). Theorising corporate social responsibility as an essentially contested concept: Is a definition necessary? *Journal of Business Ethics, 89*(4), 613–627.

Okoye, A. (2012). Exploring the relationship between corporate social responsibility, law and development in an African context: Should government be responsible for ensuring corporate responsibility? *International Journal of Law and Management, 54*(5), 364–378.

Olatunji, J. S. (2018). Tribune newspaper 'reps pass bill to prescribe CSR for companies' July 5, 2018. https://tribuneonlineng.com/153855/. Accessed 5 June 2019.

Onyekwere, J. (2019). (Guardian Newspaper Nigeria) 'Koyinsola Ajayi interview: There is need to harmonise business facilitation laws in Nigeria' (7 May, 2019) https://guardian.ng/features/konyinsola-ajayi-there-is-need-to-harmonise-business-facilitation-laws-in-nigeria/. Accessed 6 June 2019.

Oraegbunam, I., & Ubanyionwu, C. J. (2019). Examining some gains and Pitfalls of Nigeria's companies and allied matters act: Urgent need for review. *International Review of Law and Jurisprudence, 1*(2), 48–57.

Onyekwere, J. (2015). (Guardian Newspaper Nigeria) 'Fabian Ajogwu interview: Legislation on corporate responsibility is taxation' (29 June 2015) https://guardian.ng/features/law/legislation-on-corporate-social-responsibility-is-taxation/. Accessed 6 June 2019.

Osemeke, L., Adegbite, S., & Adegbite, E. (2016). Corporate social responsibility initiatives in Nigeria. In *Key initiatives in corporate social responsibility* (pp. 357–375). Springer, Cham.

Oserogho & Associates. (2014). Corporate Responsibility Bill' (5 January 2014) https://www.oseroghoassociates.com/articles/75-corporate-social-responsibility-bill. Accessed 6 June 2019.

Osuji, O. K. (2015). Corporate social responsibility, juridification and globalisation: 'inventive interventionism' for a 'paradox.' *International Journal of Law in Context, 11*(3), 265–298.

Osuji, O. K., & Obibuaku, U. L. (2016). Rights and corporate social responsibility: Competing or complementary approaches to poverty reduction and socioeconomic rights? *Journal of Business Ethics, 136*(2), 329–347.

Owoeye, O. (2015). Corporate social accountability in developing countries: A recipe for the socially responsible corporation. *International Journal of Account Bus Manage, 1*(1), 1–7.

Pillay, R. (2014). The Limits to self-regulation and voluntarism: From Corporate social responsibility to corporate accountability. *Amicus Curiae, 19,* 10–13.

Pokhrel, K. (2017). Constitutional frameworks for corporate social responsibility in context of Nepal. NJA LJ, 11, 281.

Renouard, C., & Lado, H. (2012). CSR and inequality in the Niger Delta (Nigeria). *Corporate Governance: The International Journal of Business in Society, 12*(4), 472–484.

Rossouw, G. J. (2005). Business ethics and corporate governance in Africa. *Business & Society, 44*(1), 94–106.

Spaces for Change. (2018). Memorandum submitted to the house of representatives adhoc committee on petroleum industry bills' (May 2018) https://www.spacesforchange.org/wp-content/uploads/2018/05/MEMORANDUM-SUBMITTED-TO-THE-HOUSE-REPS-COMMITTEE-ON-PETROLEUM-INDUSTRY-BILL.pdf

Stone, C. (2012). Where the law ends: The social control of corporate behaviour. Harper & Row.

Tuodolo, F. (2007). *Corporate social responsibility, local communities and TNCs in the oil and gas sector of Nigeria.* Unpublished PhD Thesis, United Kingdom : University of Liverpool.

Tuodolo, F. (2009). Corporate social responsibility: Between civil society and the oil industry in the developing world. *ACME: An International E-Journal for Critical Geographies, 8*(3), 530–541.

Umoru, G. (2015). Advancing citizenship for corporate entities in Nigeria. *University of Benin Law Journal, 16*(1), 247–265.

UNEP. (2011). Environmental assessment of Ogoni Land Report. Available at https://www.unenvironment.org/explore-topics/disasters-conflicts/where-we-work/nigeria/environmental-assessment-ogoniland-report. Accessed 6 June 2019.

Ventures. (2013). Shell back at Ogoni after two decades of absence. Available at https://venturesafrica.com/shell-back-at-ogoniland-after-absence-of-two-decade/. Accessed 6 June 2019.

Waddock, S. (2004). Parallel universes: Companies, academics, and the progress of corporate citizenship. *Business and Society Review, 109*(1), 5–42.

Zhao, J. (2017). Promoting more socially responsible corporations through a corporate law regulatory framework. *Legal Studies, 37*(1), 103–136.

An Overview of Corporate Social Responsibility in Kenya's Extractives Sector

Angela Khanali Mutsotso

Abstract Kenya's oil & gas sector is quickly developing with industry players predicting the date of first oil to be 2024 once the upstream and pipeline projects are developed. Kenya has a robust mining sector with rare earth and soda ash accounting for the largest mineral exports. Kenya is increasing investment in its extractives sector and angling to position itself as a technical leader in the region by building its technical expertise. Companies operating in the extractives sector often negotiate their economic and financial terms with governments gaining a legal right to operate and separately negotiate with host communities for a social licence to operate. Companies must obtain both licences in order to run a successful operation. In this chapter, we view corporate social responsibility (CSR) as a method for the company to obtain a social licence based on the practice in Kenya where the national government was previously not directly involved in informing and consulting communities. We shall first conceptualise CSR in the Kenyan context, examine two case studies, identity a few instances of the best and worst CSR experiences and conclude with an assessment of the challenges companies face when engaging communities. The chapter aims to inform the reader of the practice in Kenya with insights from the author's interviews and interaction with host community members, companies and government officials, and finally, this chapter is intended to contribute to the literature on CSR in the extractives industry in Kenya.

Keywords Corporate social responsibility · Base Titanium Kenya · Tullow Kenya · Turkana County · Kwale County · Sustainability

---

A. K. Mutsotso (✉)
Kenya Civil Society Platform on Oil & Gas, Nairobi, Kenya
e-mail: angie.mutsotso@gmail.com

E. G. Pereira et al. (eds.), *Sovereign Wealth Funds, Local Content Policies and CSR*, CSR, Sustainability, Ethics & Governance,
https://doi.org/10.1007/978-3-030-56092-8_26

1 Introduction

In Kenya's legal framework, there is no specific requirement for companies to carry out corporate social responsibility (CSR) activities. The laws governing the sector are seen to encourage companies to act in a socially responsible manner. Companies are however expected to carry out social projects with host communities. Both case studies examined in this chapter are of projects in marginalised areas, Kwale and Turkana Counties.

In this chapter, we adopt a definition of CSR as projects carried out by companies that do not directly result in company profits, but are carried out for the economic, environmental and social benefit of resource host nations and project host communities (What is CSR? | UNIDO, n.d.). This chapter describes the approaches adopted by companies based on demand from communities and under the current legal framework.

In this section of the chapter, we briefly traverse CSR approaches and requirements under the Kenyan legal framework, and in the next, turn to the practical examples by describing approaches adopted within the two example projects. We shall conclude this chapter with a review of the best and works experiences in CSR in the Kenyan context and a review of the challenges that companies face when implementing CSR. Much of the information about the two projects comes from information provided on the websites produced by the two extractives companies and other desk-based research.

1.1 CSR Features and Approaches in the Kenyan Legal Space

In the Kenyan extractives space, the legal framework highlights the conventional approaches to CSR, which include establishing economic linkages, training and technology transfer and infrastructural and social development through the usage of collaborative plans and programmes.

In this part, we identify four approaches and features of CSR as provided for under the Petroleum Act, 2019 and the Mining Act, 2016. We conclude this section by discussing the sources of financing and enforcement of CSR requirements in Kenya.

1.1.1 Company Centric Development Plans

When applying for licences, companies are now required to present internally developed overarching CSR plans; these detail their intended CSR approaches and activities to be carried out during the life of the licence. In Kenya, these plans as shown in the extractives sector are not always called CSR plans but have an element of voluntariness, corporate responsible behaviour and provide a road map of all intended

CSR activities for the particular project within the licence period. These plans are the government's way to inform themselves, and the public of intended programmes and ideally should allow for planning and collaboration between national and county governments and companies, and communities and companies.

The Mining Act, 2016 at Section 47 requires holders of mineral rights to carry out businesses in a socially responsible manner. Further, Section 101 (j) of the Mining Act, 2016 requires holders of large-scale mining rights to submit Social Responsible Investment Plans that they will undertake during their operations to the benefit of host communities.

The Petroleum Act, 2019 does not require the companies to develop a CSR plan, but it is envisioned in Section 125 that the company will involve the community in their CSR programmes.

1.1.2 Establishing Economic Linkages: Local Content

Both the Petroleum and Mining Laws contain local content provisions. This is a common feature of corporate responsible behaviour by companies in the extractives sector and was practised before the current legal frameworks as a result of citizen and political demand. Companies operating in Kenya are required to give preference to Kenyan employees, both skilled and semi-skilled, preference to Kenyan service providers and contractors, and Kenyan manufactured goods.

Section 50 of the Petroleum Act, 2019 details the different goods and services that Kenyans and Kenyan-owned businesses should be prioritised on. Further Section 50 requires contractors to submit Long Term and Annual Local Content Plans to the Energy and Petroleum Regulatory Authority (EPRA). The Model Production Sharing Contract within the Petroleum Act, 2019 requires companies to give preference to Kenyan employees and service providers.

The Mining Act, 2016 at Section 47 requires that Kenyans should be given preference of employment, and where this is not possible, the company is allowed to engage expatriates. In such instances where expatriates are contracted, the companies are expected to work towards replacing the technical non-Kenyan staff with Kenyan employees.

1.1.3 Technology Transfer and Trainings

Compared to the economic linkages feature, this is a novel feature especially and is intended to build the country's technical capacity. Section 47 of the Mining Act, 2016 contains provisions on partnership with universities and training of employees. Mining companies are required to form partnerships with universities for research and environmental protection purposes. Additionally, the companies should conduct training and capacity building of employees regularly. It is important to note that the Mining Act does not establish or envision a training fund as the Petroleum Act does, but this function of training non-employees is left to the discretion of the company.

The Petroleum Act, 2019 at Section 52 establishes a petroleum training fund where all licence holders are required to contribute each year. Mutai (2019) notes that as of 2019, this fund had a net worth of approximately 9 Million US dollars. Through their monetary contributions, set in the different Petroleum Agreements, companies contribute to the training and the development of technical capacity in the upstream petroleum sector.

Clause 23 of the Model Production Sharing Contract (Model PSC) requires contractors to develop a Technology Transfer Programme in line with the Energy and Petroleum Policies. These programmes should build the technical capacity of Kenyans. The Model PSC provides details on the contents of this programme.

1.1.4 Collaborative Plans and Approaches: Infrastructural and Social Development

To effect other projects outside Local Content and Training and Development, the legal framework foresees the adoption of collaborative plans developed by companies, county governments and communities. These plans allow for community infrastructure development, which strengthens the education, social, health and other sectors to directly benefit the host community.

Section 109(j) of the Mining Act, 2016 established Community Development Agreements (CDAs). These are signed between large-scale mining companies and host communities. Large-scale licence holders are required to invest one per cent of their gross annual revenues (Mining Act, 2016 (Community Development Agreements) Regulations, 2017). Financing CDAs reduces their tax burden; however, there is no indication on the recoverability of CSR costs in the mining sector. CDAs conceptualise and action development projects to the benefit of local communities. Decisions are made at the CDA Committee, which has members from the local community, national and county governments and the mining company. Currently, communities living around the Kwale Mineral Sands Project operated by Base Titanium and those around Tata Chemicals Magadi Soda Ash Mining operations are negotiating their first CDAs. It is vital to note that this collaborative and participatory CSR approach with specified amounts of investment is unique to the mining sector and is not legally required in the upstream petroleum sector.

Tullow Oil who are developing Kenya's maiden oil fields to allow for production and export of crude oil through their South Lokichar Foundation Stage Development (FSD) project released a 2019 Draft ESIA Report (Golder and Ecologics Consultants Ltd 2019). In the draft report, they proposed the formation of a community development plan (CDP). The CDPs are to be developed for each county and will guide their community projects as well as facilitate community consultation and involvement regarding project impacts. The CDP will mainly address issues of water, access to electricity and other impacts and benefits from the South Lokichar FSD. The CDP will be managed through a consultative process incorporating community, government and company; the CDPs described above are contained in a draft environmental and social impact assessment (ESIA). Thus, there is no final decision on their existence by the company or government, and they will be company initiated.

1.2 Financing of CSR Projects

CSR projects are traditionally voluntary and financed by companies. In practice, they are accounted for as operational costs borne by the company and are generally classified as non-recoverable costs in natural resource contracts (Vermaelen 2011). As the contracts of extractives companies are not publicly available in Kenya, we rely on the provisions of the Model PSC for the petroleum sector and the mining legislation to discuss the financing of CSR projects.

The Model PSC in the Petroleum Act, 2019 at clause 3.14.1.24 classifies *corporate social responsibility costs or social infrastructure costs* as non-recoverable. As this classification is not contained in the Petroleum Act, 2019 and Kenya's upstream petroleum contracts are not publicly available, it is difficult to assess whether the costs from the CSR projects that will be described below run by Tullow and Africa Oil are non-recoverable.

1.3 Enforcement

Enforcement of CSR in Kenya is done voluntarily by the companies as it is in their best interest as it helps to secure a social licences to operate and harmonious community and political relations. Additionally, most of the institutions tasked with enforcement are still being developed, and the sector has not evolved to a size that the government can justify hiring staff to track CSR projects specifically. In both the petroleum and mining sector, the licensing authorities are required to ensure enforcement of multiple requirements ranging from technical standards to environmental standards and coordinate the overall development of the sector; thus, CSR monitoring is often overlooked.[1] Despite the legal underpinnings of different CSR approaches examined in this first part of the chapter, extractives companies often proactively design and implement CSR projects.

This section laid the ground for our exploration of CSR projects in Kenya. In the next section, we shall review the CSR approaches adopted by companies operating in the Kwale Mineral Sands Project and the South Lokichar Upstream Development. These two case studies represent the most significant and most relevant upstream petroleum and mining projects in Kenya presently.

[1]Reference can be made to the Energy Act, 2019 Section 10 on the functions of Energy and Petroleum Regulatory Authority and the Mining Act, 2016 Section 20 on the functions of the Director of Mines.

You will note that CSR projects in both case studies began before the implementation of the laws discussed in this section; however, companies are redesigning their projects to incorporate the structures within the current laws.

2 Overview of Companies Operating in Kenya's Extractives Sector

This section describes the CSR projects of two examples within the upstream petroleum sector and mining sector. We shall first review the Kwale Mineral Sands Project operated by Base Titanium as it is Kenya's largest extractives project, which contributes 108 million US dollars annually to the Kenyan economy ('Base Titanium—Project Economic Contribution', n.d.). We then turn our attention to the South Lokichar Upstream Development Project; Kenya's maiden upstream development. Exploration began in 2008, and first oil is expected in 2024. This is Kenyan's longest-running upstream project and will be the country's maiden upstream petroleum-producing project.

2.1 Kwale Mineral Sands Project (KMSP): Base Titanium (BT)

The KMSP of Kenya Special Mining Lease No. 23 began in 1998 when Tiomin conducted exploration and found commercially recoverable quantities (Ong'olo 2001). Between 1998 and 2009 Tiomin attempted to raise financing for the project but due to constant disputes, which in some cases resulted in legal battles with the community over land and disagreements with the government, were unable to continue with the project (Njiraini 2009). Base Titanium (BT) acquired the project in 2011 and were able to successfully develop the KMSP and have facilities in Kwale and Mombasa counties (Kihara 2011). This section describes BT's approach in two parts, first between 2011 and 2012 when BT embarked on finalising Tiomin's CSR projects, and secondly from 2012 when BT was adopting their own CSR approach marked by the adoption of community and environmental policies.

2.1.1 Finalising Tiomin's Programmes: 2011–2012

After the 2011 acquisition of KMSP, BT inherited about 13 CSR projects and commitments from Tiomin. Below is a table of Tiomin's commitment and their completion status by the time of acquisition:

According to Odongo et al. (2019), Base Titanium embarked on their own CSR initiatives and commitments, launching and completing 18 new CSR programmes.

Their CSR approach was broader and moved away from being fully infrastructural development and covered health care, agriculture and animal husbandry, trainings, scholarships for tertiary education (Odongo et al. 2019). The geographical scope grew to cover Kwale (mine site) and Mombasa County (Likoni export site). The change in approach to making commitments to the community can be deduced to have been for two purposes—first to quell anxiety due to the transition, and secondly to ensure the community perceived BT as a benevolent player.

Due to the underlying issues that arose in the 2000s, the most significant being the land question, the CSR programmes were well received by a section of the community. A larger demographic, however, was still unsatisfied with the compensation given for the land where KMSP sits and demanded more from the company. According to Opiyo (2006), Tiomin faced a lot of backlash over land access issues with the local community, which put great strain on the operations.[2]

2.1.2 *Base Titanium's Novel Approach 2012–2020*[3]

Base Titanium focuses on CSR programmes that support their engagement with communities. This section relies on the information from Base Titanium's website. The company runs a regularly updated and informative website, somewhat unique within Kenya's mining sector. In this sub-section, we shall review Base Titanium's CSR programme in three categories: (i) community engagement; (ii) environmental management and (iii) economic linkages.

2.2 *Community Engagement*

Base Titanium has a policy that sets out their principles for community engagement (*Base Titanium Communities Policy* 2012). In the policy, BT commits to meaningfully engaging and empowering communities, working with the government, establishing a social management system based on the Equator Principles and the IFC's Sustainability Framework, which includes the Performance Standards and minimising the footprint of their activities. To operationalise their Communities Policy, BT runs an Environment and Community Affairs Department. The department effects the policy through the following programmes and processes.

Social Impact Assessment: BT periodically carries out studies on social and health impact assessments as well as runs monitoring and management programmes. According to Base's website, (Base Titanium—Social Impact Assessment, n.d.) Base

[2]There was numerous court cases including the Said Abdalla Budzo and 36 others versus Tiomin (K) Limited and Another [2007] eKLR, 2007, where land compensation issues were brought before the court.

[3]Time frame is based on the signing of Communities and Environmental Policies in December 2012; thus, Base Titanium's approach to CSR was markedly different from 2013.

has entered into Community Development Management Plans (CDMPs) with its host communities. These will be replaced by the Community Development Agreements (CDAs) established under the Mining Act, 2016. The CDMPs enable BT to work with communities and government to fashion CSR projects and create an open line of communication.

Resettlement: In 2000 when the KMSP started, Kenyan law did not adequately address resettlement and compensation issues. To resolve legacy land issues, BT crafted several resettlement action plans (RAPs) and used them with communities affected by the: (i) Special Mining Lease; (ii) Mukumudzi Dam; and (iii) water, road and electricity lines access routes. Further, they reinterred 289 graves with community consultation. While this is the responsibility of the company and the IFC Performance Standards provide for the same, Kenyan law failed to appropriately address land issues in the 2000s when BT operationalised its RAPs. Thus, their work on resettlement can be credited to them as BT displayed corporate social responsibility through their leadership to abide by international standards during the resettlement process.

Community Engagement: BT has a Stakeholder Engagement Plan (SEP) that guides their work and established the Affected Stakeholders Committee and the Sub-County Liaison Committee, both established in 2013 (Base Titanium—Community Engagement n.d.). BT has established 11 Affected Stakeholder Committees covering the mine site, resettled communities as well as communities living in the export area (Likoni) who represent various demographics. The three Sub-County Liaison Committees are formed by affected communities, religious leaders and county and national government officials. The Msambweni Committee represents communities in the mining area, the Likoni Committee covers communities in the Likoni ship-loading facilities, and the Matuga Committee links those living in the Ukunda-Likoni transport corridor (Base Titanium—Community Engagement, n.d.). BT has established a communication system with affected communities; whether these systems assist in communication or shut out dissenting voices is disputed, as was noted by Muchai (2019).

Community Programmes: Based on their communication and engagement systems, BT has established four pillars for community programming. The first is community infrastructure, which involves the construction of schools and healthcare facilities (Base Titanium—Community Infrastructure, n.d.). The second is livelihood upliftment programmes, the key of which is the Pavi Cooperative Society, which assists local farmers who grow cotton and potatoes in Kwale County. Through this programme, farmers access technical training and financial training to enable them to access markets (Base Titanium—Projects & Livelihood Programmes, n.d.; see Spencer 2019). The third is Community Health, where BT supports health authorities, provides hospital supplies, runs health campaigns and conducts a HIV control programme, prevention of mother to child transmission (PMTCT) (Base Titanium—Community Health, n.d.). Fourth is their scholarship programme where BT partners with institutions to give students access to secondary and tertiary institutions (Base Titanium—Scholarships, n.d.).

2.3 Environmental Management

BT has a 2012 Environmental Policy, which is effected by its Environmental and Social Management System (ESMS) (Base Titanium—Our Approach, n.d.). Their 2012 Environmental Policy provides a framework for environmental protection through biodiversity protection, environmental leadership and measures to combat the effects of climate change. Further, their management plan and monitoring programmes ensure they meet the objectives of their ESMS.

BT runs several environmental programmes to lessen their ecological footprint. These range from biodiversity conservation programmes to research on rare and endangered species and wetlands restoration (Base Titanium—Environmental Programmes, n.d.). Key outcomes to note from their environmental management programmes include:

The Coastal Forests of Eastern Africa Biodiversity Hotspot: The KMSP lies close to this hotspot, and the biodiversity has been affected by the mining activity and the Mukurumidzi Dam. BT, together with National Museums of Kenya, the Kenya Wildlife Service and the Kenya Forest Service, conduct regular surveys and support research on the status of various endangered species. These species include the Shimba Hills Reed Frog, the *Gigasiphon macrosiphon* (a leguminous tree), the Changamwe Caecilian (a worm-like soil burrowing amphibian), and the *Euphorbia tanaensis* (a rare tree). In addition to the surveys, BT runs various nurseries to propagate these endangered species (Base Titanium—Biodiversity & Conservation, n.d.).

Kwale Mine Arboretum: BT runs an indigenous tree nursery, which as of 2019 hosted 275 species of indigenous trees. This contributes to promoting biodiversity in the KMSP by contributing to revegetation of the area (Base Titanium—Rare & Endangered Flora, n.d.).

Rehabilitation and Restoration: BT conducts rehabilitation on the Mukurumudzi Dam embarkment and areas affected by the dam. Additionally, in 2017, BT commenced rehabilitation of the tailings storage facility's outer wall (Base Titanium—Rehabilitation & Restoration, n.d.).

2.4 Economic Linkages

BT reports on the economic contribution of their projects on their website. In this chapter, we rely on their reporting to discuss their CSR programmes. Their public profile and information sharing is not a requirement under Kenyan law.

In 2019, BT invested 16 million US Dollars in the local community guided by their Community Development Management Plan. Finally, BT has legacy projects, which it intends to hand over to the local authorities at the end of mine-life. These include the Mukurumudzi Dam and Msambweni boreholes, electricity transmission lines: access roads and port facilities (Base Titanium—Project Economic Contribution, n.d.).

2.5 *South Lokichar Upstream Petroleum*

The South Lokichar Upstream Petroleum project is situated in Turkana County. It is run by Kenya Joint Venture (KJV) Partners and Tullow Kenya BV (operator with 50% equity stake), Africa Oil (25% equity) and Total (25% equity) (Africa Oil Corp 2020). In 2012, the Government of Kenya announced the commercial discovery of crude oil and the KJV has subsequently conducted exploration and appraisal and is geared towards full-field development (Tullow Oil East Africa Operations, n.d.). Both Tullow and Africa Oil have implemented CSR projects, as shall be elaborated below.

2.5.1 Tullow Kenya BV

Tullow Oil Plc is a multinational Irish oil and gas exploration company with operations in 15 countries. In Kenya Tullow Kenya BV a subsidiary of Tullow Oil registered in the Netherlands as part of KJV. Tullow's CSR journey in Turkana is unique. In 2010, it revolved around their land acquisition, but subsequently the scope has broadened to include health, education and economic empowerment, among other programmes.

2.6 *Community Land and CSR*

In 2010, when the KJV partners obtained their exploration licences to work in Turkana County, they first had to acquire land for exploration. At the time, untitled land was regarded as government land; thus by virtue of holding an exploration licence, the KJV had rights over the land. The KJV faced a challenge in Turkana as the land was occupied and recognised under the 2010 Constitution as belonging to the indigenous Turkana people. As a result of the discrepancy in land ownership, Tullow had to negotiate with the landowners for land to drill exploration wells and for even more acreage during the initial appraisal stage (Mullins and Wambayi 2017).

According to Oxfam's 2014 report, Tullow would negotiate with the villages living around the well-pad site, get permission to work on the land and in return would agree on monetary compensation (Mullins and Wambayi 2017). The first record of this is the compensation of approximately 7 million Kenyan shillings. The agreement with the community was to deposit the money into an account and get the community's input on how the money should be utilised—schools, health care, animal care and the like (Mullins and Wambayi 2017).

2.7 *Formalisation of Well-Pad Committees*

As Tullow's activities gained permanence, and the feasibility of the project increased, Tullow changed its approach to land acquisition. Tullow, the Turkana County Government and the host communities came together to form well-pad committees who would handle the compensation and consistent social investment by Tullow to the affected villages and the broader Turkana community. These committees enabled Tullow to consult the community when designing their CSR projects. During this period, Tullow and Africa Oil ramped up their CSR and designed programmes that were focusing on education, water provision and social infrastructure.

2.8 *Tullow's Broader CSR Activities*

Tullow has a four-tiered approach to CSR, and on their website, they refer to this as part of their sustainability. We analyse Tullow's approach to CSR as they are the operating partner in the KJV and have a significant presence in Turkana County and Nairobi County. To support their approach to CSR, Tullow relies on three policies; first, their Human Rights Policy, which states their commitment to implement the UN Guiding Principles on Business and Human Rights and the Voluntary Principles on Security and Human Rights. From a reading of their policy, it is evident that they have a do-no-harm approach and intend to work with host communities to their benefit and to secure a social licence to operate (Tullow Oil PLC Human Rights Policy Statement, 2017). Secondly, the Safe and Sustainable Operations Policy speaks to the company's work environment and internal environmental and social risks the project poses. This policy contains both internal and external measures such as not exploring in World Heritage Sites (Tullow Oil PLC Policy Statement: Safe and Sustainable Operations, 2017). Finally, the Code of Ethical Conduct touches on Grievance Management Mechanisms as well as a section on host communities and stakeholders. In the Tullow Oil PLC Code of Ethical Conduct, 2018, Tullow pledged to respect host communities, manage their activities sustainably and contribute to the economic and social growth of these communities.

Drawing for these three policies, we can make a direct connection with their four-tiered sustainability strategy as the policies speak to economic, social and environmental development and so do their core areas.

Below is an overview of Tullow's CSR operations based on the webpage (Tullow Oil: Sustainability, n.d.):

Shared Prosperity: This includes local content, skills development and social investment. With regards to education, Tullow offers approximately 120 scholarships to the Lodwar Vocational Technical Centre (LVTC) in addition to this Tullow invests in various education projects, including building dormitories and providing lab equipment. Tullow has worked with the Turkana County Government and other development agencies to drill boreholes, provide reticulation systems and run experimental demonstration farms in Kapese to promote food security. According to a media release by Tullow entitled 'Water Has Been a Game Changer' (2019), some of their projects include the Kataboi Community Water Project (2014, Turkana North). Nakukulas-Lokicheda Reticulation also resulted in the formation of the Kochodin Water Resource Users Association (2014, Turkana South & East). Tullow also offers business incubation support and training services through its contractors and requires some international contracts such as the security firm it employs to train and allow job shadowing by Turkana owned and staffed companies as described in Tullow's media release entitled 'Guarding Our Future' (2019).

Environmental Stewardship: Tullow's activities are aligned to fostering climate resilience and sustainable ecosystems. Tullow notes that their ESIAs are done in line with IFC standards in Kenya. Tullow has indicated that it will conduct its Foundation Stage Development ESIA in line with the IFC requirements, which are more rigorous than the national requirements (Golder and Ecologics Consultants Ltd 2019).

Responsible Operations: This involves an internal component of safety and wellness and an external component of responsible production. When building production facilities for the Early Oil Pilot Scheme (a scheme that tests the feasibility of production), Tullow hired a local company to work with an international company from the UAE. As a result of this partnership, 58 Kenyans were trained in the UAE and assisted in building the Degassing and Early Production Facilities (Developing production operators in Turkana, 2019). Through this initiative, Tullow is contributing to skills and technology transfer.

Equality and Transparency: This encompasses good governance and promoting equality. Tullow pledges to adhere to the Voluntary Principles on Security and Human Rights and the Extractive Industries Transparency Initiative (Tullow Oil Equality and Transparency, n.d.). In Kenya, Tullow publishes its government payment data but fails to publish the payments for the entire KJV, which makes this insufficient.

2.8.1 Africa Oil Corporation

Africa Oil Corporation (AOC) is a Canadian company with producing and development assets in Kenya and Nigeria and an exploration/appraisal portfolio in Africa and Guyana (Africa Oil Announces the Closing of the Acquisition of Producing Assets in Deepwater Nigeria 2020). AOC holds a 25% working interest in the KJV (Africa Oil 2019). In this chapter, we discuss AOC's CSR initiatives for two reasons; first, it runs geographically broader CSR projects—compared to Tullow. Secondly, AOC publishes Extractives Sector Transparency Measures Act, a 2015 Canadian Law[4] this is a legal requirement in Canada but not in Kenya and this significantly contributes to the transparency of the KJV's activities (Africa Oil Corp 2019). AOC's payment data

---

[4](ESTMA) reports on their website.

is unique as it states the payments made by the three KJV partners to government as opposed to Tullow, who only reports on the payments they made.[5]

As a member of the Lundin Group, AOC operates CSR projects in Turkana, Isiolo and Marsabit Counties regularly tapping into the expertise of the Lundin Foundation (Lundin Group: About Us, n.d.). The Foundation relies on member contributions to make high impact investments within the host community and region of operations and in Sub-Saharan Africa runs projects in Kenya, Ethiopia and Somalia (Puntland) (Africa Oil Corp.—Corporate Responsibility—Creating shared Prosperity: Community Development Highlights, n.d.). Their internal corporate responsibility policies guide AOC's CSR projects.

Africa Oil has a range of corporate responsibility policies and statements on their website.[6] A majority of these policies reinforce the ideas of community participation, environmental, health and security issues as well as policies that promote local economic growth.

AOC's CSR statement touches on the environment, ethics, dialogue and engagement, sustainable social and economic benefit, transparency and good governance as well as upholding of human rights (Africa Oil Corp.—Corporate Responsibility n.d.). Additionally, the *Community and Economic Development Policy* states that AOC adopts a three-tiered approach to their community development programmes, which are community infrastructure and sustainable livelihoods and economic development (Africa Oil Corp.—Corporate Responsibility—Community & Economic Development Policy, n.d.).

AOC's CSR activities are run in four thematic areas (Africa Oil Corp.—Corporate Responsibility—Kenya, n.d.):

Community Health: AOC constructed the Laisamis District Hospital and sanitation facilities. Additionally, AOC runs Water Sanitation and Health Programmes (WASH) in the broader Marsabit County.

Education and Skills: This involves the construction of classrooms, provision of school supplies and granting of scholarships and bursaries in Isiolo, Turkana and Marsabit Counties. In 2015, AOC signed a Memorandum of Understanding with the Turkana County Government for the reviving of the Lodwar Youth Polytechnic, which as of 2020 is functional and offers training to local youth (Turkana County Government 2015). At the national level, AOC together with the Lundin Foundation carried out a scoping study of education and skills needed of Kenya's upstream oil and gas sector. In essence, this is useful when designing CSR initiatives but also for

[5]For instance, the AOCs ESTMA Report for the year ending 31 Dec 2017 puts the payments by the three companies at 167,000 US Dollars for fees for Blocks 10BB and 13T but in Tullow's Extractive Transparency Disclosure within their Annual Report and Accounts for the year ended 31 December 2017 at pg 174 they state licence fees for Block 10BB and 13T to be 112,000 US Dollars.

[6]These include (i) Corporate Social Responsibility Commitment; (ii) Health, Safety & Environmental Policy; (iii) Security Policy; (iv) Ethics Policy; (v) Community Relations Policy; (vi) Community & Economic Development Policy; (vii) Financial Transparency & Good Governance Policy; (viii) ESTMA Reports; (ix) Human Rights Policy; (x) Key Health, Safety, Environmental and Community Documents; (xi) Creating shared Prosperity: Community Development Highlights; (xii) Country Overview; (xiii) Statements and (xiv) Feedback.

planning by the national government. Finally, AOC supports staff from the Ministry of Petroleum and the National Oil Corporation to attend Petrad Training (run by the Norwegian Government).

Access to Energy: In 2012, the Lundin Foundation invested in M-Kopa, a mobile technology company that enables rural customers to access energy products. The products include home solar lighting and phone charging systems.

Sustainable Livelihood and Economic Development: In a bid to leave a lasting footprint into the economic development of the local community, AOC facilitates solar companies to set up local sales agents, occasional veterinary clinics in Kaisut, and develop local sourcing chains to boost the economic participation of local businesses in the upstream oil and gas sector.

3 Experiences with Community Engagement

Based on the examples in the Kwale Mineral Sands Project and the South Lokichar Upstream Development Projects in the previous section, we shall now highlight two outstanding examples of successful CSR approaches. In Kwale County, BT's collaborative approach is noted as a sustainable path for the community and the company internally. In Turkana County, we review the KJV's activities relating to education and economic development and its impact on the community beyond the upstream petroleum industry.

3.1 KMSP: Formation of Committees

When Base Titanium took over the KMSP in 2011, they continued the trend of implementing CSR projects focusing on education and healthcare infrastructure. In essence, this was done mainly to ensure completion of Tiomin's commitments shown in Table 1. During this period, BT adopted their predecessor's approach of internally designing projects and then presenting these to the community, as was noted earlier, this was not very efficient as the company did not adequately cater to the needs of the community.

As their operations developed, BT's approach evolved, they adopted environmental and community policies in December 2012 and established an Environment and Community Affairs Department. A key feature of their new approach was the organised consultation of communities and the county government through the formation of committees. In 2013, BT established the Affected Stakeholders Committee and the Sub-County Liaison Committee as part of the implementation of their Stakeholder Engagement Plan (SEP).

The Affected Stakeholder Committees (ASCs) are formed at a sub-county level and are used by different affected groups. The ASC's focus on communities within the project footprint and members is selected by the communities in the specific area.

Table 1 Tiomin's commitments and progress by 2011 when Base Titanium acquired the project

Tiomin commitments (Abuya 2016)	Progress by 2011 (Odongo et al. 2019)	Status
Construct two primary schools	One primary school, Bwiti primary school	Partial fulfilment
Construct two secondary schools	One secondary school completed (Kiruku secondary school)	Partial fulfilment
Construct two dispensaries	One dispensary (Bwiti dispensary) at the resettlement site	Partial fulfilment
Construct one health clinic (at the mining site)	Constructed	Fulfilled
Provide water supply at each resettlement village	One borehole at Mrima Bwiti	Partial fulfilment
Construct two social halls	One social hall at Mrima Bwiti (the Bwiti community social hall)	Partial fulfilment
Reconstruct Churches and Mosques destroyed/lost	Compensated residents for their Churches and Mosques	Fulfilled

There are 11 ASCs, five of these are formed based on the location of communities around the projects including those living at the host resettlement site; north and south of the mine site; around the Mukurumudzi Dam; and the port in Likoni. Three of the ASCs focus on the usage of shared resources such as roads or the ocean (fisherfolk). Finally, a special ASC is set up by the Kaya elders, which contributes to the protection of culture, and the final ASCs focus on conservation and security.

BT incorporates three Sub-County Liaison Committees covering the following sub-counties: Msambweni (KMSP project area); Likoni (export port location); and Matiga (communities in the transport corridor between Ukunda and Likoni). These Sub-County Liaison Committees are bigger than the ASCs and consist of affected stakeholders, women community leaders, members of the farmer cooperative, and national and county government officials. From their composition, the Sub-County Committees are similar to CDA Committees established under the Mining Act, 2016. Once CDA committees are set up in Kwale, BT may dissolve their Sub-County Committees and focus on working with the gazetted CDA committees.

These committees are vital in establishing and preserving a communication mechanism for the community to input into CSR and raise concerns over the company's operations. On the other hand, the company has an avenue to communicate and consistently engage the host community. BT's approach of working through committees (commencing in 2013) enriches their CSR projects as they receive direct input and feedback, with the potential to create a sense of ownership by the community. Arguably, the approach builds the social fabric of affected communities in a context where extractives projects can be very divisive as sections of the community benefit while others demand more from the company. However, within these committee structures, there is space for dissenting views.

To cement their work with the community and civil society, BT participates in the Kwale Mining Alliance Working Group (KMAWG) (Ngei, n.d.). The working group was set up by civil society and in July 2019, the community-based organisation launched as a way to coordinate their interventions in Kwale and to give community members a forum to raise concerns. In some instances, when concerns are raised, the working group requests BT to respond directly with Base Titanium utilising the working group as an additional platform to engage affected communities (Ngei 2020).

BT's approach is proving to be efficient allowing for proper community consultation, thereby forming a strong basis to ease into the era of CDA Committees envisioned in the Mining Act, 2016.

3.2 South Lokichar Upstream Project

The Kenya Joint Venture Partners, Africa Oil and Tullow, have been running CSR projects focusing on education and building livelihoods. Tullow officials have, on numerous occasions, emphasised that the upstream sector will not provide many jobs to the local communities and even for those the jobs will only be available during exploration and appraisal. In February 2020, Tullow announced a massive layoff programme as a result of poor performance globally but in Kenya, this is directly related to the fact that they are moving to full-field development (Okoth 2020). This state of affairs illustrates that the upstream sector ultimately requires very little labour and cannot be entirely relied on by host communities to provide employment.

Projects such as Africa Oil's work on refurbishing the Lodwar Youth Polytechnic and Tullow offering scholarships to the Lodwar Vocational Technical Centre (LVTC) are vital in building the technical capacity of youth from the local community. These institutions allow for the youth to develop technical skills that can be used in Turkana but even outside in a multitude of industries.

AOC's sustainable livelihood programmes, which include facilitating solar companies to set up local sales agents and developing local sourcing chains to boost the economic participation of local businesses, are vital in building the local economies. Such projects ensure that communities develop as a result of the upstream sector but outside the sector sustainably.

The approach adopted by Tullow and Africa Oil will if adequately run in collaboration with the county government and local community go beyond the upstream sector and allow the local economy to thrive. Such programmes help shift the focus of communities from benefiting directly from the upstream sector allowing them to build on linkages and create sustainable industries within themselves.

4 The Challenge of Defining Affected Communities and Its Impact on CSR Practices and Community Involvement

As the governance of Kenya's extractives sector is still growing and evolving, we shall discuss the challenge for government and companies of defining the affected communities. Defining these groups is vital as it dictates who should be consulted when designing CSR programmes and who should be prioritised to benefit. The law attempts to define who affected communities are defining them as either host or local communities. These definitions are to be used when distributing extractives revenue at the community level and further when designing and implementing CSR programmes. We shall delve into the issue faced by companies and government when defining community and using examples from the situations in the two case studies to illustrate how this lack of a definition is straining the extractives projects. We shall then propose some recommendations.

4.1 Mapping Stakeholders

As companies work towards securing a social licence to operate, they must map out the different stakeholders and their power and influence over the project. In our discussion in this chapter, we identified four classes of actors, as illustrated in Table 2. Power, in this case, denotes political or administrative power over the project and influence denotes the level of impact from the project occasioned to different groups in society.

National governments of the extractives host countries and those of the company home country have high power over the project as they can impose restrictions on companies or create good working environments through regulatory and legal means. In some instances, they can enter into bilateral treaties for the benefit of the extractives companies. These governments, however, have very little influence over the company's behaviours beyond incentivising them and also very little influence over the affected communities and their politicians who hold the social licence to operate and are directly impacted by the project.

Table 2 Stakeholder matrix of power and influence

Influence	Power	
	High	Low
High	Affected communities Local members of parliament	County government Civil society
Low	Host national government Home government	General public International institutions

County governments have low power over the project as the large-scale extractives addressed here are under the purview of the national government. Companies involve the county governments because they speak directly to the people and in some instance have a higher sway with the people compared to the national government giving them high influence over the project. Most CSR activities incorporate county government in the planning as this is efficient and easily gels in with the county development plans. This approach allows the company to gain favour with the national government and also the local communities. Civil society organisations working within the extractives sector have low power as they are not duty-bearers in any way. However, because of their objectives, track record and understanding with the community have very high interest over the project. Communities can easily interact with county governments and civil society as these institutions listen to them and incorporate their issues in their programming.

The general public and international institutions are stated as being low power and influence stakeholders, but in some instances, can quickly gain power and influence over the project. These are mapped as stakeholders because they can quickly impact the project and it is vital to keep them informed of the project.

Arguably, in the Kenyan contexts, the affected communities, including their political leaders, have the greatest power and influence over the project. The geographical proximity to the project results in them being directly impacted as they give up their land and have to adjust to the social changes occasioned by the project. However, it is difficult for the company to accurately define who the affected communities are. In Turkana, this is due to the nomadic nature of Turkana Communities, and in Kwale, this is due to the resettlements that have taken place since 1998 when Tiomin began exploration. There is a temporal and geographical nature to these communities that ought to be addressed in such a manner that the company is not overburdened and that the communities affected the most are given higher priority when it comes to beneficiation.

4.2 *Who Should CSR Projects Prioritise?*

While many thematic challenges can be found within the KMSP and the South Lokichar Upstream Projects, the greatest of these is the fact that the definition of affected communities does not cater to all that may be affected by the project. The main focus of CSR is on communities living near the extractives project, but as shall be illustrated below the definition and mapping of the local community must be expanded to suit project-specific contexts.

4.2.1 The South Lokichar Upstream Development

The Petroleum Act, 2019 adopts a geographic definition defining affected communities as local communities who live within a particular sub-county. Section 2 of the Petroleum Act, 2019 defined local community as 'people living in a sub-county within which a petroleum resource under this Act is situated and are affected by the exploitation of that petroleum resource'. This definition is employed when referring to participation once a licence is applied for, with regard to revenue sharing, and community rights. From the usage of the phrase 'local community' in the Act, it is deduced that the contractor shall engage the local community in various ways, including prioritising them in their CSR programmes. In line with the Act's definition, Tullow and Africa Oil's projects were initially focused around well-pads (from 2010) and the adjacent villages. Later Tullow and Africa Oil projects evolved to cover the Turkana East Constituency and have now grown to cover Turkana, Marsabit and Isiolo Counties. Specific projects only focus on Turkana County such as the water provision, but education and economic empowerment projects focus on all three counties. From their approach, it is evident that the community closest to the upstream project are prioritised and rightfully so as they bear the most significant social and resource burden when designing CSR programmes.

Tullow and Africa Oil's multi-county approach is progressive, as it ensures regional development. As the project moves on to full-field development and water from River Turkwel will be diverted to the project, it is essential that the CSR approach is expanded to include communities who rely on the river and will be affected by the abstraction as another class of local communities (Golder and Ecologics Consultants Ltd 2019). It is vital to include communities, in West Pokot and Turkana Counties, whose ecosystems will change as a result of changes to water quantity to be catered for in the same manner as communities who lose access to their land for the project. The CSR approach employed by both the governments and company should cater for both as water and land resources are vital to their livelihoods.

From the growth of the South Lokichar Upstream Project, it is evident that the definition of the local community for the purposes of CSR and greater benefit-sharing should also cover communities who lose access to resources aside from land, in this case, water.

4.2.2 The Kwale Mineral Sands Project

The Mining Act, 2016 has a robust definition of community accounting for geographical proximity and resettled people. Section 2 of the Mining Act, 2016 defines community as 'a group of people living around an exploration and mining operations area; or a group of people who may be displaced from land intended for exploration and mining operations'. The term community is used throughout the Act with regards to consent for companies to gain licences, community development plans, employment preference, community land rights, community investments and royalty sharing.

Base Titanium's programmes are designed to cover the mine site, resettlement site, transport corridor, export port and supplemental infrastructure such as the Mukurumidzi Dam. This approach covers the entirety of the project, and its focus on auxiliary project infrastructure ensures the project affected people are catered for. This approach is equitable as all those impacted by the project are considered in the CSR design. This efficiency can be credited to the fact that CSR projects in the KMSP have run for the past two decades and also the progressive definition in the Mining Act, which encourages the company to cater for all affected persons.

4.3 Resource Focused Definition of Local Community

The definitions in both the petroleum and mining laws are interpreted in practice to include communities living next to the project site and those affected by the development. The Petroleum Act's definition is broader and covers all affected by the exploitation, but in practice, projects for the 'local community' are focused on those neighbouring the exploration sites. The Mining Act's definition covers those in close proximity and those displaced due to the operations. In practice, Base Titanium's approach is comprehensive and covers most communities affected by the project affected by land loss. Going forward, it is vital that companies and government in interpreting the definition of local community account for communities who lose land and other natural resources, including water. Communities are affected during projects as they lose access to resources vital to their culture and livelihood. As illustrated in Lamu Port, communities who lose access to other resources aside from land should also be prioritised and not seen as secondary groups.[7] In the Lamu Port Case, communities who were displaced from their land to make way for the Lamu Port had been compensated. However, fisherfolk whose yield would reduce as a result of the port construction had not been compensated. The High Court awarded the fisherfolk compensation of approximately one million US dollars.

5 Conclusion

At the beginning of this chapter, we settled on a definition of CSR, which incorporated aspects of CSR from Kenyan extractives law. We also discussed sources of financing for CSR projects and enforcement mechanisms. It was determined that the CSR projects in the Kenyan extractives sector are company driven and it is in the best interest of these companies to design, implement and complete these projects. We used two case studies, the Kwale Mineral Sands Project operated by Base Titanium and the South Lokichar Upstream Development operated by Tullow and Africa Oil. We established that Base Titanium has had the benefit of running CSR programmes for a decade and building on their predecessor's decade learning of CSR and community engagement. BT's approach is advanced as it incorporated committees consisting of the local community, politicians and local government before this was a legal

[7]Mohamed Ali Baadi and others v Attorney General & 11 others [2018] eKLR is referred to as the Lamu Port Case.

requirement. Tullow and Africa Oil started their project in 2010, and their approach evolved from using CSR for land-related issues to a more holistic programme. The CSR approach adopted by South Lokichar Upstream project is arguably forward-looking and works towards moving community attention from seeking employment within the petroleum sector while developing the linking sectors. The CSR activities focus on education, training and livelihoods. Tullow and Africa Oil's programmes are in line with legal learning to develop capacity through the upstream sector.

We looked at the outstanding examples of executing CSR from both case studies. From Kwale, the approach of using committees is vital in ensuring communities and county governments input into the CSR programme; his way, the company is informed of what projects to undertake. This also goes a long way in building community decision making structures, essential in other sectors and other parts of community life. However, there are still many concerns within the community over the committee members whose concerns are legitimate. This chapter in no way claims that Base Titanium's CSR activities are adequate or the community as a whole is satisfied; such a conclusion needs field research and data (see Spencer 2019). From the upstream petroleum sector, the projects established that the focus on training and economic development by Tullow and Africa Oil is a sustainable approach. Their projects help the local community build skills that they can use in other sectors without relying on the petroleum sector, given the sector needs minimal human resources during production. For this CSR approach to be genuinely sustainable and have a lasting impact, it is vital that the county and national governments partner with the company to sustain the programmes once the company leaves and to help the local economies grow, thus absorb the skilled individuals.

We identified the major challenge relating to CSR in the sector as the usage of the geographical definition of local community. We advanced arguments on why this must be expanded to encompass communities that are directly impacted as a result of disturbances to their communal resources, including land and water resources. The definition should include all resources touching on land adjacent to the extractives site and include other resources users such as water users. This way, when designing CSR programmes, companies are more likely to be issued with a social licence to operate by all impacted communities.

References

Abuya, W. (2016). Mining conflicts and Corporate Social Responsibility: Titanium mining in Kwale Kenya. *The Extractive Industries and Society, 3*(2), 489. https://doi.org/10.1016/j.exis.2015.12.008

Africa Oil Corp. (2020). *Africa Oil Corp report to shareholders December 31, 2019* (p. 3). Retrieved from https://www.africaoilcorp.com/i/pdf/2019-Q4-and-YE-Report.pdf

Africa Oil. (2019). *Africa Oil Corporation, 2018 annual report and accounts* (p. A-7). Retrieved from https://www.africaoilcorp.com/i/pdf/2019-02-28-AOI-AIF.pdf

Africa Oil announces the closing of the acquisition of producing assets in Deepwater Nigeria. (2020). Retrieved from https://news.africaoilcorp.com/releases/entry/122771

Africa Oil Corp.—Corporate responsibility—Community & Economic Development Policy. Africaoilcorp.com. Retrieved March 8, 2020, from https://www.africaoilcorp.com/s/corporate-responsibility.asp?ReportID=584655

Africa Oil Corp.—Corporate responsibility—Creating shared prosperity: Community development highlights. Africaoilcorp.com. Retrieved from https://www.africaoilcorp.com/s/corporate-respon sibility.asp?ReportID=598181

Africa Oil Corp.—Corporate responsibility—Kenya. Africaoilcorp.com. Retrieved March 8, 2020, from https://www.africaoilcorp.com/s/corporate-responsibility.asp?ReportID=598182

Africa Oil Corp.—Corporate responsibility. Africaoilcorp.com. Retrieved March 8, 2020, from https://www.africaoilcorp.com/s/corporate-responsibility.asp

Africa Oil Corp. (2019). *Extractive Sector Transparency Measures Act—2018 annual report*. Retrieved from https://www.africaoilcorp.com/i/pdf/2018-aoi-estma.pdf

Base Titanium—Biodiversity & conservation. Basetitanium.com. Retrieved March 6, 2020, from https://basetitanium.com/environment/environmental-programmes/biodiversity-conservation

Base Titanium Communities policy. (2012). Retrieved from https://basetitanium.com/governance/company-policies?download=8:communities-policy

Base Titanium—Community engagement. Basetitanium.com. Retrieved March 2, 2020, from https://basetitanium.com/community/engagement

Base Titanium—Community health. Basetitanium.com. Retrieved March 7, 2020, from https://bas etitanium.com/community/community-programmes/community-health

Base Titanium—Community infrastructure. Basetitanium.com. Retrieved March 4, 2020, from https://basetitanium.com/community/community-programmes/community-infrastructure

Base Titanium environmental policy. (2012). Retrieved from https://basetitanium.com/governance/company-policies?download=11:environmental-policy

Base Titanium—Environmental programmes. Basetitanium.com. Retrieved March 3, 2020, from https://basetitanium.com/environment/environmental-programmes

Base Titanium—Our approach. Basetitanium.com. Retrieved March 3, 2020, from https://basetitan ium.com/environment/our-approach

Base Titanium—Project economic contribution. Basetitanium.com. Retrieved March 3, 2020, from https://basetitanium.com/kwale-project/project-economic-contribution

Base Titanium—Projects & livelihood programmes. Basetitanium.com. Retrieved March 3, 2020, from https://basetitanium.com/community/community-programmes/community-projects-livelihood

Base Titanium—Rare & endangered flora. Basetitanium.com. Retrieved March 3, 2020, from https://basetitanium.com/environment/environmental-programmes/rare-endangered-flora

Base Titanium—Rehabilitation & restoration. Basetitanium.com. Retrieved March 4, 2020, from https://basetitanium.com/environment/environmental-programmes/rehabilitation-restoration

Base Titanium—Scholarships. Basetitanium.com. Retrieved March 3, 2020, from https://basetitan ium.com/community/community-programmes/scholarships

Base Titanium—Social impact assessment. Basetitanium.com. Retrieved March 3, 2020, from https://basetitanium.com/community/our-approach/sia

Developing production operators in Turkana. (2019). Tullowoil.com. Retrieved March 7, 2020, from https://www.tullowoil.com/media/case-studies/developing-production-operators-in-turkana

Golder and Ecologics Consultants Ltd. (2019). *Draft environmental and social impact assessment report for the foundation stage of the South Lokichar development for the upstream oil production in South Lokichar*. Nairobi: Tullow BV Kenya. Retrieved from https://www.tul lowoil.com/Media/docs/default-source/operations/kenya-eia/esia-report_foundation-stage-of-south-lokichar-development.pdf?sfvrsn=2

Government Printers. Mining Act, 2016 (Community Development Agreements) Regulations (2017). Nairobi.

Guarding Our Future. (2019), (2), 40–41. Retrieved March 7, 2020, from https://www.tullowoil.com/Media/docs/default-source/operations/ustawi/tullow-kenya_ustawi-_march-2019.pdf?sfv rsn=2

Kihara, G. (2011). Titanium mining firm to begin work in six months. *Daily Nation*. Retrieved March 8, 2020, from https://www.nation.co.ke/business/Titanium-mining-firm-to-begin-work-in-six-months/996-1095300-v2s7t5z/index.html

Lundin group: About us. The Lundin Group. Retrieved March 8, 2020, from https://www.thelundingroup.com/about-us/

Muchai, J. (2019). Base Titanium denies sidelining communities in mineral exploration. *Kenya News Agency*. Retrieved March 3, 2020, from https://www.kenyanews.go.ke/base-titanium-denies-sidelining-communities-in-mineral-exploration/

Mullins, D., & Wambayi, J. (2017). *Testing community consent: Tullow oil project in Kenya* (p. 23). Oxford: Oxfam GB. Retrieved from https://cng-cdn.oxfam.org/kenya.oxfam.org/s3fs-public/file_attachments/FPIC%20Report-November%202017.pdf

Mutai, E. (2019). Sh900m lies unused in petroleum training fund. *Business Daily*. Retrieved March 1, 2020, from https://www.businessdailyafrica.com/economy/Sh900m-lies-unused-in-petroleum/3946234-5197276-4kas6dz/index.html

Ngei, S. *Kwale Mining Alliance Working Group* [PDF]. Transparency International—Kenya. Retrieved March 8, 2020, from email communication

Ngei, S. (2020). *Research on the impact of Base Titanium's activities in Kwale County from the perspective of civil society* [in person]. Nairobi.

Njiraini, J. (2009). Tiomin caught between a rock and hard place. *Standard Digital*. Retrieved March 8, 2020, from https://www.standardmedia.co.ke/article/1144023414/tiomin-caught-between-a-rock-and-hard-place

Odongo, G., Abuya, W. O., Masese, E., & Jamin, M. M. (2019). The fate of CSR projects when ownership changes takes place from One Extracting Company to another. *Journal of Sociology, 1*(2), 14-28.

Okoth, E. (2020). Tullow job cuts in Kenya deepen oil export doubts. *The East African*. Retrieved March 8, 2020, from https://www.theeastafrican.co.ke/business/Tullow-job-cuts-deepen-oil-export-doubts/2560-5449274-kqgd1r/index.html

Ong'olo, D. (2001). *International investment and environmental issues: The case of Kenya's Kwale mineral sands project* (p. 2). Jaipur: OECD. Retrieved March 3, 2020, from https://www.oecd.org/env/2075490.pdf

Opiyo, K. (2006). Tiomin threatens to quit over land access. *The East African*. Retrieved March 4, 2020, from https://www.theeastafrican.co.ke/business/2560-252130-uvibli/index.html

Said Abdalla Budzo & 36 others V Tiomin (K) Limited & Another [2007] eKLR (High Court 2007).

Spencer, R. (2019). Resourcing rural livelihoods in Kenya: Intersections between mining, agriculture and development, CASE—Reference no. 719-0064-1. Retrieved from March 10, 2020, from https://www.thecasecentre.org/educators/products/view?id=166276

Turkana County Government. (2015). *Turkana County Government enters a partnership with Lundin Foundation*. Retrieved from https://www.turkana.go.ke/index.php/2016/10/24/press-release-lundin-foundation-tcg/

Tullow Oil East Africa Operations. Tullowoil.com. Retrieved March 8, 2020, from https://www.tullowoil.com/operations/east-africa/kenya

Tullow oil equality and transparency. Tullowoil.com. Retrieved March 7, 2020, from https://www.tullowoil.com/sustainability/equality-and-transparency

Tullow oil PLC code of ethical conduct. (2018). Retrieved from https://www.tullowoil.com/Media/docs/default-source/1_About_us/2018-tullow-oil-code-of-ethical-conduct.pdf?sfvrsn=9

Tullow oil PLC human rights policy statement. (2017). Retrieved from https://www.tullowoil.com/Media/docs/default-source/1_About_us/policies/tullow-oil_human-rights-policy-2017.pdf?sfvrsn=14

Tullow oil PLC policy statement: Safe and sustainable operations. (2017). Retrieved from https://www.tullowoil.com/Media/docs/default-source/1_About_us/policies/tullow-oil_safe-andsustainible-operations-policy.pdf?sfvrsn=16

Tullow oil: Sustainability. Tullowoil.com. Retrieved March 2, 2020, from https://www.tullowoil.com/sustainability

Vermaelen, T. (2011, June). An innovative approach to funding CSR projects. *Harvard Business Review*. Retrieved March 8, 2020, from https://hbr.org/2011/06/an-innovative-approach-to-funding-csr-projects

Water Has Been a Game Changer. (2019), (2), 18–25. Retrieved March 7, 2020, from https://www.tullowoil.com/Media/docs/default-source/operations/ustawi/tullow-kenya_ustawi-_march-2019.pdf?sfvrsn=2

What is CSR? | UNIDO. Unido.org. Retrieved March 8, 2020, from https://www.unido.org/our-focus/advancing-economic-competitiveness/competitive-trade-capacities-and-corporate-responsibility/corporate-social-responsibility-market-integration/what-csr

Practice of Corporate Social Responsibility (CSR) in Extractives Sector in Indonesia

Yudo Anggoro, Adita Pritasari, Rivana Mezaya, Dematria Pringgabayu, and Dany Muhammad Athory Ramdlany

Abstract Corporate social responsibility (CSR) has been receiving wide attention as this practice implies the efforts of corporations and industries to contribute to society. While the practice of CSR is expected to promote economic and social development, CSR practices in extractives sector, especially in developing countries, invites debates from stakeholders. The debates are mostly related to the negative image of the extractives sectors in harming the environment and exploiting local people. From the corporation side, debates are related to what kinds of CSR the community desires and how much money corporations for their CSR activities spend. This chapter explores the practice of CSR in the extractives sector in Indonesia, where the extractives sector has been the backbone of the economy for years. It starts with some reviews on the extractives sector and its operations in Indonesia, then the best and worst experiences in CSR practices in the extractives sector, and finally provides the main challenges that the extractives sector is facing and strategies to overcome those challenges. This chapter concludes that CSR practices in the extractives sector are sometimes performed to minimize the negative image of the industries. Often corporations perform CSR activities in order to be perceived as having good business practices by the community.

Y. Anggoro (✉) · A. Pritasari · D. Pringgabayu · D. M. A. Ramdlany
School of Business and Management, Institut Teknologi Bandung, Bandung, Indonesia
e-mail: yudo.anggoro@sbm-itb.ac.id

A. Pritasari
e-mail: adita.pritasari@sbm-itb.ac.id

D. Pringgabayu
e-mail: dematria@sbm-itb.ac.id

D. M. A. Ramdlany
e-mail: dany.athory@sbm-itb.ac.id

R. Mezaya
Business Ethics, Law and Sustainability, School of Business and Management,
Institut Teknologi Bandung, Bandung, Indonesia
e-mail: rivana.mezaya@sbm-itb.ac.id

E. G. Pereira et al. (eds.), *Sovereign Wealth Funds, Local Content Policies and CSR*, CSR, Sustainability, Ethics & Governance,
https://doi.org/10.1007/978-3-030-56092-8_27

Keywords Corporate social responsibility (CSR) · Extractives sector · Indonesia · Sustainable development · Community · Developing countries

1 Introduction

As one of the growing economies in the world, Indonesia has various industry sectors that generate economic activities. Among those sectors, companies that operate in extractives sector make significant contributions to the economy through taxes. For example, in 2017, PT Adaro Indonesia, one of the major coal mining companies in the country, has paid taxes 39,309 million USD or around IDR 5.4 trillion. This placed PT Adaro Indonesia as the number one taxpayer from the corporation category in Indonesia. From the private category, some top individual taxpayers also come from the extractives sector such as Arifin Panigoro who owns PT. Medco Energy International, Garibaldi Thohir (Adaro Energy) and T.P. Rachmat (Triputra Group). These companies mostly operate in the coal and mining sector in Indonesia.

Due to those facts, the extractives sector has an important role to develop the economy in Indonesia. However, even though the extractives sector has a huge impact on state income in Indonesia, the public has heavily criticized those companies. Most of the critiques address the environmental impact of the exploration area. Other negative view is directed to the minimum contribution of this sector to society.

In order to turn around these negative sentiments, the extractives sector puts serious effort to develop CSR. CSR practices of the extractives sector have the main objectives to benefit people, either within the company or within people outside of the company. Through CSR activities, the overall goal of sustainable development is the long-term stability of the economy and environment (Sigam and Garcia 2012).

In Indonesia, some extractives sector companies consistently conduct CSR for the community. In order to increase the awareness of society about the importance of CSR activity, some CSR awards have been granted to companies that have made extra efforts in developing the community where the company operates. An example of this CSR award is the Indonesia's Best Corporate Practices award that gives appreciation to the extractives sector in the category of corporate philanthropy and responsible business practice. The winner is selected because their CSR programs are effective to improve the community life and to preserve the environment. In this award, there are three basic aspects to process evaluation of the TOP CSR award, which are ISO 26000, Creating Shared Value (CSV) and Good Corporate Governance (GCG).

The best experience with community involvement can be observed from the local communities themselves when they accept the activities of extractives sector in their community. On the contrary, the bad experiences with community involvement might indicate that the extractives sector is unsuccessful to implement sustainable development principles, and the local communities feel aggrieved and harmed by the presence of extractives sector and their activities.

There are some examples where extractives sector creates negative impact on the surrounding community. Some notable examples include the case of conflagration in

Balikpapan, the deadly abandoned mining pits in Samarinda and Kutai Kertanegara, and the illegal drilling in Jambi, which caused the deaths of some people in those incidents. Another negative example of the extractives sector is the poor conditions in Papua province because of a prominent mining company that has been exploiting local environments in large scale for decades. The Indigenous people in Papua, called Amungme and Kamoro, explain that the mining activities in their hometown do not provide any benefits but environmental disaster.

Because of the negative reputations of the extractives sector, the companies are necessary to build business sustainability through CSR (Suastha 2018; Slack 2012). Hence, there are challenges that the extractives companies are facing about CSR practices and community involvement such as assessments of project costs and benefits, project and technology selection, respect for community consent and performance incentive structures. The companies have to recognize that CSR is more than a way for companies to manage their business processes to produce an overall positive impact on the community at large. That communities sit adjacent to the extractives sector activities must be the priority of CSR, the companies need to provide sustainable development to them as "development that meets the needs of the present without compromising the ability of future generations to meet their own needs" (Brundtland Report in Our Common Future 1987).

There are some extractives sector companies in Indonesia that have successfully implemented CSR activities to local communities, including, for example, Indika Energy, Medco Energy International, Pertamina and Vale Indonesia. The CSR practices of these companies can be viewed as best practices in Indonesia on how the extractives sector can perform CSR programs that successfully engaging local community in developing their areas. The success of this CSR program will affect the positive images of the companies and help them to achieve business sustainability.

2 Extractives Sector

The extractives sector can be defined as processes involving various activities that lead to the extraction of raw materials from the earth (such as oil, metals, minerals and aggregates), processing and utilization by consumers (Gilberthorpe and Banks 2015). This process occurs in the host country and the country of origin of the company, and the global consumer market. The global value chain produced and the distribution of values along the production chain and stakeholders can be influenced by local policies, which are increasingly becoming strategic factors in investment decisions and public policy formulation for the development of this extractives sector. Understanding these impacts requires an understanding of the structure of the extractives sector (oil, gas and mining), which consists of activities ranging from exploration to sales to end consumers (Buur 2013).

The extractives sector has significant potential to change an environment, society and economy (Renwick et al. 2018). But not infrequently, the transformation in the extractives sector can actually lead to conflicts or disputes between resources originating from developers and the local community, which results in hindering or giving disruption to company licenses to operate—especially those related to operational costs for companies, local communities and the wider community (Acuña 2015).

The extractives sector is important instruments for creating wealth for developing countries. For some countries that are rich in natural resources, the extractives sector contributes more than 50% of GDP (Renwick et al. 2018) However, various benefits that can be obtained by resource owners will be limited due to many factors, including the characteristics of the industry and capital-intensive requirements to develop the industry. The scope of the positive and negative impacts of the extractives sector is very specific in terms of the context and differs from extractive activities, location, economic environment and the quality of governance. These impacts can be influenced at any value chain level by local policies adopted in the host country. Likewise, it is very important to understand these impacts and their dynamics in order to draw up the right design rules and policies.

The impacts of the extractives sector itself are very well known. Since the 1990s, there have been a multitude of case studies and critical analysis of the impacts of the extractives sector. The biggest risks are borne by those in places where natural resources are taken (communities and countries), while the greatest benefits are enjoyed by those who take them (companies and investors) (Lohde et al. 2015).

In general, the economic condition of an area that is the location of extractives sector follows the extractives sector phase: exploration, construction, operation, post-operation. During the exploration phase, the economic benefits that arise are very small because the company is only looking for natural resources to be utilized. There is not necessarily a discovery of resources—an oil and gas exploration usually have a success factor of under 10%—and if found it is not necessarily technically and financially feasible to proceed to the next phase (Davis and Franks 2011). Because of this situation, the company rarely provides any economic benefits to the region/country where exploration is carried out.

However, if discovered natural resources are sought, and technically may be exploited, and exploitation is calculated to bring economic benefits, the company will enter the next stage, namely construction. Of course, this is carried out, at least ideally, when the licenses whether from the Government and from the society around the company's operational places have been granted. In this phase, the economic impact of the extractives sector will increase. Employment opportunities increase rapidly, before falling back in the operating phase. Likewise, local business opportunities to become suppliers of materials and services are needed. However, in aggregate, the economic value of this phase is not large due to the short period (Canel et al. 2010).

The biggest economic benefits for the country are obtained during the period of operation or exploitation, through taxes and non-tax state revenues. For certain groups of people who are able to meet labour requirements and business contracts, this phase also provides great opportunities. However, groups of people who can fulfil it are usually very few. Thus, the dominance of labour and contracts in this phase is usually non-local. Even if the local number of labours can look big, they usually occupy a low level.

In recent years, the economic benefits of the extractives sector in Indonesia have been increasing for many extractives sector that recognize the importance of carrying out CSR, especially in the form of community development. Through this community development, the company deliberately targets local communities to receive benefits through employment opportunities, business development and projects for the community (Slack 2012).

Enter the next phase, namely post-operation, the economic benefits also shrink. Worse than just shrinking, many of the extractives sector that leave the area with all valuable resources are exhausted, while there are no other resources or replacements that can be utilized by the local community. If the majority of the people are also migrants to find work in the extractives sector, they will immediately leave the location once extraction is complete. If the Government does not allocate financial benefits during the operation to ensure that life can proceed afterwards, it is almost certain that post-operation cities will decline economically. The ghost town phenomenon is commonly found throughout the world (Graves et al. 2009).

In the exploration phase, social impacts can be minimal, unless the company fails to approach the community around their operational area. The problem is that public acceptance in this period will determine whether the company can get support in the next phase. Most companies that think that Government licenses (legal and actuarial licenses to operate) are adequate without community support (social license to operate) will experience continuous problems. The best practices teach that social licenses are at least as important as legal licenses, if not more important (Graves et al. 2009).

In the next phase, negative social and environmental impacts continue to grow. The construction phase does provide employment and business opportunities, but it is also known as a trigger for labour conflicts and business struggles. The influx of other workers in large numbers also increases the incidence of conflict. Environmentally, this phase also has a significant impact. Although the site of the extractives sector is usually much smaller than the agricultural industry (e.g. palm oil plantations), the environmental impact far exceeds site size.

Nevertheless, of course, the greatest social and environmental impacts come in the phase of operations or exploitation. Tensions and conflicts with the community because of land acquisition by means of coercion and deception occur far too often. The shrinking employment opportunities make it difficult for people to fulfil their needs because at the same time the prices of goods soar around any extractives sector's operating area. Most human rights violations occur in this phase too, because protests from the public have not been responded to with an approach that honours human rights. Meanwhile, loss of biodiversity; water, land and air pollution; other natural resource depletion occurs at a high speed, and often is not reversed.

3 The Extractives Sector in Indonesia

Indonesia possesses is endowed with abundant natural resources providing a strategic extractives sector. The gas industry in Indonesia is the most abundant gas supplier in Southeast Asia with a total export of around 45% of production (Publish What You Pay Indonesia 2018). For the oil and gas sector, the Ministry of Finance noted that oil had contributed IDR 72,665.72 Billion, around 34% of the realization of non-tax state revenues (PNBP) from January to July 2018. The results exceeded the target of the state budget realization of IDR 59,582.70 Billion (around 121.94%) (Pahlevi 2018). On the other hand, mining is also one of the strategic extractives sector by contributing to non-tax state revenues (PNBP) reaching IDR 33.5 Trillion in September 2018 and is expected to reach IDR 43 Trillion by the end of 2018 (Suhartadi 2018). Although, according to Suhariyanto, Head of National Statistics Agency, the mining and quarrying sector in the previous year contracted negatively 0.49% because two large companies, namely PT Freeport Indonesia (PTFI) and PT Amman Mineral Nusa Tenggara (AMNT) ex-PT Newmont Nusa Tenggara, had experienced production decline up to 60% since January 2017 along with the expiration of concentrate export licenses for companies. These conditions also have an impact on weakening economic growth in the provinces of Papua and West Nusa Tenggara (NTB) (Fauzi 2017).

Despite this, the extractives sector still faces stigma. The community largely perceives that the extractives sector harms people and the environment rather than provides benefits. According to *Komnas HAM* (the National Commission on Human Rights), the extractives sector in East Kalimantan has violated the human rights and harmed the environment. Until June 2016, 243 people had died as a result of drowning in ex-coal mine holes, 22 of which were children: in Samarinda (fifteen children), Kutai Kertanegara (eight children) and Pasir Panajem Utara (one child). East Kalimantan is the leading producer of coal. There are 1488 Mining Business License (*Izin Usaha Pertambangan* or "IUP") issued by the Regional Governments, Provincial Governments and District/City Governments. A large number of these permits and mines in densely populated areas have left many ex-mine holes with toxic water and heavy metal contamination. In 2016, Samarinda's green open space was only around 5%, while Law No. 26/2007 on Spatial Planning mandated a green open space of at least 30% of the city area. This condition shows how mining as part of the extractives sector has a negative impact on society and the environment. Even though Law Number 4 of 2009 concerning Mining, Minerals and Coal has regulated the holders of IUP and Special Mining Business License (*Izin Usaha Pertambangan Khusus* or IUPK) must guarantee the application of environmental quality standards in accordance with the characteristics of an area. These companies are also obliged to preserve the function and carrying capacity of the water resources in question by the provisions of legislation (Komnas HAM 2016). If this continues, the environmental crisis cannot be avoided.

The environmental damage is not the only impact of the extractives sector. Conflicts often occur because of the different perceptions among stakeholders and the community. Many findings highlight the negligence and lack of coordination by the management of extractives sector companies that have been the cause of conflicts between companies and local communities (Bebbington 2010; Bracking 2009; Davis and Franks 2011; Maconachie and Menzies 2015). The village officials in each region are expected to be representatives of the community in such conflicts; however, they are often considered the puppets of mining companies. The situation is aggravated by the existence of gaps rooted in injustices of the investment returns distribution. Conflicts could also lead to human rights violations. According to Zeid Ra'ad Al Hussein, the United Nations High Commissioner for Human Rights, Indonesia's extractives sector is full of human rights violations. His statement is based on complaints from civil society from Sumatra to Papua. The mining and logging of large companies have become the main causes of human rights violations, such as farmers, workers and Indigenous peoples (Suastha 2018). Morowali District in Central Sulawesi is one example of a place where the conflict has occurred. Morowali is a fertile area in nickel minerals, so this district is nicknamed "Nickel Land". However, the majority of Morowali people who are farmers are not provided enough training opportunities to become workers in mining companies, and even the wages are not feasible for those who can work. The mining companies in Morowali prefers to recruit foreign workers or local workers from other cities, whereas the communities expect that mining companies should be able to improve the welfare of farmers, labourers and communities living around the mine (Sabintoe 2016).

This conflict can worsen when mining occurs on customary land. The existence of differences in paradigms, perceptions, and the meaning of local wisdom values are the main triggers for conflict. Mining companies consider the land and natural resources in an economic and business perspective, while Indigenous people view land as their source of life and their culture, therefore they must maintain it. For instance, the conflict that occurred in Central Kalimantan, PT AGL, is considered to have annexed the Dayak tribal land. The Indigenous community has been relying for years on agricultural land, which has now turned into a mining pit or palm oil plantation. The land was a source of livelihood for Dayak people because it provides welfare and virtue. The creation of various forms of local wisdom across generations is due to the strong agrarian culture that has been practiced for generations by villagers. In 2012, residents of Murung Raya District sacrificed a pig by conducting a traditional ritual called *hinting pali* to exclude the mining companies that destroyed sacred areas in the Tanah Siang Selatan District. *Hinting* means boundary, while *pali* means prohibition or taboo. The ropes installed by residents in palm oil plantations showed the boundaries of the area claimed by them. Referring to traditional consensus, those who violate or damage the boundary will be subject to customary sanctions from ancestral spirits. Like the "police line", *hinting pali* was installed with the intention of begging the ancestors to guard the boundary area. In 2016, residents of Tumbang Mantuhe, Gunung Mas Regency, also erected *hinting pali* in palm oil plantation areas (Luthfy 2018).

The above examples illustrate the need for the extractives sector in Indonesia to develop the country's economy from upstream to downstream under the mandate outlined by the 1945 Constitution Article 33, Paragraph 3 that the nation has to manage natural resources effectively to maximize its social welfare benefits. The different perceptions that lead to this conflict need to develop anticipatory steps as soon as possible. Comprehensive anticipation efforts to prepare a problem-solving plan based on a win–win solution and future-oriented. One step that can be conducted is to run a participatory CSR program.

4 CSR in Indonesia

CSR is a common concept for an extractives sector in Indonesia today. CSR indicates that a company's responsibility is not only the economic objectives and outcomes but also the social and environmental objectives and outcomes for the wider community that the company is situated within. Consequently, a company must not only contribute to its shareholder base, as in generating profits, it also has to satisfy the needs of its other stakeholders. These stakeholders include all elements of a community that the company has relationships with, such as suppliers, customers, employees, Government, surrounding communities and broader society.

CSR generally refers to transparent business practices that are based on ethical values, compliance with legal requirements, and respect for people, communities, and the environment (Chandler 2001). In the business context, a company needs to be a "good neighbour" within its host community, as intended by the CSR concept.

In this technology era, globalization has blurred country borders. Fierce competition is increasing between skilled employees, consumer loyalty and investors, align with the company's effort to protect their brand reputation. In order to make its business sustainable, every company must pay attention to the relationships with its workers, customers and also its host communities. CSR management in the global context has moved fast from the phase of planning and implementation to evaluation and control phase. Since November 2010, the International Standard Organization (ISO) has officially issued ISO 26000 as guidance for executing CSR planning and implementation. Meanwhile, to help CSR evaluation and control, many companies publish their CSR Report based on a standard issued by the Global Reporting Initiative (GRI).

Indonesia is the largest economy in Southeast Asia and currently is the only country in the region that became a member of G-20 (a group of 20 of the world's largest economies). Indonesia also has many corporations operating in natural resources-related business, such as mining, forestry and plantations. Thus, the need for companies to become sustainable and to optimize economic contribution, environmental performance and social responsibility is high in Indonesia.

While the international context of CSR provides many guidelines and definitions (e.g. ISO 26000, OECD Guidelines for Multinational Enterprises, Global Reporting Initiative, Sustainable Development Goals, etc.), each country has different CSR implementation according to their cultural context. With Indonesia currently being one of the fastest growing economies in the world, the national CSR agenda represents a number of challenges that highly differ from those of, for example, Europe or the USA. Hence, Visser in Country Scan CSR in Indonesia (2009:474) defines CSR in emerging economies as "the formal and informal ways in which business makes a contribution to improving the governance, social, ethical, labour and environmental conditions of the emerging economies in which they operate, while remaining sensitive to prevailing religious, historical and cultural contexts".

Initially, CSR was not familiar for Indonesian companies, including state-owned companies, even though charities and philanthropic activities were a common practice. However, in 1998, after the Reformation era, the increase in Government involvement aided the CSR movement in Indonesia. Public participation has been possible in the decision-making system, including monitoring if the companies show irresponsible actions towards environment and society. The Ministry of SoE issued a Minister Decree No. Kep-216/M-PBUMN/1999 on 28 September 1999, followed by the enactment of Law No. 19/2003 concerning SoEs and the issuance of the SoE Ministry Regulation No. Per-05/MBU/2007 concerning SoE's partnership with small enterprises and environmental management programs. All law and regulations stated that every SoE had to allocate 4% of its net profit on the partnership with small and medium enterprises and environmental management programs, equally 2% each for partnership programs and for environmental management programs.

In 2017, the Government and the parliament of the Republic of Indonesia passed Law No. 40/2007 regarding Corporation. Article 74 of the Law said that all companies operating in and/or related to natural resources have to follow social and environmental responsibilities. Different from previous laws and regulations that were applicable specifically to SoEs, this law applied to all companies either Government, private domestic or foreign-owned companies. There was also an initiative from the National Center for Sustainability Reporting, a joint effort of several non-Governmental organizations and professional associations started in 2005, to conduct the Indonesia Sustainability Reporting Award (ISRA) to give an award to companies that had published sustainability reporting.

For many Indonesian companies, economic success is the main baseline requirement needed to be achieved. The voluntary CSR adoption is usually led by national and multinational companies; they have many programs that help the objective of Government, to achieve better welfare of society in Indonesia. They fulfil their philanthropic responsibilities by setting aside budgets for partnerships and community development. These funds are then distributed to different branches of the company across the countries to address common societal issues including public health, access to education, economic empowerment, infrastructure, etc. For example, companies from oil and gas sectors such as Exxon Mobil, Pertamina, ConocoPhillips and Chevron have long-term community development programs that are aligned with their project strategy including the national/local content strategy.

Nevertheless, there are still several key issues in Indonesia as reported by CSR Netherlands, stated as follow:

4.1 People

Labour condition of CSR in Indonesia is still recognized having some issues such as child labour, unfair minimum wages per sector, incompliance of health and safety, discrimination towards ethnic minorities, and human rights. The Indonesian population, in general, has a strong hierarchical structure driven by status and wealth causing a gap between rich and poor.

4.2 Planet

There are some issues related to *Land Use* in extractives sector. Deforestation or burning forests to create land for, among others, the palm oil industry causes smouldering peat and carbonized soil. Furthermore, it leads to air pollution. This has a great impact on the health of neighbouring communities. Deforestation also causes friction with local communities, and *Environment*is placing increasing pressure on its natural resources as we can see that Indonesia is one of the world's richest countries in natural resources and biodiversity, but the growing population and economy. One of the main causes can be found in agriculture, mining and fishing. CSR possesses the responsibility of companies for their impact on society, especially in environmental impacts. It must provide community projects or nature reserves and create public policy dialogue and institutions towards more sustainable development and land use, respectively.

4.3 Profit

Corruption as a highly organized and widespread phenomenon in Indonesia. It remains difficult to completely eradicate corruption, even more, it happens in all stages of the bureaucracy process: design, implementation and supervision. The Indonesian market is tightly protected. There is a rule that within international companies for each foreign employee, 10 Indonesians must be employed as a counterbalance. Market disruption and false competition generally occur through price fixing. The *Komisi Pengawas Persaingan Usaha* (KPPU) was installed as a supervisory body to control fair competition. However, currently, the KPPU is now also subjected to discussions regarding corruption.

5 Evidence of CSR Practices in the Extractives Sector in Indonesia

In order to gain a deeper perspective on the CSR practices in the extractives sector in Indonesia, some interviews have been performed with several key stakeholders in the industry. The respondents come from industry, academia, Government and non-Government organization (NGO) backgrounds. For the industry, respondents come from one of the largest copper mining companies in Indonesia and one large state-owned oil and gas company. From the Government side, respondents include Government officials in the Ministry of Energy and Mineral Resources who are in charge of supervising community development activities managed by private sectors.

Program of community development is one form of CSR actualization that focuses on the development of human resource. The presence of the extractives sector in the midst of community life with its various activities raises socio-economic inequalities of local community members, so community development could improve the competitiveness and independence of local communities. The Government regulation requires corporations, especially in extractive resource industries, to spend some proportions of their operational budget to perform community development activities. The Government refers to this as the corporate responsibility for society and environment. The Government expects that one of the goals of this community development activity can be used to recover the exploration site to meet environment standards.

Based on the Government regulations, corporations are required to spend a minimum of 2% of their operational expenses towards their community development programs. Some big companies in Indonesia are renowned for their generosity in spending more for community development projects. These projects can range from providing employment for local people, developing infrastructure, building schools and roads, to providing scholarships for students in the local area. For the mechanism to allocate funds for community development activity, the company's management needs to include this activity in the company's annual plan and budgeting. Once the commissioners approve the plan and budgeting, the activity can be performed.

Some notable companies that are often doing community development projects in big scale for example are Freeport Indonesia in Papua province, Vale in South Sulawesi province and Pertamina, the state-owned oil and gas company. Freeport Indonesia (formerly is a subsidiary of Freeport-McMoran Inc., but since December 2018 the Indonesian Government bought 51.23% of the shares), one of the largest copper and gold mining companies in the world, has been operating in Indonesia since 1967. Its main operation for copper and gold mining is in Tembagapura highland (Grasberg mine), Papua province. Since its first operation in Indonesia, Freeport operation has sparked controversies from many people. The controversies are mostly related to the exploitation of environment in the mining site and the continuous conflict between the company and the local people. For example, Freeport has been accused of damaging the environment in Papua. The environmental damage includes the water pollution of Ajkwa and Otomona rivers due to tailing waste, and deforestation of Papua rainforest.

The local people in Papua often demand Freeport to contribute more to develop the local community since the community where Freeport located is considered under-developed. This situation is quite common in Indonesia, where most of the operation sites of the mining companies are located in rural areas, which have poor infrastructure facilities and a lack of accessibility. As a result, local communities expect the extractives sector to generate more positive spillover effects to their surroundings.

For the case of Freeport Indonesia, in order to have more alignment with the local community, the company has attempted to create a community development program that is based on sustainability principles. Especially after the Government of Indonesia became the majority owner of the company in December 2018, the company has committed to develop smelter for further processing of the copper ore. This smelter is expected to have a processing capacity of 2–2.6 million tons of copper ore per year. With this amount of processed copper ore, Freeport Indonesia can increase the value of its output. On the other hand, the commitment of Freeport to develop a smelter in Indonesia is to meet Government regulation in banning exports of unprocessed raw materials. This regulation has forced mining companies to process their mineral ore further before exporting it to other countries.

Another commitment from Freeport Indonesia is to invest US$20 billion in the Grasberg copper and gold mine until 2041, which is the end of the Freeport contract in Papua. In order to ensure the development of local communities, from the 51.23% of the Government of Indonesia's share at Freeport Indonesia, 10% of it is given to Papua provincial Government. For developing local communities in Papua, Freeport Indonesia has spent more than US$15 billion in various projects including education, gender equality, infrastructure, health and environment protection. It is now the largest private sector employer in Papua province. In the education sector, Freeport provided scholarships to more than 11,000 students to study at all education levels, from elementary to doctoral programs.

The objectives of all these CSR activities initiated by Freeport Indonesia are to ensure that the company meet the sustainable development principles especially in the field of education (number 4), health (number 3), economics (number 8) and cultural development (number 16) as well as to have alignment with the values of local people in the community (PT Freeport Indonesia 2016). The company expects that all these CSR initiatives might increase the image of the company to the local community, and to the public in general. The approach taken by the company to ensure the success of their CSR programs is mostly by making contact with top Government officials so that the Government supports the programs and maintains a network with community leaders. Having a close relationship with community leaders is important since Indonesian society is a communal society where patronage still plays an important role in collective decision-making.

Another example of CSR practice in the extractives sector in Indonesia is CSR implementation performed by Vale Indonesia. Established in 1968 as PT International Nickel Indonesia and since renamed to PT Vale Indonesia, it is the biggest nickel producer in Indonesia. Vale Indonesia produces about 75,000 tons of nickel per year, contributing 5% of the world's nickel supply. However, the huge success of Vale Indonesia has brought about enormous consequences for the environment where the mining operations are. For example, air pollution is often cited by residents complaining their house roofs get dirty and the alteration of water hydrology caused by changes in weather.

Those issues are the trigger of CSR practices at PT Vale Indonesia. The company has commitment to preserve the earth through environmentally friendly technology. First is hydroelectric power plant (PLTA), Vale Indonesia operates three PLTA with the total capacity 365 Megawatt located at Larona, Balambano and Karebbe. The function of these hydroelectric power plants is to operate furnaces that process and smelter nickel seeds, thereby minimizing the use of fossil fuels.

Second is land reclamation as post-mining rehabilitation including nursery and biodiversity programs. The technique of land reclamation is to use backfilling systems as a soil exfoliation process. Vale Indonesia rehabilitates more than 100 ha of lands per year and produces about 700,000 seeds. Until 2017, the total rehabilitated lands is 4.154 ha. Third is the building of a baghouse to absorb dust emissions and electrostatic precipitator (ESP) as dust sedimentations. Through this facility, in 2019 Vale Indonesia plans to decrease SO_2 emissions from 0.86 kg SO_2/kg Ni to 0.80 kg SO_2/kg Ni. Fourth is Lamella Gravity Settler (LGS). It integrates 17 settling ponds with the capacity of 16 million cubic metres in order to process water waste effectiveness and safeness to water flows.

6 Conclusion

The main challenges that the extractives sector are facing related to CSR practices in Indonesia are mostly around public acceptance and difficulties in dealing with the Government—social and political licenses to operate. Many people in Indonesia believe that the presence of the extractives sector is purely about the exploitation of resources while giving little benefits to the society. The public demand more contributions from extractives sector such as providing job opportunities for local people, developing local communities and providing basic infrastructure in the local areas surrounding mining operations. Related to the challenges with the Government, unclear regulation is still the main factor for extractives sector in dealing with the Government. There are many regulations in the extractives sector that cause inefficiency in governance. Other problems at the Government level include a rent-seeking mentality in some sections of Government highlighting that corruption is rampant.

Even though some challenges prevail in the extractives sector, some major corporations are still willing to expand their operations in Indonesia. These companies range from multinational, local and state-owned companies. For the multinational companies, the fact that there are many opportunities in this sector in Indonesia is the main factor they invest in the country. For the state-owned companies, their reason to operate is mainly to fulfil the obligation from the State to manage the strategic industries in the country. Some of the major companies that are still operating in Indonesia include Freeport Indonesia (copper mining), Vale Indonesia (nickel), Pertamina (oil and gas) and Chevron Indonesia (oil and gas).

From the community side, despite the fact that some people still oppose the operation of extractives sector in their community, this sector still has some lucrative offerings. An opportunity to work in the company and to learn new things in the sector still attracts many local people to work in the industry. From the Government side, much work needs to be completed in this sector. Firm regulation is the key to maintain the healthy business climate of the industry. Therefore, the Government needs to work hard to reform the regulations in energy and extractives sector so that companies will be attracted to invest their money and other resources in Indonesia.

References

Acuña, R. M. (2015). The politics of extractive governance: Indigenous peoples and socio-environmental conflicts. *The Extractives Industries and Society, 2*(1), 85–92.

Bebbington, A. (2010). The extractives sector and stunted states: Conflict, responsibility and institutional change in the Andes. In *Corporate social responsibility: Discourses, practices and perspectives* (pp. 97–115).

Bracking, S. (2009). *Hiding conflict over industry returns: A stakeholder analysis of the extractives sector transparency initiative*. Brooks World Poverty Institute: University of Manchester.

Buur, L., Therkildsen, O., Hansen, M. W., & Kjær, M. (2013). Extractive natural resource development: governance, linkages and aid (No. 2013: 28). DIIS Report.

Canel, E., Idemudia, U., & North, L. L. (2010). Rethinking extractive industry: Regulation, dispossession, and emerging claims. *Canadian Journal of Development Studies/Revue canadienne d'études du développement, 30*(1–2), 5–25.

Chandler, S. G. (2001). *Introduction: Defining corporate social responsibility, ethical performance best practice.*

Davis, R., & Franks, D. M. (2011). The costs of conflict with local communities in the extractive industry. In *Proceedings of the First International Seminar on Social Responsibility in Mining, Santiago, Chile* (Vol. 30, pp. 7576–7581).

Fauzi, Y. (2017). *Ekonomi Tumbuh, Industri Tambang Malah Melorot*. Retrieved January 30, 2019, from CNN Indonesia. https://www.cnnindonesia.com/ekonomi/20170505144748-85-212583/ekonomi-tumbuh-industri-tambang-malah-melorot.

Graves, P. E., Weiler, S., & Tynon, E. E. (2009). The economics of ghost towns. *Journal of Regional Analysis and Policy, 39*(2), 131–140.

Gilberthorpe, E., & Papyrakis, E. (2015). The extractives industries and development: The resource curse at the micro, meso and macro levels. *The Extractives Sector and Society, 2*(2), 381–390.

Lohde, L., Armstrong, C., & Nyhan Jones, V. (2015). *The art and science of benefit sharing in the natural resources sector*. Washington, DC: International Finance Corporation.

Luthfy, R. M. (2018). *Adat dan Konflik Pertanahan*. Retrieved February 19, 2019, from Koran Jakarta https://www.koran-jakarta.com/adat-dan-konflik-pertanahan/.

Maconachie, R., Srinivasan, R., & Menzies, N. (2015). *Natural resources, extractive industry investment, and social conflict*.

Pahlevi, R. (2018). *Mengenai Produksi Minyak Bumi Indonesia, Ini Faktanya*. Retrieved February 16, 2019, from Liputan 6. https://www.liputan6.com/bisnis/read/3633451/mengenai-produksi-minyak-bumi-indonesia-ini-faktanya.

PT Freeport Indonesia (October 2016). Ajang Perkenalan Program Sosial Perusahaan. Retrieved from PT Freeport Indonesia. https://ptfi.co.id/id/news/detail/promoting-company-social-programs.

Publish What You Pay Indonesia. (2018). *Penelitian: Peta Masalah dan Tata Kelola Industri Gas Indonesia, di Sepanjang Rantai Nilai Industri Ekstraktif*. Retrieved from PWYP Indonesia. https://pwypindonesia.org/id/penelitian-peta-masalah-dan-tata-kelola-industri-gas-indonesia-di-sepanjang-rantai-nilai-industri-ekstraktif.

Renwick, N., Gu, J., & Hong, S. (2018). China and African Governance in the Extractives Industries. In *International Development Policy Revue internationale de politique de développement, 10*(10.1).

Sabintoe, M. (2016). *Dampak Kehadiran Perusahaan Tambang Terhadap Kehidupan Petani di Morowali*. Retrieved February 19, 2019, from Partai rakyat Pekerja. https://www.prp-indonesia.org/2016/dampak-kehadiran-perusahaan-tambang-terhadap-kehidupan-petani-di-morowali.

Sigam, C. & Garcia, L. (2012). *Extractive Industries: Optimizing value retention in host countries*. Retrieved from https://unctad.org/en/PublicationsLibrary/suc2012d1_en.pdf.

Suastha, R. D. (2018). *PBB Sebut Industri Ekstraktif Indonesia Sarat Pelanggaran HAM*. Retrieved February 2, 2019, from CNN. https://www.cnnindonesia.com/internasional/20180208021554-106-274696/pbb-sebut-industri-ekstraktif-indonesia-sarat-pelanggaran-ham.

Suhartadi, I. (2018). *Peluang dan Tantangan Sektor Pertambangan pada 2019*. Retrieved February 22, 2019, from Berita Satu. https://id.beritasatu.com/energy/peluang-dan-tantangan-sektor-per tambangan-pada-2019/182758.

Slack, K. (2012). Mission impossible? Adopting a CSR-based business model for extractives industries in developing countries. *Resources Policy, 37*(2), 179–184.

The UN Interagency Framework Team for Preventive Action. In *Toolkit and guidance for preventing and managing land and natural resources conflict: Extractive industries and conflict*. Retrieved from https://postconflict.unep.ch/publications/GN_Extractive_Consultation_ES.pdf.

Visser, W. (2009). *Corporate social responsibility in developing countries*. The Oxford Handbook of Corporate Social Responsibility.https://doi.org/10.1093/oxfordhb/9780199211593.003.0021.

World Commission on Environment and Development (1987). *Our common future: The world commission on environment and development*. Oxford University Press.

Corporate Social Responsibility: Brazilian Case

Silvia Andrea Cupertino, Hirdan Katarina de Medeiros Costa, and Marcia Regina Konrad

Abstract The purpose of this study was to analyse the related regulations: special purpose company in the transport of natural gas through pipelines, noting that Brazilian legislation is stumbling, with no specific regulations in the gas sector. We sought to analyse the corporative governance applied to such types of corporate models, pointing to the need for improvement in the gas sector, understanding that there is a legal loophole to its application in this sector.

Keywords Specific purpose company · Corporate governance · Natural gas

1 Introduction

The Brazilian administrative reforms of the 1990s saw the State direct its infrastructure activity to the regulatory area, departing from its role as entrepreneurial State. In view of this change in State activity, it is necessary to insert control instruments, so that the public service is effectively provided to the population. Thus, we have seen a shift in Brazilian public management that focuses on public governance tools.

The gas sector in Brazil is strongly regulated and oriented by standards. Its governance focuses on the actors involved and available decision-making and implementation structures by a public agency or a private entity. It is a governmental institutional arrangement that seeks innovative solutions to the deepening of democracy through the articulation of the economic-financial, institutional-administrative and socio-political dimensions, establishing partnerships with civil society and the private

S. A. Cupertino (✉) · H. K. de M. Costa · M. R. Konrad
University of São Paulo, São Paulo, Brazil
e-mail: silviaan@senado.gov.br

H. K. de M. Costa
e-mail: hirdan@usp.br

M. R. Konrad
e-mail: marcia.konrad@usp.br

E. G. Pereira et al. (eds.), *Sovereign Wealth Funds, Local Content Policies and CSR*, CSR, Sustainability, Ethics & Governance,
https://doi.org/10.1007/978-3-030-56092-8_28

market. Thus, it presupposes a democratic project insofar as it considers the importance of civil society in the formulation of public policies (Ronconi 2011). The performance in promoting the delivery of public services to the Company should be based on efficiency and transparency instruments for greater social control and reduction of information asymmetries.

Thus, the strategy adopted by the national regulation in Brazil was the use of the Specific Purpose Company (SPC) model in the formatting of public service concessions. Its benefit is the segregation of the equity of the corporate persons and the partners who performed the activity. In such a model, there may be a close partnership between the State and the private sector, and in the case of public service concessions, between private individuals and between state and private companies, with the regulation implemented by the regulatory body. For Féres (2004), the SPC can be classified as a joint venture, creating a person distinct from the partners to achieve a common purpose. It points out that the Public–Private Partnership Law states that it can assume any corporate type. In the SPC's negotiating scheme, the State is subject to the control of the individual.

Therefore, the application of the SPC model in the Brazilian gas sector is a tool to ensure insertion of best management practices in utilities companies, which conciliates strategic objectives of the organisation and the engagement among its members, leading to the implementation of the principle of transparency and public governance as a type of state management. An important aspect for the analysis of Special Purpose Companies[1] is the scope of corporate governance as an instrument to confer management transparency and to attribute reliability to foreign investment in partnerships in the area of natural gas transportation. Thus, a more in-depth study is required of this Brazilian model that is currently used by gas transportation concessionaires through pipelines in order to point to ways for more efficient regulation that complies with corporate governance rules.

According to Petrobras (nd), this model was used in 2005 when created by 'Transportadora Gasene S/A', a finance project in which the company was created for a specific goal: raise funds to implement a project and to individualise costs, revenues and results, allowing a clear view of the business and the perception of risks to partners and potential funders. Since the late 1990s, Petrobras has been using this same SPC model for the development of its E&P, gas transportation and refining projects, such as Marlim 1999; Cabiúnas 2000; Espadarte-Voador-Marimbá 2000; Barracuda/Caratinga 2000; Albacora Japan 2000; Albacora Petros 2001; New Marlim 2001; Snapper-Carapeba-Garoupa-Cherne-Congro 2002; Knitwear Project 2002; Project Urucu-Manaus 2004; Marlim Leste 2004; Mexilhão 2005; REVAP 2006, Petrobras 2015).

At the same time, as Lantos (2001) points out, the development of the corporate social responsibility (CSR) concept involves four components: economic, legal, ethical and altruistic duties; proposing that ethical CSR, grounded in the concept of ethical duties and responsibilities, is mandatory and good for business and society. Corporate social-environmental responsibility comes from the notion that

[1] A 'Special Purpose Company' is created for a specific goal and has a determined time of existence.

firms should be accountable to society's long run needs and wants by minimising harms (ethical CSR) and promoting benefits for society (altruistic and strategic CSR) where social responsibility should orient various stakeholders in achieving economic performance, ethical performance and social performance.

Moir (2001) reviewed the literature on CSR noting that it "has strong divides between normative or ethical actions and instrumental activities" and revealed an increasing focus both by business on CSR and by society on the actions of the private sector. This led to the development of methods to identify the responsibility of companies to society, their structures, how these structures impact rights to well-being and adequate models to fulfil legal and ethical needs by stakeholders and their engagement as a way of determining precise activities. We consider the Specific Purpose Company (SPC) as a legal and ethical tool in order to reach a reliability to foreign investment in partnerships in the area of natural gas transportation.

The inclusion of best management practices in utility companies addresses the organisation's strategic objectives and engagement among its members. This leads to the implementation of the principle of transparency, where timely and reliable information should be disclosed to the general public, rather than only to regulated stakeholders. Best management practices also lead to the implementation of the principles of equity (where fair and appropriate treatment is given to all stakeholders) and accountability (full responsibility of the directors with respect to their actions during their periods of operation).

The model adopted by the Brazilian Constitution of attributing to third parties the provision of certain essential public services requires the adoption of control practices and mechanisms for measuring the quality of the activities developed by the concessionaires. In this sense, good governance practices have the scope of adding value to society, providing access to the capital market, contributing to the permanence of companies in the face of positive results (Bernardino 2014), which generates greater funding, resources and conversion of such gains into tariff modality.

The viability in implementing the infrastructure of gas transportation depends on a proper legal structure in order to create a robust environment to allow large private capital to invest in construction and operation of natural gas transportation plants. Therefore, rules oriented to specific characteristics of this activity are necessary to provide greater stability and means to view returns and reasonable risk for private investment (Prisco 2010). As a result of the national expansion of the natural gas sector, the logistics of transportation of such a product in order to meet the needs of the country stand out, taking into account the need to expand the national gas network. In this chapter, we study natural gas industry in Brazil as a case, which is expecting to increase its contributions to the Brazilian energy mix, particularly, after pre-salt discoveries.

2 Brazilian Case: Natural Gas Sector and Public Governance (PG)

Public governance focuses on the actors involved and available decision-making and implementation structures by a public agency or a private entity. Therefore, under the application of the regulatory impact analysis instrument, regulatory governance is directed to the means and instruments used by the regulator to achieve efficiency, transparency and legitimacy (Dantas and Maciel 2019).

For Bernardino (2014), good governance practices have the scope of adding value to society, providing access to the capital market, contributing to the permanence of companies in the face of positive results, which generates greater fundraising and conversion of such gains into tariff modality. The Brazilian Federal Audit Court (2014) understands that governance is directed, monitored and encouraged to organisations and involves relationships between society, senior management and employees or employees and control bodies. It should aim to gain and preserve society's trust through an efficient set of mechanisms to ensure that actions taken are always aligned with the public interest. Therefore, Federal Audit Court Ruling No. 2261 of 2011, among others, sets out as good practice the implementation of minimum transparency requirements in the agency's decision-making process.

In this sense, Decree No. 6.062 of March 16, 2007, which established the Programme for Strengthening Institutional Capacity for Regulatory Management—PRO-REG, in its Article 2, gave PRO-REG the means to formulate and implement integrated measures aimed at strengthening a regulatory system to facilitate the full exercise of functions by all actors, and the ability to formulate and analyse public policies in regulated sectors. It also established compliance with improvements in coordination and strategic alignment between sectoral policies and the regulatory process; strengthening the autonomy, transparency and performance of regulatory agencies; and the development and improvement of mechanisms for the exercise of social control and transparency within the regulatory process.

Brazilian policies are implemented by law. The current regulatory model proposed by Law 11.909/2009 is not enough to delineate many aspects of the legal regime for the concession of gas transportation by pipelines of general interest, this activity still depends on the detailed discipline of the matter, which will arise from rules of contracts to be signed between Administration and individuals. Even so, it is possible to discern in Law 11.909/2009 the basic outline of the legal-economic structure of this adjustment and its regulatory apparatus (Prisco 2010).

According to Article 4, paragraph 2 of Law 11.909/2009, the federal legislation of Public–Private Partnerships (Law 11.079/2004) is applicable to the natural gas transportation pipeline model, as well as to be able to apply the General Law of Concessions (Article 3, paragraph 1 of Law 8.987/1995). Under this aspect, it is necessary to study the ways that Specific Purpose Society can be used, in the way that it is determined in Article 9 of the Public–Private Partnership Act.

It should be noted that the term 'Specific Purpose Company' (SPC) appeared in the legal order of the country in Law 11.079, of December 30 of 2004, which established the public–private partnerships regime, in Article 9, which requires the constitution of such company before the conclusion of the agreement, with the need to comply with corporate governance standards.

Its rule is also given by Article 981 of the Brazilian Civil Code, which explains that it is possible to set up a business company aiming at the development of a very restricted activity and in some cases may have a certain term of existence, usually used to isolate the financial risk of the activity developed. As a rule, it has a specific goal to be achieved (Article 50, item XVI, of Law 11.101/2005).[2]

As a business model where there is segregation of funding for project implementation and individualisation of costs, revenues and results, it is verified that there is no specific legislation for SPCs in the field of gas transportation, but it is widely used in concessions that seek to provide public services, such as the gas transportation service.

The transportation of gas by pipeline requires high investment and presents low flexibility, its cost being determined by its length, route and volume to be transported and 50–60% of the total costs of assembly and expropriation costs. However, with the falls seen since 1985, pipeline transport has become more competitive even for distances over 5000 km (Piquet and Miranda 2009).

In this sense, investment in such a model is possible, but there is a need to improve the business exploitation model, in order to adapt the Specific Purpose Company to the gas market and to enable the segregation of the explorer of activity from activity itself. It is considered that the SPC model benefits transparency because it gives the regulator greater possibility to carry out a more adequate inspection, and in case of return of the activity to the granting power, the equity is ready for that purpose.

When integrating operation in the gas logistic field, it is essential that an SPC develops a project that is presented in a clear and objective way, in addition to its operations plan and its financial plan, which must have total transparency on fundraising, a level of compliance that meets regulatory requirements.

The natural gas transportation logistics should primarily aim to guarantee the supply of natural gas to the plants belonging to the Priority Thermoelectricity Program (PPT). According to Technical Note 002/03/SCG and based on the Petroleum Law, it should be carried out by independent enterprises that possess administrative, technical and operational capacities, as well as technical elements of their own support, decision-making autonomy and advanced level of compliance. This will result in periodic publications of their financial statements in the most appropriate and transparent way possible, which makes the importance of corporate governance evident, given the explicit context, since this is based on the dialogue between regulators, companies, market and society.

---

[2]Law 11.101/ 2005. Art. 50. The following are among the means of judicial recovery, observing the legislation pertinent to each case, among others: (...) XVI—constitution of a special purpose company to adjudicate, in payment of credits, the assets of the debtor.".

The structuring of projects in gas-nets must fully comply with technical-financial requirements of the National Petroleum Agency (NPA) and of the legislation in force through observation of applicable norms, regulations and objectives and principles established by law without disregarding ways to encourage the development of the natural gas market in Brazil.

From this perspective, the design of the natural gas distribution sector implies the consideration of possible contractual mechanisms to stimulate the competition or the maintenance of more closed contracts from the competitive point of view of the capital needs to build a network at a stage very nascent (or practically non-existent) (Costa 2006: 27).

The proposal of a gas transportation model must be fully compatible between its structure and the organisation model of the natural gas sector according to Law 9.478/97 and Law 11.909/2009. The project must foresee flexibility when strategic adaptations are needed, however, without losing focus on its main objective and its characteristics. It is important to note that financial engineering characteristics of those involved in SPCs should always be subject to critical attention, through continuous analysis and monitoring that ratify their solidity and suitability.

The project should present as fundamental to the integration among the members involved, so that there is a common vision about all its implications, such as the development of the natural gas transport infrastructure, the national interest, the generation of employment and income, the technological development, relations between financial gains and competitiveness, and, of course respecting the regulatory model established by the Oil Law and the Gas Law.

Therefore, the SPC's employment is adequate, but the model that uses it needs legal improvement. It is necessary to develop studies related to the regulation of SPCs in the natural gas transportation sector through gas pipelines and its relation to rules of public governance. In this way, as a guide to an investigative and analytical view, this theoretical study and its theoretical-conceptual model establish the current legislative framework. It builds on the work of authors such as Álvares et al. (2008), Sabbag (2015), Santos (1999, 2011), Steinberg (2003).

Another aspect of public governance is corporate governance because it is applied in companies that are in regulated sectors, such as the gas sector. This sector, which is still nascent in Brazil, requires professionalisation and insertion of robust and effective management practices. This means that corporate governance (CG) is not only applicable but is an extremely desirable practice because what makes its adoption imperative is that the system represents "competitive and innovative postures, which do not constitute extravagances of the favourable times, but necessities for survival in moments of transition and global changes" (Santos 1999: 37).

3 Gas Logistics and Corporate Governance

Good corporate governance practices were instituted in Brazil by Decree 6.021 of January 22, 2007 and conceptualised as a set of practices involving, among others, relationships between shareholders or quota holders, boards of directors and auditors, or equivalent function bodies, senior management and independent auditors, with the purpose of optimising company performance and protecting the rights of all stakeholders with a view to maximising the economic and social results of the performance of state-owned federal enterprises. In this sense, corporate governance is optimised through transparent management (Papariello 2008). According to Álvares, Giacometti and Gusso (2008), the organisational structure of corporate governance is composed of the Board of Directors, Committees of the Board of Directors, Fiscal Council, Family Council, Executive Management and Holdings.

The Brazilian Institute of Corporate Governance (IBGC) guides CG as a management system that incorporates organisations, their monitoring and the relationship between stakeholders, shareholders and control bodies in search of continuous improvements aligned with their interests, preserving and optimising its organisational value and consequently achieving its economic efficiency and permanence. As a management system, corporate governance presents four principles: transparency, fairness, accountability and corporate responsibility. These principles are supported by an organisational structure responsible for decision-making processes related to the strategic orientation of the organisation that is supported by solid and dynamic structural conditions.

The SPC should provide business management with emphasis on the four principles of corporate governance as this will allow the viability of new investments, maximisation of the use of existing infrastructure and (of gas production to Brazil and abroad) tariff planning that results in a clear plan reflecting the real costs involved in the associated logistics processes and strengthening the role of the transporter.

It is necessary to discuss on the alternatives for raising funds, as well as the total and unrestricted attendance to the regulatory events so as not to impede the introduction of new competitors in the gas sector in the country. It is not enough to recognise the economic and financial importance of the project. The national commercial interests and the regulatory objectives of the NPA and other relevant legislations need to be allied with them.

4 Corporate Social-Environmental Responsibility

Due to the intensification of social and environmental issues, natural gas companies seek to improve their public image via CSR initiatives. Improvements in gas industry production processes are required to minimise environmental impacts and risks. Due to this necessity of market and environmental legislation demanded by society, several tools, techniques and methodologies of environmental management emerge.

Over the years, countries have progressively adopted different approaches to tackling social problems, environmental degradation and pollution, ranging from simply ignoring the problem to diluting or dispersing pollution, so that its effects are less harmful, to apparent pollution control using end-of-pipe treatment, as well as avoiding pollution and waste at source through a clean production approach.

Cleaner production is defined as the continued application of integrated environmental prevention strategies applied to processes, products and services to increase overall efficiency and reduce risks to humans and the environment (UNIDO/UNEP 1995). Development of products involves reduction of negative impacts throughout its life cycle, from raw material extraction to final disposal, while for service industries, cleaner production involves the incorporation of environmental considerations into the design and provision of services.

In addition, a change of attitude on the part of the directors of companies, managers and employees is fundamental to gain the maximum in cleaner production. It is important to emphasise that cleaner production not only involves a change of attitudes, but also a technological change. In many cases, the most significant benefits of this practice can be achieved through lateral thinking (De Bono 2000).

Corporate social-environmental responsibility and sustainable development are no longer restricted to organisational matters. Companies adopt practices with a view to sustainable development through environmental actions and strategies, in which companies must contribute to economic, environmental and social progress in order to ensure sustainable development (Lomes and Azevedo 2019; OECD 2000). Nowadays, companies in the gas sector are aware of the need for less polluting technologies applied in their production processes as a tool to implement CSR.

Brazilian companies in the natural gas sector are aware of the need to implement practices of cleaner production, with the intention of applying preventive and integrated environmental action plans to the processes, products and services, in order to increase efficiency and reduce risks to humans and the environment (Moraes et al. 2007).

The largest producer, Petrobras, developed projects in socio-environmental responsibility. As reported by its Social Responsibility Policy (2019), a project was developed to prevent, eliminate, control or mitigate the impacts directly or indirectly caused by its operations, by verifying the changes in the social and environmental milieus, being established during the whole life cycle of its production process. It also develops social and environmental programmes and projects, contributing to the communities where they operate and collaborate on conservation of the environment and improvement of living conditions.

Another initiative was developed by Shell Brazil. It has developed the Sustainable Development Diagram, a set of sanitation, safety, environmental and social performance standards, under which the company establishes and maintains commitments that consider social and environmental issues as well as the involvement of the communities impacted by it (Shell 2016).

Parnaíba Gás Natural maintains corporate management of sustainability, responsible for the monitoring of all socio-environmental actions, which acts in accordance with current legislation and adoption of environmental and social quality standards to constantly improve its indicators on health, safety, environment and social responsibility (Eneva 2016).

The know-how of the natural gas sector reflects improved efficiency as a result of its experience adopting best management techniques and improvement of companies' practices and procedures. Usually, the application of technical know-how results in the optimisation of existing processes (Christie, Rolfe and Legard 1995). The gradual progression from ignoring rather than avoiding has culminated in the realisation by gas companies that it is possible to achieve cost savings in industry and, in addition, to improve the environment for society, which characterises the objective of cleaner production.

Consequently, with the low price of oil, there is a direct effect on the amount of investments and activities in this sector. In short, less investments means less skilled labour. When the amount of work increases, there is a shortage of skilled labour available. This vicious cycle creates a complex environment to work in as the price of oil affects the amount of work and investments. But in the scenario when a skilled workforce is needed, it is not easily available as the investment was not made and prepared for.

5 Conclusion

The development of an appropriate legal structure that will enable the entrance of large amounts of private capital required for the construction and operation of the natural gas transmission facilities is the basis for making feasible the gas transport infrastructure, since the investment depends on stable and attractive regulatory frameworks in terms that could encourage the entrepreneur to see a possible and attractive economic return. The SPC should be based on the four principles of corporate governance because this will allow the viability of new investments and the maximisation of usage of existing and future infrastructure. In corporate governance, CSR is where many SPCs in the gas sector are developing projects in environment and social areas in order to meet market expectations, international mandates for responsibility in promoting development and public governance rules. Brazil has environmental legislation, such as the National Solid Waste Policy, Law 12.305/10, Technical Note CGPEG/DILIC/IBAMA nº 01/11, among others, in conjunction with the cleaner and more stable industries in Brazil that seek to minimise the generation of waste, establish temporary targets and increase the range of direct and indirect job creation opportunities in its many thousands of processes, as well as the decision-making processes for actions and maintenance of projects in the gas sector.

Companies that adopt cleaner production methodologies benefit the social and natural environment, reflecting lower costs through the efficient use of both raw materials and energy and by reduced spending on environmental liabilities, as well as assisting local development. The increasingly fierce competition among natural gas companies makes them seek the necessary adaptations to meet consumers in their increasingly social and environmental demands and the legal obligations imposed by the control bodies. To do so, companies devise strategies that fit these requirements, in addition to promoting their competitiveness and continuity in the market. The adoption by gas companies of a conscious posture of their social and environmental actions, through the use of appropriate concepts and techniques, lead to the adoption of a new paradigm and the development of a new market position.

As mentioned above, an SPC is the result of joining efforts to achieve a specific enterprise, so it is imperative to adopt clear and objective rules, where there is complete transparency in their actions and high commitment to organisational compliance. The cases chosen above show that there are many companies focusing attention on actions of social and environmental responsibility. This initiative relates to internationally established norms and agreements and is also aligned to Brazilian legislation and the social expectations identified by the society. However, there are not many experiences in Brazil in the natural gas sector.

Acknowledgements We acknowledge the support of the RCGI Research Centre for Gas Innovation, sponsored by FAPESP (2014/50279-4) and Shell. In addition, we acknowledge the support of CAPES and National Agency of Petroleum, Natural Gas and Biofuels through Human Resources Program n. 04.

References

Álvares, E., Giacometti, C., & Gusso, E. (2008). *Governança Corporativa: um modelo brasileiro*. Rio de Janeiro: Elsevier.

Brasil. MME. Matriz Energética Nacional 2030.

Brasil. (2014). Tribunal de Contas da União. Referencial básico de governança aplicável a órgãos e entidades da administração pública/Tribunal de Contas da União. Versão 2 - Brasília: TCU, Secretaria de Planejamento, Governança e Gestão. 80 p.

Brazil Chevron. (2019). Meio Ambiente. São Paulo: Brazil Chevron, s.d. Available in https://brazil.chevron.com/-/media/brazil/our-businesses/documents/Brazil_fact-sheet-6-20.pdf. Access on February 17, 2019.

BRASIL. Lei n.° 11.909, de 04 de março de 2009. Regulates activies related to the transportation of natural gas, referred to in art. 177 of the Federal Constitution, as well as on the activities of treatment, processing, storage, liquefaction, regasification and commercialization of natural gas; amends Law No. 9,478, of August 6, 1997.

BRASIL. Law n° 11.079, of december 30th of 2004. Institutes general rules for bidding and contracting public-private partnerships within the scope of public administration.

BRASIL. Law n.° 8987, of february 13th of 1995, that regulates Provides for the concession and permission regime for the provision of public services provided for in art. 175 of the Federal Constitution. Available on http://www.planalto.gov.br/ccivil_03/leis/L8987compilada.htm. Access on Oct 08 2020.

BRASIL. Law n.° 12.305, of August 2nd of 2010, tha institutes the National Solid Waste Policy. Available on http://www.planalto.gov.br/ccivil_03/_ato2007-2010/2010/lei/l12305.htm. Access on Oct 08 2020.

BRASIL. Law n.° 10.406, of january 10th of 2002. Insitutes Brazilian Civil Code. Available on http://www.planalto.gov.br/ccivil_03/leis/2002/L10406compilada.htm. Access on Oct 08 2020.

BRASIL. Decrete n.° 6062, of march 16th of 2007. Institutes the Institutional Capacity Building Program fo. Regulatory Management - PRO-REG. Available on http://www.planalto.gov.br/ccivil_03/_Ato2007-2010/2007/Decreto/D6062.htm. Access on Oct 08 2020.

BRASIL. National oil, Natural gas and Biofuel Agency. Technical Note n.° 002/03/SCG, march 18th of 2003. Available on http://www.anp.gov.br/images/movimentacao-estocagem-comercializacao/transporte-gas-natural/estudos-notas-tecnicas/nota-tecnica-002-2003.pdf. Access on Oct 10 2020.

BRAZIL. Brazilian Institute of Environment and Renewable Natural Resources. Technical Note CGPEG/DILIC/IBAMA N.º 1, 2011. Available in http://www.oads.org.br/leis/3546.pdf. Access on Oct 10 2020.
Costa, H. K. M. (2006). A regulação do livre acesso na distribuição de gás natural canalizado: o caso de São Paulo. Dissertação de Mestrado. Instituto de Energia, Universidade de São Paulo, São Paulo.
Eneva. (2016). Parnaíba Gás Natural. Sustentabilidade. Rio de Janeiro: Eneva. Available in http://ir.eneva.com.br/conteudo_pt.asp?idioma=0&conta=28&tipo=57773. Access on February 19, 2019.
Féres, M. A. (2004). As sociedades de propósito específico (SPEs) no âmbito das parcerias público-privadas (PPPs): Algumas observações de direito comercial sobre o art. 9º da lei n. 11.079, de 30 de dezembro de 2004- Revista Jurídica da Presidência.
Lomes, J. S., & Azevedo, T. C. (2019, April). Corporate social responsibility: dissemination of bahian companies in the global compact. REBRAE, [S.l.], 12(2), 171–188. ISSN 2175-3350. Available at https://periodicos.pucpr.br/index.php/REBRAE/article/view/25270. Date accessed December 9, 2019. http://dx.doi.org/10.7213/rebrae.v12i2.25270.
Moir, L. (2001). What do we mean by corporate social responsibility? *Corporate Governance, 1*(2), 16–22. https://doi.org/10.1108/EUM0000000005486, http://dx.doi.org/10.1590/S1519-70772010000100004. Access on December 10, 2019.
Moraes, C. A. M., Gaspar, R., Rocha, L. K., Brehm, F. A., & Garcia, A. C. (2007, July–September). Aplicação de ferramentas do programa de produção mais limpa na gestão de resíduos. *Revista Tecnologia em Metalurgia e Materiais,* 4(1), 59–63.
OCDE. (2019). Declaração sobre o Investimento Internacional e as Empresas Multinacionais. Disponível em https://www.oecd.org/corporate/mne/38110590.pdf. Acessado em 10 dez 2019.
Papariello, B. M. (2008). Governança Corporativa em Agência Reguladora: a Experiência Australiana Comparada com a Brasileira.Brasília, 18p. Trabalho de conclusão de curso de Especialização em Mercado Financeiro e Investimentos (artigo)—Universidade de Brasília.
Princípios da Boa Governança para as Universidades: Um olhar do Canadá. IBGC em foco, nº 54, p. 17–20, Jan./Fev./Mar. 2011. Disponível em: http://www.ibgc.org.br/download/newsletter/IBGC54_Final.pdf. Acesso em 01 de agosto de 2016.
Petrobrás. (2016). Petróleo Brasileiro. Termo de Compromisso: comprometimentos assumidos pela Petrobras no sentido de aproximar-se do modelo regulatório previsto na Lei nº 9.478/97. Rio de Janeiro: Petrobr. Disponível em http://www.anp.gov.br/?dw=2035. Acesso em 01 de agosto de 2016.
Petrobras. (2019a). Política de Responsabilidade Social. RJ: Petrobras, s.d. Available in http://www.petrobras.com.br/lumis/portal/file/fileDownload.jsp?fileId=8A245DA25ADA9EA2015BABC2A0330546. Access on February 19, 2019.
Petrobras. (2019b). Esclarecimento sobre Sociedade de Propósito Específico do gasoduto Gasene. Disponível em http://www.petrobras.com.br/fatos-e-dados/esclarecimento-sobre-sociedade-de-proposito-especifico-do-gasoduto-gasene.htm. Acessado em 09 dez 2019.
Petrorio. (2019). Meio Ambiente. Rio de Janeiro: Petrorio, s.d. Available in http://www.petroriosa.com.br/seguranca-meio-ambiente-e-responsabilidade-social/meio-ambiente. Access on February 17, 2019.
PETROBRAS. (2020). Social responsibility policy. Available in https://petrobras.com.br/data/files/8A/14/5A/E3/CDB03710A3C3FE272418E9C2/Politica_de_Responsabilidade%20Social_RS.pdf. Access on Oct 08 2020.
Prisco, A. V. (2010). A concessão da atividade de transporte dutoviário de gás natural: notas ao regime jurídico básico da Lei n º 11.909/2009. *Revista de Direito Público da Economia.*
Ronconi. L. (2011). Emancipação, 2011-revistas.apps.uepg.br.
Sabbag, E. M. (2015). *Direito tributário essencial* (3ª edição). São Paulo: Saraiva.
Santos, N. M. B. F. (1999). Fatores-Chave de Sucesso no Processo de Internacionalização das Escolas de Administração. Administração em Diálogo - Programa de Estudos Pós-Graduados em Administração PUC-SP, São Paulo, 1(1), 23–37. Disponível em http://revistas.pucsp.br/index.php/rad/article/view/12505. Acesso em 01 de agosto de 2016.

Shell. (2016). *Sustainability Report 2016*. Hague: Shell Global. Available in http://reports.shell.com/sustainability-report/2016/introduction/sustainability-governance.html. Access on 19. Fev. 2019.

Steinberg, H. (2003). A dimensão humana da governança corporativa: pessoas criam as melhores e piores práticas. São Paulo: Gente.

UNIDO/UNEP. (1995). *Manual de avaliação de P + L.*. Porto Alegre: CNTL/SENAI.

Villacorta-Hernandez, M. A. (2014, June). Abandono legislativo del objetivo de alcanzar la independencia en la auditoría de cuentas. *Cuadernos de Contabilidad, 15*(37), 215–237. Available from http://www.scielo.org.co/scielo.php?script=sci_arttext&pid=S0123-14722014000100009&lng=en&nrm=iso. Access on September 19, 2016.

Corporate and Social Responsibility in Iran

Zoha Abdolalizadeh and Elham Beygi

Keywords Corporate Social Responsibility · Local communities · Mining Industry · Environmental Protection Programs · Law and Regulations

1 Introduction

The energy business comes with societal fallout. Gas and oil companies more than ever need to work within complex and regulated institutions to better balance social norms with the expectations of governments and societies. Corporate social responsibility (CSR) is a mechanism by which businesses pursue self-interest without overlooking—and by even promoting—social benefits to stakeholders, as well as the interests of all engaged parties. These include employees, suppliers, customers and local communities, and this strategy results in increasing stakeholders' utility function. In this light, efficiency gains a new definition where all concerned benefit, and the knock-on effect is felt when money is spent or donated to improve the utility of others. The notion of responsibility is a very important element in CSR, which sets it apart from other kinds of finite chained efficiencies. Companies bear a responsibility to acknowledge a society's expectations and to respect its social norms.

This chapter is concerned with CSR in the context of the oil and gas industry in Iran. Mine Social Responsivity as a branch of CSR is discussed first, followed by examples of several active projects that have been somewhat successful from a CSR perspective. Then, the discussion and significance of regulations and policies will be

Z. Abdolalizadeh (✉)
Dana Energy Company, Tehran, Iran
e-mail: zoha.abd@gmail.com

E. Beygi
McMaster University, Hamilton, Canada

E. G. Pereira et al. (eds.), *Sovereign Wealth Funds, Local Content Policies and CSR*, CSR, Sustainability, Ethics & Governance,
https://doi.org/10.1007/978-3-030-56092-8_29

addressed by analysing existing legal enforcing mechanisms in the national domain of gas and oil industry in Iran. In identifying the problems of almost each and every piece of relevant regulatory policies, this chapter concludes that Iran's oil and gas industry falls far behind developed existing schemes of social responsibility. Society should encourage extra-legal and governmental measures, because the governing body creates many obstacles to an ethical and sustainable future.

Here, we discuss the concept of CSR contextuality, defined as "every corporation in a specific country is under the influence of a prevailing institutional framework that has been formed during history",[1] and thus, the issues that define social responsibility vary, depending on the particular setting and circumstances.

Many believes that Iran's socio-economic stand is not ready for implementing western concept of CSR,[2] while there is an assumption that asserts that CSR initiatives are different in every society depending on their cultural, social and political context.[3]

In Iran, CSR has traditionally been manifested in religion, and moral norms of society such as the institution of Waghf (an Islamic concept where loans from employers must be made interest-free), Zakat (or "alms giving", one of the Five Pillars of Islam, giving of 2.5% of one's possessions (surplus wealth) to the poor and needy[4]) and Khoms (according to Shia jurisprudence, one-fifth of certain items which a person acquires as wealth must be paid as an Islamic tax). However, these kinds of religious or moral responsibilities have been limited to donating money, building schools, mosques and holy shrines.[5] Also, employers have played a patriarchal role in relation to their employees and helped them in times of illness, as well as providing funds for marriage and house-buying. In the absence of a social security system, these provisions have played an important role.

In fact, during Iran's early years of industrialisation, the vibrant role of religion and tradition served as the main driving force behind widespread social support on the part of employers. Given the dominant left-Islamic-leaning paradigm of Iran's economic and political environment after the Islamic revolution, industrial owners and investors began to be viewed skeptically. They were shrouded under an atmosphere of insecurity, going so far as to being fully marginalised for almost 20 years. Many financiers were even held accountable for their investments. This scepticism cast a shadow over the traditional social responsibility business owners had provided.

Iran's first conference on corporate social responsibility—organised by some industry associations, such as the Chamber of Commerce, the Confederation of Industry, several private companies, as well as NGOs—was the starting point for discussing the concept and literature of this subject in Iran. The intensifying competitive business environment led many of the country's top companies to consider this concept. With the emergence of environmental quality certifications and organisational excellence certificates, the concept became incorporated into corporate

[1]Chapardar and Khanlari (2011).

[2]Modabber (2013).

[3]Chapardar and Khanlari (2011).

[4]The Shia double this to 5% of one's possessions.

[5]Omidvar (2011).

management processes. Gaining these certificates, in addition to leading companies to greater productivity, also provided businesses with a competitive advantage. In addition, the need Iranian industries to find external partners was one major impetus to embrace the modern concept of CSR.[6]

A questionnaire[7] revealed that although Iranian industrial owners have always complained about the government's interventionist policies, most (75%) of the respondents agreed with the formulation of laws and regulations on corporate social responsibility; 92% saw the promotion of corporate social responsibility to be the principle role of government and governmental policies; 56% considered the main role of government to promote corporate social responsibility, and 37% held that the responsibility of monitoring business activities in CSR fell under the Institute for Standardisation and Industrial Research Organisation.

The impetus behind Iranian companies moving towards corporate social responsibility can be summarised and categorised as follows[8]:

Improving the public opinion of the company's external image, as well as that of consumers, employees and investors;

Targeting and integrating the company's current scattered activities in the field of corporate social responsibility in order to link and synergize CSR activities to other processes within the organisation;

Promoting organisational excellence.

2 Mine Social Responsibility

Mine Social Responsibility (MSR) refers to the steps mining companies take to reduce the negative impacts of their industry, and to improve the living conditions (economic, social and environmental) of the beneficiaries. These include personnel, local communities, local and state organisations, corporate investors, and generally everyone affected by the project. These actions sometimes go beyond legal obligations, contracts and agreements, and are usually investments in infrastructure as well as social and human capital. Each MSR program must be individually tailored to the conditions of the mining project and the needs of those who will benefit, an aspect continually evaluated for effectiveness. With the ultimate goal of achieving sustainable development, most MSR projects focus on three main areas: the environment, the community and the economy.[9]

At the same time, MSR programs require mining companies to submit periodic and regular activity reports on social, economic and environmental issues. These help to raise awareness among local communities, stakeholders, and the government

---

[6] Yoshanloey and Johar (2013).

[7] Omidvar (2011).

[8] Omidvar (2008).

[9] "Mining Social Responsibility" Published by Intelligent Mine Monitoring Organisation.

about the company's positive impacts and constructive actions, and also enhances the dialogue between the mining company and its beneficiaries.

The principles and objectives of implementing mining social responsibility programs are: acceptance of responsibility and management; applying ethical business practices; respect for human rights; commitment to project risk assessment and evaluation; the employment of people in the host community and other stakeholders; protecting the health and safety of workers and the local population; helping the development of community and social welfare; and protection of the environment.

In order to achieve these goals, mining companies must consider the following policies in their strategies and programs: to make their entire business approach based on social responsibility; to consider commitment and participation in addressing the needs of society, government, personnel and the environment beyond the life of the mine; making social responsibility part of their organisation's identity; and taking steps towards achieving sustainable development goals.

Implementing social responsibility plans, as well as reducing or eliminating numerous risks (such as local conflicts, environmental problems, labor protests and mining closures), offers additional benefits. These include improving the mining company's brand reputation, increasing sales and customer loyalty, reducing operating costs, better financial performance, the ability to attract talent and retain employees, organisational growth and access to capital. Top mining organisations operate in remote areas and are therefore some of the location's economic determinants, providing jobs for local people and improving their welfare. For this reason, the mining industry should pay particular attention and promote social accountability by making the sustainability of current and future ecosystems a priority.

Last year, IMIDRO's executive chief rendered a bylaw in which he asked all the organisation's subsidiaries, offices and projects to outline their social responsibility plans for the communities around them. Here are a few examples of active mining companies in the area of CSR, along with their activities.[10]

2.1 Gol Gohar Mining and Industrial Company

Gol Gohar mine is located in Sirjan, in the southern region of Iran. It is the largest known iron ore mine in the Middle East, with a geological reserve of 1019 million tons, a definitive reserve of 1000 million tons and a potential reserve of 219 million tons. Its mining products are iron ore pellet, iron ore concentrate and granulated rock. The Gol Gohar mining company is one of the most active and successful in CSR's realm, running a range of health, cultural and sports projects in the area, with US$5 billion in social responsibility invested in the region, according to Gol Gohar's CEO.[11]

[10]Gol Gohar, Chador Malu and Sarcheshmeh companies are among the biggest and wealthiest companies in mining industry of Iran.

[11]http://www.geg.ir/.

Some of the main programs and their local impacts are[12]:

1. Gol Gohar Hospital: the company plans to build a specialised hospital. It began initial studies in September 2016. This project is designed to have on-site departments of cardiology and angiography, radiotherapy, radiology, physiotherapy, orthopedics, obstetrics, gynecology, infertility, paediatrics, as well as general and internal surgery. The hospital is predicted to serve a unique role in the city's health services and surrounding areas.
2. Green Belt of Sirjan-Gol Gohar Road: the goal of this project is to beautify and improve the quality of this landscape by expanding the tree and shrub cover of this route, to control dust, optimise the use of water from rainfall and floods, and prevent soil erosion. The Gol Gohar mining and industrial area and its access road are located in Iran's desert areas. Because of recent droughts in the region, environmental measures are necessary to prevent soil erosion. Gol Gohar undertook research studies that showed the most successful way to deal with wind erosion and desertification was to establish sustainable vegetation. Desert area surveys found seedling operations to be the most effective option. Planting trees and shrubs along the roadside improved the environment, and protected it from floods and winds. The Green Belt project launched one-way roads in the first phase, with an approximate length of 25 km, and a working width of 20 m. It should be noted that Tehran University's International Wildlife Research Centre acted as a consultant to this project.
3. Gol Gohar Football Club: the company established a sports club in 1988. It runs different divisions, and the football team became famous when it was promoted to the Persian Gulf Pro League last season. Of note is the Gol Gohar Sport Complex, the club's home venue, and a well-equipped complex that includes a stadium with a 3200-seat capacity, several sports halls and an indoor swimming pool.
4. Charitable Activities: with the arrival of the holy month of Ramadan, a basket of foodstuffs valued at approximately US$37,000 was distributed among the beneficiaries of the Imam Khomeini Relief Foundation and Welfare Committee of Sirjan. In addition, the company helps survivors of natural disasters. For instance, it provided survivors of the Zemestan-Yurt winter coal mine disaster with support amounting to US$285,000, and victims of the Kermanshah province earthquake with US$150,000. In cooperation with the Imam Khomeini Relief foundation, the company also launched the Gol Gohar garment supply centre to provide jobs for women who are head of households.
5. Sanitation: Gol Gohar has made significant contributions to providing sanitation facilities in the region. It played a key role in setting up the municipal wastewater organisation and participated with other companies in the area to establish a wastewater network. The company also partnered in the construction of the Sirjan drinking water treatment plant.

[12]"Take a look at the list of Gol Gohar's investments in social responsibility; The Significant Role of Gohar Gohar in Industrial and Social Development of Region and Country" Report Published by Gohar Press, 2018.

6. Cultural and Social infrastructures: the organisation carried out various cultural measures, such as helping with construction in Masjid, restoring and upgrading the only theatre hall in Sirjan, upgrading a care centre for people with disabilities, active participation in festivals, conferences and various cultural events, as well as supporting Sirjan's scientific community.
7. Environmental Protection: the company runs several schemes to reduce its environmental damage. The most important are the pelletizing plant degassing project; the production of ammonium sulphate fertilizer at a rate of 200,000 tons per year; dewatering of tailings in two phases (which results in a water return of 200 litres per second); and creating a green belt of 5000 hectares around Gol Gohar.

2.2 Chador Malu Mining and Industrial Company

Chador Malu Mining and Industrial Company started operating in 1992 to explore, extract and exploit Chador Malu iron ore mines located in Yazd Province. The company also produces concentrate, pellet and granulated iron ore (coarse-grained and fine-grained) and apatite concentrate. This company works extensively in the field of CSR, collaborating with various institutions and organisations. According to the company's statistics, it has invested more than US$1 billion so far in the city of Ardakan, and employed more than 8000 people. This accounts for 98% of the native workers, as well as creating 35,000 indirect jobs.[13] It has thus become one of the most influential industrial units in Yazd Province, undeniably improving the quality of life for its citizens. In addition, the company has taken a proactive approach towards environmental protection, including a recently signed agreement with the Department of Environment to take part in plans to protect a rare species of Asian cheetah.

Some other significant initiatives are[14]:

1. Environmental Measures: the company undertakes multiple actions to protect the environment and reduce the impact of its contaminating industrial activities as follows:

   - Planting more than 300 hectares of fruit orchards in Chadermo Industrial Complex in Ardakan, Yazd;
   - 218 hectares of arboriculture of resistant species in desert conditions in the Chadermo Industrial Complex;
   - Using dust collector equipment and bag filters in the Ardakan pelletizing plant to reduce environmental pollutants;
   - Creating a waterfall for animals and wildlife in the desert;
   - Contributing to the provision of park rangers to the area;

[13] http://www.chadormalu.com/.

[14] Ibid.

- Constructing buildings for the park rangers of the Fig Valley (Darre Anjir) wildlife refuge;
- Buying and installing equipment to run an air pollution monitoring station in Ardakan;
- Use of a wet system in iron ore processing plants to reduce dust particles entering the environment; and
- The installation of auxiliary equipment to prevent the emission of noise, air, sewage and industrial pollutants from the pelletizing plant.

2. Health and Sanitations Infrastructures:

- Construction of a sewage treatment plant in the city of Ardakan;
- Industrial water supply from domestic wastewater;
- Implementation of household waste separation program;
- Controlling plant wastewater and environmental factors with periodic environmental monitoring; and
- Washing wastewater treatment tanks to control their sludge and to prevent air pollution and unpleasant sludge odour.

2.3 Sarcheshmeh Copper Complex

Sarcheshme copper mine, located in Rafsanjan county in Kerman province, is the second largest copper reserve in the world, and the complex is the biggest in Iran. This mine is run by the National Iranian Copper Industries Company (NICICO), which takes a progressive approach towards CSR, focusing on cultural and research activities.[15]

The main cultural measures are:

3. Holding various art competitions in the region for photography, painting and Quran reading;
4. Providing digital applications to reduce paper consumption;
5. Donating books (more than 300 were given to the Rafsanjan prison library and medical centres last year);
6. Holding different art courses to teach copper craft skills. The complex provides this opportunity for local people to better their earning potential; and
7. The company has an agreement with Vali e Asr University to fund research projects.

The complex also provides great sport facilities in the region. Along with construction of the Sarcheshmeh Copper Sports Club in 1975, sports facilities were added for football, cricket, tennis, golf, volleyball, swimming and ping pong, and wrestling was also added after the revolution in 1979. There are currently four separate buildings under the supervision of 100 professional coaches, with 26 different sports provided in the men's section and 14 in the women's section.

[15]https://sarcheshmeh.nicico.com/.

2.4 Environmental Impact

Although these examples show the commitment of major companies to CSR, there are other examples that illustrate the neglect and inadvertency of CSR by many others. Micro-mines have both degraded nature and disrupted the rural lifecycle, leaving nothing left except lifeless bush. There has been severe harm to the environment, the ignoring of local requirements and conflicts with landholders. Some of the instances are reviewed here to identify the root of the problem and to point out a range of causes:

1. The protected area of Mount Beyrami, with its rich diversity of flora and fauna, and numerous deep gorges and springs, sits at the highest elevation of Bushehr province. The variable altitude of the area has created three distinct climatic segments, making it a unique and rare natural phenomenon. This mountain also supplies drinking water to several cities, both large and small, in the region. The Government issued a marble mining permit which is still the subject of discussion, while mining-related activities, such as road construction, were prevented by park rangers from the Department of Environment. Environmental activists who oppose the mining project believe that the proposed operations will harm the unique ecosystem that plays a cultural and economic role in the lives of local people.[16]
2. When the permit was issued to exploit the Khonar silica mine, local people and officials tried to prevent it from going ahead. Khonar mine is located in a conservation area of Semnan province. The Governor of the city of Sorkhe indicated that only 5,000 hectares throughout the area remain intact, and mining permits have been issued all around, making Khonar the only pristine local area. Its wildlife has been protected by a hunting ban for 20 years. A number of local people gathered in the surroundings of the Khonar conservation area to protest the mining licensing, as they were concerned about the possibility of water contamination affecting neighbourhood villages.[17] As a result of the Department of Environment protests and public pressure, the permit was revoked.[18]

---

[16]https://www.irna.ir/news/83392064/%D9%BE%D9%88%DB%8C%D8%B4-%D9%86%D8%AC%D8%A7%D8%AA-%D8%AE%D9%88%D8%A7%D8%B3%D8%AA%D8%A7%D8%B1-%D9%84%D8%BA%D9%88-%D9%85%D8%AC%D9%88%D8%B2-%D9%85%D8%B9%D8%AF%D9%86-%D8%B3%D9%86%DA%AF-%D9%85%D8%B1%D9%85%D8%B1%DB%8C%D8%AA-%D8%AF%D8%B4%D8%AA%DB%8C-%D8%B4%D8%AF%D9%86%D8%AF, Published on 07/13/2019.

[17]https://www.isna.ir/news/98081810916/%D8%AA%D8%AC%D9%85%D8%B9-%D8%A7%D8%B9%D8%AA%D8%B1%D8%A7%D8%B6%DB%8C-%D8%A8%D9%87-%D9%81%D8%B9%D8%A7%D9%84%DB%8C%D8%AA-%DB%8C%DA%A9-%D9%85%D8%B9%D8%AF%D9%86-%D8%AF%D8%B1-%D8%AA%D9%86%D9%87%D8%A7-%D9%85%D9%86%D8%B7%D9%82%D9%87-%D8%A8%DA%A9%D8%B1-%D8%A8%D8%A7%D9%82%DB%8C%D9%85%D8%A7%D9%86%D8%AF%D9%87-%D8%B3%D8%B1%D8%AE%D9%87, Published on 11/18/2019.

[18]https://www.doe.ir/.

3. A video of the Imam of Joshaghan Ghali village protesting the inauguration of a new mine in the mountains around the city was posted on social media and went viral. The Imam threatened officials that he would close Friday prayers if the mine began operation. Joshaghan Ghali is a village on the hillside of Karkas Mountain in Isfahan province. Mining activities have been going on for several years, but recently came very close to the urban area, and the nature of the surrounding area has been extremely affected. Some village youngsters edited a documentary where they illustrated the problems that the miners' profitability had brought to the city's natural resources and environment. Mining had badly damaged the local economy, as the locals are mostly ranchers who raise livestock. Rangeland restoration regulations that prevent damage of the natural environment were disregarded in the area and resulted in livestock permits in the region being revoked. The film showed the impact on livestock production, the failure to meet requirements for criminalisation of environmental degradation, the lack of enforcement of environmental laws, the reason for supporting wholesale raw materials, the signs of mining development activities and the evidence of over-limit extractions.[19]
4. Even areas near and close to Iran's capital, Tehran, have not been protected from harmful mining operations. Shemiranat, which has the best countryside near the city, suffers from the effects of mining activities. Mountain damage, noise pollution, soil erosion, environmental pollution and unsafe roads leading to mines are among some of the destructive impacts. The local economy is dependent on tourism; in the past, more than 10,000 visitors came to the area, but these numbers are rapidly declining.[20]

The review of these examples are consequential since they illustrate the most important practices and themes of CSR in Iran. In fact, it shows that Iran has significant distance with the modern concept of CSR identified legally as an independent kind of responsibility alongside with civil and criminal liability.[21] Moreover, current Iranian CSR has not been institutionalised at the community level, for individuals and companies, and it is not seen as a targeted interaction between the company and the community which is consciously established to operate as an independent social system. Besides, the problem of lack of transparency as well as limited access to the information exists among Iranian companies which makes a serious barrier in the way of development of CSR concept in Iran. Companies perform CSR as a right

[19]https://snn.ir/fa/news/788583/%D8%B5%D8%AF%D8%A7%DB%8C-%D8%A7%D8%B9%D8%AA%D8%B1%D8%A7%D8%B6-%D9%85%D8%B1%D8%AF%D9%85-%D8%AC%D9%88%D8%B4%D9%82%D8%A7%D9%86-%D9%82%D8%A7%D9%84%DB%8C-%D8%B1%D8%A7-%D8%A8%D8%B4%D9%86%D9%88%DB%8C%D8%AF-%D8%AA%D8%AE%D8%B1%DB%8C%D8%A8-%D9%85%D8%AD%DB%8C%D8%B7-%D8%B2%DB%8C%D8%B3%D8%AA-%D9%88-%D8%A8%DB%8C%E2%80%8C%D8%A7%D8%B9%D8%AA%D9%86%D8%A7%DB%8C%DB%8C-%D8%B4%D8%B1%DA%A9%D8%AA%E2%80%8C%D9%87%D8%A7%DB%8C-%D9%85%D8%B9%D8%AF%D9%86%DB%8C, Published on 09/03/2019.

[20]Kianoushrad (2018.

[21]Nouri and Saeed (2015).

thing to do to compensate their damage or in the best case to share some small part of their benefits with the people who are their host and give them their resources.[22] while the concept of opportunity for better life or access to better living conditions for people is not usually considered as the main purpose of CSR. Thus, the things that are performed as CSR by companies have been kept their traditional essence of moral-religious origin which cannot fulfil the criteria of the modern legal CSR concept. The important characteristics of this historical content are that these activities have been performed as benevolent activities without any obligation and most of them have been carried out in secret. These features are still relevant today and this is why talking about CSR as an independent legal institution in Iran is not justified. Although the instances above have various aspects including environmental, cultural and social, the traits of these activities are the same as traditional ones.

3 Legislation and Regulations Relevant to CSR in Iran

Corporate Social Responsibility might at first seem to be a belief system, or even a beneficial practice to ensure a humanitarian approach is combined with profitable practices. But it is not difficult to trace the disastrous impacts of activities by giant companies in underdeveloped communities, such as child labour in Indonesia and East Timor.[23] In the modern profit-driven era, CSR practices need to be enforced by mechanisms to secure their execution, and links to many legal areas exist, including international, corporate and contract law. Regulatory measures within each domain of governance significantly contribute to the development of CSR. In this section, the most important legal aspects of CRS will be examined in the national context of Iran:

3.1 Guidelines to report CSR according to GRI standard published by Ministry of Petroleum[24]

The purpose of this guideline is to:

1. Strengthen a properly functioning infrastructure through relevant executive organisations in populated centres, such as cities and villages close to operation sites;
2. Provide opportunities to increase employees' interests to work and live close to the operation sites and to grow healthy interactions with nearby residents;

[22]Bazrafkan and Taghi (2018).

[23]https://unevoc.unesco.org/fileadmin/user_upload/pubs/AB4_ChildLabour.pdf

[24]https://csr.mop.ir/portal/file/?301400/https://csr.mop.ir/portal/file/?301400/راهنماي-گزارش-دهي-مسئوليت-اجتماعي-شركت.pdf.

3. Shape or support sustainable livelihood of local communities;
4. Contextualize cooperation between people, governing body and the oil industry;
5. Use the potential that active NGOs bring in various realms such as social, cultural and environmental activities;
6. Prevent exploitation of local resources and stop taking advantage of local communities;
7. Provide a context for constructive interaction between local and non-local subcultures;
8. Maintain natural resources and vulnerable species; and finally,
9. Attract foreign investment in the oil industry.

These seemingly promising goals lack specification and clarity regarding significant issues such as human rights and the inequality of power. Although these guidelines are a good first step in ensuring equality and protection of people and planet, they are not nearly sufficient. The execution of such policies in Iran are still vague, overlooked and complicated. However, based on evidence from the oil and gas sector, many changes have taken place over the last decade in terms of acknowledging and even using the capabilities of CSR to meet increasing business expectations.

3.2 CSR Manuals Published by Ministry of Petroleum[25]

As global demand for fossil fuels increases, so does the investment in oil and gas exploration and production facilities. The Ministry of Petroleum published four manuals aimed at the four main principal companies affiliated to the Ministry of Petroleum: National Iranian Gas Company (NIGC), National Iranian Oil Company (NIOC), National Iranian Petrochemical Company (NPC) and National Iranian Oil Refining and Distribution Company (NIORDC).

A map of Iran's oil and gas reserves shows that most of these sites are located in rural areas with inadequate civil development and pre-established infrastructure. These manuals are designed for the specifics discussed at the councils of Ministry of Petroleum with respect to the environmental effects of the oil industry. They aim to influence public thought and build public trust to better manage interaction with local communities and to improve work quality in the oil sector.

Nine articles are covered in the four manuals. The general language does not take the specifications of each site into consideration, and no concrete and clear steps are provided on how to ensure the mentioned goals are to be realised. Analysing the first article, we see that only a vague definition of local communities and empowerment is given. For instance, Article 1.1 defines local communities as "people with distinct identity who reside in various sites where oil industry's activities take place and whose actions have direct or indirect impact on oil industry's activities". It is unclear what characterises a distinct identity, or to what extent and how their activities might

[25] https://csr.mop.ir/portal/home/?generaltext/287500/287987/301781/.

or might not influence the oil industry. Or, how we can distinguish between direct and indirect effects.

Aside from the vague definitions briefly laid out in the first article, others are also conveyed in ambiguous and unclear language. For instance, Article 7 requires all relevant corporations to report their method of CSR to the Supreme Council of Ministry of Petroleum, the CSR section. It turns out there is no such distinctive sector of the ministry, but is instead the same Council of the Ministry of Petroleum whose members include the CEOs of the four principal companies. It is clear that serious measures to ensure the quality of these reports have not taken place. But this circular structure is not obvious until we take a closer look at the underlying and potential interpretations of the articles.

These manuals may indeed provide something beneficial; as examples in the previous section revealed, we have witnessed considerable improvement in shaping new infrastructures and civil developments through CSR in Iran's oil and gas industry. But CSR has a long way to go to advance sustainable consumption and production.

3.3 The Relationship Between Tax Laws and CSR

The relationship between corporate social responsibility and tax laws is significant in order to understand the function of CSR in Iran's oil and gas industry. More specifically, the relationship between CSR and possible harmful tax practices—that create problems in the execution of proper tax laws—should be discussed. Iran's tax system has numerous issues that would require a lengthy discussion irrelevant to the scope of this chapter, so the focus here is to examine whether the corporate tax regime is aligned with the requirements of CSR in the oil and gas industry.

According to the 2018 Global Oil and Gas Tax Guide, the standard corporate income tax rate for resident companies is 25%.[26] To contextualize the importance of the relationship between tax laws and CSR, the specificities of oil and gas companies should be taken into account. This industry deals with several kinds of hazards, including environmental, safety, health-related and ultimately financial risk. The production process burdens society, and more directly imposes nearby communities with costs such as air pollution, oil spills and ultimately habitat degradation and destruction, known as "the oil or resource curse".[27] It is not surprising that the public has ambivalent feelings about energy development, and that these accomplishments are accompanied by some sort of legal mechanism to ensure their execution. While requiring CSR from gas and oil companies to be reasonable, we need to see how this requirement is related to tax laws.

---

[26] https://www.ey.com/Publication/vwLUAssets/ey-global-oil-and-gas-tax-guide/$FILE/ey-global-oil-and-gas-tax-guide.pdf.

[27] See e.g., Paul Collier, *Essay: Laws and Codes for the Resource Curse*, 11 YALE HUM. RTS. &DEV. L.J. 9 (2008).

There are two faces of the tax coin; it can be considered one of the many expenditures companies must pay, or it can be seen as an economic contribution to society. Companies like to minimise their costs, and if a consistent and meticulous system is in place where its business is conducted, then companies will try to maximise their corporate taxes.

A strict legal approach is not advocated here, but a consistent and open tax system suggested instead, with an apparatus in place to evaluate the execution of the principles of CSR within the parameters of each company's needs and plans. This is lacking in Iran's tax legislation. Despite efforts in 2000 to reform some of these laws, all the practical measures brought minimal, constructive influence in promoting CSR within the oil and gas sector. This was due to inconsiderate and negligent policy makers who took an inconsistent approach. Also, many projects that were supposed to be run and supported by people were now budgeted by the government, and became dependent on governance bodies. In addition, many cultural or charity-run organisations reviewed the Ministry of Intelligence guidelines which created many inconsistent and non-systemic limitations. Depending on the political party to which the Ministry's decision-makers at the time belonged, and their own belief system, they might agree or disagree with a business plan. Companies are therefore unable to clearly calculate, foresee and plan their programs.

3.4 The Mining Act

Another regulatory instance of attempting to acknowledge CSR in the oil and gas industry can be found in Chapter Three of the Mining Act which refers to Exploitation, and Note 5 of its Article 14 which says: "holders of exploitation permits who endeavour for optimal exploitation and conservation of mineral deposits, enhancing productivity, research, development and exploration and preservation of environment in their mines shall be exempted up to twenty per cent (20%) from payment of royalty upon confirmation of the High Mine Council".[28]

This note attempts to take a small step towards promoting CSR disciplines. But there are still internal obstacles to executing Note 5. The most noticeable is that it only pertains to environmental preservation, the most traditional, yet very important element of CSR. This Act is silent about non-environmental aspects.

Various Bylaws

As this chapter intends to give a fair picture of the CSR endeavours taking place in the context of Iran's oil and gas industry, it also addresses some other scattered bylaws that in one way or another contribute to inclusion of corporate social responsibility requirements.

[28] Iran's Mining Act, Accessible on http://en.tccim.ir/Content/media/image/2016/06/110_orig.pdf?t=636025349930009690.

3.4.1 The Agreement Between the Iranian Department of Environment and the National Iranian Association of Gas (NIA) 2016

This agreement was made to benefit the nation by providing sustainable economic systems, increase job opportunities and remove obstacles for employment with the motto of green exploitation. Under it, the NIA is responsible to provide a thorough estimation of the costs it will impose on local communities in sites where they conduct business and evaluate the social benefits as well as stakeholders' profit.

3.4.2 Outlook of Iranian Mines and Mining Industries Development and Renovation Organisation (IMIDRO)[29]

This document intended to convey the government's outlook of sustainable development in the mining industry. It has nine statements that are too general, vague and unconstructive to even be evaluated. For instance, the document's first statement wishes to have: "a developed country in 2026" or the ninth state writes: "entrepreneur and having sustainable employment".

There are, of course, more policies, agreements or documents that can be interpreted as an endeavour to include CSR principles, but none are close to being sufficient, let alone the fact that corporations in developed countries go beyond CSR to address local and societal concerns as they conduct their business.

4 Conclusion

This chapter examined the extent to which CSR principles were applied in the context of Iran's oil and gas industry. Studying various companies revealed that setting aside regional livelihoods and environmental concerns—and despite various limitations—some attempts have been successful in providing services for local affected communities, and even those at large. The Gol Gohar Mining and Industrial Company, for instance, runs several projects, from environmental protection programs to developing cultural and social infrastructure, to building the Gol Gohar Football Club and Gol Gohar Hospital. Chador Malu Mining and Industrial Company and Sarcheshmeh Copper Complex are also examples of corporations who have successfully implemented CSR requirements into their business practices.

Not all established firms have been successful or even considered their social responsibility. Overlooking the duty owed to local communities and the environment in the protected areas of Mount Beyrami, Khonar Silica Mine, and Shemiranat has resulted in catastrophic and harmful consequences, as well as devastating costs for society at large. Mediation by the Department of Environment and consistent protests by local communities in Khonar were able to revoke mining permits.

[29] Accessible on http://www.imidro.gov.ir/general_content/157- چشم-انداز.html.

After studying examples on both sides of Iran's Mining industry spectrum, a closer look at existing regulations, laws and policies was taken to see what might encourage the inclusion of CSR. Although several legislations and regulations that could potentially promote corporate social responsibility were found, they are mostly ignored in practice. Various reasons were listed as to why legal procedures have been unsuccessful in robustly promoting CSR. The wording was vague and ambiguous, such as in the guidelines to report CSR according to GRI standards (published by Ministry of Petroleum); CSR Manuals published by the Ministry of Petroleum; the limited and restrained requirements in The Mining Act; the lack of specifics by the Outlook of Iranian Mines and Mining Industries Development and Renovation Organisation (IMIDRO); and insufficient clarity and inconsistency within the tax laws.

Legal as well as practical attempts have unsuccessfully addressed the requirements of CSR policies in Iran. Unlike developed countries, the Iranian business scene has not established corporate social responsibility as a matter of mandatory practice, and extra-legal measures might be helpful in rectifying this. Perhaps a widening range of control methods set by the public sector, rather than the political and legal domain, can bring about a new corporate responsibility. This chapter concludes by considering the possibility that institutionalised corporate social accountability could possibly be achieved by Iran's collective civil society, outside the sphere of governmental or business sectors.

References

Bazrafkan, M. I., & Taghi, M. (2018). A critical study of corporate social responsibility and development (an ethnographic study in Assaluyeh region). *Social Development Quarterly*, 11(4), 37–70.

Chapardar, H., & Khanlari, R. (2011). Iranian corporations and corporate social responsibility: An overview to adoption of CSR themes. *SAGE Open* 1–15, *1*(3), 2.

Kianoushrad, S. (2018, September 23). Destruction of the Mountain in favor of the Mine. *Hamshahri Newsletter*.

Modabber, F. (2013). Implementing corporate social responsibility policies in a developing country—A study of Iran, Durham theses, Durham University. Available at Durham E-Theses Online: http://etheses.dur.ac.uk/8457/.

Omidvar, A. (2008). Government policy in promoting corporate social responsibility. *Journal of Corporate Social Responsibility*. (Strategic Research Center).

Omidvar, A. (2011, August 19). Corporate social responsibility of companies in Iran. *Donyaye Eghtesad* Newsletter, No. 2485.

Nouri, & Saeed, J. J. (2015). The concept of corporate social responsibility in the Laws of Germany, France, England and Iran: with a Comparative Approach. *Comparative Law Studies, 5*(2), 603–627.

Yoshanloey, & Saeed, J. J. (2013). The concept of corporate social responsibility in the Laws of Germany, France, England and Iran with a comparative approach. *Comparative Law Studies Journal, 5*(2), 603–627.

The Approach to Corporate Social Responsibility in the Extractive Industries in Trinidad and Tobago and Guyana

Alicia Elias-Roberts

Abstract This chapter outlines the legal framework of corporate law and the approach to CSR in Guyana and Trinidad and Tobago. It discusses ExxonMobil Guyana's general approach towards CSR. It also discusses the worst experience with community involvement in the extractive resources industries in Guyana and one of the best experiences of CSR in the Caribbean, namely Trinidad and Tobago's Mayaro Initiative for Private Enterprises Development (MIPED) programme. Additionally, an analysis of the main challenges that the extractive resources industries are facing with respect to CSR practices is highlighted. The conclusion analyses the compatibility of Guyana's Green State Development Strategy and sustainable development under the CSR and evaluates Trinidad and Tobago's NSCSRP.

Keywords Corporate social responsibilities · Trinidad and Tobago · Guyana · Green state · Oil and gas law

1 Introduction

Corporate social responsibility (CSR) can be conceptualized as a mechanism through which companies align with and shape sustainable development in a country where they are doing business.[1] In recent years, the Government of Guyana developed a *Green State Development Strategy: Vision 2040*,[2] which has the potential to significantly affect CSR and investments in the private sector in that country. The *Green*

[1] Shah et al. (2016) at 237.

[2] Government of the Cooperative Republic of Guyana (2019).

[3] Republic of Trinidad and Tobago, Ministry of Trade, Industry, Investment and Communications (2015).

A. Elias-Roberts (✉)
Faculty of Law, University of the West Indies, St. Augustine, Trinidad and Tobago
e-mail: aeliasroberts@gmail.com

E. G. Pereira et al. (eds.), *Sovereign Wealth Funds, Local Content Policies and CSR*, CSR, Sustainability, Ethics & Governance,
https://doi.org/10.1007/978-3-030-56092-8_30

State Development Strategy: Vision 2040 outlines an approach to CSR, which incorporates sustainable development. The Government of Trinidad and Tobago has also taken a step in the right direction and developed a National Strategic Corporate Social Responsibility Policy (NSCSRP), which serves as a facilitative framework to encourage Strategic CSR practices in Trinidad and Tobago.[3] The NSCSRP outlines the major national priorities or focal areas for CSR and identifies strategies that companies can adopt to address these gaps in Trinidad and Tobago. While Guyana has a general policy document and Trinidad and Tobago has a more specialized policy document concerning CSR, the current legal framework in both jurisdictions does not incorporate sustainable development as an element for investors or directors of companies to consider when discharging their duties.

This chapter will focus on the petroleum extractive industries in both Guyana and Trinidad and Tobago. Guyana has over 60 years of experience in extractive industries,[4] with experience in bauxite, stone, sand, gold and diamond mining. However, Guyana is a new frontier oil and gas country.[5] On 15 May 2015, the ExxonMobil operator found oil at a depth of 18,730 feet (5700 feet of water), 120 miles off the coast of Guyana and the first major announcement of a significant discovery of oil was made in Guyana. The well, named Liza 1, has since been followed by several other successful discoveries. The initial reports stated that the discovery had about 1.8 billion barrels of high-quality crude oil.[6] However, the total recoverable gross resources for the block as of August 2019 is estimated to be approximately 10 billion barrels of oil equivalent.[7] All the successful wells in the Stabroek deep water exploration block contain recoverable high-quality or 'sweet' crude, which fetches a premium because it is less costly to refine.[8] While Guyana has great potential to become a major petroleum producer in the Caribbean region, Trinidad and Tobago is currently the largest oil and natural gas producer of the Caribbean islands and territories. The first discovery of oil deposits occurred in 1867 and since then Trinidad and Tobago has established a good record in offshore commercial oil and

---

[4]GYEITI (2019).

[5]'ExxonMobil Announces Significant Oil Discovery Offshore Guyana', press release, 20 May 2015 at https://news.exxonmobil.com/press-release/exxonmobil-announces-significant-oil-discovery-offshore-guyana accessed 19 February 2019.

[6]See 'More Oil Found Offshore Guyana' *Guyana Times* (26 July 2017) http://guyanatimesgy.com/more-oil-found-offshore-guyana/ accessed 19 February 2019; Brooks and Schipani (2017).

[7]Myers (2018).

[8]'More Oil Found Offshore Guyana' *Guyana Times, supra* n. 6.

gas.[9] Apart from the petroleum sector Trinidad's economy includes mineral extraction, agriculture, tourism and services. However, the petroleum industry remains the main economic driver of the country.

In this chapter, I will begin by outlining the legal framework of corporate law and the approach to CSR in Guyana and Trinidad and Tobago. Next, I will critically discuss ExxonMobil Guyana's general approach towards CSR. ExxonMobil Guyana is the main oil company operating in Guyana. This will be followed by what I consider to be the worst experience with community involvement in the extractive resources industries in Guyana. I will then discuss one of the best experiences of CSR in the Caribbean, namely Trinidad and Tobago's Mayaro Initiative for Private Enterprises Development (MIPED) programme. Next, an analysis of the main challenges that the extractive resources industries are facing with respect to CSR practices will be highlighted. The conclusion will analyse the compatibility of Guyana's Green State Development Strategy and sustainable development under the CSR and evaluate Trinidad and Tobago's NSCSRP.

2 Development of Corporate Law and the Approach to CSR in Guyana and Trinidad and Tobago

Up until the 1980s, company law in Guyana and Trinidad and Tobago, as well as the rest of English-speaking Commonwealth Caribbean[10] evolved concomitantly with the UK company law. The various Companies Acts in the region up until this time represented a replica of the corresponding UK statute. Momentum for change began in 1971 when the 8th Meeting of the Council of Ministers of the Caribbean Free Trade Association (CARIFTA) resolved to initiate a project for the reform and harmonization of the aging company laws within the Member States. A Working Party was created to execute the project and was reconstituted under the Caribbean

[9]The cumulative production is said to total over 3 billion barrels of oil. Pawan G Patil, John Virdin, Sylvia Michele Diez, Julian Roberts, Asha Singh, 'Toward A Blue Economy: A Promise for Sustainable Growth in the Caribbean; An Overview' (Report No: AUS16344 The World Bank 2016) 70. See also, The Oil & Gas Year Report, 'Trinidad and Tobago 2019: The Oil & Gas Year: Information is power,' at https://www.theoilandgasyear.com/market/trinidad-and-tobago-2/ accessed 19 February 2019; and see Government of the Republic of Trinidad and Tobago Ministry of Energy and Energy Industries, 'Historical facts on the Petroleum Industry of Trinidad and Tobago' http://www.energy.gov.tt/historical-facts-petroleum/ accessed 19 February 2019.

[10]The Commonwealth Caribbean includes the following independent island-nations; Antigua and Barbuda, The Bahamas, Barbados, Dominica, Grenada, Jamaica, Saint Kitts and Nevis, Saint Lucia, Saint Vincent and the Grenadines, and Trinidad and Tobago; the following mainland countries: Belize, once known as British Honduras and Guyana; and the following British Overseas Territories: Anguilla, Bermuda, British Virgin Islands, Cayman Islands, Montserrat and the Turks and Caicos Islands.

Community (CARICOM).[11] Out of the Working Party was the CARICOM Report on the Harmonization of Company Law, which was published in 1979 along with a model legislation incorporating all the reforms proposed in a report submitted by the group.

Barbados was the first to use the CARICOM model Companies Act, along with the Canadian Dickerson Report of 1971[12] and the (Draft) Canada *Business Corporations Act* (CBCA) 1971 to draft and enact the Barbadian *Companies Act* 1985.[13] The Barbados Act was to be the first of the new generation of Companies Acts in the Commonwealth Caribbean region which were, for the first time, no longer based on the UK legislation but instead upon the more progressive Canadian legislation. The new Act introduced a number of changes and new concepts to Caribbean company law. Amongst these were: the incorporation of a company by a single person was now permitted; there was no longer any requirement to file multiple constitutive documents, such as articles and memoranda of association, instead, incorporation was effected by simply filing 'articles of incorporation'; shares could be issued with no par value; a company now automatically had all the powers and capacity of a natural person and there was no more need to specifically list the object and powers of the company; duties and liabilities of directors were expressly stated; and greater rights and more effective remedies given to minority shareholders.

The reforms created by the Barbados legislation proved popular in the Caribbean and was followed a few years later by similar pieces of legislation in Guyana[14] in 1991 and Trinidad and Tobago in 1995.[15] The development of Guyana's corporate law adopted the principle that directors have a duty of care and in exercising these powers and discharging duty he must think about what is in the best interest of the company, the company's employees and its shareholders. To be precise, the Act provides:

> (1) Every director and officer of a company in exercising his powers and discharging his duties must -...
>
> (2) in determining what are the best interest of the company, a director must have regard to the interests of the company's employees in general as well as the interests of its shareholders.[16]

A similar provision is provided for the Trinidad and Tobago's *Companies Act*, 1995, Section 99. This statutory position is a little more extensive than the common law position, which provides that a director owes duties to the company alone. This company law principle is part of the common law and was enunciated over 100 years

---

[11] See *Revised Treaty of Chaguaramas establishing the Caribbean Community including the CARICOM Single Market and Economy*, 2259 UNTS 293 (Nassau, Bahamas 05/07/2001); and see also, *CARICOM Caribbean Community: Who We Are,* https://caricom.org/about-caricom/who-we-are, accessed 14 August 2019.

[12] Dickerson et al. (1971).

[13] *Companies Act*, Cap 308 (1985) Law of Barbados.

[14] *Companies Act* (no. 29 of 1991), Cap. 89:01, Laws of Guyana.

[15] *Companies Act* (no. 35 of 1995), Cap. 81:01, Laws of Trinidad and Tobago.

[16] *Companies Act* (Guyana) *supra* n. 12, s. 96(2).

ago in the English case of *Percival v Wright*.[17] The Canadian Supreme Court also followed this principle in *BCE Inc. v 1976 Debentureholders* where the court said:

> The fiduciary duty of the directors to the corporation originated in the common law. It is a duty to act in the best interests of the corporation. Often the interests of shareholders and stakeholders are coextensive with the interests of the corporation. But if they conflict, the directors' duty is clear – it is to the corporation [as stated in] *People's Department Stores*.[18]

The impact of this judgment and the common law position established is that directors do not owe a duty to any individual member or the shareholders or any other party other than the company alone. However, the wide interpretation given to the meaning of the 'interests of the corporation' in the Canadian case is that stakeholders and shareholders were included. In contrast to this position, the statutory corporate law position in Guyana and Trinidad and Tobago is that directors should have regard to certain listed categories of persons only, namely company employees, shareholders as well as what is in the best interest of the company, when discharging their duties.

The old common law position emanating from *Percival v Wright* is akin to the shareholder primacy theory. This theory underpins US corporate jurisprudence, as seen in cases such as *Dodge v Ford Motor Co.*[19] The shareholder primacy theory advances that the corporation's business decisions should be made with a view to benefit the shareholders only, to the exclusion of all other stakeholders. This is a reasonable contention if the company is seen principally as nothing more than a vehicle to provide a return on the capital invested by its members. Milton Friedman in *Capitalism and Freedom*[20] supported this view when he opined: 'There is one and only one social responsibility of business – to use its resources and engage in activities designed to increase its profits so long as it stays within the rules of the game…' According to Van der Weide this theory also has an economic basis. He submits that the shareholder value theory allows the company to achieve operational efficiencies and generate greater returns for its equity investors.[21] However, this can also have the negative effect of putting pressure on directors to achieve short-term gains to satisfy shareholders at the expense of the company's long-term, wider interests.

In contrast to the shareholder primacy theory, some jurisdictions, including Canada, have developed a stakeholder approach. In *People's Department Stores Inc. (Trustees of) v Wise*[22] the Canadian Supreme Court enumerated a list of categories of persons for directors to be mindful of while discharging their duties to their company. As stated by Major and Deschamps JJ:

> We accept as an accurate statement of law that in determining whether they are acting with a view to the best interests of the corporation it may be legitimate, given all the circumstances

[17] [1902] 2 Ch 421.

[18] [2008] 3 SCR 560 at para 37.

[19] 170 MW 668 (Mich 1919).

[20] Friedman (1962).

[21] Van Der Weide (1996).

[22] [2004] SCR 461 SCC, para. 42.

> of a given case, for the board of directors to consider, *inter alia*, the interests of shareholders, employees, suppliers, creditors, consumers, governments and the environment.[23]

This list of various categories of persons and things that should be considered when the director exercises his duty of care was also confirmed by the Supreme Court in 2009 in the case of *BCE Inc. v 1976 Debentureholders*.[24] It is also notable that the Supreme Court prefaced this list with the words '*inter alia*'. Therefore, this indicates that this is not an exhaustive list and the set of parties mentioned therein is not intended to be closed but may be further expanded upon in subsequent cases. In this way, the Canadian courts can be said to embrace the concept of stakeholder primacy. This is a theory that places equal importance on the interests of other stakeholders in the company, as is placed on the interests of the shareholders. The theory has its origins in Professor R. Edward Freeman, a Professor of Business Administration at the Darden School of Business at University of Virginia, USA. Professor Freeman wrote a book, *Strategic Management: A Stakeholder Approach,*[25] in which he defines a stakeholder as 'any group or individual who can affect or is affected by the achievement of the firm's objectives'.[26] He lists a variety of constituencies such as owners, employees, suppliers, government, consumer and other activist groups, trade unions and associations, political groups and competitors who are stakeholders. According to Taylor, in so doing, Professor Freeman confirms the status of the corporation as an essential part of society and the national economy.[27]

A third approach, which is like a compromise between the stakeholder and shareholder primacy theories, is the enlightened shareholder theory. This theory was adopted in the UK *Companies Act* 2006[28] on the recommendation of the UK's Department of Trade and Industry's (DTI) Company Law Reform (CLR), instead of taking a more pluralistic approach such as adopting the stakeholder primacy view, for instance.[29] One reason for this choice was that it was felt that the old common law position risks leaving directors accountable to no one, since there is no clear yardstick for judging their performance.[30] Section 172 (1) (a) *Companies Act* 2006 (UK) explicitly requires directors to consider a long-term view of the company, rather than focusing on achieving the short-term gains to appease shareholders. The section reads as follows:

> A director of a company must act in the way he considers, in good faith, would be most likely to promote the success of the company for the benefit of its members as a whole, and in doing so have regard (amongst other matters) to—

---

[23] Major and Deschamps JJ at para 42.

[24] [2008] 3 SCR 560, at para 39.

[25] Edward Freeman (1984).

[26] Ibid., 25.

[27] Taylor (2010).

[28] *Companies Act* 2006, Cap. 46 (United Kingdom), http://www.legislation.gov.uk/ukpga/2006/46/contents, accessed 5 August 2019.

[29] See Company Law Reform Bill—White Paper 2005, para. 3.3; CLR, *Modern Company Law for a Competitive Economy: A Strategic Framework*, at para 5.

[30] UK Committee on Corporate Governance, *Final Report*, para 1.17; See also: [1903] 2 Ch 506.

(a) the likely consequences of any decision in the long term,
(b) the interests of the company's employees,
(c) the need to foster company's business relationships with suppliers, customers and others,
(d) the impact of the company's operations on the community and the environment,
(e) the desirability of the company maintaining a reputation for high standards of business conduct, and
(f) the need to act fairly as between members of the company.[31]

There are a few key principles that can be distilled about the 'enlightened shareholder approach'. In particular, the duty is on the directors to 'promote the success of the company for the benefit of its members as a whole'. This phrase was very carefully and deliberately worded so that the 'company' would be the subject of the sentence, and its 'members' would be the object. According to the Company Law Reform this was done because '[w]e believe there is value in inserting a reference to the success of the company, since what is in view is not the individual interests of members, but their interests as members of an association with the purposes and the mutual arrangements embodied in the constitution'.[32] According to Sealy and Worthington, this means that the primacy of the company is paramount. The focus is first and foremost on the company, and if the interests of the company as a separate entity are in conflict with the interests of the members as a whole, or at least some of them, it would appear that the interests of the company should be preferred.[33] Therefore, while the UK law requires the company to have regard to other interests such as employees, business relationships, community and the environment, amongst others, unlike the stakeholder theory, in the enlightened shareholder view, these interests are secondary to the company and its members.

Section 172 (1) of the UK *Companies Act* 2006 codifies the 'enlightened shareholder approach' as previously explained. This is essentially a compromise between the shareholder primacy theory and the stakeholder approach. The reference to director discharging their duties and taking into consideration 'the impact of the company's operations on the community and the environment' is a welcomed addition to incorporating sustainable development goals as part of the corporate jurisprudence in the UK. There is no similar provision in the *Companies Acts* in Guyana and Trinidad and Tobago. It is respectfully suggested that *Companies Acts* in both jurisdictions should be amended to incorporate the enlightened shareholder approach as formulated in the UK statute. This is a preferable position to that which currently obtains in Guyana and Trinidad and in which the legal framework provides that the directors are to have regard only to the interests of the shareholders and employers.

---

[31] Section 172 (1) (a) *Companies Act* 2006 (UK), *supra* n. 28.

[32] White Paper, Company Law Reform (Cm6456, 2005) para 3.3; CLR, *Modern Company law for a Competitive Economy: Developing the Framework* (URN 00/656), para 3.51; See also, Monk (2000).

[33] Sealy and Worthington (2008).

3 ExxonMobil Guyana's Approach Towards CSR

While the list of major companies operating in Guyana includes several banks, beverage companies, manufactures, gold and diamond business, etc., ExxonMobil Guyana, a new entrant in the business scene, is predicated to soon become the largest company operating in Guyana. It is predicated that ExxonMobil Guyana's operations will transform the wealth of the country through revenues to be collected from their oil and gas exploration and development in Guyana. Hence, this major extractive company's approach to CSR is a worthy candidate for review.

According to the ExxonMobil Guyana website,[34] the company seeks to contribute to the social and economic progress of Guyana and the local communities because they are operating in the country. The company's CSR statement provides as follows:

> We believe that maintaining a fundamental respect for human rights, responsibly managing our impacts on communities and making valued social investments are integral to the success and sustainability of our business.
>
> We strive to establish meaningful relationships that benefit communities and the company for the long-term. Our focus areas include STEM Education; Youth, Women and Community Empowerment; and Environmental Sustainability.[35]

During the period 2018–2019, the company reports that over GYD$550 million in grants were given. This is equivalent to US$2.637 million dollars.[36] The contributions included: GYD$400 million given to Conservation International Guyana for a programme to advance Guyana's sustainable economy through education, research, sustainable management and conservation; GYD$120 million given to Iwokrama for the Centre's Science Programme; and GYD$31 million given to Volunteer Youth Corps for Science, Technology, Engineering and Mathematics (STEM) programmes, including robotics.

Other activities included: Over GYD$6 million given to the Civil Defence Commission's Voluntary Emergency Response Team (VERT) training, supporting volunteers from Regions 3, 4, 5, 6, 8, 9 and, 10; GYD$5 million given to Youth Challenge Guyana (YCG) supporting a programme to enhance the agricultural food production skills of at least 200 Guyanese in Region 4; and GYD$4 million given to STEM Guyana to support the implementation of a 'Reading & Robotics' Program in libraries in Regions 3, 4, 5, 6 and 9.

The above approach to CSR by ExxonMobil Guyana is in line with the 'enlightened shareholder approach' as previously explained. The company's statement regarding its contributions towards community and social responsibility makes reference to environmental sustainability and managing their impact on communities and making valued social investments. As the figures above demonstrate, the company has gone beyond what the law requires and what they are contractually obligated to

---

[34] ExxonMobil Guyana, 'Community and social responsibility in Guyana', https://corporate.exxonmobil.com/locations/guyana/community-and-social-responsibility-in-guyana, accessed 23 August 2019.

[35] Ibid.

[36] For currency conversion see https://www.xe.com/, accessed 23 August 2019.

do. This is a move in a positive direction that the investors and director are discharging their duties and taking into consideration the impact of the company's operations on the community and the environment. While it is commendable that the company has decided to take this approach, there is nothing binding them to continue with this as there is no legal requirement. Since there is no guarantee to ensure that the company continues with an enlightened shareholder approach there is the possibility that the investors can decide to take a different course of action if the price of oil declines. Hence, it is respectfully recommended that the Guyana *Companies Acts* should be amended to incorporate the enlightened shareholder approach as a binding legal requirement.

4 Trinidad and Tobago's Mayaro Initiative for Private Enterprises Development (MIPED) Programme

The Inter-American Development Bank (IDB) has constantly advocated for resource rich nations, such as Guyana and Trinidad and Tobago, to ensure oil companies provide ample compensation in every sphere of their operation. In a report released in 2018, the IDB said that Guyana must move quickly to put a national corporate social responsibility strategy in place for all companies operating in the oil and gas industry. In the eyes of the IDB, time is running out for Guyana to get this in order or else, it runs the risk of being cheated in this area.[37]

In the 2018 IDB report, it was noted that for CSR to be relevant they must address the most pertinent pillars of the economy. Generally, most of the money that international oil companies in the petroleum sector generates tend to repatriate, and because of this trend many have argued that they need to follow the Hartwick rule to enhance their reputation as good corporate citizens. The report referenced the study by Genasci and Pray who argued that 'building a town hall or health clinic in a mining community where the most apparent impact of the mining activity is large-scale environmental degradation or forced relocation of villages might be viewed as simply an attempt to buy off the community and create space to go on with business as usual'.[38] With regard to this issue, the IDB stressed that CSR must be relevant to the needs of a country's communities.

To further solidify the importance of CSR being relevant, the Bank gave the example of a successful case study from Trinidad and Tobago called the Mayaro Initiative for Private Enterprises Development (MIPED) programme. The programme was set up in 2002 in Mayaro, a small village located in the southeast coast of Trinidad. The Bank stated that this programme was envisioned by Trinidad's government and has helped to facilitate the growth of small businesses in that area. The Trinidadian government ensured that MIPED programme was started in 2003 by British Petroleum of Trinidad and Tobago (bpTT) in the Mayaro community after

[37]Wenner et al. (2018) and Wilburg (2018).

[38]Genasci and Pray (2008).

recognizing the challenges of small businesses and their access to finance. The IDB said that MIPED began operations with a TT$7 million-dollar investment from bpTT to develop Mayaro and its environs into a model community, creating self-sustaining employment, improving training and establishing business opportunities to help build self-esteem and improve the quality of life. According to the report, since its inception, the MIPED programme has created over 3000 jobs and at the time of reporting it has an asset base in excess of TT$70m.[39]

In 2015 the Government of Trinidad and Tobago released the National Strategic Corporate Social Responsibility Policy (NSCSRP). It is expected that the NSCSRP will assist in fostering a culture of Strategic CSR activities in Trinidad and Tobago.[40] In an earlier report on CSR Mapping in Trinidad and Tobago, which was prepared by the UNDP in collaboration with the then South Trinidad and Tobago Chamber of Industry and Commerce (STCIC) it was observed that the domestic landscape in Trinidad is littered with CSR activities; however the broader and far reaching impacts of these activities were viewed to be marginal and tended to be philanthropic in nature.

According to figures provided by UNDP and STCIC in the CSR Mapping Report (2007), for the period 2001–2006, 68 companies disclosed spending a combined total of TT$ 54 million on external social and environmental programmes. However, even though a large sum was reportedly spent on CSR initiatives, there were many issues that were identified, which continues to minimize the impact of CSR activities in the island. In 2012 UNDP released another report on CSR activities in Trinidad and Tobago, which found that there was little evidence of a substantial shift towards more strategic forms of CSR in 2012 since the 2007 report; that there was continuous strong focus on public relations and reputation management as the main drivers of CSR; and that unfortunately, there was evidence of poor reporting and disclosure practices on CSR activities. The issues identified in the UNDP's 2012 report served as an impetus for the Ministry of Trade, Industry, Investment and Communications to develop the NSCSRP to guide companies in Trinidad and Tobago to promote more Strategic CSR practices.

5 The Hartwick Rule

As stated above the IDB report also contends that CSR must be informed by the Hartwick rule. The Hartwick rule ensures that while oil companies are allowed to extract a country's resource, meaningful investment must be aligned to the development of a country's human and social development. The Hartwick rule is also intended to help companies to strengthen their legitimacy and, with the right type of intervention, foster greater genuine savings in host communities. Essentially, the Hartwick rule argues that the most important assets to which resource rents can be placed are human capital, produced capital and financial capital. If CSR activities

---

[39] IDB Report, *supra* n. 37 at 51.

[40] NSCSRP Report, *supra* n. 3 at 3.

focused on other areas, there is the possibility that it could impede the long-run growth potential of the host community.

For CSR to be relevant, it must address the most germane externality.[41] In this regard, the IDB notes that Christmas parties and hampers alongside Carnival programmes for children would not cut it. It said that CSR that does not seek to transform the country's human capital only compromises the growth of the communities within the country. The IDB said, 'CSR initiatives have often been conceived by the "helpers" in the air-conditioned offices of oil companies and consultancies rather than through ongoing participation with the beneficiaries; again, that approach follows the logic of CSR serving corporate objectives. Where oil companies have consulted local communities, the consultation exercises have usually been superficial and grossly inadequate'.[42]

While international oil companies in the energy sector will have to provide an adequate return to their head offices and shareholders, unfortunately, in many developing countries, this usually takes place while the host communities in which they operate are experiencing serious and severe environmental degradation and poverty of various forms including deprivations in human, physical and social capital.

It should be highlighted that the Hartwick rule emphasizes a replacement of the extracted natural capital with an appropriate intervention by the international oil companies to provide human, social and infrastructural capital in the host community. This is recommended because in many regards, as an economy's stock of physical capital declines, the only way to maintain the productive capacity is to replace the depleting natural resource capital (and in the case of hydrocarbons, non-renewable resources) with human capital and physical capital. Interestingly, one source cites that for an economy, the 'transformation of an exhaustible natural resource stock into a reproducible stock of capital manages to keep the level of production and thus consumption, constant'[43]

According to the IDB report, if the government is not skilful in developing a strong CSR strategy then, oil companies can end up aiming for 'soft targets' that can be seen as a form of hush money and may even help to precipitate a type of dependency mentality in some host communities of the country. It is said that this often takes the form of monetary donations to clubs, government entities, etc. In light of the foregoing, the Bank stressed, 'Meaningful CSR must be in sync with the core developmental needs of the country from a Hartwick perspective. In this regard, Guyana must put in place an appropriate national CSR strategy to promote the avenues through which large firms, both national and multinational, are guided to link their CSR strategy with the pertinent Hartwick rule needs of communities'.[44] This recommendation is strongly supported. It is critical for Guyana to urgently put in place an appropriate national CSR strategy. This will help to clarify the government's approach to CSR, and to identify the avenues through which international oil

[41] Frynas (2005).

[42] IDB Report, *supra* n. 38 at 52; See also, Frynas (2000), at 580.

[43] See Van der Ploeg (2008), at 7.

[44] IDB Report, *supra* n. 38 at 52.

companies as well as national companies, which are investing in the country, can operate and will be guided to link their CSR strategy with the pertinent Hartwick rule needs of the community. According to Article 28 of the 2016 Petroleum Agreement between the Government of Guyana and ExxonMobil, the Government and the Contractor shall establish a program of financial support for environmental and social projects to be funded by the Contractor. The CSR from ExxonMobil Guyana and its partners will amount to US$300,000 per calendar year, and any unspent funds will be taken over into the ensuing year.[45]

6 The Worst Experience with Community Involvement in the Extractive Resources in Guyana: Omai Gold Mine Environmental Disaster

The Omai gold mine was one of the largest gold mines in Guyana. The mine was located in the north-west of the country in Cuyuni-Mazaruni area. On 19 August 1995, the mining company had a major failure of its tailings system causing some 4.2 billion liters of cyanide-laced waste to flow into a tributary of the Essequibo River, the main water source in the country. The incident lasted over a period of five days and marine life in the river was essentially killed off. Eighty kilometres of the Essequibo River were declared an environmental disaster zone.[46] Operation at the mine was halted for several months while the spill was investigated, and the mine was subsequently forced to close. The principal mine owners were Cambior Inc., a Canadian based company; Golden Star Resources Inc., a company based in Colorado, USA; and the government of Guyana. Cambior owned 65%, Golden Star 30%, and the Government of Guyana 5% of the mine.

Attempts were made in Guyana and in Canada to sue the principal company. Some 23,000 victims of the spill filed suit in Quebec against Cambior. The Guyana case sought $2 billion in damages. These cases were dismissed in Canada in 1998, and in Guyana in 2002 and 2006.[47] In 1998, the Quebec Superior Court refused to take jurisdiction over the legal action in which it was alleged that Cambior was negligent with respect to the tailings dam collapse at its mine in Guyana. The Guyanese plaintiffs sought to have the Canadian courts exercise jurisdiction over alleged human rights abuses and environmental damages committed in Guyana. The tailings dam collapse allegedly contaminated the water supply of thousands of

[45]Petroleum Agreement between the Government of Guyana and Esso Exploration and Production Guyana Limited, CNOOC Nexen Petroleum Guyana Limited and Hess Guyana Exploration Limited (2016), Article 28.7.

[46]See Mineral Policy Institute (2014); see also Davidson (1995).

[47]*Recherches Internationales Quebec v. Cambior Inc.* [1998] Q.J. No. 2554 (S.C.J.); See also Business and Human Rights Resource Centre, *Cambior Lawsuit (re Guyana),* https://www.business-humanrights.org/fr/node/86220?page=1, accessed 27 August 2019.

Guyanese. While the Quebec Court determined that there were some connections to Quebec, it ultimately determined that Guyana was the appropriate forum.[48]

The cyanide spill affected loggers, indigenous villagers/native Indians, wildlife and fish along the Essequibo river. Those who live along the banks of the river rely on it for their drinking water and the river is renowned for its fishing. It was reported that the remedial action undertaken by Cambior included sending Canadian mining engineers who tried to build a new dam on the Essequibo river within a week after the accident to stem the flow. The company also sent out helicopters and foot patrols to distribute drinking water and warn indigenous villagers and others along the banks not to drink, fish or bathe in the affected rivers. While the government lifted the environmental alert on the river after one-week, indigenous villagers on the river were still using alternative water sources, at considerable inconvenience, seven years after the spill.[49]

This incident is one of the worst cases of CSR in Guyana's extractive industry experience. The remedial action taken by the company was woefully inadequate. Sadly, there was no CSR activities by Omai to align the companies' objectives and investment with broader goals towards sustainable development and development of the country's human and social development. Apart from the Omai environmental disaster incident, the Bai Shan Lin controversy is another case of poor CSR within the extractive industries in Guyana. It was reported in the local newspapers in Guyana that Bai Shan Lin, a large logging company operating in Guyana, was doing nothing more than capitalizing on an opportunity for its own corporate gain.[50] It was alleged that the company's operations were not 'socially, ethically, and environmentally' responsible or accountable to the people of Guyana. There were many articles published about the company, which alleged their absolute disregard for Guyana law and process. A Stabroek News article published in 2013 stated: 'The Chinese logging company Bai Shan Lin Forest Development Inc. has been accused of carrying out unlawful works at Moblissa, Linden, and refusing to sign a Cease Work Order (CWO) served on it by the Guyana Geology and Mines Commission (GGMC)'.[51]

The Omai environmental disaster and the Bai Shan Lin controversy demonstrate an important element for companies in their approach to CSR. Mining, logging or any company operating in the extractive industries need to realize that it is no longer just about having individuals with good finance or the other relevant expertise, but good social conscience is now a necessary ingredient in conducting meaningful investments. Companies operating in the extractive industries need to be socially responsible. Governments who regulate the industries ought to ensure that along with granting an operating license, CSR activities become an integral part of the daily operations of companies operating within their jurisdictions.

---

[48] Ibid.

[49] Reuters (1995).

[50] Henry-Williams (2013).

[51] Thomas (2013).

7 The Main Challenges that the Extractive Resources Industries Are Facing with Regards to CSR Practices

As highlighted above, the main challenges that the extractive resources industries are facing with regard to CSR practices is the need to incorporate sustainable development. Incorporating sustainable development in CSR practices will help to ensure that companies conduct meaningful investment that is aligned to the development of the country's human and social development goals. Another challenge is that the *Companies Acts* in both Guyana and Trinidad and Tobago should be amended to include an enlightened shareholder approach towards directors and investors' duties. The current legal framework in Trinidad and Tobago and Guyana provides that the directors are to have regard only to the interests of the shareholders and employers. However, various other laws and policy documents released by the Governments in both states mention sustainable development as an important goal, which should guide investors. It is important for the legal and regulatory framework to be compatible with other strategies, such as the Green State Development Strategy and the NSCSRP released by the governments.

Turning to the *Green State Development Strategy: Vision 2040 and community involvement in Guyana*, this strategy is Guyana's twenty-year, national development policy that reflects the guiding vision and principles of the 'green agenda'. According to the information about this strategy shared by the government, the aim is for 'an inclusive and prosperous Guyana that provides a good quality of life for all its citizens based on sound education and social protection, low-carbon and resilient development, providing new economic opportunities, justice and political empowerment'.[52] The central objective of the Green State Development Strategy is stated to be development that provides a better quality of life for all Guyanese derived from the country's natural wealth—its diversity of people and abundant natural resources (land, water, forests, mineral and aggregates, biodiversity). The vision of the 'green agenda' is centred on principles of a green economy defined by sustainable, low-carbon and resilient development that uses its resources efficiently and sustained over generations. According to the Strategy document, the development philosophy emphasizes the importance of a more cohesive society based on principles of equity and tolerance between ethnic groups—recognizing that diversity of culture and heritage is the underlying strength of the country's human capital. So the overall picture generated by the strategy is to promote development objectives that seek to improve the health, education and overall well-being of Guyanese citizens, to lift people out of poverty through an economy that generates decent jobs and that provides opportunities for sustaining livelihoods over the long term.

While the principle of 'sustainable development' is often repeated in the Strategy document it is not defined in the document. According to the Brundtland Commission on Environment and Development,[53] sustainable development is: 'development

[52] *Green State Development Strategy, Vision 2040, supra* n. 2 at 1.

[53] Brundtland Commission on Environment and Development (1987).

that meets the needs of the present without compromising the ability of future generations to meet their own needs'. This definition will be adopted here. Note that the principle sustainable development has the status of a legal norm and is evident in the Rio Principles 4 and 25, which provides the concepts of integration and interdependence are an integral part of achieving peace and development.[54] Also, one of the Millennium Development Goals, number 7, to 'Ensure environmental sustainability', provides that governments must 'integrate the principles of sustainable development into country policies and programmes…'[55] Guyana has incorporated sustainable development into several Acts related to the environment. However, it is time for the government to consider incorporating sustainable development into corporate laws, such as the *Companies* Act.

Regarding Trinidad and Tobago, as mentioned above, notwithstanding the fact that a large sum of money was reportedly spent on CSR initiatives, there were many issues that were identified, which continues to minimize the impact of CSR activities in the country. A major challenge is to shift CSR activities towards more strategic forms of CSR. The Government of Trinidad and Tobago indicated that it will lead by example and actively participate in and promote the NSCSRP. The overall aim of the NSCSRP is to shift the focus from mainly charitable activities by companies to more strategic forms of CSR. One of the benefits of the Government's participation in Strategic CSR, and as a leader in the practice of Strategic CSR, is that it will encourage Foreign Direct Investments (FDIs). Foreign investors are usually attracted to nations that operate responsibly, support good governance, anti-corruption and a robust, responsible business sector.[56]

The Government of Trinidad and Tobago intends to lead by example in adhering to CSR principles and practices through the following four main channels:

(i) Promoting policies that provide support to CSR efforts (e.g. endorsement of CSR labels, publication of best practices, support to civil society transparency initiatives).
(ii) Partnering of public and private CSR efforts (e.g. public–private partnerships, ensuring stakeholder consultation in the development of public policy and laws).
(iii) Facilitating measures to enable CSR efforts (e.g. awareness raising campaigns, incentives, tax rebates, procurement policies, capacity building, supporting the development of sustainable and green businesses); and
(iv) Mandating frameworks to enforce and guide CSR activities in its departments and state enterprises (e.g. developing standards and codes of conduct, reporting guidelines, corporate governance codes).[57]

---

[54] United Nations Conference on Environment and Development. *Agenda 21, Rio Declaration, Forest Principles* (New York: United Nations, 1992).

[55] UN Office of the High Commissioner for Human Rights (OHCHR) 2008.

[56] NSCSRP Report, *supra* n. 3 at 15.

[57] Ibid., 16.

The Government of Trinidad and Tobago also reported that it will make it mandatory for all Government departments and State Enterprises to publicly report on an annual basis all activities and expenditures relating to CSR. This bold initiative should strengthen reporting and disclosure practices on CSR activities and help to promote more Strategic CSR practices in Trinidad and Tobago.

8 Conclusion

Since late 2001, many corporate failures, scandals and wrongdoing that have come to light have increased the global drive for CSR. The abuses at Enron, Tyco, Global Crossing, Adelphia and WorldCom in the USA, and Shell in Europe, have severely impacted investor and other stakeholder confidence in the integrity of those charged with the supervision and management of large companies. The incidents in North America and Europe have severely impacted the Caribbean. In Trinidad and Tobago and Guyana there is the view that corporate corruption is widespread and there is a general lack of transparency and accountability.

Companies operating in the extractive industries in Trinidad and Tobago and Guyana should develop clear CSR. While the primary role of corporate businesses is to maximize shareholder value, the global marketplace, where reputations matter deeply, dictates that shareholder value increasingly depends on corporate values. Corporate leaders understand that practising corporate responsibility affects their corporate reputation and brand image. Business managers are becoming more aware that socially responsible investors and activist shareholders can impact the bottom line. It is recommended that companies operating in the extractive industries should develop clear CSR policies to strengthen their ties with local communities through sustainable development programmes and corporate codes of conduct. By conveying clear values and principles, and accepting responsibility for workplaces and workplace conduct, companies can not only build trust and mutual understanding with stakeholders, they can also support the role of governments and the communities in which they operate. Also, corporate responsibility can provide a competitive advantage to companies and company executives can motivate employees by their labour practices.

It is recommended that Guyana develop a National CSR policy like Trinidad and Tobago's NSCSRP, to encourage foreign companies operating in the extractive industries and large domestic companies to strategically give back to the economy, and specifically to host communities. Also, in line with the Green State Development Strategy, the Government of Guyana must aggressively promote sustainable development as part of investment. Additionally, it is recommended that the Government of Trinidad and Tobago continue to map CSR activities to assess whether the initiatives under the NSCSRP to maximize the impact of CSR activities in the country, and shift CSR activities towards more strategic forms of CSR. Finally, it is recommended that the *Companies Acts* in both Guyana and Trinidad and Tobago be amended to incorporate the enlightened shareholder approach as formulated in the UK statute.

Of the various approaches discussed above, this approach provides the best situation to investors while still accommodating elements of Corporate Social Responsibility.

References

Brooks, E., & Schipani, A. (2017, March 19). Guyana oil prospects stir friction between Venezuela and ExxonMobil. *Financial Times*. https://www.ft.com/content/013bfd26-0a8e-11e7-ac5a-903b21361b43?mhq5j=e1. Accessed July 19, 2019.

Brundtland Commission on Environment and Development. (1987). *Our common future*. New York: Oxford University Press.

Davidson, P. (1995, August 23). Cyanide from mine threatens Guyana river. *The Independent*. https://www.independent.co.uk/news/world/cyanide-from-mine-threatens-guyana-river-1597531.html. Accessed August 27, 2019.

Dickerson, R. W. V., et al. (1971). *Proposal for a new business corporations law for Canada*. Ottawa: Information Canada.

Edward Freeman, R. (1984). *Strategic management: A stakeholder approach*. Pitman Publishing.

Friedman, M. (1962). *Capitalism and freedom: Fortieth anniversary edition* (p. 133). University of Chicago Press.

Frynas, G. (2000). *Oil in Nigeria: Conflict and litigation between oil companies and village communities*. Hamburg, Germany and New Brunswick, NJ: LIT Verlag and Transaction Publishers.

Frynas, G. (2005). The false developmental promise of corporate social responsibility: Evidence from multinational oil companies. *International Affairs, 81*(3), 581–598.

Genasci, M., & Pray, S. (2008). Extracting accountability: The implications of the resource curse for CSR theory and practice. *Yale Human Rights and Development Journal*, 11(1), Article 4.

Government of the Cooperative Republic of Guyana. (2019, May). *Green state development strategy, vision 2040*. https://www.doe.gov.gy/published/document/5cd1d69fe5569929a69b35b0. Accessed August 23, 2019.

GYEITI. (2019, April). *Guyana Extractive Industries Transparency Initiative (GYEITI) report for 2017 fiscal year*.

Henry-Williams, T. (2013, November 2). Customer service must not be at the cost of insulting public intelligence. *Kaieteur News*. https://www.kaieteurnewsonline.com/2013/11/02/customer-service-must-not-be-at-the-cost-of-insulting-public-intelligence/. Accessed August 25, 2019.

Mineral Policy Institute. (2014, August 7). *Chronology of major tailings dam failures*. http://www.mpi.org.au/2014/08/chronology-of-major-tailings-dam-failures/. Accessed June 23, 2019.

Monk, R. (2000). Modern company law for a competitive economy: The strategic framework. *Corporate Governance: An International Review, 8*(1), 16–24.

Myers, K. (2018, September 5). Latest Guyana discovery opens the way to a new 10 billion barrel oil province and transformation for one of South America's smallest countries. Westwood Global Energy Group. *Westwood Insight*. https://www.westwoodenergy.com/news/westwood-insight/latest-guyana-discovery-opens-the-way/. Accessed August 14, 2019.

Republic of Trinidad and Tobago, Ministry of Trade, Industry, Investment and Communications. (2015, August). *National strategic corporate social responsibility policy of Trinidad and Tobago*.

Reuters. (1995, August 25). World news briefs: cyanide spill is plugged along river in Guyana. *The New York Times*. https://www.nytimes.com/1995/08/25/world/world-news-briefs-cyanide-spill-is-plugged-along-river-in-guyana.html. Accessed August 27, 2019.

Sealy, L., & Worthington, S. (2008). *Cases and materials in company law* (p. 293). Oxford University Press.

Shah, K. U., Arjoon, S., & Rambocas, M. (2016). Aligning corporate social responsibility with green economy development pathways in developing countries. *Sustainable Development, 24,* 237–253.

Taylor, P. N. (2010). *Enlightened shareholder value and the companies act 2006* (Ph.D. thesis). Birkbeck College.

Thomas, M. (2013, April 18). Chinese logging co accused of illegal Moblissa excavation—Operations continuing despite GGMC cease orders. *Stabroek News.* https://www.stabroeknews.com/2013/news/guyana/04/18/chinese-logging-co-accused-of-illegal-moblissa-excavation/. Accessed August 29, 2019.

UN Office of the High Commissioner for Human Rights (OHCHR). (2008). *Claiming the millennium development goals: A human rights approach.* HR/PUB/08/3. https://www.refworld.org/docid/49fac1162.html. Accessed September 7, 2019.

Van der Ploeg, F. (2008). *Voracious transformation of a common natural resource in productive capital* (OxCarre Research Paper No. 2008-02). Oxford, UK: Oxford Centre for the Analysis of Resource-Rich Economies, Oxford University.

Van Der Weide, M. (1996). Against fiduciary duties to corporate stakeholders. *Delaware Journal of Corporate Law, 21*(27), 56.

Wenner, M., Bollers, E., & Hosein, R. (2018, July). The Dutch disease phenomenon and lessons for Guyana: Trinidad and Tobago's experience. IDB report. *Country Department Caribbean Group, Technical Note No. IDB-TN-1470.*

Wilburg, K. (2018, October 3). Time is running out; get appropriate CSR strategy for oil companies—IDB urges. *Kaieteur News.* https://www.kaieteurnewsonline.com/2018/10/03/time-is-running-out-get-appropriate-csr-strategy-for-oil-companies-idb-urges/. Accessed August 23, 2019.

Corporate and Environmental Responsibility Among Russian Oil Companies

Nina Poussenkova and Indra Overland

Abstract Corporate social responsibility (CSR) and commitment to sustainable development have become integral elements of long-term corporate strategies across sectors and countries. The CSR policies of Russian vertically integrated oil companies (VIOCs) are of particular interest because these companies share some characteristics of international oil companies (IOCs) and the national oil companies (NOCs) of other petroleum-producing countries.

1 Between IOCs and NOCs

Corporate social responsibility (CSR) and commitment to sustainable development have become integral elements of long-term corporate strategies across sectors and countries. The CSR policies of Russian vertically integrated oil companies (VIOCs) are of particular interest because these companies share some characteristics of international oil companies (IOCs) and the national oil companies (NOCs) of other petroleum-producing countries.

In general, NOCs bear a heavier social burden than Russian VIOCs since the former usually have a "national mission" and perform political and social tasks that distract them from the task of profit maximisation.[1] In return, they receive additional

[1] National Oil Companies and Value Creation. Silvana Tordo with Brandon S. Tracy and Noora Arfaa.

[2] Poussenkova (2012).

N. Poussenkova (✉)
Institute of World Economy and International Relations, Russian Academy of Sciences; ENERPO Centre of the European University at St. Petersburg, Moscow and St. Petersburg, Russia
e-mail: npoussenkova@imemo.ru

I. Overland
Centre for Energy Research, Norwegian Institute of International Affairs (NUPI), Oslo, Norway
e-mail: INO@nupi.no

E. G. Pereira et al. (eds.), *Sovereign Wealth Funds, Local Content Policies and CSR*, CSR, Sustainability, Ethics & Governance,
https://doi.org/10.1007/978-3-030-56092-8_31

benefits from the state. In practice, one of the heaviest social burdens for NOCs is their obligation to provide petroleum products to the population and other sectors of the economy at subsidised prices.[2] Russian oil companies do not have such an obligation, although the Russian government does occasionally ask them to freeze gasoline prices. Thus, in February 2011, Prime Minister Vladimir Putin ordered the VIOCs to change price tags at fuel stations: "The end consumer should sense that the decrease of fuel prices benefits his pocket, and it should be visible at fuel stations"; the oilmen rushed to obey. In early 2012, before the presidential elections, the Ministry of Energy reached an agreement with the VIOCs not to raise retail prices until March.[3] This is an unusual combination of political and social objectives achieved at the expense of the Russian oil companies, and definitely transcends traditional Western CSR—but is quite similar to non-commercial obligations of NOCs.

There are also certain specifics deriving from the Russian context, particularly the complicated relations between the State and the oil sector that makes CSR in the Russian petroleum sector comparable to the non-commercial obligations of NOCs. NOCs often have to help improve living standards in the poorer areas of their home country. In Russia, sometimes the state "offers" the oil oligarchs a chance to resolve social issues in Russia's backward regions by spending some of the fortunes they amassed during the 1990s—an offer they can hardly refuse.[4] For example, in 2000, the oligarch Roman Abramovich was tasked by the Kremlin with taking over the governorship of Chukotka, an impoverished and remote region in the East of Russia. He saved the region's finances by registering his companies in Chukotka, where they supplied around 60% of Chukotka's revenues as long as Abramovich was Governor. His personal income tax also went to the province.[5] Besides, Abramovich channelled money to Chukotka through two charitable organisations. As a result, during the period 2000–2004, salaries grew by 350%, and quality of life radically improved in the region.

Despite their shared characteristics with national oil companies, Russian vertically integrated oil companies now seek to emulate international oil companies in terms of CSR. The transformation is particularly striking in the case of Rosneft, Russia's largest oil company. During the 1990s, the then fully state-owned Rosneft was in a weak position, particularly vis-a-vis private VIOCs. During that period, it appeared to welcome non-commercial responsibilities, perhaps because it wanted to show that it was useful to the Russian state. However, as the company returned to strength during the rule of Vladimir Putin, it changed its stance and started behaving more like an IOC than an NOC and increasingly came to see its social engagements as an onus that stood in the way of profitability. Meanwhile, it continued to see the perks of being a national oil company as its birth right.

Rosneft's activities in the Republic of Chechnya exemplify its changing role. After the two wars in Chechnya, Rosneft was tasked with rebuilding the Chechen oil

[3] Predvybornoye sderzhivaniye tsen na benzin mozhet byt oplacheno iz byudzheta, Vedomosti, 30.01.12.

[4] Neft I Capital, 2004, Roman Abramovich.

[5] Anna Nikolayeva. Odin za Vsekh Oligarkhov. Vedomosti, 13.09.05.

sector in the early 2000s—in the national interest. Thus, in cooperation with the regional government of Chechnya, Rosneft created the subsidiary Grozneftegaz. It rebuilt 256 pieces of infrastructure and raised oil production in the Republic from 0.7 million tonnes in 2001 to 2 million tonnes in 2004.[6] But this surge in oil output was unsustainable: By 2011, oil production had declined again to 0.8 million tonnes, and by 2018, it had fallen all the way to 0.3 million tonnes. A refinery that Rosneft was to build in Grozny experienced severe delays. In 2013, work on the refinery appeared to accelerate when Vladimir Putin intervened to push it forward. However, in 2016, it was put on hold until an unspecified date. In 2017, Rosneft decided that the refinery was not commercially viable and suggested building a cheaper bitumen facility instead.[7] Also this was shelved later on.[8] Rosneft's reluctance to invest in Chechnya, and especially in the refinery, was driven by entirely financial considerations related to the oil price and market conditions for refined petroleum products in and near Chechnya.[9]

Ramzan Kadyrov, leader of the Republic of Chechnya, made a public statement that the Chief Executive Officer of Rosneft, Igor Sechin, as a "big politician" who came to the oil sector from a "top-level position", should take into account the Chechen Republic's status as a former war zone in need of support and rebuilding.[10] Rosneft's response says a lot about its new self-image. "*We have a profound sympathy with the fact that Ramzan Akhmatovich [Kadyrov] refers to the social problems of Chechnya, but we do not have the right to discount the value of assets to the detriment of the interests of our shareholders*".[11] This statement is clearly far removed from the perception of Rosneft as the faithful vassal of the state, ready to undertake any strategic business that the Kremlin entrusted it with—and there is hardly a more strategic business in Russia than Chechnya.

2 "Good Old Days" of Soviet Petroleum Social Responsibility?

Partly, this similarity of Russian oil companies to NOCs in the social sphere originated in the socialist past. Many older generation oilmen nostalgically recall the good old days of the "Soviet social responsibility". This psychological phenomenon stems from the evolution of the Soviet oil industry. It was born before the 1917 revolution, in Azerbaijan and the North Caucasus. These territories have a fairly mild climate and relatively well-developed social infrastructure. After WWII, petroleum activities

---

[6] *Neft i Kapital* 2004: 37.

[7] http://www.rbc.ru/business/30/03/2017/58dd00469a794728782e107c.

[8] https://ria.ru/economy/20170817/1500541945.html.

[9] https://oilcapital.ru/news/companies/04-05-2017/chechnya-i-rosneft-avantyurnyy-roman-s-otkrytym-finalom.

[10] http://www.rbc.ru/business/30/03/2017/58dd00469a794728782e107c.

[11] https://www.rbc.ru/business/25/04/2017/58fdccad9a79478ad5714d5a.

shifted to Bashkiria and Tatarstan, two densely populated regions in the Volga-Urals area. Their economy was reasonably diversified, so there was no need for a special "petroleum" social responsibility.

However, in the early 1960s, the oilmen moved to West Siberia where giant oil fields were discovered in the Ob river basin. It is a swampy, sparsely populated region with a harsh climate. Overnight, tens of thousands of workers from all over the USSR arrived to explore and extract "black gold" in an area with practically no roads and very few houses; in a case of typical Soviet mismanagement on a grand scale and with blatant disregard for the people whose welfare was sacrificed to political and ideological considerations. Soviet energy policy was driven by the desire to produce as much oil as quickly as possible mainly to provide financing for Soviet geopolitical goals.[12] The Central Committee of the Communist Party of the Soviet Union (CPSU) and the government relentlessly raised oil production targets, and they did not much care how the West Siberian oilmen achieved them. Thus, given the low productivity of labour in the USSR, production associations (PAs) had to hire numbers of personnel that were hugely excessive by Western standards.

Thus, in 1964, only seven enterprises operated in Surgut, a tiny West Siberian town where just 5 thousand people lived. A year later, over 60 new petroleum enterprises were established, and their personnel had to be somehow housed and fed.[13] The regional Communist Party committee tried to cope with these challenges, but failed. So, oil PAs came to rescue—they needed able-bodied workers to fulfil tough five-year plans.

Many new arrivals were lured to West Siberia by higher "Northern" salaries and/or the chance for rapid promotion. But money alone was not enough to retain workers in the mid-Ob area. Harsh living conditions were, probably, the main reasons for high-personnel turnover (which was detrimental for efficiency of operations): initially, many oilmen had to live in tents with winter temperature falling to −35–40°. So, PAs rapidly erected wooden barracks or converted storage facilities to house the newcomers… They also had to build cinemas and recreation centres to prevent potential conflicts among the young workers who had nothing to do in their free time.

Lev Tchourilov, the last USSR Minister of the Petroleum Industry, recalls that when he was appointed Director of the Nefteyugansk Oil Production Unit, there was only one small bakery (built by geologists) in the town of Nefteyugansk. However, its capacity was insufficient for the rapidly growing oil-producing personnel. Tchourilov ordered two mobile military ovens, but they kept breaking down. One unlucky day both of them broke and the population of Nefteyugansk was left without bread. Oilmen rushed to build a new powerful bakery for their own needs—and meanwhile bread was delivered by airplanes to Nefteyugansk from Surgut, Nizhnevartovsk and

---

[12]See in detail Elena Nikitina, Nina Poussenkova, Petroleum CSR in Russia: Affordable Luxury or Basic Necessity, Russian Analytical Digest, # 181, 2016.

[13]Krol (1995).

Khanty-Mansiisk.[14] It was a typical example of oil enterprises being more efficient and socially oriented (by dire necessity) than the government.

Yet, with application of the state-of-the-art technologies and higher labour productivity, the West Siberian oil industry would not have needed the excessive number of personnel, and the social burden of the oil PAs could have been reduced. As Thane Gustafson remarked in his "Crisis Amidst Plenty", "much of the housing problem, therefore, was a symptom rather than a cause."[15]

Consequently, oil PAs had to keep on their balance sheets non-core social assets (schools, cinemas, hospitals, agricultural farms, etc.) necessary to ensure reasonable living standards for their employees. Thus, Vagit Alekperov, who was appointed head of Kogalymneftegas PA in 1986, had not only to deal with oil production issues, but also to organise adequate life for his workers; because at that time the town of Kogalym was just being established (it is noteworthy that Alekperov insisted from the start on building high-quality houses for the oilmen rather than the traditional temporary barracks).[16] These non-core social activities consumed much of the time and energy of the Soviet oil "generals", sometimes interfering with their key job, as was often the case with NOCs from other petrostates.

Thus, the 1985 oil production plan in the Soviet Union was fulfilled only by 94% and West Siberia accounted for the lion's share of the shortfall. As Vagit Alekperov wrote in his book "Oil of Russia", the CPSU Central Committee and the government admitted the following main reasons of this failure: *insufficient focus on providing housing and social and cultural infrastructure to the workers*, low rates of new field development, shortage of state-of-the-art equipment, etc.[17]

Therefore, this much-admired "petroleum" Soviet social responsibility resulted from several objective and subjective factors: the harsh West Siberian climate; an unbalanced and controversial State energy policy; and low labour productivity. And directors of oil PAs had to cope with adverse nature, challenging crude output plans, relentless pressure from Moscow, and, in addition had to be "founding fathers" of oil towns. The USSR used petrodollars earned by exporting hydrocarbons that were produced with such hardships to import basic food and low-quality consumer goods, as well as to pursue the geopolitical ambitions of the establishment, rather than to modernise the economy and the oil industry itself.

Besides, though the circumstances forced Siberian oil PAs to strongly focus on providing social services to their workers, they were unable (or unwilling) to spend time and efforts on minimising environmental impact of their activities. Lev Tchourilov recalled with bitterness what he termed "environmental barbarism" practiced by the oilmen in West Siberia since they had to fulfil extremely demanding five-year oil production plans regardless of the inevitable damage to nature.[18] A similar cavalier attitude prevailed in mature oil regions of Russia. Representatives of

[14]Tchourilov et al. (1996).

[15]Thane Gustafson, Crisis amidst Plenty, p. 116.

[16]Slavkina M. Rossiiskaya Dobycha, Moscow, 2014, p. 328.

[17]Alekperov (2011).

[18]See in detail Lifeblood of the Empire.

Tatneft say: "… unprecedented rates of drilling, development, and production under the tough dictate of the centre—"Oil at any cost", and continuously rising output targets of "His Majesty the Plan" had a negative impact on the environmental situation in the region."[19] And currently the Russian oil companies have to cope with this legacy environmental damage.

When oil PAs were privatised in the early 1990s, many of them transferred their social assets to the local municipalities in order to raise corporate efficiency. Still, it is unclear whether their efficiency improved due to this divestment, and whether the quality of social services provided by the municipalities to the residents remained adequate.

The 1990s were characterised by economic, social and political crises, low oil prices, declining crude production, galloping inflation and rampant non-payments in Russia. Oil companies were forced to reduce investments, shut down wells, and delay payment of salaries. Clearly, CSR and sustainability were not their top priority. At the same time, the oil companies were globalising, hiring foreign advisors, establishing international alliances, entering world capital markets, and learning the rules of the (international) game, including in the CSR sphere.

In the 2000s, the economic and political situation in Russia was stabilising, oil prices were rising, and crude production was increasing. The Timan-Pechora and East Siberia petroleum provinces were launched, but they are much smaller than West Siberia and require considerably less "imported" manpower. Oil companies now do not have to provide comprehensive "Soviet" social packages to their employees, but their social policy began to embrace new stakeholders, e.g. indigenous peoples.[20] So their "sotsialnaya otvetstvennost" (social responsibility) became closer to Western notions of CSR, particularly in companies with a strong foreign influence such as Rosneft (Table 1).

The five biggest oil companies in Russia—Gazprom neft, LUKOIL, Rosneft, Surgutneftegas and Tatneft—perform fairly similar CSR activities, though each one has different priorities. Compared to its West Siberian peers, Tatneft, which mainly operates in the Central part of Russia, has not had to build oil towns from scratch, nor does it interact with northern indigenous peoples. Still, it has a greater visibility because it functions in densely populated areas, and any environmental accident is likely to draw immediate large-scale attention.

In the CSR sphere, Russian VIOCs are basically guided by the provisions of the relevant Russian legislation and of the E&P licences that they receive and which contain certain social obligations—and by their own social and environmental policies.

---

[19] https://www.tatneft.ru/ekologiya/ohrana-okruzhayushchey-sredi-i-obespechenie-ekologicheskoy-bezopasnosti/?lang=ru.

[20] See in detail Nikitina, Poussenkova.

Table 1 Five Largest Oil Companies in Russia

	Gazprom neft	LUKOIL	Rosneft	Surgutneftegas	Tatneft
Oil production in Russia in 2018 (mt)	62.9	82.0	230.2	61.0	29.5
President	Alexander Dyukov	Vagit Alekperov	Igor Sechin	Vladimir Bogdanov	Nayl Maganov
HQ	St. Petersburg	Moscow	Moscow	Surgut	Almetievsk
Main regions of E&P operations in Russia	West Siberia, Tomsk, Omsk and Orenburg regions	West Siberia, Timan-Pechora, Perm Region, Volga-Urals Region, Caspian and Baltic Seas	West Siberia, East Siberia, Volga-Urals Region, Far East, Timan-Pechora, Krasnodar Krai, continental shelf	West Siberia, East Siberia	Tatarstan, Samara and Orenburg regions, Nenetsk Autonomous District
State stake	Yes	No	Yes	No	Yes
Foreign shareholders	No	No	Yes	No	No
Overseas operations	Yes	Yes	Yes	No	Yes
Foreigners in BoD/Managing Board	No/no	Yes/no	Yes/yes	No/no	Yes/no
Listing at international stock exchanges	Yes	Yes	Yes	Yes	Yes

Sources Company data

3 Social and Environmental Policies of Russian Oil Companies

All petroleum companies in Russia proclaim their firm commitment to CSR. Many of them incorporate these issues into their long-term strategies. Thus, in 2018, Rosneft included three new social initiatives in its strategy up to 2022: *Social Medicine* (expansion of the network of industrial health centres, development of tele-medical technologies and introduction of a new system of regular checkups), *Active Longevity* (indexation of corporate pensions on the basis of investment income of Rosneft's pension fund) and *Accessible Housing* (opportunities for employees to improve their living conditions through dedicated interest-free loans of Rosneft and cheaper mortgage provided by partner banks).[21]

Most VIOCs publish regular sustainability reports; the more internationalised companies, such as Rosneft or LUKOIL, produce them in accordance with the requirements of the Global Reporting Initiative, the UN Global Compact and other national and international sustainability reporting guidelines. Thus, LUKOIL's reports comply with Business Reporting on SDGs, GRI, the UN Global Compact and the Social Chart of the Russian business.[22] In contrast, Surgutneftegas, which is resolutely Russian-based, has been producing since 2004 an annual environmental report that resembles Soviet style documents.

Over time, the proclaimed perceptions of CSR by most Russian oil companies are evolving: They increasingly take into account internationally recognised principles, such as the UN Sustainable Development Goals-2030, probably prompted by their investors and shareholders. For example, the contents of LUKOIL's sustainability reports exemplify the changes in LUKOIL's approach to corporate responsibility. Vagit Alekperov's introductory words to the 2012 issue are about LUKOIL's financial successes and technological innovations.[23] Subsequent prefaces are increasingly about the environment and local communities. Thus, in his preface to the 2017 issue of the report, Alekperov write that "The Company has always sought to work not only to benefit its shareholders and employees, but also society as a whole. We are convinced that our successful development is only possible if we take into account the interests of the communities of the countries where we operate'.[24] This evolution shows that there has been a changed understanding—at least in LUKOIL's communications department—to see that a company's social responsibility goes beyond simply paying taxes.

By now, most Russian VIOCs admit the importance of non-financial indicators. In his introductory message to Gazprom neft's 2018 sustainability report, Alexander Dyukov states: "Gazprom neft is consistently based on the principles of sustainable development in its activities. *We measure success of the company not only by financial and production indicators.* Our key priorities are care about the environment and prudent use of natural resources, safety, high technological level and systemic improvement of living standards in the regions where the company operates."[25]

Currently, the UN Sustainable Development Goals (SDGs) are recognised by many Russian oil companies as viable guidelines. Thus, Rosneft says that an integral element of its Rosneft-2022 strategy is the commitment to the UN SDGs. In line with the practice of international majors, it identified five main goals to pursue in its core activities: (1) good health and wellbeing, (2) affordable and clean energy, (3) worthy

[21] https://www.rosneft.ru/Development/social/m-0.

[22] http://www.lukoil.ru/InvestorAndShareholderCenter/ReportsAndPresentations/SustainabilityReport.

[23] LUKOIL 2012 Sustainability Report.

[24] LUKOIL 2017 Sustainability Report.

[25] Gazprom neft 2018 Sustainability Report, p. 7.

work and economic growth, (4) climate change mitigation and (5) partnership in the interests of sustainable development.[26]

Most companies now realise that adherence to sustainability principles results in their greater attractiveness for investors. Thus, Nayl Maganov said in Tatneft's BoD report to the AGM held in June 2019: "We implement target corporate programmes aimed at support of health protection, science, education, preservation of spiritual heritage, culture and sport in the form of social partnership and social investments. *This contributes to social stability in the regions of our operations, which simultaneously enhances investment attractiveness of the company*".[27]

It is noteworthy that one of the most environmentally and socially responsible Russian oil companies, Surgutneftegas, firmly believes that actions speak louder than words. In the 2018 Environmental Report of Surgutneftegas, the introductory message of the first deputy director Anatoliy Nuryaev simply says that "the JSC understand how important it is to maintain the balance ensuring not only high economic indicators of Surgutneftegas' activities, but also reliability and environmental aspects of all production processes, as well as social responsibility of business".[28]

3.1 Environmental Protection

All Russian VIOCs proclaim their commitment to environmental protection. Thus, Gazprom Neft states in its 2018 Sustainability Report "Goal—zero: no harm to people, environment and assets in the process of activities".[29]

Many of them had to deal with the Soviet "legacy pollution." Thus, LUKOIL that bought KomiTEK in the late 1990s had to eliminate the consequences of the major oil spill from Komineft's feeder pipelines that happened in the mid-1990s. In a similar vein, Gazprom neft recycles waste generated over several decades. After five years of such clean-up activities, dumps that were accumulated before 1991 at its Moscow refinery were completely liquidated. More than 180,000 tonnes of oil-containing waste was handled, and 15 hectares of the territory re-cultivated.

However, given the specifics of the contemporary Russia, several controversial factors affect environmental activities of VIOCs, particularly given their considerable lobbying potential. On the one hand, a popular Russian saying, "the strictness of our laws is offset by the fact that it is not necessary to observe them" is fully applicable to the petroleum sector. Sometimes, oil companies find it easier to pay environmental fines and penalties than to install expensive purification equipment—or they successfully lobby delays in introduction of stricter environmental regulations.

[26]https://www.rosneft.ru/Development/.

[27]https://www.tatneft.ru/press-tsentr/press-relizi/more/6483?lang=ru.

[28]Surgutneftegas Environmental Report for 2018, p. 3.

[29]https://www.gazprom-neft.ru/social/ecology/.

Table 2 Dynamics of LUKOIL's environmental expenditures, 2012–2018, bln rubles

	2012	2013	2014	2015	2016	2017	2018
Environmental expenditures	23.4	42.1	59.2	48.4	53.3	42.4	35.5

Source http://www.lukoil.ru/Responsibility/SafetyAndEnvironment/Ecology/Results

Also, a serious controversy stems from the dual nature of the Russian Ministry of Natural Resources and Environment that plays a role simultaneously of a gamekeeper and poacher, and there were numerous cases when the poacher defeated the gamekeeper in the process of issuing E&P licences in environmentally vulnerable areas. The much-publicised conflicts around Surgutneftegas' operations in Numto Nature Reserve and the sacred Imlor Lake areas are a good example of this dual role of the Ministry. Despite protests by local inhabitants from 2010 onwards against petroleum activity in the area, in 2012 Surgutneftegas reached an agreement with the Governor of Khanty-Mansi Autonomous District. In the ensuing years, almost all indigenous people moved out of the area. Regarding the Numto Nature Reserve, 36,000 letters were written to officials requesting them to stop oil drilling in the area. Nonetheless, in 2016, the federal government decided to allow drilling to move forwards.[30]

On the other hand, the oil companies realise that environmental offences are a convenient lever for the authorities to put pressure on business in certain cases, as the conflict around Sakhalin Energy demonstrated. Then, Gazprom and federal agencies pressured shareholders of Sakhalin-2 consortium accusing them, among other things, of environmental violations, and as a result Gazprom managed to join Sakhalin Energy as a majority shareholder on very attractive terms in 2006.[31] Also, Russian oil companies are now increasingly influenced by their shareholders, investors and consumers of petroleum products, all demanding greater environmental responsibility.

The Russian VIOCs report considerable outlays on environment protection. Thus, in 2018, Gazprom Neft's investments in environment protection rose to 19.0 bln rubles[32]; Surgutneftegas spent 17.4 bln rubles on environmental protection; payments for negative impact on the environment amounted to 75.4 mln rubles, including 173 thousand rubles for impact above the established limits.[33] LUKOIL allocated 35.5 bln rubles for capital and operating environmental expenditures in 2018, i.e. less than in 2017, mainly because it completed construction of several APG (associated petroleum gas) utilisation facilities. Key directions of its expenditures were as follows: 41% for prevention and elimination of accidents, 31% for protection of atmospheric air, 10% for waste utilisation, 9% for protection and rational use of water resources and 9% for other directions (Table 2).[34]

---

[30] https://regnum.ru/news/2197112.html.

[31] Sakhalin perestal byt goryachei tochkoi. Kommersant, 22.12.2006.

[32] https://www.gazprom-neft.ru/social/ecology/.

[33] https://www.surgutneftegas.ru/responsibility/ecology/prirodookhrannye-aspekty-khozyaystvennoy-deyatelnosti/osnovnye-napravleniya-prirodookhrannoy-deyatelnosti/.

[34] LUKOIL 2018 Annual Report, p. 71.

In general, all five companies focus on traditional directions of environmental protection: air, water, land, biodiversity, waste utilisation, environmental monitoring, prevention of accidents, R&D, education and training of personnel.[35] Here are some specific examples of these activities. In 2006, Rosneft introduced the integrated system of HSE management (which complies with ISO 14001 and OHSAS 18001), and target HSE indicators are incorporated in Rosneft-2022 Strategy. Its main HSE goal is to be included in the first quartile of the relevant rating of the global energy companies by end of 2022.[36] In January 2018, TANECO, Tatneft's refinery, became the first enterprise in Russia to introduce automatic monitoring of pollutant emissions to the atmosphere.[37] In 2017, Gazprom neft commissioned the 9 bln rubles Biosphera treatment facilities at the Moscow Oil Refinery that purify 99.9% of the refinery's wastewater. A similar installation is also being introduced at its Omsknefteorgsyntez.[38]

Gazprom neft implements a Clean Territory project which envisages, among other things, re-cultivation of land; it also takes preventive measures to reduce the number of accidents at pipelines; in 2018, 37 thousand km of pipelines were protected by inhibitors of corrosion. Surgutneftegas places a special focus on maintaining biodiversity, particularly in environmentally sensitive areas. In Numto, Surgutneftegas implements measures aimed at minimising environmental risks of oil production in swamps, as well as at preventing death of birds at electricity facilities.[39] Rosneft launched a programme of protecting polar bears: since 2013 it has been a guardian of 34 polar bears that are exhibited in 16 Russian zoos, and it implements a programme of protection of polar bears in their natural environment, for example, saves orphaned bear cubs. Since 2014, Rosneft has been studying polar bears in the regions of future oil developments.[40]

Breaking with the Soviet traditions of environmental secrecy, Russian oilmen provide fairly detailed information about their environmental challenges. Thus, Surgutneftegas, normally the least transparent company among its peers, posts a surprising amount of environmental data on its website, even a detailed analysis of environmental incidents in a given year. Thus, in 2019, it reported three serious accidents: one was fire at a well in Rogozhnikovskiy licensing plot (the reasons are being investigated); another was an oil spill from a pipeline caused by an attempt at oil theft by an unknown person; and the third was an unauthorized spill of oil-containing liquid by an unknown person in the area of Zapadno-Surgutskoye field.[41]

---

[35]https://www.surgutneftegas.ru/responsibility/ecology/prirodookhrannye-aspekty-khozyaystvennoy-deyatelnosti/osnovnye-napravleniya-prirodookhrannoy-deyatelnosti/.

[36]https://www.rosneft.ru/Development/HealthSafetyandEnvironment/.

[37]https://www.comnews.ru/digital-economy/content/111407/2018-01-24/taneko-avtomatiziruet-kontrol-promyshlennyh-vybrosov.

[38]Gazprom Neft 2017 Sustainability Report, p. 7.

[39]https://www.surgutneftegas.ru/responsibility/ecology/osobo-okhranyaemye-prirodnye-territorii/prirodnyy-park-numto/.

[40]https://www.rosneft.ru/press/news/item/197505/.

[41]https://www.surgutneftegas.ru/upload/iblock/009/Сведения%20об%20инцидентах%20на%20нефтепромыслах%20по%20ПАО%20Сургутнефтегаз%202019.pdf.

Table 3 Surgutneftegas environmental indicators

Indicator	Unit	2016	2017	2018
Gross emissions of pollutants into atmosphere	Thous. tons	185,120	176,707	139,485
Level of APG utilisation	%	99.34	99.32	99.56
Water consumption	Thous. cub. m	10,155,104	10,110,556	9,996,563
Generation of waste	Thous. tons	714.0	797.3	798.6
Utilisation of waste in own production	Thous. tons	417.1	462.3	480.5

Source https://www.surgutneftegas.ru/upload/iblock/009/Показатели%20воздействия%2011102019.pdf

In general, their environmental policies bring desirable results. Thus, Surgutneftegas was rated number 1 in terms of environmental management, and number 5 in the total 2018 rating of WWF and Kreon[42] reported the following improvements of its environmental indicators (Table 3):

Oil companies increasingly apply state-of-the-art environment protection technologies and approaches in their activities. Thus, in 2018, Gazprom neft launched a Green Seismic 2.0 project that introduces environmentally friendly seismic technologies (while traditional seismic requires clearing wide paths in forests so that heavy vehicles could pass, Green Seismic uses registering equipment that can be installed with light vehicles. Therefore, it reduces cutting of trees, consumption of fuel and atmospheric emissions).

All these are traditional directions that were implemented even in the USSR. However, recently, a new area of environmental protection emerged connected with the commencement of offshore crude exploration and production by Gazprom neft, LUKOIL and Rosneft, including in the Arctic, and it is particularly important for them to demonstrate to the global community their environmental reliability and responsibility.

Thus, after Greenpeace's activists protested in 2012–2013 against environmentally unsafe and economically unfeasible oil production in the Arctic and boarded twice the Prirazlomnaya oil platform of Gazprom neft in the Pechora Sea, the company places a special focus on environmental aspects of the Prirazlomnoye field development. It stresses that specialised ice-class vessels equipped with the state-of-the-art complexes for gathering spilt oil are on constant duty in the vicinity of the platform. According to the company, Prirazlomnaya platform operates under a "zero-discharge" principle: the drilling muds, sludge and other technological wastes are pumped into a special absorption well.[43]

---

[42]https://wwf.ru/what-we-do/green-economy/ekologicheskiy-reyting-neftegazovykh-kompaniy-rf-sovmestnyy-proekt-wwf-i-kreon/.

[43]https://shelf.gazprom-neft.com/upload/iblock/109/spravka_o_proekte_prirazlomnoe.pdf.

LUKOIL that produces oil in the Caspian Sea and the Baltic Sea also emphasises environmental aspects of its offshore activities. Vagit Alekperov says in 2018 Sustainability Report that environmental protection and occupational safety technologies and standards that the company uses often exceed the current international practice. "For example, LUKOIL applies "zero-discharge" principles at all its offshore projects, and this approach was recognised as exemplary by the Helsinki Commission on the Protection of the Marine Environment in the Baltic Sea (HELCOM)".[44] It is noteworthy that in 2017, the Year of Ecology in Russia, LUKOIL and Gazprom neft conducted joint trainings to clean up oil spills in the Arkhangelsk region in the vicinity of Varandei terminal of LUKOIL and Prirazlomnaya platform of Gazprom neft.[45]

In 2012 and 2013, Rosneft published joint declarations on environmental protection in the Arctic with its international partner companies ExxonMobil, Equinor and Eni. In these declarations, they committed to protection of the Arctic environment, including keeping their impact on indigenous peoples and climate change to a minimum.

3.2 *Climate Change*

The most important challenge faced by oil companies—Russian or international—during the coming decades will be that of climate policy. Russian vertically integrated oil companies have a nonchalant approach to climate change; however, there are differences among them.

Although Rosneft has made climate change, one of its five main goals, Igor Sechin's attitude indicates that the company's climate engagement is at best shallow.

> I can tell you I am not a big expert in that area, but I know a few things. First of all, we are subject to global climate change cycles... those cycles repeat every 30 million years, so everything is normal. The human effect on the environment is less than any volcano. A volcanic eruption produces more CO_2 than any human activity. The rotting of algae in the ocean significantly exceeds any human-made effect.[46]

LUKOIL appears to take climate policy a bit more seriously and has been part of the International Carbon Disclosure Project since 2013. In 2016, LUKOIL achieved a climate score of D on a scale from A (best) to F (worst). By contrast, BP, Eni, Statoil (Equinor) and Total all achieved A-.[47]

Russian VIOCs' main contribution to climate change mitigation is the introduction of energy efficiency programmes (thus, Surgutneftegas has been implementing the Programme of Energy Saving and Enhancing Energy Efficiency of Facilities since

---

[44]LUKOIL 2018 Sustainability Report, p. 2.

[45]http://www.lukoil.ru/Responsibility/SafetyAndEnvironment/Ecology/AccidentElimination.

[46]https://www.independent.co.uk/news/business/analysis-and-features/igor-sechin-the-oil-man-at-the-heart-of-putins-kremlin-10043230.html.

[47]LUKOIL 2016 Sustainability Report, p. 67.

Table 4 Rational utilisation of APG by companies in 2018

Company	Level (%)
Surgutneftegas	99.5
LUKOIL	97.4
Tatneft	96.2
Rosneft	84.4
Gazprom neft	78.4

Source Company data

1997)[48] and limiting flaring of associated petroleum gas (APG), which is required by the Russian authorities—95% rational utilisation of APG is the declared official target (Table 4).

It is noteworthy that the two Russian privately owned companies, Surgut and LUKOIL, achieved a higher level of APG utilisation than companies that are partly owned by the State. Thus, while Surgutneftegas has avoided making statements about climate change and climate policy, the company has taken the lead on reducing the flaring of associated petroleum gas. It has advanced facilities for associated petroleum gas utilisation.[49]

Gazprom neft has reported the lowest utilisation of associated petroleum gas of the Russian oil companies, but the company has been taking consistent and innovative measures to limit gas flaring. In 2012, the company agreed with SIBUR to create the Noyabrsk Integrated Project for the Vynagapurovskaya group of fields in the Yamal-Nenets Autonomous District. Associated petroleum gas from these fields is supplied to SIBUR's Vynagapurovskiy plant.[50] Extending this trend, in 2018 Gazprom neft started injecting associated petroleum gas from the East Messoyakha field into the West Messoyakha field, thus avoiding having to flare the gas and saving it for the future at the same time.[51] In June 2017, LUKOIL became the first Russian oil company to support the World Bank initiative "Full Elimination of Regular Gas Flaring by 2030", and intends to take all necessary measures to implement this initiative both in Russia and in its overseas operations.[52] Tatneft's situation with APG utilisation is somewhat different because it mainly produces oil in mature regions with well-developed gas gathering infrastructure. Its interest in climate issues is growing: it plans to apply the latest international standards of the ISO 14064 "Greenhouse Gases".[53]

---

48 https://www.surgutneftegas.ru/responsibility/ecology/prirodookhrannye-aspekty-khozyaystvennoy-deyatelnosti/energoeffektivnost-i-resursosberezhenie/.

49 https://www.surgutneftegas.ru/responsibility/ecology/prirodookhrannye-meropriyatiya/meropriyatiya-po-okhrane-atmosfernogo-vozdukha/.

50 https://www.gazprom-neft.ru/press-center/sibneft-online/archive/2012-october/1103828/.

51 https://www.gazprom-neft.ru/press-center/news/messoyakhaneftegaz-realizuet-unikalnyy-proekt-po-utilizatsii-poputnogo-neftyanogo-gaza/.

52 http://www.lukoil.ru/Responsibility/SafetyAndEnvironment/Ecology/Results.

53 http://2018.tatneft.ru/climate/company-position/.

Reforestation is another climate change mitigation measure taken by VIOCs. Thus, since 2008, Tatneft planted over 9 million trees, i.e. some 5000 hectares of forests, trying to compensate GHG emissions through sinking capacity of woods. Tatneft plants trees along highways, creates "green fuel stations", establishes "green belts" around industrial facilities, develops parks in cities and towns, as well as improves "green infrastructure" of social facilities.[54]

The relative indifference of Russian oilmen to climate change goes hand in hand with their scepticism about renewables. Among the Russian oil companies, LUKOIL has shown the greatest interest in renewable energy, with an electricity generation capacity 390 MW in 2016 (including hydropower). In 2009, LUKOIL installed solar panels at three petrol stations in Serbia and Russia. The company has also commissioned solar panels at its refineries in Romania and Bulgaria. LUKOIL's joint venture with the Italian ERG SpA acquired four Bulgarian wind power firms.[55]

According to Gazprom Neft, oil and gas will continue to be important and supply some 90% of the world's primary energy supply for the foreseeable future. Sergei Vakulenko, Head of the Department of Strategy and Innovations stated in a recent interview: "If we speak about alternative energy: solar, air and water, electric vehicles and so on, we certainly take them into account in our strategy but in a much longer-term perspective".[56]

Nonetheless, Gazprom Neft started dabbling in renewables through its Serbian subsidiary NIS. Geothermal units have been installed near 11 oil wells owned by NIS, with another 20 wells under construction. In 2013, NIS started building a wind farm in Plandishte with a capacity of 100 MW, consisting of 40 wind generators.[57] Gazprom Neft is also involved in some minor wind and solar power projects on the Yamal Peninsula, where electricity in remote locations is generated by a combination of wind, solar and diesel.

Tatneft is taking its first tentative steps towards renewables. In 2018, 0.24% of the total power generated by Tatneft was produced on their basis: 99.9% by generating heat by pellet-fired boilers, and 0.01% by solar panels.[58]

There are also leaders and laggards among the five companies with respect to the use of electric vehicles. At the 2017 Saint Petersburg Forum, Igor Sechin discussed the outlook for electric cars, stating that they would have a niche in the market but not on the scale anticipated by some actors. He repeated this forecast in even stronger negative terms at the 2019 Forum.[59] By contrast, Tatneft has involved itself actively in electric vehicle developments by installing charging points for them. In 2016, the company set up its first charging point in Khimgrad Industrial Park in Kazan. Tatar President Minnikhanov took part in the opening and stated that electric vehicles were

[54] http://2018.tatneft.ru/climate/measures/.

[55] http://www.lukoil.ru/Responsibility/SafetyAndEnvironment/Ecology/RenewableEnergy.

[56] https://www.gazprom-neft.ru/press-center/lib/1642212/.

[57] https://www.nis.eu/en/presscenter/news/nis-started-construction-first-wind-farm-serbia-plandiste.

[58] http://2018.tatneft.ru/climate/measures/.

[59] https://www.kommersant.ru/doc/3992564.

environmentally safe and reliable, and that Tatneft should do more work in this area. In June 2017, a second charging point was unveiled at a Tatneft petrol station in the city of Almetievsk.[60]

Thus, Gazprom Neft and LUKOIL are most transparent in reporting their greenhouse gas emissions and have taken some first steps into the renewable energy sector, mostly outside Russia. In sum, the Russian vertically integrated oil companies are similar to their international peers, with a delay of some years.[61]

3.3 Agreements on Socio-economic Cooperation with the Regions

Russian oil companies normally sign agreements on socio-economic cooperation with the heads of the regions where they operate. Such agreements are particularly important in the Russian high North and Siberia, where CSR programmes of the oil companies are still essential for ensuring higher living standards of the population—and, subsequently, social and economic stability. Thus, in September 2019, Rosneft and Yamal-Nenetsk Autonomous District signed additional agreements on cooperation in the social and environmental spheres, the basic agreement being effective since 2015. Under them, a number of infrastructural projects will be implemented in the region with Rosneft's support. The company will assist in medical treatment and rehabilitation of handicapped children, will provide support to organizations of veterans of wars and armed forces, as well as indigenous people of the North. Rosneft will also take measures to restore biodiversity in water bodies of the district.[62]

Also, during the 2019 Saint Petersburg International Economic Forum, Rosneft and the Samara region signed an additional agreement on cooperation in social development. Under it, Rosneft will finance construction of a swimming pool in Samara, repair of a school building in Syzran, reconstruction of Children's School of Art building in the town of Otradniy and of the kindergarten in Novokuibyshevsk.[63]

3.4 Stakeholder Engagement

Stakeholder engagement is a fairly recent phenomenon in the Russian oil business. Russian oilmen usually stress that they strive to involve stakeholders in their decision-making, particularly on environmental issues. However, it is difficult to say whether the general public in Russia can make a real impact on operating decisions of the

[60] https://tender.tatneft.ru/news/2017/na-azs-tatnefti-otkrylas-eshche-odna-stantsiya-dlya-zaryadki-elektromobiley/.

[61] See in detail: Challenge of Change, Indra Overland, Nina Poussenkova.

[62] https://www.rosneft.ru/press/releases/item/196909/.

[63] https://www.rosneft.ru/press/releases/item/195559/.

powerful oil companies. Thus, in late 2015, a series of public hearings were organised by LUKOIL-Komi in the town of Ukhta concerning the Yareg field development. Usually, from 11 to 26 people registered for such hearings, and it is unclear whether these were merely formalities or real discussions that could have made a difference.[64]

Rosneft's sustainability reports explain its involvement of stakeholders, with the main types of stakeholder engagement being public consultations about environmental impact, and roundtables the company has held regularly since 2007 in the regions where it operates. Rosneft reported 127 public awareness measures in 2017; over 50 of these were consultations about offshore exploration; and 15 were roundtables.[65] Rosneft diligently publishes the results of such hearings on its website, usually stating that the participants of the hearings wholeheartedly support its projects. Thus, in February 2016 RN-Shelf-Arctic, subsidiary of Rosneft, held public hearings in Arkhangelsk where it presented a programme of maritime geochemical studies to be held in 2016-2020 in the Barents Sea, detailing the schedule of activities and environment protection measures. Rosneft's press release said that the participants approved the programme, including the environmental impact assessment.[66] Given the power of Rosneft, it might have been difficult for the participants to take a negative stance during the hearing.

3.5 *Relations with Indigenous Peoples*

Interaction with northern indigenous peoples became an important new direction of CSR activities of Russian oil companies, with four out of five companies introducing this element in their strategy (Tatneft is less involved because of the geography of its operations). Most of Russia's vertically integrated oil companies have formalised policies with the northern indigenous peoples. Gazprom neft approved its Policy for Interaction with Indigenous Minorities of the North, Siberia and the Far East in 2017.[67] LUKOIL's approach to indigenous peoples is set out in its Social Code (Section 2.4, Preserving National and Cultural Identity). The company also seeks to regulate the behaviour of its subsidiaries in this area. LUKOIL-Western Siberia, for example, is subject to Order 262, "On Measures to Limit Access to Territories of Communal Lands". Subsidiaries have Departments on Interaction with Indigenous peoples to handle complaints and proposals.[68] Surgutneftegas has departments on relations with indigenous people at management level as well as further down in the

[64] Indra Overland, Nina Poussenkova. "Russia: Public Debate and the Petroleum Sector", in Public Brainpower.

[65] Rosneft, 2017a, p. 48–9.

[66] https://www.rosneft.ru/press/news/item/180545/.

[67] Gazprom Neft, 2017n, p. 124.

[68] LUKOIL 2015-16: 92–93.

company hierarchy.[69] Rosneft applies Convention 169 of the International Labour Organisation, as well as the UN Declaration on Rights of Indigenous Peoples.[70]

One of LUKOIL's initiatives to deliver medical assistance to the Northern indigenous people is particularly noteworthy. The Red Tent (Krasniy Chum) project was launched in 2002 in the Nenetsk Autonomous District by LUKOIL-Komi, the district administration, Yasavei public movement and Total E&P Russia. Red Tent ensures accessible medical aid to the nomad tribes in the remote Northern regions. The project also supplies deer-breeding crews with necessary medicine and teaches them principles of providing first medical aid.[71]

Surgutneftegas signs and implements cooperation agreements with the administrations of the regions where it operates and enters into direct agreements with indigenous families. The company pays the families' transportation expenses, reimburses expenditure for medical treatment and care, and supports children's education. Surgutneftegas organises seminars for its own and contractors' personnel to discuss norms of behaviour towards indigenous people, as well as briefings before employees are allowed to work within the habitation areas of indigenous people.[72] Rosneft's approach to interaction with indigenous people is an element of its Sustainable Development Policy. Rosneft helps them to buy vehicles to drive children to school, trucks, fuel and lubricants, communication equipment, finances repair and procurement of medical equipment for district hospitals, and assists in creating jobs for indigenous people.[73] Gazprom neft also grants financial assistance to families who suffered in emergency situations. To preserve national identity of Northern people, Gazprom neft supports celebration of Days of Deer-herder in Polar towns and settlements.[74]

All of the Russian companies claim to involve indigenous peoples in decision-making on future activities. LUKOIL coordinates its exploration schedule with indigenous people and states that it takes into consideration their interests already at the planning stage. But despite the official policies on indigenous peoples, conflicts do occur. In 2014, there was a high-profile scandal in the Izhma District of the Komi Republic when the local inhabitants voted unanimously to terminate LUKOIL-Komi's activities in the region.[75]

---

[69] https://www.surgutneftegas.ru/responsibility/ecology/vzaimootnoshenie-s-kmns/politika-v-oblasti-vzaimootnosheniy-s-kmns/.

[70] https://www.rosneft.ru/Development/cooperation/Podderzhka_korennih_malochislennih_narodov_Severa/.

[71] http://www.lukoil.ru/Responsibility/SocialInvestment/SocialInitiatives/Theredtent.

[72] https://www.surgutneftegas.ru/responsibility/ecology/vzaimootnoshenie-s-kmns/politika-v-oblasti-vzaimootnosheniy-s-kmns/.

[73] https://www.rosneft.ru/Development/cooperation/Podderzhka_korennih_malochislennih_narodov_Severa/.

[74] https://www.gazprom-neft.ru/social/indigenous-people/.

[75] https://www.greenpeace.org/russia/ru/news/blogs/green-planet/-/blog/49459/.

3.6 Charity

Charity is a traditional element of oil companies' social policy. Thus, LUKOIL established a charity fund in 1993, one of the first of its kind in contemporary Russia. LUKOIL's position in this respect is quite clear: charity should not generate social parasitism. Therefore, LUKOIL, alongside with the traditional forms of charity, uses programmes of strategic charity and social investments. LUKOIL and its charity fund have been organising contests of social and cultural projects in host regions since 2002. Project applications are submitted in three main nominations: Environment; Spirituality and Culture; and Sport. Additional nomination is approved every year: in 2016, in honour of the 25th anniversary of LUKOIL, a nomination Energy for the Benefit of People was approved; in 2018, the Year of Volunteer in Russia, a nomination Youth Initiatives.[76]

To raise efficiency of social investments, in 2018 Tatneft united its various charitable funds in a single Charitable Fund while retaining all its main long-term programmes such as Mercy, Talented Children, Ruhiyat and Tazalyk. The annual charitable programme of Tatneft is approved by the Board of the Fund, which is chaired by the General Director of the company. In 2018, the Fund approved grants to 44 projects. The initiative Talented Children, for example, provides grants for gifted children. As a new trend, in 2018, financial support was allocated for the "green fitness" activities.[77] In a first for a Russian oil company, Tatneft started leasing bicycles to the residents of Almetievsk at subsidised rates in 2017.[78]

Gazprom neft's social programme is called 'Native Towns'. Projects are selected every year, taking into account the views of stakeholders. The programme included over 2000 projects in 35 regions in 2017.[79] Kustendorf CLASSIC is one of the key projects of the Native Cities aiming at helping young musicians and developing cultural ties between Russia and Serbia.[80]

3.7 Sponsorship of Education and Health Protection

These spheres have traditionally been important CSR foci for all Russian oil companies. Rosneft supports education, focusing on nine institutions of higher education. The company helped set up a school connected with Moscow State University and finances the naval higher education establishments, for example, at the Saint Petersburg State Naval Technical University and Makarov State University of Sea and River

---

[76]http://www.lukoil.ru/Responsibility/SocialInvestment/SocialProjectsCompetition.

[77]http://2018.tatneft.ru/socialpartnership/charity/.

[78]http://almetievsk-ru.ru/news/goryachie-novosti/v-almetevske-zapustyat-sotsialnyiy-proekt-po-proka.

[79]https://www.gazprom-neft.ru/social/regions/.

[80]https://www.gazprom-neft.ru/social/fest/about/.

Fleet. At Tatneft, Nayl Maganov emphasises support for medical services. Projects have included support for an oncology centre and a children's clinic and various other medical facilities.[81]

3.8 Support of Entrepreneurship

Support of entrepreneurship is a relatively new direction of CSR activities for Russian oilmen, which resembles traditional activities of IOCs. LUKOIL's CEO Vagit Alekperov created the foundation "Our Future", which promotes entrepreneurship. "Our Future" helps people to set up new businesses, with all-Russian contests for project proposals held twice a year. In 2014, LUKOIL and "Our Future" established the initiative "More than a Purchase". It involves the selling of goods produced by social entrepreneurs at 106 LUKOIL petrol stations across the country.[82] Gazprom neft focuses on supporting local industry to serve the petroleum sector. In 2017, the company entered into agreements with seven Russian regions to replace the imported lubricants and fluids.[83]

Tatneft launched an initiative of cooperation with stakeholders in support of small and mid-size business. Tatneft is ready to give SMEs an opportunity to produce new goods or provide services by leasing or selling them its redundant production facilities. In 2004, Tatneft approved a Special Corporate Project to Support Development of SMEs in order to enhance business activity in the South-East of Tatarstan, create new jobs and ensure social stability in the host regions.[84]

In June 2019, Rosneft and the Kurgan region signed an agreement on cooperation in promoting import substitution. According to Rosneft, it is aimed at maximising use of Kurgan's industrial potential by the company. The parties plan to develop proposals on cooperation with regional industrial and R&D organisations, as well as to compile a list of equipment and technologies to be used at Rosneft's facilities.[85] On the part of Rosneft, this move was, probably, prompted by sectoral sanctions against Russian VIOCs, as they began to focus on import substitution in response.[86] From a social perspective, this initiative will help to create jobs in the Kurgan region and will contribute to improvement of living standards.

[81] http://zt116.ru/news/1/nail-maganov-my-zdes-zhivem-dyshim-odnim-vozdukhom-so-vsemi.

[82] http://www.lukoil.ru/Responsibility/SocialInvestment/SocialInitiatives/JointprogramswiththeFundOurFuture.

[83] Gazprom Neft 2017 Annual Report, p. 124.

[84] https://www.tatneft.ru/sotsialnaya-politika/podderzhka-malogo-i-srednego-biznesa?lang=ru.

[85] https://www.rosneft.ru/press/releases/item/195551/.

[86] Fjaertoft and Overland (2015).

4 Conclusion

Some of the CSR activities of the five oil companies reflect the traditional Soviet approach (support for health facilities and cultural festivals) while others resemble more Western notions of CSR (investment in the supply chain capacities of local enterprises). Agreements with local indigenous communities combine elements of the Soviet, Russian and international standards and values. In general, Russian VIOCs initially used to closely resemble NOCs in the social sphere, since they inherited the traits of the "Soviet social responsibility" from their Socialist predecessors. As it usually happens with NOCs, this dual role of being both a commercial entity and an instrument of the "national mission" often resulted in the lower business efficiency of oil companies and undermined their investment potential. Currently, VIOCs try to evade non-commercial obligations and purely social projects, considering them unprofitable, and tend to evolve towards the CSR models of IOCs, largely driven by the demands of their investors, shareholders and consumers. In contrast to the non-commercial functions of NOCs, corporate social responsibility policies and commitment to sustainable development help to enhance investment attractiveness of Russian oil companies.[87]

Acknowledgement This chapter was funded by the Research Council of Norway, grant number 287937.

References

Alekperov, V. (2011). *Oil of Russia: Past, present and future* (pp. 333–334). Moscow: Creative Economy.

Fjaertoft, D., & Overland, I. (2015). Financial sanctions impact Russian Oil, Equipment Export Ban's Effects Limited. *Oil and Gas Journal, 113*(8), 66–72. https://www.researchgate.net/publication/281776234.

Krol, M. (Ed.). (1995). *Oil epopee of West Siberia* (p. 33). Moscow: Publishing House Neftyannik.

Poussenkova, N. (2012). *Noviye Zvezdy Mirovoi Neftyanki*. Moscow.

Tchourilov, L., Gorst, I., & Poussenkova, N. (1996). *Lifeblood of empire. A personal history of the rise and fall of the Soviet oil industry*. PIW Publications.

[87] This chapter is a product of the project "Is this Russia's Kodak Moment", which is financed by the Research Council of Norway, grant number 287937.

Alaska's Corporate Social Responsibility: The Economics of the Corruption Case of VECO

Douglas B. Reynolds

Abstract During the early 2000s, Alaska tried to develop its natural gas industry just as natural gas prices in the US market were high, highlighting an interesting case of Corporate Social Responsibility (CSR). In order to develop an Alaskan natural gas industry, then, a two thousand mile, multi-billion dollar pipeline would be needed, and such a pipeline would require a tax contract between the state of Alaska and the Alaskan North Slope oil producers, which would create tough negotiations and a strenuous relationship between Alaska and the international oil companies (IOCs). Usually, governments choose petroleum taxes, regulations, and government support within the petroleum industry based on maximizing social welfare, but not based on maximizing corporate profits. The corporations though needed a certain level of profits in order to be able to invest in a new petroleum development. Therefore, there is a natural incentive for corporations to try to change government policy in order to reduce their risk and assure profitability. However, there is one other factor to consider when trying to get such a gigantic pipeline built and a new natural gas industry started, which are the risks involved. Since investors are taking huge risks to build natural gas infrastructure, then they need favorable taxes to get the job done or they cannot make the investment pay off, and sometimes, they use corrupt means to get a good deal, in which case their Corporate Social Responsibility can be lost. Still, it is not always possible to know the absolute best tax rates or the perfect government terms and conditions that will maximize social value. Therefore, if social welfare includes not just the best terms possible of a tax contract but also whether a project is completed or not, then even if taxes are low, the state could still receive jobs and maximize social welfare compared to not having a new industry at all. Thus, the Alaska VECO corruption case analysis also shows an intricate economic calculation of the expected costs and benefits of the project, after the fact, of how things could have turned out.

D. B. Reynolds (✉)
Department of Economics, School of Management, University of Alaska Fairbanks, Fairbanks, USA
e-mail: dbreynolds@alaska.edu

E. G. Pereira et al. (eds.), *Sovereign Wealth Funds, Local Content Policies and CSR*, CSR, Sustainability, Ethics & Governance,
https://doi.org/10.1007/978-3-030-56092-8_32

Keywords Resource development · Corruption · Negotiating terms

1 Introduction

Almost inevitably the idea of corporate social responsibility (CSR) is about how corporations are going to help various social causes or reduce pollution or in some way help the citizens of the countries where they work. What can be more interesting, though, is the interaction of the International Oil Companies (IOCs) with the governments they deal with over terms and condition of their business relationship with the government. Thus, we have (1) the government that represents society, (2) the IOCs that represent investor interests in developing the petroleum resources for a profit, and (3) the interaction between the IOCs and the government. It is this interaction then that can be the most interesting thing about CSR since within this relationship cases of corruption can emerge, and where CSR can be lost. For example, in Alaska, it is relatively cheap to give money to specific community projects and causes within the state since the population is so small. The more pertinent issue then is not so much local giving or local environmental protection, but rather the oil taxes and the government take themselves which allow the state to gain value from the petroleum industry. Indeed, more often than not the workers in the industry do not live in Alaska but come into the state for two weeks of work every month and then live the other two weeks outside of the state. This makes the most pertinent issue in CSR not so much corporate giving to help local causes, but rather the state's oil taxes on those corporations where the amount of revenues to the state determines how much the state itself can pursue social welfare programs. However, there is more to gaining social welfare than mere government take because the real tension is between society's desire to get the most benefit out of its petroleum resources and the corporate investor's desire to make a fair return on the money that petroleum companies invest in Alaska, especially since if a given project has a risk of losing money, then the investors could see a sizable loss.

> I was seldom able to see an opportunity until it has ceased to be one.
>
> A man cannot be comfortable without his own approval.
>
> Let us live so that when we come to die even the undertaker will be sorry.
>
> Mark Twain.

In Alaska, one of the most controversial petroleum projects in terms of creating social welfare and also in terms of CSR is a long sought after natural gas pipeline, which in the years before the new millennium was supposed to encompass a pipeline all the way from the North Slope of Alaska down to Alberta, Canada, a roughly 2000 mile (3000 km) distance depending on which route was to be used. The idea of this pipeline project was that it would develop Alaska's stranded, but massive, natural gas reserves that exist on the Alaskan North Slope and sell them to Alberta, Canada, which is connected by pipeline to Chicago and the Lower-48 natural gas markets.

Considering such a massive multi-billion dollar project, if, on the one hand, the project came in under cost and the natural gas prices held up, then the project could have made a healthy return such that investors would have been compensated and the state government would have obtained healthy revenues. On the other hand, if the costs where to have blown out and the price of US natural gas declined, as they in fact did, then the whole project could have lost money and the investments would make a loss. Considering the high probability for a loss, the project clearly needed a substantial return for the natural gas investors.

But on the other side of the deal, if there had been very high profits for such a project, then the people of the state of Alaska would have felt that they had been unfairly swindled. It is rather like negotiating the price of a single used car where no other cars are available. Both sides of the negotiation cannot be sure if they are getting a good deal and so they haggle over and over again and never eventually have a transaction. The problem with a natural gas pipeline project though is that you can lose a window of opportunity to get a natural gas pipeline put in place, in which case both sides can lose.

The stakes for the project, then, were so high that the negotiations between the IOCs and the government were bound to be contentious. This is especially so in Alaska where there is no state-operated petroleum company and where the private producer companies, the IOCs, need to gauge whether they will do such a project depending on how good the terms the state can provide. In the case of Alaska developing its stranded natural gas reserves, which the oil producers had lease ownership over, the project depended on obtaining favorable state government tax rates, and it was the state legislature that needed to approve it.

2 The Nature of Natural Gas

In order to understand Alaska's failed natural gas project more thoroughly, you have to understand some of the economics of natural gas, which can be quite different than the economics of oil. The greatest difference between petroleum oil (which includes hydro carbons above C5, pentane) and natural gas (particularly methane C1), and where natural gas liquids (NGLs) like ethane (C2), propane (C3) and butane (C4) are somewhere in the middle, is that oil is dense and natural gas is light weight. What this means is that natural gas is about a magnitude more expensive per joule (or British Thermal Unit, BTU) than oil to store. That storage issue is important because in order to make the economics of natural gas cost effective, you have to use it as soon as you produce it. So, any time you extract natural gas, it needs to be sent by pipelines or by Liquefied Natural Gas (LNG), a super cooled liquid (minus 260° Fahrenheit, minus 160 °C), to the consumer to use almost immediately, which makes the natural gas market rather a lot like the electric power market, where the natural gas consumer and the natural gas wellhead producer work in synchronous tandem.

The other problem with natural gas is that due to it being less dense and needing to be squeezed under high pressures to reduce its bulk, the transport system is going to be cheaper the larger it is, i.e. natural gas has immense economies of scale. Whenever you have such huge economies of scale there is tendency for a monopoly to buy out natural gas infrastructure in order to create Rockefeller-esqe high profits. The US history of natural gas pipelines and regulations shows that natural gas has always been a problematic fuel even compared to coal or nuclear power because of the difficulty in storing natural gas, in regulating the natural gas pipeline system and in assuring high enough profits for the producers.

In the US throughout the twentieth century with plenty of oil supplies and associated natural gas supplies surrounding those oil supplies, there was quickly established a natural gas pipeline system from major natural gas producing hubs, like in Texas, to major natural gas consumer hubs, such as in Chicago. But every time, a new natural gas pipeline was built to a new city outside of the main hubs, that particular pipeline created for itself a monopoly to raise consumer prices and lower sales value. Therefore, the then Federal Power Commission (FPC) was instituted in order to regulate all pipelines. However, even with regulations in place, the pipeline system did not keep up with demand and new needs for energy, so by the 1970s, the US regulations for pipelines were deregulated and changed to finally allow natural gas markets to be established. Then, natural gas prices started to rise at the wellhead to allow new exploration and development to occur.

The US supplies of natural gas then were adequate throughout the 1980s and 1990s for most regions in the USA and indeed natural gas and oil often were substituted for each other in the US electric utility market. Starting after the year 2000, though, the US could not keep up with the greater demand for natural gas that the de-regulation created in the 1980s and 1990s, and the reserves of conventional natural gas started to go into decline. Then more trouble hit the US economy as natural gas production declined in 2004 and 2005, which caused natural gas prices to boom well above oil prices in the US, which had never happened before to such an extent. While oil was at $30 per barrel, about $5 per mm Btu ($5 per million British Thermal Unit, or about $5 per Giga Joule) natural gas was twice that price at $10 per mm Btu, which was an unheard of price inversion between oil and natural gas in America's petroleum industry history. Usually natural gas is plentiful and cheap and indeed more often than not it is a nuisance rather than a valuable commodity, but in 2005, it was extremely valuable.

Therefore, partly due to an emerging crisis that hit California due to a natural gas pipeline explosion and due to a drought that reduced available hydro-power, and also due to concerns for US supplies of natural gas, the US Federal Government passed a law, the Alaska Natural Gas Pipeline Act of 2004 (ANGPA) that would guarantee a loan for a large natural gas pipeline from Alaska to the US Lower-48,[1] or if need be a natural gas pipeline from the North Slope to Valdez, Alaska and an LNG project to the Lower-48. The US Federal Government loan guarantee meant that if the entire project went bust, the USA would pay back the loan, although usually such a large

[1]Note, the USA has 48 contiguous states on the North America continent but where the state of Alaska is separated by Canada from those 48 states. Therefore, Alaskans often call those 48 states, the Lower-48 as they are south of Alaska, i.e. lower on the map.

project would have take-or-pay contracts by consumer entities, like Chicago, and so the US Federal Government loan guarantee is more of a subsidy than a full payback. Nevertheless, this loan guarantee would help immensely in getting a pipeline built, if the State of Alaska could also help. And so, with the US loan guarantee and the high natural gas prices, the major oil companies at that time started looking at the possibility of developing Alaska's large natural gas resources and connecting them to the Lower-48.

3 The Push for Alaska Natural Gas

Alaska like most petroleum regions not only wants to develop its oil production but also its natural gas production. Since oil and gas often occur hand in hand in most petroleum producing places, then it only makes sense to produce both valuable commodities. The problem with developing natural gas though is that it is less valuable than oil and much more problematic to transport than oil. So natural gas development often requires a much more involved government to industry collaboration than does oil development such as lower natural gas production taxes, subsidized infrastructure or out and out state investment into the development of the natural gas resources. And that kind of help was going to be needed in Alaska, particularly since Alaska's natural gas was stranded, i.e. it was thousands of miles away from any consumer markets, which means it was going to require a huge project with favorable taxes to be able to cost effectively get that natural gas to market. This means the development of natural gas created an even more intertwined relationship between corporations, government tax and business policy than normal, which can lead to corruption. Alaska's VECO example is a striking example.

VECO which stands for Veltri Enterprises and renamed VE Construction or VECO for short, was never a major oil and gas company in its own right. Nevertheless, it became a large sub-contractor in Alaska's oil and gas industry working under the IOCs. More than that, VECO understood that if a large natural gas project ever came to fruition, then it would gain immeasurable profits from all the contracting work that would occur.

Now Alaska had been trying to develop and export its natural gas from the very beginning of the North Slope development in the 1970s. One of the early plans for getting that natural gas to market was to actually build a natural gas pipeline and locate it right on top of the trans-Alaska (oil) pipeline system (TAPS), which was mostly an above ground pipeline, due to permafrost issues,[2] whereby the natural gas would go to Alaska's south coast and be transformed into LNG and then shipped for consumption in Japan. Indeed, some of the design behind TAPS pipeline was done

[2]Permafrost is soil that is permanently frozen due to cold winters and cool summers and where the summer time grasses keep the soil insulated from the sun so that they remain frozen. A hot oil pipeline could melt that permafrost and cause certain plant and animal species to be disturbed or killed.

for the deliberate purpose of putting a natural gas pipeline just over the top of the oil pipeline. Also TAPS was further able to have sliding capabilities so that it could withstand an earthquake. Nevertheless, the above ground design meant that TAPS would have been a relatively easy structure to have a natural gas pipeline fit higher up above the oil pipeline.

However, it was thought that if a natural gas pipeline were on top of the oil pipeline that it would be too easy to have some sort of gas explosion similar to what happened to the natural gas pipeline from Texas to California in 2000, but instead of blowing up the natural gas by itself, such an explosion would stop up the oil as well and create a double hazard. So when that first idea for a natural gas pipeline and an LNG facility in South Central Alaska was rejected, then other plans kept surfacing mostly surrounding the idea of building a large natural gas pipeline, well away from TAPS, all the way to Valdez and then to sell the natural gas to Japan using LNG. But none of those proposals gained traction.

Finally, though, in 2005, the price of natural gas in the Lower-48 skyrocketed, and with the new ANGPA loan guarantee from the US Federal Government, the talk of an Alaskan natural gas pipeline was heating up, and gave everyone the idea that Alaskan natural gas could be routed to the US through northern Canada to Alberta instead of by LNG to Japan. So, the State of Alaska started to promote the idea of getting involved to help push a natural gas development and build such a natural gas pipeline, but the producers, who owned the natural gas leases, needed help. The oil companies insisted that they needed "fiscal stability", which meant a guaranteed low Alaska tax rate on petroleum for 40 years. One idea was to change the rate of taxes on Prudhoe Bay oil and gas and elsewhere on the North Slope so that the natural gas would be more valuable to develop. However, the state's constitution did not allow tax contracts, but based on court cases, the state was allowed to institute a tax subsidy for about 10 years which would have been a de facto tax contract, but that needed legislative approval.

4 The Crucial Year 2005

One aspect of CSR is that corporations can actively engage the government to change laws or rules in the corporations' favor. Such engagements can be considered responsible if the project gives society an overall gain, or irresponsible if it gives corporations much more gain than consumers or than local citizens' gain within a producing region. The economics of natural gas in 2005 and surrounding VECO show the balance that is needed for CSR.

The way the VECO scandal played out was as follows. While a huge Alaska Highway pipeline was certainly looking feasible by 2005 and was extensively looked into, nevertheless, it would have been a very challenging project economically. Such a large pipeline could have had cost overruns and the natural gas prices in the Lower-48 could have gone down, as they in fact did do, and indeed a current analysis of past proposed pipelines shows scenarios where such a pipeline could have lost money. But at the time, everyone figured that such a pipeline was set to make money. By 2005 with natural gas prices so high and natural gas supplies in the Lower-48 actually in decline, and a Federal loan guarantee on offer, it was thought that such a long pipeline was assured to happen. Note, this was just before fracking and shale-gas technologies came into wide-spread use. Once shale-gas and fracking got going, the natural gas price went down and Alaska's natural gas probably would not have been economic. Although, even without the new fracking technologies, the Alaska pipeline economics were going to be a challenge.

The only way to make such a proposed pipeline economically viable, even then, would have been not just hoping for high natural gas prices in the Lower-48 but also by having a tax contract between the State of Alaska and the oil and natural gas producers, the IOCs, whereby the contract would set Alaska's natural gas production taxes at a low level for at least 10 years into the future. The producers wanted a 40-year contract, but the state constitution only allowed about a 10-year contract, which still might have worked. Granted, such a project would have helped both the state of Alaska and the producers, but in any negotiation when both sides see great opportunity and a sure bet, the perception is that you can gain even more depending on how hard you negotiate the final outcome. The end result was that a negotiated outcome was going to have to be voted on by the legislators of the State of Alaska, so legislators were going to get involved.

What happened then, was that VECO, which was not an IOC, but which worked as a contractor for IOCs, wanted the State of Alaska to make a deal fast, and with low taxes so that the oil companies would build the pipeline and develop the natural gas. Then, in that case, VECO would gain immense wealth through contracts it would obtain from all aspects of the pipeline construction and the natural gas development. The way VECO tried to influence the legislators was by bribing individual legislators. The evidence came to light during a US Federal Government initiated FBI and US Justice Department probe into bribery. VECO's head, Bill Allen, and several legislators were either charged with crimes or found guilty of them, or in some cases allowed to go free by handing over state evidence on the case.

While the details are clouded in courtroom anonymity, nevertheless, this shows that one way for IOCs to affect corporate social responsibility is by either accepting the government's rules or by trying to change them by means of bribes or corruption. Alaska, i.e. the United States of America, is no stranger to such occurrences. While the IOCs were not involved, VECO did subcontract to them and so it would have hurt their reputations had this come out later during the construction of a potential project had such a project been sanctioned. Interestingly, one of Alaska's own US Senators, Ted Stevens, who was in the US Federal Government's Congress for decades was accused of corruption and accepting a bribe in the form of goods in kind. The charges were dropped on account of failed trial processes, but there was a report of a tape recording of Stevens actually accepting a bribe. So, even the US can have corruption, and even Alaskan legislators can be involved in corruption. However, this also shows that one aspect of how Alaska was able to deal with bribery and corruption was through the fact that the US Federal Government can always come into Alaska and investigate any problems based on tips or other information. This shows how Alaska in general is able to have a higher standard of CSR and a higher level of law enforcement than it might ordinarily have based on being part of the US where the rule of law is held to a high standard. Nevertheless, Alaska is not immune to corruption and it shows that corruption in general can happen to change laws and rules in favor of the oil and gas or other industries.

5 A Post-VECO Analysis

One question that is often asked is what would have happened had this gigantic pipeline project gone forward. The rough estimates are that the project would have cost close to $20 billion in 2005 (about $40 billion in 2019 with normal inflation) with a possible $3 per MCF (one thousand cubic feet) tariff, plus at least a $1 per MCF wellhead price, although that could have been more. Since US natural gas prices in 2005 were as high as $8/MCF, and the since projected prices were expected to stay that high, then based on that projecting there would have been good profits on the project.

However, cost estimates were increased due to high inflation and fuel costs to drive trucks that would put the pipeline in place. Plus the permafrost ground required a lot of gravel that would be expensive to put in place, so the costs might have been higher. Even though in 2008, there was another natural gas price spike, nevertheless, by 2011 the price of natural gas was down to about $3 in Alberta and going down further. So given these de facto prices and costs, the tariff on the pipeline would have climbed above $3 per MCF and the price at the Canadian end of the pipeline could have been lower than $3 per MCF at times, which would have made the pipeline uneconomic. Although, since there was an ANGPA loan guarantee, that may have been invoked and could have caused the US Federal Government to be forced to pay some of the losses.

Interestingly, during the heady years around 2005 with natural gas prices as high as they were, the concern by US Federal Government Regulators was not if the pipeline would make money or not, since everyone assumed it would, but rather if the pipeline owners would allow independent natural gas producers on the North Slope to transport their natural gas production through that pipeline. You see, on the one hand, the IOCs were taking a great risk just to build such a huge pipeline and so they would want to fill it with only their own natural gas in order to profit from both the pipeline and the production. On the other hand, the US Federal Government, with its loan guarantee, wanted to require that any independent producer could send natural gas through that pipeline. So there was a subtle conflict between the major IOCs on the North Slope and the US Federal Government over allowing all natural gas resources to be able to go through the pipeline. This also was a point of contention. This point alone might have made an actual project more problematic had it actually been built since it can ruin the cost to benefits ratio.

Still, everyone was blinded by the natural gas price forecasts and the potential big benefits that would roll into the state with this project. There would be benefits to the small independent petroleum producers, i.e. the small oil companies, on the North Slope due to the Federal requirement that they could have access to the pipeline. There would be benefits to the Lower-48 if the pipeline were to bring in 4.5 billion cubic feet (BCF) per day or, if expanded, 6 BCF per day to cut Lower-48 Americans' energy costs. There would be massive benefits to the state of Alaska with wellhead natural gas production taxes and cheaper energy within the State of Alaska. Few thought the project would not happen and so many advocated negotiating longer and harder, so that Alaska could gain more benefits than the deal being proposed.

It is theoretically possible, though, that had a tax deal and contract been struck fast enough and regulation been minimal enough, and the US Federal and Canadian regulatory process smooth enough, and with the loan guarantee, that indeed Alaska could have gotten a pipeline built to Alberta, Canada, and Lower-48 markets by 2010. And even if such a pipeline had all those loss making years, from about 2011 to about 2020, it might have broken even with the loan guarantee, although it probably would have lost money due to the shale-gas supplies. Still, in a break-even case, it is conceivable the pipeline would have provided the state with badly needed revenues, jobs, and economic activity. So, many will argue that had the bribes been accepted to a few legislators of a few tens of thousands of dollars, and not found out about, then the State of Alaska as a whole would have gained billions of dollars in value compared to it not happening. So, it is easy to see how social welfare can be gained, or lost, on one deal, which means that corporate social responsibility is important.

But there is a tough balancing act. In this particular case, had the bribes and the deal gone through, and assuming the pipeline were sanctioned and built, a big if, you can argue that the State of Alaska would have gained immensely from this project, at the expense of the US Federal Government and the IOCs, who would have lost. Still, had news of the bribes come out later, and a project sanctioned and built, the IOCs would have lost all of their credibility and lawsuits would have happened, and they might have lost many millions on the whole deal, even if the actual project could have broken even. Nobody could have predicted the bleak picture for Lower-48 natural gas prices, and how the shale-gas revolution would bring those prices down later in the decade, and so the VECO scandal coming out as it did probably helped the IOCs and the US Federal Government the most.

Since that project, though, Alaska has seen many more planned natural gas projects, usually encompassing a pipeline from the North Slope to Anchorage, Alaska and from there a LNG facility to export natural gas to Japan and China. New deals with China or Japan have been proposed and tax breaks for such projects have been passed around. But to date, no project has happened. In the end, the natural gas on the North Slope is still there and can be produced for value to the state and to society eventually. It is just that it will have to wait for a better market, or more practical minds.

6 CSR Versus Oil Taxes

In general, corporate social responsibility in Alaska is alive and well. Consider environmental issues. While certainly there have been some bad environmental issues in Alaska such as the Exxon Valdez oil tanker crash and spillage, the BP North Slope oil pipeline leak and the Williams Alaska Petroleum Inc., North Pole, Alaska refinery water pollution case, in general Alaska's environment is in very good shape and the petroleum industry does a good job of taking care of the environment.

Take the Exxon Valdez oil spill for example. It was one of the worst ever disasters in the USA to have occurred in one concentrated area. While such a catastrophe should not have happened with proper over sight, nevertheless most Alaskans would rather have an oil industry than not considering the jobs and economic growth the industry has created. However, if we consider the entire territory of Alaska in terms of the Exxon Valdez oil spill, that amounts to 30,000 tons of oil spilled in Prince William Sound in 1989[3] compared to 660,000 square miles (1.7 million square kilometers) of Alaska.

For comparison, the Seattle area is say 200 square miles and there are roughly 200 tons of motor oil spillages from all the cars in that area every year, or about 1 ton per square mile. Exxon Valdez in all of Alaska is only 0.05 tons per square mile and it only happened the one time. So most of Alaska is environmentally cleaner than most of the Seattle area. This does not mean that the oil industry in Alaska should always do no harm, but if Alaska had an industrial economy as dense per square mile as Seattle has, then there would be much more environmental problems in the state. People not living in Alaska tend to exaggerate environmental problems in Alaska when most of the state is fairly clean.

If anything, the environmental regulations on the oil producers are too onerous such as making oil producers wait too long for the winter operating season to start their North Slope drilling, such that they must delay building ice roads to get to the oil fields for exploration and development well drilling. The state's regulators probably wait too long to open up the tundra.[4] In addition, the US Federal regulators are often too quick to shut down areas of oil exploration to save a small pond or a lake on the North Slope for the sake of single specie which in turn reduces oil and gas exploration. When these regulators cause the companies to wait longer, it creates less ability to drill exploration and other wells and reduces Alaska' economic well being.

As far as philanthropic giving is concerned, the petroleum industry does give millions of dollars to organizations in Alaska such as Alaska's university and other groups. The real issue is not the amount of philanthropic giving, the real issue is that since Alaska's population is so small, it takes very little philanthropic giving to gain a high level of good public relation perceptions for the average Alaskan citizen. In the meantime, the producers who give millions are also earning billions on the oil they produce, so the government of Alaska could conceivably tax the oil companies

---

[3]For a full account see McBeath et al. (2008).

[4]See Wall (2005).

at a higher rate than they do and then use those revenues for government sponsored social welfare.

As the VECO case shows, Alaska has had legislative hearings with public involvement surrounding oil and gas projects. However, partly due to the fairly small population of Alaska, its large geographic area and the state of Alaska's relatively small sized government, there is still plenty of opportunity for corporations to use corruption or political pressure as a means to an end to side step public involvement.

For example, when the oil tax change also known as Senate Bill 21 (SB 21) reduced taxes and gave tax credits to oil producers, the issue went to a public vote. Clearly, SB 21 was more lucrative for Alaskan producers than the average oil tax for other states.[5] With tax credit of $8 per barrel for oil prices under $80 per barrel, Alaska has reduced its normal oil take by over a billion a year to the oil companies' advantage, with Alaska's government take lower than for other states.[6] Yet when the vote to rescind SB 21 happened in 2015, it was narrowly rejected, i.e. SB 21 was kept, partly due to the CSR giving to so many community groups.

What you get with such a small population is many opportunities to give money to local community causes to illicit goodwill while in the meantime, your company can advocate a lower government take for the petroleum industry in general and make more profits. On the surface and as far as the public is concerned, the corporations have a high degree of CSR in terms of their giving but since Alaska's population is so small it takes less giving to appear to have a high CSR and that can in turn helps politically to put in place a reduced government take for the corporations.

7 Conclusion

Not many could believe that corruption and bribing could happen in an advanced well-policed country like the USA, but when so much money is on the line, not only do politics become crazy electric, but debates become hardened along polarized lines, such as trying to extract as much value as you can from a project. On the state citizen side, you want to obtain the highest possible tax revenues and job prospects. On the IOC side, you want the best deal possible not just for a normal return on a given investment but also because of the fear factor that indeed, as actually happened, profits from low natural gas prices and high cost overruns and additional taxes on a natural gas project, the pipeline could look at a total loss. The IOCs concerns are especially well founded since the project requires such a long time frame of stable prices, costs and taxes, and because the project would have been one of the biggest construction projects ever in the world since the building of the Great Wall of China.

During the dizzying heights of the natural gas bubble of 2005 and 2008, you could not have convinced many state residents that the project would not have happened, just as the IOCs may have been afraid of losing money. Then, when one of the North

[5]See Nebert (2019).

[6]See Paskavan (2019).

Slope oil service contracting firms, VECO, decided to help smooth the negotiations with actual bribes to legislators, no one would be surprised. While, on the one hand, there is a chance that such bribes would have allowed the project to move forward and benefit the state by many billions of dollars in revenue, jobs, and new industries. On the other hand, had that project gone forward and actually have broken even, although looking back in hindsight it probably would have lost money unless the Federal loan guarantee were invoked, and had the citizens then found out about the bribes, then the trust of the public would have been lost and the IOCs would have forever been under the microscope of both the US Federal authorities and State Alaskan authorities. The IOCs never would have been trusted again, making any new project difficult if not impossible to move forward.

Nevertheless, in general, Alaska's environmental quality is high, and some of the environmental regulations are too onerous. Also, giving philanthropically is fine, but at a cost of a politically induced lower government take.

References

McBeath, M. Berman, J. R., & Mary F. E. (2008). *The Political Economy of Oil in Alaska: Multinationals vs. the State*, Lynne Rienner Publishers, Boulder, Co.

Nebert, D. (2019). An Oil Tax Fix, Fairbanks, Daily-News-Miner, Sept 1 2019, p. C5.

Paskavan, J. (2019). Alaskans should stop giving costly oil tax credits, Fairbanks, Daily-News-Miner, Mar 17, 2019, p. C5.

Wall, S. (2005). *The Economic Implications of the Climate-Change Induced Shortened Oil Exploration Season on Alaska's North Slope*, Master's Project In Fulfillment of M.S. Resource and Applied Economics Requirement, Apr 30 2005.

Corporate Social Responsibility in the Mining Sector in Canada

Jocelyn Fraser and Andre Xavier

Companies working in the extractives sector have become increasingly adept at managing legal, financial and operating risk. They have also become proficient at developing expertise in order to mitigate threats to project production schedules and budgets, and ensuring that their shareholders profit from their investments. However, in recent decades, two fundamental changes have rattled the business-as-usual approach and have adversely affected the success of mining and oil and gas projects. The first is a change in the definition of business success. Milton Friedman's belief that the "business of business is business" has largely been superseded as companies endorse a new standard of corporate performance—one in which a growing emphasis is placed on social imperatives as well as financial performance, and where profits benefit stakeholders, not just shareholders. The second change involves an increase, both in number and in cost to companies, of incidents of company–community conflict. From minor disagreements to sustained and violent conflict, the failure to earn stakeholder approval has emerged as one of the leading causes of project delays and a key strategic risk. From 2008—the peak of the super cycle—until 2020, earning a "social license to operate" has been ranked by industry executives as one of the top business risks faced by the extractive sector. In other words, for companies whose projects can only be built where deposit exists, and where the life of an operating facility can extend for several decades, generating value for both the company and the community is becoming a strategic imperative and

J. Fraser · A. Xavier (✉)
Norman B. Keevil Institute of Mining Engineering, University of British Columbia, Vancouver, Canada
e-mail: andre.xavier@ubc.ca

J. Fraser
e-mail: jocelyn.fraser@ubc.ca

E. G. Pereira et al. (eds.), *Sovereign Wealth Funds, Local Content Policies and CSR*, CSR, Sustainability, Ethics & Governance,
https://doi.org/10.1007/978-3-030-56092-8_33

has given rise to initiatives designed to demonstrate corporate social responsibility (CSR).

Numerous authors and organizations have proposed definitions of CSR. Driven by the question, "What responsibilities to society may business reasonably be expected to assume?" Howard Bowen, in his seminal 1953 book *Social Responsibilities of the Businessman,* was reluctant to provide a definition. He believed that "the way to greater responsiveness of businessmen toward their social obligations lies in the processes of broadly-based discussion and individual soul-searching on the part of actual participants—not in the spelling out of 'answers' by outside observers" (p. xi). Bowen went on to interpret social responsibilities as the obligations that business entrepreneurs have to pursuing policies, decision making, and implementing actions that are desirable in terms of the objectives and values of society (1953: p. 6).

Bowen's contemporary, Theodore Levitt warned about the dangers of social responsibility, espousing the more pragmatic view that the goal of for-profit organizations must be to maximize profits to shareholders while respecting the "rules of the game". In 1970, American economist, Friedman (1970) famously declared that "the one and only one social responsibility of business [is] to use its resources and engage in activities designed to increase its profits". Friedman's article in the New York Times was one trigger for a more fulsome discussion amongst scholars and practitioners about the idea of social responsibility. Archie Carrol's view that corporate social responsibility "encompasses the economic, legal, ethical, and discretionary (philanthropic) expectations that society has of organizations at a given point in time" (Carroll 1979, 1991, 2016; Carroll and Shabana 2010) gained popularity and created a platform for scholars to discuss questions related to social performance, stakeholder engagement and the importance of value creation. As such, questions regarding the ways in which a company can work to leave customers, suppliers, communities, employees and financiers better off (Freeman et al. 2006a, b: 5) took on greater levels of importance.

In the Canadian context, the first national CSR strategy, *Building the Canadian Advantage* (2009), which was developed by Global Affairs Canada (GAC), conceptualized CSR as "the voluntary activities undertaken by a company to operate in an economically, socially, and environmentally sustainable manner" (p. 2). A few years later, a new definition was proposed in the report entitled *Doing Business the Canadian Way*, in which GAC defined CSR "as the voluntary activities undertaken by a company, over and above legal requirement, to operate in an economically, socially and environmentally suitable manner" (GAC 2014: 3).

In the mining sector, more relevant reflection and discussion about corporate social responsibility was initiated during the 1992 Sustainable Development conference in Rio de Janeiro, Brazil. These conversations gained momentum in 1999 in Davos, Switzerland, and at the Rio +10, in 2002 in Johannesburg, South Africa, influenced by concern that there was a growing disconnect between mining and minerals-related practices and the needs and values of society (MMSD 7 Questions 2002: iv). In this context, and in the face of growing social opposition, industry formalized its commitment to responsible mining with the formation of the Metals,

Mining, and Sustainable Development (MMSD) initiative, which, in 2001, gave rise to the International Council for Metals and Mining (ICMM).

In this chapter, CSR is defined as the initiatives that go "beyond philanthropy and compliance to address the way companies manage their economic, social, and environmental impacts and their stakeholder relationships in all their key spheres of influence: the workplace, the marketplace, the supply chain, the community and the public policy realm" (Kytle and Ruggie 2005: 9).

1 CSR and the Mining Sector in Canada

Canada is recognized as one of the world's leading mining nations. The industry provides direct employment to more than 400,000 and creates an additional 200,000 indirect jobs. There are exploration or mining activities taking place in every province and territory in the country, with mineral production contributing $97 billion to the country's gross domestic product (GDP) in 2017. Vancouver, home to more than 700 exploration companies, is recognized as a global centre for mineral exploration. Toronto is considered a global centre for mine finance: the Toronto Stock Exchange has the largest number of listed mining companies of any stock exchange globally.

Over the years, Canadian companies have made significant progress in terms of environmental and social performance. The concept of a social license to operate (SLO)[1] was introduced in 1997 by Canadian mining executive James Cooney in response to a growing need to engage with communities and establish ongoing, positive relationships (Cooney 2017). Cooney's Canadian colleagues Robert Boutilier, Susan Joyce and Ian Thompson worked to take the idea of SLO from metaphor to management tool—an effort that has been built upon by many other scholars.

Mining companies in Canada were the first in the world to develop an externally verified performance system for sustainable mining practices with the creation of the Mining Association of Canada's (MAC) Towards Sustainable Mining (TSM) tool. Established in 2004, TSM is built on three pillars—communities and people, environmental stewardship, and energy efficiency. TSM focuses on providing the tools to enable mining companies to meet society's needs for minerals, metals and energy products in the most socially, economically and environmentally responsible way (MAC 2019). Participation in TSM is mandatory for all MAC members. TSM reports, produced annually and verified by third parties, provide the public with an overview of individual company performance in key environmental and social areas. This award-winning initiative has been adopted in other resource-rich countries including Argentina, the Philippines, Finland, Botswana and Spain, with a number of additional countries considering implementation.

The Government of Canada has been clear that it expects Canadian mining companies operating both at home and abroad to adhere to the highest standards of social

[1] SLO can be defined as having the tacit permission of governments and communities to undertake exploration and mining activities.

responsibility. In 2009, the government produced a CSR strategy for the international extractive sector entitled *Building the Canadian Advantage*. This was followed in 2015 with new legislation—the Extractive Sector Transparency Measures Act—and by the creation, in January 2018, of the Canadian Office for Responsible Enterprise (CORE). In addition, Canada was one of the first countries to call for extractive companies to negotiate impact benefit agreements (or participation agreements) with indigenous communities in order to provide compensation for the use of their traditional lands.

Impact benefit agreements (IBAs) are binding agreements signed between mining companies and indigenous communities whose traditional territories are impacted by extractive exploration and operations, and are seen as a way to move towards a more equitable and sustainable approach to mineral development. In Canada, IBAs are a mandatory requirement under the Nunavut Land Claims Agreement. In the other provinces and territories, IBAs are not legally required but are normatively expected.

The primary purpose of IBAs is to address the adverse effects of mining activity on local communities and their environment, and to ensure that aboriginal peoples benefit from the development of mineral resources (O'Faircheallaigh and Corbett 2005; Hitch 2006; Tuulentie et al. 2019). IBAs were previously confidential documents; however, new federal legislation requires disclosure of any payment of over $100,000, and it is expected that transparency will strengthen the negotiation process and associated outcomes.

IBAs can cover a range of issues: employment and business contracting opportunities, training and education (including apprenticeships and scholarships), equity participation, revenue sharing, cash compensation, social and environmental monitoring and/or mitigation measures, archaeological site preservation, access to facilities and infrastructure, information exchange, agreement management and dispute resolution mechanisms. Recently negotiated IBAs are exploring new ways to support sustainable development: the IBA concluded in early 2019 between the Nisga'a and Ascot Gold for the Red Mountain project in British Columbia includes a commitment to create a Business Opportunities Committee to identify, plan and build capacity for local procurement. When strategically implemented and effectively monitored, IBAs can be used to promote resource development in a manner that contributes to the sustainability of the local environment and economy, and the social and cultural fabric of affected communities.

At the provincial level, royalty revenue sharing agreements, such as those in use in British Columbia, Ontario and the Yukon, provide another means for the financial benefits of mining to be distributed back to communities impacted by the industry. As well, many companies fund their own social responsibility and community investment programs, and work with stakeholders and rights holders to identify programs and initiatives with which companies can become engaged.

These initiatives are both important and relevant but do not address all concerns. In 2019, three Canadian companies were the subject of lawsuits brought by claimants from abroad that allege human rights abuses by the mining sector, and a number of Canadian-owned mining operations have been the focus of mining-community conflict, sometimes referred to as social risk.

2 Social Risk and Social Responsibility

As with other terms used in this chapter, the definition of social risk has been contested. The definition most frequently endorsed by the extractive industry is that of Kytle and Ruggie (2005) who write, "from a company's perspective, social risk arises when an empowered stakeholder takes up a social issue area and applies pressure on the corporation (exploiting a vulnerability in the earnings drivers—e.g. reputation, corporate image), so that the company will change policies or approaches in the marketplace" (p. 6). Those who endorse this definition advocate for using CSR and stakeholder engagement to collect intelligence then integrate the information into the organization's strategic risk paradigm. In other words, they advocate for aligning social risk with other forms of risk management such as technical, legal, operating, financial and political risk.

Others (for example, Lapalme 2003; Schafrik and Kazakidis 2011) have suggested that this definition is not appropriate and argue that social risk is the risk that specific operations pose to local communities. There is a growing movement that advocates for the differentiation of social risk and business risk (Brereton and Parmenter 2006; Graetz and Franks 2015; Kemp et al. 2016). These scholars suggest that business risk and social risk are conflated and that confusion over the definitions of these terms has the potential to lead to greater conflict. They propose that social risk should be defined as "the perceived or expected potential future threats to, and unwanted impact on, individuals and groups of individuals arising from the processes of social change precipitated by development interventions and the decisions of external actors" (Graetz and Franks 2016: 587). This approach posits that threats and unwanted impacts on a company's operations, reputational capital arising from operational decisions and strategies, and the exogenous response of other actors to these decisions and strategies, should be labelled as business risk. Scholars (e.g. Kemp et al. 2016) have suggested that the mining industry is equating social risk with social acceptability risk, a term introduced by Miller and Lessard in 2001 that stresses the importance of effective stakeholder engagement for reducing community opposition to large engineering projects.

In this chapter, we endorse the position that regardless of how social risk is defined—as a risk to people or as a risk to business—CSR represents an excellent mechanism through which to address these challenges across the business enterprise provided that it is an extension of global operations policies, and that CSR programs are not discretionary expenditures or the target of cost-cutting activities, and that CSR is strategically linked to core business. In other words, CSR needs to be embedded as a strategic business imperative and a core element of risk management. This idea raises a question to be considered in this chapter: which approaches to CSR are most effective and how is success measured?

There are operational, financial, moral and ethical reasons for companies to create processes to proactively engage with their communities of interests. Communities and governments have high expectations that mining projects will deliver socio-economic benefits to the regions where mining operations take place (Rooke 2016;

Fraser 2018). Conflicts and opposition to mining projects are often triggered when communities feel that the engagement process is not adequate or that the local benefits are limited or unfair (Kapelus 2002; Calvano 2008; Franks et al. 2014; Andrews et al. 2017; Mercer-Mapstone et al. 2019). Domestic and international experiences have taught mining companies the importance of consulting and engaging with local communities of interest. Studies show that the cost of conflict for a company with a capital expenditure of between US$3 and US$5 billion can go up to $20 million per week (Franks et al. 2014).

In recent years, access to financing has become one more compelling reason for project proponents to put robust community engagement strategies in place. Through the International Finance Corporation (IFC), the World Bank Group was one of the first organizations to establish performance standards with which companies seeking project financing with IFC must comply.[2] The performance standards align with the IFC mission of promoting responsible investment in developing countries, reducing poverty and improving people's lives, as well as ensuring that projects are socially and environmentally responsible.

In addition to the financial risks that can arise for a company for not having engaged adequately with the communities with which they work, there are human rights risks and impacts that can adversely affect communities of interest, trigger community discontent and result in negative impacts on businesses. The United Nations Guiding Principles on Business and Human Rights (UNDRIP), which was endorsed by the UN Human Rights Council in 2011, provides a global standard for preventing and addressing these risks and impacts. A specific right pertaining to indigenous peoples, known as Free Prior and Informed Consent (FPIC), is embodied within UNDRIP and is important for extractive companies to factor into their engagement and social responsibility planning.

In Canada, national and provincial governments and industry organizations have published tools and guidelines to promote proactive engagement with communities of interest. These guidelines cover the entire mine life cycle from exploration to closure. Some focus specifically on how to successfully undertake meaningful community engagement with First Nations groups such as the *Exploration and Mine Guide for Aboriginal Communities* published in 2008 by Natural Resources Canada in partnership with the Prospectors and Development Association of Canada (PDAC), the Mining Association of Canada (MAC) and the Canadian Aboriginal Minerals Association.

Over the years, academia, governments, communities and industry have enhanced their understanding and improved their practices regarding community engagement in mining projects (Kapelus 2002; Kemp et al. 2006; Lin et al. 2015a, b; Delannon et al. 2016). It is acknowledged as common practice that project proponents should

[2]Other relevant CSR standards include the OECD Guidelines for multinational enterprises, UN Guiding Principles, Global Reporting Initiative, ISO 26000—Social responsibility, Voluntary Principles, United Nations Global Compact, Equator Principles, Extractive Industries Transparency Initiative, UN Principles for Responsible Investment.

engage local communities as early as possible (Zandvliet and Anderson 2009). Similarly, it is expected that national and local governments inform, consult with, and secure spaces for community participation in the early days of exploration and mining activities. In addition to "when" community engagement takes place, another critical piece of the community engagement puzzle is the "how" the engagement takes place (Rooke 2016).

Once again, there are a variety of tools and guidelines available to support engagement planning. Most agree that the first step for project proponents is to build an understanding of the communities in which they hope to operate. Publicly available documents, such as official community plans, council minutes and municipal budgets, can be reviewed to learn about community-identified issues and priorities. It is also important to be able to identify who to talk to within the communities of interest, to determine how best to conduct those talks, and to plan an engagement timeline that recognizes that building relationships takes longer than building a mine or pipeline. A strategic approach to community engagement should be viewed as an opportunity to develop trust, meaningful relationships and partnerships with communities of interest.

The "how" component recognizes that there is a process to be followed in order for engagement to take place in a proper way. The process may be preceded by observing cultural rituals, traditions and practices that can be overlooked if a project proponent's representatives are driven by a set of short-term performance indicators when community engagement is approached as simply another box to be checked on the company's engagement checklist (Delannon et al. 2016).

Leading practices with respect to successful engagement suggest that the process should be co-designed and constructed together with the communities of interests (Bowen et al. 2010). Furthermore, "Co-management relationships are now emerging as one form of combining community, government, and companies in order to achieve greater adaptive capacity for resource management" (Rooke 2016: 40). Ultimately, early, proactive and meaningful engagement with communities contributes to reducing financial and social risks to companies and mining projects, and facilitates access to financing and supporting companies' social license to operate.

An example of co-management is the Elk Valley Environmental Monitoring Committee. The committee is comprised of representatives from Teck, the British Columbia Ministry of Environment and Climate Change Strategy (ENV), the Ministry of Energy, Mines, and Petroleum Resources (EMPR) and representatives from the Ktunaxa Nation Council (KNC) as well as the Interior Health Authority (IHA). An independent scientist and a professional facilitator are also on the committee, which monitors selenium concentrations in the biota. The Environmental Monitoring Committee (EMC) was created in 2015 as a requirement of the Environmental Management Act (EMA) permit issued to Teck. The committee reviews the monitoring and reporting submissions that are required under the permit and provides technical advice to Teck and the director of the Ministry of Environment and Climate Change Strategy.

3 CSR Challenges Facing Extractives Industries

The issue of trust is a concern for extractive projects around the world. Trust is based on many factors, including respect, inclusivity, transparency and consistency, and is an essential requirement for local communities' acceptance of mining projects. Unfortunately, many communities, both in Canada and abroad, distrust mining companies and their related governmental institutions (Horowitz 2010; Andrews et al. 2017; Conde and Le Billon 2017; Moffat et al. 2015). This lack of trust has been identified as a trigger for conflict and a factor in escalating confrontation (Muradian et al. 2003; Calvano 2008).

The Edelman Trust Barometer, which surveys more than 30,000 people annually across 27 countries consistently finds that trust in business, industry and government, is low. In Canada, for example, trust inequality is at a record high with more than 50% of participants holding the belief that the system is failing them, and with CEOs and government officials being seen as the least trusted spokespeople amongst the survey participants (Trust Barometer 2019).

Andrews et al. (2017) identified weak government capacity, corruption and lack of transparency as key factors that can lead to distrust in government representatives, and which contribute to negative impressions of companies that appear to be working with government personnel. Moreover, Slack (2012) argues that companies that do not integrate CSR into their business models end up creating a contradiction between CSR rhetoric and its implementation. This is problematic because while trust is difficult to develop, it can be compromised or lost with one unethical or untrustworthy action. In today's hyper-connected world, incidents of deficient performance or perceived corporate malfeasance can become known well beyond immediately impacted stakeholders. Furthermore, while business expects an increase in trust amongst stakeholders and rights holders by delivering CSR-related activities, its impacts are limited by the fact that communities often feel that they have not participated in the process (Conde and Le Billion 2017; Majer 2013).

A shift towards community inclusion in natural resource governance processes began a few decades ago and manifests today in the form of co-management or collaborative governance, where the extractive companies establish joint decision-making processes with communities, rights holders and government (Bowen et al. 2010; Delannon et al. 2016). Successful partnerships and practices of community inclusion "encompasses mutual influence, with a careful balance between synergy and individual autonomy, which incorporates mutual respect, equal participation in decision making, mutual accountability and transparency" (Zurba 2017: 6).

CSR-related activities can result in trust building, provided these activities are inclusive and participatory. To be effective, CSR needs to deliver results both in the short term and contribute to the sustainable development of the region, which often takes time and requires a long-term perspective. Traditionally, many companies have focused their CSR programs on philanthropic investments and have responded to community requests for the company to support short-term projects with direct cash contributions or by providing funding to improve infrastructure or to build roads, hospitals or schools (Frynas 2005; Jenkins and Obara 2006) While philanthropic investment can improve quality of life for local residents over the short term, many mining companies, governments and communities have struggled with the fact that the maintenance of these facilities and the continuation of the services that they provide are jeopardized when the mine closes either temporarily, as can happen when commodity prices go down, or permanently, when the resources are exhausted or are no longer worth extracting (Xavier 2013; Xavier et al. 2015).

An unintended consequence of this traditional approach is that communities can start to rely on philanthropic investment and look for short-term solutions to the detriment of long-term solutions that are oriented to contributing to the sustainable development of the region (Frynas 2005; Bainton and Holcombe 2018). As such, that which may have made business sense in the early days of a project turns into a problem with unfulfilled expectations on the part of all parties involved (Taarup-Esbensen 2019). This situation can become increasingly challenging for companies since communities and governments frequently expect that the private sector is responsible for resolving the problem.

Such a situation can lead to resentment, a withdrawal of interest from collaboration, and the erosion of trust between the parties, which could ultimately impede the parties' ability to think collaboratively about the design and implementation of transformative activities and plans for contributing to the sustainable development of the region.

It is recognized that both old and new mining jurisdictions expect transactional short-term, philanthropic, easy cash solutions to address community issues, and it would be unwise for a project proponent to dismiss this situation. Nevertheless, leading practices in CSR are thorough and purposeful with respect to striving to find a balance between transactional initiatives and strategies and programs that lead to transformative outcomes (Zvarivadza 2018; Tuulentie et al. 2019).

4 Approaches to CSR: From Transactional to Transitional to Transformative

As noted earlier, mining companies are increasingly recognizing the importance of engaging all stakeholders, not just shareholders or institutional investors, and of building collaborative relationships with communities proximal to resource deposits.

Different companies employ a variety of CSR strategies at various stages of the mine life cycle. Approaches are often dictated by the financial and personnel resources that are available for supporting social responsibility and by the needs of the communities of interest. Three common categorizations of CSR are illustrated in Fig. 1.

Transactional engagement provides information or resources to a community through arm's length transactions. Philanthropy, employee volunteerism and community investment are three examples of transactional, or one-way, engagement. This

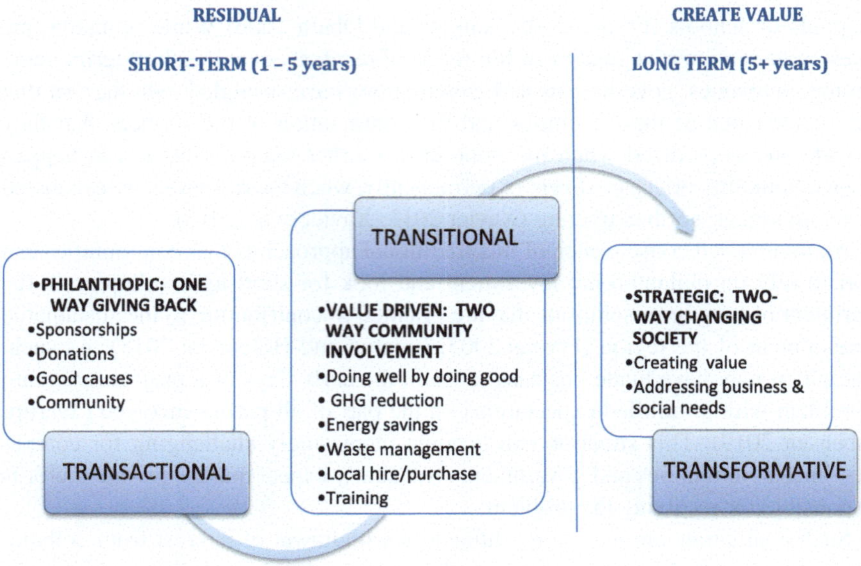

Fig. 1 CSR continuum: moving from transactional engagement through transitional engagement and arriving at transformative CSR. Adapted from "*Strategy and Society: The link between competitive advantage and corporate social responsibility*"—Michael Porter & Mark Kramer, HBR, Jan–Feb 2011 and Network for Business Sustainability Engaging the Community: A Systematic Review Sept 2008

approach is best described as traditional CSR, whereby companies redistribute wealth by investing in initiatives and infrastructure that community stakeholders view as important but that has little relevance to core business. For example, opportunities for this type of philanthropy include swimming pools, daycare centres, medical equipment and schools. Investing in transactional CSR has a strong appeal: opportunities are brought forward by the community, the short-term focus (typically one to five years) allows the transaction to be quickly executed and disclosed to stakeholders, and the common objective of shareholders who are looking for short-term financial reports is also met.

One challenge with respect to transactional CSR that is relevant to the mining sector is the belief that companies should not redistribute wealth created via irresponsible practices. In other words, compensating communities for environmental and social impacts that have arisen from deficient performance is not aligned with socially responsible behaviour. Other challenges regarding investments that are based on community wish lists are that a company may experience that they are chasing a moving target; the investment may only fulfil the expectations of a small minority; and philanthropy can create situations where the community cannot afford the annual operating costs of the facility built by corporate sponsorship (Rodríguez et al. 2014). Finally, attempting to build relationships based upon transactional CSR may increase the risk of conflict both because stakeholders compete for funding, and because the more vocal and vociferous the community, the more attention and money it may receive from company officials, who tend to respond more immediately to aggressive demands and threats than to polite requests (Zandvliet 2004). Nevertheless, transactional CSR has an important role to play. Communities may expect that wealth from resource extraction should be redistributed via philanthropy and may be unwilling to abandon it, especially before trust is built with company representatives.

Transitional CSR, the second stage in the continuum, is characterized by two-way communication, consultation and collaboration and sees companies investing in projects that yield benefits to both the business and the community. This type of CSR, sometimes described as "doing well by doing good" (Aguilera et al. 2007), is seen as a more strategic approach than pure philanthropy since there is often a link between the mining company's business and the communities in which it operates. A criticism of the transitional approach is that companies may report progress on issues such as greenhouse gas reduction (GHG), recycling, local procurement, local hiring, energy conservation and community engagement, without changing the underlying business practices that cause environmental and social degradation (Stubbs and Cocklin 2008). Despite these shortcoming, as with transactional engagement, there is a time and place in the mining cycle when transitional CSR activities can add value. One avenue to explore involves the ways in which the CSR activities of mining and oil and gas can align and contribute to communities, and municipal and regional development plans. Implementing CSR activities that respond to these plans demonstrates commitment to local issues and avoids the risks of long wish lists that, in many cases, are based on wants rather than needs.

The third approach is the concept of transformative CSR in which value is created for both the mining company and the community. Since transformational CSR is anchored in strategy, results can take longer to achieve. This is due to the need to prioritize projects aligned with core business, and finding and recruiting partners to deliver the projects requires trust. Mining projects, where entirely different teams of people are often used to find, then build, then operate the mine, are particularly susceptible to changes in the social chain of custody.[3] Changes in company personnel can erode the trust held by community stakeholders, and can also cause CSR initiatives to be re-evaluated, adjusted, or abandoned. An example of how such an issue can be addressed along the CSR continuum is offered in Table 1.

As discussed, there is an expectation that project proponents will engage and develop relationships with their communities of interests. Many companies purposefully dedicate time and resources to meeting such expectations and create channels and ongoing spaces for community voices de be heard. Some of these spaces include participatory processes that allow for the joint oversight of activities, results and the impacts of mining operations, both with respect to the environment and communities. Participatory monitoring involves the collaborative process of collecting and analysing data, as well as communicating results with the aim of identifying and solving problems together (CAO 2008). As such, participatory environmental

---

[3]Social chain of custody refers to efforts to maintain reputation capital and social commitments made throughout the life cycle of a project.

Table 1 CSR approaches within the continuum. Example illustrating how companies adopting different CSR approaches might approach dealing with infrastructure issues

CSR approaches using the example of infrastructure		
Transactional	Transitional	Transformative
Donation to capital campaign to build infrastructure, for example, a recreational facility for the community	Augment mine infrastructure to provide community benefits: e.g. installing a cell tower to provide service to the mine and adjacent communities, or road building and maintenance to service the mine and adjacent communities	Collaborative partnership to design shared infrastructure: e.g. a mine waste water treatment facility that provides potable water to adjacent communities
No stakeholder engagement	Limited stakeholder engagement	Integrated stakeholder engagement/program ownership
Limited strategic relevance for business	Strategic relevance recognized	CSR is strategic imperative
No correlation between money spent and business success	Little correlation between money spent and business success	Addresses macro-economic conditions to benefit business and society

monitoring programs, as well as socio-economic monitoring programs, have been increasingly adopted as mechanisms through which to foster and secure community participation and engagement in the decision-making process.

From the company's point of view, in addition to allowing it to respond to the concerns of the community, participatory environmental monitoring can also lead to earning the community's approval. Furthermore, the participatory spaces that are created can facilitate dialogue and create trust between the company and its communities of interests. This approach can also provide community members with learning opportunities, skill and training in new technologies (Pareja et al. 2017). Environmental monitoring programs are concerned with issues related to the fauna and flora of a region, and most often focus on water quality and quantity, dust, animal migration, biodiversity and greenhouse gas emissions.

Following a similar pattern, companies are setting up joint programs through which to monitor and evaluate the socio-economic benefits that a mining project brings to a region. In common with participatory environmental monitoring, the aim of socio-economic monitoring is to understand a project's contributions to the local people, to identify opportunities through which to maximize benefits, and to find and address issues that may be limiting the communities' full potential to benefit from the existence of a mine operation. Socio-economic monitoring programs often examine issues related to local employment, training, education, indigenous employment and local businesses development.

In general, participatory monitoring programs can help companies, governments and communities with the early identification of concerns so that they can be corrected and mitigated in a timely manner. Such programs can also help establish baseline information and allow communities to participate in the decision-making process related to their local environment and socio-economic future. However, if these participatory monitoring programs are perceived as token exercises without relevant participation from the communities of interest, it can lead to discontent and mistrust (Himley 2014).

There are many examples of CSR approaches across the continuum described in Fig. 1. It is important to recognize that different strategies can be used at different stages of the life cycle of a mine—from exploration through to closure. To illustrate how transformative CSR can be employed throughout the life of a mining project, three applied examples are offered.

4.1 Exploration: Early Engagement to Creating and Sharing Value

The exploration phase is the time when communities have their first contact with prospective miners. These first impressions can make or break trust, which is a critical attribute for negotiating company–community relationships. Erdene Resource Development Corporation, a Canadian exploration company, is working to advance a gold discovery towards a mine development in Mongolia's Gobi Desert. The nearest town is about 90 km away from the exploration site and was built in an area with no access to potable water. Due to Erdene's community engagement, company personnel were aware of the town's interest in improving its access to potable water. In 2017, when Erdene commissioned hydrogeological work to source a water supply for the potential mine, the company offered to expand the area for the water investigation to include the town. This led to the discovery of a potable water source three kilometres from the town centre. Local municipal officials and Erdene personnel then worked together to drill the new water well, which now provides drinking water to local residents and replaces the need to truck potable water from the provincial capital, more than 200 km to the north.

4.2 Operations: First Nations and Local Procurement Practices in Saskatchewan's Uranium Industry

Cameco's mining operations are located in the northern part of the Canadian province of Saskatchewan. The company was one of the first in Canada to sign collaboration agreements with communities near its operations, and its corporate social responsibility strategy aims to build relationships, strengthen partnerships and secure the support of nearby communities. Cameco's strategy is focused on workforce and business development, community engagement and investment, as well as environmental stewardship. Through its Northern Preferred Supplier Program, Cameco provides guidelines to develop northerners' capacity to provide safe, high-quality and cost-effective goods and services. Since 2004, $3.76 billion worth of goods have been procured locally for the mine. In 2018, 89% of the mine's goods and services including catering, transportation, general construction and maintenance support, were supplied by local businesses. In addition to contributing to local business development in the region, Cameco recognizes that local suppliers reduced the company's operational risks (Partnerships in Procurement—understanding aboriginal business engagement in the Canadian mining industry (2016) https://www.ccab.com/research/partnerships-in-procurement-2016-ccab-and-msv/).

4.3 Closure: New Gold's Cerro San Pedro Mine

The impending closure of New Gold's gold mine was of concern to the 4000 residents living in the municipality of Cerro de San Pedro in central Mexico: the mine was the principal employer in the region and a key economic contributor to the community. As the end of mining approached, the importance of ensuring that site workers and the surrounding community would be able to transition to a post-mining economy was recognized as a core planning need, and provided the impetus for the development of the Cerro San Pedro Integrated Mine Closure Program. The program was co-designed with the community with the aim of empowering local citizens to play an integral role in designing the closure plan and to establish a mechanism through which to support economic diversification. This community engagement strategy was recognized with a 2019 TSM Award of Excellence for community engagement from the Mining Association of Canada.

5 The Way Forward

Corporate social responsibility in the extractives sector has advanced significantly in the last 20 years. Canadian companies and governments have made contributions towards this progress by establishing concepts such as social license to operate, a term that is perhaps even more relevant today than it was when introduced in the late 1990s. The importance of a social license is evidenced by the fact that more than 50% of mining executives surveyed by Ernst Young in 2018 predicted that SLO would be the most significant business risk facing the extractives sector in the next two to three years. As noted earlier in this chapter, made-in-Canada toolkits and guidance initiatives are available to support companies wishing to demonstrate their commitment to social responsibility. Industry's use of the TSM reporting protocol provides further evidence of the sector's broad acceptance of the idea that extractives must deliver benefits to stakeholders and rights holders, not just shareholders.

Nevertheless, there remains room for improvement. Media coverage and social networks illustrate that extractive companies—regardless of country of origin—continue to fail to meet social expectations in multiple jurisdictions around the world.

One part of the challenge for oil and gas and mining companies is that they need to better understand social risk. Some have argued that industry's tendency to view social risk as a business risk is problematic (Graetz and Franks 2014; Kemp et al. 2016). These authors suggest that there are in fact two separate risks—risks to people, and risks to the project, and that conflating the two can lead to problems. For mining companies to be able to properly assess "social" risk, we believe that it is important for CSR to be considered as a core and strategic business function. To accomplish this objective, companies will need to break down entrenched organizational silos, increase collaboration and provide training for cross-functional teams. Assembling an integrated team to address social issues should result in stronger internal alignment and create synergy across business lines. This alignment should then enable more effective social risk management and support the identification of opportunities for company–community collaboration to create value.

The idea of creating value is an important one. Extractives contribute much to society that is valued, including materials required to transition to a low-carbon economy. Despite this, the industry is not commonly viewed positively by resource-rich communities and governments. Transformative approaches to CSR offer an opportunity through which the extractives sector can be realigned with the values of society, as well as the potential to rebuild trust through a new social contract. An economic management strategy, called creating shared value (CSV), is one approach companies could embrace to ensure the extractive sector contributes to the long-term sustainable development of the communities and countries in which they operate.

The CSV terminology was introduced by Porter and Kramer (2011) and establishes a framework for identifying opportunities to improve socio-economic outcomes and related core business performance (e.g. decreased operational costs, enhanced productivity, and/or a predictable and stable business environment). The objective of CSV is to find the points of intersection between the needs of business and society, and then build collaborative partnerships to address complex social problems that are at the root of market failures—situations in which socio-economic conditions prevent conventional business models from succeeding (Kramer and Pfister 2016). In the mining sector, points of intersection between what local communities need and what miners need to be successful are not hard to find. For example, both benefit from an educated workforce and healthy communities, both need access to water, both require energy and infrastructure. Creating shared value does not replace corporate social responsibility. Instead, it brings business strategy into the equation, creating the opportunity for more relevant and impactful contributions from companies.

Efforts to prove the connection between financial performance and performance on social issues have proven problematic for many companies with CSR frequently viewed as a sunk cost, one that offers little, or a questionable, return on investment. In this chapter we have argued that CSR needs to be reframed as a strategic business imperative. The objective of a strategic approach to CSR is for extractive companies to build more positive relationships with communities that host operations, reduce conflict and increase the probability that projects can earn the approval required to build and operate facilities with impacts on society and the environment. We believe that, moving forward, the best-practice examples of CSR will share three features. First, CSR will be business strategy with parallel goals: improving company performance while delivering economic value to host communities. Second, CSR programs will focus on issues affecting both company and community, recognizing that the more closely tied a social issue is to a company's business, the greater the opportunity to leverage the firm's resources and benefit society. Third, collaboration with stakeholders will improve business performance while delivering tangible social benefits.

List of Resources, Best Practice Documents and Websites

Amor-Esteban, V., Galindo-Villardón M-P, García-Sánchez I.-M., & David, F. (2019). An extension of the industrial corporate social responsibility practices index: New information for stakeholder engagement under a multivariate approach. *Corporate Social Responsibility and Environmental Management, 2019*(26), 127–140.

Amor-Esteban, V., & Galindo-Villardón M-P, García-Sánchez I.-M. (2018). Useful information for stakeholder engagement: A multivariate proposal of an Industrial Corporate Social Responsibility Practices Index. *Sustainable Development, 2018*(26), 620–637.

Bodruzic, D. (2015). Promoting international development through corporate social responsibility: The Canadian government's partnership with Canadian mining companies. *Canadian Foreign Policy Journal, 21*(2), 129–145.

Böhling, K., Murguía, D. I., & Godfrid, J. (2019). Sustainability reporting in the mining sector: Exploring its symbolic nature. *Business and Society, 58*(1), 191–225.

Buchanan, S., & Marques, J. C. (2018). How home country industry associations influence MNE international CSR practices: Evidence from the Canadian mining industry. *Journal of World Business, 53*(1), 63–74.

Chollet, P., & Sandwidi, B. W. (2018). CSR engagement and financial risk: A virtuous circle? International evidence. *Global Finance Journal, 38,* 65–81.

Conde, M. (2017). Resistance to mining. A review. *Ecological Economics, 132,* 80–90.

Conde, M., & Le Billon, P. (2017). Why do some communities resist mining projects while others do not? *The Extractive Industries and Society, 4*(3), 681–697.

Corrigan, C. C. (2018). Corporate social responsibility and local context: The case of mining in Southern Africa. *Resources Policy, 55*(C), 233–243.

Craik, N., Gardner, H., & McCarthy, D. (2017). Indigenous-corporate private governance and legitimacy: Lessons learned from impact and benefit agreements. *Resources Policy, 52,* 379–388.

Dashwood, H. (2007). Canadian mining companies and corporate social responsibility: Weighing the impact of global norms. *Canadian Journal of Political Science, 40*(1), 129–156.

Dashwood, H. S. (2014). Sustainable development and industry self-regulation: Developments in the global mining sector. *Business and Society, 53*(4), 551–582.

Davis, G. A., & Manzano, O. (2018). The extractive industries in Central America. *The Extractive Industries and Society, 5*(3), 215–217.

de Lange, W., de Wet, B., Haywood, L., Stafford, W., Musvoto, C., & Watson, I. (2018). Mining at the crossroads: Sectoral diversification to safeguard sustainable mining? *The Extractive Industries and Society, 5*(3), 269–273.

de Loë, R. C., Murray, D., & Brisbois, M. C. (2016). Perspectives of natural resource sector firms on collaborative approaches to governance for water. *Journal of Cleaner Production, 135,* 1117–1128.

Devenin, V. (2018). Collaborative community development in mining regions: The Calama Plus and Creo Antofagasta programs in Chile. *Resources Policy.*

Devenin, V., & Bianchi, C. (2018). Soccer fields? What for? Effectiveness of corporate social responsibility initiatives in the mining industry. *Corporate Social Responsibility and Environmental Management, 25*(5), 866–879.

Diavik Diamond Mines Inc. (1999). *Diavik Diamonds Project Socio-Economic Monitoring Agreement* (Rep.)

Dilling, P. (2016). Reporting on long-term value creation—The example of public canadian energy and mining companies. *Sustainability, 8*(9), 938.

Engineers Without Borders Canada and Canadian Council for Aboriginal Business. *Partnerships in procurement: Understanding aboriginal business engagement in the Canadian mining industry* (Rep.).

Fitzpatrick, P., Fonseca, A., & McAllister, M. L. (2011). From the whitehorse mining initiative towards sustainable mining: Lessons learned. *Journal of Cleaner Production, 19*(4), 376–384.

Fontana, E. (2018). Corporate social responsibility as stakeholder engagement: Firm–NGO collaboration in Sweden. *Corporate Social Responsibility and Environmental Management, 25*(4), 327–338.

Frederiksen, T. (2018). Corporate social responsibility, risk and development in the mining industry. *Resources Policy, 59,* 495–505.

Frederiksen, T. (2019). Political settlements, the mining industry and corporate social responsibility in developing countries. *The Extractive Industries and Society, 6*(1), 162–170.

Freeman, R. E., & Dmytriyev, S. (2017). Corporate social responsibility and stakeholder theory: Learning from each other. *Symphonya. Emerging Issues in Management, 1,* 7–15.

Freeman, R. E., Velamuri, S. R., & Moriarty, B. (2006). Company stakeholder responsibility: A new approach to CSR. *Business Roundtable Institute for Corporate Ethics, 19.*

Gibson, G., & Klinck, J. (2005). Canada's resilient North: The impact of mining on aboriginal communities. *Pimatisiwin: A Journal of Aboriginal and Indigenous Community Health, 3*(1), 116–139.

Gorman, M. R., & Dzombak, D. A. (2018). A review of sustainable mining and resource management: Transitioning from the life cycle of the mine to the life cycle of the mineral. *Resources, Conservation and Recycling, 137,* 281–291.

Government of the Northwest Territories. (2014). *Diavik socio-economic monitoring agreement* (Rep.).

Graetz, G. (2015). Energy for whom? Uranium mining, indigenous people, and navigating risk and rights in Australia. *Energy Research & Social Science, 8,* 113–126.

Gustafsson, M. T., & Scurrah, M. (2019). Unpacking the extractivist state: The role of weak state agencies in promoting institutional change in Peru. *The Extractive Industries and Society, 6*(1), 206–214.

Haslam, P. A., Tanimoune, N. A., & Razeq, Z. M. (2018). Do Canadian mining firms behave worse than other companies? Quantitative evidence from Latin America. *Canadian Journal of Political Science/Revue canadienne de science politique, 51*(3), 521–551.

Heisler, K., & Markey, S. (2013). Scales of benefit: Political Leverage in the negotiation of corporate social responsibility in mineral exploration and mining in Rural British Columbia, Canada. *Society & Natural Resources, 26*(4), 386–401.

Horowitz, L. S. (2010). "Twenty years is yesterday": Science, multinational mining, and the political ecology of trust in New Caledonia. *Geoforum, 41*(4), 617–626.

Idemudia, U., & Kwakyewah, C. (2018). Analysis of the Canadian national corporate social responsibility strategy: Insights and implications. *Corporate Social Responsibility and Environmental Management, 25*(5), 928–938.

Intergovernmental Working Group on the Mineral Industry. (2016). *Good practices in community engagement and readiness* (Rep.).

Jeswiet, J. (2017). Including towards sustainable mining in evaluating mining impacts. *Procedia CIRP, 62,* 494–499.

Karakaya, E., & Nuur, C. (2018). Social sciences and the mining sector: Some insights into recent research trends. *Resources Policy, 58,* 257–267.

Kasimba, S. A., & Lujala, P. (2019). There is no one amongst us with them! Transparency and participation in local natural resource revenue management. *The Extractive Industries and Society, 6*(1), 198–205.

Khalid, S. M., Atkins, J., & Barone, E. (2019). Sartrean bad-faith? Site-specific social, ethical and environmental disclosures by multinational mining companies. *Accounting, Auditing & Accountability Journal, 32*(1), 55–74.

Kivinen, S., Vartiainen, K., & Kumpula, T. (2018). People and post-mining environments: PPGIS mapping of landscape values, knowledge needs, and future perspectives in Northern Finland. *Land, 7*(4), 151.

Knierzinger, J., & Sopelle, I. T. I. (2019). Mine closure from below: Transformative movements in two shrinking West African mining towns. *The Extractive Industries and Society, 6*(1), 145–153.

Laplante, J. P., & Nolin, C. (2014). Consultas and socially responsible investing in Guatemala: A case study examining Maya perspectives on the indigenous right to free, prior, and informed consent. *Society & Natural Resources, 27*(3), 231–248.

Larsen, R. K., Österlin, C., & Guia, L. (2018). Do voluntary corporate actions improve cumulative effects assessment? Mining companies' performance on Sami lands. *The Extractive Industries and Society, 5*(3), 375–383.

Lawrence, R., & Moritz, S. (2019). Mining industry perspectives on indigenous rights: Corporate complacency and political uncertainty. *The Extractive Industries and Society, 6*(1), 41–49.

Li, X., Wang, S. S., & Wang, X. (2017). Trust and stock price crash risk: Evidence from China. *Journal of Banking & Finance, 76,* 74–91.

Lin, P. T., Li, B., & Bu, D. (2015a). The relationship between corporate governance and community engagement: Evidence from the Australian Mining Companies. *Resources Policy, 43,* 28–39.

Littlewood, D. (2014). 'Cursed' communities? Corporate social responsibility (CSR), company towns and the mining industry in Namibia. *Journal of Business Ethics, 120*(1), 39–63.

Louw, H., & Marais, L. (2018). Mining and municipal finance in Kathu, an open mining town in South Africa. *The Extractive Industries and Society, 5*(3), 278–283.

Lujala, P. (2018). An analysis of the extractive industry transparency initiative implementation process. *World Development, 107*(2018), 358–381.

Majer, M. (2013). The practice of mining companies in building relationships with local communities in the context of CSR formula. *Journal of Sustainable Mining, 12*(3), 38–47.

Manning, S. M. (2016). Intersectionality in resource extraction: A case study of sexual violence at the Porgera mine in papua New Guinea. *International Feminist Journal of Politics, 18*(4), 574–589.

McLennan, S., & Banks, G. (2019). Reversing the lens: Why corporate social responsibility is not community development. *Corporate Social Responsibility and Environmental Management, 26*(1), 117–126.

Nicholls, A., & Dees, J. G. (2015). Social Innovation. *International encyclopedia of the social & behavioral sciences* (2nd ed., pp. 355–361)

O'Faircheallaigh, C. (2013). Extractive industries and indigenous peoples: A changing dynamic? *Journal of Rural Studies, 30,* 20–30.

Owen, J. R., & Kemp, D. (2014). 'Free prior and informed consent', social complexity and the mining industry: Establishing a knowledge base. *Resources Policy, 41,* 91–100.

Papillon, M., & Rodon, T. (2017). Proponent-indigenous agreements and the implementation of the right to free, prior, and informed consent in Canada. *Environmental Impact Assessment Review, 62,* 216–224.

Saenz, C. (2018). The context in mining projects influences the corporate social responsibility strategy to earn a social licence to operate: A case study in Peru. *Corporate Social Responsibility and Environmental Management, 25*(4), 554–564.

Selmier, W. T., II, Newenham-Kahindi, A., & Oh, C. H. (2015). "Understanding the words of relationships": Language as an essential tool to manage CSR in communities of place. *Journal of International Business Studies, 46*(2), 153–179.

Slack, K. (2012). Mission impossible?: Adopting a CSR-based business model for extractive industries in developing Countries. *Resources Policy, 37*(2), 179–184.

St-Laurent, G. P., & Le Billon, P. (2015). Staking claims and shaking hands: Impact and benefit agreements as a technology of government in the mining sector. *The Extractive Industries and Society, 2*(3), 590–602.

Stapenhurst, F., Karakas, F., Sarigöllü, E., Jo, M. S., & Draman, R. (2017). The supply and demand sides of corruption: Canadian extractive companies in Africa. *Canadian Foreign Policy Journal, 23*(1), 60–76.

Vivoda, V., & Kemp, D. (2019). How do national mining industry associations compare on sustainable development? *The Extractive Industries and Society, 6*(1), 22–28.

Wang, J., Huang, X., Hu, K., & Li, X. (2018). Evaluation on community development programs in mining industry: A case study of small and medium enterprise in China. *Resources Policy, 59,* 516–524.

Wasylycia-Leis, J., Fitzpatrick, P., & Fonseca, A. (2014). Mining communities from a resilience perspective: Managing disturbance and vulnerability in Itabira, Brazil. *Environmental management, 53*(3), 481–495.

References

Aguilera, R. V., Rupp, D. E., Williams, C. A., & Ganapathi, J. (2007). Putting the S back in corporate social responsibility: A multilevel theory of social change in organizations. *Academy of Management Review, 32*(3), 836–863.

Bainton, N., & Holcombe, S. (2018). A critical review of the social aspects of mine closure. *Resources Policy, 59,* 468–478.

Barnett, M. L. (2007). Stakeholder influence capacity and the variability of financial returns to corporate social responsibility. *Academy of Management Review, 32*(3), 794–816.

Barnett, M. L. (2019). The business case for corporate social responsibility: A critique and an indirect path forward. *Business & Society, 58*(1), 167–190. https://doi.org/10.1177/0007650316660044.

Barnett, M. L., & Salomon, R. M. (2012). Does it pay to be really good? Addressing the shape of the relationship between social and financial performance. *Strategic Management Journal, 33*(11), 1304–1320.

Blowfield, M., & Frynas, J. G. (2005). Editorial setting new agendas: critical perspectives on corporate social responsibility in the developing world. *International affairs, 81*(3), 499–513.

Bowen, F., Newenham-Kahindi, A., & Herremans, I. (2008). Engaging the community: A systematic review. *Research Network for Business Sustainability, 1*(1).

Bowen, F., Newenham-Kahindi, A., & Herremans, I. (2008b). *Engaging the community: A systematic review. A synthesis of academic and proactionner knowledge on best practices in community engagement*. London, Ontario: Research Network for Business Sustainability.

Bowen, F., Newenham-Kahindi, A., & Herremans, I. (2010). When suits meet roots: The antecedents and consequences of community engagement strategy. *Journal of Business Ethics, 95*(2), 297–318.

Brereton, D., & Parmenter, J. (2006). Water, communities and mineral resource development—understanding the risks and opportunities. *Proceedings of Water in Mining*, 14–16

Calvano, L. (2008). Multinational corporations and local communities: A critical analysis of conflict. *Journal of Business Ethics, 82*(4), 793–805.

CAO, J. (2008). *Participatory water monitoring; A guide for preventing and managing conflict*. The Office of the Compliance Advisor/Ombudsman. http://www.caoombudsman.org/howwework/advisor/documents/watermoneng.pdf. Accessed on April 16, 2019.

Carroll, A. B. (1979). A three-dimensional conceptual model of corporate performance. *Academy of Management Review, 4*(4), 497–505.

Carroll, A. B. (1991). The pyramid of corporate social responsibility: Toward the moral management of organizational stakeholders. *Business Horizons, 34*(4), 39–48.

Carroll, A. B. (2016). Carroll's pyramid of CSR: Taking another look. *International Journal of Corporate Social Responsibility, 1*(1), 3.

Carroll, A. B., & Shabana, K. M. (2010). The business case for corporate social responsibility: A review of concepts, research and practice. *International Journal of Management Review, 12*(1), 85–105.

Cooney, J. (2017). Reflections on the 20th anniversary of the term 'social license'. *Journal of Energy & Natural Resources Law*, 1–4. https://doi.org/10.1080/02646811.2016.1269472

Delannon, N., Raufflet, E., & Baba, S. (2016). Corporate community engagement strategies and organizational arrangements: A multiple case study in Canada. *Journal of Cleaner Production, 129,* 714–723.

Edelman, Inc. (2019). *Edelman trust barometer* (Rep.)

Engineers Without Borders Canada and Canadian Council for Aboriginal Business. *Partnerships in procurement: Understanding aboriginal business engagement in the Canadian mining industry* (Rep.).

Foreign Affairs Trade and Development Canada. (2018, January 17). *Doing business the canadian way: A strategy to advance corporate social responsibility in Canada's extractive sector abroad*. Retrieved from https://www.international.gc.ca/trade-agreements-accords-commerciaux/topics-domaines/other-autre/csr-strat-rse.aspx?lang=eng

Franks, D., Davis, R., Bebbington, A., Ali, S., Kemp, D., & Scurrah, M. (2014). Conflict translates environmental and social risk into business costs. *Proceedings of the National Academy of Sciences, 111*(21), 7576–7581. https://doi.org/10.1073/pnas.1405135111.

Fraser, J. (2018). Mining companies and communities: Collaborative approaches to reduce social risk and advance sustainable development. *Resources Policy*. https://doi.org/10.1016/j.resourpol.2018.02.003. In press.

Freeman, R. E., Velamuri, S. R., & Moriarty, B. (2006b). *Company stakeholder responsibility: A new approach to CSR*. Institute for Corporate Ethics Bridge Paper: Business Roundtable.

Friedman, M. (1970, September 13). The social responsibility of business is to increase its profits. *New York Times Magazine*.

Frynas, G. J. (2005). The false developmental promise of CSR: Evidence from multinational oil companies. *International Affairs 81*(3) (Critical Perspectives on Corporate Social Responsibility).

Graetz, G., & Franks, D. (2016). Conceptualising social risk and business risk associated with private sector development projects. *Journal of Risk Research, 19*(5), 581–601.

Himley, M. (2014). Monitoring the impacts of extraction: Science and participation in the governance of mining in Peru. *Environment and Planning A, 46*(5), 1069–1087.

Hitch, M. (2006). *Impact and Benefit Agreements and the political ecology of mineral development in Nunavut*.

Jenkins, H., & Obara, L. (2006, September). Corporate social responsibility (CSR) in the mining industry-the risk of community dependency. In *Corporate Responsibility Research Conference* (pp. 4–5), Dublin.

Kapelus, P. (2002). Mining, corporate social responsibility and the "community": The case of Rio Tinto, Richards Bay Minerals and the Mbonambi. *Journal of Business Ethics, 39*(3), 275–296.

Kemp, D., Boele, R., & Brereton, D. (2006). Community relations management systems in the minerals industry: Combining conventional and stakeholder-driven approaches. *International Journal of Sustainable Development, 9*(4), 390–403.

Kemp, D., Worden, S., & Owen, J. (2016). Differentiated social risk: Rebound dynamics and sustainability performance in mining. *Resources Policy, 50,* 19–26.

Kramer, M., & Pfister, M. (2016). The ecosystem of shared value. *Harvard Business Review, 94*(10), 80–89.

Kytle, B., & Ruggie, J. G. (2005). *Corporate social responsibility as risk management: A model for multinational companies.* Harvard. Cambridge, MA: John F Kennedy School of Government. Retrieved March 16, 2015, from https://www.ksg.harvard.edu/m-rcbg/CSRI/.../workingpaper_10_kytle_ruggie.pdf

Lapalme, L. A. (2003). *The social dimension of sustainable development and the mining industry.* Ottawa: Natural Resources Canada.

Lapierre, D. M., & Bradshaw, B. (2008, May). Why mining firms care: Determining corporate rationales for negotiating impact and benefit agreements. In *Canadian institute of mining, metallurgy and petroleum conference*, Edmonton.

Levitt, T. (1958). The dangers of social-responsibility. *Harvard Business Review, 36*(5), 41–50.

Lin, P. T., Li, B., & Bu, D. (2015b). The relationship between corporate governance and community engagement: Evidence from the Australian mining companies. *Resources Policy, 43,* 28–39.

Mercer-Mapstone, L., Rifkin, W., Louis, W., & Moffat, K. (2019). Power, participation, and exclusion through dialogue in the extractive industries: Who guets a seat at the table? *Resources Policy, 61,* 190–199.

Mining, Minerals and Sustainable Development North America. (2002). *Seven questions to sustainability* (Rep.).

O'Faircheallaigh, C., & Corbett, T. (2005). Indigenous participation in environmental management of mining projects: The role of negotiated agreements. *Environmental Politics, 14*(5), 629–647.

Porter, M., & Kramer, M. (2011). Creating shared value. *Harvard Business Review, 89*(1–2), 62–77.

Prno, J., & Bradshaw, B. (2007). *Assessing the effectiveness of impact and benefit agreements from the perspective of their Aboriginal signatories.* Canadian Association of Geographers.

Rodríguez, L., Montiel, I., & Ozuna, T. (2014). A conceptualization of how firms engage in corporate responsibility based on country risk. *Business and Society, 53*(5), 625–651.

Rooke, R. (2016). *Investigating relationships: How mining companies and Aboriginal communities can improve impact mitigation for terrestrial wildlife and traditional harvesting practices in the Canadian Arctic.* Doctoral dissertation.

Schafrik, S., & Kazakidis, V. (2011). Due diligence in mine feasibility studies for the assessment of social risk. *International Journal of Mining, Reclamation and Environment, 25*(1), 86–101.

Stubbs, W., & Cocklin, C. (2008). Conceptualizing a "sustainability business model". *Organization & Environment, 21*(2), 103–127. https://doi.org/10.1177/1086026608318042.

Taarup-Esbensen, J. (2019). Managing risk through dependency: how do mining MNEs strategize to legitimize business continuity? *The Extractives Industries and Society, 6,* 489–497.

Tuulentie, S., Halseth, G., Kietavainen, A., Ryser, L., & Simila, J. (2019). Local community participation in mining in Finnish Lapland and Northern British Columbia, Canada—Practical applications of CSR and SLO. *Resources Policy, 61,* 99–107.

Xavier, A. (2013). *Socio-Economic Mine Closure (SEMC) framework: A comprehensive approach for addressing the socio-economic challenges of mine closure.* Doctoral dissertation, University of British Columbia. Retrieved from https://circle.ubc.ca/handle/2429/45595

Xavier, A. M., Veiga, M., & van Zyl, D. (2015). Introduction and assessment of a socio-economic mine clsoure framework. *Journal of Management and Sustainability, 5*(1), 38. https://doi.org/10.5539/jms.v5n1p38.

Zandvliet, L. (2004). *Redefining corporate social risk mitigation strategies.* Washington: World Bank. Retrieved April 28, 2017, from http://documents.worldbank.org/curated/en/200261468778821090/Redefining-corporate-social-risk-mitigation-strategies

Zandvliet, L., & Anderson, M. (2017). *Getting it right: Making corporate-community relations work.* London: Routledge.

Zurba, M. (2017). Long-term and meaningful community leadership and engagement in natural resources and environmental governance through university access programs. *Journal of Adult and Continuing Education, 23*(1), 5–14. https://doi.org/10.1177/1477971416686582.

Zvarivadza, T. (2018). Large scale miners—Communities partnerhsips: A plausible option for communities furvival beyond mine closure. *Resources Policy, 56,* 87–95. https://doi.org/10.1016/j.resourpol.2017.12.005.

Corporate Social Responsibility in Australia

Martin Brueckner

Abstract Corporate Social Responsibility (CSR) in Australia has had a slow start and—while gaining momentum—continues to lag behind international trends and developments. This chapter seeks to describe the state, and characterise the nature of CSR 'Down Under', offering explanations for the somewhat lacklustre approach by business and government to operationalise and regulate CSR, respectively. Answers provided, based on select industry examples, will point to Australia's political economy of CSR and prevalent ideologies among corporate and political decision-makers as drivers of a kind of CSR that is largely reactive and based on economic legitimacy.

1 Civic Virtue "Down Under"

It is widely recognised that different economic systems give rise to different forms of CSR (Matten and Moon 2008; Gjølberg 2009) and Australia in this regard is no exception. Since white occupation in the late eighteenth century, the country has been pursuing an aggressive development agenda, evolving into the kind of 'stock market capitalism' (Dore 2000) found in other Anglo-Saxon countries such as the USA and the UK. The size of the country and its small population base have fuelled fears since colonisation of economic underdevelopment of successive governments and may help explain Australia's developmentalist ambitions, especially from the 1960s onwards; particularly, in resource-rich states such as Queensland and Western Australia (Layman 1982; Kellow and Niemeyer 1999). Developmentalism (following Thurbon 2012: 275) can largely be understood as viewing national economic prosperity and security as a primary priority of government thus granting governing elites an active role in facilitating the transformation and upgrading of the national techno-industrial infrastructure. In some state jurisdictions, governments even had a

M. Brueckner (✉)
Centre for Responsible Citizenship and Sustainability, Murdoch University, Perth, WA, Australia
e-mail: m.brueckner@murdoch.edu.au

E. G. Pereira et al. (eds.), *Sovereign Wealth Funds, Local Content Policies and CSR*, CSR, Sustainability, Ethics & Governance,
https://doi.org/10.1007/978-3-030-56092-8_34

central role in "attracting, facilitating and supporting development through providing key infrastructure, and being prepared occasionally to underwrite projects through bearing or sharing risk" (Phillimore 2014: 42). Unsurprisingly, the moral standing of development in Australia, as observed by Trigger (1997: 164) in relation to Australia's extractive sector, tends to be promoted routinely by industry and governments alike; an equation of economic development with moral progress.

The country's strong pro-development stance has had a bearing on the evolution, extent and nature of CSR in Australia. In the international comparison, a little over a decade ago, the uptake of CSR among Australian companies was still very limited, evidenced, for example, by the relatively small number of non-financial reports published (20% of top 100 listed companies); disclosure was largely restricted to multinational, foreign-owned or government-owned entities (Australian Government Department of the Environment and Heritage 2004). A national survey as recent as 2014 described Australian firms' progress on CSR as slow and insufficient (Australian Centre for Corporate Social Responsibility 2014), with higher levels of CSR maturity within Australian companies not recorded before 2017 (Australian Centre for Corporate Social Responsibility 2017). While recent progress in this space has elevated the 'Australian CSR brand' internationally (ROBECOSAM 2017), other measures of responsible business conduct generally do not rate Australian business enterprises. Forbes (2017) and the Corporate Knights Index (Corporate Knights Inc. and Innovest Strategic Value Advisors Inc. 2007), for example, only rank two Australian firms within their respective top 100 most sustainable[1] companies; ironically, both companies had their CSR credentials publicly questioned in recent years and been the subject of a Royal Commission[2] investigation (Gluyas 2016; Schmulow et al. 2019). Further, Australian mining companies regularly stand in the media crossfire over claims of irresponsible business conduct (Saunders et al. 2015; Campbell 2017), along with aged care, financial advice services and the superannuation industry, only to mention a few, are other sectors in the Australian economy with their CSR reputation tarnished in recent years. This state of play reflects poorly on the CSR performance of Australian firms and makes understandable the diminished public trust in the private sector, which today stands at an all-time low (Humpage 2014; Edelman 2017).

For the purposes of this chapter, the state of play in Australian CSR sketched above will be seen in the context of the country's development approach and economic system. In what follows, it will be shown how a particular political-cum-commercial CSR discourse has enabled a particular CSR framing that sees development itself as an expression of CSR. This framing serves to reduce CSR to a matter of economic legitimacy and explains the gulf between corporate performance and societal expectations on company conduct. Select industry cases will be presented to illustrate the

[1]Differences between CSR and sustainability are acknowledged here. Both concepts within the business context fall within what is referred to as firms' environmental, social and governance (ESG) performance.

[2]In Australia, Royal Commissions are appointed by the government to look into matters of great importance and usually controversy. Commission findings and recommendations are then to be responded to by government.

key points of the argument. An ensuing discussion will seek to problematise the Australian approach to CSR.

2 Econo-Political Constructions of CSR

As suggested earlier, Australia's development approach has helped to give rise to a particular kind of CSR that is rooted in economic legitimacy. The country's development focus, so the argument here, has helped foreground companies' economic contributions in terms of employment and wealth generation; frequently at the expense of broader social and environmental objectives (Brueckner and Mamun 2010; Pini et al. 2010; Wesley and MacCallum 2014; Luke 2017). The prevailing CSR discourses in Australia are further explored below for they hold clues about how CSR is being constructed by econo-political decision-makers and how these discourses help shape particular company practices. Of particular note are discourse features around the issue of voluntarism and wealth creation and the way in which these features are reflected in the treatment of CSR by the authorities. These issues will be addressed in turn.

The CSR discourse globally has long been favouring non-mandatory approaches to CSR. Despite mounting critiques on self-regulation (Blowfield 2005; Doane 2005; Banerjee 2006; Epstein 2007), a seeming trend towards mandated CSR in countries such as India, Indonesia and Mauritius (Waagstein 2011; Pillay 2016; Subramaniam et al. 2017) and tightening company disclosure regulations across Europe (Szabó and Sørensen 2015), indirect forms of regulation and a range of internal CSR instruments in the form of voluntary corporate codes of conduct, as well as strategic stakeholder partnerships are still widely considered adequate substitutes for 'hard' regulatory approaches (Shamir 2004, 2010; Enoch 2007). Especially, in countries committed to neoliberal economics, a regulatory approach is largely being avoided[3] (Hanlon 2011). Unsurprisingly, thus, in Australia, with its strong neoliberal leanings since the 1980s (Connell 2010) successive governments have championed industry self-regulation, a policy position unlikely to shift in the near future. To illustrate, a review by the Parliamentary Joint Committee on Corporations and Financial Services (2006) explored the extent to which organisational decision-makers (should) have an existing regard for the interests of stakeholders (other than shareholders) and the broader community. The review, which involved government, industry, research bodies, consultancies and industry coalitions, resulted in the reaffirmation of companies' 'enlightened self-interest' as a sufficient guarantor for adequate industry self-regulation amidst fears expressed by the business community about the implications of tighter rules imposed by government. Chiefly among them were concerns that regulation would be heavy-handed, stifle economic growth and overly constrain the entrepreneurial, innovative forces of business. Industry groups argued that CSR activities were more likely to

[3]Central to this neoliberal consensus are corporate desires to avoid government regulation and governments' intent to have companies fulfil duties that were abandoned by the state.

be embedded within corporations' mindsets if these were voluntary and governed through corporate self-regulation (Parliamentary Joint Committee on Corporations and Financial Services 2006; Wesley 2014).

Central to the industry arguments listed above are the purported risks any form of government regulation might pose to companies' financial success, and wealth creation is frequently invoked as the key welfare contribution of Australian companies. The argument takes the form of a perversity thesis (after Hirschman 1991), for it suggests that any kind of government regulation would end up hurting the very people it seeks to protect. In other words, society would suffer the consequences of any government intervention in the CSR space. The welfare contribution of business is highlighted routinely in public statements by industry. For example, the Business Council of Australia, a key lobby group and industry association that comprises the chief executives of more than 100 of Australia's biggest corporations, states in its submission to the Parliamentary Joint Committee on Corporations and Financial Services (2006: 10) that "it is important to note that … the greatest social contribution made by corporations is through the goods and services they provide, the wealth they create and the employment they generate". In a similar vein, mining production company Alcoa suggests that its "investment has provided essential infrastructure and supported the growth of regional communities" (Parliamentary Joint Committee on Corporations and Financial Services 2006: 10), again invoking the notion of development as a moral good.

Industry sentiments such as these are often echoed in Australia's political realm. At the national political level, a firm commitment can be observed to "corporate self-determination" and the governance of CSR through "liberal and indirect means of steering" (Vallentin and Murillo 2012: 827) reflected in policies that merely encourage and facilitate "socially responsible activities, via communication, education, incentives (financial) and regulatory trade-offs for undertaking responsible business practice" (Wesley 2014: 124). Over the years and irrespective of political persuasion, numerous federal politicians have been vocal about the win–win logic of CSR, advocating how it can "contribute to long term profitability of corporate Australia and deliver benefits to the community" (Sherry 2009).

The win–win logic is at the very heart of the widely promoted 'business case for CSR' (World Business Council for Sustainable Development 2000), which rests on the assumed interdependencies between business and society for the improvement of living standards and social conditions and for the creation of jobs and wealth (Hoque 1985). This mutual dependence between business and society enables the equation of community well-being and companies' economic interests, overcoming "entrenched perceptions of an incompatibility between economic and social goals" (Brueckner and Mamun 2010: 328). The integration of CSR into business (Lee 2008) is said to enable firms to align their capabilities with the opportunities their social environment presents and to deliver both societal benefits and competitive advantage (Munilla and Miles 2005; Miles et al. 2006). While the critical CSR literature raises concerns about areas where the interests of business do not merge with the interests of society (e.g. environmental impacts, social inequality, etc.) (e.g. Newell 2001; Banerjee 2007), in Australia, the virtues of economic development seem to trump these fears. This sentiment is expressed, for example, by the Chair of Western Australia's Regulatory Impact Assessment Process Review who suggests in connection with environmental impacts from resource development that "the environment is an important consideration, but it is not always—or even often—the only one. A weak economy is a far greater threat to the environment than is responsible mining" (Government of Western Australia 2009: 1–2). This stance is reflective of a particular ideology that gives primacy to economic development, implicitly treating it—as suggested earlier—as a moral end in itself and as the very foundation of nation building.

The above illustrates how in Australia both industry and government frame a particular CSR discourse, effectively blending corporate self-interest within the mantra of responsibility (Selznick 2002). The economic responsibility of the firm is being foregrounded and in a Friedmanite (Friedman 1970) tradition and elevated to the all-encompassing societal contribution of business. It thus follows that direct regulatory responses by the Australian government are largely absent or decidedly light touch as profit-making and the protection of shareholder interests are given first priority (Madsen and Ulhøi 2001; Farrar 2008). While the Australian Corporations Act (2001) compels company directors to have regard for the interests of stakeholders other than shareholders, it leaves open questions about the extent to which this fosters corporate practice beyond strategic, relational CSR (Brueckner and Mamun 2010). When taking CSR reporting systems of Australian companies as a proxy, these are seen to be primarily aimed at serving the interests of equity holders (Chen and Bouvain 2009: 304).

Below, a number of case examples are presented with a view to illustrate the impacts of Australia's discursive CSR landscape on companies' CSR practices, the broader implication of which will be discussed subsequently.

3 CSR at Work in Australia: Case Examples

The industry cases detailed below shed light on the ways in which both Australian state and federal governments tend to protect industry interests, generally, by way of reference to how business secures, and contributes to, the well-being of the country and its citizens. Even in situations of public disquiet over widely publicised corporate wrong-doing or unacceptable environmental risks, governments are often found to be in support of the implicated business entities and/or to be defending the adequacy of existing industry regulation.

3.1 Banking

Australia's banking industry self-describes as "solid, safe and efficient" and to be "at the heart of [the Australian] economy and [...] community. Banks [purportedly] allow [...] Australian families to realise their dreams [...], help Australian businesses start and grow, be profitable and well governed, and create jobs" (Australian Bankers' Association Inc. 2015: 1). In short, strong banks = strong Australia. This stout sense of responsibility is also reflected in Australia's major banks' value and mission statements. The Commonwealth Bank, for example, speaks of a "vision [...] to excel at securing and enhancing the financial well-being of people, businesses and communities" and a commitment to maintain the highest professional standards and [to] act with integrity during the course of [their] business activities" (Commonwealth Bank 2019). Similarly, National Australia Bank (2019) states to have Corporate Responsibility at the heart of their approach and to be seeking to "make a positive and sustainable impact on the lives of [their] customers, people, shareholders, communities, and on the environment in which [they] operate, while Australia and New Zealand Banking Group (2019) speaks of seeking "to build strong and lasting relationships with [their] customers" and to provide "banking that [...] is delivered in a responsible manner by [their] people and in accordance with the highest standards of integrity".

Australia's major banks have long been seen in a similar light in the political realm—mostly by politicians of a conservative persuasion—where they have been described as "among the best-run, the most prudentially supervised, and the most well-capitalised in the world" (Anon 2017a). Also, the former Treasurer and current Prime Minister suggest that "[w]e want a banking system that is strong" for "healthy and profitable banks [are] a 'pillar' of the economy" (Kehoe 2016).[4]

Despite the accolades, the banking industry has been plagued by numerous scandals over the years (Verrender 2017), having faced allegations of fraud, deception, and money laundering, among various other crimes (Vedelago 2018). In recent years, calls for an inquiry into the banks became louder, "emanating from a broad section of the community — from farmers, small business and households, jaded and disillusioned with the industry's rampant profiteering, fee gouging and blatant disregard for the law" (Verrender 2017). Notwithstanding the seriousness of the allegations, members of government and the conservative liberal party continued to defend the banking sector, suggesting the alleged scandal was "all a conspiracy theory, whipped up by populists for cheap political gain" and describing the called-for Royal Commission as "rank socialism" (Anon 2017a). The banking inquiry was decried as a "populist whinge" and as "reckless political games with one of the core pillars of our economy" that could "[...] undermine confidence in the banking and finance system" (Knott 2016). Again, a conflation of corporate success and national economic well-being become visible here.

[4]The view that Australia's banks are indispensable for Australia's economic well-being (too big to fail) may explain the political protection banks have received over the years. For example, banks benefited from the government's deposit guarantee scheme during the Global Financial Crisis in 2008 (Australian Government 2008), which sought to maintain the banks' credit ratings and their access to lucrative overseas loans. While scaled back since 2008, the ongoing guarantee scheme, which critics have likened to 'agrarian socialism' (Verrender 2017), has cost the taxpayer AUS$5 billion in 2017 alone in bank subsidies (Joye 2017).

Due to a series of media revelations about misconduct and mounting pressure from the public and the political opposition the government reluctantly announced a Royal Commission into the banking sector in late 2017. In its seven rounds of public hearings, the inquiry unearthed systematic breaches of corporate law by the banks, including money laundering for drug syndicates and terrorism financing as well as failures to comply with statutory reporting responsibilities and impropriety in foreign exchange trading. Other breaches included fees without service, inappropriate advice and conduct among others in the "pursuit of short-term profit at the expense of basic standards of honesty" (Hayne 2018, p. xix). Lack of government oversight was also identified as an enabling factor and the lack of regulatory intervention by the relevant government authorities. As Commission's Interim Report (2018: xix) suggests:

> The conduct regulator, ASIC, rarely went to court to seek public denunciation of and punishment for misconduct. The prudential regulator, APRA, never went to court. Much more often than not, when misconduct was revealed, little happened beyond apology from the entity, a drawn out remediation programme and protracted negotiation with ASIC of a media release, an infringement notice, or an enforceable undertaking that acknowledged no more than that ASIC had reasonable 'concerns' about the entity's conduct.

The Commission handed down its final report (Hayne 2019) in early 2019, spelling out 76 separate recommendations, targeting banks' corporate culture and governance as well as lending and advisory practices. While the government has promptly accepted most of the Commission's recommendations (Murphy 2019), at the time of writing it was still considering its full policy response to the inquiry, flagging that the implementation of the recommendations would not commence prior to the next federal election scheduled for May 2019 (Remeikis 2019). At the same time, the government warned that it "would balance action on a tougher regulatory framework against the need to ensure there was a free flow of credit", which it described as the "lifeblood of the economy" (Coorey 2019). There is widespread concern, given the high level of political support the industry enjoys and the government's reluctance to commission the inquiry in the first place that it will affect its willingness to implement the Commission's recommendations, which some commentators saw as too lenient on the banks (Keane and Dyer 2019a; b). As such, media speculation may be warranted that another banking inquiry will be a matter of course within a few years (Linden and Staples 2019).

3.2 *Coal Mining*

Australia is resource-rich and many minerals and metals are some of the country's key export commodities, including coal (Connor et al. 2009). Climate change concerns, however, makes the extraction and combustion of coal increasingly untenable amidst moves globally towards decarbonising industrial processes and power generation (International Energy Agency 2017). Thus, unsurprisingly, the proposed Carmichael coal mine in the north of the Galilee Basin in Central Queensland is currently one of Australia's most contentious econo-political and environmental issues (Brueckner and Eabrasu 2018). The AUS$16.5 billion project proposed by Adani Mining Pty Ltd., involves six open-cut and five underground coal mines. The Carmichael coal mine is expected to yield 60 million tons of thermal coal per annum, amounting to 2.3 billion tonnes over its expected 60-year life (Slezak 2017b). The mine proves controversial on environmental, economic and socio-cultural grounds, with public opinion polls showing growing opposition to the project (Slezak 2017a; Massola 2018).

With regard to environmental matters, as detailed by Brueckner and Eabrasu (2018), the coal extracted is projected to produce emissions of about 77 million tonnes of CO_2-e per annum and produce a total of 4.6 billion tonnes CO_2-e over the expected lifetime of the mine (Meinshausen 2015). Moreover, the Queensland government's granting of unlimited access to groundwater from the Great Artesian Basin until the year 2077 is a key concern to water-poor Queensland farming communities. While the mine is likely to require 12 billion litres of water per year, Adani will merely be required to monitor and report the amount of water extracted (Hannam 2017). Finally, the mine poses a risk to the Great Barrier Reef, one of Australia's environmental icons and UNESCO-listed heritage areas, indirectly through coral bleaching due to rising ocean temperatures consequent to climate change (Hughes et al. 2017) but also more directly due to the transport of coal to export markets via ship, increasing the dispersion of coal dust and risks of collisions and spills (Climate Council 2017a; Sparrow 2017). Adani's poor track record on environmental stewardship (Environmental Justice Australia 2015,2017) has further heightened concerns about the mine's potential to harm the environment and local communities.

In economic terms, the mine promises to deliver 10,000 jobs for Queensland where the unemployment rate is currently at 6.4%, with even higher rates experienced in some regional areas (Australian Government 2018). While the promise of job creation explains strong support of the project by the Queensland government (Robertson 2017), which echoes Adani's job figures (Queensland Government 2018), only 1464 jobs are considered likely to result from the project (Fahrer 2015). At the same time, the Carmichael mine may also create job losses in coal producing states such as New South Wales (NSW) (Long 2017) but also in the tourism sector due to impacts on Great Barrier Reef-based tourism, which employs (both directly and indirectly) around 64,000 people (Pearce 2017). The economic case for the mine is also considered weak, evidenced by 19 major banks refusing to underwrite the project owing to environmental, reputational and viability concerns (Climate Council 2017b). The project is also deemed "unbankable" (Cox 2015; Institute for Energy Economics and Financial Analysis 2017) because of Adani's high level of debt, unclear corporate structure and use of offshore entities as well as diminished future prospects for coal in a carbon constrained global economy (International Energy Agency 2017).

There is also considerable opposition to the mine on cultural grounds as the proposed mine is to be built on traditional Wangan and Jagalingou land (W&J). The W&J Traditional Owners Family Council opposes the Adani Carmichael coal mine and has mounted a Federal Court challenge against the project. After having rejected a land rights deal with Adani on four separate occasions since 2012 (Wangan and Jagalingou Family Council 2018), the Council rejects the legitimacy of the Indigenous Land Use Agreement (ILUA) Adani has struck with traditional owners of the land, alleging the use of duress and bribery payments (Anon 2017c).

The case description shows that Adani has been able to secure both political and legal licences for the development of the Carmichael mine, albeit with considerable legal challenges and immense public outcry over the political support given to the project. While public anti-coal sentiments are on the rise, the proposed mine enjoys political support at state and federal level. Despite various court challenges by conservation groups, the mine received Federal Government approval (with conditions) under the Environment Protection and Biodiversity Conservation Act 1999 in 2015 as well as conditional approval by the Queensland Department of Environment and Heritage Protection in 2016 (Queensland Government 2016), helping Adani to secure its legal licenses for the project.

The coal industry has long been enjoying a close relationship with, and bilateral support from, Australian state and federal governments (Connor et al. 2009). This is what Baer (2016) describes as the state/coal industry nexus, with the corporate sector being the dominant player and governments ideological wedded to coal. This is evidenced by governments continuing to do the bidding on behalf of the increasingly besieged coal industry as shown in statements by former Australian Prime Minister Tony Abbott: "Coal is good for humanity, coal is good for prosperity, coal is an essential part of our economic future" (Anon 2014). Even in the face of growing calls for a phase out of coal-fired power generation and a push towards decarbonisation, Australia's former Treasurer and current Prime Minister Scott Morrison brandished a lump of coal in front of parliament, mocking his political opponents' alleged aversion to coal: "This is coal—don't be afraid, don't be scared ", suggesting that coal gave Australia "an energy competitive advantage for more than 100 years, delivering prosperity to businesses and ensuring industry had been able to remain competitive in a global market" (Anon 2017b).

The case illustrates how the government's pro-industry discourse offers political protection to a sunset industry that faces public opposition on environmental, economic and cultural grounds. The promise of jobs and wealth generation (despite the case for it being rather weak) seems to override the broad range of CSR concerns levelled at Adani and its Carmichael project.

3.3 *Unconventional Gas*

The resource-rich state of Western Australia is believed to have some of the country's largest reserves of offshore conventional gas as well as onshore unconventional gas (Geoscience Australia & Australian Bureau of Agricultural and Resource Economics 2010), a potential that the industry has been eager to tap in recent years.[5] (Committee

---

[5]The state of the unconventional gas industry in Western Australia is described in detail by Luke et al. (2018).

for Economic Development of Australia 2012; Upstream Petroleum Resources Working Group Report to COAG Energy Council 2015). The industry's enthusiasm for the exploration of unconventional gas (Australian Petroleum Production & Exploration Association 2015) has been matched by that of successive Western Australian state governments, which have been actively supporting gas development in recent years (Australian Mining 2013; Government of Western Australia 2015). Support was provided by way of public information campaigns about the purported safety of the extraction of unconventional gas (see Department of Mines and Petroleum 2017) and public statements made about the economic benefits the industry will deliver to the state (Verstegen 2018).

Notwithstanding, the arrival of the unconventional gas industry in Western Australia was met with considerable public opposition, especially in relation to fracking.[6] Protests took the form of organised community action groups (e.g. Frack Free WA and Lock the Gate WA), anti-fracking campaigns by environmental groups (The Wilderness Society 2018) as well as a number of communities declaring themselves 'Gasfield Free' to pre-empt the establishment of the industry (e.g. Christian 2015; de Poloni 2017). The industry is being opposed for a number of reasons, ranging from concerns about health impacts and fugitive emissions to groundwater contamination and the disclosure of chemical substances. Community concern was the trigger for a number of independent studies into the impacts of unconventional gas exploration (Vogwill 2017; Hare et al. 2018) and the establishment of a parliamentary inquiry into the implications of fracking for unconventional gas.

In its final report, the inquiry (Western Australian Standing Committee of Environment and Public Health Affairs 2015: 170), while vindicating community concerns and highlighting the need for ongoing community consultation, overall found the issues surrounding the industry to be manageable. An independent review of the inquiry (Haswell 2017), however, pointed to failures in the assessment of potential risks and benefits of the industry to human health, urging 'the Western Australia government to conduct an updated and fully comprehensive review of the potential direct and indirect impacts of proceeding with an unconventional gas industry on human health and well-being' (p. 3).

Ensuing community agitation against fracking (Smith 2017) prompted a newly elected state government to introduce a ban in 2017 on fracking for existing and future petroleum titles in the South-West, Peel and Perth metropolitan regions and a moratorium on fracking throughout the rest of Western Australia (Government of Western Australia 2017). The government set up an inquiry into fracking to "determine whether the moratorium may be lifted in locations or circumstances where potential risks can be mitigated to an acceptable level" (Government of Western Australia 2017). The year 2018 saw considerable industry lobbying and reportedly pressure applied by the federal government, threatening states with the reduction of their share of federal funding from the Goods and Services Tax (ABC 2018) unless state-imposed bans on fracking were lifted. In late 2018, following the release

[6]Fracking stands for hydraulic fracturing using high pressure to inject liquids into subterranean rocks to open existing fissures for the extraction of oil or gas.

of the report from the 12-month inquiry, the Western Australian state government announced the lifting of its moratorium on fracking but maintained a fracking ban in the state's socio-economically strong and politically well-connected South-West. While Australia's oil and gas industry welcomed the government's decision to lift the hydraulic fracturing moratorium for onshore gas projects (Oil and Gas Australia 2018), industry opponents saw in the decision proof that the government was "more supportive of big mining, oil and gas companies than they are about the environment and the future generations of Western Australians" (Jenkins 2018). Critics also feared that the lifting of the moratorium would enable the expansion of the unconventional gas industry and with it trigger a raft of adverse environmental, social, cultural and economic impacts (Verstegen 2018).[7] Independent analysis of the industry's economic impact on Western Australia showed little potential for economic gain and job creation but instead a large potential for negative community impacts and higher gas prices (Campbell et al. 2018).

In light of the industry's 'political licence' (after Brueckner et al. 2014) provided by the state government, its future growth in Western Australia seems certain. Industry-community conflicts centred on unconventional gas are thus likely to intensify in years to come as communities are vowing to fight the industry every step of the way (Farcic 2018).

4 Discussion and Concluding Comments

The data presented in this chapter highlight the strong economic focus on the side of industry and government in connection with CSR, one that is restricted largely to the generation of jobs, income and government revenue. This economic logic was shown to contrast starkly with community expectations concerning company conduct and acceptable levels of impact as shown in the cases of banking, coal mining and unconventional gas development. Australia's dominant development narrative has produced narrowly construed CSR discourses and practices, which frequently collide with communities' social, cultural and environmental interests as evidenced by a large number of industry-community conflicts around the country (see Higginbotham et al. 2010; Cleary 2012; Duus 2013; Scambary 2013; White 2013; Brueckner 2014; Brueckner and Eabrasu 2018; Luke et al. 2018); it shows that the CSR agenda extends far beyond mere questions of employment and income.

The combination of developmentalism and voluntary CSR was found to protect industry interests by way of promoting an approach to CSR that closely ties corporate success to the national interest. It is tantamount to a transformation of CSR from an attempt at reducing, mitigating or eliminating negative impacts associated with

[7] In a similar vein, attempts by the Western Australian Environmental Protection Authority (EPA) in early 2019 to require industry projects with direct carbon emissions over 100,000 tons per annum to create carbon offsets not only faced fierce industry opposition but also pressure from the federal resources minister and the Western Australian government, prompting the EPA to withdraw its guidelines (Hastie and Latimer 2019).

company conduct or doing good for society to the mere demonstration of economic benefits to company stakeholders. CSR, originally rooted in critiques of orthodox economics and laissez faire capitalism (e.g. Clark 1916), strikes in the Australian context as having wealth creation as its central leitmotif (Windsor 2001). The civic virtues of corporate self-interest and rent-seeking, however, are highly questionable for a CSR limited to a narrow business case is unlikely to deliver corporate contributions to social welfare beyond arguments of economic efficiency (Kok et al. 2001). Such a CSR is also likely to prove ineffective in addressing "the very social dilemmas it is meant to solve" and prone to overlook "critical structural and procedural aspects of industry-community conflicts as they relate to issues such as power and stakeholder dissent" (Brueckner and Mamun 2010: 328).

While company mission statements, as those of Australian banks, are employing the kind of high-minded rhetoric Matten and Moon (2008) would describe as an explicit form of CSR, stating steadfast commitments to, and responsibility for, communities, company actions and public reactions towards them often demonstrate their failure to effectively balance the 'efficiency-legitimacy dichotomy' (after Banerjee 2006). The expectation for companies in an unregulated CSR environment to get this balance right also strikes as wishful as suggested by Reich (2008) even though proponents of voluntarism continue to invest faith in their ability to do so.

With Australia's development orthodoxies well entrenched there is little hope at present that political pressure will be brought to bear on companies with a view to improve their CSR performance. Community disquiet in response to adverse industry impacts or misconduct continues to be routinely overridden by a pro-development discourse that backs business-as-usual approaches under the façade of progress and nation building with systemic blind spots towards issues at the heart of community grievances. In this regard, Australia appears to be years behind international 'best' practice in relation to CSR and is likely to remain a CSR laggard for years to come unless communities find ways of increasing commercial and political risks for decision-makers that would render CSR concerns too precarious to ignore. Yet, years of neoliberal atomisation of society and rising levels of political apathy (Clark 2018) coupled with a limited and partisan media landscape with overt pro-industry leanings (Lidberg 2019) augur poorly for such changes to occur as these barriers effectively militate against political mobilisation. At the same time, "citizens who take their responsibility to democracy (and/or social and environmental matters] seriously" are often seen to "pose a dire threat to corporate [and political] power" (Giroux 2019) and thus frequently face vilification and disparagement in the media and in political debate, as was seen in connection with anti-mining and climate justice protests in recent years (e.g. Anon 2018). In part, this is why some political commentators in the media even raise the spectre of a 'lost decade' (Keane 2019), referring to the current era in Australian politics that is marked by a political laziness that panders to special interests, blocks reforms and maintains the status quo. As such, in terms of future outlook for CSR in Australia, it seems the stakes would need to increase significantly for public demands to affect changes in corporate CSR practices or to prompt intervention by government.

References

Anon. (2014, October 13). Coal 'good for humanity', Prime Minister Tony Abbott says at $3.9b Queensland mine opening. *ABC News*. Retrieved from https://www.abc.net.au/news/2014-10-13/coal-is-good-for-humanity-pm-tony-abbott-says/5810244

Anon. (2017a, November 23). Bank royal commission 'rank socialism': Howard. *Perthnow*. Retrieved from https://www.perthnow.com.au/politics/bank-royal-commission-rank-socialism-howard-ng-fb2e65449dc582adb1697179546e0bb3

Anon. (2017b, February 9). Don't be afraid, it's only coal: MPs told. *SBS News*. Retrieved from https://www.sbs.com.au/news/don-t-be-afraid-it-s-only-coal-mps-told

Anon. (2017c, December 4). Traditional owners file Adani injunction. *Sunshine Coast Daily*. Retrieved from https://www.sunshinecoastdaily.com.au/news/traditional-owners-file-adani-inj unction/3281604/

Anon. (2018, November 30). Climate change protest will lead to dole queue, minister tells students. *The Sydney Morning Herald*. Retrieved from https://www.smh.com.au/politics/federal/climate-change-protest-will-lead-to-dole-queue-minister-tells-students-20181130-p50jbt.html

Australia and New Zealand Banking Group. (2019). *Corporate sustainability*. Retrieved from https://www.anz.com/about-us/corporate-sustainability/customers/

Australian Bankers' Association Inc. (2015). *Australia's banks: An integral part of the economy and the community. Economic report*. Sydney.

Australian Government. (2008). *Guarantee scheme for large deposits and wholesale funding*. Retrieved from https://www.guaranteescheme.gov.au

Australian Government. (2018). *Unemployment rate by state and territory*. Retrieved from https://lmip.gov.au/default.aspx?LMIP/LFR_SAFOUR/LFR_UnemploymentRate

Australian Government Department of the Environment and Heritage. (2004). *The state of sustainability reporting in Australia*. Canberra.

Australian Mining. (2013). *WA Government throws support behind fracking*. Retrieved from https://www.miningaustralia.com.au/news/wa-government-throws-support-behind-fracking

Australian Petroleum Production & Exploration Association. (2015). *The shale gas opportunity*. Retrieved from https://www.appea.com.au/oil-gas-explained/benefits/the-shale-gas-opportunity/

Baer, H. A. (2016). The nexus of the coal industry and the state in Australia: Historical dimensions and contemporary challenges. *Energy Policy., 99*, 194–202.

Banerjee, B. S. (2007). *Corporate social responsibility: The good, the bad and the ugly*. Cheltenham, UK: Edward Elgar.

Banerjee, S. B. (2006). The problem with corporate social responsibility. In S. Clegg & C. Rhodes (Eds.), *Management ethics: What is to be done?* (pp. 55–76). New York: Palgrave.

Blowfield, M. (2005). Corporate social responsibility - the failing discipline and why it matters for international relations. *International Relations, 19*(2), 173–191.

Brueckner, M. (2014). On the social sustainability of development in Western Australia: A community perspective. In M. Brueckner, A. Durey, R. Mayes, & C. Pforr (Eds.), *Resource curse or cure? On the sustainability of development in Western Australia* (pp. 239–255). Heidelberg: Springer.

Brueckner, M., Durey, A., Pforr, C., & Mayes, R. (2014). The civic virtue of developmentalism: On the mining industry's political licence to develop Western Australia. *Impact Assessment and Project Appraisal, 32*(4), 315–326.

Brueckner, M., & Eabrasu, M. (2018). Pinning down the social license to operate (SLO): The problem of normative complexity. *Resources Policy, 59*, 217–226.

Brueckner, M., & Mamun, M. A. (2010). Living downwind from CSR: A community perspective on corporate practice. *Business Ethics: A European Review, 19*(4), 326–348.

Campbell, E. (2017, March 9). World's largest sovereign wealth fund threatens corrupt and polluting firms with divestment. *ABC News*. Retrieved from https://www.abc.net.au/news/2017-03-09/nor way-sovereign-wealth-fund-threatens-unethical-firms/8341372

Campbell, R., Ogge, M., & Murray, C. (2018). *The economic impacts of unconventional gas in Western Australia*. Canberra.

Chen, S., & Bouvain, P. (2009). Is Corporate Responsibility Converging? A Comparison of Corporate Responsibility Reporting in the USA, UK, Australia, and Germany. *Journal of Business Ethics, 87*(1), 299–317.

Christian, B. (2015). WA town of Irwin declares itself gas field-free. *ABC News*. Retrieved from https://www.abc.net.au/news/2015-06-14/wa-town-goes-frack-free/6545188

Clark, J. E. S. (2018). *Network homogenisation & party disengagement: The political sociology of post-industrial democracies: An Australian case study*. Doctoral Thesis, The University of Adelaide.

Clark, J. M. (1916). The changing basis of economic responsibility. *the Journal of Political Economy, 24*(3), 209–229.

Cleary, P. (2012). *Mine-Field. The dark side of Australia's resources rush*. Collingwood: Black Inc.

Climate Council. (2017a). New report: Adani monster mine could damage climate, health, tourism. Retrieved from https://www.climatecouncil.org.au/carmichaelmediarelease

Climate Council. (2017b). *Risky business: Health, climate and economic risks of the Carmichael coalmine*.

Committee for Economic Development of Australia. (2012). *Australia's unconventional energy options*. Melbourne.

Commonwealth Bank. (2019*). How we implement sustainable business practices*. Retrieved from https://www.commbank.com.au/about-us/opportunity-initiatives/opportunity-from-good-business-practice/sustainable-business-practices.html#Corporate%2520governance

Connell, R. (2010). Understanding neoliberalism. In S. Braedley & M. Luxton (Eds.), *Neoliberalism and everyday life* (pp. 22–36). Montreal: McGill-Queen's University Press.

Connor, L., Freeman, S., & Higginbotham, N. (2009). Not just a coalmine: Shifting grounds of community opposition to coal mining in Southeastern Australia. *Ethnos, 74*(4), 490–513.

Coorey, P. (2019, 4 February). Banking royal commission final report: Coalition pulls punches on broker crackdown. *Australian Financial Review*. Retrieved from https://www.afr.com/business/banking-and-finance/banking-royal-commission-coalition-rules-out-broker-crackdown-20190203-h1aszv

Cox, L. (2015, 30 June). Adani's Carmichael mine is unbankable says Queensland Treasury. *The Sydney Morning Herald*. Retrieved from https://www.smh.com.au/business/companies/adanis-carmichael-mine-is-unbankable-says-queensland-treasury-20150630-gi1l37.html

de Poloni, G. (2017, 6 November). Gas exploration proposal draws fire in Western Australia's key south-west agricultural region. *ABC News*. Retrieved from https://www.abc.net.au/news/2017-11-06/wa-south-west-farmers-say-no-to-gas-exploration/9120728

Department of Mines and Petroleum. (2017). *The facts about natural gas and fracture stimulation in Western Australia*. Perth.

Doane, D. (2005). Beyond corporate social responsibility: Minnows, mammoths and markets. *Future, 37*, 215–229.

Dore, R. P. (2000). *Stock market capitalism: Welfare capitalism: Japan and Germany versus the Anglo-Saxons*. Oxford: Oxford University Press.

Duus, S. (2013). Coal contestations: Learning from a long, broad view. *Rural Society, 22*(2), 96–110.

Edelman. (2017). *2017 Edelman trust barometer—Global results*. Retrieved from https://www.edelman.com

Enoch, S. (2007). A greener potemkin village? Corporate social responsibility and the limits of growth. *Capitalism Nature Socialism, 18*(2), 79–90.

Environmental Justice Australia. (2015). *A review of the Adani group's environmental history in the context of the Carmichael coal mine approval*. Carlton, Victoria.

Environmental Justice Australia. (2017). *The Adani brief. What governments and financiers need to know about the Adani Group's record overseas*. Carlton, Victoria.

Epstein, E. M. (2007). The good company: Rhetoric or reality - corporate social responsibility and business ethics redux. *Amercian Business Law Journal, 44*(Summer), 207–222.

Fahrer, J. G. (2015). *Indiviidual report to the Land Court of Queensland*. Melbourne.

Farcic, E. (2018, 3 December). Fracking protesters pledge to fight on *The West Australian*. Retrieved from https://www.pressreader.com

Farrar, J. (2008). *Corporate governance. Theories, principles and practice* (3rd ed.). Sydney: Oxford University Press.

Friedman, M. (1970, 13 September). The social responsibility of business is to increase its profits. *New York Times Magazine,* pp. 122–126.

Geoscience Australia, & Australian Bureau of Agricultural and Resource Economics. (2010). *Australian gas resource assessment 2012.* Canberra.

Giroux, H. A. (2019). Beyond dystopian education in a neoliberal society. *Fast Capitalism, 10*(1).

Gjølberg, M. (2009). Measuring the immeasurable? Constructing an index of CSR practices and CSR performance in 20 countries. *Scandinavian Journal of Management, 25,* 10–22.

Gluyas, R. (2016, 18 April). Are Australia's banks really behaving badly? *The Australian.* Retrieved from https://www.theaustralian.com.au/news/inquirer/are-australias-banks-really-behaving-badly/news-story/f575d114197c1785d0b922530df97abb

Government of Western Australia. (2009). *Regulatory impact assessment process review.* Perth.

Government of Western Australia. (2015, 8 June). Focus on future of WA's emerging gas industry. *Media Release.* Retrieved from https://www.mediastatements.wa.gov.au/Pages/Barnett/2015/06/Focus-on-future-of-WAs-emerging-gas-industry.aspx

Government of Western Australia. (2017). *The State Government has stopped the use of hydraulic fracture stimulation (fracking) for onshore petroleum exploration or development in Western Australia.* Perth.

Hanlon, R. J. (2011). Engineering corporate social responsibility: Elite stakeholders, states and the resilience of neoliberalism. *Contemporary Politics, 17*(1), 71–87.

Hannam, P. (2017, 4 April). 'Barbaric': Farmers rattled as Adani coal mine granted unlimited water access. *The Sydney Morning Herald.* Retrieved from https://www.smh.com.au/environment/barbaric-farmers-rattled-as-adani-coal-mine-granted-unlimited-water-access-20170404-gvdk5v.html

Hare, B., Roming, N., Fuentes Hutfilter, U., Schaeffer, M., & Beer, M. (2018). *Western Australia's gas gamble.* Berlin, Germany.

Hastie, H., & Latimer, C. (2019, 14 March). EPA buckles under pressure from WA government, oil and gas sector. *The Sydney Morning Herlad.* Retrieved from https://www.smh.com.au/business/the-economy/epa-buckles-under-pressure-from-wa-government-oil-and-gas-sector-20190314-p5144l.html \

Haswell, M. (2017). *Health concerns associated with unconventional gas mining in Western Australia: A critical review.* Brisbane.

Hayne, K. (2018). *Interim report—royal commission into misconduct in the banking.* Canberra: Superannuation and Financial Services Industry.

Hayne, K. (2019). *Final report—royal commission into misconduct in the banking, superannuation and financial services industry.* Canberra.

Higginbotham, N., Freeman, S., Connor, L., & Albrecht, G. (2010). Environmental injustice and air pollution in coal affected communities, Hunter Valley, Australia. *Health & Place, 16*(2), 259–266.

Hirschman, A. O. (1991). *The rhetoric of reaction.* Cambridge: Harvard University Press.

Hoque, K. B. (1985). Social responsibilities of business: Myth or reality? *The Dhaka University Studies, 42*(1 (Part A)), 119–127.

Hughes, T. P., Kerry, J. T., Álvarez-Noriega, M., Álvarez-Romero, J. G., Anderson, K. D., Baird, A. H. … Berkelmans, R. (2017). Global warming and recurrent mass bleaching of corals. *Nature, 543*(7645), 373

Humpage, L. (2014). *Policy change, public attitudes and social citizenship: Does neoliberalism matter?* Bristol, UK: Policy Press.

Institute for Energy Economics and Financial Analysis. (2017). *IEEFA Update: An increasingly cursed Australian coal project.* Ohio: Cleveland.

International Energy Agency. (2017). *Market report series: Coal 2017.* Paris.

Jenkins, C. (2018, 30 November). WA Labor opens the gate to fracking. *Greenleft Weekly*. Retrieved from https://www.greenleft.org.au/content/wa-labor-opens-gate-fracking

Joye, C. (2017, 19 May). Big bank subsidy worth more than $5 billion annually. *Australian Financial Review*. Retrieved from https://www.afr.com/markets/equity-markets/big-bank-subsidy-worth-more-than-5-billion-annually-20170518-gw8491

Keane, B., & Dyer, G. (2019, 6 February). What Hayne should have done. *Crikey*. Retrieved from https://www.crikey.com.au/2019/02/06/hayne-royal-commission-recommendations

Keane, B., & Dyer, G. (2019, 9 October). *A stagnant government has created a stagnant Australia. Crikey*. Retrieved from https://www.crikey.com.au/2019/10/08/a-stagnant-government-has-created-a-stagnant-australia/

Kehoe, J. (2016, 7 October). Treasurer Scott Morrison defends 'strong banks' in US. *Australian Financial Review*. Retrieved from https://www.afr.com/news/politics/treasurer-scott-morrison-defends-strong-banks-in-us-20161006-grw2tf

Kellow, A., & Niemeyer, S. (1999). The development of environmental administration in Queensland and Western Australia: Why are they different? *Australian Journal of Political Science, 34*(2), 205–222.

Knott, M. (2016, 7 August). Scott Morrison slams banking royal commission proposal as nothing more than a 'populist whinge'. *Sydeny Morning Herald*. Retrieved from https://www.smh.com.au/politics/federal/scott-morrison-slams-banking-royal-commission-proposal-as-nothing-more-than-a-populist-whinge-20160807-gqms5y.html

Kok, P., van der Wiele, T., McKenna, R., & Brown, A. (2001). A corporate social responsibility audit within a quality management framework. *Journal of Business Ethics, 31*(4), 285–297.

Layman, L. (1982). Changing resource development policy in Western Australia, 1930s to 1960s. In E. J. Harman & B. W. Head (Eds.), *State, capital and resources in the North and West of Australia* (pp. 149–165). Nedlands, WA: UWA Press.

Lee, M.-D.P. (2008). A review of the theories of corporate social responsibility: Its evolutionary path and the road ahead. *International Journal of Management Reviews, 10*(1), 53–73.

Lidberg, J. (2019). The distortion of the Australian public sphere: Media ownership concentration in Australia. *AQ: Australian Quarterly, 90*(1), 12–20.

Linden, A., & Staples, W. (2019, 5 February). Banking inquiry's omissions mean we'll need another royal commission. *INDaily*. Retrieved from https://indaily.com.au/opinion/2019/02/05/banking-inquirys-omissions-mean-well-need-another-royal-commission/

Long, S. (2017, July 6). Galilee Basin mines will slash coal output, jobs elsewhere, Wood Mackenzie says. *ABC News*. Retrieved from https://www.abc.net.au/news/2017-07-06/galilee-basin-mining-project-will-reduce-coal-output:-research/8682164

Luke, H. (2017). Social resistance to coal seam gas development in the Northern Rivers region of Eastern Australia: Proposing a diamond model of social license to operate. *Land Use Policy, 69*, 266–280.

Luke, H., Brueckner, M., & Emmanouil, N. (2018). Unconventional gas development in Australia: A critical review of its social license. *The Extractive Industries and Society*.

Madsen, H., & Ulhøi, J. P. (2001). Integrating environmental and stakeholder management. *Business Strategy and the Environment, 10*, 77–88.

Massola, J. (2018, February 1). Big surge in opposition to Adani, new polling reveals. *The Sydney Morning Herald*. Retrieved from https://www.smh.com.au/politics/federal/big-surge-in-opposition-to-adani-new-polling-reveals-20180131-p4yz4o.html

Matten, D., & Moon, J. (2008). 'Implicit' and 'explicit' CSR: A conceptual framework for a comparative understanding of corporate social responsibility. *Academy of Management Review, 33*(2), 404–424.

Meinshausen, M. (2015). *Individual report to the Land Court of Queensland on "Climate Change—Emissions"*. Melbourne.

Miles, M., Munilla, L., & Darroch, J. (2006). The Role of Strategic Conversations with Stakeholders in the Formation of Corporate Social Responsibility Strategy. *Journal of Business Ethics, 69*(2), 195–205.

Munilla, L. S., & Miles, M. P. (2005). The corporate social responsibility continuum as a component of stakeholder theory. *Business and Society Review, 110*(4), 371–387.

Murphy, K. (2019, February 4). Yes, Kenneth: the Coalition response to the banking inquiry is pure politics. *The Guardian*. Retrieved from https://www.theguardian.com/australia-news/2019/feb/04/coalition-response-banking-inquiry-pure-politics

National Australia Bank. (2019). *We're more than money*. Retrieved from https://www.nab.com.au/about-us/corporate-responsibility

Newell, P. (2001). Citizenship, accountability and community: The limits of the CSR agenda. *Journal of International Development, 13*(3), 907–919.

Oil & Gas Australia. (2018). *Oil and gas industry welcomes lifting of WA fracking moratorium*. Retrieved from https://webcache.googleusercontent.com/search?q=cache:x4oJ-Oz5JGwJ:www.oilandgasaustralia.com.au/oil-gas-industry-welcomes-lifting-wa-fracking-moratorium/+&cd=16&hl=en&ct=clnk&gl=au&client=safari

Parliamentary Joint Committee on Corporations and Financial Services. (2006). *Corporate responsibility: Managing risk and creating value*. Canberra.

Pearce, L. (2017, 26 June). The Great Barrier Reef is worth $56 billion to our economy. Here's the proof. *Huffington Post (Australian Edition)*. https://www.huffingtonpost.com.au/2017/06/26/the-great-barrier-reef-is-worth-56-billion-to-our-economy-here_a_23001277/

Phillimore, J. (2014). The politics of resource development in Western Australia. In M. Brueckner, A. Durey, R. Mayes, & C. Pforr (Eds.), *Resource curse or cure? On the sustainability of development in Western Australia* (pp. 25–40). Heidelberg: Springer.

Pillay, R. G. (2016). The limits to self-regulation and voluntarism: From corporate social responsibility to corporate accountability. *Amicus Curiae, 2014*(99), 10–13.

Pini, B., Mayes, R., & McDonald, P. (2010). The emotional geography of a mine closure: A study of the Ravensthorpe nickel mine in Western Australia. *Social and Cultural Geography, 11*(6), 559–574.

Queensland Government. (2016). *Adani's Carmichael coal mine environmental authority approved*. Brisbane.

Queensland Government. (2018). *Carmichael coal mine and rail project*. Brisbane.

Reich, R. (2008). *Supercapitalism. The transformation of business, democracy and everyday life*. Melbourne: Scribe.

Remeikis, A. (2019, February 10). Banking reform won't happen until after the election, Christopher Pyne says. Retrieved from https://www.theguardian.com/australia-news/2019/feb/10/banking-reform-wont-happen-until-after-the-election-christopher-pyne-says

ROBECOSAM. (2017). *Country sustainability ranking*.

Robertson, J. (2017, June 6). Adani gives 'green light' to $16bn Carmichael coal mine. *The Guardian*. Retrieved from https://www.theguardian.com/environment/2017/jun/06/adani-gives-green-light-to-16bn-carmichael-coal-mine

Saunders, A., Timson, L., & Desloires, V. (2015, November 11). Report shows safety issues at BHP Billiton's failed Brazilian dam were highlighted two years ago. *The Sydney Morning Herald*. Retrieved from https://www.smh.com.au/business/mining-and-resources/report-shows-safety-issues-at-bhp-billitons-failed-brazilian-dam-were-highlighted-two-years-ago-20151111-gkwb5w.html

Scambary, B. (2013). *My country, mine country: Indigenous people, mining and development contestation in remote Australia*. Canberra: ANU E Press.

Schmulow, A., Fairweather, K., & Tarrant, J. (2019). Restoring Confidence in Consumer Financial Protection Regulation in Australia: A Sisyphean Task? *Federal Law Review, 47*(1), 91–120.

Selznick, P. (2002). *The communitarian persuasion*. Baltimore, MD: Johns Hopkins University Press.

Shamir, R. (2004). The de-radicalisation of corporate social responsibility. *Critical Sociology, 30*(3), 669–689.

Shamir, R. (2010). Capitalism, governance, and authority: The case of corporate social responsibility. *Annual Review of Law and Social Science, 6*(1), 531–553.

Sherry, N. (2009, February 5). *Government to provide assistance for establishment of responsible investment academy*. Retrieved from https://ministers.treasury.gov.au/

Slezak, M. (2017a, October 7). Most Australians oppose Adani mine, poll shows, amid national protests. *The Guardian*. Retrieved from https://www.theguardian.com/environment/2017/oct/07/most-australians-oppose-adani-mine-poll-shows-amid-national-protests

Slezak, M. (2017b, 16 August). Why Adani's planned Carmichael coalmine matters to Australia—and the world. *The Guardian*. Retrieved from https://www.theguardian.com/business/2017/aug/16/why-adanis-planned-carmichael-coalmine-matters-to-australia-and-the-world

Smith, K. (2017, June 6). Anger in WA as Labor reneges on gas promises. *Green Left Weekly*, p. 6.

Sparrow, J. (2017, April 7). It's either Adani or the Great Barrier Reef. Are we willing to fight for a wonder of the world? *The Guardian*. Retrieved from https://www.theguardian.com/commentisfree/2017/apr/07/its-either-adani-or-the-great-barrier-reef-are-we-willing-to-fight-for-a-wonder-of-the-world

Subramaniam, N., Kansal, M., & Babu, S. (2017). Governance of mandated corporate social responsibility: Evidence from Indian government-owned firms. *Journal of Business Ethics, 143*(3), 543–563.

Szabó, D. G., & Sørensen, K. E. (2015). New EU directive on the disclosure of non-financial information (CSR). *Nordic & European Company Law Working Paper*(15–01).

The Wilderness Society. (2018). The truth about fracking and the environment. Retrieved from https://wilderness.org/article/truth-about-fracking-and-environment

Thurbon, E. (2012). From Developmentalism to Neoliberalism and Back Again? Governing the Market in Australia from the 1980s to the Present. In C. Kyung-Sup, B. Fine, & L. Weiss (Eds.), *Developmental Politics in Transition: The Neoliberal Era and Beyond* (pp. 274–295). London: Palgrave Macmillan UK.

Trigger, D. S. (1997). Mining, landscape and the culture of development ideology in Australia. *Ecumene, 4*(2), 161–180.

Upstream Petroleum Resources Working Group Report to COAG Energy Council. (2015). *Coal seam, shale and tight gas in Australia: Resources assessment and operation overview.* . Caberra.

Vallentin, S., & Murillo, D. (2012). Governmentality and the politics of CSR. *Organization, 19*(6), 825–843.

Vedelago, C. (2018, May 11). Cocaine, girls, million-dollar fraud: What the Commonwealth Bank lender did next. *The Sydney Morning Herald*. Retrieved from https://www.smh.com.au/business/banking-and-finance/cocaine-girls-million-dollar-fraud-what-the-commonwealth-bank-lender-did-next-20180511-p4zeqt.html

Verrender, I. (2017, November 27). Royal commission into banking: The inquiry that could kill the Government. *ABC News*. Retrieved from https://www.abc.net.au/news/2017-11-27/how-bad-are-our-banks/9194984

Verstegen, P. (2018, December 18). Western Australia is being opened to fracking on misleading and ill-founded grounds *The Guardian*. Retrieved from https://www.theguardian.com/commentisfree/2018/dec/18/western-australia-is-being-opened-to-fracking-on-misleading-and-ill-founded-grounds

Vogwill, R. (2017). *Western Australia's tight gas industry - A review of groundwater and environmental risks*. Perth.

Waagstein, P. R. (2011). The mandatory corporate social responsibility in Indonesia: Problems and implications. *Journal of Business Ethics, 98*(3), 455–466.

Wangan and Jagalingou Family Council. (2018). *Our fight*. Retrieved from https://wanganjagalingou.com.au/our-fight/

Wesley, A. (2014). *The socio-political construction and experience of corporate social responsibility (CSR): An investigation into the conflict surrounding the James Price Point LNG precinct, Kimberley, Western Australia*. Ph.D., Curtin University, Perth.

Wesley, A., & MacCallum, D. (2014). The political economy of corporate social responsibility in the resource sector in Western Australia. A case study of the proposed James Price Point LNG precinct. In M. Brueckner, A. Durey, R. Mayes, & C. Pforr (Eds.), *Resource curse or cure? On the sustainability of development in Western Australia* (pp. 59–73). Heidelberg: Springer.

Western Australian Standing Committee of Environment and Public Health Affairs. (2015). *Implications for Western Australia of hydraulic fracturing for natural gas. Report 42*. Perth.

White, R. (2013). Resource extraction leaves something behind: Environmental justice and mining. *International Journal of Crime and Justice, 2*(1), 50–64.

Windsor, D. (2001). The future of corporate social responsibility. *the International Journal of Organizational Analysis, 9*(3), 225–256.

World Business Council for Sustainable Development. (2000). *Corporate social responsibility: making good business sense*. Geneva.

CSR in the Norwegian Context

Siri Granum Carson and Heidi Rapp Nilsen

Abstract In this chapter, we sketch the rise of explicit CSR in the Norwegian context by focusing on the extractive industries' entry into a global market and the resulting legitimacy challenges arising from this transition. Explicit CSR, in the sense of expressing social and environmental responsibility and voluntarily committing to promote societal benefits, can be viewed as a strategy by which the companies attempt to fill the governance gaps of global capitalism. We review two major Norwegian companies, Hydro and Statoil/Equinor, and argue that while their CSR strategy has been quite successful, it is challenging to reconcile the role of a socially and environmentally responsible company with being an actor in the global extractive industries.

Keywords Implicit CSR · Explicit CSR · Globalisation · Hydro · Statoil/Equinor

The extractive industries have significant impact on the environment and on the societies in which they operate, and should arguably take on environmental and social responsibilities accordingly. Corporate Social Responsibility (CSR) is a concept suggesting that private companies have obligations towards society beyond making a profit. Following internationalisation and increasing expectations from stakeholders, CSR has become a central management concept in the extractive industries also in Norway.

An important driver for CSR in the Norwegian context has been the perceived governance gap in the global economy, and official Norwegian CSR policy was primarily directed towards business operations abroad. The need for explicit CSR

S. G. Carson (✉)
Department of Philosophy and Religious Studies, Norwegian University of Science and Technology (NTNU), Trondheim, Norway
e-mail: siri.granum.carson@ntnu.no

H. R. Nilsen
Department of Industrial Economics and Technology Management, Norwegian University of Science and Technology (NTNU), Trondheim, Norway
e-mail: heidi.r.nilsen@ntnu.no

E. G. Pereira et al. (eds.), *Sovereign Wealth Funds, Local Content Policies and CSR*, CSR, Sustainability, Ethics & Governance,
https://doi.org/10.1007/978-3-030-56092-8_35

in the relatively highly regulated domestic market, where labour rights and environmental performance is secured through laws, regulation and tripartite agreements, was initially viewed as low. But eventually, globalisation also affects the domestic socio-cultural and political environment of business challenging the traditional ideals of the Nordic countries, among other things through a higher level of industry's self-regulation.

In this chapter, we look into how Hydro and Equinor, two large and partly state-owned companies of the extractive industries in Norway, have addressed some of these issues. Both companies operate domestically and abroad. Both companies have a history of high ambitions of CSR and corporate sustainability, among other things through their involvement in initiatives such as Extractive Industries Transparency Initiative (EITI). In the Hydro example, we explore the historic development of an explicit CSR policy in response to the cross-pressure of internationalisation. In the Equinor example, we look into some specific challenges of a national oil company in terms of political and environmental responsibility.

1 From Implicit to Explicit CSR in Norway[1]

Matten and Moon (2008) introduce a distinction between two kinds of CSR. Depending on the nature of the overall business-society relations, CSR may be characterised as either explicit, i.e. codified as corporate policies explicitly formulated by (or for) the companies, or implicit, i.e. codified as institutional frameworks implicitly assumed by the companies. Although European companies traditionally have kept a lower profile on the topic of CSR than American companies, Matten and Moon argue that no evidence suggests that European companies in general are less socially responsible. The difference between American and European companies is, thus, not necessarily a matter of more or less socially responsible business, but rather a difference between what they call "explicit and implicit forms of CSR". In other words: American companies have typically developed company-specific and explicit policies on social and environmental responsibility, while European companies have not—rather, corporate responsibility has been implied in the overall structures of the business-society relations. Explicit CSR has, however, gained ground, also in the Western and Northern parts of Europe (Matten and Moon 2008). Matten and Moon specify business-society relations using Whitley's comparative business system framework, which identifies four key elements: The political system (power and engagement of the state), the financial system (form and distribution of corporate ownership), education and labour systems, and the cultural system (broad assumption about society, business and government) (cf. Whitley 1997). They employ this framework in order to understand the historic differences between the US and Europe when it comes to CSR:

[1] This section and the next section with the Hydro example draws heavily on Carson et al. (2015).

> U.S.-style CSR has been embedded in a system that leaves more incentive and opportunity for corporations to take comparatively explicit responsibility. European CSR has been implied in systems of wider organisational responsibility that have yielded comparatively narrow incentives and opportunities for corporations to take explicit responsibility (Matten and Moon 2008: 409).

Whitley's theory of national business systems explains how American and European forms of CSR are different. Further, Matten and Moon refer to neo-institutionalism in order to explain why these differences are gradually dissolving. The neo-institutionalist thesis of *homogenisation* suggests that organisations become increasingly similar because popular ideas and recipes for organisational management—such as CSR—gradually become standard criteria for the legitimacy of an organisation and therefore becomes institutionalised all over the world. Traditionally, European companies did not articulate and promote their contributions to the enhancement of social and environmental goods as "CSR". Changes within the national business systems of Europe has, however, increasingly led companies to adopt explicit CSR strategies, even in the Nordic context (cf. Strand 2013; Vallentin and Murillo 2010; Røvik 2007). Explicit CSR, thus, increasingly becomes a strategic tool for the enhancement of corporate reputation in order to build public trust. Carson et al. (2015) analyses the transition from implicit to explicit CSR in the Norwegian context, and identifies two drivers for this development: Re-legitimising and organisational expressiveness, which can both be viewed as strategic adjustments to the ongoing globalisation processes changing both the market-and production conditions for Scandinavian companies.

The development of the Nordic welfare states from World War II up until the 1990s has historically conditioned implicit CSR. Much of what has been framed as the social responsibilities of U.S. companies—specifically in relation to labour rights such as workers' pay, benefits and safe working conditions—has in the Nordic context typically been negotiated and institutionalised through political compromises and tripartite agreements. A consensual political culture, a strong social-democratic welfare state, and well-functioning partnerships between business, government and labour organisations can be viewed as characteristic of a "Nordic" or "Scandinavian" model of implicit CSR (cf. e.g. Vallentin and Murillo 2010; Gjølberg 2010; Midttun, Gautesen and Gjølberg 2006). In this perspective, offering safe working conditions and social benefits is primarily a matter of complying with laws and agreements. The same is arguably the case for environmental issues. Midttun et al. (2006) argue that Nordic countries have strong traditions of regulation and social embeddedness of business behaviour, which give them certain advantages when it comes to institutionalising CSR. Scandinavian companies have strong traditions for stakeholder engagement beyond what the law requires (cf. Rhenman 1964, 1968; Näsi 1995), which can be framed as a "cooperative advantage" of these companies in a global market (Strand 2009; Strand/Freeman 2013).

Gjølberg (2010) argues that CSR in the Nordic countries builds on a certain "Nordic normative legacy" in the way solidarity, decency and respect is institutionalised (Gjølberg 2010). This leads up to a Nordic consensus that CSR is more or less superfluous in the domestic context, but could nevertheless be important for

Nordic companies when they get involved in operations abroad, in the absence of a global, multilateral framework. Among the Nordic countries, the specific Norwegian approach to CSR is characterised by the substantial state ownership in large companies, notably in the extractive industries operating in high-risk markets like Nigeria, Angola and Azerbaijan (Equinor), Qatar and Vietnam (Hydro). As a major operator in these countries, the Norwegian government has perceived a rising need to explicitly address questions regarding business and human rights, corruption and environmental hazards, and to develop an official policy on CSR. Official Norwegian CSR policy has thus, almost exclusively focused on business operations abroad, illustrated by the fact that the CSR initiatives of the Norwegian government from the beginning were coordinated from the Ministry of Foreign Affairs. The governmental "White Paper" on CSR (Norwegian Ministry of Foreign Affairs 2008/2009) also reflects this focus concerning impact and responsibility of Norwegian companies involved in activities abroad:

> Increasing internationalisation means that Norwegian companies are operating to a greater extent than before in countries where there is little respect for human rights. Working conditions are often unacceptable and child labour is used in production processes. There is discrimination in the workplace on the basis of gender, religious belief or ethnic background. Too little account is taken of environmental impacts; corruption may be widespread. Often these conditions are related to deficiencies in legislation, weak enforcement or a lack of sanctions. In many cases, the individual company thus faces a number of fundamental questions. Should we, or should we not, become engaged in the country? What can reasonably be required or expected of our company, and how can we meet these expectations?[2]

This view was also supported by the employers' association the Confederation of Norwegian Enterprise (NHO), as well as by the Norwegian Labour Organisation (LO). CSR was perceived as mainly voluntary, versus national questions of labour rights, which should be kept within the framework of binding tripartite agreements. Trygstad and Lismoen (2008) argue that the Norwegian, highly institutionalised and law-regulated labour situation represents the paradigm of implicit CSR. This general agreement concerning the business–society relation is the reason why the CSR debate in Norway primarily has centred on Norwegian companies' activities abroad (cf. Trygstad and Lismoen 2008). Gjølberg points out that the perceived governance gap in the global economy is an important driver behind Norwegian CSR, thus "the Norwegian approach to CSR remains decidedly global, and even explicitly non-domestic, in its orientation" (Gjølberg 2010).

Globalisation affects the socio-cultural and political environment of business within the Norwegian labour situation as well, and contributes to the rise of explicit CSR. The traditional ideal of labour rights being negotiated through tripartite agreements and secured at a national level is challenged by a competing ideal of industry's self-regulation of labour standards. In the 1997 reform of the safety, health and environment (SHE) systems in Norway called "internal control", the direct control of SHE conditions was delegated from the government to the enterprises (cf. Hovden 1998; Saksvik and Quinlan 2003). When individual companies become responsible

[2] https://www.regjeringen.no/en/dokumenter/report-no.-10-to-the-storting-2008-2009/.

for implementing SHE, this promotes explicit statements from the companies on how they address these issues, and can thus be seen as an early driver of explicit CSR in the Norwegian context.

The explication of CSR in Scandinavia might be seen as a rise in organisational expressiveness" (Schultz et al. 2000), with increasing attention to the image, reputation, identity, and values of products and organisations. This tendency is especially striking when it comes to social and environmental issues (Røvik 2007). Globalisation and the resulting increased competition lead to more focus on the expanded product concept in order to differentiate products, brands and companies. Organisations seek to be associated with positive values, and environmental and social performance becomes strategically important for branding and image building (Hagen 2009). CSR as a popular management concept (Røvik 2007) and a "corporate megatrend" (Midttun 2013), is the ultimate expression of this expressiveness. CSR blurs the boundaries between marketing and public relations, e.g. when a petroleum company address environmental threats with the voice of a worried "corporate citizen", but at the same time might be seen to act as a reputation-building commercial actor. The move towards explicit CSR, where Norwegian companies become more expressive regarding social and environmental values, can be seen as a strategic move by which the companies attempt to manage corporate legitimacy in the complex situation of a global market (Carson 2019).

2 "The Hydro Way"

Hydro is a Norwegian aluminium producer with activity in over 50 countries, employing 35,000 people. Its headquarters is in Oslo, nearly two-thirds of the shares of the company are in Norwegian hands, and the Norwegian government is the major shareholder with 34% of the shares. The company was established by engineer Sam Eyde in 1905 as an innovative producer of fertiliser based on Norwegian hydropower (Andersen 2005). Later, the company developed into a conglomerate with diverse activities: fertilisers, aluminium, oil and gas, and aquaculture. Aluminium eventually became the core activity. Aluminium production is a very energy-demanding activity, and Norway's abundance of hydropower created favourable natural preconditions for this industry. Hydro first established aluminium works at Karmøy (1967), and later bought several works in Norway, among others Årdal og Sunndal Verk (ÅSV), with factories in Årdal, Sunndalsøra and Høyanger—industrial locations where the aluminium works were the cornerstones of the respective local communities (Frøland and Karlsen 2008). After 2000, Hydro abandoned the diversification strategy[3] and became a pure aluminium company with global ambitions. Wholesale purchase of German and French aluminium works, building of a large aluminium

[3]The fertilizer production was spun off as a separate business under the name of Yara in 2004, while the oil and gas activity was sold and merged with Statoil in 2007.

factory in Qatar, and buying up smelting works and raw material producers in Brazil were among the moves to secure this position (Røyrvik 2008).

In terms of social responsibility, Hydro tends to point to its historic role as a cornerstone company in vulnerable local communities along the Norwegian coastline. Hydro's position in these communities may be conceptualised as "implicit CSR", with a perceived consistency between the needs of the company and the local community. Hydro's initiatives towards e.g. building schools, roads and railways, arose out of a kind of "enlightened self-interest", constituting a win-win situation for company and local community (cf. Reiten 2007). In the 1960s and -70 s, however, the debate concerning Hydro's obligations towards society changed character. The relocation of business operations, specifically the move of the nitrate production from Rjukan to Herøya in the 1960s (Sagafoss 2005), was controversial. These were economically motivated decisions, which went against the benefit of the local community, thus the established assumption that what was good for Hydro was good for the community as well did not hold anymore. The growing environmental concerns also created controversy for Hydro. The narrow fjord arms of Høyanger and Årdal were for a period deeply affected by fluoride emissions from the electrolysis (Andersen and Yttri 1997). Since the end of the 1980s, these emissions decreased considerably as a result of investments in new technology, but in general, the environmental challenges of Hydro remained high on the agenda. Civil society protests and demonstrations pointed out major shortcomings in the environmental policy of Hydro.

In the beginning, Hydro dealt with these challenges reactively, almost reluctantly. Throughout the 1980s and 1990s, social and environmental issues were primarily addressed in response to external pressure. Eventually, this policy changed and Hydro aimed to become more proactive, for example, by releasing environmental information before the protesters or the media did—even if the information was negative. From 1989, the company published an environmental report in addition to the annual report. At the time, this was seen as pioneering work that set a standard for all companies. In the 1990s, Hydro managers participated in the establishment of the World Business Council for Sustainable Development (Sagafoss 2005). In 1999, Hydro hosted an international seminar resulting in the publication of "Invitation to Dialogue" in 2000. This was the first of several Hydro publication on CSR.

The background for this strategic move was the internationalisation processes that Hydro was involved in, and the increasing competition the company faced as a result of this. Hydro closed down several of their operations along the coastline of Norway, among them Årdal and Høyanger, with significant social and economic impacts as these plants constituted the cornerstones of their respective communities (Sagafoss 2005). At the same time, Hydro invested heavily in large, new plants abroad. These investments generated novel challenges for the company with regard to both social and environmental responsibility. In December 2001, Hydro withdrew from a controversial mining project in Orissa in India, following protests and critique from local community and environmental and human rights organisations. The protests were based on fear of losing access to water resources and arable land. In addition, in the area where the mining operations were to take place, there was a mountain held to be holy by one of the local tribes. Hydro allocated considerable resources to solving the conflicts in the area, among other things by organising stakeholder dialogues. Eventually, they decided that the conflicts ran too deep and risks were too high, and that the prudent thing was to withdraw from the whole project. That did not put a complete stop to the protest against the company, however. A Norwegian NGO argued that Hydro should stay involved in order to take responsibility for the continuation of the process, rather than choosing an exit strategy.[4]

Another example illustrating Hydro's new dilemmas is their involvement in the building of a gigantic aluminium smelting works in Qatar in collaboration with the Qatari national petroleum company. The factory opened in 2011, and the project was heavily criticised for a number of reasons: for jeopardising Norwegian jobs, for accepting CO_2 emissions way above Norwegian standards, and for disregarding basic human rights. As regards the latter point: Unionising is against the law in Qatar, and Hydro was forced to accept this condition for a large majority of the workers on the project. In addition, Hydro had to accept considerable differences in salaries based on the worker's nationality in their aluminium factory in Qatar.[5]

In February 2018, Hydro faced serious accusations of contaminating drinking water supply in Brazil's metropolitan area of Para's capital in the Amazon region. Heavy rains led to the overflow of several basins, and samples taken by the Brazilian ministry's technicians found high levels of lead, aluminium, sodium, and other substances harmful to human and animal health (The Rio Times 2018). Hydro was accused that their plant Alunorte, in the affected area, was the cause of this contamination by the heavy rain triggering the leaking of these harmful substances. Hydro denied this and said that they cooperated with the relevant authorities to address both the environmental situation and the accusations, and as of January 2019 the production embargo on Alunorte was lifted. Hydro has in this same period strengthened their cooperation and engagement with the local community, including handing out drinking water to the local inhabitants (Hydro 2019).

These examples illustrate the cross-pressure that constitutes the basis of Hydro's CSR policy, between old domestic expectations and new demands abroad. In response to this, the company develops an explicit CSR strategy which they term "The Hydro Way", framed as the explication of the already existing core values of the company, cf. their annual report from 2004:

> Hydro has always taken special pride in doing things that help build a better society. When Hydro was founded 100 years ago, we helped build a country, not just a company. That sense of responsibility remains with us today. We have developed our businesses in ways that contribute more over time, not just to customers and shareholders, but also to society in general (Hydro 2004: 5).

This quote points to certain implicit values of the company, which they now explicate in order to emphasise that the company should still be perceived as a "community builder". Hydro claims a history as community builder in the Norwegian context, and on this basis point forward towards taking on global challenges. While

---

[4]Cf. Carson and Skauge (2019: 196).

[5]Dagens Næringsliv 21.11.2011.

social responsibility used to be viewed as a commitment towards small, vulnerable local communities in Norway, Hydro today frames social responsibility more in terms of global challenges such as sustainable development and energy efficiency.

3 The Political and Environmental Responsibility of Equinor

Equinor, formerly Statoil, is the National Oil Company (NOC) in Norway. It is one of the largest producers of petroleum in the world, and the Norwegian state is the majority owner. Since the 1990s, the company has moved from extracting and producing only on the Norwegian continental shelf, towards internationalisation in both extraction and production. Today, it is represented in more than 30 countries among which some, such as Angola and Azerbaijan, are estimated to be among the most corrupt regimes in the world.[6]

In 2018, Equinor changed their brand, among other things in order to communicate a commitment to be part of the transition towards a post-carbon society. However, the main production is still, and will in the foreseeable future, be carbon-based. As an international petroleum company, Equinor is unavoidably involved in environmentally degrading activity, and operates in several countries marked by poverty, inequality, and political oppression. Thus, one might expect that it would have a hard time passing as a socially responsible company. Nevertheless, the company has performed very well on international ratings of socially responsible companies. In 2011, Equinor (then Statoil) ranked as the number one petroleum company on Fortune Magazine's "World's Most Admired Companies" ranking, and the company has for a number of years scored top marks on Dow Jones Sustainability Index. The company has participated actively in the shaping of CSR initiatives regarding transparency and the protection of human rights, and is award-winning for its community projects. From the company's own perspective, these initiatives are justified in terms of «enlightened self-interest»: Maintaining good relations to its surroundings is part of a strategy to achieve competitive advantages, and their good results on these areas could be seen as indications that this is a successful strategy. However, this strategy also indicate limits when it comes to the social responsibilities of the company, in the sense that the commitment to promote social and environmental good only go as far as what benefits the company.

As we saw above, the Norwegian government's white paper on CSR brought up the question whether Norwegian companies should be involved in countries where there is little respect for human rights, and whether they should use their influence in these situations. The white paper expresses high expectations that Norwegian

[6]These are ranked at, respectively, number 165 (Angola) and 152 (Azerbajan) out of 180 on Transparency International's Corruption Perception Index of 2018, ranking 180 countries and territories by their perceived level of public sector corruption according to experts and businesspeople, cf. https://www.transparency.org/cpi2018, accessed Sept. 9, 2019.

companies should promote good societal and environmental values in the countries where they operate, and specifically under conditions where the standard and level of regulation is low. It is reasonable to expect that a company such as Equinor, where the state is majority owner, will be among the companies that meet the highest expectations in terms of upholding high standards. Representatives from Equinor express such high standards, but at the same time take care to communicate the limits of their responsibility. This is typically the case when questions about the human rights situation in the countries where they operate are raised, for example here in a statement from the information director in 2013:

> We as a company must follow rules and human rights, while the government should promote human rights. We take care that our activities contribute to value creation, both for the society and the shareholders [...] This may in itself contribute to a positive development, but we cannot take on a political responsibility. This is an obligation for the political authorities, for which we do not have a mandate.[7]

From one perspective, parts of Equinor's activity seem to run counter to explicit political goals of the Norwegian government. In a collection of case studies of the role of National Oil Companies in global markets, Richard Gordon and Thomas Stenvoll (2007) write:

> [Norway] has taken a proactive stand in the involvement of ethics in foreign policy as a contributor of aid and supporter of peace processes. On the other hand, as the primary owner of Statoil, the Norwegian government is involved in the extraction of resource wealth and the implicit subsidy of regimes that go against the very principles that Norwegian foreign policy is trying to work against.

The authors mention Angola and Azerbaijan as core examples of this apparent contradiction, where the activities of the Norwegian NOC help create and enforce the very same resource curse that Norwegian foreign policy seeks to counteract in these countries.

Equinor's CSR strategy in these countries is to support and build initiatives in three key areas: transparency, labour and human rights, and community projects. These initiatives are, however, explicitly formulated within the limits of a so-called *business case* for CSR, where building good relation to stakeholders is part of a strategy to gain a competitive advantage.[8] In other words: The CSR strategy is only successful to the extent that Equinor achieves improved economic results in these areas, compared to other companies that do not to the same extent promote transparency human rights and community development (cf. Carson 2015).

Even though there has been a development towards exercising political influence in the countries where they operate (cf. Gulbrandsen and Moe 2005), one could argue that the CSR initiatives of Equinor only scratch the surface of the problems with doing business in a country such as Azerbaijan. Azerbaijan is one of the most

[7]Petromedia, May 21, 2013, http://petro.no/statoil-vi-kan-ikke-ta-over-et-politisk-ansvar/20003, Accessed Aug 27, 2015.

[8]Cf. e.g. 2013 Sustainability Report (Statoil 2014), p. 1:"The approach is based on our fundamental belief in the business case for sustainability—efficiency in resources and therefore costs, a long-term social license to operate and technology that will secure future business opportunities".

corrupt regimes of the world, and a country where a small, ruling elite enjoys the wealth created by the petroleum industry, while these riches do not benefit the poor and politically oppressed majority of the people. Arguably, by conducting business under these circumstances, Equinor serves the enriching and further strengthening of the ruling elite, thus stalling a potential development in the direction of civil rights and political freedom (Gordon and Stenvoll 2007). Is Equinor as a company responsible for this situation in the sense that they ought to take measures towards changing it (e.g. by attempting to influence the Azeri government or by supporting its opponents), or are such measures beyond the sphere of corporate social responsibility? In this particular example, the general questions concerning the limits of corporate responsibility are further complicated by the state ownership, in the sense that it could be argued that the Norwegian state as a responsible owner is obligated by certain principles of foreign policy.

The increasing severe situation with climate change (IPCC 2018) constitutes another challenge to Equinor's stated ambition to be a socially and environmentally responsible company while at the same time, remaining a producer of fossil fuels. The Norwegian Government Pension Fund Global (GPFG) is built on revenues from the petroleum extraction on the Norwegian continental shelf, primarily executed by Statoil/Equínor. Recently, GPFG made a decision to exclude oil companies from its portfolio, in order to reduce the risk for a permanent decline in the oil price towards the transition to a post-carbon society (The Norwegian Government 2019). A nuance in this decision is to not exclude companies which also engage in production of renewable energy, and Equinor are among the companies that do so, although by only 5 per cent of their total investments (Aftenposten 2018). However, the domestic policy of the Norwegian government's petroleum policy remains unchanged, with no limitations on exploration and production of oil from the Norwegian continental shelf, including for Equinor's operations also elsewhere in the world. Still, there is no doubt that the decision of excluding oil companies from the portfolio puts Norway's continuous exploration of oil in a paradox, or even ironic situation as suggested by a foreign reporter (Canadian public radio 2019). For Equinor, the decision by GPFG constitutes a challenge when it comes to the objective to portray themselves as a socially responsible company.

4 Conclusion

The development of the Nordic welfare states has historically conditioned implicit CSR. Much of what has been framed as voluntary initiatives—and explicitly as CSR—in the USA has in the Nordic context been negotiated and institutionalised through political compromises and tripartite agreements. Hence, the "Nordic" model of CSR has largely been that of implicit CSR. However, an increasing, globalised business world with a perceived governance gap with regard to environmental standards and human rights has spurred Nordic and Norwegian firms to explicate CSR, especially linked to their business operations abroad.

In this chapter, we have sketched the rise of explicit CSR in the Norwegian context by focusing on the extractive industries' entry into a global market and the resulting legitimacy challenges arising from this transition. Explicit CSR, in the sense of expressing social and environmental responsibility and voluntarily committing to promote societal benefits, can be viewed as a strategy by which the extractive companies attempt to fill the governance gaps of global capitalism. Our brief review of Hydro and Statoil/Equinor indicates that this strategy has been relatively successful for these two major Norwegian companies, but it also shows that the role of a socially and environmentally responsible company can be hard to reconcile with being an actor in the global extractive industries. For Hydro, this is especially challenging with regard to portraying themselves as being a "community builder" while facing strong criticisms from the very same communities, both at home and abroad. For Statoil/Equinor, there is an increasing discrepancy between their proportionally minor contribution to production of renewable energy and their claim to be perceived as a socially responsible energy company in the age of a mounting climate crisis.

References

Aftenposten. (2018). Statoil skryter av sin grønne satsing men bruker mer enn 95% av investeringene på olje og gass. Downloaded Mar 2019 from: https://www.aftenposten.no/okonomi/i/a27oqd/Statoil-skryter-av-sin-gronne-satsing–men-bruker-mer-enn-95-prosent-av-investeringene-pa-olje-og-gass.

Andersen, K. G. (2005). Flaggskip i fremmed eie. Hydro 1905–1945 [Flagship in foreign hands. Hydro 1905–1945]. Oslo: Pax Forlag.

Andersen, K. G., & Yttri, G. (1997). Et forsøk verdt: Forskning og utvikling i Norsk Hydro gjennom 90 år [Worth a trial. Research and development in Hydro through 90 years]. Oslo: Universitetsforlaget.

Canadian public radio's national evening show, 8 Mar 9 (2019). Downloaded 15 March 2019 from: https://www.cbc.ca/radio/asithappens/as-it-happens-friday-edition-1.5048766.

Carson, S. (2015). Bedrifters politiske ansvar (The political responsibility of corporations). *Magma Tidsskrift for Økonomi og Ledelse, 6*, 22–29.

Carson, S. G. (2019). The corporate legitimacy matrix–A framework to analyze complex business-society relations. *Philosophy of Management, 18*(2),169–187.

Carson, S. G., Hagen, Ø., & Sethi, S. P. (2015). From implicit to explicit CSR in a Scandinavian context: The cases of HÅG and Hydro. *Journal of Business Ethics, 127*(1), 17–31.

Carson, S. G., & Skauge, T. (2019). *Etikk for beslutningstakere [Ethics for decision makers]*. Oslo: Cappelen Damm Akademisk.

Frøland, H. O., & Karlsen, A. (2008). Innledning: Globalisering gjennom et århundre. Langsiktige trekk ved norsk aluminiumsindustri [Globalization through a century. Long-term features of Norwegian aluminum industry]. In J. Henden, H. O. Frøland, & A. Karlsen (Eds.), *Globalisering gjennom et århundre. Norsk aluminiumsindustri 1908–2008* (pp. 7–32). Bergen: Fagbokforlaget.

Gjølberg, M. (2010). Varieties of corporate social responsibility (CSR): CSR meets the "Nordic Model". *Regulation & Governance, 4*(2), 203–229.

Gordon, R., & Stenvoll, T. (2007). Statoil: A study in political entrepeneurship. *The Changing Role of National Oil Companies in International Energy Markets*. The James A. Baker III Institute for Public Policy, Rice University.

Gulbrandsen, L. H. & Moe, A. (2005). Oil Company CSR Collaboration in 'New' Petro-States. *Journal of Corporate Citizenship, 20*(12), 53–64.

Hagen, Ø. (2009). *Do socially responsible brands lead to socially responsible companies? Understanding change in expressive organizations*. PhD Thesis, Norwegian University of Science and Technology, Trondheim.

Hovden, J. (1998). The ambiguity of contents and results in the Norwegian internal control of safety, health and environment reform. *Reliability Engineering and System Safety, 60*(2), 133–141.

Hydro. (2004). Corporate social responsibility—Invitation to action. Oslo: Hydro. Retrieved April 10, 2012, from http://www.hydro.com/pagefiles/7233/CSR_en.pdf.

Hydro. (2019). Strengthening our community engagement. Downloaded 17 Mar 2019 from: https://www.hydro.com/en/press-room/the-situation-in-brasil/timeline-of-key-events/strengthening-our-community-engagement/.

IPCC. (2018). Global Warming of 1.5°C, an IPCC special report on the impacts of global warming of 1.5°C above pre-industrial levels and related global greenhouse gas emission pathways, in the context of strengthening the global response to the threat of climate change, sustainable development, and efforts to eradicate poverty. Downloaded Oct 2018 from http://www.ipcc.ch/report/sr15/.

Matten, D., & Moon, J. (2008). 'Implicit' and 'explicit' CSR: A conceptual framework for a comparative understanding of corporate social responsibility. *Academy of Management Review, 33*(2), 404–424.

Midttun, A. (Ed.). (2013). *CSR and beyond. A Nordic perspective*. Oslo: Cappelen Damm Akedemisk.

Midttun, A., Gautesen, K., & Gjølberg, M. (2006). The political economy of CSR in Western Europe. *Corporate Governance, 6*(4), 369–385.

Näsi, J. (1995). A Scandinavian approach to stakeholder thinking: An analysis of its theoretical and practical uses, 1964–1980. In J. Näsi (Ed.), *Understanding stakeholder thinking* (pp. 97–115). Helsinki: LSR-Julkaisut Oy.

Norwegian Ministry of Foreign Affairs. (2008/2009). *Corporate Social Responsibility in a Global Economy*. Report No. 10 (2008-2009) to the Storting.

Reiten, E. (2007). Askeladden og den globaliserte økonomien—Hydro i endring. [Askeladden (The rising star) and the global economy—Hydro changing]. Kristofer Lehmkuhl Lecture, Norges Handelshøyskole. Retrieved April 10, 2012, from http://brage.bibsys.no/nhh/bitstream/URN:NBN:no-bibsys_brage_24761/1/Lehmkuhl_2007.pdf.

Rhenman, E. (1964). *Foeretagsdemokrati och foeretagsorganisation [Industrial democracy and industrial management]*. Stockholm: Thule.

Rhenman, E. (1968). *Industrial democracy and industrial management*. London: Tavistock.

Røvik, A. (2007). *Trender og Translasjoner: Ideer som former det 21. århundrets organisasjon.* [Trends and Translations: Ideas Shaping the 21st Century's Organization]. Universitetsforlaget, Oslo.

Røyrvik, E.A. (2008). *Directors of creation: an anthropology of capitalist conjunctures in the contemporary.* PhD Thesis, Norwegian University of Science and Technology, Trondheim.

Sagafos, O.J. (2005). *Progress of a different nature. Hydro 1905-2005.* Oslo, Pax Forlag.

Saksvik, P. Ø., & Quinlan, M. (2003). Regulating systematic occupational health and safety management: Comparing the Norwegian and Australian experience. *Industrial Relations, 58*(1), 33–56.

Schultz, M., Hatch, M. J., & Larsen, M. H. (2000). *The expressive organization: Linking identity, reputation, and the corporate brand*. Oxford: Oxford University Press.

Strand, R. (2009). Corporate responsibility in Scandinavian supply chains. *Journal of Business Ethics, 85,* 179–185.

Strand, R. (2013). The chief officer of corporate social responsibility: A study of its presence in top management teams. *Journal of Business Ethics, 112*(4), 721–734.

Strand, R., & Freeman, R.E. (2013). Scandinavian cooperative advantage: The theory and practice of stakeholder engagement in scandinavia….

The Norwegian Government (2019). Excludes exploration and production companies from the Government Pension Fund Global. Press release 8 March 2019. Downloaded Mar 2019 from: https://www.regjeringen.no/en/id4/.

The Rio Times (2018). Downloaded 17 Mar 2019 from: https://riotimesonline.com/brazil-news/rio-politics/toxic-water-seeps-from-norwegian-mining-site-in-brazils-amazon/.

Trygstad, S., & Lismoen, H. (2008). Fagbevegelsen og CSR [The trade union movement and CSR]. Oslo: Fafo.

Vallentin, S., & Murillo, D. (2010). Government, governance and collaborative social responsibility. In A. Tencati & L. Zsolnai (Eds.), *The Collaborative Enterprise*. Bern: Peter Lang.

Whitley, R. (1997). Business systems. In A. Sorge & M. Warner (Eds.), *The IEBM handbook of organizational behaviour* (pp. 173–186). London: International Thomson Business Press.

Conclusion

Jonathon W. Moses, Eduardo G. Pereira, and Rochelle Spencer

Abstract This concluding chapter surveys the importance of the extractive industries across twelve countries in terms of their level of (national) dependence on those industries and their role as global providers. We then consider some of the lessons generated by comparing the results offered in the preceding 35 chapters, in terms of their implementation of sovereign wealth funds, local content policies and corporate social responsibility.

Keywords Resource dependence · Sovereign wealth funds · Local content policies · Corporate social responsibility

It is not easy to generalise about how our 12 different countries manage their extractives industries, as their national contexts, industries, interests, needs and strategies vary so significantly. Such a smorgasbord of difference is the attraction of an anthology of the sort we have collected. It should not be surprising, then, to learn that these cases also differ in the ways that they employ the three main instruments we have studied more carefully.

In this concluding chapter, we take a step back and consider what larger patterns can be deciphered in the anthology. To do so, this chapter is divided into four parts

J. W. Moses (✉)
Department of Sociology and Political Science, Norwegian University of Science and Technology, Trondheim, Norway
e-mail: jonathon.moses@ntnu.no

E. G. Pereira
Siberian Federal University, Krasnoyarsk, Russia
e-mail: egp.portugal@gmail.com

The University of West Indies, St. Augustine Campus, St. Augustine, Trinidad and Tobago

University of São Paulo, São Paulo, Brazil

R. Spencer
Centre for Responsible Citizenship and Sustainability, Murdoch University, Perth, WA, Australia
e-mail: rochelle.spencer@murdoch.edu.au

E. G. Pereira et al. (eds.), *Sovereign Wealth Funds, Local Content Policies and CSR*, CSR, Sustainability, Ethics & Governance,
https://doi.org/10.1007/978-3-030-56092-8_36

and a conclusion. The first provides some general country characteristics, as gleaned from reading all of the case contributions. This contextual comparison provides a foundation and reminds us of the varied nature of our selection of cases. We then move to discuss the findings vis-à-vis each of the three central tools under investigation: sovereign wealth funds, local content policies and corporate social responsibility, before ending with a conclusion.

1 Contextual Comparisons

As we saw in the introduction, the cases in this collection come from a wide variety of states, at different levels of economic development and political aggregation. In an effort to impose some order to this variation, we have introduced each case by their national level of human development, using the HDI as an organising operator, and as explained in the introductory chapter.

But there are other important ways in which these states differ. Some of the more interesting differences concern their approach to ownership and Contractual regimes. As regards to the first, the case contributions showcase an important difference: in some cases, it is possible to have private ownership of subnational petroleum and mineral rights. In most other cases, however, it is the state itself (local or federal) that owns the resource. For those of us working in a single national/cultural context, it is important to bear this variation in mind. It is easy, but often incorrect, to assume that legal conditions elsewhere are similar to those at home. With respect to contract/licensing forms, we find even more variation. Several states rely on concession or royalty-based regimes (as are common in North America), but other cases rely on production sharing contracts (PSCs), service contracts, or a combination of different contract forms.

One important difference separating our cases is the degree to which they, as a country or a province/state, depend upon the extractives sector. In countries that are heavily reliant on natural resources, the cost of mismanaging those resources can be particularly high. This level of dependence is difficult to compare across our cases, as the cases rely upon different extractives sectors (mining, oil, gas, etc.), but also because many of the cases are subnational in form (states or provinces in federal arrangements), making it difficult to find comparable statistics. Table 1 provides a couple of grounds for making comparisons along these lines. Here, there are two patterns worth noting.

The first concerns level of energy dependence. The left-hand side of Table 1 compares the degree of reliance, across cases, as captured by two main indicators. The first compares the size of the natural resource rents, as a percentage of GDP, using the World Bank's indicator.[1] This gives us an idea of how important the natural resource is, in any given case, as a share of the national economy. Here, we find

[1] World Bank (2020a: [NY.GDP.TOTL.RT.ZS]). This indicator captures "... the sum of oil rents, natural gas rents, coal rents (hard and soft), mineral rents, and forest rents". The World Bank calculates rents as the difference between the value of production at world prices and the total costs of production. This is made clear when it calculates oil and gas rents (see, e.g. World Bank 2019a, b).

Table 1 Energy Reliance across our 12 countries

Nation	Subnational	Regional share of energy (%) 2018	Total natural resource rents/GDP (%) 2017	Fuel exports (% merchandise exports) 2018	% of World Energy Production (2018)			
					Oil	Gas	Coal	Rare earth metals
Australia			7.2	24.4	0.3	3.4	6.1	11.1
	Western Australia	56						
Brazil			3.5	12.5	3.1	0.7	–	0.6
Canada		8	1.7	24.7	5.7	4.8	0.7	
	British Columbia	20.1						
	Northwest Territories	4.7						
	Quebec	19.6						
	Alberta	5.6						
Guyana			25.3	0	–	–	–	–
Indonesia			3.5	23.2	0.9	1.9	6.8	–
Iran			17.8	70.8	4.9	6.2	–	–
Kenya			2.5	1.0	–	–	–	–
Nigeria			8.7	94	2.2	1.3	–	–
Norway			5.9	62.3	1.9	3.1	–	–
Russian Federation			10.7	52	12.6	17.3	5.5	
Trinidad and Tobago			7.7	48.9	0.1	0.9	–	–
United States			0.5	13.7	15	21.5	8.6	9
	Alaska	1.7						

Note The regional share of energy is determined differently in each country. In Australia, the share covers both mineral and petroleum production. In Canada, we look at just mineral shares. It is for this reason that Alberta scores so low. In terms of crude oil production, Alberta's share is much larger, about 80%. In the USA, we look at total energy production. See the relevant references for more information. The "Fuel Exports" data uses the World Bank indicator, where "fuels" is operationalised at the commodities in SITC section 3 (mineral fuels, lubricants and related materials) and measured as a percentage of merchandise exports. In this category, the Trinidad and Tobago data are from 2015; the Iran data are from 2017

Sources World Energy Shares: BP (2019); Natural Resource Rents: World Bank (2020a); Fuel Exports: World Bank (2020b); Regional Shares: GWA (2018: 43); Chap. 11; and EIA (2017)

that the resource value in Guyana is extremely high (25.3% of GDP) and still very high in countries like Iran and the Russian Federation. As we do not have a similar indicator at the subnational level, it is difficult to compare these cases, but Western Australia seems particularly exposed to the natural resource rent, considering that more than half of Australia's mineral and petroleum production comes from that state. Indeed, in the Chapter "Local Content Policies for Regional Economic Development in Western Australia", we learn that mining accounts for 30% of the Gross State Product in Western Australia. In Canada, mining seems to be more evenly spread across different provinces, but the majority of extractive resources (especially for oil and gas) remain in Alberta. In the US case, Alaska provides a surprisingly small share of total US energy production. Of course, Alaska (or these other states/provinces) can still be very dependent upon petroleum resources, even if the USA is not very dependent upon Alaska.

Another indicator of reliance can be seen in the column titled "Fuel Exports" as a share of merchandise exports. This column also relies on a World Bank (2020b) measure to capture cases of more extreme natural resource dependence (e.g. Nigeria, but with Iran and Norway not far behind). As Guyana has just begun exporting the energy it has in reserves, it displays a very low score on this indicator, and Kenya is clearly not a major fuel exporter. For these cases, the extractives industry is an important, but not necessarily the most important source of exports or GDP.

In addition, we should consider how dependent the rest of the world is on these cases for supplying their energy needs. This is done with the columns on the right-hand side of Table 1. Here too, we find that the cases in this anthology are important, but not particularly dominant in the world market for oil, gas, coal and rare earth metals. From our own selection of cases, the most important market players seem to be Russia (oil and gas), the USA (gas) and Australia (rare earth metals).

We now have a better foundation for understanding the role these states play in the global energy economy and their degree of dependence on the extractives sectors. All in all then, our sample of cases offers a very good picture of the challenges facing most states with access to natural resources wealth. We can now examine the most significant patterns in each of our three case groupings: sovereign wealth funds; local content policies and corporate social responsibility.

2 Sovereign Wealth Funds

The chapters included in the sovereign wealth fund (SWF) section offer a very good place to start thinking about national strategies for managing resource wealth. This is so for at least two reasons. First, most of these chapters cover more than the creation and management of their SWFs: they also provide important background information about the nature of the resource that each case relies on (oil/minerals/hydro…); the national legislation and licensing/contract forms used to bring that resource to market; different ways that governments capture the revenue streams from their extractive industry; and how these streams are channelled into (or not) their SWFs. In short, there is a lot of conceptual territory covered in these chapters.

Second, the focus in this section is on the development and use of SWFs, and we find a remarkable variation across the sample set. Our cases include some of the oldest SWFs in the world (Alberta and Alaska), alongside some of the youngest (e.g. Guyana). We include the world's largest SWF (Norway) and two cases (Indonesia and Kenya) that have yet to adopt SWFs. The current allure of SWFs can be seen in the case of Guyana, which has gone to the trouble of creating legislation for a SWF, without yet accessing the resource, whose wealth it hopes will fill it. Guyana's priorities seem to differ significantly from what we see in the Norwegian and Indonesian cases (for example), where petroleum revenues were first ploughed back into the domestic economy before there was any thought of setting them aside in a SWF (in Norway it came much later—27 years after commercially viable oil was discovered; in Indonesia, that day has yet to come). The reason for this may be Guyana's potential to become extremely dependent on the resource, once it hits the market, as seen in Table 1 (Natural Resource Rents/GDP). Most of the money collected in these funds is used to support their respective government budgets. In other cases, the government distributes its SWF earnings directly to the people (e.g. Alaska). Still other cases keep the SWF money trapped offshore (e.g. Norway), to minimise its effect on the economy. In this section, our aim is to try and make some sense from this remarkable variation.

While our original case selection covers 12 countries, the SWF section introduced 16 different sovereign wealth funds, as several of our cases employ more than one SWF. This variance, and the list of SWFs covered, can be seen in Table 2. It is very difficult to evaluate the utility or success of these SWFs, given the varied contexts in which they are found, but we can say something about common tendencies.

Let us begin with the last column, on the right-hand side, of Table 2. Here, we provide an estimate of the value of these SWFs, converted to US dollars. The range in value is not particularly surprising, considering the range in size and level of development across our sundry cases. After all, our collection stretches from the largest SWF in the world (the Norwegian GPFG), which is worth over a trillion US dollars, to several cases where published values are simply not available. The fact that governments are not making these (value) figures easily available is both remarkable and frightening. The smallest fund in our selection (of those for whom we could find figures) was the Western Australia Futures Fund, at $923,287,000. The HSF in Trinidad and Tobago is perhaps the most surprising SWF, given the large amount of money it has amassed (over $6 billion) and its relatively small population (ca 1.4 million).

Other than value, we have noted four important differences separating these cases: degree of transparency/accountability; the type of fund; its source of revenue; and how the money can be spent (including expenditure conditionality).

Table 2 SWFs described in this collection

Acronym	Fund name	Case	Year founded	Type	Revenue source	Withdrawal conditions	Value 2019 USD
FGF	Future Generations Fund	Nigeria	2011	Savings	Budget surplus	No	1,690,440,000
NIF	Nigerian Infrastructure Fund	Nigeria	2011	Savings	Budget surplus	No	
NSF	Stabilisation Fund	Nigeria	2011	Stabilisation	Budget surplus	No	
–	–	Kenya	–	–	–	–	–
–	–	Indonesia	–	–	–	–	–
NDF	National Development Fund of Iran	Iran	2011	Savings	Mostly oil/gas	Some	91,000,000,000
BSF	Social Fund	Brazil	2010	Mostly savings	Mostly oil/gas	No	4,123,940,000
HSF	Heritage and Stabilisation Fund	Trinidad and Tobago	2007	Both	Budget surplus	No	6,255,349,599
NRF	Natural Resource Fund	Guyana	2019	Both	Petro + mining and forest	Some	0
NWF	National Welfare Fund	Russia	2008	Both	Budget surplus	Unclear	124,140,000,000
PF	Permanent Fund	Alaska	1976	Savings	Petroleum	No	67,208,500,000
BCPF	British Columbia Prosperity Fund	Canada	2013	Savings	Natural gas	No	Unclear
NWHF	Northwest Territories Heritage Fund	Canada	2012	Savings	Non-renewable resources	Yes	13,098,200,000
WGF	Quebec's Generation Fund	Canada	2006	Savings	Hydro, mining, alcohol	No	6,866,220,000
AHF	Alberta Heritage Fund	Canada	1976	Savings	Petroleum	No	86,289,300,000

(continued)

Table 2 (continued)

Acronym	Fund name	Case	Year founded	Type	Revenue source	Withdrawal conditions	Value 2019 USD
FF	Future Fund	Australia	2006	Savings	Budget surplus	No	112,019,000,000
WAFF	Western Australia Futures Fund	Australia	2012	Savings	Minerals	Yes (until 2032)	923,287,000
GPFG	Government Pension Fund Global	Norway	1990	Both	Petroleum	Yes	1,098,820,000,000

Note All information is gained from the chapter contributions, except the estimated value (right-hand column). The estimated values come mostly from SWFI (2019) except for the data on Brazil (InfoRoyalties 2020); the NWT Heritage Fund (CBC 2018) and Quebec's Generations Fund (Finances Quebec 2019). The NWT Heritage Fund data is from 2018. See original data sources for more information

2.1 Independence and Transparency

When we consider the role of SWFs in managing resource wealth generated by the extractives sector, one of the most important issues to address is the degree to which that wealth is managed in a professional, transparent way, devoid of corruption. In doing so, we are forced to ask a number of questions, including: How does the SWF interact with the government's regular budgetary functions? Are the objectives of the SWF made clear and explicit? Can the money be used to support domestic political projects, or are there constraints on how it can be used? Are their expectations on how the money should be invested (e.g. maximise returns?) Who decides what is to be deposited in the SWF, and what is withdrawn? Is it possible to hold the relevant decision makers responsible?

To our surprise, we found that it is astonishingly difficult to get information on many of these SWFs—on how they are managed, on the limits of the authority, even their current market value (as we saw in Table 2). This is especially surprising in the case of several of the subnational Canadian funds (although not Alberta), where we would expect to see much more transparency.

It is difficult to compare answers across such a wide array of questions, but their content might be boiled down to three main (and related) concerns: legal autonomy (or independence), transparency and accountability. Citizens are concerned that their SWF will be raided by corrupt officials or squandered on poor investments. Investors are concerned that these funds may be puppets of governments. For both groups, one of the most important questions that we can ask is: How transparent and accountable are these SWFs?

Questions about autonomy and transparency go to the heart of governance and remind us of the opening epigraph in the introductory chapter by Edmund Burke: that the revenues of the state are the state. When states create a SWF, we should wonder about their intent: What exactly do they hope to achieve with this fund? Why is a SWF an appropriate vehicle for those investments? After all, governments have a long history of savings and investments and have a variety of instruments at their disposal. Most central banks have experience with managing their reserves: they invest in safe government paper, with relatively low (but low-risk) returns. Why not place the money in a government deposit account, or with the central bank, in safe government bonds, where it remains in open political space? Norway did this early on, and the Russian National Welfare Fund and the NW Territories Heritage Fund in Canada continues to do so today.

From where do we get this new appetite, among states, for higher risks and returns? Are the people aware of the risks that are involved? If the money is to be ploughed back into the domestic economy, why not use traditional budget instruments? Is the SWF being used to hide money from public scrutiny? These are the sort of questions we are left with after examining the SWFs employed by our cases.

Once a SWF is chosen as a sort of savings or spending vehicle, the next set of questions to ask are: How can the people keep policymakers accountable for the investment decisions and returns on this money? How do we know that the money is being invested wisely (and not in pet political projects that will deliver less certain returns)? How will we know that policymakers cannot access these funds in the middle of the night and whisk them away to secret and private foreign bank accounts?

Perhaps the easiest way to check if a SWF is legally autonomous, accountable and transparent is to turn to the International Forum on Sovereign Wealth Funds (IFSWF) and their so-called Santiago Principles (IWGSWF 2008). The Santiago Principles are a list of 24 generally accepted principles, in which IFSWF members voluntarily endorse to demonstrate that they comply with applicable regulatory and disclosure requirements and embrace transparent government structures. The assessment is voluntary and self-imposed, and in 2019, there were 36 member funds, three of which were associate members (see Table 3). Thus, IFSWF membership functions as a sort of gold standard for investors—signalling which SWFs are at least trying to look legitimate and attractive.

Among the total list of IFSWF members, we find only six of our case countries (listed in italics), but only five of our 16 funds. The Alaskan Permanent Fund, Australia's Future Fund, the National Development Fund of Iran, the Nigeria Sovereign Investment Authority and The Heritage and the Stabilisation Fund in Trinidad and Tobago are all signatories to the Santiago Principles, and readers can turn to the IFSWF homepage to find their self-reported assessment of how they meet those principles. The Russian Federation also has a fund that has agreed to the Santiago Principles (the Russian Direct Investment Fund, or RDIF), but it is not among our cases.

These differences can be explained, in part, by the nature of the funds concerned. While investors and markets are looking for funds that are de-coupled from politics, the people in each country may be looking for a fund that responds more directly to their concerns. In short, the interest of international investors and domestic citizens does not always coincide. Countries that are concerned about their reputations and those which hope to attract foreign investors use the Santiago Principles as a way to demonstrate that they have tied their political hands (whether they have actually done so, is a matter of discussion, as we saw in the Nigerian case).

This is why the Russian Direct Investment Fund (RDIF) is a member of the IFSWF. The RDIF was created in 2011 as a "catalyst for direct investment in Russia" (RDIF 2020). It aims to attract market actors who may be otherwise afraid of investing in Russia. For this reason, it needs to present itself as a technical, non-political and investment machine that can promise top-shelf returns. But the RDIF does not invest the wealth generated from Russia's extractives sector—it relies on a global set of investors to contribute money. For this reason, it was not included among our SWFs.

What is perhaps more revealing, however, are two of the cases in our selection that are *not* members of the IFSWF: Alberta and Norway. Alberta is an interesting example; in that, the Alberta Heritage Savings Trust is the largest of the Canadian SWFs and one of the world's oldest. Although its objectives have changed significantly over the years, it enjoys a high level of transparency and independent oversight over the fund (see Chapter "Non Renewable Resource Revenue Savings and Distribution in Canada: Alberta"). Still, it is not a member of the IFSWF, for reasons that are not entirely clear to us.

Table 3 IFSWF members, 2019

Fund name	Country	Membership
Abu Dhabi Investment Authority	United Arab Emirates	Full
Agaciro Development Fund	Rwanda	Full
Alaska Permanent Fund	*USA*	*Full*
BpiFrance	France	Full
Budgetary Income Stabilisation Fund	Mexico	Full
CDP Equity SpA	Italy	Full
China Investment Corporation	China	Full
Fonde de Ahorro de Panamá	Panama	Full
Fonds Souverain d'Investissements Stratégiques S.A.	Senegal	Full
Fundo Soberano de Angola	Angola	Full
Future Fund	*Australia*	*Full*
GIC Private Limited	Singapore	Full
Intergenerational Trust Fund for the People of the Republic of Nauru	Nauru	Full
Ireland Strategic Investment Fund	Ireland	Full
Ithmar Capital	Morocco	Full
JSC National Investment Corporation of the National Bank of Kazakhstan	Kazakhstan	Full
JSC Samruk-Kazyna	Kazakhstan	Full
Khazanah Nasional Berhad	Malaysia	Full
Korea Investment Corporation	Republic of Korea	Full
Kuwait Investment Authority	Kuwait	Full
Libyan Investment Authority	Libya	Full
National Development Fund of Iran	*Iran*	*Full*
New Zealand Superannuation Fund	New Zealand	Full
Nigeria Sovereign Investment Authority	*Nigeria*	*Full*
Palestine Investment Fund	State of Palestine	Full
Qatar Investment Authority	Qatar	Full
Russian Direct Investment Fund	*Russia*	*Full*
State General Reserve Fund	Oman	Full
State Oil Fund of the Republic of Azerbaijan	Azerbaijan	Full
The Heritage and Stabilisation Fund	*Trinidad and Tobago*	*Full*
The Pula Fund	Botswana	Full
Timor-Leste Petroleum Fund	Timor-Leste	Full
Turkey Wealth Fund	Turkey	Full
Future Heritage Fund	Mongolia	Associate
National Infrastructure Investment Fund	India	Associate

(continued)

Table 3 (continued)

Fund name	Country	Membership
National Investment Fund	Cyprus	Associate

Source IFSWF (2020)

The absence of the Norwegian case is perhaps more informative, in at least two regards. First, the Norwegian fund (GPFG) is not included on the list of members—even though Norway was an active player in launching the IFSWF. As was the case in Alberta, it is not exactly clear to us why this is the case, but it may be that the inclusion of many rather questionable funds and the use of voluntary assessments have raised some Norwegian eyebrows. As the State Secretary of the Norwegian Ministry of Finance, Tom Vamraak, noted:

> It is important for the Norwegian government to encourage and ensure transparency about the management of sovereign wealth funds, including objectives, governance framework, investments and risk management. The IFSWF has not met our expectations as an organisation with sufficiently strong progress in the implementation of these principles. Therefore, we decided to discontinue our membership in the organisation in 2016 (IFSWF 2018: 46).

But the Norwegian case is even more interesting because the Norwegian authorities have been unwilling to grant legal autonomy to the GPFG: they want to ensure that the Norwegian people, through their parliament, can still influence how that money is being invested. In short, Norway does not believe that financial technocrats, who feign political neutrality in making investment decisions, are best suited to decide how to invest the Norwegian people's money. They want to channel this technical expertise within set parameters that are explicitly political. The Norwegian authorities maintain strong political guidelines on how the GPFG can spend its money, including a unique "Council on Ethics", and it is this ability to respond to political pressure that is arguably one important reason why the Norwegian SWF has been so successful (Moses 2020).

After all, if you want your government to invest its money in a way that is politically and ethically accountable, you cannot simply shoot for the highest returns. You need to place political and ethical constraints on how that money should, or should not, be used. At the very least, this should be part of the political discussion when launching a SWF.

2.2 *Type of Fund*

The second important difference separating our cases is the decision of political authorities about whether the fund should be used as a large savings or spending account—either for future generations or for current needs (such as infrastructure, health and pensions)—or whether it should be used as an active buffer fund. It is often difficult to distinguish between the two roles, as a savings fund can be created in ways that also buffer the economy from the volatility of the extractive industries' production and price levels. It is only when we look closer at the particular rules of how deposits are made to the fund, and how withdrawals can be taken, that we can begin to see the true nature of the funds being compared.

As a general rule, it has not been very easy to see how states manage their SWFs. Some states use laws to declare whether the SWF should be designed as a savings or buffer fund. But other cases are less clear or explicit. Our effort to distinguish between these two main types of SWFs is shown in Table 2. The ranking there is informal and based on the information provided by the individual case studies included in this anthology.

Here, it is possible to see how the case studies have explained the need for a fund in terms of savings or stabilisation. A majority of the SWFs in our sample were designed as savings funds. Among these, only one was specifically designed as a stabilisation fund (Nigeria), and the Nigerian chapter (Chapter "Public Wealth Management and Distribution in the Extractive Industry in Nigeria") suggests that this is not a particularly effective fund, given the differences between federal and state officials as to how to divide the monies.[2] Four cases (Trinidad and Tobago, Guyana, Russia and Norway) use their funds as both savings and stabilisation accounts. In the Guyana case (Chapter "The Experiences of Managing the Heritage and Stabilisation Fund in Trinidad and Tobago and the Sovereign Wealth Fund Guyana"), we saw that this (duplicate objective) has created some tension and discussion about whether there should be, in fact, two different funds.

To really understand the role that these funds can play, however, we need to follow the money more closely: How are they funded (the nature of the deposits going into the fund) and how are they spent (withdrawals)?

2.3 Deposits

For our purposes, one of the most important questions concerns that the source of (and the route by which) the people's money gets deposited in the SWF. As we have made clear, we are interested in countries that are using SWFs as part of their strategies for managing the wealth generated from the extractives industries. This is why, for example, we look at the Russian National Welfare Fund and not the Russian Direct Investment Fund. But beyond this common attribute, what patterns are revealed in our cases?

Our cases receive their deposits from a variety of different extractives industries, and some even draw further than that. As we can see in Table 2, some funds are directly linked to revenues generated by the petroleum sector (in, e.g. Alaska, Alberta and Norway), and others reflect the case's reliance on the mining sector (e.g.

[2] As we learned in the Russian case from the Chapter "Russian Sovereign Wealth Fund", Russia also had a dedicated stabilisation fund (established in 2004), but it was superseded by the National Welfare Fund in 2008.

Western Australia), or a broader mix of natural resource rents (Guyana and Northwest Territories in Canada) and even hydroelectric power (Quebec Generations Fund).

More importantly, from a management perspective, is how these revenues make their way into the fund. This matters. If the government has the freedom to choose how much of its natural resource revenues are placed in the fund, it will prove tempting to spend the revenues of the resource directly, rather than place them in the fund. In a little less than half of our cases, the funds are filled with a sort of budgetary surplus: when the funding comes from the residual once the actual earnings prove to be larger than the budgeted earnings (see, e.g. Nigeria; Trinidad and Tobago; and Australia's Future Fund). Other states require that a certain share of revenues from the sector (e.g. a percentage of royalty, or signing bonus) goes directly into the fund, without going through the government budget. The Brazilian case is a particularly good example of this, because we can see (Chapter "The Social Fund: A Brazilian Sovereign Wealth Fund") an explicit accounting of where the money comes from: a share of the signing bonus, a share of royalties, income from the sale of oil/gas, returns on prior investments and money diverted from other funds.

If the fund is based on a budgetary surplus, as it is in 6 of our 16 cases, it basically functions as a savings account for the government. That savings account can be accessed by the government, but it will be difficult to use it as a buffer against the general economy: as the full force of the business cycle (production levels and price swings) will then flow through the government budget. For these (savings-based) funds to function as a buffer, we need to consider *where* the investments are being made. Among our cases, we can clearly see two different strategies being employed. In the Alaskan case, we see a willingness to invest in the "Lower 48" economies as a buffer for the Alaskan economy; in that, the returns on these investments are unlikely to be correlated with the fortunes of the Alaskan oil and gas industry. For Alaska (Chapter "Alaska's Petroleum Industry, Institutions and Sovereign Wealth Fund"), the buffer protection is aimed at the business cycle. Norway uses a very different strategy, the result of enjoying a sovereign currency of its own: it invests the money offshore (and is explicitly forbidden from investing in Norway) in an effort to avoid appreciation pressures (Dutch Disease). In contrast to Alaska, Norway (Chapter "Norway's Sovereign Wealth Fund") is buffering its economy from the appreciation effect of an overheated economy.

2.4 Withdrawals

Finally, it is interesting to compare how the money is being used. In a majority of our cases, the governments have more-or-less free access to their respective funds, and this access has—at times—proven costly. This problem was seen in both the Alaskan and Albertan cases, where money was siphoned off to pay down the deficit or to invest in some rather dubious assets (such as a golf course). In other cases, there are explicit requirements on how the money should be spent: e.g. on infrastructure (Nigerian Infrastructure Fund, BC Prosperity Fund); on education (e.g. Brazil);

on pensions (Russian National Welfare Fund, Australian Future Fund); debt reduction (BC Prosperity Fund; Russian National Welfare Fund); as a supplement to the national budget (Norway); or not until some future point in time (e.g. NWT Heritage Fund; Western Australia Future Fund).

To get an idea of how this money is used in our different cases, Fig. 1 provides a rough outline. As should be evident, states can employ strategies on both the left- and right-hand side of the figure.

Of those cases that use a SWF as an alternative vehicle for state spending, we can distinguish between two main types. In a very few cases, the money is distributed directly to the people, by way of a citizen dividend (in Alaska, the Permanent Fund Dividend; in Alberta, earlier, "Ralph Bucks"). In most of the other cases, the government spends the money on specific (e.g. infrastructure) or general needs. These cases populate the left-hand side of Fig. 1.

Other cases think of their SWF as an investment vehicle: as a means to increase future value for the country. These cases find themselves on the right-hand side of Fig. 1. Here, the money accumulates, rather than being spent. The largest fund in our sample is the one that is perhaps tapped the least: the Norwegian GPFG. Here, only 3–4% of the returns on investments are repatriated back into the national economy through the government budget. This allows the fund to grow quite large.

The main difference between states that think of their SWF as an investment vehicle lies between those that want to encourage domestic economic growth and those that do not. The first group has to proceed very carefully, as they will need to consider the political ramifications of their investment decisions. For example: why should the government support one type of commercial activity over another (i.e. can it pick winners?); and should it invest enough to be able to control the activities associated with that investment (i.e. secure a controlling share). The remaining cases move their investments outside the domestic economy—either as a buffer to the local business cycles (as we saw in the Alaskan case) or as a buffer to Dutch Disease (e.g. the Norwegian case). In any case, there are no straightforward answers to what the best choice is as it depends on a case-by-case analysis of needs and aspirations within each country and society.

3 Local Content Policies

Our second selection of cases deals with how these states employ (or do not employ) so-called local content policies (LCPs). In this section, even more than the first, we find very different understandings of what constitutes local content, whether LCPs are needed, and even what they should be used for. This variance reflects the very different contexts in which our states find themselves: national or subnational; liberal or more interventionist; developing or developed.... For example, in the Alaskan contribution in Chapter "Alaska's Tug of War on Land Rights", the discussion of local content policy revolves around land-use rights. In that case, one could make

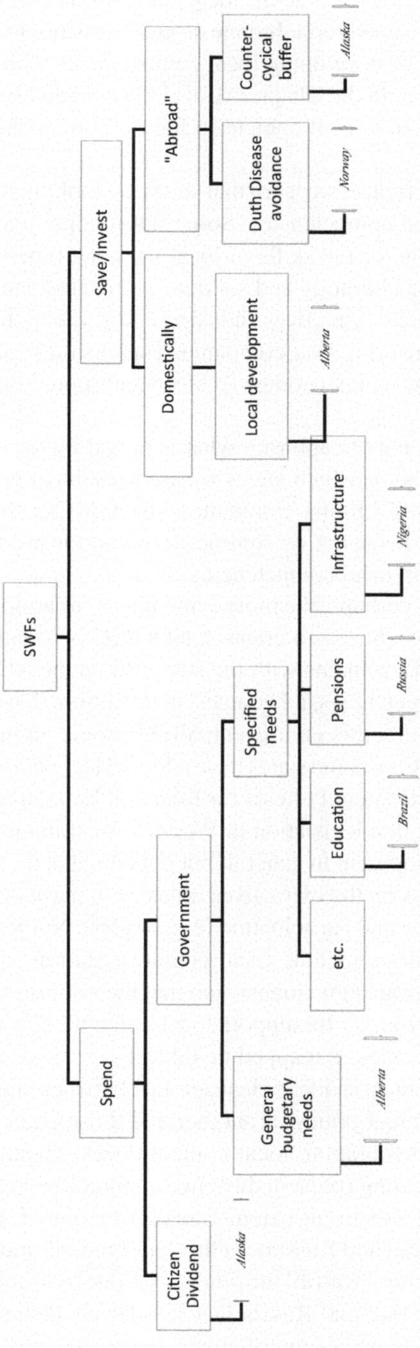

Fig. 1 SWF strategies employed

the argument that LCP comes in the form of its citizen dividend, where petroleum-based revenues are distributed directly to the local population in the form of an annual cheque. In other countries (e.g. Nigeria, Indonesia and Norway), local content policy is a multi-layered strategy for developing local competencies with the assistance of powerful multinational firms. In the Chapter "Local Content and Extractive Industry in Brazil", we see that Brazil's local content is focused on technology, goods and services.

From this bounty of experience, we learn that states are looking for different things when trying to develop local competencies. Some states hope to use the extractives sectors to boost employment, or the skills of local workers; others hope to develop a local mining equipment, technology and services sector that can help service and supply the more international extractives industries; still others hope to maximise the government take, by ensuring that companies establish a legal foothold in the country (and with it a claim on tax revenues). More ambitious states want to do all this, and more!

Despite the variety of ways of capturing what is meant by local content, we can distinguish three main tools by which states secure their local content policy: the use of National Oil, Gas and Mining Companies (NOGMCs); the use of targeted legislation; and the use of licensing or contract terms to ensure that international companies will help develop local competencies.

The use of NOGMCs is, perhaps, the most evident tool for building local content. The main reason that states choose to create a NOGMC is to ensure that a larger share of the government take remains with the state. When the NOGMC is allowed to develop operational competencies, it becomes an institution that can develop local expertise, hire local subcontractors and (eventually) become an international actor, competitive abroad. In addition, some states have adopted legislation—either broadly (e.g. in the Chapter "Local Content Policies for Regional Economic Development in Western Australia", we see that legislation in Western Australia is aimed at encouraging local economic development in general, but not aimed at the extractives sector in particular) or with a focus on the extractives industry in particular, to support and prioritise local value creation and participation (e.g. Nigeria, Norway and Indonesia). Finally, a number of countries include local content requirements of one sort or another, as part of their contract agreements—so that the allocation of a licence will depend upon an explicit promise to support local content. The variation in tools employed, by the different cases, is mapped in Table 4.

From this variance, the most evident cleavage lies between those states that are using one form of local content policy or another and those cases that do not seem particularly interested in encouraging local content development. The reasons for the latter can be many, including (presumably) free-market ideological reasons (e.g. Alaska, Western Australia and to an extent Canada) or out of a desire for more international contact (e.g. Iran and Russia). In this latter regard, Iran appears to be an extreme outlier. Iran is different from all the other cases due to its international (political) isolation. In addition, Iran and Russia face similar challenges with sanctions, which might prevent the presence of international companies and foreign technologies. For this reason, the discussion about local content in the Iranian and Russian

Table 4 Local content instruments

Country	NOMC	Legislation	Contract
Nigeria	Nigerian National Petroleum Corporation (NNPC)	Yes, especially Oil and Gas Industry Local Content Development Act (2010)	Included in Model Service Contracts and used to distribute Marginal Fields contracts
Kenya	National Oil Corporation of Kenya	No overarching policy, but supportive regulatory frameworks	Vague commitments made in Model PSA
Indonesia	Pertamina	Yes, several relevant laws and regulations	Gross-split arrangements consider local content provision when determining profit sharing
Brazil	Yes, especially Petrobras; less so with Vale do Rio Doce Compane (VALE)	Yes, especially the National Programme for Oil and Natural Gas Industry Mobilisation (PROMINP)	Local content clause in contracts
Iran	National Iranian Oil Company (NIOC)	Unclear	Buy-back contracts, pre 2014; Iranian Petroleum Contract (IPC)
Trinidad and Tobago	Petrotrin	Yes, but outdated	Model PSA provides for LC
Guyana	N/A	Yes, but outdated	N/A
Russian Federation	Roseneft and Gasprom	None	Licences only for legal Russian entities
Alaska	None	None	None
Canada	N/A	None	None
Western Australia	N/A	None	Included in some State Agreements
Norway	Statoil/Equinor	Technological agreements and regulatory regime	Extensive

cases (see the Chapters “Local Content Requirements in Iran” and “Local Content Within Extractive Resources Industry in the Russian Federation”, respectively) is somewhat reversed as they encourage foreign technology in order to enhance their existing technology and increase their production whereas most of the other chapters (to varying degrees) tend to recognise the need to build up (or deepen) local competencies in the face of more powerful international market actors.

In addition, we would like to highlight the importance of strategic planning and careful consideration before setting targets and goals for LCPs. What is doable to be implemented today? What are the goals for the short-, mid- and long-term targets?

Consideration needs to be given to how a country can transition in a reasonable progression from the current reality towards the ultimate goals. If a country does not conduct a thoughtful and detailed consideration before establishing such policies, it could face serious challenges from a lack of compliance for unrealistic targets. This lack of compliance could lead to potentially huge fines, which might provide an "unfriendly" environment that ultimately could deter future investors and lead to greater corruption.

Alternatively, if nothing is going to be done locally, foreign companies will dominate this domain as has happened in the past. Balance is needed, and a critical assessment is required to understand each case. In Brazil and Nigeria (Chapters "Local Content and Extractive Industry in Brazil" and "Local Content in the Extractive Resource Industry in Nigeria", respectively), the main concerns centre on technology, goods and services rather than human resources, so that jobs given there may exclude many qualified Brazilians and Nigerians within the extractives sector. At the same time, however, Brazil and Nigeria highlight the negative outcomes of setting unrealistic targets of LCPs. These LCPs can bring about higher costs to the industry, delay investments and fail to improve local content standards—quite the opposite of what the governments had anticipated. In Alaska, no LCP was needed to develop local competency (see Chapter "Alaska's Tug of War on Land Rights"). Guyana (Chapter "The Development and Implementation of Local Content in the Extractive Industries in Trinidad and Tobago and Guyana"), on the other hand, might be more concerned with developing their national competencies for goods, services and technology, as well as their human resources. In short, the nature of LCPs reflects the context within which they are implemented as needs and priorities will differ. In Norway, LCPs came to an end with the EU regulations they agreed to when joining the European Economic Area. So, the Norwegian case (Chapter "Norwegian Local Content Policy") poses a relevant question: Should LCPs have an expiration date? Is it fair to expect local companies and industries to be able to compete on an equal footing at some point?

Another key factor for LCPs is the approach that any country might take with regards to the compliance of their LCPs. Unfortunately, the common practice tends to be the "comply toolbox" mentality versus a "punishment" approach, which by and large is followed by most countries. In many cases, countries focus on pure compliance on the given terms imposing sanctions and penalties for any breach; but they do not provide any real benefit for companies that try to exceed such terms. However, there are a few countries that recognise they should take a different approach with direct incentives to increase local content compliance. Indonesia is one of the few examples that recently changed its legal system and provides direct and fiscal incentives to increase local content performance (see Chapter "Local Content Policy in Indonesia Oil and Gas Industry"). The higher local content performance that a contract has, the higher the profit split might be. Therefore, LCPs are not just a good "marketing" initiative or a means of "reducing costs" or ticking off the "compliance check box"; in this case, companies might get an additional financial benefit to increase local content performance.

4 Corporate Social Responsibility

The third and final collection of chapters considers the way that our cases deal with issues of corporate social responsibility (CSR). As with the local content section, we find a remarkable variety of approaches in the case studies. To try and create some order out of this variety, we consider two ways of characterising the sundry chapters in the CSR section.

The first characterisation concerns the sort of commitment states are willing to make to support CSR activities. Figure 2 presents a continuum of attitudes, stretching from relative indifference or inaction about CSR (e.g. Iran, Alaska and Australia), to states that wish to encourage it by developing explicit strategies (e.g. Canada, Guyana and Trinidad and Tobago), to those states—at the top of the figure—who go so far as to introduce legislation and regulations that require firms to behave in a particular manner, in addition to maximising shareholder value (e.g. Norway, Nigeria and Indonesia).

The second characterisation concerns the direction in which this commitment is aimed. In other words, are the case studies looking at the motives of foreign firms active in the case context, or are they interested in case-based firms, as they act abroad? Following Gourevitch (1978), loosely, we might say that the first group is employing an outside-in approach, whereas the latter group uses an inside-out approach. For the vast majority of the cases in our study, the approach is outside-in: the focus is on recognising how the impetus for CSR often comes from multinational corporations and is foreign to the domestic context (e.g. Trinidad and Tobago, Guyana, Australia). Other outside-in approaches look at how governments are actively trying to encourage IOGMCs to behave in a more responsible manner when extracting local resources (e.g. Nigeria, Indonesia and to a lesser extent Brazil).

Russia is an interesting outlier; in that, one can see remnants of earlier Soviet attitudes influencing Russian firms and the way they are willing to take on larger, community interests, even when it incurs financial costs. Curiously, this sort of path dependency seems to erode with the international successes of the companies in

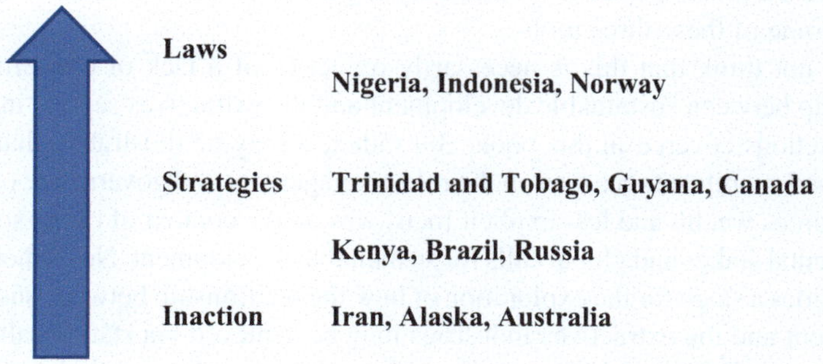

Fig. 2 CSR strategies

question, as seen in the example of Roseneft, who welcomed non-commercial obligations when it was a smaller, weaker company, but increasingly shuns them as it has grown more powerful.

A similar pattern can be seen in what is arguably the most unique case in our SWF sample: Norway. Here, the government is concerned that Norwegian firms will shun their corporate social responsibility as they go abroad in search of new economic rewards. The Norwegian chapter is very different from the others; in that, its focus is on ensuring that Norwegian companies, when working abroad, take with them the sort of CSR attitude that is already legally and culturally engrained in Norway. Unlike all the other cases (with the possible exception of Canada), the Norwegian chapter looked at how Norwegian multinational firms are behaving beyond Norway—rather than looking at how IOGMCs are expected to behave in Norway.

4.1 Sustainable Development and Extractives

In the current context of Agenda 2030, the foregrounding of the role of the private sector in progressing the Sustainable Development Goals, the threat of global climate disruption (Intergovernmental Panel on Climate Change 2018) and the commitment of nations (including the extractives sectors in many nations) to the Paris Agreement, it is surprising that sustainability in mining continues to be a contentious and somewhat vexed issue (Rajaram et al. 2005; Horowitz 2006; Aaron 2012; Spencer 2018). In the proposal for this anthology, contributors were invited to reflect on the state's management of extractives resources wealth as a key component of the sustainable development of a nation and how the extractives resources sector can support sustainable development of local industry and communities throughout the lifespan of their projects using local content policies and corporate social responsibility activities. We asked contributors to consider some of the challenges and contradictions their chapters unearth relating to the use of state wealth management and distribution, local content policies and CSR practices for sustainable development. Interestingly, most contributors shied away from explicitly engaging with sustainable development as an outcome of these three tools.

We do not think that this is necessarily reflective of a lack of concern for the relationship between sustainable development and the extractives sectors in most of the jurisdictions covered in this book. But rather, it may be illustrative that several contributors' manifest focus was on the role, management and governance of extractives resources wealth and less explicit focus was on the context of rising social and environmental stakes and global calls for sustainable development. Nevertheless, this book provides a stage for the exploration of how the relationship between sustainable development and the extractives industries may be achieved via state wealth funds, local content policies and CSR practices. The value of such an anthology is that it brings together the diverse experiences and challenges encountered in different parts of the world that might inform equitable and sustainable development, particularly for the communities directly adjacent to extractives operations but more broadly for the wider society and environment.

5 Conclusion

Each of these points of variation makes it difficult to generalise across cases and challenges us to recognise the complexity of the task at hand. This is the most important lesson one takes home after reading the cases in this anthology. This is especially true with respect to the contributions in the LCP and CSR sections, where case authors demonstrate a remarkable variety of approaches, attitudes and views in how they approach the subject matter. Frankly, local content policies and corporate social responsibility mean very different things in our many different contexts. The section on sovereign wealth funds demonstrates a little more common ground; in that, SWFs were used in every case, save two (Indonesia and Kenya). SWFs are used in different ways and responsible to different types of needs and actors, but most of our resource-wealthy states have found it worthwhile to start a SWF.

Perhaps the one commonality we find across the three sections is that each has a dominant outlier: a single state that behaves very different from the others in that section. These outliers might help us to generalise; in that, they can be seen as providing an exception to the rule in each section. As already mentioned, Indonesia is unique among the cases under consideration in the SWF section in its unwillingness to start a SWF (see the Chapter "Overview of Extractive Resources Management in Indonesia"). Kenya has been considering SWF legislation for some time but has yet to adopt it. This resistance makes us realise how quickly and widely the SWF fad has spread in recent decades, prompting us to wonder if it is not time to ask about what this pattern says about the nature of the relationship between democracy, technocratic authority and government transparency.

In the LCP section, Iran (the Chapter "Local Content Requirements in Iran") followed by Russia (the Chapter "Local Content Within Extractive Resources Industry in the Russian Federation") were the clear outliers. Every other case implicitly or explicitly recognised the need to protect local interests from powerful external and multinational factors and used different strategies to protect those local interests. This trend seems to reflect a familar theme in the extractives sector: where it is common to discover an imbalance of power between underdeveloped political institutions and powerful multinational corporate interests. Given Iran's international isolation, its perspective on these matters is very different: it seems more aware of the limits to autarky and is actively seeking more international influences to help it become more efficient. Iran's outlier position makes us realise how important it is to strike a balance between domestic and international influences, so that the people in each state can ensure that their resources are being managed in the most efficient and just manner. But it also reminds us of the serious and draconian implications that any nation can suffer from international and/or certain unilateral sanctions.

Finally, in the CSR section, the Norwegian contribution in the Chapter "CSR in the Norwegian Context" was remarkably different from the others; in that, the case study was concerned about the behaviour of Norwegian firms abroad, rather than foreign firms acting in Norway. Here, it would seem that a long history of strong legislative and political traditions has been useful in ensuring that foreign firms in Norway will obey strong CSR norms. The challenge has been to get Norwegian firms to continue acting in a socially responsible manner when competing in less transparent and more corrupt contexts elsewhere. This seems to suggest that there is a great amount of work still to be done in the international arena, and in particular across the extractives sectors, to ensure that all firms are held to the highest ethical and social standards.

These comparative examples from around the world provide real-world examples and insights into the ways in which policy, administrative and technical instruments can be harnessed to ethically and responsibly manage the revenue from extracting natural resources. Such structures are intended to provide governments and communities with the means to redress poverty and inequality through sustainable development.

References

Aaron, K. (2012). New Corporate social responsibility models for oil companies in Nigeria's Delta region: What challenges for sustainability? *Progress in Development Studies, 12*(4), 259–273.

BP. (2019). *BP Statistical Review of World Energy 2019*. Online at https://www.bp.com/en/global/corporate/energy-economics/statistical-review-of-world-energy.html. Accessed January 5, 2020.

CBC. (2018). N.W.T. Heritage Fund not keeping up with inflation, gov't borrows money to feed it. *CBC News*. July 5. https://www.cbc.ca/news/canada/north/nwt-heritage-fund-inflation-1.4734937. Accessed January 14, 2020.

EIA. (2017). Rankings: Total Energy Production, 2017 (trillion Btu). Online at https://www.eia.gov/state/rankings/?sid=AK#series/101. Accessed January 13, 2020.

Finances Quebec. (2019). The Generations Fund. November update. Online at http://www.budget.finances.gouv.qc.ca/fondsdesgenerations/index_en.asp. Accessed January 14, 2020.

Gourevitch, P. A. (1978). The second image reversed: The international sources of domestic politics. *International Organization, 32*(4), 881–912.

GWA [Government of Western Australia]. (2018). *Western Australian mineral and petroleum statistics digest 2017–18*. Perth, WA: Department of Mines, Industry Regulation and Safety, Government of Western Australia. Online at www.dmp.wa.gov.au/Documents/About-Us-Careers/Stats_Digest_2017-18.pdf. Retrieved April 3, 2019.

Horowitz, L. (2006). Mining and sustainable development. *Journal of Cleaner Production, 14*(3–4), 307–308.

IFSWF. (2018). The origin of Santiago Principles. Experiences from the past; guidance for the future. https://www.ifswf.org/sites/default/files/IFSWF_Santiago_Principles_book.pdf. Accessed November 11, 2019.

IFSWF. (2020). Our Members. Online at https://ifswf.org/our-members. Accessed January 9, 2020.

InfoRoyalties. (2020). Royalties + Participações Especiais em valores reais, corrigidos pelo IGP-DI. A partir de Agência Nacional do Petróleo e Fundação Getúlio Vargas. http://inforoyalties.ucam-campos.br/informativo.php. Accessed March 23, 2020.

Intergovernmental Panel on Climate Change. (2018). *Global warming of 1.5 °C. An IPCC Special Report on the impacts of global warming of 1.5 °C above pre-industrial levels and related global greenhouse gas emission pathways, in the context of strengthening the global response to the threat of climate change, sustainable development, and efforts to eradicate poverty*. Geneva, Switzerland: IPCC.

IWGSWF [International working Group of Sovereign Wealth Funds]. (2008, October). Sovereign Wealth Funds. Generally accepted principles and practices. 'Santiago Principles'. https://www.ifswf.org/sites/default/files/santiagoprinciples_0_0.pdf. Accessed January 13, 2020.

Moses, J. W. (2020). A Sovereign Wealth Fund: Norway's Government Pension Fund, Global. Forthcoming book chapter in E. Okpanachi & R. Tremblay (Eds.), *The political economy of natural resource funds*. Palgrave Macmillan.

Rajaram, V., Dutta, S., & Parameswaran, K. (2005). *Sustainable mining practices: A global perspective* (London: Taylor & Francis Group).

RDIF. (2020). Overview. Online at https://rdif.ru/Eng_About/. Accessed January 14, 2020.

Spencer, R. (2018). Corporate social responsibility as sustainable development: A critical anthropology of development perspective. In M. Brueckner, R. Spencer, & M. Paull (Eds.), *Disciplining the undisciplined: Perspectives on responsible citizenship, corporate social responsibility and sustainability*, pp. 73–88 (UK: Springer).

SWFI. (2019). Top 88 Largest Sovereign Wealth Fund Rankings by Total Assets. Online at https://www.swfinstitute.org/fund-rankings/sovereign-wealth-fund. Accessed January, 14 2020.

World Bank. (2019a). Natural gas rents (% of GDP). Online at https://data.worldbank.org/indicator/NY.GDP.NGAS.RT.ZS. Accessed on December 20, 2019.

World Bank. (2019b). Oil rents (% of GDP). Online at https://data.worldbank.org/indicator/NY.GDP.PETR.RT.ZS. Accessed on 20 December 2019.

World Bank. (2020a). Total natural resources rents (%of GDP). World Development Indicators. Data Bank. Online at https://databank.worldbank.org/reports.aspx?source=2&type=metadata&series=NY.GDP.TOTL.RT.ZS#. Accessed January 13, 2020.

World Bank. (2020b). Fuel exports (% of merchandise exports). Databank. Online at https://databank.worldbank.org/reports.aspx?source=2&series=TX.VAL.FUEL.ZS.UN&country=#. Accessed January 14, 2020.

Author Index

E. G. Pereira et al. (eds.), *Sovereign Wealth Funds, Local Content Policies and CSR*, CSR, Sustainability, Ethics & Governance,
https://doi.org/10.1007/978-3-030-56092-8

The manufacturer's authorised representative in the EU is Springer Nature Customer Service Centre GmbH, Europaplatz 3, 69115 Heidelberg, Germany. If you have any concerns regarding our products, please contact ProductSafety@springernature.com

Printed and bound by CPI Group (UK) Ltd, Croydon, CR0 4YY

30/09/2026

02225462-0019